Lecture Notes in Artificial Intelligence 1955

Subseries of Lecture Notes in Computer Science
Edited by J. G. Carbonell and J. Siekmann

Lecture Notes in Computer Science
Edited by G. Goos, J. Hartmanis and J. van Leeuwen

Springer

Berlin
Heidelberg
New York
Barcelona
Hong Kong
London
Milan
Paris
Singapore
Tokyo

Michel Parigot   Andrei Voronkov (Eds.)

Logic
for Programming and
Automated Reasoning

7th International Conference, LPAR 2000
Reunion Island, France, November 6-10, 2000
Proceedings

Springer

Series Editors

Jaime G. Carbonell, Carnegie Mellon University, Pittsburgh, PA, USA
Jörg Siekmann, University of Saarland, Saabrücken, Germany

Volume Editors

Michel Parigot
CNRS - Université de Paris 7
case 7012, 2 place Jussieu, 75251 Paris Cedex 05, France
E-mail: parigot@logique.jussieu.fr

Andrei Voronkov
University of Manchester, Computer Science Department
Oxford Rd, Manchester M13 9PL, United Kingdom
E-mail: voronkov@cs.man.ac.uk

Cataloging-in-Publication Data applied for

Die Deutsche Bibliozthek - CIP-Einheitsaufnahme

Logic for programming and automated reasoning : 7th international
conference ; proceedings / LPAR 2000, Reunion Island, France,
November 6 - 10, 2000. Michel Parigot ; Andrei Voronkov (ed.). - Berlin ;
Heidelberg ; New York ; Barcelona ; Hong Kong ; London ; Milan ; Paris ;
Singapore ; Tokyo : Springer, 2000
   (Lecture notes in computer science ; Vol. 1955 : Lecture notes in
   artificial intelligence)
   ISBN 3-540-41285-9

CR Subject Classification (1998): I.2.3, F.3, F.4.1

ISBN 3-540-41285-9 Springer-Verlag Berlin Heidelberg New York

Springer-Verlag Berlin Heidelberg New York
a member of BertelsmannSpringer Science+Business Media GmbH
© Springer-Verlag Berlin Heidelberg 2000
Printed in Germany

Typesetting: Camera-ready by author
Printed on acid-free paper   SPIN: 10781844   06/3142   5 4 3 2 1 0

Preface

This volume contains the papers presented at the Seventh International Conference on Logic for Programming and Automated Reasoning (LPAR 2000) held on Reunion Island, France, 6–10 November 2000, followed by the Reunion Workshop on Implementation of Logic.

Sixty-five papers were submitted to LPAR 2000 of which twenty-six papers were accepted. Submissions by the program committee members were not allowed. There was a special category of experimental papers intended to describe implementations of systems, to report experiments with implemented systems, or to compare implemented systems. Each of the submissions was reviewed by at least three program committee members and an electronic program committee meeting was held via the Internet.

In addition to the refereed papers, this volume contains full papers by two of the four invited speakers, Georg Gottlob and Michaël Rusinowitch, along with an extended abstract of Bruno Courcelle's invited lecture and an abstract of Erich Grädel's invited lecture.

We would like to thank the many people who have made LPAR 2000 possible. We are grateful to the following groups and individuals: the program and organizing committees; the additional referees; the local arrangements chair Teodor Knapik; Pascal Manoury, who was in charge of accommodation; Konstantin Korovin, who maintained the program committee Web page; and Bill McCune, who implemented the program committee management software.

September 2000

Michel Parigot
Andrei Voronkov

Conference Organization

Program Chairs

Michel Parigot (University of Paris VII)
Andrei Voronkov (University of Manchester)

Assistant to Program Chairs

Konstantin Korovin

Program Committee

Stefano Berardi (Università di Torino)
Manfred Broy (Technische Universität München)
Maurice Bruynooghe (Catholic University of Leuven)
Hubert Comon (Ecole Normale Supérieure de Cachan)
Gilles Dowek (INRIA Rocquencourt)
Harald Ganzinger (Max-Planck Institut, Saarbrücken)
Mike Gordon (University of Cambridge)
Yuri Gurevich (Microsoft Research)
Pascal van Hentenryck (Brown University)
Neil Jones (DIKU University of Copenhagen)
Teodor Knapik (Université de la Réunion)
Yves Lafont (Université de la Méditerrannée)
Daniel Leivant (Indiana University)
Maurizio Lenzerini (Universitè di Roma)
Giorgio Levi (Pisa University)
Leonid Libkin (Bell Laboratories)
Patrick Lincoln (SRI International)
David McAllester (AT&T Labs Research)
Robert Nieuwenhuis (Technical University of Catalonia)
Mitsuhiro Okada (Keio University)
Leszek Pacholski (University of Wroclaw)
Catuscia Palamidessi (Pennsylvania State University)
Frank Pfenning (Carnegie Mellon University)
Helmut Schwichtenberg (Ludwig-Maximilian Universität)
Jan Smith (Chalmers University)
Wolfgang Thomas (RWTH Aachen)

Local Organization

Teodor Knapik (Université de la Réunion)
Pascal Manoury (University of Paris VI)

List of Referees

Klaus Aehlig
Ofer Arieli
Philippe Audebaud
Roberto Bagnara
Mike Barnett
Chantal Berline
Michel Bidoit
Thierry Boy de la Tour
Max Breitling
Francisco Bueno
Diego Calvanese
Felice Cardone
Roberto di Cosmo
Ferruccio Damiani
Norman Danner
Juergen Dix
Norbert Eisinger
Sandro Etalle
Gianluigi Ferrari
Michael Gelfond
Paola Giannini
Roberta Gori
Jean Goubault-Larrecq
Colin Hirsch
Steffen Hölldobler
Ullrich Hustadt
Felix Joachimski
Emanuel Kieronski
Claude Kirchner
Konstantin Korovin
Ingolf Krüger
Ugo de'Liguoro
Denis Lugiez
Ian Mackie
Pascal Manoury

Wiktor Marek
Dale Miller
Marcin Mlotkowski
Christophe Morvan
Hans de Nivelle
Damian Niwinski
David von Oheimb
Etienne Payet
Olivier Ridoux
Riccardo Rosati
Dean Rosenzweig
Francesca Rossi
Albert Rubio
Thomas Rudlof
Paul Ruet
Salvatore Ruggieri
Pawel Rychlikowski
Bernhard Schätz
Philippe Schnoebelen
Francesca Scozzari
Jens Peter Secher
Alexander Serebrenik
Natarajan Shankar
Katharina Spies
Robert Staerk
Ralf Steinbrüggen
Karl Stroetmann
Peter Stuckey
David Toman
Tomasz Truderung
Moshe Vardi
Margus Veanes
Klaus Weich
Emil Weydert
Enea Zaffanella

Conferences preceding LPAR 2000

RCLP'90, Irkutsk, Soviet Union, 1990
RCLP'91, Leningrad, Soviet Union, aboard the ship "Michail Lomonosov", 1991
LPAR'92, St. Petersburg, Russia, aboard the ship "Michail Lomonosov", 1992
LPAR'93, St. Petersburg, Russia, 1993
LPAR'94, Kiev, Ukraine, aboard the ship "Marshal Koshevoi", 1994
LPAR'99, Tbilisi, Republic of Georgia, 1999

Table of Contents

Session 6. Logic programming and CLP

Session 7. Nonclassical logics and lambda calculus

Session 8. Logic and databases

Session 9. Program analysis

Session 10. Mu-calculus

Session 11. Planning and reasoning about actions

On the Complexity of Theory Curbing

Thomas Eiter and Georg Gottlob

Institut für Informationssysteme, TU Wien
Favoritenstraße 9–11, A-1040 Wien, Austria
eiter@kr.tuwien.ac.at, gottlob@dbai.tuwien.ac.at

Abstract. In this paper, we determine the complexity of propositional theory curbing. Theory Curbing is a nonmonotonic technique of common sense reasoning that is based on model minimality but unlike circumscription treats disjunction inclusively. In an earlier paper, theory curbing was shown to be feasible in PSPACE, but the precise complexity was left open. In the present paper we prove it to be PSPACE-complete. In particular, we show that both the model checking and the inferencing problem under curbed theories are PSPACE complete. We also study relevant cases where the complexity of theory curbing is located – just as for plain propositional circumscription – at the second level of the polynomial hierarchy and is thus presumably easier than PSPACE.

1 Introduction

Circumscription [15] is a well-known technique of nonmonotonic reasoning based on model-minimality. The (total) circumscription $Circ(T)$ of a theory T, which is a finite set of sentences, consists of a formula whose set of models is equal to the set of all *minimal* models of T. For various variants of circumscription, see [14].

As noted by various authors [5,6,17,18,19,20], reasoning under minimal models runs into problems in connection with disjunctive information. The minimality principle of circumscription often enforces the *exclusive* interpretation of a disjunction $a \lor b$ by adopting the models in which either a or b is true but not both. There are many situations in which an *inclusive* interpretation is desired and seems more natural (for examples, see Section 2).

To redress this problem, and to be able to handle inclusive disjunctions of positive information properly, the method of *theory curbing* was introduced in [8]. This method is based on the notion of a *good model* of a theory. Roughly, a good model of a theory T is either a minimal model, or a model of T that constitutes a minimal upper bound of a set of good models of T. The sentence $Curb(T)$ has as its model precisely the good models of T. When T is a first-order theory, $Curb(T)$ is most naturally expressed as a third-order formula. However, in [8], it was shown that $Curb(T)$ is expressible in second-order logic.

Circumscription is usually not applied to *all* predicates of a theory, but only to the members of a list **p** of predicates, where the predicates from a list **z** disjoint with **p**, called the *floating* predicates, may be selected such that the predicates in **p** become as small as possible; the remaining predicates not occurring in **p** and **z** (called *fixed* predicates) are treated classically. In analogy to this, in [8], formulas of the form

M. Parigot and A. Voronkov (Eds.): LPAR 2000, LNAI 1955, pp. 1–19, 2000.

$Curb(T; \mathbf{p}, \mathbf{z})$ are defined, where curbing is applied to the predicates in list $\mathbf{p}$ only, while those from list $\mathbf{z}$ (the floating predicates) are interpreted in the standard way. In the propositional case, the lists $\mathbf{p}$ and $\mathbf{q}$ of predicate symbols are lists of propositional variables (corresponding to zero-ary predicates).

Since its introduction in [8], the curbing technique has been used and studied in a number of other papers. For instance, Scarcello, Leone, and Palopoli [21], provide a fixpoint semantics for propositional curbing and derive complexity results for curbing Krom theories, i.e., clausal theories where each clause contains at most two literals. Liberatore [11,12] bases a belief update operator on a restricted version of curbing. Note that curbing is a purely model-theoretic and thus syntax-independent method. In particular, for two logically equivalent theories T and T', it holds that $Curb(T)$ is logically equivalent to $Curb(T')$. Curbing can be applied to arbitrary logical theories and not just to logic programs. In the context of disjunctive logic programming, various syntax-dependent methods of reasoning that do not treat disjunction exclusively were defined in [5,18,17,19,20,6].

In [8], the following two major reasoning problems under curbing where shown to be in PSPACE:

Curb Model Checking: Given a propositional theory T, an interpretation M of T, and disjoint lists $\mathbf{p}$ and $\mathbf{z}$ of propositional variables, decide whether M is a good model of T w.r.t. $\mathbf{p}$ and $\mathbf{z}$ (i.e., decide whether M is a model of $Curb(T; \mathbf{p}, \mathbf{z})$).

Curb Inference : Given a propositional theory T, disjoint lists $\mathbf{p}$ and $\mathbf{z}$ of propositional variables, and a propositional formula G, decide whether $Curb(T; \mathbf{p}, \mathbf{z}) \models G$.

The precise complexity of curbing, for both model checking and inferencing, was left open in [8]. Note that model checking for propositional circumscription is coNP complete [3] and inferencing under propositional circumscription is Π_2^P complete [7]. It was conjectured in [21,11] that curbing is of higher complexity than circumscription. This is intuitively supported by a result of Bodenstorfer [2] stating that in an explicitly given set of models, witnessing that some particular model is good may involve an exponential number of smaller good models (for a formal statement of this result, see Section 3).

The main result of this paper answers the above questions. We prove that Curb Model Checking and Curb Inference are PSPACE-complete. Both problems remain PSPACE-hard even in case of *total* curbing, i.e., when curbing is applied to *all* propositional variables, and thus the list $\mathbf{z}$ of floating propositional variables is empty and no propositional variables are fixed. The proof takes Bodenstorfer's construction as a starting point and shows how to reduce the evaluation of quantified Boolean formulas to theory curbing.

The PSPACE-completeness result strongly indicates that curbing is a much more powerful reasoning method than circumscription, and that it can not be reduced in polynomial time to circumscription. Thus, circumscriptive theorem provers can not be efficiently used for curb reasoning. On the other hand, a curb theorem prover could be based on a QBF solver (see [10,4,16,1,9]).

After proving our main result, we identify classes of theories for which the complexity of curbing is located at a lower complexity level. Specifically, we show that if a theory T has the *lub property*, that is, every set of good models of T has a *least* (unique minimal) upper bound, then propositional Curb Model Checking is in Σ_2^P, while Curb Inference is feasible in Π_2^p. Note that relevant classes of theories have this property. For example, as shown by Scarcello, Leone, and Palopoli [21], Krom theories enjoy the lub property. More specifically, in [21] it is shown that the *union* of any pair of good models of a Krom theory is a good model, too. This is clearly a special case of the lub property; in in [21], this special property is used to show that Curb Model Checking for propositional Krom theories is in Σ_2^P. The lub property can be further generalized. We show that following less restrictive *weak least upper bound property (weak lub property)* also leads to complexity results at the second level of the polynomial hierarchy: T has the weak least upper bound (weak lub) property, if every non-minimal good model of φ is the lub of *some* collection $\mathcal{M}$ of good models of T. The lub and the weak lub property are of interest not only in the case of propositional circumscription, but also in case of predicate logic. We therefore discuss these properties in the general setting.

The rest of this paper is organized as follows. In the next Section 2, we review some examples from [8] and give a formal definition of curbing. We then prove in Section 3 the main result stating that propositional Curb Model Checking and Curb Inference are both PSPACE-complete. In Section 4 we discuss the lub property, and the final Section 5 the weak lub property.

2   Review of Curbing

In this section, we review the concept of "good model" and give a formal definition of curbing. The presentation follows very closely the exposition in [8]; the reader familiar with [8] may skip the rest of this section.

2.1   Good Models

Let us first describe two scenarios in which an inclusive interpretation of disjunction is desirable. Models are represented by their positive atoms.

Example 1: Suppose there is a man in a room with a painting, which he hangs on the wall if he has a hammer and a nail. It is known that the man has a hammer or a nail or both. This scenario is represented by the theory T_1 in Figure 1. The desired models are h, n, and hnp, which are encircled. Circumscribing T_1 by minimizing all variables yields the two minimal models h and n (see Figure 1). Since p is false in the minimal models, circumscription tells us that the man does not hang the painting up. One might argue that the variable p should not be minimized but fixed when applying circumscription. However, starting with the model of T_1 where h, n and p are all true and then circumscribing with respect to h and p while keeping p true, we obtain the models hp and np, which are not very intuitive. If we allow p to vary in minimizing h and n, the outcome is the same as for minimizing all variables. On the other hand, the model hnp seems plausible. This model corresponds to the inclusive interpretation of the disjunction $h \vee n$.   □

Example 2: Suppose you have invited some friends to a party. You know for certain that one of Alice, Bob, and Chris will come, but you don't know whether Doug will

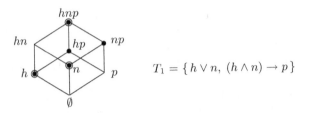

$$T_1 = \{\, h \vee n,\ (h \wedge n) \to p \,\}$$

Fig. 1. The hammer-nail-painting example

come. You know in addition the following habits of your friends. If Alice and Bob go to a party, then Chris or Doug will also come; if Bob and Chris go, then Alice or Doug will go. Furthermore, if Alice and Chris go, then Bob will also go. This is represented by theory T_2 in Figure 2. Now what can you say about who will come to the party? Look

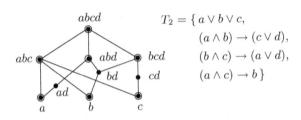

$$T_2 = \{\, a \vee b \vee c,$$
$$(a \wedge b) \to (c \vee d),$$
$$(b \wedge c) \to (a \vee d),$$
$$(a \wedge c) \to b \,\}$$

Fig. 2. The party example

at the models of T_2 in Figure 2. Circumscription yields the minimal models a, b, and c, which interpret the clause $a \vee b \vee c$ exclusively in the sense that it is minimally satisfied. However, there are other plausible models. For example, abc. This model embodies an inclusive interpretation of a and b within $a \vee b \vee c$; it is also minimal in this respect. abd is another model of this property. Similarly, bcd is a minimal model for an inclusive interpretation of b and c. The models ad, bd, and cd are not plausible, however, since a scenario in which Doug and only one of Alice, Bob or Chris are present does not seem well-supported. □

In the light of these examples, the question arises how circumscription can be extended to work satisfactory. An important insight is that such an extension must take disjunctions of positive events seriously and allow inclusive (hence non-minimal) models, even if such models contain positive information that is not contained in any minimal model. On the other hand, the fruitful principle of minimality should not be abandoned by adopting models that are intuitively not concise. The idea of curbing is based on the synthesis of both: adopt the minimal inclusive models. That is, adopt for minimal models M_1, M_2 any model M which includes both M_1 and M_2 and is a minimal such model; in other words, M is a *minimal upper bound* (*mub*) for M_1 and M_2.

To illustrate, in Example 1 hnp is a mub for h and n (notice that hn is not a model), and in Example 2 abc is a mub for a and c; abd is another one, so several mub's can exist. In order to capture general inclusive interpretations, mub's of arbitrary collections $M_1, M_2, M_3, \ldots$ of minimal models are adopted.

It appears that in general not all "good" models are obtainable as mub's of collections of minimal models. The good model $abcd$ in Example 2 shows this. It is, however, a mub of the good models a and bcd (as well as of abc and abd). This suggests that not only mub's of collections of minimal models, but mub's of any collection of good models should also be good models.

The curbing approach to extend circumscription for inclusive interpretation of disjunctions is thus the following: adopt as good models the least set of models which contains all circumscriptive (i.e. minimal) models and which is closed under including mub's. Notice that this approach yields in Examples 1 and 2 the sets of intuitively good models, which are encircled in Figs. 1 and 2.

2.2 Formal Definition of Curbing

In this section we state the formal semantical definition of good models of a first-order sentence as defined in [8].

As for circumscription, we need a language of higher-order logic (cf. [22]) over a set of predicate and function symbols, i.e. variables and constants of finite arity $n \geq 0$ of suitable type. Recall that 0-ary predicate symbols are identified with propositional symbols.

A sentence is a formula φ in which no variable occurs free; it is of order $n + 1$ if the order of any quantified symbol occurring in it is $\leq n$ [22]. We use set notation for predicate membership and inclusion. A *theory* T is a finite set of sentences. As usual, we identify a theory T with the sentence φ_T which is the conjunction $\bigwedge_{\varphi \in T} \varphi$ of all sentences in T.

A structure M consists of a nonempty set $|M|$ and an assignment $\mathcal{I}(M)$ of predicates, i.e. relations (resp. functions), of suitable type over $|M|$ to the predicate (resp. function) constants. The object assigned to constant C, i.e. the extension of C in M, is denoted by $[\![C]\!]_M$ or simply C if this is clear from the context. Equality is interpreted as identity. A model for a sentence φ is any structure M such that φ is true in M (in symbols, $M \models \varphi$). $\mathcal{M}[\varphi]$ denotes all models of φ.

Let $\mathbf{p} = p_1, \ldots, p_n$ be a list of first-order predicate constants and $\mathbf{z} = z_1, \ldots, z_m$ a list of first-order predicate or function constants disjoint with $\mathbf{p}$. For any structure M, let $\mathcal{M}_{\mathbf{p};\mathbf{z}}^M$ be the class of structures M' such that $|M| = |M'|$, and $[\![C]\!]_M = [\![C]\!]_{M'}$ for every constant C not occurring in $\mathbf{p}$ or $\mathbf{z}$. The pre-order $\leq_{\mathbf{p};\mathbf{z}}^M$ on $\mathcal{M}_{\mathbf{p};\mathbf{z}}^M$ is defined by $M_1 \leq_{\mathbf{p};\mathbf{z}}^M M_2$ iff $[\![p_i]\!]_{M_1} \subseteq [\![p_i]\!]_{M_2}$ for all $1 \leq i \leq n$. The pre-order $\leq_{\mathbf{p};\mathbf{z}}$ is the union of all $\leq_{\mathbf{p};\mathbf{z}}^M$ over all structures. We write $\mathcal{M}_{\mathbf{p}}^M$ etc. if $\mathbf{z}$ is empty; $\leq_{\mathbf{p}}^M$ and $\leq_{\mathbf{p}}$ are partial orders on $\mathcal{M}_{\mathbf{p}}^M$ resp. all structures.

The circumscription of $\mathbf{p}$ in a first-order sentence $\varphi(\mathbf{p}, \mathbf{z})$ with $\mathbf{z}$ floating is the second-order sentence [13]

$$\varphi(\mathbf{p}, \mathbf{z}) \wedge \neg \exists \mathbf{p}', \mathbf{z}'(\varphi(\mathbf{p}', \mathbf{z}') \wedge (\mathbf{p}' \subset \mathbf{p}))$$

which will be denoted by $Circ(\varphi(\mathbf{p}, \mathbf{z}))$ ($\mathbf{p}$ and $\mathbf{z}$ will be always presupposed). Here $\mathbf{p}'$, $\mathbf{z}'$ are lists of predicate and function variables matching $\mathbf{p}$ and $\mathbf{z}$ and $\mathbf{p} \subset \mathbf{p}'$ stands for

$(\mathbf{p}' \subseteq \mathbf{p}) \wedge (\mathbf{p}' \neq \mathbf{p})$, where $(\mathbf{p}' \subseteq \mathbf{p})$ is the conjunction of all $(p_i' \subseteq p_i)$, $1 \leq i \leq n$. The following is a straightforward consequence of the definitions.

Proposition 2.1. [13] $M \models Circ(\varphi(\mathbf{p}, \mathbf{z}))$ *iff* M *is* $\leq_{\mathbf{p};\mathbf{z}}$-*minimal among the models of* $\varphi(\mathbf{p}, \mathbf{z})$.

We formally define the concept of a "good" model as follows. First define the property that a set of models is closed under minimal upper bounds.

Definition 2.1. *Let* $\varphi(\mathbf{p}, \mathbf{z})$ *be a first-order sentence. A set* $\mathcal{M}$ *of models of* $\varphi(\mathbf{p}, \mathbf{z})$ *is* $\leq_{\mathbf{p};\mathbf{z}}$-*closed iff, for every* $\mathcal{M}' \subseteq \mathcal{M}$ *and any model* M *of* $\varphi(\mathbf{p}, \mathbf{z})$, *if* M *is* $\leq_{\mathbf{p};\mathbf{z}}$-*minimal among the models of* $\varphi(\mathbf{p}, \mathbf{z})$ *which satisfy* $M' \leq_{\mathbf{p};\mathbf{z}} M$ *for all* $M' \in \mathcal{M}'$ *then* $M \in \mathcal{M}$.

Clearly the set of all models is closed. Further, every closed set must contain all $\leq_{\mathbf{p};\mathbf{z}}$-minimal models of $\varphi(\mathbf{p}, \mathbf{z})$ (let $\mathcal{M}' = \emptyset$); the empty set is closed iff $\varphi(\mathbf{p}, \mathbf{z})$ has no minimal model. We define goodness as follows.

Definition 2.2. *A model* M *of* $\varphi(\mathbf{p}, \mathbf{z})$ *is good with respect to* $\mathbf{p}; \mathbf{z}$ *iff* M *belongs to the least* $\mathbf{p}; \mathbf{z}$-*closed set of models of* $\varphi(\mathbf{p}, \mathbf{z})$.

Notice that good models only exist if a unique smallest closed set exists. The latter is immediately evident from the following characterization of goodness.

Proposition 2.2 ([8]). *A model* M *of* $\varphi(\mathbf{p}, \mathbf{z})$ *is good with respect to* $\mathbf{p}; \mathbf{z}$ *iff* M *belongs to the intersection of all* $\mathbf{p}; \mathbf{z}$-*closed sets.*

In [8], it was shown how to capture goodness by a sentence $Curb(\varphi(\mathbf{p}, \mathbf{z}); \mathbf{p}, \mathbf{z})$ whose models are precisely the good models of $\varphi(\mathbf{p}, \mathbf{z})$. Similar to circumscription, $\mathbf{p}$ are the minimized predicates (here under the *inclusive* interpretation of disjunction), $\mathbf{z}$ are the floating predicates, and all other predicates are fixed. Curbing is naturally formalized as a sentence of third-order logic, given that the definition of the set of good models of a theory involves sets of sets of models. However, in [8] it was also shown that curbing can be formalized in second-order logic.

In the present paper we do not need the formal definitions of $Curb(\varphi(\mathbf{p}, \mathbf{z}); \mathbf{p}, \mathbf{z})$ in third or second order logic, but we are interested in the problems Curb Inference and Curb Model Checking as defined in the introduction.

2.3 Previous Complexity Results on Propositional Curbing

Recall that in the propositional case, a structure M is a truth-value assignment to the propositional variables. The problems *Curb Model Checking* and *Curb Inference* were described in the introduction. In [8] it was shown that both problems are in PSPACE, and in fact can be solved in quadratic space.

Two possibilities to approximate the full set of good models by a subset are discussed in [8]. The first approximation is to limit iterated inclusion of minimal upper bounds. Let us define the notion of α-goodness for ordinals α.

Definition 2.3. *A model M of $\varphi(\mathbf{p}, \mathbf{z})$ is 0-good with respect to $\mathbf{p}$ and $\mathbf{z}$, if M is $\leq_{\mathbf{p};\mathbf{z}}$-minimal among the models of φ.*

A model M of $\varphi(\mathbf{p}, \mathbf{z})$ is α-good with respect to $\mathbf{p}$ and $\mathbf{z}$, if M is a $\leq_{\mathbf{p};\mathbf{z}}$ minimal upper bound of a set of models $\mathcal{M}$ of φ, such that for each model $M' \in \mathcal{M}$ there exists an ordinal $\beta < \alpha$ such that M' is β-good w.r.t. $\mathbf{p}$ and $\mathbf{z}$.

Informally, in the approximation, one chooses only the models that are α-good for some α such that $\|\alpha\| \leq \|\delta\|$, where the ordinal δ is a limit on the depth in building minimal upper bounds. The operator corresponding to such a restricted version of curbing is denoted by $Curb^{\delta}$. Notice that circumscription appears as the case $\delta = 0$, i.e. $Curb^{0}(\varphi(\mathbf{p}, \mathbf{z}); \mathbf{p}, \mathbf{z})$ is equivalent to $Circ(\varphi(\mathbf{p}, \mathbf{z}); \mathbf{p}, \mathbf{z})$.

Concerning the computational complexity, the following was shown in [8]:

Theorem 2.1. *For $Curb^{\delta}$ (with fixed constant δ) the model checking problem is Σ_2^P complete, while inferencing is Π_2^P complete.*

Thus, the inference problem is in the propositional case for finite constant δ as easy (and as hard) as circumscription.

Another potential approximation to curbing studied in [8] is to limit the cardinality of model sets from which minimal upper bounds are formed. Intuitively, this corresponds to limiting the number of inclusively interpreted disjuncts by a cardinal $\kappa > 0$. The concept of closed$_\kappa$ set is defined by adding in the definition of closed set the condition "$\|\mathcal{M}'\| \leq \kappa$"; goodness$_\kappa$ is the relative notion of goodness.

Clearly, goodness$_1$ is equivalent to circumscription. For $\kappa \geq 2$, (i.e. $|M|$ is finite) the following result was proven:

Theorem 2.2 ([8]). *Over finite structures, for every $\kappa \geq 2$ a model of $\varphi(\mathbf{p}, \mathbf{z})$ is good$_\kappa$ with respect to $\mathbf{p}; \mathbf{z}$ iff it is good with respect to $\mathbf{p}; \mathbf{z}$.*

This result, which fails for arbitrary structures, implies a dichotomy result on the expressivity of κ-bounded disjuncts: Either we get only the minimal models, or all models obtainable by unbounded disjuncts. Thus the method of bounded disjunction is not a really useful approximation.

3   Main Result: PSPACE Completeness of Theory Curbing

In this section, we shall prove that inference as well as model checking under curbing is PSPACE-complete. Intuitively, the problems have this high complexity since checking whether a model is good requests a "proof", given by a proper collection of models, which may have non-polynomial size in general.

That such large proofs are necessary has been shown by Bodenstorfer [2]. A *support* of a model M in a collection $\mathcal{F}$ of models is a subset $\mathcal{F}' \subseteq \mathcal{F}$ containing M such that every $M' \in \mathcal{F}'$ is in $\mathcal{F}$ a mub of some models $\mathcal{M} \subseteq \mathcal{F}' \setminus \{M'\}$. Note that every minimal model $M \in \mathcal{F}$ has a support $\{M\}$ and that all models in a support are good models. Furthermore, every good model of $\mathcal{F}$ has some support.

Bodenstorfer has defined a family $\mathcal{F}_n$, $n \geq 0$, of sets of models on an alphabet of $O(n)$ propositional atoms, such that $\mathcal{F}_n$ contains exponentially many models (in n), and

$\mathcal{F}_n$ itself is the only support of the unique maximal model M_n of $\mathcal{F}_n$. Informally, $\mathcal{F}_0 = \{\{a_0\}\}$, and the family $\mathcal{F}_n$ is constructed inductively by cloning $\mathcal{F}_{n-1}$ and adding some sets which ensure that the ¡maximal model needs all models for a proof of goodness (see Figure 3).

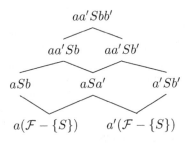

Fig. 3. Cloning a family $\mathcal{F}$ with unique maximal model S

3.1 Describing the exponential support family $\mathcal{F}_n$

We describe Bodenstorfer's family $\mathcal{F}_n$ by a formula Φ_n, such that $\mathcal{F}_n = mod(\Phi_n)$. The letters we use are $At_n = \{a_i, a_i', b_i, b_i' \mid 1 \le i \le n\} \cup \{a_0\}$. We define the formula Φ_n inductively, where we set $\Phi_0 = a_0$ and $M_0 = \{a_0\}$, and for $n > 1$:

$$\Phi_n = (M_{n-1} \wedge \gamma_n) \vee (\neg M_{n-1} \wedge \Phi_{n-1} \wedge (a_n \leftrightarrow \neg a_n') \wedge \neg b_n \wedge \neg b_n'),$$

where

$$\gamma_n = (a_n \wedge b_n \wedge \neg a_n' \wedge \neg b_n') \vee (a_n \wedge a_n' \wedge \neg b_n \wedge \neg b_n') \vee$$
$$(a_n' \wedge b_n' \wedge \neg a_n \wedge \neg b_n) \vee (a_n \wedge b_n \wedge a_n' \wedge \neg b_n') \vee$$
$$(a_n' \wedge b_n' \wedge a_n \wedge \neg b_n) \vee (a_n \wedge b_n \wedge a_n' \wedge b_n');$$

$$M_n = M_{n-1} \cup \{a_n, a_n', b_n, b_n'\}.$$

Note that the left disjunct of Φ_n gives rise to six models, which extend M_{n-1} by the following sets of atoms:
$A_{n,1} = \{a_n, b_n\}$, $A_{n,0} = \{a_n', b_n'\}$, $B_n = \{a_n, a_n'\}$, $C_{n,1} = \{a_n, a_n', b_n\}$, $C_{n,0} = \{a_n, a_n', b_n'\}$, and $D_n = \{a_n, a_n', b_n, b_n'\}$.

Informally, $A_{n,1}$ (resp., $A_{n,0}$) represents the assignment of true (resp., false) to the atom a_n. The right disjunct of Φ_n generates recursively assignments to the other atoms $a_{n-1}, \ldots, a_1$, such that certain minimal models of Φ_n represent truth assignments to the atoms $a_1, \ldots, a_n$ (see Figure 4).

Note that $M_n = M_{n-1} \cup D_n$ (i.e., all atoms are true) is, as easily seen, the unique maximal model of the formula Φ_n. The set of models of Φ_n over At_n, $mod(\Phi_n)$, defines the family $\mathcal{F}_n$ as described in [2]. Thus, each model $M \in mod(\Phi_n)$ is good, and M_n requires an exponential size support.

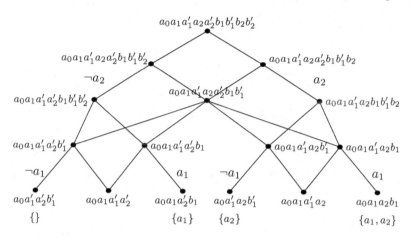

Fig. 4. The set of models $mod(\Phi_2)$

3.2   Evaluating a quantified Boolean formula on $mod(\Phi_n)$

We now show that a quantified Boolean formula (QBF)

$$F = Q_n a_n Q_{n-1} a_{n-1} \cdots Q_1 a_1 \varphi,$$

where each $Q_i \in \{\forall, \exists\}$ and φ is a Boolean formula over atoms $a_1, \ldots, a_n$, can be "evaluated" on the collection $mod(\Phi_n)$ of good models exploiting the curbing principle.

Roughly, the idea is as follows: $mod(\Phi_n)$ can be layered into n overlapping layers of models, where each layer i contains the models which are recursively generated by the left disjunct of the formula Φ_i. In each layer we have three levels of models. Neighbored layers i and $i - 1$ overlap such that the bottom level of i is the top level of $i-1$ (see Figure 5). The minimal models in $mod(\Phi_n)$ are the bottom models of layer 1, and might be considered as the top model of an artificial layer 0. Similarly, the maximal model M_n in $mod(\Phi_n)$ might be viewed as a bottom model of an artificial layer $n + 1$.

In order to "evaluate" the QBF F, we will obtain a formula $\Psi(F)$ from F by adding conjunctively a set of formulas $\Gamma(F)$ to Φ_n. Thus $\Psi(F) = \Phi_n \wedge \Gamma(F)$. The formulas in Γ will be chosen such that the overall structure of the set of good models of $\Psi(F)$ does not differ from the one of the set of models of Φ_n. In particular, each model M of Φ_n will correspond to some good model $f(M)$ of $\Psi(F)$ which augments M by certain atoms that describe the truth status of subformulas of F.

By adjoining $\Gamma(F)$ to Φ_n, we "adorn" the models in $mod(\Phi_n)$ with additional atoms which help us in evaluating the formula F along the layers. At a layer i in $mod(\Phi_n)$, we have fixed an assignment to the variables $a_{i+1}, \ldots, a_n$ already, where a_j is true if a_j occurs in the model, and a_j is false if a'_j occurs in the model, for all $j \geq i + 1$ (there are some ill-defined assignments in top elements of layer i, in which both a_{i+1} and a'_{i+1} occur; these assignment will be ignored). Then, at two sets at the bottom of the layer i which correspond to the possible extensions of the assignment to $a_{i+1}, \ldots, a_n$ by setting a_i either true (effected by the set $A_{i,1}$) or to false (by $A_{i,0}$),

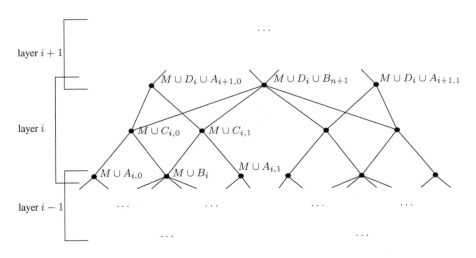

layer $i + 1$

$M \cup D_i \cup A_{i+1,0}$ $M \cup D_i \cup B_{n+1}$ $M \cup D_i \cup A_{i+1,1}$

layer i

$M \cup C_{i,0}$ $M \cup C_{i,1}$

$M \cup A_{i,0}$ $M \cup B_i$ $M \cup A_{i,1}$

layer $i - 1$

Fig. 5. Layers in $mod(\Phi_n)$

we "evaluate" the formula $Q_{i-1}a_{i-1} \cdots Q_1 a_1 \varphi(a_i, a_{i+1}, \ldots, a_n)$ where the variables $a_i, \ldots, a_n$ are fixed to the assignment. If that formula evaluates to true, then if a_i is true an atom v_i is included (resp., if a_i is false an atom v_i') at this bottom element. The quantifier Q_i is then evaluated by including in the top element "above" the two bottom sets an atom t_i if, in case of $Q_i = \exists$, either v_i or v_i' occurs in one of the two bottom elements, and in case of $Q_i = \forall$, v_i resp. v_i' occur in the bottom elements. The top element is itself a bottom element at the next layer $i + 1$, and the atom t_i is used there to see whether the formula $Q_i a_i \cdots Q_1 a_1 \varphi(a_{i+1}, \ldots, a_n)$ evaluates to true.

In what follows, we formalize this intuition. We introduce a set of new atoms $At_n' = \{v_i, v_i', t_i \mid 1 \leq i \leq n\} \cup \{t_0\}$.

The following formulas are convenient for our purpose:

$$ass_i = a_i \leftrightarrow \neg a_i', \quad 1 \leq i \leq n;$$
$$\lambda_i = (\neg b_{i+1} \vee \neg b_{i+1}') \wedge (a_{i+1} \wedge a_{i+1}' \rightarrow \neg b_{i+1} \wedge \neg b_{i+1}'), \quad 1 \leq i \leq n;$$
$$\Lambda_i = \lambda_i \wedge \neg \lambda_{i-1}, \quad 2 \leq i \leq n;$$
$$\Lambda_1 = \lambda_1.$$

Informally, ass_i tells whether the model considered assigns the atom a_i legally a truth value. The formula λ_i says that the model is at layer i or below. The formula Λ_i says that the model is at layer i. The models at the bottom of layer i which are of interest to us are those in which ass_i is true; all other models of the entire layer violate ass_i.

At layer $i \geq 1$, we evaluate the formula using the following formulas:

$$\Lambda_i \wedge ass_i \wedge t_{i-1} \wedge a_i \rightarrow v_i$$

$$\Lambda_i \wedge ass_i \wedge t_{i-1} \wedge a_i' \rightarrow v_i'$$

For $i = 1$, we add

$$\varphi \to t_0,$$

which under curbing evaluates the quantifier-free part after assigning all variables. Depending on the quantifier Q_i, we add a clause as follows. If $Q_i = \exists$, then we add

$$\Lambda_i \wedge (v_i \vee v_i') \to t_i;$$

otherwise, if $Q_i = \forall$, then we add

$$\Lambda_i \wedge v_i \wedge v_i' \to t_i.$$

For "garbage collection" of the new atoms used at lower layers, we use a formulas $trap_i$ which adds all values v_j, v_j', t_j' of lower layers to all elements of layer i which correspond to an illegal assignment to a_i:

$$trap_i = \Lambda_i \wedge \neg ass_i \to t_0 \wedge \bigwedge_{j=1}^{i-1} v_j \wedge v_j' \wedge t_j.$$

Informally, models corresponding to different extensions of an assignment will always have a mub which is upper bounded by the bottom model at layer i which is an illegal assignment.

Let the conjunction of all formulas introduced for layer i, where $1 \leq i \leq n$, be Γ_i, and let $\Gamma(F) = \bigwedge_{i=1}^{n} \Gamma_i$. Then we define

$$\Psi(F) = \Phi_n \wedge \Gamma(F).$$

Note that $\Phi(F)$ has a unique maximal model M_F, which is given by $M_F = M_n \cup \{v_i, v_i', t_i \mid 1 \leq i \leq n\}$ (i.e., all atoms are true).

Let us call a model $M \in mod(\Psi(F))$ an *assignment model*, if either $M \cap At_n = M_n$, or (b) $M \models \Lambda_i \wedge ass_i$, i.e., either M extends the maximal model of Φ_n or M is at the bottom of layer i and assigns a_i a unique truth value. In case (a), we view M at the bottom of an artificial layer $n + 1$. M represents a (partial) assignment σ_M to $a_i, \ldots, a_n$ defined by $\sigma_M(a_j) = $ true if $a_j \in M$ and $\sigma_M(a_j) = $ false if $a_j' \in M$, for all $j = i, \ldots, n$.

We show the following

Lemma 3.1. *For each model $M \in mod(\Phi_n)$, there exists a good model $f(M)$ of $mod(\Psi(F))$, such that:*

1. *$f(M) \cap At_n = M$ (i.e., $f(M)$ coincides with M on the atoms of Φ_n);*
2. *if M is an assignment model at layer $i \in \{1, \ldots, n+1\}$, then $f(M)$ contains t_{i-1} iff the formula*

$$F_i = Q_{i-1}a_{i-1}Q_{i-2}a_2 \cdots Q_1a_1\varphi(a_1, \ldots, a_{i-1}, \sigma_M(a_i), \ldots, \sigma_M(a_n))$$

is true

3. *If M is at layer $i \in \{1, \ldots, n\}$ but not an assignment model, then*

$$f(M) = \begin{cases} M \cup At'_{i-1}, & \text{if } M = M_{n-1} \cup B_n; \\ f(M_{n-1} \cup A_{n,k}) \cup f(M_{n-1} \cup B_n), & \text{if } M = M_{n-1} \cup C_{n,k}, \ k \in \{0, 1\}. \end{cases}$$

4. *$f(M_n)$ is the unique maximal good model of $\Psi(F)$, and if $Q_n = \forall$, then $t_n \in f(M_n)$ iff $f(M_n) = At_n \cup At'_n$.*

An example of the construction of $f(\cdot)$ for the formula $F = \forall a_2 \exists a_1 (a_2 \rightarrow a_1)$ is shown in Figure 6.

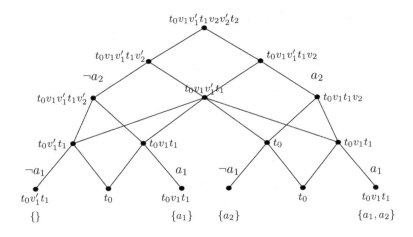

Fig. 6. Evaluating $F = \forall a_2 \exists a_1 (a_2 \rightarrow a_1)$: Extending M to $f(M) = M \cup X$ (X shown)

Proof. We first note that each model M' of $\Psi(F)$ is of the form $M \cup S$, where $M \in mod(\Phi_n)$ and $S \subseteq At'_n$, and each $M \in mod(\Phi_n)$ gives rise to at least one such M' (just add At'_n to M).

We prove the lemma showing by induction on $n \geq 0$ how to construct such a correspondence $f(M)$.

The base case $n = 0$ (in which F contains no variables and is either truth or falsity) is easy: $mod(\Phi_0) = \{\{a_0\}\}$ and, if F is truth, then $mod(\Psi(F)) = \{\{a_0, t_0\}\}$ and $f(\{a_0\}) = \{a_0, t_0\}$, and if F is falsity, then $mod(\Psi(F)) = \{\{a_0\}, \{a_0, t_0\}\}$ and $f(\{a_0\}) = \{a_0\}$.

Consider the case $n > 1$ and suppose the statement holds for $n - 1$. Let $M \in mod(\Phi_n)$. We consider two cases.

(1) $M \models \lambda_{n-1}$ and $M \not\models a_n a'_n$. Then, $M \models a_n \leftrightarrow \neg a'_n$, and either M is an assignment model at the bottom of layer n (in which case, M satisfies the left disjunct of Φ_n) or some model not at layer n (in which case M satisfies the right disjunct of M). In any case, $N = M \setminus \{a_n, a'_n, b_n, b'_n\}$ is a model of Φ_{n-1}. By the induction hypothesis, it follows that for N we have a good model $\hat{f}(N)$ of $\Psi(F')$, where $F' =$

$Q_{n-1}a_{n-1}\cdots Q_1 a_1 \varphi'$ and $\varphi' = \varphi[a_n/\top]$ (where $\top$ is truth) if $a_n \in M$ and $\varphi' = \varphi[a_n/\bot]$ (where $\bot$ is falsity) if $a'_n \in M$ (i.e., $a_n \notin M$), such that $\hat{f}(N)$ fulfills the items in the lemma. We define $f(M)$ as follows. If $N \subset M_{n-1}$, then $f(M) := M \cup \hat{f}(N)$; otherwise, if $N = M_{n-1}$, then $f(M) = M \cup f(N) \cup S_M$, where

$$
S_M = \begin{cases}
\emptyset, & \text{if } t_{i-1} \notin \hat{f}(N); \\
\{v_n, t_n\}, & \text{if } t_{i-1} \in \hat{f}(N), Q_n = \exists, \text{ and } a_i \in M; \\
\{v'_n, t_n\}, & \text{if } t_{i-1} \in \hat{f}(N), Q_n = \exists, \text{ and } a'_i \in M; \\
\{v_n\}, & \text{if } t_{i-1} \in \hat{f}(N), Q_n = \forall, \text{ and } a_i \in M; \\
\{v'_n\}, & \text{if } t_{i-1} \in \hat{f}(N), Q_n = \forall, \text{ and } a'_i \in M.
\end{cases}
$$

As easily checked, $f(M)$ is a model of $\Psi(F)$. Furthermore, $f(M)$ is either a minimal model of $\Psi(F)$ (if $n = 1$), or the mub of good models $f(M_1)$ and $f(M_2)$ such that $M_1, M_2 \in mod(\Phi_{n-1})$, $M_1, M_2 \subset M$, and M is a mub of M_1, M_2 in $mod(\Phi_{n-1})$. (If not, then $\hat{f}(N)$ were not a mub of $\hat{f}(N_1), \hat{f}(N_2)$ in $mod(\Psi(F'))$, which is a contradiction.) We can see that $f(M)$ fulfills the items 1-3 in the lemma.

(2) $M \not\models \lambda_{n-1}$ or $M \models a_n a'_n$, i.e., M is at layer n but not an assignment model at its bottom. We consider the following possible cases for M:

(2.1) $M = M_{n-1} \cup B_n$: If $n = 1$, then M is a minimal model of Φ_n, and $f(M) = M \cup \{t_0\}$ is a minimal model of $\Psi(F)$, thus $f(M)$ is a good model of $\Psi(F)$; otherwise (i.e., $n > 2$), M is a mub of any arbitrary models $M_1, M_2 \in mod(\Phi_n)$ such that M_1 contains a_n and M_2 contains a'_n, respectively, and $M_i \setminus \{a_n, a'_n, b_n, b'_n\} \subset M_{n-1}$, for $i \in \{1, 2\}$. Since, by construction, $\hat{f}(M_i) \subseteq M_{n-1} \cup At'_{n-1} =: f(M)$, this set is an upper bound of $f(M_1)$ and $f(M_2)$ in $mod(\Psi(F))$; from formula $trap_{n-1}$ it follows that $f(M)$ is a mub of $f(M_1), f(M_2)$. Thus, $f(M)$ is a good model of $\Psi(F)$.

(2.2) $M = M_{n-1} \cup C_{n,k}, k \in \{0, 1\}$: As easily checked, $f(M) = f(M_{n-1} \cup A_{n,k}) \cup f(M_{n-1} \cup B_n) (= M_{n-1} \cup B_n \cup S_{M_{n-1} \cup A_{n,k}})$ is a model of $\Psi(F)$. Since, as already shown, both $f(M_{n-1} \cup A_{n,k})$ and $f(M_{n-1} \cup B_n)$ are good models of $\Psi(F)$, clearly $f(M)$ is a mub of them and thus a good model of $\Psi(F)$.

(2.3) $M = M_n$: We define

$$
f(M) = f(M_{n-1} \cup C_{n,0}) \cup f(M_{n-1} \cup C_{n,1}) \cup \begin{cases} \{t_n\}, & \text{if } Q_n = \forall \text{ and } v_n, v'_n \in X; \\ \emptyset, & \text{otherwise.} \end{cases}
$$

Observe that $f(M) = M_n \cup At'_{n-1} \cup X$, where $X \subseteq \{v_n, v'_n, t_n\}$. Then, as easily checked, $f(M)$ is a model of $\Psi(F)$. Clearly, $f(M)$ is a mub of $f(M_{n-1} \cup C_{n,0})$ and $f(M_{n-1} \cup C_{n,1})$, and thus, $f(M)$ is a good model of $\Psi(F)$.

We now show that $f(M)$ in (2.1)–(2.3) satisfies items 1-3 in the lemma. Obviously, this is true for (2.1) and (2.2). For the case (2.3), from the definitions of $f(\cdot)$ in (1) and (2.1)–(2.2) it follows that $t_n \in f(M)$ if and only if $t_{n-1} \in f(M_{n-1} \cup A_{n,k})$ holds for for some $k \in \{0, 1\}$ if $Q_n = \exists$ and for both $k \in \{0, 1\}$ if $Q_n = \forall$. By the induction hypothesis, $t_{n-1} \in f(M_{n-1} \cup A_{n,k})$ is true iff the QBF $Q_{n-1}a_{n-1}\cdots Q_1 a_1 \varphi'$, where $\varphi' = \varphi[a_n/\top]$ if $k = 1$ and $\varphi' = \varphi[a_n/\bot]$ if $k = 0$, is true. Thus, $t_n \in f(M)$ iff the QBF F is true. Hence, $f(M)$ satisfies items 1-3 of the lemma.

As for property 4, Furthermore, in the case where $Q_n = \forall$, we have by definition of $f(M)$ that $t_n \in f(M)$ iff $f(M) = M_n \cup At'_{n-1} \cup \{v_n, v'_n, t_n\} = At_n \cup At'_n$.

Finally, it remains to show that $f(M_n)$ is the unique maximal good model of $\Psi(F)$. As easily seen, every finite propositional theory which has a unique maximal model has a unique maximal good model, thus $\Psi(F)$ has a unique maximal good model M'. From the induction hypothesis, it follows that $M_k = f(M_{n-1} \cup A_{n,k})$ is the unique maximal good model M'_k of $\Psi(F)$ such that $M' \cap At_n \subseteq M_{n-1} \cup A_{n,k}$, for $k \in \{0, 1\}$. Since $M_2 = f(M_{n-1}B_n)$ is the unique maximal good model N of $\Psi(F)$ such that $N \cap At_n \subseteq M_{n-1} \cup B_n$, we conclude from the structure of layer n, which has the lub property (see Section 4), that M' is a mub of M_0, M_1, M_2. Since, by construction, $f(M)$ is an upper bound of M_1, M_2, M_3, it follows $M' = f(M)$.

This proves that the claimed statement holds for n, and completes the induction. $\square$

We thus obtain the following result.

Theorem 3.1.   *1. Given a propositional formula G and a model M of G, deciding whether M is a good model of G is PSPACE-hard.*
   2. Given a propositional formula G and an atom p, deciding whether $Curb(G) \models p$ is PSPACE-hard.

Proof. By items 2 and 4 in Lemma 3.1, $M = At_n \cup At'_n$ is a good model of $\Psi(F)$ for a QBF $F = \forall a_n Q_{n-1} a_{n-1} \cdots Q_1 a_1 \varphi$ iff F is true. Furthermore, F is false if and only if no good model of $\Psi(F)$ contains t_n. Deciding whether any given QBF of this form is true (resp. false) is clearly PSPACE-hard, and the formula $\Psi(F)$ is easily constructed in polynomial time from F. This proves the result. $\square$

Combined with the previous results [8] that Curb Inference and Curb Model Checking are in PSPACE, we obtain the main result of this section.

Theorem 3.2.   *1. Curb Model Checking, i.e., given a propositional theory T and sets $\mathbf{p}$, $\mathbf{z}$ of propositional letters, deciding whether M is a $\mathbf{p}; \mathbf{z}$-good model of T is PSPACE-complete.*
   2. Curb-Inference, i.e., given a propositional theory T, sets $\mathbf{p}; \mathbf{z}$ of propositional letters, and a propositional formula G, deciding whether $Curb(T; \mathbf{p}, \mathbf{z}) \models G$ is PSPACE-complete.

4   The Lub Property

While curbing of general theories is PSPACE-complete, it is possible to identify specific classes of theories on which curbing has lower complexity. In this section, we identify a relevant fragment of propositional logic for which curb-inference is in Π_2^P.

Definition 4.1. *A theory T has the lub property iff every nonempty set S of good models has a least upper bound (lub) M.*

Lemma 4.1. *Let S_1, S_2 be nonempty sets of good models of theory T such that $S_1 \subseteq S_2$, and let M_1, M_2 be mubs of S_1 and S_2, respectively. If M_1 is the lub of S_1, then $M_1 \leq M_2$.*

Theorem 4.1. *If theory T has the lub property, then a model is good iff it is 1-good.*

Proof. Prove by induction on α that if model M is α-good, then it is 1-good. Obvious for $\alpha \leq 1$. Assume $\alpha > 1$. Then, M is a mub of $\mathcal{S} = \{M' : (< \alpha)\text{-good}(M'), M' \leq M\}$. Now, by the hypothesis, each $M' \in \mathcal{S}$ is the mub of some $\mathcal{S}' \subseteq \mathcal{S}$ which contains only minimal models. Let $\mathcal{S}_m$ be the minimal models from $\mathcal{S}$. If $\mathcal{S}_m = \emptyset$, then M is a minimal model and the statement holds. Else $\mathcal{S}_m$ has a lub M_m. From the unique mub property and Lemma 4.1, it follows that $M' \leq M_m$ for each $M' \in \mathcal{S}$. Thus M_m is an upper bound of $\mathcal{S}$, hence $M \leq M_m$. On the other hand, since $\mathcal{S}_m \subseteq \mathcal{S}$, it follows from Lemma 4.1 that $M_m \leq M$. Since $\leq$ is a partial order, it follows $M_m = M$. Thus M is 1-good and the statement holds. $\square$

Corollary 4.1. *For propositional theories T having the lub property,* Curb Inference *is in Π_2^P, and* Curb Model Checking *is in Σ_2^P.*

Proof. To show $Curb(T) \not\models F$, guess a model M of $Curb(T)$ such that $M \not\models F$. To verify M, guess k from $\{0, \ldots, |V|\}$, where V is the variable set, and minimal models $M_1, \ldots, M_k$ of T such that M is a mub of them. Use an NP oracle for testing whether M_i is minimal (is in coNP) and for testing if M is a mub of the M_i (is in coNP). $\square$

Notice the following characterization of lub theories.

Definition 4.2. *A theory T is mub-compact over a domain iff every good model is a mub of a finite set of good models.*

Theorem 4.2. *Let T be a mub-compact theory over some domain. Then T has the lub property iff every pair of good models has a lub.*

Proof. (Sketch) To show the *if* direction, demonstrate by induction on finite cardinality κ that every set $\mathcal{S}$ such that $\|\mathcal{S}\| \leq \kappa$ has a lub. For $\kappa \leq 2$, this is obvious. For $\kappa > 2$, let $M \in \mathcal{S}$ be a maximal element in $\mathcal{S}$. By the hypothesis, $\mathcal{S} - \{M\}$ has a lub M'. M and M' have a lub M'', which must (Lemma 4.1) be the lub of $\mathcal{S}$. $\square$

Corollary 4.2. *If the domain is finite and the models of T form an upper semi-lattice, then T has the lub property and a model is* good *iff it is 1-good.*

As already mentioned in the introduction, Scarcello, Leone, and Palopoli [21] derived complexity results for curbing Krom theories, i.e., clausal theories where each clause contains at most two literals. They showed that Curb Model Checking for propositional Krom theories is in Σ_2^P. To establish this result, they showed that the *union* of any pair of good models of a propositional Krom theory is also a good model. From this it clearly follows that propositional Krom theories enjoy the (more general) lub property. Hence their Σ_2^P upper bound and, in addition, a Π_2^P upper bound for curb inferencing can also be derived via our more general results.

5   Good Models and Least Upper Bounds

The lub property defined in Section 4 requires that *all* nonempty collections of good models of a theory have a lub. Let us weaken this property by requiring merely that for every non-minimal good model M there exists a collection of models whose lub is M.

Definition 5.1. *A theor T has the weak least upper bound (weak lub) property, if every non-minimal good model of T is the lub of some collection $\mathcal{M}$ of good models of T.*

Notice that the lub property implies the weak lub property, but not vice versa. This is shown by the following example.

Example 5.1. Suppose the models of a propositional theory T are the ones shown in Figure 7. All models are good, and $M_1 = \{a, b, c\}$, $M_2 = \{b, c, d\}$ are the lubs of

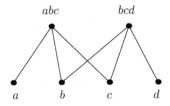

Fig. 7. The weak lub property does not imply the lub property

the collections $\{\{a\}, \{b\}, \{c\}\}$ and $\{\{b\}, \{c\}, \{d\}\}$, respectively. However, the good models $\{b\}$ and $\{c\}$ do not have a lub; thus, the theory satisfies the weak lub property but not the lub property.

Intuitively, if a theory satisfies the weak lub property, then any good model M in a collection $\mathcal{M}$ of good models can be replaced by a collection $\mathcal{M}'$ of good models whose lub is M, without affecting the mubs of the collection, i.e., $\mathcal{M}$ has the same mubs as $\mathcal{M} \setminus \{M\} \cup \mathcal{M}'$. By repeating this replacement, $\mathcal{M}$ can be replaced by a collection $\mathcal{M}^*$ of minimal models that has the same mubs as $\mathcal{M}$. This is actually the case, provided that the collection of good models has the following property.

Definition 5.2. *The collection of good models of a sentence φ is well-founded if every decreasing chain $M_0 \supseteq M_1 \supseteq \cdots$ of good models has a smallest element.*

Notice that in the context of circumscription, theories were sometimes called well-founded if every model M of a sentence φ includes a minimal model of φ [14]. That notion of well-foundedness is different from the one employed here.

The collection of good models of a theory is not necessarily well-founded, as shown by the following example.

Example 5.2. Consider the theory T on the domain $\mathbb{Z}$ of all integers:

$$\varphi = (\forall x)(p(x) \longleftrightarrow x < 0)) \vee (\exists x \geq 0)(\forall y)(p(y) \longleftrightarrow \neg(1 \leq y \leq x)) \vee$$
$$(\exists x \geq 0)(\forall y)(p(y) \longleftrightarrow (y > x) \vee (-x \leq y \leq 0))$$

Informally, T says that the numbers having property p are either all the negative numbers ($\mathbb{Z}^- = \{-1, -2, \ldots\}$), all numbers except some interval $[1, 2, \ldots, k]$, $k \geq 0$, or all nonnegative numbers where the interval $[0, k]$, $k \geq 0$, is replaced by the interval $[-k, -0]$. All models of T are good. The minimal models are $\mathbb{Z}^-$ and $N_k = (N_0 \setminus [0, k]) \cup [-k, -0]$, $k \geq 0$; every model $M_k = \mathbb{Z} \setminus [1, k]$, $k \geq 0$, is a mub of the models $\mathbb{Z}^-$ and N_k (see Figure 8). Clearly, $M_0 \supseteq M_1 \supseteq \cdots \supseteq M_i \supseteq \cdots$, $i \in \omega$, forms

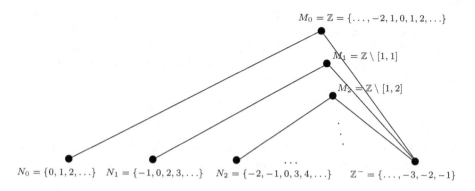

Fig. 8. A collection of good models that is not well-founded.

a decreasing chain of good models. This chain has no smallest element, and hence the collection of good models of T is not well-founded.   □

Theorem 5.1. *Let φ be a first-order sentence such that the collection of good models of φ is well-founded. If φ hast the weak lub property, then every good model is either minimal or the lub of some collection of minimal models.*

Proof. We show this by contradiction. Assume the contrary holds. Let $\mathcal{B}$ be the set of good models which are not the lub of some collection of minimal models; note that $\mathcal{B}$ is not empty. Since the collection of good models is well-founded, $\mathcal{B}$ must have a minimal element M. (To obtain such an M, construct a maximal chain in $\mathcal{B}$, and take the unique minimal element of this chain, which must exist). Since φ has the weak lub property, M is the mub of some collection $\mathcal{S}$ of good models. The definition of $\mathcal{B}$ and the weak lub property of φ imply that every $M' \in \mathcal{S}$ is the lub of a collection $\mathcal{S}_{M'}$ of minimal models. Let $\mathcal{S}'$ be the union of all these $\mathcal{S}_{M'}$. We show that M is the lub of $\mathcal{S}'$. Clearly, M is an upper bound of $\mathcal{S}'$. Assume then that M is not a minimal. Then there exists a good model $M' < M$ which is an upper bound of $\mathcal{S}'$. But this M' is also an upper bound of $\mathcal{S}$. This means that M is not a mub of $\mathcal{S}$, which is a contradiction. It follows that M is a mub of $\mathcal{S}'$. On the other hand, every upper bound M' of $\mathcal{S}'$ must satisfy $M \leq M'$.

Therefore, M is the unique mub of $\mathcal{S}'$. Consequently, M is the unique minimal upper bound of a collection of minimal models. By definition, this means $M \notin \mathcal{B}$. This is a (global) contradiction. □

The converse of this theorem (which is equivalent to the statement that a theory, if every good model is either minimal or the lub of some collection of minimal models, is well-founded) is not true. This is shown by Example 5.2. Furthermore, this theorem does not hold if the collection of good models is arbitrary. This is shown by the following example.

Example 5.3. Replace in Example 5.2 every model M_i, $i \in \omega$, by the two models $M_i^a = M_i \cup \{a\}$ and $M_i^b = M_i \cup \{b\}$ and extend the domain with the new elements a and b.

In the resulting collection of models, which is clearly axiomatizable by a first-order sentence φ, every model is good and the lub of some collection of good models (M_i^a is the lub of $\{N_i, M_{i+1}^a\}$, and M_i^b of $\{N_i, M_{i+1}^b\}$; all other models are minimal). However, no M_i^a is the lub of a collection of minimal models. Notice that each good model is the lub of two good models and 1-good. □

From Theorems 5.1 and 2.1, we immediately get the following complexity results for propositional theories.

Theorem 5.2. *For propositional theories which enjoy the weak lub property, the problem* Curb Model Checking *is in* Σ_2^P, *while the problem* Curb Inference *is in* Π_2^P.

A possible attempt to strengthen the weak-lub property is to use ordinals. Say that the collection of good models of a theory has the *inductive weak-lub property*, if every non-minimal α-good model is the lub of a collection of $(< \alpha)$-good models. Notice that collection of good models in Example 5.2 has the inductive weak-lub property (which, as a consequence, does not imply well-foundedness). However, the following result is an easy consequence of our results from above.

Theorem 5.3. *Let* φ *be a first-order sentence whose collection of good models is well-founded. Then, it has the inductive weak-lub property if and only if it has the weak-lub property.*

Proof. The *only if* direction is trivial. The *if* direction follows from Theorem 5.1. □

Acknowledgments. This work was supported by the Austrian Science Fund (FWF) Project N. Z29-INF.

References

1. bddlib. http://www.cs.cmu.edu/ modelcheck/bdd.html. 2
2. B. Bodenstorfer. How Many Minimal Upper Bounds of Minimal Upper Bounds. *Computing*, 56:171–178, 1996. 2, 7, 8
3. M. Cadoli. The Complexity of Model Checking for Circumscriptive Formulae. *Information Processing Letters*, 44:113–118, 1992. 2

4. M. Cadoli, A. Giovanardi, and M. Schaerf. An Algorithm to Evaluate Quantified Boolean Formulae. In *Proc. AAAI/IAAI-98*, pp. 262–267, 1998. 2

5. E. Chan. A Possible Worlds Semantics for Disjunctive Databases. *IEEE Trans. Knowledge and Data Engineering*, 5(2):282–292, 1993. 1, 2

6. H. Decker and J. C. Casamayor. Sustained Models and Sustained Answers in First-Order Databases. In A. Olivé, editor, *Proc. 4th International Workshop on the Deductive Approach to Information Systems and Databases (DAISD 1993), 1993, Lloret de Mar, Catalonia*, pp. 267–286, Report de recerca, LSI/93-25-R, Departament de Llenguatges i Sistemes Informatics, Universitat Politecnica de Catalunya, 1990. 1, 2

7. T. Eiter and G. Gottlob. Propositional Circumscription and Extended Closed World Reasoning are Π_2^P-complete. *Theoretical Computer Science*, 114(2):231–245, 1993. Addendum 118:315. 2

8. T. Eiter, G. Gottlob, and Y. Gurevich. Curb Your Theory! A circumscriptive approach for inclusive interpretation of disjunctive information. In R. Bajcsy, editor, *Proc. 13th Intl. Joint Conference on Artificial Intelligence (IJCAI-93)*, pp. 634–639. Morgan Kaufmann, 1993. 1, 1, 1, 2, 2, 2, 3, 3, 3, 5, 6, 6, 6, 6, 6, 7, 7, 7, 14

9. R. Feldmann, B. Monien, and S. Schamberger. A Distributed Algorithm to Evaluate Quantified Boolean Formulae. In *Proc. National Conference on AI (AAAI'00)*, Austin, Texas, 2000. AAAI Press. 2

10. A. Flögel, M. Karpinski, and H. Kleine Büning. Resolution for Quantified Boolean Formulas. *Information and Computation*, 117:12–18, 1995. 2

11. P. Liberatore. The Complexity of Iterated Belief Revision. In F. Afrati and P. Kolaitis, editors, *Proc. 6th Intl. Conference on Database Theory (ICDT-97)*, LNCS 1186, pp. 276–290, Springer, 1997. 2, 2

12. P. Liberatore. The Complexity of Belief Update. *Artificial Intelligence*, 119(1-2):141–190, 2000. 2

13. V. Lifschitz. Computing Circumscription. In *Proc. International Joint Conference on Artificial Intelligence (IJCAI-85)*, pp. 121–127, 1985. 5, 6

14. V. Lifschitz. Circumscription. In D. Gabbay, C. Hogger, and J. Robinson, editors, *Handbook of Logic in Artificial Intelligence and Logic Programming*, volume III, pages 297–352. Clarendon Press, Oxford, 1994. 1, 16

15. J. McCarthy. Circumscription – A Form of Non-Monotonic Reasoning. *Artificial Intelligence*, 13:27–39, 1980. 1

16. J. Rintanen. Improvements to the Evaluation of Quantified Boolean Formulae. In *Proc. IJCAI '99*, pp. 1192–1197. AAAI Press, 1999. 2

17. K. Ross. The Well-Founded Semantics for Disjunctive Logic Programs. In W. Kim, J.-M. Nicholas, and S. Nishio, editors, *Proc. First Intl. Conf. on Deductive and Object-Oriented Databases (DOOD-89)*, pp. 352–369. Elsevier Science Pub., 1990. 1, 2

18. K. Ross and R. Topor. Inferring Negative Information From Disjunctive Databases. *Journal of Automated Reasoning*, 4(2):397–424, 1988. 1, 2

19. C. Sakama. Possible Model Semantics for Disjunctive Databases. In W. Kim, J.-M. Nicholas, and S. Nishio, editors, *Proc. First Intl. Conf. on Deductive and Object-Oriented Databases (DOOD-89)*, pp. 337–351. Elsevier Science Pub., 1990. 1, 2

20. C. Sakama and K. Inoue. Negation in Disjunctive Logic Programs. In *Proc. ICLP-93*, Budapest, Hungary, June 1993. MIT-Press. 1, 2

21. F. Scarcello, N. Leone, and L. Palopoli. Curbing Theories: Fixpoint Semantics and Complexity Issues. In M. Alpuente and M. I. Sessa, editors, *Proc. 1995 Joint Conference on Declarative Programming (GULP-PRODE'95)*, pp. 545–554. Palladio Press, 1995. 2, 2, 3, 3, 3, 15

22. J. van Benthem and K. Doets. Higher Order Logic. In D. Gabbay and F. Guenthner, editors, *Handbook of Philosophical Logic, Vol.I*, chapter I.4, pp. 275–329. D. Reidel Pub., 1983. 5, 5

Graph Operations and Monadic Second-Order Logic: A Survey

Bruno Courcelle*

LaBRI (CNRS, UMR 5800), Université Bordeaux-I,
351 cours de la Libération,
33405 Talence, France,
courcell@labri.u-bordeaux.fr
http://dept-info.labri.u-bordeaux.fr/~courcell/ActSci.html

We handle finite graphs in two ways, as relational structures on the one hand, and as algebraic objects, i.e., as elements of algebras, based on graph operations on the other.

Graphs as relational structures

By considering a graph as a relational structure (consisting typically, of the set of vertices as domain and of a binary relation representing the edges), one can express graph properties in logical languages like First-Order Logic or fragments of Second-Order Logic. The purpose of *Descriptive Complexity* is to relate the complexity of graph properties (or more generally of properties of finite relational structures) with the syntax of their logical expressions, and to characterize complexity classes in logical terms, independently of computation models like Turing machines.

The logical expression of graph properties raises also *satisfiability problems* for specific classes of graph, namely the problems of deciding whether a given formula of a certain logical language is satisfiable by some graph belonging to a fixed class.

Monadic Second-Order Logic

As main logical language, we will consider *Monadic Second-Order Logic*, i.e., the extension of First-Order Logic with variables denoting sets of elements of the considered structures. Despite the fact that it does not correspond exactly to any complexity class, this language enjoys a number of interesting properties.

First, it is rich enough to express nontrivial graph properties like planarity, k-vertex colorability (for fixed k), connectivity, and many others (that are not expressible in First-Order Logic).

Second, it is an essential tool for studying context-free graph grammars. In particular, certain graph transformations expressible by Monadic Second-Order formulas behave very much like Rational Transductions (or Tree Transductions) used in the Theory of Formal Languages.

* This research is supported by the European Community Training and Mobility in Research network GETGRATS.

M. Parigot and A. Voronkov (Eds.): LPAR 2000, LNAI 1955, pp. 20–24, 2000.
© Springer-Verlag Berlin Heidelberg 2000

Third, the verification, optimization and counting problems, expressible in Monadic Second-Order logic are efficiently solvable for certain classes of "hierarchically structured graphs" i.e., of graphs built from finite sets of graphs by means of finitely many graph operations. (This the case of the well-known class of partial k-trees, equivalently, of graphs of tree width at most k). We will discuss this second way of handling graphs shortly. Let us precise here that a *verification* (or *model checking*) *problem* consists in testing whether a given graph from a certain class is a model of a fixed closed logical formula (here a Monadic Second-Order formula). An *optimization problem* consists in computing for a given graph (from a certain class), the minimum (or maximum) cardinality of a set of vertices satisfying a fixed formula with one free set variable, again of Monadic Second-Order Logic. The length of a shortest path between two specified vertices and the maximum size of a planar induced subgraph in a given graph are expresible in this way. A *counting problem* consists in counting the number of sets satisfying a given formula. The number of paths between two specified vertices is of this form.

Graph operations

The algebraic approach to graphs pertains to the extension to sets of finite (and even countably infinite) graphs of several notions of Formal Language Theory based on the monoid structure of words. Two basic such notions are *context-freeness* and *recognizability* (defined in terms of finite congruences). The graph operations we will consider can be seen as generalizations of the concatenation of words, or of the construction of a tree from smaller trees connected by a new root.

From an algebra of finite graphs, generated by finitely many graph operations and basic graphs, one obtains:

- a specification of its graphs by algebraic terms: this yields a linear notation for these graphs, and also a background for inductive definitions and proofs,
- a notion of *context-free graph grammar* formalized in terms of *systems of recursive set equations* having least solutions: this is, by far, the easiest and most general way to handle context-free graph grammars,
- an *algebraic notion of recognizability*, defined in terms of finite congruences; an algebraic notion is useful because there is no appropriate notion of finite-state automaton for graphs except in very special cases; however, algebraic recognizability yields finite-state tree automata processing the syntax trees of the considered graphs.

Provided the graph operations are *compatible with Monadic Second-order logic* (this notion has a precise definition), we also obtain linear algorithms for every verification, optimization or counting problem expressed in Monadic Second-Order Logic, on the graphs of the corresponding algebras. These algorithms are based on tree automata traversing the syntax trees of the given graphs, and these automata exist because Monadic Second-Order Logic is equivalent to recognizability on finite graphs.

A drawback of this theory is that we cannot handle the class of all finite graphs as a single finitely generated algebra based on operations compatible with Monadic Second-Order Logic. But this is unavoidable, unless P = NP.

Context-free graph grammars and Monadic Second-Order Logic

There exist only two classes of context-free graph grammars. They are called the *HR Grammars* (HR stands for *hyperedge replacement*) and the *VR Grammars* (VR stands for *vertex replacement*). The corresponding classes of sets of graphs are closed under graph transformations expressible in Monadic Second-Order Logic, and are generated from the set of binary trees by such transformations. Hence, the classes HR and VR are robust and have characterizations independent of any choice of graph operations. This establishes a strong connection between context-free graph grammars and Monadic Second-Order Logic, and, more generally, between the two ways we handle graphs.

Graph operations compatible with Monadic Second-Order Logic

The graph operations we know enjoying the desired "compatibility properties" are the following ones, dealing with k-graphs, i.e., with graphs the vertices of which are colored with k colors (neighbour vertices may have the same color):

(a) disjoint union of two k-graphs,
(b) uniform change of color (i.e., all vertices of the input k-graph colored by p are then colored by q), for fixed colors p, q,
(c) redefinition of the (binary) edge relation of the structure representing the input k-graph by a fixed quantifier-free formula using possibly the k color predicates,
(d) an operation that fuses all vertices of the input k-graph having color p into a single one, for fixed color p.

The edge complement is an example of an operation of type (c) (its definition needs no color predicate).

For generating graphs, the operations of the forms (c) and (d) can be replaced by operations of the restricted form:

(c') addition of edges between any vertex colored by p and any vertex colored by q,

at the cost of using (many) more colors. (See [6]).

The *clique-width* of a graph G is defined as the minimal number k of colors that can be used in an algebraic expression denoting this graph and built from one-vertex graphs and operations of the forms (a), (b), (c') (using only these k colors). This complexity measure is comparable to tree-width, but stronger in the sense that, for a set of finite graphs, bounded tree-width implies bounded clique-width, but not vice versa. (See [1,6,7]).

The HR context-free graph grammars are those defined as systems of equations using operations (a), (b) and (d). The VR context-free graph grammars are those defined as systems of equations using operations of all four types. (See [6]).

Summary of the lecture

In a first part, we will review these notions, we will give various examples of graph operations and of Monadic Second-Order graph properties ([2,5,6,7]). In a second part, we will focus our attention on the "compatibility" conditions mentioned above and on operations of type (c) and (d): we will review results from [5]. In a third part, we will present the following open problems:

1. *The parsing problem*: What is the complexity of deciding whether the clique-width of a given graph is at most k? It is polynomial for k at most 3, NP otherwise ([1]). Is it NP complete for fixed values of k? For the algorithmic applications, one needs an algorithm producing not only a "yes/no" answer but also an algebraic expression in case of "yes" answers.
2. *Alternative complexity measure*: Can one define a complexity measure *equivalent* to clique-width (equivalent in the sense that the same sets of finite graphs have bounded "width"), such that the corresponding parsing problem is polynomial for each value k?
3. *Countable graphs*: From infinite expressions using the operation of types (a), (b), (c'), one can define the clique-width of a countable graph G. It may be finite but strictly larger than the maximum clique-width of the finite induced subgraphs of G. How large can be the gap? Is there an equivalent complexity measure for which there is no gap? Preliminary results can be found in [4].
4. *An open conjecture by Seese*: If a set of finite or countable graphs has a decidable satisfiability problem for Monadic Second-Order Formulas, then it has bounded clique-width. The result of [4] reduces this conjecture to the case of sets of finite graphs, but the hard part remains; partial results have been obtained in [3].

More open problems can be found from: http://dept-info.labri.u-bordeaux.fr/~courcell/ActSci.html. A list of results on clique-width is maintained on the page: http://www.laria.u-picardie.fr/~vanherpe/cwd/cwd.html

References:

[1] **D. Corneil, M. Habib, J.M. Lanlignel, B. Reed, U. Rotics**, Polynomial time recognition of clique-width at most 3 graphs, Conference LATIN 2000, to appear.

[2] **B. Courcelle**: The expression of graph properties and graph transformations in monadic second-order logic, Chapter 5 of the "Handbook of graph grammars and computing by graph transformations, Vol. 1 : Foundations", G. Rozenberg ed., World Scientific (New-Jersey, London), 1997, pp. 313-400.

[3] **B. Courcelle**: The monadic second-order logic of graphs XIV: Uniformly sparse graphs and edge set quantifications, 2000, submitted.

[4] **B. Courcelle**: Clique-width of countable graphs: a compactness property. International Conference of Graph Theory, Marseille, 2000.

[5] **B. Courcelle, J. Makowsky**: Operations on relational structures and their compatibility with monadic second-order logic, 2000, submitted.

[6] **B. Courcelle, J. Makowsky, U. Rotics**: Linear time solvable optimization problems on certain structured graph families Theory of Computer Science (formerly Mathematical Systems Theory) **33** (2000) 125-150.

[7] **B. Courcelle, S. Olariu**: Upper bounds to the clique-width of graphs, Discrete Applied Mathematics **101** (2000) 77-114.

Full texts of submitted or unpublished papers and other references can be found from: http://dept-info.labri.u-bordeaux.fr/~courcell/ActSci.html.

Efficient First Order Functional Program Interpreter with Time Bound Certifications

Jean-Yves Marion and J.-Y. Moyen

Loria, Calligramme project, B.P. 239, 54506 Vandœuvre-lès-Nancy Cedex, France,
{marionjy,moyen}@loria.fr

Abstract. We demonstrate that the class of functions computed by first order functional programs over lists which terminate by multiset path ordering and admit a polynomial quasi-interpretation, is exactly the class of function computable in polynomial time. The interest of this result lies on (i) the simplicity of the conditions on programs to certify their complexity, (ii) the fact that an important class of natural programs is captured, (iii) potential applications for program optimisation.

1 Introduction

This paper is part of a general investigation on the implicit complexity of a specification. To illustrate what we mean, we write below the recursive rules that compute the longest common subsequence of two words. More precisely, given two strings $u = u_1 \cdots u_m$ and $v = v_1 \cdots v_n$ of $\{0,1\}^*$, a common subsequence of length k is defined by two sequences of indices $i_1 < \cdots < i_k$ and $j_1 < \cdots < j_k$ satisfying $u_{i_q} = v_{j_q}$.

$$\mathtt{lcs}(\epsilon, y) \to \mathbf{0}$$
$$\mathtt{lcs}(x, \epsilon) \to \mathbf{0}$$
$$\mathtt{lcs}(\mathbf{i}(x), \mathbf{i}(y)) \to \mathtt{lcs}(x, y) + 1$$
$$\mathtt{lcs}(\mathbf{i}(x), \mathbf{j}(y)) \to \max(\mathtt{lcs}(x, \mathbf{j}(y)), \mathtt{lcs}(\mathbf{i}(x), y)) \qquad \mathbf{i} \neq \mathbf{j}$$

The number of recursive calls is exponential because of the recomputing values. So, the execution of this specification would be unrealistic, and moreover, it is well-known that this problem is efficiently solved by a dynamic programming algorithm. Our purpose is to analyse a source program, such as the specification above, to determine the computational complexity of the function computed. The next step is to generate an efficient implementation of the function denoted by the source program. Practically, we address the question of the efficiency of a program which is crucial, in particular in the context of semi-automatic program constructions. For example, Benzinger [5] has developed a prototype which analyses the complexity of a program extracted from a Nuprl proof. The same problem motivates Crary and Weirich in [13] who introduce a type system to take into account the resources involved in a computation, see also [19]. Our approach is different because the complexity analysis must diagnose the intentional behaviour of data, and thereby help us to interpret programs more efficiently.

M. Parigot and A. Voronkov (Eds.): LPAR 2000, LNAI 1955, pp. 25–42, 2000.

In this paper, we present an interpreter which carries out online memoizing and runs in polynomial time on a non-trivial class of first-order functional programs over lists. The class of programs involved consists of programs which (i) terminate by *multiset path ordering* and (ii) admit a polynomial quasi-interpretation.

This interpreter is similar to the interpreter of cons-free WHILE programs defined by Jones [21,22,23]. (See also the recent work of Liu and Stoller [29]) This technique was initiated by Cook in [12] to demonstrate that languages which are recognised by 2-way pushdown automata are exactly polynomial time languages.

The program analysis combines and depends on one the hand on *the predicative analysis of recursion* and on the other hand on *term orderings*.

The predicative analysis of recursion was initiated by Bellantoni and Cook [3] and by Leivant [27,26] who characterised polynomial time computable functions (PTIME). The analysis permits us to separate the active data in a recursion from the others, by assigning tiers to data. In essence, those characterisations are extensional. So many polynomial time algorithms are ruled out by the mechanism of data tiering. These intentional gaps were first coped by Caseiro in [7] and then by K. Aehlig and H. Schwichtenberg [1], Bellantoni, Niggl and Schwichtenberg [4], Hofmann [17,18], and Leivant [28] in their studies on characterisations of PTIME by mean of higher type recursions.

Term orderings play a central role in proving the termination of term rewriting systems, and of the programs that we shall consider. The *Multiset path ordering* (MPO) was introduced by Plaisted [32] and Dershowitz [14]. A predicative analysis of recursion leads in [30] to the introduction of the term ordering *light multiset path ordering* ($LMPO$). It was established that $LMPO$ captures a significant class of algorithms. However, because of the underlying data-tiering, it was not permitted to iterate non-size increasing functions [17], for example.

The solution proposed in this paper is the outgrowth of the work [30] on $LMPO$[1] and includes several improvements. Firstly, the new approach does not refer to data-tiering but uses it to build a predicative analysis which leads to a time resource bound certification. Secondly, the definition of the class of programs is much simpler and is closer to programming practice than the definition of $LMPO$. Thirdly, the class of programs delineated allows iteration on non-size increasing functions and recursive templates like `lcs` above. Finally, we could use this result to analyse proof search based on ordered resolution as proposed by Basin and Ganzinger [2], for example.

In the next section we give the definition of the first order functional programming language. In Section 3, we define the multiset path ordering, MPO, and a variant called MPO'. We see that the class of functions which terminate by MPO and MPO' is the class of primitive recursive functions. In Section 4, we present the light multiset path ordering, $LMPO$. It turns out that the class of functions which terminate by $LMPO$ is exactly the class of functions which are computable in polynomial time. We sketch the proof and we present call by value interpreter which uses a cache. In Section 5, we introduce quasi-interpretation

---

[1] We presuppose no familiarity with [30]

and we state our main result. (Section 5 might be read independently from Section 4.) The last section is devoted to the proof of the main result. The proof consists to transform a program, which terminates by MPO' and admits a quasi-polynomial interpretation, into a $LMPO$-program.

2   First Order Functional Programming

We shall consider first order programs over constructors which resemble ELAN programs, or a first-order restriction of HASKELL or ML programs. Throughout the following discussion, we consider four disjoint sets $\mathcal{X}, \mathcal{F}, \mathcal{C}, \mathcal{S}$ of variables, function symbols, constructors and sorts.

2.1   Syntax of Programs

Definition 1. *The sets of terms, patterns and function rules are defined in the following way:*

$$
\begin{array}{llll}
(Constructor\ terms) & \mathcal{T}(\mathcal{C}) \ni u & ::= \mathbf{c}(u_1, \cdots, u_n) \\
(Ground\ terms) & \mathcal{T}(\mathcal{C}, \mathcal{F}) \ni s & ::= \mathbf{c}(s_1, \cdots, s_n) \mid f(s_1, \cdots, s_n) \\
(terms) & \mathcal{T}(\mathcal{C}, \mathcal{F}, \mathcal{X}) \ni t & ::= x \mid \mathbf{c}(t_1, \cdots, t_n) \mid f(t_1, \cdots, t_n) \\
(patterns) & \mathcal{P} \ni p & ::= \mathbf{c}(p_1, \cdots, p_n) \mid x \\
(rules) & \mathcal{D} \ni d & ::= f(p_1, \cdots, p_n) \to t
\end{array}
$$

where $x \in \mathcal{X}$, $f \in \mathcal{F}$, and $\mathbf{c} \in \mathcal{C}$.[2]

Definition 2. *An untyped program is a quadruplet* $\mathtt{main} = \langle \mathcal{X}, \mathcal{C}, \mathcal{F}, \mathcal{E} \rangle$ *such that:*

- $\mathcal{E}$ *is a set of $\mathcal{D}$-rules.*
- *Each variable in the right-hand side of a rule also appears in the left hand side of the same rule.*
- $\mathtt{main}$ *is the main function symbol of* $\mathtt{main}$.

The size $|t|$ of a term t is the number of symbols in t. It is defined by $|\mathbf{c}| = 1$ and $|f(t_1, \cdots, t_n)| = 1 + \sum_{i=1}^{n} |t_i|$, where $f \in \mathcal{C} \cup \mathcal{F}$.

2.2   Typing Programs

Definition 3. *Let $\mathcal{S} = \{\mathbf{s}_1, \cdots, \mathbf{s}_n\}$ be a set of sorts. The set of types built from $\mathcal{S}$ is Types $\ni \tau ::= \mathbf{s} \to \tau \mid \mathbf{s}$*

Notice that each type is of the form $\mathbf{s}_1 \to (\cdots ((\mathbf{s}_n) \to \mathbf{s}) \cdots)$ and so denotes a function. We write it as $(\mathbf{s}_1, \cdots, \mathbf{s}_n) \to \mathbf{s}$.

---

[2] We shall use type writer font for function symbol and bold face font for constructors.

Definition 4. *A typing environment is a mapping Γ from $\mathcal{X} \bigcup \mathcal{C} \bigcup \mathcal{F}$ to Types which assigns to each symbol f of arity $n > 0$ a type $\Gamma(f) = (s_1, \cdots, s_n) \rightarrow s$, and, to each symbol of arity 0 a sort.*

Definition 5. *Let Γ be a typing environment. A term $t \in T(\mathcal{C}, \mathcal{F}, \mathcal{X})$ is of type s, noted $t : s$, iff:*

- *t is a symbol of arity 0 and $\Gamma(t) = s$.*
- *$t = f(t_1, \cdots, t_n)$, $t_i : s_i$ and $\Gamma(f) = (s_1, \cdots, s_n) \rightarrow s$.*

Definition 6. *A typed program is a sextuplet main $= \langle \mathcal{X}, \mathcal{C}, \mathcal{F}, \mathcal{E}, \mathcal{S}, \Gamma \rangle$ such that $\langle \mathcal{X}, \mathcal{C}, \mathcal{F}, \mathcal{E} \rangle$ is an untyped program and for each rule $f(p_1, \cdots, p_n) \rightarrow t$, the type of $f(p_1, \cdots, p_n)$ and of t are the same with respect to Γ.*

Throughout the paper, we assume that we are talking about a typed program called *main*. We shall not explicitly mention $\langle \mathcal{X}, \mathcal{C}, \mathcal{F}, \mathcal{E}, \mathcal{S}, \Gamma \rangle$. Also, $f : \Gamma(f)$ is an abbreviation for f is of type $\Gamma(f)$.

Example 1. Given a list *list* of type List(Nat), sort(*list*) sorts the elements of *list*. The algorithm is the insertion sort. The sorts are $\mathcal{S} = \{\text{bool}, \text{Nat}, \text{List}(\alpha)\}$ where α is a type variable, *i.e.* can be instantiated by any type $s \in \mathcal{S}$. (This construction does not belong to our programming language but is admissible.) Constructors are $\mathcal{C} = \{\text{tt} : \text{bool}, \text{ff} : \text{bool}, 0 : \text{Nat}, \text{suc} : \text{Nat} \rightarrow \text{Nat}, \text{nil} : \text{List}(\alpha), \text{cons} : \alpha \times \text{List}(\alpha) \rightarrow \text{List}(\alpha)\}$.

if_then_else : bool, $\alpha, \alpha \rightarrow \alpha$

$$\text{if } \mathbf{tt} \text{ then } x \text{ else } y \rightarrow x$$
$$\text{if } \mathbf{ff} \text{ then } x \text{ else } y \rightarrow y$$

_<_ : Nat, Nat $\rightarrow$ bool

$$0 < \mathbf{suc}(y) \rightarrow \mathbf{tt}$$
$$x < \mathbf{0} \rightarrow \mathbf{ff}$$
$$\mathbf{suc}(x) < \mathbf{suc}(y) \rightarrow x < y$$

insert : Nat, List(Nat) $\rightarrow$ List(Nat)

$$\text{insert}(a, \mathbf{nil}) \rightarrow \mathbf{cons}(a, \mathbf{nil})$$
$$\text{insert}(a, \mathbf{cons}(b, l)) \rightarrow \text{if } a < b \text{ then } \mathbf{cons}(a, \mathbf{cons}(b, l))$$
$$\text{else } \mathbf{cons}(b, \text{insert}(a, l))$$

sort : List(Nat) $\rightarrow$ List(Nat)

$$\text{sort}(\mathbf{nil}) \rightarrow \mathbf{nil}$$
$$\text{sort}(\mathbf{cons}(a, l)) \rightarrow \text{insert}(a, \text{sort}(l))$$

2.3   Semantics

The set $\mathcal{E}$ of rules induces a rewrite system. We recall briefly some basic notions of rewriting theory. For further details, one might consult Dershowitz and Jouannaud's survey [15]. The rewriting relation $\rightarrow$ induced by a program *main* is defined as follows. $t \rightarrow s$ if s is obtained from t by applying one of the rules of $\mathcal{E}$. The relation $\xrightarrow{+}$ ($\xrightarrow{*}$) is the transitive (reflexive-transitive) closure of $\rightarrow$. Lastly, $t \xrightarrow{!} s$ means that $t \xrightarrow{*} s$ and s is in normal form, *i.e.* no other rule may be applied. A ground (resp. constructor) substitution is a substitution from $\mathcal{X}$ to $\mathcal{T}(\mathcal{C}, \mathcal{F})$ (resp. $\mathcal{T}(\mathcal{C})$).

Since our intention is to interpret a program by a function, an important property is that each term have a unique normal form, if it exists. Henceforth, we just consider *confluent* programs, that is programs for which the induced relation $\rightarrow$ is confluent. In particular, our definition includes orthogonal programs which are confluent [20]. A program is orthogonal if in addition there are no overlapping rules.

We now give the semantics of (confluent) programs, based on the "standard" interpretation. The domain of the interpretation is the constructor algebra $\mathcal{T}(\mathcal{C})$. Hence, we only consider normal forms in $\mathcal{T}(\mathcal{C})$ as being defined. Closed terms are interpreted by elements of the algebra $\mathcal{T}(\mathcal{C})/\mathcal{E}$, that is the initial algebra of the class of models of the set of program rules $\mathcal{E}$.

Definition 7. *The semantics of a type* s *is the set of constructor terms whose type is* s. *That is* $[\![\mathsf{s}]\!] = \{t : \mathsf{s} \mid t \in \mathcal{T}(\mathcal{C})\}$.

Definition 8. *Let main be a program and* $\Gamma(main) = (\mathsf{s}_1, \cdots, \mathsf{s}_n) \rightarrow \mathsf{s}$. *The function computed by main is* $[\![main]\!] : [\![\mathsf{s}_1]\!] \times \cdots \times [\![\mathsf{s}_n]\!] \mapsto [\![\mathsf{s}]\!]$ *which is defined as follows. For all* $u_i \in [\![\mathsf{s}_i]\!]$, $[\![main]\!](u_1, \cdots, u_n) = v$ *iff* $main(u_1, \cdots, u_n) \xrightarrow{!} v$ *and* $v \in [\![\mathsf{s}]\!]$, *otherwise* $[\![main]\!](u_1, \cdots, u_n)$ *is undefined.*

3   Termination Orderings

3.1   Multiset Path Ordering

The *Multiset Path Ordering* (MPO) is a syntactic termination ordering which was introduced by Plaisted [32] and Dershowitz [14]. MPO is widely employed to prove program terminations. We briefly describe it together with some basic notions which we shall use later on.

A multiset M is a finite mapping $M : \mathcal{T}(\mathcal{C}, \mathcal{F}, \mathcal{X}) \mapsto \mathbb{N}$ which associates to each term t the number $M(t)$ of occurrences of terms t in M. An ordering $\prec$ on terms induces an ordering $\prec^m$ on multisets.

Definition 9. $M \prec^m N$ *iff there is a term* s *such that* $M(s) > N(s)$ *then there is a term* t *such that* $s \prec t$ *and* $M(t) < N(t)$.

A precedence $\preceq_{\mathcal{F}}$ (strict precedence $\prec_{\mathcal{F}}$) is quasi-ordering (total ordering) on the set $\mathcal{F}$ of function symbols. Define the equivalence relation $\approx_{\mathcal{F}}$ as $f \approx_{\mathcal{F}} g$ iff $f \preceq_{\mathcal{F}} g$ and $g \preceq_{\mathcal{F}} f$. It is worth noting that we shall consider constructors as minimal symbols with respect to $\preceq_{\mathcal{F}}$ and all variables are incomparable.

From $\approx_{\mathcal{F}}$, we define the *permutative congruence* $\approx$ as the smallest equivalence relation on terms which satisfied $f(t_1, \cdots, t_n) \approx g(s_1, \cdots, s_n)$ if $f \approx_{\mathcal{F}} g$ and $t_i \approx s_{\pi(i)}$ for some permutation π over $\{1, \cdots, n\}$.

Definition 10. *The multiset path ordering* $\prec_{mpo}$ *is defined recursively by*

1. $s \prec_{mpo} f(\cdots, t_i, \cdots)$ *if* $f \in \mathcal{C} \cup \mathcal{F}$ *and* $s \preceq_{mpo} t_i$
2. $\mathbf{c}(s_1, \cdots, s_n) \prec_{mpo} f(t_1, \cdots, t_m)$ *if* $s_i \prec_{mpo} f(t_1, \cdots, t_m)$, *for* $i \leq n$, *and* $\mathbf{c} \in \mathcal{C}$, $f \in \mathcal{F}$. *Note that* $\mathbf{c}$ *can be a 0-ary.*
3. $g(s_1, \cdots, s_m) \prec_{mpo} f(t_1, \cdots, t_n)$, *if* $s_i \prec_{mpo} f(t_1, \cdots, t_n)$, *for* $i \leq m$, *and* $g \prec_{\mathcal{F}} f$.
4. $g(s_1, \cdots, s_m) \prec_{mpo} f(t_1, \cdots, t_n)$ *if* $\{s_1, \cdots, s_m\} \prec_{mpo}^m \{t_1, \cdots, t_n\}$ *and* $g \approx_{\mathcal{F}} f$.

where $\preceq_{mpo} = \prec_{mpo} \bigcup \approx$.

3.2   A Restriction of MPO

We introduce a termination ordering MPO' by restricting the multiset ordering of MPO. We shall see that both orderings define the class of primitive recursive functions.

Definition 11. *Let* $M = \{m_1, \cdots, m_p\}$ *and* $N = \{n_1, \cdots, n_p\}$ *be two multisets with the same number of elements,* $M \prec^{m'} N$ *iff* $M \neq N$ *and there exists a permutation* π *such that:*

- $\exists i \leq p$ *such that* $m_i \prec n_{\pi(i)}$,
- $\forall j \leq p, m_j \prec n_{\pi(j)}$ *or* $m_j \approx n_{\pi(j)}$.

Definition 12. *The ordering* MPO' *is defined by replacing* $\prec^m$ *by* $\prec^{m'}$ *in Definition 10.*

Proposition 1. $\prec_{mpo}$ *is an extension of* $\prec_{mpo'}$, *that is* $s \prec_{mpo'} t$ *implies* $s \prec_{mpo} t$.

Proof. Immediate by observing that $\prec^m$ is an extension of $\prec^{m'}$.

A program is terminating by MPO' (resp. MPO) if there is a precedence on $\mathcal{F}$ such that for each rule $l \rightarrow r$ of $\mathcal{E}$, we have $r \prec_{mpo'} l$ (resp. $\prec_{mpo}$).

Example 2.

1. By putting if_then_else $\prec_{\mathcal{F}}$ _<_ $\prec_{\mathcal{F}}$ insert $\prec_{\mathcal{F}}$ sort, we see that the program written in Example 1 terminates by MPO'.

2. The following program computes the exponential. It is convenient to identify **0** with 0 and **suc**(n) with $n + 1$. Henceforth, we write **n** instead of **suc**$^n(\mathbf{0})$. In the same way, **n** + **m** stands for **suc**$^{n+m}(\mathbf{0})$. The program is ordered by MPO' with d $\prec_{\mathcal{F}}$ exp.

d : Nat $\to$ Nat and $[\![\mathrm{d}]\!](\mathbf{n}) = \mathbf{n} + \mathbf{n}$

$$\mathbf{d(0)} \to \mathbf{0}$$
$$\mathbf{d(suc}(x)) \to \mathbf{suc(suc(d}(x)))$$

exp : Nat $\to$ Nat and $[\![\exp]\!](\mathbf{n}) = \mathbf{2^n}$

$$\mathbf{exp(0)} \to \mathbf{suc(0)}$$
$$\mathbf{exp(suc}(x)) \to \mathbf{d(exp}(x))$$

Define $\mathrm{Prog}(MPO')$ as the set of programs which terminate by MPO'. A program of $\mathrm{Prog}(MPO')$ is constructed following some templates which are defined by the ordering MPO'. It is a particular case of "un-nested multiple recursion". Typical examples of such templates are illustrated by `inf` in Example 1 and by `lcs` in Example 3(3).

Now, define SPR as the following class of programs.
A program $\langle \mathcal{X}, \mathcal{C}, \mathcal{F}, \mathcal{E} \rangle$ is in SPR if (i) there is a strict precedence $\prec_{\mathcal{F}}$ on the set $\mathcal{F}$ of defined function symbols, and (ii) each rule is of the following form:

Explicit definition assuming that for each symbol h in t, we have h $\prec_{\mathcal{F}}$ f

$$\mathbf{f}(p_1, \cdots, p_n) \to t$$

Primitive recursion assuming that g, h $\prec_{\mathcal{F}}$ f

$$\mathbf{f(0}, x_1, \cdots, x_{n-1}) \to \mathbf{g}(x_1, \cdots, x_{n-1})$$
$$\mathbf{f(suc}(t), x_1, \cdots, x_{n-1}) \to \mathbf{h}(t, x_1, \cdots, x_{n-1}, \mathbf{f}(t, x_1, \cdots, x_{n-1}))$$

It is not difficult to see that the class of functions $\mathbb{F}(SPR)$ computed by SPR is exactly the class of primitive recursive functions.

On the other hand, it is routine to check that each SPR rule is terminating by MPO'. Next, by proposition 1, each program which terminates by MPO' also terminates by MPO. Hofbauer in [16] demonstrated that functions which terminate by MPO are exactly primitive recursive functions. So as a corollary, we state

Corollary 1. *The class of functions* $\mathbb{F}(MPO')$ *computed by a confluent program which terminates by* MPO' *is exactly the class of the primitive recursive functions.*

So, we have two sets of programs, $\mathrm{Prog}(MPO')$ and $\mathrm{Prog}(SPR)$, which are extensionally equivalent, *i.e.* $\mathbb{F}(MPO') = \mathbb{F}(SPR)$. But, there are more programs in $\mathrm{Prog}(MPO')$ than in $\mathrm{Prog}(SPR)$. More importantly, Colson showed, see [11], that there is no program in SPR which computes the minimum of two integers n and m in $\inf(n, m)$ steps. However, $\_<\_$ defined in Example 1 provides such an algorithm. So, $\mathrm{Prog}(MPO')$ contains more "good" programs than SPR because more algorithmic patterns are allowed. This remark raises the question of "intentional completeness" with respect to the class of primitive recursion functions. This problems was studied by Péter in her monograph [31] by showing that many recursion schema stay within the realm of primitive recursion. Simmons [34] studied what could be the rationale behind such results. A new approach to those questions, is suggested by Cichon and Weiermann in [8], and which contains also other references.

To summarise the discussion above, termination orderings, like MPO', are a starting point for the study of intentional properties of classes of programs and the resource bounds for computing a function denoted by such a program.

4 Ordering and Feasible Computation

4.1 Data Tiering

Bellantoni and Cook [3] and Leivant [27] characterised polynomial-time functions in syntactic way, unlike characterisations *à la* Cobham [10] and Ritchie [33]. (More information may be found in Clote's survey [9].)

Their approaches are based on the predicative analysis of recursion. The rationale for this analysis is intuitively explained from the Example 2(2). The existence of the function E denoted by exp is demonstrated by an induction on natural numbers. At some point of the proof, one has to assume that $E(n)$ is defined, in order to show the existence of $E(n + 1)$. That is, one has to assume that the term exp(n) can be reduced to 2^n. This reasoning is legitimate as long as one does not care about computations and resource bounds. The predicative analysis of recursion indicates that this hypothesis is the source of trouble in studying feasible computations. For this reason, data ramification, or data tiering, is introduced to control recursion by assigning to the result of a recursion a tier which is strictly lower than the tier of the recursion parameter. Thus, it is not possible to use the result of a recursion as a recursion parameter because of the tier difference. And so, primitive recursion is tame.

4.2 Light Multiset Path Ordering

The term ordering Light multiset path ordering (LMPO) was introduced in [30] to encompass a broad class of algorithms delineating polynomial time functions. $LMPO$ is a restriction of MPO' based on the predicative analysis of recursion. For this, each argument of a function symbol has a valency which predicts its intentional use.

Definition 13. *A valency of a function symbol f of arity n is a mapping $\nu(f)$:* $\{1, \cdots, n\} \mapsto \{0, 1\}$.

The valency of a function symbol f indicates how to compare arguments of f with respect to $LMPO$. Technically, valencies are like the notion of status which was suggested by Kamin and Levy [24].

Definition 14. *Let ν be a valency function on the set $\mathcal{F}$ of function symbols. A permutation π over $\{1, \cdots, n\}$ respects the valency of f and g if the arity of f and g is n and $\nu(g, i) = \nu(f, \pi(i))$, for all $i \leq n$.*

Definition 15. *Let $\approx_{\mathcal{F}}$ be as equivalence relation on $\mathcal{F}$. A permutative congruence $\approx$ which respects the valency ν is the smallest equivalence relation which satisfies:*

1. $c(s_1, \cdots, s_n) \approx c(t_1, \cdots, t_n)$ *if* $c \in \mathcal{C}$ *and* $s_i \approx t_i$ *for all* $i \leq n$.
2. $f(t_1, \cdots, t_n) \approx g(s_1, \cdots, s_n)$ *if* $f \approx_{\mathcal{F}} g$, $t_i \approx s_{\pi(i)}$ *for some permutation* π *which respects the valency of f and g.*

Definition 16. *Let $\prec_0$ and $\prec_1$ be two term orderings. These orderings are lifted to an ordering over n-tuples of terms which respects the valency function ν on $\mathcal{F}$ as follows. $\{s_1, \cdots, s_n\} \prec^{\nu}_{g,f} \{t_1, \cdots, t_n\}$ iff there is a permutation π which respects the valency of f and g and which satisfies the condition that there is $j \leq n$ such that $\nu(g, j) = 1$ and $s_j \prec_1 t_{\pi(j)}$, and for all $i \leq n$, $s_i \prec_{\nu(g,i)} t_{\pi(i)}$ or $s_i \approx_{\mathcal{F}} t_{\pi(i)}$.*

Definition 17. *Let $\preceq_{\mathcal{F}}$ be a precedence and ν be a valency function on $\mathcal{F}$. The light multiset path ordering is a pair $(\prec_k)_{k=0,1}$ of orderings which is recursively defined on $\mathcal{T}(\mathcal{C}, \mathcal{F}, \mathcal{X})$ as follows:*

1. $s \prec_k c(\cdots, t_i, \cdots)$ *if* $s \preceq_k t_i$.
2. $s \prec_k f(\cdots, t_i, \cdots)$ *if* $s \preceq_{\nu(f,i)} t_i$ *and* $k \leq \nu(f, i)$.
3. $c(s_1, \cdots, s_n) \prec_k f(t_1, \cdots, t_m)$ *if* $s_i \prec_k f(t_1, \cdots, t_m)$, *for each* $i \leq n$, *Note that c can be a 0-ary.*
4. $g(s_1, \cdots, s_n) \prec_k f(t_1, \cdots, t_m)$ *if* $s_i \prec_{\max(k, \nu(g,i))} f(t_1, \cdots, t_m)$, *for each* $i \leq n$ *and if* $g \prec_{\mathcal{F}} f$.
5. $g(s_1, \cdots, s_n) \prec_0 f(t_1, \cdots, t_n)$ *if* $\{s_1, \cdots, s_n\} \prec^{\nu}_{g,f} \{t_1, \cdots, t_n\}$, *and* $g \approx_{\mathcal{F}} f$.

Where $\preceq_k = \prec_k \bigcup \approx$, $c \in \mathcal{C}$, *and* $f \in \mathcal{F}$.

Definition 18. *A program main is terminating by $LMPO$ if there is a valency function ν such that for each rule $(l \rightarrow r) \in \mathcal{E}$, we have $r \prec_0 l$. A $LMPO$-program is a program which terminates by $LMPO$.*

Example 3. Throughout, we use the following convention to indicate the valency of a function argument. The arguments of valency 1 are the first arguments of f and the last ones are of valency 0. There are separated by ';', like the normal/safe arguments of [3].

1. The following program is terminating using either MPO, MPO' or $LMPO$ by setting $\text{add} \prec_{\mathcal{F}} \text{mult}$.

$\text{add} : \text{Nat}_1, \text{Nat}_0 \rightarrow \text{Nat}$ and $[\![\text{add}]\!](\mathbf{x}, \mathbf{y}) = \mathbf{x} + \mathbf{y}$

$$\text{add}(\mathbf{0}; y) \rightarrow y$$
$$\text{add}(\mathbf{suc}(x); y) \rightarrow \mathbf{suc}(\text{add}(x; y))$$

$\text{mult} : \text{Nat}_1, \text{Nat}_1 \rightarrow \text{Nat}$ and $[\![\text{mult}]\!](\mathbf{x}, \mathbf{y}) = \mathbf{x} \times \mathbf{y}$

$$\text{mult}(\mathbf{0}, y;) \rightarrow \mathbf{0}$$
$$\text{mult}(\mathbf{suc}(x), y;) \rightarrow \text{add}(y; \text{mult}(x, y;))$$

2. The exponential program given in Example 2(2) is terminating using either MPO or MPO' but not with LMPO. Indeed, the second equation will force $\nu(\mathbf{d}, 1) = 1$ and so the fourth equation cannot be ordered because $\exp(x) \not\prec_1 \exp(\mathbf{suc}(x))$.

3. Take again the problem of computing the longest common subsequence of two words. To write a $LMPO$-program to solve this problem, we encode binary words by the sort word generated by the constructors $\{\epsilon : \text{word}, \mathbf{a} : \text{word} \rightarrow \text{word}, \mathbf{b} : \text{word} \rightarrow \text{word}\}$. Then, we write the recursive solution of the problem.

$\text{max} : \text{word}_1, \text{Nat}_0, \text{Nat}_0 \rightarrow \text{Nat}$

$$\text{max}(x; n, \mathbf{0}) \rightarrow n$$
$$\text{max}(x; \mathbf{0}, m) \rightarrow m$$
$$\text{max}(\mathbf{i}(x); \mathbf{suc}(n), \mathbf{suc}(m)) \rightarrow \mathbf{suc}(\text{max}(x; n, m)) \qquad \mathbf{i} \in \{\mathbf{a}, \mathbf{b}\}$$

$\text{lcs} : \text{word}_1, \text{word}_1 \rightarrow \text{Nat}$

$$\text{lcs}(x, \epsilon;) \rightarrow \mathbf{0}$$
$$\text{lcs}(\epsilon, y;) \rightarrow \mathbf{0}$$
$$\text{lcs}(\mathbf{i}(x), \mathbf{i}(y);) \rightarrow \mathbf{suc}(\text{lcs}(x, y;))$$
$$\text{lcs}(\mathbf{i}(x), \mathbf{j}(y);) \rightarrow \text{max}(\mathbf{i}(x); \text{lcs}(x, \mathbf{j}(y);), \text{lcs}(\mathbf{i}(x), y;)) \quad \mathbf{i} \neq \mathbf{j}$$

The program is ordered by $\prec_0$ by putting $\text{max} \prec_{\mathcal{F}} \text{lcs}$. It is worth noting that it is necessary that max has an extra argument of valency 1. The definition is unnatural. Similar defect is observable with the insertion sort in Example 1. The reason was clearly analysed by Caseiro [7], Hofmann [17] and K. Aehlig and H. Schwichtenberg [1]. Data-tiering prevents iteration on a result (*i.e.* on the value $\text{lcs}(x, \mathbf{j}(y);)$ of the example above) previously obtained by iteration, even if the function (*i.e.* max) does not increase the size of the output. We remedy this defect in Section 5.

4.3 LMPO-Interpreter

Definition 19. *A sort* $s \in \mathcal{S}$ *is simple if for each constructor* $c \in \mathcal{C}$ *of type* $(s_1, \cdots, s_n) \rightarrow s$, *the sort* s *appears at most once in* $(s_1, \cdots, s_n)$. *A LMPO-program over a simple constructor signature is a LMPO-program in which each sort is simple.*

The sorts `bool`, `Nat`, and `List`(α) where α is simple, are simple. The sort of binary trees is not simple.

Theorem 1. *Each LMPO-program over a simple constructor signature is computable in polynomial time, and conversely each polynomial time function is computed by a LMPO-program over a simple constructor signature.*

Proof (Sketch of proof). The detailed proof can be found in [30]. The second implication is proved by programming each function of the class B of Bellantoni and Cook [3], by merely identifying safe (normal) arguments with arguments of valency 1 (resp. 0).

Conversely, a $LMPO$-program is evaluated by call-by-value interpreter with a cache memory. Whenever the interpreter wants the value of a call $f(t_1, \cdots, t_n)$, it looks first in the cache. If $f(t_1, \cdots, t_n)$ is not in the cache, then it computes $f(t_1, \cdots, t_n)$ and stores the result in the cache, so that the next call to $f(t_1, \cdots, t_n)$ will be in the cache. The algorithm of the interpreter is presented in Figure 1.

The time-bound of the interpreter is obtained by establishing that the number of recursive calls is bounded by a polynomial in the height of the inputs. Since inputs are of simple sorts, the size of an input is at most quadratic in the height. Consequently, the run-time of the interpreter is bounded by a polynomial in the size of the inputs.

It is necessary to memorise intermediate values because the length of a derivation can be exponential. Indeed, the length of a derivation of `lcs`, in Example 3(3), is exponential. Consequently, the interpreter can transform a program into an exponentially faster one.

5   Quasi-Interpretation and MPO'

5.1 Polynomial Quasi-Interpretation

Definition 20. *Let* $f \in \mathcal{F} \bigcup \mathcal{C}$ *be either a function symbol or a constructor of arity* n. *A quasi-interpretation of* f *is a mapping* $\{f\} : \mathbb{N}^n \rightarrow \mathbb{N}$ *which satisfies*

1. $\{f\}$ *is (non-strictly) increasing with respect to each argument.*
2. $\{f\}(X_1, \cdots, X_n) \geq X_i$, *for each* $1 \leq i \leq n$.
3. $\{c\} > 0$ *for each 0-ary symbol* c.

We extend a quasi interpretation $\{\bullet\}$ to terms as follows :

$$\{f(t_1, \cdots, t_n)\} = \{f\}(\{t_1\}, \cdots, \{t_n\})$$

$$\frac{\sigma(x) = v}{\mathcal{E}, \sigma \vdash \langle C, x \rangle \to \langle C, v \rangle}$$

$$\frac{\mathbf{c} \in \mathcal{C} \quad \mathcal{E}, \sigma \vdash \langle C_{i-1}, t_i \rangle \to \langle C_i, v_i \rangle}{\mathcal{E}, \sigma \vdash \langle C_0, \mathbf{c}(t_1, \cdots, t_n) \rangle \to \langle C_n, \mathbf{c}(v_1, \cdots, v_n) \rangle}$$

$$\frac{\mathbf{f} \in \mathcal{F} \quad \mathcal{E}, \sigma \vdash \langle C_{i-1}, t_i \rangle \to \langle C_i, v_i \rangle \quad (\mathbf{f}(v_1, \cdots, v_n), v) \in C_n}{\mathcal{E}, \sigma \vdash \langle C_0, \mathbf{f}(t_1, \cdots, t_n) \rangle \to \langle C_n, v \rangle}$$

$$\frac{\mathcal{E}, \sigma \vdash \langle C_{i-1}, t_i \rangle \to \langle C_i, v_i \rangle \quad \mathbf{f}(p) \to r \in \mathcal{E} \quad p_i \sigma' = v_i \quad \mathcal{E}, \sigma' \vdash \langle C_n, r \rangle \to \langle C, v \rangle}{\mathcal{E}, \sigma \vdash \langle C_0, \mathbf{f}(t_1, \cdots, t_n) \rangle \to \langle C \cup (\mathbf{f}(v_1, \cdots, v_n), v), v \rangle}$$

Fig. 1. Call by value $LMPO$-interpreter with a cache.

Given a set of equations $\mathcal{E}$ and a ground substitution σ, $t \xrightarrow{!} v$ *using the cache C and obtaining the cache C' is written:*

$$\mathcal{E}, \sigma \vdash \langle C, t \rangle \to \langle C', v \rangle$$

$\mathcal{E}, \emptyset \vdash \langle \emptyset, \mathbf{f}(t_1, \cdots, t_n) \rangle \to \langle C, v \rangle$ means that $[\![\mathbf{f}]\!](t_1, \cdots, t_n) = v$

Definition 21. $\{\bullet\}$ *is a quasi-interpretation of a program main if for each rule $l \to r \in \mathcal{E}$ and for each constructor substitution σ, $\{r\sigma\} \leq \{l\sigma\}$.*

Definition 22. *A program main admits a polynomial quasi interpretation $\{\bullet\}$, if $\{\bullet\}$ is bounded by a polynomial and each constructor $\mathbf{c} \in \mathcal{C}$ is interpreted by $\{\mathbf{c}\}(X_1, \cdots, X_n) = \sum_{i=1}^{n} X_i + a$ where $a > 0$.*

Clearly, a quasi-interpretation is not sufficient to assure the termination of a program. For example, take the rule $\mathbf{f}(x) \to \mathbf{f}(x)$ which has a quasi-interpretation but does not terminate. It is worth noting that a quasi-interpretation does not bound the size of terms involved in a derivation.

Remark 1. Termination proofs by polynomial interpretations were proposed by Lankford [25]. Termination is assured by requiring that $\{\bullet\}$ is strictly increasing and also that for each rule, $\{r\sigma\} < \{l\sigma\}$. Bonfante, Cichon, Marion and Touzet in [6] established that the complexity of functions computed by programs admitting a polynomial interpretation termination proof, depends on constructor interpretations. In particular, when constructors are interpreted by $\{\mathbf{c}\}(X_1, \cdots, X_n) = \sum_{i=1}^{n} X_i + a$ (as we do), the functions computed by such systems are exactly the functions computed in polynomial-time.

Let us end with a Lemma which we shall use later on.

Lemma 1. *Let t and s be two constructor terms of $\mathcal{T}(\mathcal{C})$. If $t \preceq_k s$ then $\{t\} \leq \{s\}$, for $k = 0, 1$.*

Proof. By induction on the size of s.

5.2   Main Result

Definition 23. *The class of functions $\mathbb{F}^{poly}(MPO')$ is the class of functions computed by a program which terminates by MPO' and admits a polynomial quasi-interpretation.*

Theorem 2. *The class $\mathbb{F}^{poly}(MPO')$ is exactly the class of functions computed in polynomial-time.*

Proof. In [3], Bellantoni and Cook demonstrated that the class of B functions is exactly the class of polynomial time functions. It is easy to see that each function of the class B is ordered by MPO'. Then, Lemma 4.1 of [3] provides a quasi-interpretation. It follows that each polynomial time function is in $\mathbb{F}^{poly}(MPO')$.

Conversely, each program is executed by the $LMPO$-interpreter described in Figure 1. Let us give an informal account on how to establish that the runtime is polynomially bounded. We lift each source program *main* to a program $\lceil main \rceil$ which is ordered by $LMPO$. Intuitively, the lifting mapping $\lceil \bullet \rceil$ is akin to a polynomial reduction from MPO'-programs to $LMPO$-programs. Each rule of the program $\lceil main \rceil$ has two new arguments. The first argument plays the role of a virtual clock. The second one is a copy of the initial setting of the clock. The semantics of the source program *main* is preserved only if the virtual clock is set to a value greater than the quasi-interpretation. We see that if the quasi-interpretation is bounded by a polynomial, then the runtime of the execution of *main* is also polynomial by Theorem 1. The details of the proof are in Section 6.

Example 4.

1. Consider again the problem which computes the longest subsequence between two words. The rules, given in the introduction, which defined `lcs` are ordered by MPO' and possess the following polynomial quasi-interpretation: $\{0\} = \{\epsilon\} = 1$, $\{\mathbf{suc}\}(X) = \{\mathbf{a}\}(X) = \{\mathbf{b}\}(X) = X + 1$, $\{\mathbf{max}\}(X, Y) = \max(X, Y)$, $\{\mathbf{lcs}\}(X, Y) = \max(X, Y)$. So, this program is evaluated in polynomial time. Finally, notice also that `lcs`-rules have no polynomial interpretation.

2. The insertion sort in Example 1 terminates by MPO' and admits the following polynomial quasi-interpretation :

$$\{\mathbf{tt}\} = \{\mathbf{ff}\} = \{\mathbf{nil}\} = \{\mathbf{0}\} = 1$$
$$\{\mathbf{suc}\}(X) = \{\mathbf{cons}\}(X) = X + 1$$
$$\{\mathbf{if\_then\_else}\}(X, Y, Z) = \max(X, Y, Z)$$
$$\{\_\mathbf{<}\_\}(X, Y) = \max(X, Y)$$
$$\{\mathbf{insert}\}(X, Y) = X + Y + 1$$
$$\{\mathbf{sort}\}(X) = X$$

Notice that $LMPO$ is not able to order the rules of the insertion sort, because of the data-tiering.

3. The definition of the exponential in Example 2(2) has no polynomial quasi-interpretation, because $\{\mathbf{d}\}(X) = 2 \times X$ and, we obtain the inequality $\{\mathtt{exp}\}(X + c) \geq 2 \times \{\mathtt{exp}\}(X)$ whose solution is exponential.

6   Runtime Analysis

Without loss of generality, we assume that each program contains the sort Nat.

6.1   Program Lifting

Let $\preceq_{\mathcal{F}}$ be a precedence over $\mathcal{F}$. We define a lift mapping $\uparrow\bullet\uparrow$ which transforms each program $main = \langle \mathcal{X}, \mathcal{C}, \mathcal{F}, \mathcal{E} \rangle$ into a program $\uparrow main\uparrow = \langle \uparrow\mathcal{X}\uparrow, \mathcal{C}, \uparrow\mathcal{F}\uparrow, \uparrow\mathcal{E}\uparrow \rangle$ where

- $\uparrow\mathcal{X}\uparrow = \mathcal{X} \cup \{X, Y\}$ where X and Y are two variables which are not in $\mathcal{X}$,
- The set of function symbols $\uparrow\mathcal{F}\uparrow$ consists of a new symbol $\mathbf{f}'$ of arity $n + 2$, for each symbol $\mathbf{f} \in \mathcal{F}$ of arity $n > 0$.
- The set of rules $\uparrow\mathcal{E}\uparrow$ contains a rule

$$\mathbf{f}'(\mathbf{suc}(X), Y; p_1, \cdots, p_n) \rightarrow \uparrow s \uparrow^{\mathtt{f}}_{X,Y}$$

for each rule $\mathbf{f}(p_1, \cdots, p_n) \rightarrow s$ of $\mathcal{E}$ and $\uparrow s \uparrow^{\mathtt{f}}_{X,Y}$ is defined below.

The lift mapping $\uparrow\bullet\uparrow^{\mathtt{f}}_{X,Y}$ over terms is defined as follows:

- $\uparrow x \uparrow^{\mathtt{f}}_{X,Y} = x$ for all $x \in \mathcal{X}$.
- $\uparrow \mathbf{c}(t_1, \cdots, t_m) \uparrow^{\mathtt{f}}_{X,Y} = \mathbf{c}(\uparrow t_1 \uparrow^{\mathtt{f}}_{X,Y}, \cdots, \uparrow t_m \uparrow^{\mathtt{f}}_{X,Y})$ if $\mathbf{c} \in \mathcal{C}$.
- $\uparrow \mathbf{g}(t_1, \cdots, t_m) \uparrow^{\mathtt{f}}_{X,Y} = \mathbf{g}'(Y, Y, \uparrow t_1 \uparrow^{\mathtt{f}}_{X,Y}, \cdots, \uparrow t_m \uparrow^{\mathtt{f}}_{X,Y})$, if $\mathbf{g} \prec_{\mathcal{F}} \mathbf{f}$
- $\uparrow \mathbf{g}(t_1, \cdots, t_n) \uparrow^{\mathtt{f}}_{X,Y} = \mathbf{g}'(X, Y, \uparrow t_1 \uparrow^{\mathtt{f}}_{X,Y}, \cdots, \uparrow t_n \uparrow^{\mathtt{f}}_{X,Y})$, if $\mathbf{g} \approx_{\mathcal{F}} \mathbf{f}$

Proposition 2. Let $main$ be a MPO'-program, $\uparrow main\uparrow$ is a $LMPO$-program.

Proof. We define the valency function ν over $\uparrow\mathcal{F}\uparrow$ by $\nu(\mathbf{f}', 1) = \nu(\mathbf{f}', 2) = 1$ and $\nu(\mathbf{f}', 3 + i) = 0$, $i < n$, for each symbol $\mathbf{f} \in \mathcal{F}$ of arity $n > 0$.

Now, consider a rule $t = \mathbf{f}(p_1, \cdots, p_n) \rightarrow s$ of $\mathcal{E}$. We show that if $s \prec_{mpo'} t$ then $\uparrow s \uparrow^{\mathtt{f}}_{X,Y} \prec_0 \mathbf{f}'(\mathbf{suc}(X), Y; p_1, \cdots, p_n) = T$. The proof goes by induction on the size of the right hand side term s.

If $s = \mathbf{c}(u_1, \cdots, u_m)$, $\mathbf{c} \in \mathcal{C}$, then $\uparrow s \uparrow^{\mathtt{f}}_{X,Y} = \mathbf{c}(\uparrow u_1 \uparrow^{\mathtt{f}}_{X,Y}, \cdots, \uparrow u_m \uparrow^{\mathtt{f}}_{X,Y})$. By the MPO' definition, $u_i \prec_{mpo'} t$ for each $1 \leq i \leq m$. So, by induction hypothesis, $\uparrow u_i \uparrow^{\mathtt{f}}_{X,Y} \prec_0 T$. So, by definition of LMPO (rule 3), $\uparrow s \uparrow^{\mathtt{f}}_{X,Y} \prec_0 T$.

If $s = \mathbf{g}(u_1, \cdots, u_m)$, where $\mathbf{g} \in \mathcal{F}$ and $\mathbf{g} \prec_{\mathcal{F}} \mathbf{f}$, then we have $\uparrow s \uparrow^{\mathtt{f}}_{X,Y} = \mathbf{g}'(Y, Y; \uparrow u_1 \uparrow^{\mathtt{f}}_{X,Y}, \cdots, \uparrow u_m \uparrow^{\mathtt{f}}_{X,Y})$. By the MPO' definition, $u_i \prec_{mpo'} t$ for each $1 \leq i \leq m$. So, by induction hypothesis, $\uparrow u_i \uparrow^{\mathtt{f}}_{X,Y} \prec_0 T$. We have $Y \prec_1 T$. It follows by definition of LMPO (rule 4) that $\uparrow s \uparrow^{\mathtt{f}}_{X,Y} \prec_0 T$.

If $s = g(u_1, \cdots, u_n)$, where $g \in \mathcal{F}$ and $g \approx_{\mathcal{F}} f$, then we have $\lceil s \rceil_{X,Y}^f = g'(X, Y; \lceil u_1 \rceil_{X,Y}^f, \cdots, \lceil u_n \rceil_{X,Y}^f)$. By definition of MPO', there exists a permutation π such that $u_i \preceq_{mpo'} p_{\pi(i)}$ for each $1 \leq i \leq n$. Since u_i's and p_i's are terms of $\mathcal{T}(\mathcal{T}(\mathcal{C}, \mathcal{X}))$, we have $\lceil u_i \rceil_{X,Y}^f \preceq_0 p_{\pi(i)}$. Next, we have $X \prec_1 \mathbf{suc}(X)$. The definition of LMPO (rule 5) gives: $\lceil s \rceil_{X,Y}^f \prec_0 T$.

Since each rule of *main* is ordered by MPO', we have shown that each rule of $\lceil main \rceil$ is ordered by $LMPO$, which leads to the conclusion.

6.2 Quasi-Interpretations are Virtual Clocks

We define $\lfloor \bullet \rfloor$ as the inverse of the term lifting $\lceil \bullet \rceil_{X,Y}^f$ as follows.

- $\lfloor x \rfloor = x$ if $x \in \mathcal{X}$.
- $\lfloor c(t_1, \cdots, t_n) \rfloor = \mathbf{c}(\lfloor t_1 \rfloor, \cdots, \lfloor t_n \rfloor)$ if $\mathbf{c} \in \mathcal{C}$.
- $\lfloor f'(s, u; t_1, \cdots, t_n) \rfloor = \mathbf{f}(\lfloor t_1 \rfloor, \cdots, \lfloor t_n \rfloor)$ if $\mathbf{f} \in \mathcal{F}$.

We see that $\lfloor \lceil t \rceil_{X,Y}^f \rfloor = t$.

Lemma 2. *Let main* $= \langle \mathcal{X}, \mathcal{C}, \mathcal{F}, \mathcal{E} \rangle$ *be a MPO'-program which admits a quasi-interpretation* $\{\bullet\}$. *Let d be the greatest arity of a function symbol in main. Let t be a ground term of* $\mathcal{T}(\mathcal{C}, \lceil \mathcal{F} \rceil)$.

Suppose that for each subterm $f'(\mathbf{h_1}, \mathbf{h_2}; t_1, \cdots, t_n)$ *of t,*
we have (i) $h_1 \geq \sum_{i=1}^n \{\lfloor t_i \rfloor\}$ and (ii) $h_2 \geq d \times \{f(\lfloor t_1 \rfloor, \cdots, \lfloor t_n \rfloor)\}$.
If $\lfloor t \rfloor \to u$ then there exists a term v such that

1. $\lfloor v \rfloor = u$ *and* $t \to v$.
2. *For each subterm* $g'(\mathbf{h'_1}, \mathbf{h'_2}; v_1, \cdots, v_m)$ *of v,*
   we have (i) $h'_1 \geq \sum_{i=1}^m \{\lfloor v_i \rfloor\}$ and (ii) $h'_2 \geq d \times \{g(\lfloor v_1 \rfloor, \cdots, \lfloor v_m \rfloor)\}$.

Proof. $\lfloor t \rfloor \to u$ means that there are a rule $e \equiv f(p_1, \cdots, p_n) \to r$ in $\mathcal{E}$, a subterm $f(\lfloor t_1 \rfloor, \cdots, \lfloor t_n \rfloor)$ of $\lfloor t \rfloor$ and a constructor substitution σ such that $f(\lfloor t_1 \rfloor, \cdots, \lfloor t_n \rfloor) = f(p_1, \cdots, p_n)\sigma$ and $u = \lfloor t \rfloor [r\sigma / f(\lfloor t_1 \rfloor, \cdots, \lfloor t_n \rfloor)]$.

The rule e is lifted to $\lceil e \rceil \equiv f'(\mathbf{suc}(X), Y; p_1, \cdots, p_n) \to \lceil r \rceil_{X,Y}^f$. There are $\mathbf{h_1}$, and $\mathbf{h_2}$ such that $\lfloor f'(\mathbf{h_1}, \mathbf{h_2}; t_1, \cdots, t_n) \rfloor = f(\lfloor t_1 \rfloor, \cdots, \lfloor t_n \rfloor)$. We have $h_1 > 0$ because quasi-interpretations are always strictly positive. It follows that there is a constructor substitution σ' which extends σ and such that $\sigma'(X) = \mathbf{h_1} - 1$ and $\sigma'(Y) = \mathbf{h_2}$. Therefore, we apply $\lceil e \rceil$ to t and we obtain $v = t[\lceil r \rceil_{X,Y}^f \sigma' / f'(\mathbf{h_1}, \mathbf{h_2}; t_1, \cdots, t_n)]$. We get $\lfloor v \rfloor = u$.

It remains to establish (2). Take a subterm $g'(\mathbf{h'_1}, \mathbf{h'_2}; v_1, \cdots, v_m)$ of v. If this subterm remains as it was before the reduction step, then (2) is satisfied. Otherwise, it is a subterm of $(\lceil r \rceil_{X,Y}^f)\sigma'$, and we have $\mathbf{h'_2} = \sigma'(Y) = \mathbf{h_2}$.

Because the quasi-interpretation is increasing and the Lemma assumptions,

$$h'_2 = h_2 \geq d \times \{f(\lfloor t_1 \rfloor, \cdots, \lfloor t_n \rfloor)\} \geq d \times \{g(\lfloor v_1 \rfloor, \cdots, \lfloor v_m \rfloor)\}$$

So the condition (ii) is satisfied.

Next, if $\mathbf{g} \prec_{\mathcal{F}} \mathbf{f}$, we have $\mathbf{h}'_1 = \sigma'(Y) = \mathbf{h_2}$. The condition (2) on the definition of a quasi-interpretation enforces

$$h'_1 = h_2 = d \times \{\mathbf{g}(\lfloor v_1 \rfloor, \cdots, \lfloor v_m \rfloor)\} \geq \sum_{i=1}^{m} \{\lfloor v_i \rfloor\}$$

Lastly, if $\mathbf{g} \approx_{\mathcal{F}} \mathbf{f}$ then $h'_1 = h_1 - 1$. Since the rule e is ordered by MPO', there is a permutation π such that:

- for all $1 \leq i \leq n$, $v_i \preceq_{mpo'} t_{\pi(i)}$.
- There exists a j such that $v_j \prec_{mpo'} t_{\pi(j)}$.

By Lemma 1, $\{\lfloor v_i \rfloor\} \leq \{\lfloor t_{\pi(i)} \rfloor\}$ for all $1 \leq i \leq n$ and $\{\lfloor v_j \rfloor\} < \{\lfloor t_{\pi(j)} \rfloor\}$. So we have, $h'_1 = h_1 - 1 \geq \sum_{i=1}^{n} \{\lfloor t_i \rfloor\} - 1 \geq \sum_{i=1}^{n} \{\lfloor v_i \rfloor\}$. So (i) is proved.

Proposition 3. *Let main be a MPO'-program which admits a quasi interpretation and let $t_1, \cdots, t_n$ be constructor terms of $\mathcal{T}(\mathcal{C})$. Let d be the greatest arity of a function symbol in $\mathcal{F}$.*

Suppose that $main(t_1, \cdots, t_n) \xrightarrow{!} s$.

Let main′ be the main symbol of $\lceil main \rceil$. If $h \geq d \times \{main(t_1, \cdots, t_n)\}$, we have $main'(\mathbf{h}, \mathbf{h}; t_1, \cdots, t_n) \xrightarrow{!} s$

Proof. By induction on the length of the derivation, using the above Lemma 2.

Theorem 3. *Let main be a MPO'-program which admits a polynomial quasi-interpretation. The function $\llbracket main \rrbracket$ is computed in polynomial time in the size of the inputs.*

Proof. The computation of $\llbracket main \rrbracket$ on inputs $t_1, \cdots, t_n$ is performed by evaluating *main* with the $LMPO$-interpreter described in Figure 1. The computation runtime is at least as efficient as the evaluation of $main'(\mathbf{h}, \mathbf{h}; t_1, \cdots, t_n)$ which is defined by the lifted program $\lceil main \rceil$ and where $h = d \times \{main(t_1, \cdots, t_n)\}$. Let q be a polynomial which bounds the quasi-interpretation $\{main\}$. So, the size of h is bounded by $q(c \times \max_{i=1}^{i=n} |t_i|)$ for some constant c which depends on constructor interpretations. Following Theorem 1, the runtime of $main'$ is bounded by a polynomial p, in the maximal size of the argument. We conclude, that the runtime of $main'(\mathbf{h}, \mathbf{h}; t_1, \cdots, t_n)$ is bounded by $p(q(c \times \max_{i=1}^{i=n} |t_i|))$.

References

1. AEHLIG, K., AND SCHWICHTENBERG, H. A syntactical analysis of non-size-increasing polynomial time computation. In *Proceedings of the Fifteenth IEEE Symposium on Logic in Computer Science (LICS '00)* (2000), pp. 84 – 91. 26, 34
2. BASIN, D., AND GANZINGER, H. Automated complexity analysis based on ordered resolution. *Journal of the ACM* (2000). To appear. 26

3. BELLANTONI, S., AND COOK, S. A new recursion-theoretic characterization of the poly-time functions. *Computational Complexity 2* (1992), 97–110. 26, 32, 33, 35, 37, 37

4. BELLANTONI, S., NIGGL, K.-H., AND SCHWICHTENBERG, H. Higher type recursion, ramification and polynomial time. *Annals of Pure and Applied Logic 104*, 1-3 (2000), 17–30. 26

5. BENZINGER, R. *Automated complexity analysis of NUPRL extracts.* PhD thesis, Cornell University, 1999. 25

6. BONFANTE, G., CICHON, A., MARION, J.-Y., AND TOUZET, H. Complexity classes and rewrite systems with polynomial interpretation. In *Computer Science Logic, 12th International Workshop, CSL'98* (1999), vol. 1584 of *Lecture Notes in Computer Science*, pp. 372–384. Full version will appear in Journal of Functional Programming. 36

7. CASEIRO, V.-H. An equational characterization of the poly-time functions on any constructor data strucure. Tech. Rep. 226, University of Oslo, Dept. of informatics, December 1996. http://www.ifi.uio.no/~ftp/publications. 26, 34

8. CICHON, A., AND WEIERMANN, A. Term rewriting therory for the primitive recursive functions. *Annals of pure and applied logic 83* (1997), 199–223. 32

9. CLOTE, P. Computational models and function algebras. In *LCC'94* (1995), D. Leivant, Ed., vol. 960 of *Lecture Notes in Computer Science*, pp. 98–130. 32

10. COBHAM, A. The intrinsic computational difficulty of functions. In *Proceedings of the International Conference on Logic, Methodology, and Philosophy of Science*, Y. Bar-Hillel, Ed. North-Holland, Amsterdam, 1962, pp. 24–30. 32

11. COLSON, L. Functions versus algorithms. *Bulletin of EATCS 65* (1998). The logic in computer science column. 32

12. COOK, S. Characterizations of pushdown machines in terms of time-bounded computers. *Journal of the ACM 18(1)* (January 1971), 4–18. 26

13. CRARY, K., AND WEIRICH, S. Ressource bound certification. In *ACM SIGPLAN-SIGACT symposium on Principles of programming languages, POPL* (2000), pp. 184 – 198. 25

14. DERSHOWITZ, N. Orderings for term-rewriting systems. *Theoretical Computer Science 17*, 3 (1982), 279–301. 26, 29

15. DERSHOWITZ, N., AND JOUANNAUD, J.-P. *Handbook of Theoretical Computer Science vol.B.* Elsevier Science Publishers B. V. (NorthHolland), 1990, ch. Rewrite systems, pp. 243–320. 29

16. HOFBAUER, D. Termination proofs with multiset path orderings imply primitive recursive derivation lengths. *Theoretical Computer Science 105*, 1 (1992), 129–140. 31

17. HOFMANN, M. Linear types and non-size-increasing polynomial time computation. In *Proceedings of the Fourteenth IEEE Symposium on Logic in Computer Science (LICS'99)* (1999), pp. 464–473. 26, 26, 34

18. HOFMANN, M. Programming languages capturing complexity classes. *SIGACT News Logic Column 9* (2000). 26

19. HOFMANN, M. A type system for bounded space and functional in-place update. In *European Symposium on Programming, ESOP'00* (2000), vol. 1782 of *Lecture Notes in Computer Science*, pp. 165–179. 25

20. HUET, G. Confluent reductions: Abstract properties and applications to term rewriting systems. *Journal of the ACM 27*, 4 (1980), 797–821. 29

21. JONES, N. *Computability and complexity, from a programming perspective.* MIT press, 1997. 26

22. JONES, N. Logspace and ptime characterized by programming languages. *Theoretical Computer Science 228* (1999), 151–174. 26

23. JONES, N. The expressive power of higher order types or, life without cons. To appear, 2000. 26

24. KAMIN, S., AND LÉVY, J.-J. Attempts for generalising the recursive path orderings. Tech. rep., Univerity of Illinois, Urbana, 1980. Unpublished note. 33

25. LANKFORD, D. On proving term rewriting systems are noetherien. Tech. Rep. MTP-3, Louisiana Technical University, 1979. 36

26. LEIVANT, D. A foundational delineation of computational feasiblity. In *Proceedings of the Sixth IEEE Symposium on Logic in Computer Science (LICS'91)* (1991). 26

27. LEIVANT, D. Predicative recurrence and computational complexity I: Word recurrence and poly-time. In *Feasible Mathematics II*, P. Clote and J. Remmel, Eds. Birkhäuser, 1994, pp. 320–343. 26, 32

28. LEIVANT, D. Applicative control and computational complexity. In *Computer Science Logic, 13th International Workshop, CSL '99* (1999), vol. 1683 of *Lecture Notes in Computer Science*, pp. 82–95. 26

29. LIU, Y. A., AND STOLLER, S. D. Dynamic programming via static incrementalization. In *Proceedings of the 8th European Symposium on Programming* (Amsterdam, The Netherlands, March 1999), Springer-Verlag, pp. 288–305. 26

30. MARION, J.-Y. Analysing the implicit complexity of programs. Tech. rep., Loria, 2000. Presented at the workshop Implicit Computational Complexity (ICC'99), FLOC, Trento, Italy. The paper is submitted and accessible from http://www.loria.fr/~marionjy. 26, 26, 26, 32, 35

31. PÉTER, R. *Rekursive Funktionen*. Akadémiai Kiadó, Budapest, 1966. English translation: *Recursive Functions*, Academic Press, New York, 1967. 32

32. PLAISTED, D. A recursively defined ordering for proving termination of term rewriting systems. Tech. Rep. R-78-943, Department of Computer Science, University of Illinois, 1978. 26, 29

33. RITCHIE, R. Classes of predictably computable functions. *Transaction of the American Mathematical Society 106* (1963), 139–173. 32

34. SIMMONS, H. The realm of primitive recursion. *Archive for Mathematical Logic 27* (1988), 177–188. 32

Encoding Temporal Logics in Executable Z: A Case Study for the **ZETA** System

Wolfgang Grieskamp and Markus Lepper

Technische Universität Berlin, FB13, Institut für Kommunikations- und Softwaretechnik, Sekr. 5–13, Franklinstr. 28/29, D–10587 Berlin, E-mail: {wg,lepper}@cs.tu-berlin.de

Abstract. The **ZETA** system is a Z-based tool environment for developing formal specifications. It contains a component for *executing* the Z language based on the implementation technique of *concurrent constraint resolution*. In this paper, we present a case-study for the environment, by providing an executable encoding of *temporal interval logics* in the Z language. As an application of this setting, test-case evaluation of trace-producing systems on the base of a formal requirements specifications is envisaged.

1 Introduction

The **ZETA** system [3] is a tool environment for developing formal specifications based on the Z notation [12]. It contains a component for *executing* the Z language, using a computation model of *concurrent constraint resolution*, described in [6]. A wide range of Z's logic can be executed within this approach, integrating the power of higher-order functional and logic computation.

In this paper, we present a case study of the system. We develop an executable encoding of *discrete temporal interval logics* (in the style of Moszkowski's logic, [9]), and illustrate it by animation in the **ZETA** system. The example demonstrates the interplay of logical search and of higher-orderness, the last one allowing us to build abstractions by passing predicates (generally represented as sets in Z) to functions and storing them in data values.

As an application of our encoding of temporal logics we briefly look at *test-case* evaluation for *safety-critical* embedded systems. Given a formal requirements specification which uses temporal logics, some input data describing a test case, and the output data from a run of the system's implementation on the given input, we check by executing the specification whether the implementation meets its requirements. This application stems from the context of a research project funded by Daimler-Chrysler.

This paper is organized as follows. In Sec. 2, we introduce the basic features of executing Z in the **ZETA** system. In Sec. 3 we develop the encoding of temporal logics, and describe the application to test-case evaluation, where we use the example of an *elevator controller*. In Sec. 4 we give a conclusion, discussing the results and related work.

M. Parigot and A. Voronkov (Eds.): LPAR 2000, LNAI 1955, pp. 43–53, 2000.

Fig. 1 ZETA's Graphical User Interface After Executing an Expression

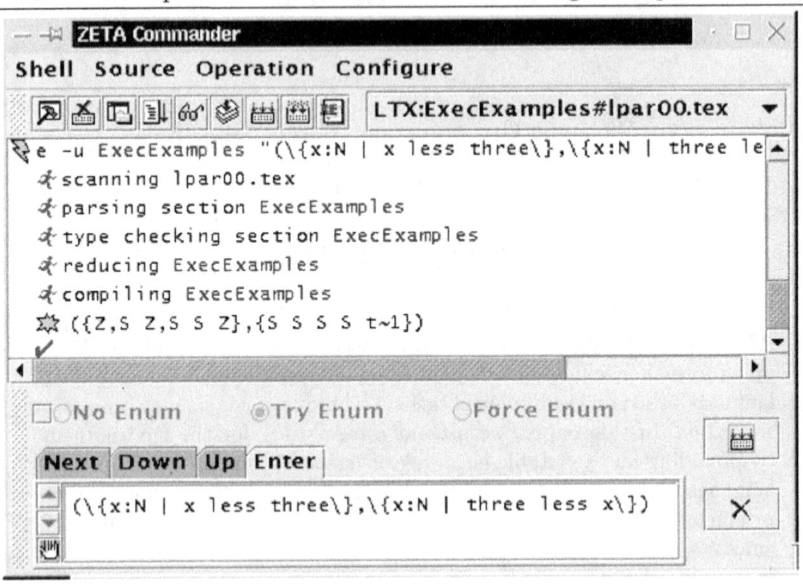

2 Executing Z under ZETA

In [5,6] a computation model based on concurrent constraint resolution has been developed for Z. A high-performance virtual machine has been derived, which is implemented as a component of the ZETA system. In this implementation, all idioms of Z which are related to *functional* and *logic* programming languages are executable. Below, we look at some examples to illustrate the basic features. We assume some knowledge of Z (see e.g. [12]; the Z implemented by ZETA actually confirms to the forthcoming Z ISO Standard [16], which, however, does not make a significant difference in our application).

As sets are paradigmatic for the specification level of Z, they are for the execution level. Set objects – relations or functions – are eventually defined by (recursive) equations, as in the following example, where we define natural numbers as a free type, addition on these numbers and an order relation:

$$N ::= Z \mid S \langle\!\langle \{x : N\} \rangle\!\rangle \qquad\qquad \mid three == S(S(S\,Z))$$

$$\mid add : \mathbb{P}((N \times N) \times N)$$

$$\mid add = \{y : N \bullet (Z, y) \mapsto y\} \cup \{x, y, z : N \mid (x, y) \mapsto z \in add \bullet (S\,x, y) \mapsto S\,z\}$$

$$\mid \_less\_ == \{x, y : N \mid (\exists\, t : N \bullet (x, S\,t) \mapsto y \in add)\}$$

A few remarks on the syntax. With ::= a free type is introduced in Z. The declaration form $n == E$ declares and defines a (non-recursive) name simultaneously. The form $x \mapsto y$ is just an alternative notation for (x, y). A set-comprehension in Z, $\{x : T \mid P \bullet E\}$, describes the values E such that P holds for the possible assignments of x; if the $\bullet$ is omitted (as in the definition for *less*, where the $\bullet$ actually belongs to the existential quantor), the set of tuples of the assignments to the variables in the declaration part is denoted (thus $\{x, y : T \mid P\} = \{x, y : T \mid P \bullet (x, y)\}$).

We may now execute under ZETA queries such as the following, where we ask for the pair of sets less and greater than *three*:

$(\{x : N \mid x\ less\ three\}, \{x : N \mid three\ less\ x\})$
$\Rightarrow (\{\texttt{Z},\texttt{S(Z)},\texttt{S(S(Z))}\}, \{\texttt{S(S(S(S(t\textasciitilde{}))))}\})$

The query as it is entered into the ZETA GUI is visualized in Fig. 1. In the sequel, however, we will use a conceptual notation as above.

As the result of the query, we get the pair of the numbers less than and greater than *three*, where the second value of the resulting pair is a singleton set containing the free variable $\texttt{t\textasciitilde{}}$ (the $\textasciitilde{}$ results from internal variable renaming). These capabilities are obviously similar to logic programming. In fact, we can give a translation from any clause-based system to a system of recursive set-equations in the style given for *add*, where we collect all clauses for the same relational symbol into a union of set-comprehensions, and map literals $R(e_1, \ldots, e_n)$ to membership tests $(e_1, \ldots, e_n) \in R$.

The functional paradigm comes into play as follows. A binary relation R can be *applied*, written as $R\ e$, which is syntactic sugar for the expression $\mu\, y : X \mid (e, y) \in R$. This expression is defined iff their exists a unique y such that the constraint is satisfied; it then delivers this y. The set *add* is a binary relation (since it is member of the set $\mathbb{P}((N \times N) \times N)$), and therefore we can for example evaluate $add(three, three) \Rightarrow \texttt{S(S(S(S(S(S(Z))))))}$.

Note the semantic difference of $(e, y) \in R$ and $y = R\ e$: the first is not satisfied if R is not defined at e, or produces several solutions for y if R is not unique at e, whereas the second is *undefined* in these cases. This difference is resembled in the implementation: application, μ-expressions, and related forms are realized by *encapsulated search*. During encapsulated search, free variables from the enclosing context are not allowed to be bound. A constraint requiring a value for such variables *residuates* until the context binds the variable. As a consequence, if we had defined the recursive path of *add* as $\{x, y, z : N \mid z = add(x, y) \bullet (S\,x, y) \mapsto S\,z\}$ (instead of using $(x, y) \mapsto z \in add$), backwards computation is not be possible:

$\{x : N \mid x\ less\ three\}$
$\Rightarrow$ unresolved constraints:
    LTX:cpinz(48.24-48.31) waiting for variable x

Here, the encapsulated search for $add(x, y)$, solving $\mu\, z : N \mid ((x, y), z) \in add$, cannot continue, since it is not allowed to produce bindings for the context variables x and y.

The elegance of the functional paradigm comes from the fact that functions are first-class citizens. In our implementation of execution for Z, sets are first-class citizens as well. For example, we define a function describing relational image as follows:

$$\boxed{\begin{array}{l} \underline{[X, Y]} \\ \_(\!\_\!) == \lambda\, R : \mathbb{P}(X \times Y);\ S : \mathbb{P}\, X \bullet \{x : X;\ y : Y \mid x \in S;\ (x, y) \in R \bullet y\} \end{array}}$$

A query for the relational image of the *add* function over the cartesian product of the numbers less then three yields in:

let $ns == \{x : N \mid (x, \mathit{three}) \in \mathit{less}\} \bullet \mathit{add}(\!ns \times ns\!)$
$\Rightarrow$ {Z,S(Z),S(S(Z)),S(S(S(Z))),S(S(S(S(Z))))}

Universal quantification is executable if it deals with finite ranges. For example, we can define the operator denoting the set of partial functions in Z, $A \nrightarrow B$, as follows:

$$\boxed{\begin{array}{l} \underline{[X, Y]} \\ \_\nrightarrow\_ == \{R : \mathbb{P}(X \times Y) \mid (\forall\, x : X \mid x \in \mathrm{dom}\, R \bullet \exists_1\, y : Y \bullet (x, y) \in R)\} \end{array}}$$

Universal and unique existential quantification are resolved by enumeration. Thus, if we try to check whether *add* is a partial function, we get in a few seconds:

$add \in N \times N \nrightarrow N$
$\Rightarrow$ still searching after 200000 steps
  gc #1 reclaimed 28674k of 32770k
  · · ·

In enumerating the domain of *add* our computation diverges. However, if we restrict *add* to a finite domain it works:

$\exists_1\, ns == \{x : N \mid (x, \mathit{three}) \in \mathit{less}\} \bullet ((ns \times ns) \triangleleft add \in N \times N \nrightarrow N)$
$\Rightarrow$ *true*

Above, $A \triangleleft R$ restricts the domain of R to the set A; the existential quantor is used to introduce a local name in the predicate.

3 Encoding of Temporal Interval Logics

Temporal interval logics [9,4] is a powerful tool for describing requirements on traces of the behavior of real-time systems. For a discrete version of this logic, related to Moszkowski's version of ITL, an embedding into Z has been described in [2]. Here, we develop an executable shallow encoding for the positive subset of this kind of ITL. The encoding supports resolution for timing and observation constraints (going behind Moszkowski's Tempura implementation), demonstrating some of the capabilities of Executable Z in the ZETA system.

3.1 The Encoding

We define temporal formulas generic over a state type Σ, such that the behaviors we look at have type $\mathrm{seq}\, \Sigma$ ($\mathrm{seq}\_$ is Z's type constructor for sequences). A *predicate* over a state binding is a unary relation, $p \in SP[\Sigma] = \mathbb{P}\, \Sigma$:

$$SP[\Sigma] == \mathbb{P} \, \Sigma$$

A *temporal formula* is encoded by a set of so-called arcs, $w \in TF[\Sigma] = \mathcal{P}ARC[\Sigma]^{1}$, which basically model a transition relation. An arc is either a proper transition, $tr(p, w)$, where p is the guard for this transition and w a followup formula, or the special arc *eot* which indicates that an interval which satisfies this formula may end at this point:

$$TF[\Sigma] == \mathcal{P}ARC[\Sigma] \qquad\qquad ARC[\Sigma] ::= eot \mid tr\langle\!\langle SP[\Sigma] \times TF[\Sigma]\rangle\!\rangle$$

$xs \in_T w$ is the satisfaction relation of this encoding of temporal formulas, and is defined as follows:

$$
\begin{array}{l}
\rule{1.5em}{0pt}[\Sigma]\rule[0.5ex]{10em}{0.4pt}\\[2pt]
\; \_ \in_T \_ : \operatorname{seq} \Sigma \leftrightarrow TF[\Sigma] \\[4pt]
\; (\_ \in_T \_) = \{w : TF[\Sigma] \mid eot \in w \bullet (\langle\rangle, w)\}\cup \\
\rule{4em}{0pt}\{x : \Sigma;\ xs : \operatorname{seq} \Sigma;\ p : SP[\Sigma];\ w, w' : TF[\Sigma] \mid \\
\rule{8em}{0pt} tr(p, w') \in w;\ x \in p;\ xs \in_T w' \bullet (\langle x\rangle \frown xs, w)\}
\end{array}
$$

Thus, if *eot* is an arc of the transition relation, then the empty interval is valid. Moreover, all intervals are valid such that their exists a transition whose predicate fulfills the head of the interval, and the tail of the interval satisfies the followup formula of this transition.

We know define the operators of our logic, which construct values of type $TF[\Sigma]$. The formula which is satisfied exactly by the empty trace is encoded by the singleton transition containing the *eot* arc. The formula $\uparrow p$ lifts a state predicate p to an interval formula which holds exactly for those intervals of length 1 containing a state which satisfies p:

$$
\begin{array}{l}
[\Sigma]\\
\mathsf{empty} == \{eot[\Sigma]\}
\end{array}
\qquad\qquad
\begin{array}{l}
[\Sigma]\\
\uparrow\; == \lambda\, p : SP[\Sigma] \bullet \{tr(p, \mathsf{empty})\}
\end{array}
$$

Next we look at disjunction, $w_1 \sqcup w_2$, and its generalized form. Disjunction is realized by simply mapping it to the union of the arc sets of both formulas:

$$
\begin{array}{l}
[\Sigma]\\
\_ \sqcup \_ == \lambda\, w_1, w_2 : TF[\Sigma] \bullet w_1 \cup w_2
\end{array}
\qquad
\begin{array}{l}
[\Sigma]\\
\bigsqcup\; == \lambda\, ws : \mathbb{P}\, TF[\Sigma] \bullet \bigcup ws
\end{array}
$$

---

[1] We use the powerset-constructor $\mathcal{P}$ which models a computable powerset *domain*. Using the general power, $\mathbb{P}$, our free type definition of ARC would be inconsistent in Z, since a free type's constructor cannot have a general powerset of the type in its domain.

Beware that the generalized disjunction operator can be used for introducing "local variables", as on $\bigsqcup\{x : T \bullet TF[x]\}$.

Conjunction, $w_1 \sqcap w_2$, constructs new arcs by pairwise combination of all arcs of w_1 and w_2 – the conjunction is recursively "pushed" through these combinations:

$$
\begin{array}{l}
\underline{\quad} \sqcap \underline{\quad} : TF[\Sigma] \times TF[\Sigma] \longrightarrow TF[\Sigma] \\
\hline
(\underline{\quad} \sqcap \underline{\quad}) = \lambda\, w_1, w_2 : TF[\Sigma] \bullet \\
\quad (\textbf{if } eot \in w_1 \wedge eot \in w_2 \textbf{ then empty else } \emptyset) \sqcup \\
\quad \{p_1, p_2 : SP[\Sigma];\ w_1', w_2' : TF[\Sigma] \\
\qquad |\ tr(p_1, w_1') \in w_1;\ tr(p_2, w_2') \in w_2 \bullet tr(p_1 \cap p_2, w_1' \sqcap w_2')\}
\end{array}
$$

$w_1 \,\fatsemi\, w_2$ is sequential composition ("chop"). The followup-formula w_2 is recursively pushed through the arcs of w_1 until eot is reached:

$$
\begin{array}{l}
\underline{\quad} \,\fatsemi\, \underline{\quad} : TF[\Sigma] \times TF[\Sigma] \longrightarrow TF[\Sigma] \\
\hline
(\underline{\quad} \,\fatsemi\, \underline{\quad}) = \lambda\, w_1, w_2 : TF[\Sigma] \bullet \\
\quad (\textbf{if } eot \in w_1 \textbf{ then } w_2 \textbf{ else } \emptyset) \sqcup \\
\quad \{p : SP[\Sigma];\ w_1' : TF[\Sigma] \mid tr(p, w_1') \in w_1 \bullet tr(p, w_1' \,\fatsemi\, w_2)\}
\end{array}
$$

w^* is the repetition of w for zero or more times, w^+ for one or more times. In the definition of $\underline{\quad}^*$, we need to embed the recursive reference to $\underline{\quad}^*$ in a set-comprehension, since our implementation of Z imposes a *strict* (eager) evaluation order. The formula skip holds for arbitrary singleton intervals. Temporal truthness, satisfied by any interval, is the repetition of skip. Temporal falsity is described by the empty set of arcs:

$$
\begin{array}{l}
\underline{\quad}^* : TF[\Sigma] \longrightarrow TF[\Sigma] \\
\hline
(\underline{\quad}^*) = \lambda\, w : TF[\Sigma] \bullet \text{empty} \sqcup ((w \setminus \text{empty}) \,\fatsemi\, \{a : ARC[\Sigma] \mid a \in w^*\})
\end{array}
$$

$$
\begin{array}{l}
\underline{\quad}^+ == \lambda\, w : TF[\Sigma] \bullet w \,\fatsemi\, w^*
\end{array}
\qquad
\begin{array}{l}
\text{skip} == \uparrow \Sigma
\end{array}
$$

$$
\begin{array}{l}
\text{true} == \text{skip}[\Sigma]^*
\end{array}
\qquad
\begin{array}{l}
\text{false} == \emptyset[ARC[\Sigma]]
\end{array}
$$

We animate the encoding of some formulas. Suppose type Σ is instantiated with Z. Recall that our observation predicates are sets, hence we can use e.g. $\{1\}$

as a predicate which is exactly true for the state value 1:

$$\uparrow\{1\} \,\mathbin{\mathring{,}}\, \mathsf{empty} \,\mathbin{\mathring{,}}\, \uparrow\{2\} \,\mathbin{\mathring{,}}\, \mathsf{empty} \Rrightarrow \{\mathsf{tr}(\{1\},\{\mathsf{tr}(\{2\},\{\mathsf{eot}\})\})\}$$
$$\uparrow\{1\}^* \qquad\qquad\qquad\quad \Rrightarrow \{\mathsf{eot},\mathsf{tr}(\{1\},\{\mathsf{eot},\mathsf{tr}(\{1\},\dots)\})\}$$
$$\uparrow\{2,3\}^* \sqcap (\uparrow\{1,2\} \,\mathbin{\mathring{,}}\, \uparrow\{3,4\}) \Rrightarrow \{\mathsf{tr}(\{2\},\{\mathsf{tr}(\{3\},\{\mathsf{eot}\})\})\}$$

The first example shows neutrality of empty on chop. The next example illustrates how the repetition operator incrementally "unrolls" its operand (the ZETA displayer has stopped unrolling after a certain depth). In the last example, the effect of conjunction is shown.

Using the satisfaction relation $t \in_T w$, we can now test whether a trace t fulfills a formula w and – provided the state predicates are finite – also generate the set of traces which satisfy a formula. Here are some examples

$$\langle 1,2,3,1,2,1 \rangle \in_T (\mathsf{true} \,\mathbin{\mathring{,}}\, \uparrow\{x : \mathsf{Z} \mid x \geq 2\}^+)^* \Rrightarrow *\mathtt{false}*$$
$$\langle 2,2,2,1,2,2 \rangle \in_T (\mathsf{true} \,\mathbin{\mathring{,}}\, \uparrow\{x : \mathsf{Z} \mid x \geq 2\}^+)^* \Rrightarrow *\mathtt{true}*$$
$$\{t : \mathsf{seq}\,\mathsf{Z} \mid t \in_T \uparrow\{1,2\}^+\} \qquad\qquad \Rrightarrow \{\mathtt{<1>},\mathtt{<2>},\mathtt{<1,1>},\mathtt{<1,2>},\dots\}$$

In the first two examples above, the formula states that the interval must be partitionable into zero or more sub-interval such that in each sub-interval, from some point only numbers greater or equal two appear. This is not satisfied by the first trace, but by the second, choosing the right partitioning. The third example shows the generation of traces.

Our encoding allows the use of free variables in state predicates. For example, we can define a formula which is satisfied by all traces which contain adjacent values. The variable can be existential quantified, or as in the example below, bound by a set comprehension to enumerate its possible bindings:

$$\{x : \mathsf{Z} \mid \langle 4,1,1,3,2,2 \rangle \in_T \mathsf{true} \,\mathbin{\mathring{,}}\, \uparrow\{x\} \,\mathbin{\mathring{,}}\, \uparrow\{x\} \,\mathbin{\mathring{,}}\, \mathsf{true}\} \Rrightarrow \{1,2\}$$

We will use this feature in the next section in order to introduce timing constraints.

3.2   Timing Constraints

Due to the higher-orderness of Z and our implementation, it is easily possible to add new temporal operators. Suppose that our state type Σ contains a duration stamp describing the time distance to the next observation[2], and that this stamp is selected by the function $getd : \Sigma \longrightarrow T$. We then can define a duration operator $DUR(getd, d)$ which holds for those intervals whose duration is d[3]:

$$\mathcal{T} == \mathsf{Z}$$

---

[2] Currently, in our implementation of Z only integral numbers are supported – hence we define time as integral numbers.

[3] Beware that we do not support an "overlapping chop"; therefore intervals which limits fall between two data samples of the given behavior are never considered.

$$
\begin{array}{l}
\underline{[\Sigma]} \\
DUR : (\Sigma \rightarrow T) \times T \rightarrow TF[\Sigma] \\
\hline
DUR = \lambda\, getd : \Sigma \rightarrow T;\ d : T \bullet \\
\quad \bigsqcup\{\sigma : \Sigma \mid getd\,\sigma = d \bullet \uparrow\{\sigma\}\} \sqcup \\
\quad \bigsqcup\{\sigma : \Sigma;\ d' : T \mid d = getd\,\sigma + d' \bullet \uparrow\{\sigma\} \,\S\, DUR(getd, d')\}
\end{array}
$$

This definition makes use of the "generalized disjunction" for temporal formulas, $\bigsqcup$ (see 3.1), to introduce local variables σ and d'. In general, the set-comprehension $\{x : T \mid P \bullet w\}$, where w is a temporal formula, denotes the set of all formulas for instances of x which satisfy P. Since a temporal formula is a set of arcs, the generalized disjunction simply collects all arcs from all formulas, by its definition $\bigsqcup = \bigcup$. The name $\bigcup$ is in turn defined in the Z standard library as $\bigcup SS = \{x : X;\ S : \mathbb{P}\,X \mid S \in SS;\ x \in S \bullet x\}$. Our implementation enumerates the solutions to $S \in SS$ symbolically; henceforth $\bigcup$ also works if SS is not finite, as in the definition of DUR.

In the definition of $DUR(getd, d)$ two cases are distinguished. Either the interval contains exactly one state with duration d, or d is the result of adding $getd\,\sigma$ of the heading state and d' for the remaining states.

As an example, we calculate the partitions of an interval with equal duration, using repetition on the duration operator (where our state contains only durations, and the identity function id selects them):

$$\{d : T \mid \langle 1,1,2,2,2\rangle \in_T DUR(\text{id}, d)^*\}$$

$$\Rrightarrow \{2,4,8\}$$

Note that the partitionings are not of equal length regarding the number of states in an interval. For the duration 2, we use $\langle 1,1\rangle$ and the remaining three $\langle 2\rangle$ partitions. For the duration 4, we have $\langle 1,1,2\rangle$ and $\langle 2,2\rangle$. For duration 8, one partition containing all states is recognized.

3.3   Application

Fig. 2 gives a very simplified example how to apply our temporal logics for requirements specification. The specification defines some aspects of the behavior of a (much simplified) elevator controller. The elevator's state is modeled by a set of sensors which are combined with a duration stamp into the system state $STATE$. The sensors are the current position of the elevator and two sets which represent the state of doors at each floor and of request buttons. Floors are modeled as a subset of positions.

Fig. 2 Elevator Requirements

$| \ POS \ == \mathbb{N}$ $| \ FLOOR \ == \{0, 20, 40, 60, 80\}$

$| \ delay \ == 15$

$\begin{array}{l} \rule{3cm}{0pt} STATE \rule{8cm}{0pt} \\ dur : T; \ pos : POS; \ open, request : \mathbb{P} \, FLOOR \end{array}$

$\begin{array}{lll}
getd & == & \lambda \sigma : STATE \bullet \sigma.dur \\
Safety & == & \uparrow[STATE \mid \forall f : FLOOR \mid f \in open \bullet pos = f]^* \\
Serve & == & \lambda f : FLOOR \bullet \\
& & \quad \uparrow[STATE \mid f \notin request] \sqcup \\
& & \quad (\uparrow[STATE \mid f \in request \wedge f \neq pos]^+ \, \S \\
& & \quad \bigsqcup\{d : T \mid d < delay \bullet \uparrow[STATE \mid f = pos]^* \sqcap DUR(getd, d)\} \, \S \\
& & \quad \uparrow[STATE \mid f \in open]^+) \\
Liveness & == & Serve(0)^* \sqcap Serve(20)^* \sqcap Serve(40)^* \sqcap Serve(60)^* \sqcap Serve(80)^* \\
Reqs & == & Safety \sqcap Liveness
\end{array}$

Our requirements are composed from the conjunction of sub-requirements:

- *Saftey*: a door must be only open if the elevator is at the floor of the door.
- *Serve*: describing the service requirements for a given floor f: Either the floor is not requested, or – if the elevator is requested at this floor – the elevator can be anywhere else. But as soon as it reaches the floor, it must stop there and open the door at least after *delay* seconds. (The specification does not handle error situations, where the elevator does not work for some reason.)
- *Liveness*: is simply the conjunction of all service requirements for all floors.

Such a specification can now be used for test-evaluation, feeding it with the concrete traces produced by an implementation of the controller. For example, let some test traces (parameterized over a duration stamp) be defined as follows:

$$\begin{array}{ll}
t_1 == \lambda \, d : T \bullet \langle\!\langle dur == d, pos == 0, open == \varnothing, request == \{20\}\rangle\!\rangle, \\
\qquad\qquad\qquad \langle\!\langle dur == d, pos == 20, open == \varnothing, request == \{20\}\rangle\!\rangle, \\
\qquad\qquad\qquad \langle\!\langle dur == d, pos == 20, open == \{20\}, request == \varnothing\rangle\!\rangle\rangle \\
t_2 == \lambda \, d : T \bullet \langle\!\langle dur == d, pos == 0, open == \varnothing[\mathbb{Z}], request == \{20\}\rangle\!\rangle, \\
\qquad\qquad\qquad \langle\!\langle dur == d, pos == 20, open == \varnothing, request == \{20\}\rangle\!\rangle, \\
\qquad\qquad\qquad \langle\!\langle dur == d, pos == 40, open == \varnothing, request == \{20\}\rangle\!\rangle\rangle
\end{array}$$

Here are some evaluation results:

$t_1 \ 10 \in_T Reqs \Rightarrow$ *true*; $t_1 \ 40 \in_T Reqs \Rightarrow$ *false*; $t_2 \ 10 \in_T Reqs \Rightarrow$ *false*

In the second case, the elevator stopped at the requested floor but did not opened the door in time. In the third case, the elevator passed a requested floor without stopping.

The performance of test-evaluation highly depends on the kind of specification. For the above specification we check traces of around thousand elements in approx. 30 seconds. However, it is possible to formulate specifications which are intractable to execution since deep backtracking is required to recognize traces. These specifications involve constructs such as $(\text{true} \mathbin{\fatsemi} decision_1) \sqcup (\text{true} \mathbin{\fatsemi} decision_2)$.

4   Conclusion and Related Work

We have presented a case study of the ZETA system, a practical, working setting for developing specifications based on the Z language, which allows for executing a subset of Z based on concurrent constraint resolution. The example of encoding temporal interval logics showed that higher-orderness is a key feature for an environment where we can add new abstractions and notations in a convenient and consistent way: in that temporal formulas are first-class citizens, we could define the operators of the logic as functions over formulas. Below, we discuss some further aspects.

Animating Z. Animation of the "imperative" part of Z is provided by the ZANS tool [8], imperative meaning Z's specification style for sequential systems using state transition schemas. This approach is highly restricted. An elaborated *functional approach* for executing Z has been described in [13], though no implementation exists today, and logic resolution is not employed. Other approaches are based on a mapping to Prolog (e.g. [14,15]), but do not support higher-orderness. The approach presented in this paper goes beyond all the others, since it allows the combination of the functional and logic aspects of Z in a higher-order setting.

Functional and Logic Programming Languages. There is a close relationship of our setting to functional logic languages such as Curry [7] or Oz [11]: in these languages it is possible to write functions which return constraints, enabling abstractions as have been used in this paper. However, our setting provides a tighter integration and has a richer predicate language as f.i. Curry, including negation and universal quantification which are treated by encapsulated search. The role of a function as a special kind of relation as a special kind of set, and of application $e\,e'$ just as an abbreviation for $\mu\,y \mid (e', y) \in e$, makes this tight integration possible.

Integrating Specific Resolution Techniques. Currently, our implementation is not very ambitious regarding the basic employed resolution techniques. Central to the computation model is not the basic solver technology (which is currently mere term unification) but the management of abstractions of constraints via sets. However, the integration of specialized solvers for arithmetic, interval and temporal constraints is required for our application to test-evaluation. The extension of the model to an architecture of cooperating basic solvers is therefore subject of future work.

References

1. H. Boley. *A Tight, Practical Integration of Relations and Functions*, volume 1712 of *Lecture Notes in Artificial Intelligence*. Springer-Verlag, 1999.
2. R. Büssow and W. Grieskamp. Combinig Z and temporal interval logics for the formalization of properties and behaviors of embedded systems. In R. K. Shyamasundar and K. Ueda, editors, *Advances in Computing Science – Asian '97*, volume 1345 of *LNCS*, pages 46–56. Springer-Verlag, 1997. 46
3. R. Büssow and W. Grieskamp. A Modular Framework for the Integration of Heterogenous Notations and Tools. In K. Araki, A. Galloway, and K. Taguchi, editors, *Proc. of the 1st Intl. Conference on Integrated Formal Methods – IFM'99*. Springer-Verlag, London, June 1999. 43
4. Z. Chaochen, C. A. R. Hoare, and A. Ravn. A calculus of durations. *Information Processing Letters*, 40(5), 1991. 46
5. W. Grieskamp. *A Set-Based Calculus and its Implementation*. PhD thesis, Technische Universität Berlin, 1999. 44
6. W. Grieskamp. A Computation Model for Z based on Concurrent Constraint Resolution. to appear in ZUM'00, January 2000. 43, 44
7. M. Hanus. Curry – an integrated functional logic language. Technical report, Internet, 1999. Language report version 0.5. 52
8. X. Jia. An approach to animating Z specifications. Internet: http://saturn.cs.depaul.edu/~fm/zans.html, 1996. 52
9. B. Moszkowski. *Executing Temporal Logic Programs*. Cambridge University Press, 1986. updated version from the authors home page. 43, 46
10. G. Nadathur and D. Miller. An overview of λProlog. In *Proc. 5th Conference on Logic Programming & 5th Symposium on Logic Programming (Seattle)*. MIT Press, 1988.
11. G. Smolka. Concurrent constraint programming based on functional programming. In C. Hankin, editor, *Programming Languages and Systems*, Lecture Notes in Computer Science, vol. 1381, pages 1–11, Lisbon, Portugal, 1998. Springer-Verlag. 52
12. J. M. Spivey. *The Z Notation: A Reference Manual*. Prentice Hall International Series in Computer Science, 2nd edition, 1992. 43, 44
13. S. Valentine. The programming language Z^{--}. *Information and Software Technology*, 37(5–6):293–301, May–June 1995. 52
14. M. M. West and B. M. Eaglestone. Software development: Two approaches to animation of Z specifications using Prolog. *IEE/BCS Software Engineering Journal*, 7(4):264–276, July 1992. 52
15. M. Winikoff, P. Dart, and E. Kazmierczak. Rapid prototyping using formal specifications. In *Proceedings of the Australasian Computer Science Conference*, 1998. 52
16. Drafts for the Z ISO standard. Ian Toyn (editor). Available via the URL http://www.cs.york.ac.uk/~ian/zstan, 1999. 44

Behavioural Constructor Implementation for Regular Algebras

Sławomir Lasota*

Institute of Informatics, Warsaw University
Banacha 2, 02-097 Warszawa, Poland
sl@mimuw.edu.pl

Abstract. We investigate *regular algebras*, admitting infinitary regular terms interpreted as least upper bounds of suitable approximation chains. The main result of this paper is an adaptation of the concept of behavioural constructor implementation, studied widely e.g. for standard algebras, to the setting of regular algebras. We formulate moreover a condition that makes proof of correctness of an implementation step tractable. In particular, we indicate when it is sufficient to consider only finitary observational contexts in the proofs of behavioural properties of regular algebras.

Keywords: Algebraic specifications, observational equivalence, regular algebras, behavioural constructor implementation, proofs of behavioural properties.

Introduction

Behavioural semantics of algebraic specifications is widely accepted to capture properly the "black box" character of data abstraction. As a nontrivial example of an algebraic framework where behavioural ideas may be applied, we consider regular algebras, differing from the standard algebras in one respect: they allow one to additionally model "infinite" datatypes, like streams. Regular algebras were introduced in [16], and then studied e.g. in [17,6]. Our starting point here is a more recent paper [4], investigating observational equivalence of regular algebras and the induced behavioural semantics of specification. Regular algebras contain properly continuous algebras [15], intended usually to model "infinite" datatypes. Unfortunately continuous algebras are not well suited for behavioural semantics of specification, as they lack some crucial algebraic properties, e.g. quotients of continuous algebras do not compose (cf. [5]).

The subject of this paper is to analyze the applicability of regular algebras as models of behavioural specifications in the process of stepwise development of software systems. A general methodology is proposed for regular algebras, as an adaptation of the *constructor behavioural implementation* [12,13]. To our knowledge the constructor implementation has not been studied in this setting so

---

* The work reported here was partially supported by the KBN grant 8 T11C 019 19.

M. Parigot and A. Voronkov (Eds.): LPAR 2000, LNAI 1955, pp. 54–69, 2000.

far. Whereas regular algebras require a separate treatment, since their structure differs substantially from standard algebras: they have partially ordered carrier sets and all term-definable mappings have fixed points, given by the least upper bounds of appropriate approximation chains.

This paper reports briefly contents of Chapter 5 of [9]. After preliminary Section 1 introducing regular algebras, observational equivalence and constructor implementation, in Section 2 we explain how to combine, roughly, regular and standard algebras. We extend the setting of regular algebras with a possibility to have carriers of some sorts essentially unordered (these sorts are called *algebraic*), which is strongly required in practical examples.

In Section 3 we prove that observational indistinguishability in a regular algebra is induced by only finitary observable contexts – surprisingly, even when non-trivial approximation chains exist in carrier sets of observable sorts. This makes proofs of behavioural properties of regular algebras substantially easier, and allows one to exploit methods known for standard algebras, like context induction [7] or results of [1,2]. Moreover, when all observable sorts are algebraic, the notion of observational equivalence can be characterized by only finitary contexts too. Hence one can prove equivalence of regular algebras e.g. using standard observational correspondences [14].

Finally, in Sections 4 and 5 we propose an adaptation of behavioural constructor implementation methodology [12,13] to the new framework. Initially, our main motivation was to enable finitary implementations of infinitary regular data structures. To this aim we introduce a method, called μ-induction, for defining fixed-points of recursors. This gives rise, at an intermediate step in constructor implementation, to *pre-regular algebras*, defined as algebras of a syntactical monad. Next we investigate proofs of correctness of such an implementation step. A property is formulated, called *behavioural consistency*, that guarantees that such proofs are tractable. In Lemma 1 and Theorem 3 we a derive sufficient condition for this property to hold. The condition formalizes a methodological paradigm to ensure behavioural consistency: when defining a pre-regular algebra in the implementation step, one should ensure that $\bot$ in each sort has the smallest observable behaviour and the observable behaviour of each least upper bound of an approximation chain is the least upper bound of the observable behaviours of approximants. Moreover, under that condition, the results from Section 3 apply and one only needs to consider finitary observational contexts while proving correctness of implementation. Our general considerations are illustrated by few examples in Section 1 and in Appendix.

Acknowledgements The author is grateful to Andrzej Tarlecki for many fruitful discussions and valuable comments during this work.

1   Preliminaries

Regular algebras Let Σ, Σ' be fixed many-sorted algebraic signatures throughout this paper. We omit introducing classical notions of standard Σ-algebra, homomorphism (the category of those is denoted Alg_Σ), subalgebra, congruence, quotient (cf. e.g. [5]). By $t_{A[v]}$ we denote the value of Σ-term t in algebra A under valuation v. For A being a standard or regular algebra, by $|A|$ we denote the many-sorted carrier set of A; by $|A|_s$ the carrier of sort s; by $|A|_S$, for a subset S of sorts of Σ, the carrier sets of sorts from S. All sets are implicitly meant to be many-sorted in the sequel.

An *ordered Σ-algebra* is a standard Σ-algebra whose carrier set on each sort s is partially ordered (let $\leq_s^A$ denote below the partial order in $|A|_s$) and has a distinguished element $\perp_s^A \in |A|_s$. However, we do not assume operations in an ordered algebra to be monotonic.

The set of *regular Σ-terms* $T_\Sigma^\mu(X)$ over X is defined inductively as usual, with the only additional case: for any $t \in T_\Sigma^\mu(X \cup \{z : s\})_s$ and a distinguished variable z of the same sort s, there is a *μ-term* $\mu z.t$ in $T_\Sigma^\mu(X)_s$. Similarly, the inductive definition of the value $t_{A[v]}$ of a term under a valuation $v : X \to |A|$ in an ordered algebra A needs one more case. For $t \in T_\Sigma^\mu(X \cup \{z : s\})_s$, put:

- $t_{A[v]}^0(\perp) = \perp_s^A$,
- for $i \in \omega$, $t_{A[v]}^{i+1}(\perp) = t_{A[v_i]}$, where $v_i : X \cup \{z : s\} \to |A|$ extends v by $v_i(z) = t_{A[v]}^i(\perp)$.

Now, $(\mu z.t)_{A[v]}$ is defined if $t_{A[v]}^i(\perp)$ are defined, for all $i \in \omega$, $t_{A[v]}^i(\perp) \leq_s^A t_{A[v]}^{i+1}(\perp)$, and the least upper bound $\bigsqcup_{i \in \omega} t_{A[v]}^i(\perp)$ exists in $|A|_s$; if so, then $(\mu z.t)_{A[v]} = \bigsqcup_{i \in \omega} t_{A[v]}^i(\perp)$. An ordered Σ-algebra A is *regular* if it satisfies the following conditions:

- *completeness*: for all $t \in T_\Sigma^\mu(X)$ and $v : X \to |A|$, the value $t_{A[v]}$ is defined,
- *continuity*: for all $t \in T_\Sigma(X \cup \{y : s\})_{s'}$, $q \in T_\Sigma^\mu(X \cup \{z : s\})_s$ and valuation $v : X \to |A|$, $t_{A[v_i]} \leq_s^A t_{A[v_{i+1}]}$, for $i \in \omega$, and $t_{A[v']} = \bigsqcup_{i \geq 0} t_{A[v_i]}$, where valuation $v' : X \cup \{y : s\} \to |A|$ extends v by $v'(y) = (\mu z.q)_{A[v]}$ and $v_i : X \cup \{y : s\} \to |A|$ extends v by $v_i(y) = q_{A[v]}^i(\perp)$, for $i \geq 0$.

Continuity is required only for *finitary* terms $t \in T_\Sigma(X \cup \{y : s\})_{s'}$, i.e., those not containing symbol μ. A seemingly stronger condition (considered in [4]), concerning all $t \in T_\Sigma^\mu(X \cup \{y : s\})_{s'}$, is equivalent to the above one – a detailed proof can be found in [9]. By completeness, $\perp_s^A$ is the least element in $|A|_s$.

Note that completeness does not imply general ω-completeness. Completeness and continuity as defined above, correspond to ω-completeness and ω-continuity w.r.t. definable ω-chains only. Moreover, operations in a regular algebra need not even be monotonic.

By a regular Σ-homomorphism $h : A \to B$ we mean any function h such that for all terms $t \in T_\Sigma^\mu(X)$ and valuations $v : X \to |A|$, $h(t_{A[v]}) = t_{B[h \circ v]}$. Regular algebras together with regular homomorphisms form a category, called $RAlg_\Sigma$ in the sequel. A regular subalgebra of A is any regular algebra B the carrier of which is a subset of A and such that for all terms $t \in T_\Sigma^\mu(X)$ and valuations $v : X \to |B|$, $t_{B[v]} = t_{A[v]}$. It is easy to see that all operations of B are restrictions of operations of A to $|B|$ and moreover $\perp_s^B = \perp_s^A$ for each sort s. For $Y \subseteq |A|$, by *the subalgebra of A generated by Y* we mean the least regular subalgebra of A whose carrier includes Y.

A relation $\sim\, \subseteq |A| \times |A|$ is a regular congruence iff $\sim\, = \sqsubseteq \cap \sqsubseteq^{-1}$, for some *pre-congruence* $\sqsubseteq$, where a pre-congruence is any pre-order on $|A|$ satisfying:

1. for all $t \in T_\Sigma(X \cup \{y : s\})_{s'}$, $q \in T_\Sigma^\mu(X \cup \{z : s\})_s$ and $v : X \to |A|$, the family $\{t_{A[v_i]}\}_{i \geq 0}$ is a chain w.r.t. $\sqsubseteq$ with a least upper bound $t_{A[v']}$, where valuation $v' : X \cup \{y : s\} \to |A|$ extends v by $v'(y) = (\mu z.q)_{A[v]}$ and valuations $v_i : X \cup \{y : s\} \to |A|$ extend v by $v_i(z) = q^i_{A[v]}(\bot)$, for $i \geq 0$; [1]

2. the equivalence $\sqsubseteq \cap \sqsubseteq^{-1}$ is preserved by the operations (i.e., is a standard congruence).

In particular, instantiating t in 1 with a single variable, we get: for all $q \in T_\Sigma^\mu(X \cup \{z : s\})_s$ and $v : X \to |A|$, $\{q^i_{A[v]}(\bot)\}_{i \geq 0}$ is a chain w.r.t. $\sqsubseteq$ with the least upper bound $(\mu z.q)_{A[v]}$. Given a regular congruence $\sim$, the quotient regular algebra $A/\sim$ is defined (for each sort s) by $|A/\sim|_s = |A|_s/\sim$, $\bot^{A/\sim}_s = [\bot^A_s]$, $[a]_\sim \leq^{A/\sim}_s [a']_\sim \Leftrightarrow a \sqsubseteq a'$ (where $\sqsubseteq$ is a pre-congruence inducing $\sim$) and $f_{A/\sim}([a_1]_\sim, \ldots, [a_n]_\sim) = [f_A(a_1, \ldots, a_n)]_\sim$. For details we refer to [4].

Observational equivalence There have been a number of different formalizations of the concept of behavioural equivalence of algebras (see [8] for an overview). In the following we concentrate on an observational equivalence induced by a subset of observable sorts; hence throughout this paper let us fix a subset *OBS* of *observable sorts* of Σ.

In the following, let X denote some *OBS*-sorted set of variables. By a Σ-*context* of sort s' on sort s we mean any term $\gamma \in T_\Sigma(X \cup \{z_s : s\})_{s'}$, where z_s is a special, distinguished variable of sort s such that $z_s \notin X$. Note that z_s, for $s \notin OBS$, is the only variable of a non-observable sort appearing in a context. A special role is played by *observable contexts*, i.e., contexts of observable sort ($s' \in OBS$). For any regular Σ-algebra A, Σ-context γ on sort s, valuation $v : X \to |A|_{OBS}$ and value $a \in |A|_s$, we will write $\gamma_{A[v]}(a)$ for $\gamma_{A[v_a]}$ where v_a extends v by $v(z_s) = a$.

For any A, let A_{OBS} denote its subalgebra generated by (carrier sets of) observable sorts; we call A_{OBS} the *observational subalgebra* of A. The regular congruence $\sim^{OBS}_A$ on A_{OBS} is defined as follows: for any $a, a' \in |A_{OBS}|_s$, $a \sim^{OBS}_A a'$ if and only if for all valuations v into carriers of observable sorts of A_{OBS} and all observable contexts γ, $\gamma_{A[v]}(a) = \gamma_{A[v]}(a')$. The congruence $\sim^{OBS}_A$ is called *observational indistinguishability* in A; $\sim^{OBS}_A$ is the greatest congruence on A_{OBS} being identity on observable sorts (cf. [4]). The quotient of A_{OBS} by $\sim^{OBS}_A$ represents the observable behaviour of A; $A/\sim^{OBS}$ is *fully abstract* in the sense that its indistinguishability is identity (in particular it equals its own observational subalgebra, i.e., $(A/\sim^{OBS})_{OBS} = A/\sim^{OBS}$). Two regular algebras are taken as equivalent when their behaviours are isomorphic:

---

[1] In [4] a stronger requirement was assumed, for all regular terms $t \in T_\Sigma^\mu(X \cup \{y : s\})_{s'}$, similarly as in continuity condition above. In the same vein as above, finitary terms are sufficient here.

Definition 1 ([4]). *Observational equivalence* $\equiv_{OBS}$ *of regular* Σ*-algebras is defined by:*

$$A \equiv_{OBS} B \quad iff \quad A_{OBS}/\sim_A^{OBS} \cong B_{OBS}/\sim_B^{OBS}.$$

$A_{OBS}/\sim_A^{OBS}$ will be written shortly $A/\sim_A^{OBS}$ or $A/\sim^{OBS}$ in the sequel.

All definitions in this subsection are still valid when standard algebras are taken into account (cf. [3,2]) – the only modification required is to restrict contexts γ to only finitary terms $T_\Sigma(X)$ in definition of indistinguishability.

Behavioural implementation A concept of observational equivalence plays a crucial role in the process of step-wise refinement. It allows one to consider possibly large class of acceptable realizations of a specification, under the only assumption that the observable behavior of the implementation conforms to the specification requirements. We recall below a formalization of these ideas by the notion of behavioural constructor implementation [12,13] in the framework of standard algebras.

Let us look at an example specification of stacks of integers, as an illustration of a behavioural approach to implementation:

```
specification STACKS extends INT, BOOL by
sorts
        stack;
operations                              axioms
    empty : stack;                         empty?(empty) = true,
    push(_,_) : int × stack → stack;       empty?(push(n, s)) = false,
    pop(_) : stack → stack;                top(push(n, s)) = n,
    top(_) : stack → int;                  pop(push(n, s)) = s.
    empty?(_) : stack → bool;
```

Consider the following candidate A for implementation of this specification, which realizes a stack as an infinite array of integers (modeled here as a function from natural numbers to integers) together with a pointer to a current position (top of the stack):

$$A_{stack} := \mathbb{N} \times \mathbb{Z}^{\mathbb{N}},$$
$$empty_A := \langle 0, \lambda i.\ 0 \rangle,$$
$$empty?_A(\langle k, f \rangle) := (k = 0),$$

$$push_A(n, \langle k, f \rangle) := \langle k + 1, \lambda i.\ \text{if}\ \ i = k\ \text{then}\ \ n\ \text{else}\ \ f(i) \rangle,$$
$$pop_A(\langle k, f \rangle) := \text{if}\ \ k = 0\ \text{then}\ \ \langle 0, f \rangle\ \text{else}\ \ \langle k - 1, f \rangle,$$
$$top_A(\langle k, f \rangle) := \text{if}\ \ k = 0\ \text{then}\ \ 0\ \text{else}\ \ f(k - 1).$$

Obviously, this algebra does not satisfy the axioms of STACKS; in particular, the last axiom does not hold. On the other hand, intuitively, this seems to be a "reasonable" realization of the datatype of stacks. The intuition behind this is as follows: although $pop(push(n, s))$ and s need not be identical, they are indistinguishable w.r.t. the observable sorts $\{bool, int\}$. This leads to the behavioural semantics of specification, according to which models of a specification $SP = \langle \Sigma, Ax \rangle$ are all algebras which *behaviourally (observationally) satisfy* its axioms:

$$BehMod(SP) = \{A \in Alg_\Sigma : A \vDash_{OBS} Ax\}.$$

Relation of the behavioural satisfaction $\vDash_{OBS}$ is defined as usual, with the only difference that equality is interpreted in an algebra A as the indistinguishability $\sim_A^{OBS}$ and variables range over the subalgebra A_{OBS}. Formally, $A \vDash_{OBS} \phi \iff A/\sim_A^{OBS} \vDash \phi$. Behavioural semantics is closely related to the observational equivalence of algebras. When only equational specifications are considered, the class of behavioural models coincides with the closure of classical models under observational equivalence (see [3,5]).

The most straightforward formalization of the concept of implementation refers to the inclusion of model classes: specification $SP' = \langle \Sigma', Ax' \rangle$ *implements* (*refines*) SP, if each model of SP' is a model of SP: $Mod(SP') \subseteq Mod(SP)$ (both SP and SP' are over the same signatures here, $\Sigma' = \Sigma$). This concept has been refined in two ways [12,13]. First, a notion of *constructor implementation* was proposed: SP' implements SP via a constructor $\kappa : Alg_{\Sigma'} \to Alg_{\Sigma}$, denoted by $SP \stackrel{\kappa}{\leadsto} SP'$, if $\bar{\kappa}(Mod(SP')) \subseteq Mod(SP)$ ($\bar{\kappa}(\_)$ is direct image of κ). Intuitively, function κ, called *constructor*, represents a parametrized program, realizing one refinement step. Second, a behavioural realization was taken into account: we say that SP' implements behaviourally SP via κ if

$$\bar{\kappa}(BehMod(SP')) \subseteq BehMod(SP). \tag{1}$$

Appendix A contains an example of such an implementation step. In general, development of a system consists of a sequence of such steps,

$$SP \stackrel{\kappa_1}{\leadsto} SP_1 \stackrel{\kappa_2}{\leadsto} \ldots \stackrel{\kappa_n}{\leadsto} SP_n,$$

which finishes when SP_n is the empty specification (e.g. in the implementation of stacks by infinite arrays above, the implementing specification is implicitly assumed to be empty).

2 Regular and Algebraic Sorts

In many practical situations we need only some sorts of a regular algebra to have upper bounds of approximation chains. Assume in the rest of this paper that the set of sorts of Σ is partitioned into two disjoint subsets of *regular* sorts and *algebraic* sorts. The aim is to simplify the work with regular algebras and not to be bothered with considering limit values in algebraic sorts.

In [9] it was argued that the best way to achieve this is to require carrier sets of algebraic sorts to be *essentially flat*. Formally, we say that the carrier set of sort s of a regular Σ-algebra is essentially flat if this algebra is isomorphic (i.e., related via a bijective regular homomorphism) to a regular algebra A whose carrier set of sort s has flat ordering $\leq$ with the least element $\perp_s^A$ (i.e., $a \leq b$ iff $a = \perp_s^A$ or $a = b$). Evidently, isomorphic regular algebras can have different orders.

In the rest of this paper we implicitly assume that some subset of algebraic sorts is distinguished and that all regular algebras considered are essentially flat on those sorts.

3 Finitary Observations

Roughly, finitary observational contexts (i.e., those not containing symbol μ) are powerful enough for indistinguishability in a regular algebra; the only infinitary regular terms really needed are to denote bottoms $\perp$. For fixed OBS, let $\mathcal{F}^{OBS}$ denote the set of all observable contexts $\gamma \in T_\Sigma^\mu(X \cup \{z_s : s\})$, such that the only possible μ-subterms of γ are of the form $\mu x.x$. In the sequel, by finitary terms we also mean those containing μ-subterms of the form $\mu x.x$.

Theorem 1 ([10]). *In a regular algebra A, the observational indistinguishability $\sim_A^{OBS}$ coincides with the contextual indistinguishability induced in A by contexts from $\mathcal{F}^{OBS}$.*

As a corollary, methods of proving behavioural properties of standard algebras (like context induction [7] or methods developed in [1,2]) can be reused in the framework of regular algebras.

Let $\Sigma(\perp)$ denote signature Σ enriched by a constant symbol $\perp_s$ in each sort s. For a regular Σ-algebra A, let $|A|_{\Sigma(\perp)}$ denote the standard $\Sigma(\perp)$-algebra with carrier sets and operations as in A and with $\perp_s$ interpreted as $\perp_s^A$ in each sort.

By Theorem 1 we conclude that observational indistinguishability in a regular algebra A coincides with the indistinguishability in $|A|_{\Sigma(\perp)}$. But observational equivalence of regular algebras is not reducible to observational equivalence of standard $\Sigma(\perp)$-algebras: it *does not* hold

$$A \equiv_{OBS} B \quad \Leftrightarrow \quad |A|_{\Sigma(\perp)} \equiv_{OBS} |B|_{\Sigma(\perp)} \tag{2}$$

(note that on the right-hand side $\equiv_{OBS}$ denotes observational equivalence of standard $\Sigma(\perp)$-algebras).

Let us find out where the difficulties are. Since $\sim_A^{OBS}$ is clearly standard $\Sigma(\perp)$-congruence, the forgetful functor $|\_|_{\Sigma(\perp)}$ commutes with observational quotient:

$$|A_{OBS}|_{\Sigma(\perp)}/_{\sim_A^{OBS}} = |A_{OBS}/_{\sim_A^{OBS}}|_{\Sigma(\perp)}. \tag{3}$$

The observational (standard, regular) subalgebras are generated by (finitary, regular) terms, hence $(|A|_{\Sigma(\perp)})_{OBS} = (|A_{OBS}|_{\Sigma(\perp)})_{OBS}$. Moreover $(|A|_{\Sigma(\perp)})_{OBS}$ may be a *proper* subalgebra of $|A_{OBS}|_{\Sigma(\perp)}$ in general – this is why we need another symbol $\approx_A^{OBS}$ to stand for the indistinguishability (by means of $\Sigma(\perp)$-contexts) in $(|A|_{\Sigma(\perp)})_{OBS}$. However, by Theorem 1 $\approx_A^{OBS}$ and $\sim_A^{OBS}$ agree on $(|A|_{\Sigma(\perp)})_{OBS}$, so the implication from left to right holds in (2). On the other hand, from $|A|_{\Sigma(\perp)} \equiv_{OBS} |B|_{\Sigma(\perp)}$, i.e. from

$$(|A|_{\Sigma(\perp)})_{OBS}/_{\approx^{OBS}} \cong (|B|_{\Sigma(\perp)})_{OBS}/_{\approx^{OBS}}$$

we cannot even conclude (e.g. using (3)) that $|A/_{\sim^{OBS}}|_{\Sigma(\perp)} \cong |B/_{\sim^{OBS}}|_{\Sigma(\perp)}$; but even if we could, this would not guarantee $A/_{\sim^{OBS}} \cong B/_{\sim^{OBS}}$ in general. Intuitively, finitary contexts are more powerful in regular algebras than in standard algebras.

It is common in practical examples that all observable sorts are intended to be essentially flat. Besides practical advantages, essentially flat carriers of observable sorts imply that finitary observational contexts (and consequently standard correspondences) are sufficient for observational equivalence of regular algebras, in contrast to the negative statement (2) above.

Theorem 2 ([10]). *For regular algebras A and B with essentially flat carrier sets of observable sorts, $A \equiv_{OBS} B \;\Leftrightarrow\; |A|_{\Sigma(\perp)} \equiv_{OBS} |B|_{\Sigma(\perp)}$.*

As a conclusion, we obtain an effective complete proof technique for observational equivalence of regular algebras:

Corollary 1. *Regular algebras with essentially flat carrier sets of observable sorts are observationally equivalent iff they are related by an observational $\Sigma(\perp)$-correspondence [14].*

The proofs omitted here can be found in [10].

4 Behavioural Implementation of Regular Algebras

In [9] it was argued that when implementation of regular algebras is considered, it is not always possible to reuse the partial order of a model of the implementing specification in a construction of a model of the implemented one. This problem especially arises when an algebraic sort is to implement a regular one, what seems to be common in practical examples. As an example, consider regular algebras as models of the specification of stacks from Section 1 – this opens the possibility to define infinite streams, e.g. $\mu x.push(1, push(0, x))$. Let us look at the following Pascal-like implementation of stacks by pointers linking dynamically allocated memory cells.

```
specification MEMORY extends INT, BOOL by
sorts
        memory, address;
operations
        initmem : memory;
        null : address;
        alloc(_) : memory → memory × address;
        avail(_,_) : memory × address → bool;
        _[_].val : memory × address → int;
        _[_].nxt : memory × address → address;
        (_[_].val := _) : memory × address × int → memory;
        (_[_].nxt := _) : memory × address × address → memory;
        _[_] ← ⟨_,_⟩ : memory × address × int × address → memory;
        copy(_,_,_) : memory × address × address → memory;
axioms
        a ≠ null ⇒ avail(initmem, a) = true,
        alloc(m) = ⟨m', a⟩  ⇒  a ≠ null ∧
                avail(m, a) = true ∧ avail(m', a) = false ∧ identical(m, m', a),
        a ≠ null ∧ m' = (m[a].val := x)  ⇒  m'[a].val = x ∧
                avail(m', a) = avail(m, a) ∧ m'[a].nxt = m[a].nxt ∧ identical(m, m', a),
        a ≠ null ∧ m' = (m[a].nxt := a'')  ⇒  m'[a].nxt = a'' ∧
                avail(m', a) = avail(m, a) ∧ m'[a].val = m[a].val ∧ identical(m, m', a),
        m[a] ← ⟨x, a'⟩ = ((m[a].val := x)[a].nxt := a'),
        copy(m, a, a') = ((m[a'].val := m[a].val)[a'].nxt := m[a].nxt),

        ( identical(m, m', a) ⟺ (∀a' ∈ address. a ≠ a' ⇒
          avail(m, a') = avail(m', a') ∧ m[a'].val = m'[a'].val ∧ m[a'].nxt = m'[a'].nxt) ).
```

For MEMORY, a set of observable sorts is $OBS = \{bool, int\}$; all sorts are intended to be essentially flat. We choose a constructor κ_{MEMORY} taking a regular algebra over the signature of MEMORY to an algebra over the signature of STACKS. It is defined by:

$$stack := memory \times address,$$
$$empty := \langle initmem, null \rangle,$$
$$top(\langle m, a \rangle) := \textbf{if} \ \ a = null \vee a = \bot_{address} \ \ \textbf{then} \ \ \bot_{int} \ \textbf{else} \ \ m[a].val,$$
$$pop(\langle m, a \rangle) := \textbf{if} \ \ a = null \vee a = \bot_{address} \ \ \textbf{then} \ \ \langle m, a \rangle \ \textbf{else} \ \ \langle m, m[a].nxt \rangle,$$
$$push(x, \langle m, a \rangle) := \textbf{let} \ \ \langle m', a' \rangle = alloc(m) \ \ \textbf{in} \ \ \langle m'[a'] \leftarrow \langle x, a \rangle, a' \rangle.$$

It implements in a straightforward manner all finite stacks, but what about infinite ones? Are they realizable in (behavioural) models of MEMORY? In this particular case a right idea is to exploit cyclic lists; however, a more universal method is needed in general.

For a given model of MEMORY, consider an algebra A yielded by $\kappa_{\mathtt{MEMORY}}$. Roughly, our idea is to define a semantical counterpart of μ-operator, that is for each function $f : |A|_{stack} \rightarrow |A|_{stack}$, a value $\mathsf{fix}_{stack}(f)$ in $|A|_{stack}$. This is enough to define values of all regular terms, since we can take $\mathsf{fix}_{stack}(f)$ as the value of a μ-term $\mu x.t$ under a valuation v, for an appropriate function f induced by t and v. Formally, we can define inductively the value of $(\mu x.t)_{A[v]}$ by: $(\mu x.t)_{A[v]} := \mathsf{fix}_{stack}(\lambda a.t_{A[v_a]})$, where v_a extends v by $v_a(x) = a$. In our example this could look like

$$\mathsf{fix}_{stack}(f) := \textbf{let} \ \ \langle m', a' \rangle = alloc(initmem), \ \langle m, a \rangle = f(\langle m', a' \rangle) \ \ \textbf{in}$$
$$\textbf{if} \ \ a \neq a' \ \textbf{then} \ \ \langle copy(m, a, a'), a' \rangle \ \textbf{else} \ \ \langle initmem, \bot_{address} \rangle.$$

This method of defining values of all μ-terms will be called μ-induction in the sequel. In general, it is sufficient to define $\bot$ in each algebraic sort and a fix operator for each regular sort. For regular sorts s, we derive $\bot_s$ for instance by $\bot_s = \mathsf{fix}_s(\mathrm{id}_{A_s})$ (e.g. $\bot_{stack} = \langle initmem, \bot_{address} \rangle$); for algebraic sorts s, one can assume that $\mathsf{fix}_s(f) = f(\bot_s)$. Hence μ-induction in an algebra A needs a family $\{\bot_s\}$ indexed by algebraic sorts s and a family of functions, $\mathsf{fix}_s : |A|_s{}^{|A|_s} \rightarrow |A|_s$, indexed by regular sorts s.

According to the terminology used below, A defined in this way is a *pre-regular algebra*, differing from a regular algebra in at least one respect: its carrier sets are not ordered and consequently it does not satisfy continuity; in particular, values of μ-terms given by μ-induction do not have to be fixed points. In what follows, we introduce pre-regular algebras and use them in behavioural implementation step of regular algebras.

4.1   Pre-Regular Algebras

Let S denote the set of sorts of a fixed signature Σ. It is reasonable to consider an abstract syntax given by the set $\mathcal{T}_\Sigma^\mu(X)$ of regular terms up to α-conversion, equal to the quotient of $T_\Sigma^\mu(X)$ by all equalities of the form: $\mu x.t(x) = \mu y.t(y)$, for all terms t and variables x, y. The mapping $X \mapsto \mathcal{T}_\Sigma^\mu(X)$ can be extended to an endofunctor $\mathcal{T}_\Sigma^\mu$ in Set^S, similarly to an endofunctor $X \mapsto T_\Sigma(X)$; moreover, in a similar way, $\mathcal{T}_\Sigma^\mu$ can be extended to a monad, i.e., equipped with unit and multiplication. By μ-induction we define in fact an algebra of this monad - $\mathcal{T}_\Sigma^\mu$-algebras are called *pre-regular algebras* in the sequel. By the very definition, a pre-regular algebra has enough structure to assign values to all regular terms, in a canonical way. Morphisms of $\mathfrak{T}_\Sigma^\mu$-algebras are precisely those functions which preserve values of all regular terms, similarly as regular homomorphisms: h is a morphism from A to B iff $h(t_{A[v]}) = t_{B[h \circ v]}$, for any regular term t and valuation v in A (this category is denoted $Pre\text{-}RAlg_\Sigma$ in the sequel).

Each regular algebra is (via the evident forgetful functor) a pre-regular algebra; in fact, category $RAlg_\Sigma$ is equivalent to a full subcategory of pre-regular algebras. Moreover, the forgetful functor has a left adjoint:

Proposition 1. $RAlg_\Sigma$ *is a reflective subcategory of* $Pre\text{-}RAlg_\Sigma$.

Proof. Observe that the notion of pre-congruence from Section 1 is meaningful in a pre-regular algebra. By the quotient of a pre-regular algebra A by a pre-congruence $\sqsubseteq$, denoted by $A/\sqsubseteq$, we mean the quotient by a congruence $\sqsubseteq \cap \sqsubseteq^{-1}$ induced by $\sqsubseteq$; $A/\sqsubseteq$ is a regular algebra, i.e., can be appropriately equipped with a partial order, similarly as in Section 1.

Exploiting these facts, the left adjoint to the forgetful functor is given by the quotient of a pre-regular algebra A by the smallest pre-congruence $\preccurlyeq_A$ in A, $A \mapsto A/\preccurlyeq_A$. This mapping is functorial. To see this, notice that every morphism $f : A \to B$ of pre-regular algebras together with a pre-congruence $\sqsubseteq$ on B induces a pre-congruence: $\{(a, a') : f(a) \sqsubseteq f(a')\}$ on A.

Since $\preccurlyeq_A$ is the smallest pre-congruence, there exists a unique morphism $A/\preccurlyeq_A \to B/\preccurlyeq_B$ making the square on the right commute (horizontal maps are quotient projections). Moreover, by the very definition, this induced morphism is a regular homomorphism.

$$
\begin{array}{ccc}
A & \xrightarrow{\;\pi_A\;} & A/\preccurlyeq_A \\
{\scriptstyle f}\downarrow & & \downarrow{\scriptstyle !} \\
B & \xrightarrow{\;\pi_B\;} & B/\preccurlyeq_B
\end{array}
$$

When B is regular, $\preccurlyeq_B$ induces identity congruence; thus, by instantiating the diagram above we immediately get a 1-1 correspondence between hom-sets $Hom_{Pre\text{-}RAlg_\Sigma}(A, B)$ and $Hom_{RAlg_\Sigma}(A/\preccurlyeq_A, B)$. Hence the quotient functor $\_/\preccurlyeq\_$ is the left adjoint to the embedding of $RAlg_\Sigma$ into $Pre\text{-}RAlg_\Sigma$; the projections π form a unit. $\qquad\square$

Notions of subalgebra, quotient, etc. can be straightforwardly extended to pre-regular algebras. The relation $\models_{OBS}$ of behavioural satisfaction can be lifted to pre-regular algebras in a natural way. By $A \models_{OBS} \phi$, for pre-regular A, we mean that $A/\sim_A^{OBS} \models \phi$, where $\sim_A^{OBS}$ denotes the indistinguishability induced by the set of *all* observable contexts (including μ-terms) in the pre-regular subalgebra A_{OBS} of A generated by OBS. (A_{OBS} contains precisely those elements that are a value of some regular term with only observable variables.) Note that it is indispensable to take all contexts into account, since values of μ-terms in a pre-regular algebra can be defined arbitrarily and we can not use continuity, as in the case of regular algebras in Theorem 1, to restrict the set of relevant contexts. Moreover, considering all regular contexts guarantees that $\sim_A^{OBS}$ is a congruence on A_{OBS} (in fact, the greatest congruence which is identity on OBS). In the following section we overload symbol $\equiv_{OBS}$ to denote also the equivalence of pre-regular algebras factorized by indistinguishabilities $\sim_A^{OBS}$ similarly as stated for regular algebras in Definition 1.

5   Constructor Implementation Step

By now, we found pre-regular algebras useful in the behavioural implementation of regular algebras. Motivated by examples and by Proposition 1, we propose an implementation methodology for regular algebras, as a slight adaptation of the behavioural constructor implementation of standard algebras presented in Section 1. By the abuse of notation, given a specification SP over Σ, by $BehMod(SP)$ we mean the class of all regular Σ-algebras which satisfy behaviourally SP and whose algebraic sorts are essentially flat.

Let SP, SP' be two specifications over Σ and Σ', respectively. Let OBS denote the subset of observable sorts in SP. In order to express the way how SP' implements (refines) SP, we define a constructor $\kappa : RAlg_{\Sigma'} \to Pre\text{-}RAlg_{\Sigma}$, for instance by μ-induction. This mapping induces a function $\kappa' : RAlg_{\Sigma'} \to RAlg_{\Sigma}$: $A \mapsto \kappa(A)/\preccurlyeq_{\kappa(A)}$. But for some technical reason, which becomes apparent in (8) below, instead of $\preccurlyeq_{\kappa(A)}$ we prefer to quotient $\kappa(A)$ by the smallest pre-congruence on the observable subobject, that is on the subobject $\kappa(A)_{OBS}$ of $\kappa(A)$ generated by the observable sorts[2]:

$$\kappa'(A) := \kappa(A)_{OBS}/\preccurlyeq_{\kappa(A)_{OBS}}.$$

Then, we say that SP' implements behaviourally SP via κ if

$$\bar{\kappa}'(BehMod(SP')) \subseteq BehMod(SP). \tag{4}$$

5.1   Behavioural Consistency

Similarly as in the case of standard algebras, for the correctness of the constructor implementation step one needs to show (4). When $SP = \langle \Sigma, Ax \rangle$ is a basic specification given by a set of axioms, (4) can be proved for instance by showing that $\kappa'(A)$ satisfies behaviourally axioms of SP, for any behavioural model A of SP'. However, this is difficult in practice, since the congruence in the construction of $\kappa'(A)$ is not given explicitly – the task would be simpler if we could consider $\kappa(A)$ instead of $\kappa'(A)$. This would be the case when $\kappa(A)$ and $\kappa'(A)$ satisfy behaviourally the same formulas:

$$\{\phi : \kappa(A) \vDash_{OBS} \phi\} = \{\phi : \kappa'(A) \vDash_{OBS} \phi\}. \tag{5}$$

We say that constructor κ is *behaviourally consistent* if (5) holds for all A in $BehMod(SP')$. A sufficient condition for (5) is:

$$\kappa(A) \equiv_{OBS} \kappa'(A), \quad \text{for each } A \in BehMod(SP'), \tag{6}$$

---

[2] $\preccurlyeq_{\kappa(A)_{OBS}}$ needs not coincide with $\preccurlyeq_{\kappa(A)}$ restricted to $\kappa(A)_{OBS}$ – the latter is coarser in general.

where $\equiv_{OBS}$ denotes the equivalence of pre-regular algebras factorized by the indistinguishabilities $\sim_A^{OBS}$. The rest of this section is devoted mainly to formulating of a condition sufficient for (6).

In the sequel let $B := \kappa(A)_{OBS}$ denote the observable subalgebra of the pre-regular algebra yielded by κ, for an arbitrary fixed $A \in BehMod(SP')$. Since $\kappa(A) \equiv_{OBS} B$, (6) is equivalent to

$$B \equiv_{OBS} B/\preccurlyeq_B. \tag{7}$$

Hence, for behavioural consistency (in fact for (7)) it is sufficient to know that $\preccurlyeq_B$ induces an observational congruence on B:

$$\preccurlyeq_B \cap \preccurlyeq_B^{-1} \subseteq \sim_B^{OBS}. \tag{8}$$

(8) states that the congruence induced by $\preccurlyeq_B$ is identity on observable sorts; in other words, imposing continuity on B does not lead to identification of observable elements. From the practical perspective this is a natural condition, since it conforms neatly to the situation when observable sorts are to support some standard datatypes, which rest unaffected during a refinement step.

For (8) to hold, $\sim_B^{OBS}$ should be necessarily a total congruence (i.e., $B = B_{OBS}$). Fortunately, it is so since B is generated by OBS (recall definition of κ' in the beginning of this section).

Evidently, $\sim_B^{OBS}$ is equal to $\lesssim_B^{OBS} \cap (\lesssim_B^{OBS})^{-1}$, where the *observational pre-order* $\lesssim_B^{OBS}$ is defined analogously to $\sim_B^{OBS}$: for $b, b' \in |B|_s$, $b \lesssim_B^{OBS} b'$ if and only if for all valuations $v : X \to |B|_{OBS}$ and all regular observable contexts γ, $\gamma_{B[v]}(b) \leq \gamma_{B[v]}(b')$; $\leq$ stands here for the flat order in all observable sorts.[3] Having this, we conclude that for (8) it is sufficient to have $\preccurlyeq_B \subseteq \lesssim_B^{OBS}$. Now, recalling that $\preccurlyeq_B$ is the smallest pre-congruence on B, we deduce:

Lemma 1. *A constructor κ is behaviourally consistent whenever $\lesssim_{\kappa(A)}^{OBS}$ is a pre-congruence on $\kappa(A)_{OBS}$, for each $A \in BehMod(SP')$.*

From now on we develop a sufficient condition for this to hold. Since B is generated by OBS, we can suitably represent each non-observable value by a term with only observable variables. This implies that operations in B are monotonic w.r.t. $\lesssim_B^{OBS}$, hence $\sim_B^{OBS}$ is preserved by operations. We need to check the other condition from the definition of pre-congruence (cf. Section 1). For any recursor $q \in |T_\Sigma^\mu(X \cup \{z : s\})|_s$ and valuation $v : X \to |B|$, let $b_i := q_{B[v]}^i(\bot)$, for $i = 0, 1, \ldots$, $b := (\mu z.q)_{B[v]}$; then we should show that

for each $t \in |T_\Sigma(X \cup \{y : s\})|_{s'}$ and $u : X \to |B|$, $\{t_{B[u_i]}\}_{i \in \omega}$ is a chain w.r.t. $\lesssim_B^{OBS}$ with a l.u.b. $t_{B[u']}$ (u_i, $i \in \omega$, and u' extend u by $u_i(y) = b_i$ and $u'(y) = b$).

As B is generated by OBS, we may replace each occurrence of a non-observable variable in t (besides y) by a term with only observable variables; hence the

---

[3] Surprisingly, at this point we could use *any* partial order instead of $\leq$.

mapping $a \mapsto t_{B[u[y \mapsto a]]}$ preserves $\lesssim_B^{OBS}$. This allows us to replace condition above with a simpler one:

$\{b_i\}_{i \in \omega}$ is a chain w.r.t. $\lesssim_B^{OBS}$ with a least upper bound b.

Moreover, the mapping $a \mapsto q_{B[v[z \mapsto a]]}$ preserves $\lesssim_B^{OBS}$ by the same argument as above, hence for the last formula it suffices that the following two conditions hold:

$$\perp_s^B \lesssim_B^{OBS} b', \quad \text{for all } b' \in |B|_s, \tag{9}$$

$$b = \bigsqcup_{i \in \omega} b_i \quad \text{w.r.t. } \lesssim_B^{OBS}. \tag{10}$$

The observable behaviour of an element of B consists of the infinite tuple of values yielded by all observable contexts (with all valuations into observable sorts) applied to it. As $\lesssim_B^{OBS}$ is the point-wise pre-order, (9) and (10) say that a bottom $\perp$ has the smallest observable behaviour and that the observable behaviour of b is the least upper bound of behaviours of b_i. This allows us to formulate a methodological paradigm:

> *A constructor is guaranteed to be behaviourally consistent whenever $\perp$ in each sort has the smallest observable behaviour and the observable behaviour of each (candidate for) least upper bound of an approximation chain is the least upper bound of the observable behaviours of approximants.*

Expanding definition of $\lesssim_B^{OBS}$ and recalling that $\leq$ was chosen to be flat, the paradigm can be formalized as follows:

Theorem 3. *A constructor κ is behaviourally consistent if for each behavioural model A of SP' it holds: for all contexts $\gamma \in T_\Sigma^\mu(X \cup \{z_s : s\})_o$ of an observable sort o and valuations w into observable carriers of $B := \kappa(A)_{OBS}$,*

- $\gamma_{B[w]}(\perp_s) \neq \perp_o \Rightarrow \forall b \in B_s. \ \gamma_{B[w]}(b) = \gamma_{B[w]}(\perp_s)$,
- *for any $q \in T_\Sigma^\mu(X \cup \{z : s\})_s$, $v : X \to |B|$, let $b_i := q_{B[v]}^i(\perp)$, $i \in \omega$ and $b := (\mu z.q)_{B[v]}$; then*
  $$\forall i \in \omega.(\gamma_{B[w]}(b_i) \neq \perp_o \Rightarrow \gamma_{B[w]}(b) = \gamma_{B[w]}(b_i)),$$
  $$(\forall i \in \omega.\gamma_{B[w]}(b_i) = \perp_o) \Rightarrow \gamma_{B[w]}(b) = \perp_o.$$

For instance, κ_{MEMORY} from the beginning of previous section is behaviourally consistent.

5.2   Finitary Contexts

Theorem 3 concerns all regular observable contexts γ. From practical perspective, especially when considering proof methods for behavioural properties, it would be useful to be able to restrict to only finitary contexts. Assume for a while that we replaced $\lesssim_B^{OBS}$ in Lemma 1 with pre-order induced by only contexts from $\mathcal{F}^{OBS}$ (similarly as in Theorem 1 in Section 3). If we prove now that this pre-order is a pre-congruence, then the induced congruence, say $\sim$, necessarily coincides with $\sim_B^{OBS}$. This follows from the fact that $\sim$ is obviously coarser than $\sim_B^{OBS}$, $\sim_B^{OBS} \subseteq \sim$, but on the other hand it is necessarily identity on OBS. Since $\sim_B^{OBS}$ is the greatest congruence being identity on OBS, we get $\sim = \sim_B^{OBS}$ (this idea of identifying a sufficient subset of "crucial" contexts comes e.g. from [1]). From this we conclude that Lemma 1 would still hold and behavioural consistency would follow.

Unfortunately, our further considerations (culminating in Theorem 3) are no longer valid when only contexts from $\mathcal{F}^{OBS}$ are taken into account. Infinitary regular terms are indispensable when valuation maps a variable to value in B not representable by a finitary term with observable variables. Consequently, we can only restrict the proof obligation stated in Theorem 3 to those contexts γ that contain no occurrence of the context variable inside a μ-term.

Surprisingly, finitary contexts are sufficient for proving correctness of implementation when we already know that κ is behaviourally consistent. Namely, by (5), every proof in $\kappa(A)$ involving observational contexts can be carried over to $\kappa'(A)$, where finitary contexts suffice by Theorem 1.

6   Final Remarks

The main result of this paper is an adaptation of the behavioural constructor implementation methodology to the framework of regular algebras. A proof obligation was given that guarantees (together with results on finitary character of observational indistinguishability) that the proof of correctness of an implementation step is feasible.

The subject needs still more studies, especially the issue of proving correctness of implementation. In particular, we did not tackle the task of proving that carriers of algebraic sorts are essentially flat in the implementation step. Moreover, an interesting topic for further research is to investigate relationship between regular algebras and coalgebraic specifications [11] and to try to apply proof methods used there. It could be probably of some relevance here that the initial regular Σ-algebra can be seen as a suitable sub-coalgebra of the final $\Sigma(\bot)$-coalgebra, hence it admits the coinduction principle.

References

1. Bidoit, M., Hennicker, R. Behavioural theories and the proof of behavioural properties. *Theoretical Computer Science* 165(1): 3-55, 1996. 55, 60, 67
2. Bidoit, M., Hennicker, R. Modular correctness proofs of behavioural implementations. *Acta Informatica* 35(11):951-1005, 1998. 55, 58, 60
3. Bidoit, M., Hennicker, R., Wirsing, M. Behavioural and abstractor specifications. *Science of Computer Programming* 25(2-3), 1995. 58, 59
4. Bidoit, M., Tarlecki, A. Regular algebras: a framework for observational specifications with recursive definitions. Report LIENS-95-12, Ecole Normale Superieure, 1995. 54, 56, 57, 57, 57, 58

5. Bidoit, M., Tarlecki A. Behavioural satisfaction and equivalence in concrete model categories. *Manuscript*. A short version appeared in *Proc. 20th Coll. on Trees in Algebra and Computing* CAAP'96, Linköping, LNCS 1059, 241-256, Springer-Verlag, 1996. 54, 56, 59

6. Guessarian, I., Parisi-Presicce, F. Iterative vs. regular factor algebras. *SIGACT News* 15(2), 32-44, 1983. 54

7. Hennicker, R. Context induction: A proof principle for behavioural abstractions and algebraic implementations. *Formal Aspects of Computing*, 3(4):326-345, 1991. 55, 60

8. Knapik, T. Specifications algebriques observationnelles modulaires: une semantique fondee sur une relation de satisfaction observationnelle. PhD Thesis, Ecole Normal Superieur, Paris, 1993. 57

9. Lasota, S. *Algebraic observational equivalence and open-maps bisimilarity*. PhD Thesis, Institute of Informatics, Warsaw University, March 2000. Accessible at http://www.mimuw.edu.pl/~sl/work/phd.ps.gz. 55, 56, 59, 61, 69

10. Lasota, S. Finitary observations in regular algebras. To appear in Proc. 27th Seminar on Current Trends in Theory and Practice of Informatics *SOFSEM'2000*, LNCS. Accessible at http://www.mimuw.edu.pl/~sl/papers. 60, 61, 61

11. Rutten, J.J.M.M. Universal coalgebra: a theory of systems. CWI Report CS-R9652, 1996. 67

12. Sannella, D., Tarlecki, A. Towards formal development of programs from algebraic specifications: implementations revisited. *Acta Informatica* 25:233-281, 1988. 54, 55, 58, 59

13. Sannella, D., Tarlecki, A. Essential Concepts of Algebraic Specification and Program Development. *Formal Aspects of Computing* 9:229-269, 1997. 54, 55, 58, 59

14. Schoett, O. *Data abstraction and correctness of modular programming*. PhD thesis, CST-42-87, Department of Computer Science, University of Edinburgh, 1987. 55, 61

15. Tarlecki, A., Wirsing, M. Continuous abstract data types. *Fundamenta Informaticae* 9(1986), 95-126. 54

16. Tiuryn, J. Fixed-points and algebras with infinitely long expressions. Part I. Regular algebras. *Fundamenta Informaticae* 2(1978), 102-128. 54

17. Tiuryn, J. Fixed-points and algebras with infinitely long expressions. Part II. Muclones of regular algebras. *Fundamenta Informaticae* 2(1979), 317-336. 54

A   An Example of Behavioural Constructor Implementation

As an example, consider a specification of queues and its behavioural implementation by pairs of stacks.

```
specification QUEUES extends INT, BOOL by
sorts
        queue;
operations
        empty_queue : queue;
        empty_queue?(_) : queue → bool;
        put(_, _) : queue × int → queue;
        get(_) : queue → int;
        rest(_) : queue → queue;
```

axioms

$$empty\_queue?(empty\_queue) = true$$
$$empty\_queue?(put(q, e)) = false$$
$$empty\_queue?(q) \Rightarrow get(put(q, e)) = e \ \wedge \ rest(put(q, e)) = q$$
$$\neg empty\_queue?(q) \Rightarrow get(put(q, e)) = get(q) \ \wedge \ rest(put(q, e)) = put(rest(q), e)$$

A constructor behavioural implementation of QUEUES by STACKS may be given by the following definitions (some ad hoc, but hopefully self-explanatory notation is used here to define a function that maps $Sig(\text{STACKS})$-algebras to $Sig(\text{QUEUES})$-algebras):

$$queue := stack \times stack$$
$$empty\_queue := \langle empty, empty \rangle$$
$$empty\_queue?(\langle s_1, s_2 \rangle) := empty?(s_1) \ \wedge \ empty?(s_2)$$
$$put(\langle s_1, s_2 \rangle, e) := \langle push(e, s_1), s_2 \rangle$$
$$get(\langle s_1, s_2 \rangle) :=$$
$$\qquad \textbf{if} \ empty?(s_2) \ \textbf{then} \quad \textbf{let} \ s_2' := reverse(\langle s_1, s_2 \rangle) \ \textbf{in} \ top(s_2')$$
$$\qquad \textbf{else} \ top(s_2)$$
$$rest(\langle s_1, s_2 \rangle) :=$$
$$\qquad \textbf{if} \ empty?(s_2) \ \textbf{then} \quad \textbf{let} \ s_2' := reverse(\langle s_1, s_2 \rangle) \ \textbf{in} \ \langle empty, pop(s_2') \rangle$$
$$\qquad \textbf{else} \ \langle s_1, pop(s_2) \rangle$$

$$reverse(\langle empty, s \rangle) = s$$
$$reverse(\langle push(n, s_1), s_2 \rangle) = reverse(\langle s_1, push(n, s_2) \rangle)$$

This is a correct implementation only when behavioural satisfaction is assumed, w.r.t. observable sorts $\{int, bool\}$.

If we intend specifications STACKS and QUEUES to describe regular algebras, we ought to extend the implementation step by a definition of fix$_{queue}$. An infinite queue behaves like an infinite stack, with $get(\_)$ and $rest(\_)$ operations corresponding to $pop(\_)$ and $top(\_)$, respectively. Moreover, putting new elements into an infinite queue has no effect. Hence we may essentially re-use the fix operator of stacks:

$$\text{fix}_{queue}(f) := \langle \ empty \ , \ \text{fix}_{stack}(\ \lambda s.reverse(\ f(\langle empty, s \rangle) \ ) \ ) \ \rangle.$$

In particular, we derive $\perp_{queue} = \langle empty, \perp_{stack} \rangle$.

Some larger examples of specification and implementation in the setting of regular algebras can be found in [9].

An Extensible Proof Text Editor

Thomas Hallgren and Aarne Ranta*

Department of Computing Science
Chalmers University of Technology
S-412 96 Göteborg, Sweden
{hallgren,aarne}@cs.chalmers.se

Abstract. The paper presents an extension of the proof editor Alfa with natural-language input and output. The basis of the new functionality is an automatic translation to syntactic structures that are closer to natural language than the type-theoretical syntax of Alfa. These syntactic structures are mapped into texts in languages such as English, French, and Swedish. In this way, every theory, definition, proposition, and proof in Alfa can be translated into a text in any of these languages. The translation is defined for incomplete proof objects as well, so that a text with "holes" (i.e. metavariables) in it can be viewed simultaneously with a formal proof constructed. The mappings into natural language also work in the parsing direction, so that input can be given to the proof editor in a natural language.

The natural-language interface is implemented using the Grammatical Framework GF, so that it is possible to change and extend the interface without recompiling the proof editor. Such extensions can be made on two dimensions: by adding new target languages, and by adding theory-specific grammatical annotations to make texts more idiomatic.

1 Introduction

Computer algebra systems, such as Mathematica [21] and Maple [14], are widely used by mathematicians and students who do not know the internals of these systems. Proof editors, such as Coq [1], LEGO [2], Isabelle [4], and ALF [15], are less widely used, and require more specialized knowledge than computer algebras. One important reason is, of course, that the structures involved in manipulating algebraic expressions are simpler and better understood than the structures of proofs, and typically much smaller. This difference is inescapable, and it may well be that formal proofs will never be as widely interesting as formal algebra. At the same time, there is one important factor of user-friendliness that can be improved: the language used for communication with the system. While computer algebras are reasonably conversant in the "ordinary language" of mathematics, that is, expressions that occur in ordinary mathematical texts, proof editors only read and write artificial languages that are designed by logicians and computer scientists but not used in mathematical texts.

---

* The authors are grateful to anonymous referees for many suggestions and corrections.

M. Parigot and A. Voronkov (Eds.): LPAR 2000, LNAI 1955, pp. 70–84, 2000.

Making proof editors conversant in the language of ordinary proofs is clearly a more difficult task than building support for algebraic expressions. There are two main reasons for this: first, ordinary algebraic symbolism is quite formal already, and reflects the underlying mathematical structures more closely than proof texts in books reflect the structure of proofs. Second, the realm of proofs is much wider than algebraic expressions, which is already shown by the fact that proofs can contain arbitrary algebraic expressions as parts and that they also contain many other things.

We are far from a situation in which it is possible to take an arbitrary mathematical text (even a self-contained one) and feed it into a proof editor so that the machine can check whether the proof is correct, or even return a list of open problems if the proof contains leaps too long for the machine to follow. What is within reach, however, is a *restricted language* at the same time intelligible to non-specialist users, formally defined, and implemented on a computer. With such a language, it is not guaranteed that the machine understands all input that the user finds meaningful, but the machine will always be able to produce output meaningful for the user.

The idea of a natural-language-like formal language of proofs was presented by de Bruijn under the title of Mathematical Vernacular [12]. Implementations of such languages have been made in connection with at least Coq [11], Mizar [3], and Isabelle [4]. Among these implementations, it is Coq that comes closest to the idea of having a *language of proofs*, in the same sense as type theory: a language in which proofs can be written, so that parts of the proof text correspond to parts of the formal proof. The other languages reflect the *proof process* rather than the *proof object*: they explain what commands the user has given to the machine, or what steps the machine has made automatically, when constructing the proof. While sometimes possibly more useful and informative than a text reflecting the proof object (because it communicates the heuristics of finding the proof), a description of the proof process is more system-dependent and less similar to ordinary proof texts than a description of the proof object.

Like the "text extraction" functionality of Coq [11], the present work aims to build a language of proofs whose structures are similar to the structures of proof objects. The scope of the present work is wider in certain respects:

- We do not only consider proofs but propositions and definitions as well.
- Our language can be used not only for output but for input as well[1].
- Our language can be extended by the user in the same way as proof editors are extended by user-defined theories.

At the same time, the present work is more modest in one respect:

- We do not study automatic optimizations of the text.

The user of our interface always gets a proof text that directly reflects the formal proof, and thus has to do some extra work on the proof (and possibly

---

[1] An extension of the Coq interface [10], however, has a reversible translation of proofs to texts.

on language extensions) to make the proof texts short. The Coq interface, in contrast, automatically performs certain abbreviating optimizations on the proof [9]. However, the optimization feature is orthogonal to the novel features of our system, and one may well consider combining the two into something yet more powerful.

The focus of this paper is on the architecture and functionalities of a natural language interface to a proof editor. Little will be said about the linguistic questions of mathematical texts; some of the linguistic background work can be found in [17,18].

2 Proof Editors, Type Theory and Functional Programming

Alfa [13] is a graphical, syntax-directed editor for the proof system Agda. Agda [7] is an implementation of *structured type theory* (STT) [8], which is based on Martin-Löf's type theory [16]. The system is implemented completely in Haskell, using the graphical user interface toolkit Fudgets [6].

Like its predecessors in the ALF family of proof editors [15], Alfa allows the user to, interactively and incrementally, define theories (axioms and inference rules), formulate theorems and construct proofs of the theorems. All steps in the proof construction are immediately checked by the system and no erroneous proofs can be constructed.

Alternatively, since Martin-Löf's type theory is a typed lambda calculus, one can view Alfa as a syntax-directed editor for a small purely functional programming language with a powerful type system.

Figure 1 gives an idea of what the system looks like.

In virtue of being based on Martin-Löf type theory, STT can draw on the Curry-Howard isomorphism and serve as a unified language for propositions and proofs, specifications and programs. This allows Alfa to be used many ways:

- As a tool for pure logic. Alfa has in fact been used in undergraduate courses, allowing the students to practice doing natural deduction style proofs in propositional logic and predicate logic. As shown in Figure 2, Alfa has a mode of editing where terms are displayed as natural deduction style proof trees.
- As a tool for functional programming with dependent types. The language STT is closely related[2] to the language Cayenne [5], a full-fledged functional programming language with dependent types.
- As a tool for programming logic. The power of the language makes it possible to express both specifications and programs and to construct the proofs that the programs meet their specifications.

---

[2] The differences are to some extent due to the fact that Cayenne was designed to be used with an ordinary text editor and a batch compiler, whereas STT is designed for use in interactive proof editors.

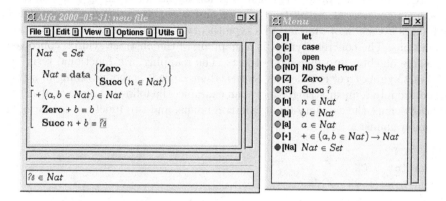

Fig. 1. A window dump of Alfa.

The user has defined the natural numbers and is working on the definition of addition. Question marks are *metavariables*, also called *place holders*, and allows the user to make a definition by starting from a skeleton and gradually refine it into a complete definition in a top down fashion. When a metavariable is selected, its type is displayed at the bottom of the window, and the menu indicates which ones of the identifiers in scope may be used to construct an expression of the required type.

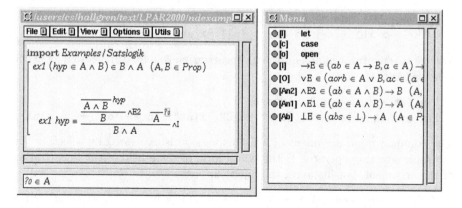

Fig. 2. A natural deduction proof in progress in Alfa.

3   The Grammatical Framework

GF (Grammatical Framework) [20] is a formalism for defining grammars. A grammar consists of an *abstract syntax* and a *concrete syntax*. The abstract syntax is a version of Martin-Löf's type theory, consisting of type and function definitions. The concrete syntax is a mapping of the abstract syntax, conceived as a free algebra, into linguistic objects. The mapping of a functional term (= abstract syntax tree) is called its *linearization*, since it is the flattening of a tree structure into a linear string. To give an example, the following piece of abstract syntax defines the category CN of common nouns, and two functions for forming common nouns:

```
cat CN
fun Int : CN
fun List : CN -> CN
```

To map this abstract syntax into English, we first define the class of linguistic objects corresponding to CN:

```
param Num = sg | pl
lincat CN = {s : Num => Str}
```

The first judgement introduces the parameter of number, with the two values the singular and the plural. The second judgement states that common nouns are records consisting of one field, whose type is a table of number-string pairs. The linearization rule for Int is an example of such a record[3]:

```
lin Int = {s = tbl {{sg} => "integer" ; {pl} => "integers"}}
```

In practice, it is useful to employ the GF facility of defining morphological operations, such as the inflection of regular common nouns:

```
oper regCN : Num => Str =
  \str -> tbl {{sg} => str ; {pl} => str + "s"}
```

We use this operator in an equivalent linearization rule for Int, as well as in the rule for List:

```
lin Int = {s = regCN "integer"}
lin List A = {s = tbl {n => regCN "list" ! n ++ "of" ++ A.s!pl}}
```

The common noun argument of a list expression is expressed by selecting (by the table selection operator !) the plural form of the s-field of the linearization of the argument. For instance, the functional term

---

[3]   GF uses the double arrow => for tables, or "finite functions", which are representable as lists of argument-value pairs. The table type is distinguished from the ordinary function type for metatheoretical reasons, such as the derivability of a parsing algorithm. A parallel distinction is made on the level of objects of these types: ordinary functions have the λ-abstract form $\backslash x \rightarrow \ldots$ whereas tables have the form tbl { ...}.

```
List (List Int)
```

is linearized into the record

```
{s = tbl {
       {sg} => ["list of lists of natural numbers"] ;
       {pl} => ["lists of lists of natural numbers"]}}
```

showing the singular and the plural forms of the complex common noun.

The concrete-syntax part of a grammar can be varied: for instance, the judgements

```
param Num = sg | pl
param Gen = masc | fem
oper regCN : Num => Str =
  \str -> tbl {{sg} => str ; {pl} => str + "s"}
oper de : Str =
  pre {"de" ; "d'"/strs {"a";"e";"i";"o";"u";"y"}}
lincat CN = {s : Num => Str ; g : Gen}
lin Int = {s = regCN "entier" ; g = masc}
lin List A =
  {s = tbl {n => regCN "liste" ! n ++ de ++ A.s ! pl ; g = fem}}
```

define a French variant of the grammar above. Notice that, unlike English, the French rules also define a gender for common nouns, as a supplementary field of the record.[4]

The class of grammars definable in GF includes all context-free grammars but also more[5]. Thus GF is applicable to a wide range of formal and natural languages. The implementation of GF includes a generic algorithm of linearization, but also of parsing, that is, translating from strings back to functional terms[6].

4   GF-Alfa: an Interface to Alfa

The GF interface to Alfa consists of two kinds of GF grammars:

– *Core grammars*, defining the translations of framework-level expressions.

---

[4] Also notice the elision of the preposition "de" in front of a vowel. An ordinary linguistic processing system might treat elision by a separate morphological analyser, but the user of a proof editor may appreciate the possibility of specifying everything in one and the same source file.

[5] The most important departure from context-free grammars is the possibility to permute, reduplicate, and suppress arguments of syntactic constructions. Rules using parameters, although conceptually non-context-free, can be interpreted as sets of context-free rules.

[6] The parsing algorithm is context-free parsing with some postprocessing. Suppressed arguments give rise to metavariables, which, in general, can only be restored interactively.

   – *Syntactic annotations*, defining translations of user-defined concepts.

The only grammar that is hard-coded in the Alfa system is the abstract syntax common to all core grammars. It is the grammar with which the normal syntax of Alfa communicates: the natural-language interface does not directly generate English or French, but expressions in this abstract syntax. The concrete syntax parts of core grammars are read from GF source when Alfa is started. Users of Alfa may thus modify them and add their own grammars for new languages[7].

   The syntactic categories of the interface are, essentially, those of the syntax of type theory used in the implementation of Alfa. The most important ones are expressions, constants (=user-defined expressions), and definitions:

```
cat Exp ; Cons ; Def
```

The category `Exp` covers a variety of natural-language categories: common nouns, sentences, proper names, and proof texts. Rather than splitting up `Exp` into all these categories, we introduce a set of corresponding parameters, and state that a given expression can be linearized into all of these forms:

```
param ExpForm = cn Num | sent | pn | text ; Num = sg | pl
lincat Exp = {s : ExpForm => Str}
```

For instance, the expression `emptySet`, which "intrinsically" is a proper name, has all of these forms, of which the `pn` form is the shortest:

```
lin emptySet = {s = tbl {
  (cn {sg}) => ["element of the empty set"] ;
  (cn {pl}) => ["elements of the empty set"] ;
  {sent}    => ["the empty set is inhabited"] ;
  {pn}      => ["the empty set"] ;
  {text}    => ["we use the empty set"]}}
```

This rule can be obtained as the result of a systematic transformation:

```
oper mkPN : Str -> {s : ExpForm => Str} = \str -> {s = tbl {
  (cn {sg}) => ["element of"] ++ str ;
  (cn {pl}) => ["elements of"] ++ str ;
  {sent}    => str ++ ["is inhabited"] ;
  {pn}      => str ;
  {text}    => ["we use"] ++ str}}
lin emptySet = mkPN ["the empty set"]
```

Such transformations can be defined for each parameter value taken as the "intrinsic" one for a constant. The user of GF-Alfa can, to a large extent, rely on these operations and need not write explicit tables and records. However, a custom-made annotation may give more idiomatic language:

---

[7] This is relatively easy: using the English core grammar as a model, the Swedish one was constructed in less than a day. It required ca. 400 lines of GF code, of which a considerable part is not used in the core grammar itself, but consists of macros that make it easier for Alfa users to write syntactic annotations.

```
lin emptySet = {s = tbl {
  (cn {sg}) => ["impossible element"] ;
  (cn {pl}) => ["impossible elements"] ;
  {sent}    => ["we have a contradiction"] ;
  {pn}      => ["the empty set"] ;
  {text}    => ["we use the empty set"]}}
```

The abstract syntax of the core grammars is extended every time the user defines a new concept in Alfa. The extension is by a function whose value type is Cons. For instance, the Alfa judgement

```
List (A::Set) :: Set = ...
```

is interpreted as a GF abstract syntax rule

```
fun List : Exp -> Cons
```

GF-Alfa automatically generates a default annotation,

```
lin List A = mkPN ("List" ++ A.s ! pn)
```

which the user may then edit to something more idiomatic for each target language: for instance,

```
lin List A =
  mkCN (tbl {n => regCN "list" ! n ++ "of" ++ A.s ! (cn pl)})
```

The reading given to proofs is not different from other type-theoretical objects. For instance, the conjunction introduction rule, which in Alfa reads

```
ConjI (A::Set)(B::Set)(a::A)(b::B) :: Conj A B = ...
```

can be given the GF annotation

```
lin ConjI A B a b = mkText (
  a.s ! text ++ "." ++ b.s ! text ++ "." ++
  "Altogether" ++ A.s ! sent ++ "and" ++ B.s ! sent)
```

The rest of natural deduction rules can be treated in a similar way, using e.g. the textual forms used in [11]. It is, of course, also possible to define *ad hoc* inference rules and give them idiomatic linearization rules.

On the top level, an Alfa theory is a sequence of definitions. Even theorems with their proofs are definitions of constants, which linguistically correspond to names of theorems. The linearization of a definition depends on whether the constant defined is intrinsically a proper name, common noun, etc. This intrinsic feature is by default proper name, but can be changed in a syntactic annotation. In the following section, examples are given of definitions of common nouns ("natural number") and proper names ("the sum of a and b"). Section 8 shows a definition of a constant conceived as the name of a theorem.

5   Natural Language Output

The primary and most basic function of GF in Alfa is to generate natural language text from code. Any definition or expression visible in the editor window can be selected and converted into one of the supported languages by using a menu command.

As an example, the default linearization of the (completed) definitions shown in Figure 1 would be as follows:

⌈ Definition. Nat is defined as follows:
  – the constructor *Zero* .
  – the constructor *Succ* applied to *n* where *n* is an element
  of Nat

⌈ Definition. Let *a* and *b* be elements of Nat. + applied to *a*
  and *b* is an element of Nat, depending on *a* as follows:
  – for the constructor *Zero* , choose *b* .
  – for the constructor *Succ* applied to *n* , choose the
  constructor *Succ* applied to + applied to *n* and *b*

By adding the following grammatical annotations,

```
Nat     = mkRegCN ["natural number"]
Zero    = mkPN "zero"
Succ n  = mkPN (["the successor of"] ++ n.s ! pn)
(+) a b = mkPN (["the sum of"] ++ a.s!pn ++ "and" ++ b.s!pn)
```

and similar grammatical annotations for Swedish and French, we obtain the following versions of the above definitions:

Definition. A natural number is defined by the following constructors:
- zero
- the successor of *n* where *n* is a natural number.
Definition. Let *a* and *b* be natural numbers. Then the sum of *a* and *b* is a natural number, defined depending on *a* as follows:
- for zero, choose *b*
- for the successor of *n*, choose the successor of the sum of *n* and *b*.

Définition. Les entiers naturels sont définis par les constructeurs suivants :
- zéro
- le successeur de *n* où *n* est un entier naturel.
Définition. Soient *a* et *b* des entiers naturels. Alors la somme de *a* et de *b* est un entier naturel, qu'on définit dépendant de *a* de la manière suivante :
- pour zéro, choisissons *b*
- pour le successeur de *n*, choisissons le successeur de la somme de *n* et de *b*.

Definition. Ett naturligt tal definieras av följande konstruerare:
- noll
- efterföljaren till *n* där *n* är ett naturligt tal.
Definition. Låt *a* och *b* vara naturliga tal. Summan av *a* och *b* är ett naturligt tal, som definieras beroende på *a* enligt följande:
- för noll, välj *b*
- för efterföljaren till *n*, välj efterföljaren till summan av *n* och *b*.

It is possible to switch between the usual syntax, different language views and multilingual views by simple menu commands.

6   Symbolic parts of natural-language expressions

Using natural language in every detail is not always desirable. A more suitable expression for addition, for instance, would often be

```
(+') a b = mkPN (a.s ! pn ++ "+" ++ b.s ! pn)
```

A problem with $+'$ is, however, that it generates bad style if one or both of its arguments are expressed in natural language:

the successor of zero + 2.

The proper rule is that all parts of a symbolic expression must themselves be symbolic. This can be controlled in GF by introducing a parameter of formality and making pn dependent on it:

```
param Formality = symbolic | verbal
param ExpForm = ... | pn Formality | ...
```

The definition of mkPN must be changed so that it takes two strings as arguments, one symbolic and one verbal. We can then rephrase the annotations:

```
Zero    = mkPN "zero" "0"
(+) a b = mkPN
  (["the sum of"] ++ a.s!(pn verbal) ++ "and" ++ b.s!(pn verbal))
  (a.s!(pn symbolic) ++ "+" ++ b.s!(pn symbolic))
```

A separate symbolic version of + now becomes unnecessary. In a text, those parts that are to be expressed symbolically, are enclosed as arguments of an operator

```
MkSymbolic A a = mkPN (a.s!(pn symbolic)) (a.s!(pn symbolic))
```

Semantically, this operation is identity: its definition in Alfa is

```
MkSymbolic (A::Type)(a::A) = a
```

This is a typical example of an identity mapping that can be used for controlling the style of the output text.

7   Natural Language Input

In addition to obtaining natural language output, you can also use the parsers automatically generated by GF to enter expressions in natural language. This way, you can make definitions without seeing any programming language syntax at all. As a simple example, suppose you want to add a definition of *one* as the successor of zero. By using the command to add a new definition, you get a skeleton:

[Definition. *one* is an element of ?o , defined as ?₁

The first hole to fill in is the type of *one*. You can use the commands "Give in English", "Give in French", "Give in Swedish":

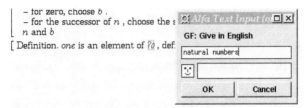

The last step is to enter the body of the definition,

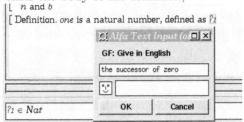

and the final result is:

> [Definition. *one* is a natural number, defined as the
> successor of zero

The parser understands only the fragment of natural languages we have defined, but can actually correct minor grammatical errors in the input. A completion mechanism helps in finding accepted words. The smiley in the input window gives feedback from the parser.

Since GF covers arbitrary context-free grammars (and more), it is possible for the concrete syntax to be ambiguous. When an ambiguous string is entered, Alfa asks the user to choose between the resulting alternative terms.

Ambiguous structures belong intimately to natural language, including the informal language of mathematics. Banning them from the proof editor interface would thus be a drastic limitation. Syntactic ambiguity is not so disastrous as one might think: careful writers use potentially ambiguous expressions only in contexts in which they can be disambiguated. The disambiguating factor is often type checking. For instance, the English sentence

for all numbers x, x is even or x is odd

has two possible syntactic analyses, corresponding to the formulas

$$(\forall x \in N)(\mathrm{Ev}(x) \vee \mathrm{Od}(x)),$$
$$(\forall x \in N)\mathrm{Ev}(x) \vee \mathrm{Od}(x).$$

Only the first reading is actually relevant, because the second reading has an unbound variable x. In this case, the GF-Alfa interface, which filters parses through type checker, would thus not even prompt the user to choose an alternative.

Since the annotation language of GF permits the user to introduce ambiguous structures, the parsing facility plays an important role even in natural language output: the question whether a text generated from a proof is ambiguous can be answered by parsing the text. Even a user who does not care about the natural language input facility of GF-Alfa may want to use the GF parser to find ambiguities in natural language output.

8   An Example: Insertion Sort

As a small, but non-trivial, example where GF and many features of Alfa are used together, we show some fragments from a correctness proof of a sorting algorithm.

We have defined insertion sort for lists of natural numbers in the typical functional programming style:

$$
\begin{bmatrix}
insert\ (x \in Nat, xs \in [Nat]) \in [Nat] \\
insert\ x\ [] \equiv x \mathbin{:} [] \\
insert\ x\ (x' \mathbin{:} xs') \equiv \text{if } x <= x' \text{ then } x \mathbin{:} xs \text{ else } x' \mathbin{:} insert\ x\ xs'
\end{bmatrix}
$$

$$
\begin{bmatrix}
sort\ (xs \in [Nat]) \in [Nat] \\
sort\ [] \equiv [] \\
sort\ (x \mathbin{:} xs') \equiv insert\ x\ (sort\ xs')
\end{bmatrix}
$$

The English translation of the definition of *sort* is

$$
\begin{bmatrix}
\text{Definition. Let } xs \text{ be a list of natural numbers. Insertion} \\
\text{sort applied to } xs \text{ is a list of natural numbers, depending on} \\
xs \text{ as follows:} \\
\text{– for the empty list, choose the empty list.} \\
\text{– for } x : xs'\ , \text{ choose } x \text{ inserted into insertion sort applied to} \\
xs'
\end{bmatrix}
$$

As a specification of the sorting problem, we use the following:

$$
\begin{bmatrix}
SortSpec\ (xs, ys \in [Nat]) \in Set \\
SortSpec\ xs\ ys \equiv xs \sim ys \wedge IsSorted\ ys
\end{bmatrix}
$$

We have chosen "*ys* is a sorted version of *xs*" as the English translation of *SortSpec xs ys*. The body of *SortSpec* translates to "*ys* is a permutation of *xs* and *ys* is sorted".

After proving some properties about permutations and the *insert* function, we can fairly easily construct the correctness proof for *sort* by induction on the list to be sorted. The proof is shown in natural deduction style in Figure 3.

The same proof can also be viewed in English. The beginning of it is:[8]

$$
\begin{bmatrix}
\text{The correctness proof for insertion sort. Let } xs \text{ be a list of} \\
\text{natural numbers. Insertion sort applied to } xs \text{ is a sorted} \\
\text{version of } xs\ . \\
\text{Proof. Use the element depending on } xs \text{ as follows: – for} \\
\text{the empty list, choose the result of the following procedure:} \\
\text{first, insertion sort applied to the empty list is a} \\
\text{permutation of the empty list: the empty list is a} \\
\text{permutation of itself. Second, insertion sort applied to the} \\
\text{empty list is sorted: trivial.} \\
\text{– for } x : xs'\ , \text{ choose the result of the following procedure:}
\end{bmatrix}
$$

Using the above proof, we can easily prove the proposition

$$
\forall xs \in [Nat] . \exists ys \in [Nat] . SortSpec\ xs\ ys
$$

The English translation of the proof is:

---

[8] We omit the rest of the proof for the time being. Some fine tuning is needed to make the text look really nice.

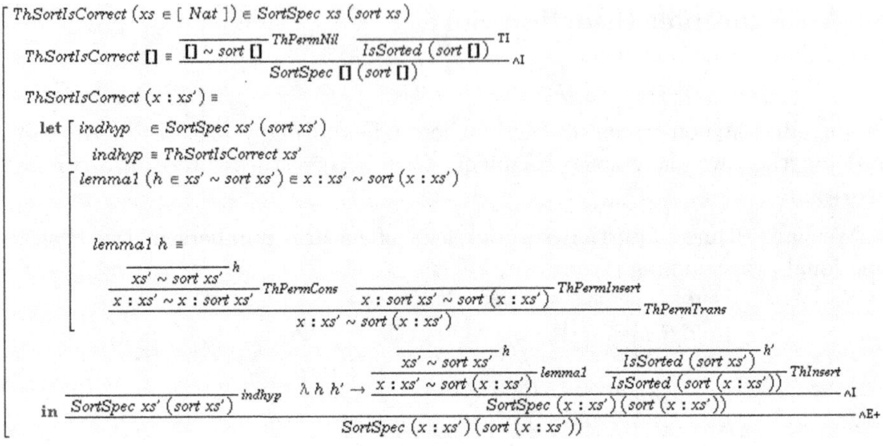

$$\left[\begin{array}{l} ThSortIsCorrect\,(xs \in [\,Nat\,]) \in SortSpec\,xs\,(sort\,xs) \\[4pt] ThSortIsCorrect\;[] \equiv \dfrac{\dfrac{}{[] \sim sort\,[]}\,ThPermNil \quad \dfrac{}{IsSorted\,(sort\,[])}\,TI}{SortSpec\,[]\,(sort\,[])}\,\wedge I \\[10pt] ThSortIsCorrect\,(x : xs') \equiv \\[4pt] \quad let \left[\begin{array}{l} indhyp \quad \in SortSpec\,xs'\,(sort\,xs') \\ indhyp \equiv ThSortIsCorrect\,xs' \end{array}\right. \\[8pt] \qquad \left[\begin{array}{l} lemma1\,(h \in xs' \sim sort\,xs') \in x : xs' \sim sort\,(x : xs') \\[10pt] lemma1\,h \equiv \\[4pt] \dfrac{\dfrac{\dfrac{}{xs' \sim sort\,xs'}\,h}{x : xs' \sim x : sort\,xs'}\,ThPermCons \quad \dfrac{}{x : sort\,xs' \sim sort\,(x : xs')}\,ThPermInsert}{x : xs' \sim sort\,(x : xs')}\,ThPermTrans \end{array}\right. \\[10pt] \quad in\;\dfrac{\dfrac{}{SortSpec\,xs'\,(sort\,xs')}\,indhyp \quad \lambda\,h\,h' \to \dfrac{\dfrac{\dfrac{}{xs' \sim sort\,xs'}\,h}{x : xs' \sim sort\,(x : xs')}\,lemma1 \quad \dfrac{\dfrac{}{IsSorted\,(sort\,xs')}\,h'}{IsSorted\,(sort\,(x : xs'))}\,ThInsert}{SortSpec\,(x : xs')\,(sort\,(x : xs'))}\,\wedge I}{SortSpec\,(x : xs')\,(sort\,(x : xs'))}\,\wedge E+ \end{array}\right.$$

Fig. 3. The correctness proof for insertion sort. See section 8.
The specification $SortSpec\ xs\ ys$ is defined to mean that ys is a permutation of xs, denoted $xs \sim ys$, and ys is sorted, denoted $IsSorted\ ys$.

> A sorting theorem. For every list of natural numbers xs, there exists a list of natural numbers ys such that ys is a sorted version of xs.
> Proof. Let xs be an arbitrary list of natural numbers. Let ys be insertion sort applied to xs. We know that ys is a sorted version of xs, since we can use the correctness proof for insertion sort. We conclude that, for every list of natural numbers xs, there exists a list of natural numbers ys such that ys is a sorted version of xs. QED

9   Conclusion

While Alfa dates back to 1995 and GF to 1998, the work on GF-Alfa only started at the end of 1999. It has been encouraging that the overall concept of integrating GF and Alfa works. Moreover, there is nothing particular to Alfa that makes this type of interface work; an earlier interface with the same architecture (i.e. core grammar + syntactic annotations) was built for the completely different formalism of extended regular expressions [19]. Similar lessons can be learnt from both systems:

- Formal structures can be mapped to natural-language structures so that arbitrarily complex expressions always give grammatically correct results. Thus it is possible to translate from formal to natural languages.
- Complex expressions are harder to understand in natural than in formal languages. Thus it is important to structure the code and break it into small units (such as lemmas), in order for the resulting text to be readable.
- It is useful to define equivalent variants of formal objects and equip them with different linearization rules. In this way stylistic variation can be included in the text. Linearization rules can also implement different degrees of information hiding.

- Natural-language input is only useful for small expressions, since entering a long expression runs the risk of falling outside the grammar.
- The interactive construction of a formal object is helped by a simultaneous view of the object as informal text.
- Ambiguities need not be forbidden in natural language, since they can be handled by interaction. Moreover, syntactic ambiguities are often automatically resolved by type checking.

The technique of improving the style of generated texts by syntactic annotations is interactive rather than automatic. Thus it fits well in the concept of interactive proof editors. This does not exclude the possibility of automatic text optimizations, e.g. factorizing parts of texts into shared parts (cf. [9]). The preferable place of such operations is on the level of abstract syntax, from where they are propagated to all target languages. Language-specific optimizations would certainly enable more elegant texts to be produced, but they would at the same time reduce the extensibility of the system.

References

1. Coq Homepage. http://pauillac.inria.fr/coq/, 1999. 70
2. The LEGO Proof Assistant. http://www.dcs.ed.ac.uk/home/lego/, 1999. 70
3. The Mizar Homepage. http://mizar.org/, 1999. 71
4. Isabelle Homepage. http://www.cl.cam.ac.uk/Research/HVG/Isabelle/, 2000. 70, 71
5. Lennart Augustsson. Cayenne — a language with dependent types. In *Proc. of the International Conference on Functional Programming (ICFP'98)*. ACM Press, September 1998. 72
6. M. Carlsson and T. Hallgren. *Fudgets — Purely Functional Processes with applications to Graphical User Interfaces*. PhD thesis, Department of Computing Science, Chalmers University of Technology, S-412 96 Göteborg, Sweden, March 1998. 72
7. C. Coquand. AGDA Homepage. http://www.cs.chalmers.se/~catarina/agda/, 1998. 72
8. C. Coquand and T. Coquand. Structured type theory. In *Workshop on Logical Frameworkds and Meta-languages*, Paris, France, Sep 1999. 72
9. Y. Coscoy. A natural language explanation of formal proofs. In C. Retoré, editor, *Logical Aspects of Computational Linguistics*, number 1328 in Lecture Notes in Artificial Intelligence, pages 149–167, Heidelberg, 1997. Springer. 72, 83
10. Y. Coscoy. *Explication textuelle de preuves pour le calcul des constructions inductives*. PhD thesis, Université de Nice-Sophia-Antipolis, 2000. 71
11. Y. Coscoy, G. Kahn, and L. Théry. Extracting text from proof. In M. Dezani and G. Plotkin, editors, *Proceedings of the International Conference on Typed Lambda Calculus and Applications (TLCA), Edinburgh*, number 902 in Lecture Notes in Computer Science. Springer-Verlag, 1996. 71, 71, 77
12. N. G. de Bruijn. Mathematical Vernacular: a Language for Mathematics with Typed Sets. In R. Nederpelt, editor, *Selected Papers on Automath*, pages 865–935. North-Holland Publishing Company, 1994. 71
13. T. Hallgren. Home Page of the Proof Editor Alfa. http://www.cs.chalmers.se/~hallgren/Alfa/, 1996-2000. 72

14. Waterloo Maple Inc. Maple Homepage. http://www.maplesof.com/, 2000. 70
15. L. Magnusson. *The Implementation of ALF - a Proof Editor based on Martin-Löf's Monomorphic Type Theory with Explicit Substitution.* PhD thesis, Department of Computing Science, Chalmers University of Technology and University of Göteborg, 1994. 70, 72
16. P. Martin-Löf. *Intuitionistic Type Theory.* Bibliopolis, Napoli, 1984. 72
17. A. Ranta. Context-relative syntactic categories and the formalization of mathematical text. In S. Berardi and M. Coppo, editors, *Types For Proofs and Programs*, number 1158 in Lecture Notes in Computer Science, pages 231–248. Springer-Verlag, 1996. 72
18. A. Ranta. Structures grammaticales dans le français mathématique. *Mathématiques, informatique et Sciences Humaines*, (138, 139):5–56, 5–36, 1997. 72
19. A. Ranta. A multilingual natural-language interface to regular expressions. In L. Karttunen and K. Oflazer, editors, *Proceedings of the International Workshop on Finite State Methods in Natural Language Processing*, pages 79–90, Ankara, 1998. Bilkent University. 82
20. A. Ranta. Grammatical Framework Homepage. http://www.cs.chalmers.se/~aarne/GF/index.html, 2000. 74
21. Inc. Wolfram Research. Mathematica Homepage. http://www.wolfram.com/products/mathematica/, 2000. 70

A Tactic Language for the System Coq

David Delahaye*

Project Coq
INRIA-Rocquencourt**

Abstract. We propose a new tactic language for the system Coq, which
is intended to enrich the current tactic combinators (tacticals). This language is based on a functional core with recursors and matching operators for Coq terms but also for proof contexts. It can be used directly
in proof scripts or in toplevel definitions (tactic definitions). We show
that the implementation of this language involves considerable changes
in the interpretation of proof scripts, essentially due to the matching operators. We give some examples which solve small proof parts locally and
some others which deal with non-trivial problems. Finally, we discuss the
status of this meta-language with respect to the Coq language and the
implementation language of Coq.

1 Introduction

In a proof[1] system, we can generally distinguish between two kinds of languages:
a proof language, which corresponds to basic or more elaborate primitives and
a tactic language, which allows the user to write his/her own proof schemes.
In this paper, we do not deal with the first kind of language which has been
already extensively studied by, for example, John Harrison in a comparative
way ([7]), Don Syme with a declarative prover ([11]) and Yann Coscoy with a
"natural" translation of proofs ([2]). Here, we focus on the tactic language which
is essentially the criterion for assessing the power of automation of a system (to
be distinguished from automation which is related to provided tactics). In some
systems, the tactic language does not exist and the automation has to be quite
powerful to compensate for this lack. For example, this is the case for PVS ([10])
where nothing is given to extend the system. Also, Mizar ([12]), one of the oldest
provers, is based on a unique tactic by and it is impossible to automate some
parts of the proofs or more generally, some logic theories.

The tactic language must be Turing-complete, which is to say that we must
be able to build proof strategies without any limitation imposed by the language
itself. Indeed, in general, this language is nothing other than the implementation
language of the prover. The choice of such a language has several consequences
that must be taken into account:

---

* David.Delahaye@inria.fr, http://coq.inria.fr/~delahaye/.
** INRIA-Rocquencourt, domaine de Voluceau, B.P. 105, 78153 Le Chesnay Cedex,
France.
[1] The word "proof" is rather overloaded and can be used in several ways. Here, we
use "proof" for a script to be presented to a machine for checking.

M. Parigot and A. Voronkov (Eds.): LPAR 2000, LNAI 1955, pp. 85–95, 2000.

- the prover developers have to provide the means to prevent possible inconsistencies arising from user tactics. This can be done in various ways. For example, in LCF ([6]) and in HOL ([5]), this is done by means of an abstract data type and only operations (which are supposed to be safe) given by this type can be used. In Coq ([1]), the tactics are not constrained, it is the type-checker which, as a Cerberus, verifies that the term, built by the tactic, is of the theorem type we want to prove.
- the user has to learn another language which is, in general, quite different from the proof language. So, it is important to consider how much time the user is ready to spend on this task which may be rather difficult or at least, tedious.
- the language must have a complete debugger because finding errors in tactic code is much harder than in proof scripts developed in the proof language, where the system is supposed to assist in locating errors.
- the proof system must have a clear and a well documented code, especially for the proof machine part. The user must be able to easily and quickly identify the necessary primitives or he/she could easily get lost in all the files and simply give up.

Thus, we can notice that writing tactics in a full programmable language involves many constraints for developers and more especially for users. In fact, we must recognize that the procedure is not really easy but we have no alternative if we want to avoid restrictions on tactics. However, we can wonder if this method is suitable for every case. For example, if we want a tactic which can solve linear equations on an Abelian field, it seems to be a non-trivial problem which requires a complete programming language. But, now suppose that we want to show that the set of natural numbers has more than two elements. This can be expressed as follows:

$$\vdash (\exists x : \mathsf{N}.\exists y : \mathsf{N}.\forall z : \mathsf{N}.x = z \lor y = z) \to \bot$$

To show this lemma, we introduce the left-hand member of the conclusion (say H) and eliminate it, then we introduce the witness (say a) and the instantiated hypothesis H (say H_a), finally, we eliminate H_a to introduce the second witness (say b) and the instantiation of H_a (say H_b). At this point, we have the following sequent:

$$..., H_b : \forall z : \mathsf{N}.a = z \lor b = z \vdash \bot$$

It remains to eliminate H_b with any three natural numbers (say 0, 1 and 2). Finally, we have three equalities (that we introduce) with a or b as the left-hand member and 0, 1 or 2 as the right-hand member. To conclude in each case, it is simply necessary to apply the transitivity of the equality between two equations with the same left-hand member, then we obtain an equality between two distinct natural numbers which validates the contradiction (depending on the prover, this last step must be detailled or not).

Of course, the length of this proof depends on the automation of the prover used. For example, in PVS, it may be imagined that applying the lemma of transitivity is quite useless and assert would solve all the goals generated by the eliminations of H_b. In Coq, the proof would be done exactly in this way and we may want to automate the last part of the proof where we use the transitivity. Unfortunately, even if this automation seems to be quite easy to realize, the current tactic combinators (tacticals) are not powerful enough to make it. So, the user has two choices: to do the proof by hand or to write his/her own tactic, in Objective Caml[2] ([8]), which will be used only for this lemma.

Thus, it is clear that a large and complete programming language is not a good choice to automate small parts of proofs. This is essentially due to the fact that the interfacing is too heavy with respect to the result the user wants to obtain. Moreover, the need for small automations must not only be seen as a lack of automation of the prover because tactics are intended to solve general problems and sometimes, user problems are too specific to be covered by primitive tactics. Thus, it seems that there is a gap between the proof language and the language used for writing tactics.

Here, we want to propose, in the context of Coq, the idea of an intermediate language, integrated in the prover and less powerful than the Turing-complete language for writing tactics, which is able to deal with small parts of proofs we may want to automate locally. This language is intended to be a kind of middle-way where it is possible to better enjoy both the usual language of Coq and some features of the full programmable language.

2   Presentation of the Language

2.1   Definition

Currently, the only way to combine the primitive tactics is to use predefined operators called tacticals. These are listed in table 1.

As seen previously, no tactical given in table 1 seems to be suitable for automating our small proof. In fact, we would like to do some pattern matchings on terms and even better, on proof contexts. So, the idea is to provide a small functional core with recursion to have some high order structures and with pattern matching operators both for terms as well as for proof contexts to handle the proof process. The syntax of this language, we call $\mathcal{L}_{tac}$, is given, using a BNF-like notation, by the entry *expr* in table 2, where the entries *nat*, *ident*, *term* and *primitive_tactic* represent respectively the natural numbers, the authorized identificators, Coq's terms and all the basic tactics. In *term*, there can be specific variables like ?n, where n is a *nat* or ?, which are metavariables for pattern matching. ?n allows us to keep instantiations and to make constraints whereas ? shows that we are not interested in what will be matched. We can also use this language in toplevel definitions (Tactic Definition) for later calls.

---

[2] This is the implementation language of Coq.

$tac_1;tac_2$	Applies tac_1 and tac_2 to all the subgoals
$tac;[tac_1\|...\|tac_i\|...\|tac_n]$	Applies tac and tac_i to the i-th subgoal
tac_1 Orelse tac_2	Applies tac_1 or tac_2 if tac_1 fails
Do n tac	Applies tac n times
Repeat tac	Applies tac until it fails
Try tac	Applies tac and does not fail if tac fails
First $[tac_1\|...\|tac_i\|...\|tac_n]$	Apply the first tac_i which does not fail
Solve $[tac_1\|...\|tac_i\|...\|tac_n]$	Apply the first tac_i which solves
Idtac	Leaves the goal unchanged
Fail	Always fails

Table 1. Coq's tacticals

2.2   Semantics

We do not wish to give a formal semantic here. It is not our main aim and would be premature. We can just say that in the context of a reduction semantics (small steps), the interpretation is almost usual. This language can give expressions which are tactics (to apply to a goal) and others which represent terms, for example. Thus, we must evaluate the expressions in an optional environment which is a possible goal. This environment is used for Match Context which makes non-linear first order unification as well as Match. Match Context has a very specific behavior. It tries to match the goal with a pattern (hypotheses are on the left of |- and conclusion is on the right) and if the right-hand member is a tactic expression which fails then it tries another matching with the same pattern. This mechanism allows powerful backtrackings and we will discuss an example of use below.

2.3   Typechecking

This language is not yet typechecked; although this might be useful in the future for at least two reasons. First, we have some ambiguities which must be solved by syntactic means and a consequence is the presence of a quote to mark the application of $\mathcal{L}_{tac}$ (see table 2). Another reason for building a typechecker is that we want to detect statically the free variables in a proof script. Experience of proof maintainability shows that proofs are quite sensitive to naming conventions and the idea is mainly to watch the names of hypotheses. Thus, typechecking will be an interesting and original feature of the language and will allow robust scripts to be built.

2.4   Implementation

To implement $\mathcal{L}_{tac}$, we had to make some choices regarding the existing code. First, we decided to keep an interpreted language. We are not really convinced

expr	::=	*expr* ; *expr*
	|	*expr* ; [*(expr |)* expr*]
	|	*atom*
atom	::=	Fun *input_fun*$^+$ -> *expr*
	|	Let *(let_clause* And*)* rec_clause* In *expr*
	|	Rec *rec_clause*
	|	Rec *(rec_clause* And*)* rec_clause* In *expr*
	|	Match Context With *(context_rule |)* context_rule*
	|	Match *term* With *(match_rule |)* match_rule*
	|	'(*expr*)
	|	'(*expr expr*$^+$)
	|	*atom* Orelse *atom*
	|	Do *(int | ident) atom*
	|	Repeat *atom*
	|	Try *atom*
	|	First [*(expr |)* expr*]
	|	Solve [*(expr |)* expr*]
	|	Idtac
	|	Fail
	|	*primitive_tactic*
	|	*arg*
input_fun	::=	*ident*
	|	()
let_clause	::=	*ident* = *expr*
rec_clause	::=	*ident input_fun*$^+$ -> *expr*
context_rule	::=	[*(context_hyps* ;*)* context_hyps* |- *term*] -> *expr*
	|	[|- *term*] -> *expr*
	|	_ -> *expr*
context_hyps	::=	*ident* : *term*
	|	_ : *term*
match_rule	::=	[*term*] -> *expr*
	|	_ -> *expr*
arg	::=	()
	|	*nat*
	|	*ident*
	|	*term*

Table 2. Definition of $\mathcal{L}_{tac}$

that we could save a significant amount of time in the execution of compiled scripts, in general run once, especially if we consider the cost of compilation time. Compared to the previous interpretation core[3], we have made great changes in the main function which executes the tactics, by, for example, adding the new structures we saw previously (see table 2). Also, to be able to deal with substitutions coming from abstracted variables (Fun) and metavariables (Match Context, Match), we interpret the tactic arguments in the main function. The tactics now take already interpreted arguments rather than AST's (Abstract Syntax Trees) coming from syntactical analysis. To be extendable, it is possible to dynamically associate interpretation functions to specific AST nodes.

3   Examples

A first natural example is the one we discussed in the introduction. We want to show that the set of natural numbers has more than two elements. With the current tactic language of Coq, the proof could look like the script given in table 3. As can be seen, after the three inductions (Elim), we have eight cases which can be solved by eight very similar instructions which are possibly different in the equality we cut and the term used to apply transitivity. As we know that this equality, say x=y, is such that there exist the equalities a=x and a=y in the hypotheses, it would be easy to automate this part provided that we can handle the proof context. This can be done by using $\mathcal{L}_{tac}$ and especially, the Match Context structure. Table 4 shows the corresponding script. We can notice that the proof is considerably shorter[4] and this is increasingly true when we add cases (with three, four , ... elements). Moreover, the work is much less tedious than in the case of the proof by hand and the script can be written without the help of the interactive toplevel loop. This results in a proof style which is much more batch mode like.

Another example, a little less trivial, is the problem of list permutation on closed lists. Indeed, we may be faced with this problem when we want to show that a list is sorted and it is quite annoying to do the proof by hand when we know it can be done automatically. To use Objective Caml[5] is certainly quite excessive compared to the difficulty of what we want to solve and $\mathcal{L}_{tac}$ seems to be much more appropriate. To do this, first, we define the permutation predicate as shown in table 5, where ^ represents the append operation on lists. Next, we can write naturally the tactic by using $\mathcal{L}_{tac}$ and the result can be seen in table 6. We can notice that we use two toplevel definitions PermutProve and Permut. The function to be called is PermutProve which is intended to solve goals of the form ...|-(permut l1 l2), where l1 and l2 are closed list expressions. PermutProve computes the lengths of the two lists and calls Permut with the length if the two lists have the same length. Permut works as expected. If the two lists are equal, it

---

[3] Of the last release V6.3.1.

[4] In this respect, we can see that the non-linear pattern matching solves the problem in one pattern instead of two successive patterns.

[5] This is the full programmable language to write tactics in Coq.

```
Lemma card_nat: ~(EX x:nat|(EX y:nat|(z:nat)(x=z)\/(y=z))).
Proof.
    Red;Intro H.
    Elim H;Intros a Ha.
    Elim Ha;Intros b Hb.
    Elim (Hb (0));Elim (Hb (1));Elim (Hb (2));Intros.
    Cut (0)=(1);[Discriminate|Apply trans_equal with a;Auto].
    Cut (0)=(1);[Discriminate|Apply trans_equal with a;Auto].
    Cut (0)=(2);[Discriminate|Apply trans_equal with a;Auto].
    Cut (1)=(2);[Discriminate|Apply trans_equal with b;Auto].
    Cut (1)=(2);[Discriminate|Apply trans_equal with a;Auto].
    Cut (0)=(2);[Discriminate|Apply trans_equal with b;Auto].
    Cut (0)=(1);[Discriminate|Apply trans_equal with b;Auto].
    Cut (0)=(1);[Discriminate|Apply trans_equal with b;Auto].
Save.
```

Table 3. A proof on cardinality of natural numbers in Coq

concludes. Otherwise, if the lists have identical first elements, it applies Permut on the tail of the lists. Finally, if the lists have different first elements, it puts the first element of one of the lists (here the second one which appears in the permut predicate) at the end if that is possible, i.e., if the new first element has been at this place previously. To verify that all rotations have been done for a list, we use the length of the list as an argument for Permut and this length is decremented for each rotation down to, but not including, 1 because for a list of length n, we can make exactly $n - 1$ rotations to generate at most n distinct lists. Here, it must be noticed that we use the natural numbers of Coq for the rotation counter. In table 2, we can see that it is possible to use usual natural numbers but they are only used as arguments for primitive tactics and they cannot be handled, in particular, we cannot make computations with them. So, a natural choice is to use Coq data structures so that Coq makes the computations (reductions) by Eval Compute in and we can get the terms back by Match.

Beyond these small examples, we discovered that $\mathcal{L}_{tac}$ is much more powerful than might have been expected and, even if it was not our initial aim, this language can deal with non-trivial problems. For example, we coded a tactic to decide intuitionnistic propositional logic, based on the contraction-free sequent calculi LJT* of Roy Dyckhoff ([4]). There was already a tactic called Tauto and written in Objective Caml by César Muñoz ([9]). We observed several significant differences. First, with $\mathcal{L}_{tac}$, we obtained a drastic reduction in size with 40 lines of code compared with 2000 lines. This can be mainly explained by the complete backtracking provided by Match Context. Moreover, we were very surprised to get a considerable increase in performance which can reach 95% in some examples. In fact, this is understandable since $\mathcal{L}_{tac}$ is a proof-dedicated language and we can suppose that some algorithms (such as Dyckhoff's) may be coded very naturally.

```
Lemma card_nat: ~(EX x:nat|(EX y:nat|(z:nat)(x=z)\/(y=z))).
Proof.
    Red;Intro H.
    Elim H;Intros a Ha.
    Elim Ha;Intros b Hb.
    Elim (Hb (0));Elim (Hb (1));Elim (Hb (2));Intros;
        Match Context With
            [_:?1=?2;_:?1=?3|-?] ->
                Cut ?2=?3;[Discriminate|Apply trans_equal with ?1;Auto].
Save.
```

Table 4. A proof on cardinality of natural numbers using $\mathcal{L}_{tac}$

```
Section Sort.

Variable A:Set.

Inductive permut:(list A)->(list A)->Prop:=
    permut_refl:(l:(list A))(permut l l)
  |permut_cons:
        (a:A)(l0,l1:(list A))(permut l0 l1)->(permut (cons a l0) (cons a l1))
  |permut_append:(a:A)(l:(list A))(permut (cons a l) (l^(cons a (nil A))))
  |permut_trans:
        (l0,l1,l2:(list A))(permut l0 l1)->(permut l1 l2)->(permut l0 l2).

End Sort.
```

Table 5. Definition of the permutation predicate

Finally, readibility has been greatly improved so that maintainability has been made much easier (even if there is no debugger for $\mathcal{L}_{tac}$ yet).

We dealt with another important example which was to verify equalities between types and modulo isomorphisms. We chose to use the isomorphisms of the simply typed λ-calculus with Cartesian product and *unit* type (see, for example, [3]). Again, the code, we wrote by using $\mathcal{L}_{tac}$, was quite short (about 80 lines with the axiomatization) and quite readable so that extensions to more elaborated λ-calculi can be easily integrated.

4   Conclusion

We have presented a language ($\mathcal{L}_{tac}$) which is intended to make a real link between the primitive tactics and the implementation language (Objective Caml) used to write large tactics. In particular, it deals with small parts of proofs that

```
Tactic Definition Permut n:=
    Match Context With
        [|-(permut ? ?1 ?1)] -> Apply permut_refl
        |[|-(permut ? (cons ?1 ?2) (cons ?1 ?3))] ->
            Let newn=Eval Compute in (length ?2)
            In
                Apply permut_cons;'(Permut newn)
        |[|-(permut ?1 (cons ?2 ?3) ?4)] ->
            '(Match Eval Compute in n With
                [(1)] -> Fail
                |_ ->
                    Let l0'=(?3^(cons ?2 (nil ?1)))
                    In
                        Apply (permut_trans ?1 (cons ?2 ?3) l0' ?4);
                            [Apply permut_append|
                            Compute;'(Permut (pred n))]).

Tactic Definition PermutProve ():=
    Match Context With
        [|-(permut ? ?1 ?2)] ->
            '(Match Eval Compute in ((length ?1)=(length ?2)) With
                [?1=?1] -> '(Permut ?1)).
```

Table 6. Permutation tactic in $\mathcal{L}_{tac}$

are to be automated. It can be seen that this language has some interesting features:

- it is in the toplevel of Coq. We do not need a compiler or any specification of the implementation of Coq to write tactics in this language. Moreover, to learn this small language would be certainly easier than tackling the manual of the implementation language. Of course, these remarks must be considered with regard to small tactics.
- the code length is, in general, quite short compared to the same proofs made by hand (see tables 3 and 4) and, even when solving non-trivial problems, we still have reductions in size, which are sometimes very impressive (as in the case of Tauto seen previously). Thus, the scripts are more compact and much simpler.
- the scripts are more readable. This is already the case with small proofs but even more so with large tactics (as with Tauto again).
- the scripts are more maintainable, as a direct consequence of the increase in readibility.

It is important to carefully define the scope of $\mathcal{L}_{tac}$ compared to Objective Caml. We must not be tempted to enrich $\mathcal{L}_{tac}$ too much in order to write tactics which are more and more complex. Even if we can at present deal with some complex examples, this must be considered as a bonus and not as a goal. We must make sure that Coq does not draw too much upon Objective Caml and, for the moment, we think that $\mathcal{L}_{tac}$ is complete enough. However, we plan to enable Objective Caml to enjoy the advantages of $\mathcal{L}_{tac}$ by a quotation or a syntax extension. With this system, we could use $\mathcal{L}_{tac}$ in Objective Caml like a true Application Programming Interface (API for short) with specific calls, as seen previously, so that we could write tactics more easily and without any limitation.

From the user point of view, it could be a tricky problem to decide which language is the most appropriate to solve his/her problem. The user must know whether the problem in hand can be coded with $\mathcal{L}_{tac}$. There is no general rule but we can identify several criteria by which Objective Caml must be used rather than $\mathcal{L}_{tac}$. First, $\mathcal{L}_{tac}$ is not suitable for tactics which handle the environment. For example, searching the global context is only possible by using Objective Caml and certain functions of Coq's code. Another indicator that $\mathcal{L}_{tac}$ is not suitable is the use of data structures. The more we use data structures, the more complex the problem is, as is the tactic to build. As shown previously with the example of list permutation (see tables 5 and 6), we can use data structures in $\mathcal{L}_{tac}$ by means of Coq's data structures[6] which can be handled by Match (and possibly Match Context) and the number of data structures we need is a good indication of the difficulty of the tactic we want to write. Moreover, if you are concerned about performances, it is better to use Objective Caml's data structures which are much more efficient than those of Coq. Finally, there are more libraries implementing usual data structures in Objective Caml than in Coq and this may be a decisive argument in some cases. Thus, in general, the use of data structures must be limited in $\mathcal{L}_{tac}$ and the user must make choices. For example, the use of natural numbers in the previous example concerning list permutation seems to be quite reasonable and we may consider that this is also the case for other data structures such as booleans or lists.

References

1. Bruno Barras et al. *The Coq Proof Assistant Reference Manual Version 6.3.1.* INRIA-Rocquencourt, May 2000.
   http://coq.inria.fr/doc-eng.html. 86
2. Yann Coscoy. A natural language explanation for formal proofs. In C. Retoré, editor, *Proceedings of Int. Conf. on Logical Aspects of Computational Liguistics (LACL), Nancy*, volume 1328. Springer-Verlag LNCS/LNAI, September 1996. 85
3. Roberto Di Cosmo. *Isomorphisms of Types: from λ-calculus to information retrieval and language design.* Progress in Theoretical Computer Science. Birkhauser, 1995. ISBN-0-8176-3763-X. 92
4. Roy Dyckhoff. Contraction-free sequent calculi for intuitionistic logic. In *The Journal of Symbolic Logic*, volume 57(3), September 1992. 91
5. M. J. C. Gordon and T. F. Melham. *Introduction to HOL: a Theorem Proving Environment for Higher Order Logic.* Cambridge University Press, 1993. 86

---

[6] This can be seen as a step towards a bootstrapped system.

6. M. J. C. Gordon, R. Milner, and C. P. Wadsworth. Edinburgh LCF: a mechanised logic of computation. In *Lectures Notes in Computer Science*, volume 78. Springer-Verlag, 1979. 86
7. John Harrison. Proof style. In Eduardo Giménez and Christine Paulin-Mohring, editors, *Types for Proofs and Programs: International Workshop TYPES'96*, volume 1512 of *LNCS*, pages 154–172, Aussois, France, 1996. Springer-Verlag. 85
8. Xavier Leroy et al. *The Objective Caml system release 3.00*. INRIA-Rocquencourt, April 2000.
   http://caml.inria.fr/ocaml/htmlman/. 87
9. César Muñoz. Démonstration automatique dans la logique propositionnelle intuitionniste. Mémoire du DEA d'informatique fondamentale, Université Paris 7, Septembre 1994. 91
10. Sam Owre, Natarajan Shankar, and John Rushby. PVS: A prototype verification system. In *Proceedings of CADE 11, Saratoga Springs, New York*, June 1992. 85
11. Don Syme. *Declarative Theorem Proving for Operational Semantics*. PhD thesis, University of Cambridge, 1998. 85
12. Andrzej Trybulec. The Mizar-QC/6000 logic information language. In *ALLC Bulletin (Association for Literary and Linguistic Computing)*, volume 6, pages 136–140, 1978. 85

Proof Simplification for Model Generation and Its Applications

Miyuki Koshimura and Ryuzo Hasegawa

Graduate School of Information Science and Electrical Engineering
Kyushu University
6-1 Kasuga-Kouen, Kasuga, Fukuoka 816-8580, Japan
{koshi, hasegawa}@ar.is.kyushu-u.ac.jp

Abstract. Proof simplification eliminates unnecessary parts from a proof leaving only essential parts in a simplified proof. This paper gives a proof simplification procedure for model generation theorem proving and its applications to proof condensation, folding-up and completeness proofs for non-Horn magic sets. These indicate that proof simplification plays a useful role in theorem proving.

1 Introduction

A theorem prover for first-order logic called SATCHMO [13] was proposed by Manthey and Bry, which is based on *model generation* and effectively utilizes logic programming technologies. SATCHMO tries to construct models for a given clause set and determines its satisfiability. The model generation method maintains a set of ground atoms called a model candidate, finds violated clauses that are not satisfied under the model candidate, extends it to satisfy them, and repeats the process until a model is found or all model candidates are rejected.

Thus, we make use of model generation not only for model finding [4,16,8] but also refutation [7]. There are two types of redundancies in model generation: One is that the same subproof tree may be generated at several descendants after a case-splitting occurs. Another is caused by unnecessary model candidate extensions.

Folding-up is a well known technique for eliminating duplicate subproofs in a tableaux framework [12]. In order to embed folding-up into model generation, we have to analyze dependency in a proof for extracting lemmas from proven subproofs. Lemmas are used for pruning other subproofs. Dependency analysis makes unnecessary parts visible because such parts are independent of essential parts in the proof. In other words, we can separate unnecessary parts from the proof according to dependency analysis.

Identifying unnecessary parts and eliminating them are considered as *proof simplification*. The computational mechanism for their elimination is essentially the same as that for *proof condensation* [17] and *level cut* [2]. Considering this, we implement not only folding-up but also proof condensation by embedding one mechanism, i.e. proof simplification, into model generation. Proof simplification

M. Parigot and A. Voronkov (Eds.): LPAR 2000, LNAI 1955, pp. 96–113, 2000.
© Springer-Verlag Berlin Heidelberg 2000

can be achieved by computing "relevant atoms", which contribute to closing subproof, during the proof.

On the other hand, we developed a method called *non-Horn magic sets* (NHM) [7] to avoid unnecessary model candidate extensions. An ideal proof by the NHM method contains no unnecessary model candidate extension. A simplified proof agrees with the ideal, that is, it contains no unnecessary model candidate extensions. This implies that we can transform a proof by model generation into one by the NHM method by modifying proof simplification. This transformation gives a new completeness proof for the NHM method in a syntactical way.

The paper is organized as follows. In Section 2, we give a model generation procedure and in Section 3, we present a proof simplification procedure. From Section 4 to Section 6, we show a model generation procedure with proof condensation, a model generation procedure with proof condensation and folding-up, and a completeness proof of the NHM method by modifying the proof simplification procedure. In Section 7, we evaluate effects of proof condensation and folding-up by proving problems taken from the TPTP problem library.

2   Model Generation

Throughout this paper, a *clause* $\neg A_1 \vee \ldots \vee \neg A_n \vee B_1 \vee \ldots \vee B_m$ is represented in implicational form: $A_1 \wedge \ldots \wedge A_n \to B_1 \vee \ldots \vee B_m$ where A_i $(1 \le i \le n)$ and B_j $(1 \le j \le m)$ are atoms; the left hand side of "$\to$" is said to be the *antecedent*; and the right hand side of "$\to$" the *consequent*.

A clause is said to be *positive* if its antecedent is *true* $(n = 0)$, and *negative* if its consequent is *false* $(m = 0)$; otherwise it is *mixed* $(n \neq 0, m \neq 0)$. A clause is said to be *violated* under a set M of ground atoms if with some ground substitution σ the following condition holds: $\forall i(1 \le i \le n)A_i\sigma \in M \wedge \forall j(1 \le j \le m)B_j\sigma \notin M$.

A model generation proof procedure is sketched in Fig. 1. The procedure MG takes a partial interpretation Mc (model candidate) and a set of clauses S to be proven, and builds an annotated (sub)proof-tree of S. An annotated proof-tree records which clauses are used for model extension or rejection. The annotation is used for proof simplification described in the next section.

A leaf labeled with $\top$ tells us that a model of S has been found as a current model candidate. If every leaf of the constructed proof-tree is labeled with $\bot$, S is unsatisfiable; otherwise S is satisfiable. In the latter case, at least one leaf is labeled with $\top$ or at least one branch grows infinitely. In this paper, we deal with only finite proof trees for simplicity.

A normal proof-tree is obtained from an annotated proof-tree by removing the annotations. Conversely, an annotated proof-tree is obtained from a normal proof-tree by adding annotations to the normal proof-tree. In this way, we consider that an annotated proof-tree is equivalent to its corresponding normal proof-tree. Therefore, we use *proof-tree* to refer to either annotated or normal proof-trees where confusion does not arise.

---

procedure $MGTP(S) : AP$; /* Input(S):Clause set,
                          Output(AP):Annotated proof-tree of S */
   return$(MG(\emptyset, S))$;

procedure $MG(Mc, S) : AP$;/* Input(Mc): Model candidate */

1. (Model rejection) If a negative clause $(A_1 \wedge \ldots \wedge A_n \rightarrow false) \in S$ is violated under Mc with a ground substitution σ,
   return $\langle \overset{!}{\bot}, A_1\sigma \wedge \ldots \wedge A_n\sigma \rangle$
2. (Model extension) If a positive or mixed clause $(A_1 \wedge \ldots \wedge A_n \rightarrow B_1 \vee \ldots \vee B_m) \in S$ is violated under Mc with a ground substitution σ,
   return an annotated proof-tree in the form depicted in Fig. 2 where $AP_i = MG(Mc \cup \{B_i\sigma\}, S)$ $(1 \leq i \leq m)$.
3. (Model finding) If neither 1 nor 2 is applicable, **return** $\langle \overset{!}{\top}, \emptyset \rangle$;

Fig. 1. Model generation procedure

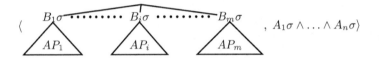

Fig. 2. Model extension

Example 1. Consider the set of clauses $S1$:

$$C1 : true \rightarrow r \quad C3 : r \rightarrow p \vee c \vee d \quad C5 : p \rightarrow false$$
$$C2 : r \rightarrow a \vee b \quad C4 : r \rightarrow p \vee q \quad\quad C6 : q \rightarrow false$$

Fig. 3 (a) shows an annotated proof-tree of $S1$ and Fig. 3 (b) shows a normal proof-tree of $S1$.

Example 2. Consider the set of clauses $S2$:

$$C1 : true \rightarrow t \vee p \quad C3 : q \rightarrow r \quad C5 : t \rightarrow p$$
$$C2 : p \rightarrow q \vee s \quad\quad C4 : s \rightarrow r \quad C6 : p \wedge r \rightarrow false$$

Fig. 4 (a) shows an annotated proof-tree of $S2$ and Fig. 4 (b) shows a normal proof-tree of $S2$.

3   Proof Simplification

In order to eliminate unnecessary model extensions from a proof-tree P, we have to make a decision which model extension can be eliminated. Relevant atoms and relevant model extensions as defined below provide a criterion for making the decision.

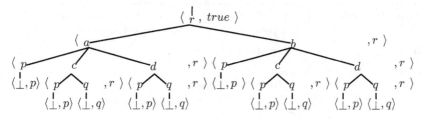

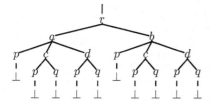

(a) An annotated proof-tree

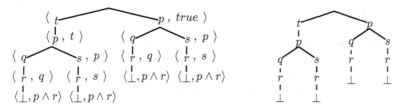

(b) A normal proof-tree

Fig. 3. Proof-trees of $S1$

(a) An annotated proof-tree   (b) A normal proof-tree

Fig. 4. Proof-trees of $S2$

Definition 1 (Relevant atom). *Let AP be an annotated finite (sub)proof-tree. A set $Rel(AP)$ of relevant atoms of AP is defined as follows:*

1. *If $AP = \langle \perp, A_1\sigma \wedge \ldots \wedge A_n\sigma \rangle$, then $Rel(AP) = \{A_1\sigma, \ldots, A_n\sigma\}$.*
2. *If $AP = \langle \top, \emptyset \rangle$, then $Rel(AP) = \emptyset$.*
3. *If AP is in the form depicted in Fig. 2 and*
    (a) $\forall i(1 \leq i \leq m)B_i\sigma \in Rel(AP_i)$, then
        $Rel(AP) = \cup_{i=1}^{m}(Rel(AP_i) \setminus \{B_i\sigma\}) \cup \{A_1\sigma, \ldots, A_n\sigma\}$
    (b) $\exists i(1 \leq i \leq m)B_i\sigma \notin Rel(AP_i)$, then $Rel(AP) = Rel(P_{i_0})$ (where i_0 is the minimal index[1] satisfying $1 \leq i_0 \leq m$ and $B_{i_0}\sigma \notin Rel(AP_{i_0})$)

---

[1]   We assume a fixed total order on indices. The order is also used by the proof simplification procedure shown in Fig. 5 and the proof-tree transformation shown in Fig. 11.

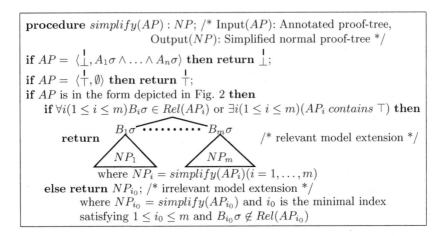

Fig. 5. Proof simplification procedure

Informally, relevant atoms of a (sub)proof-tree P are atoms which contribute to building P and appear as ancestors of P if P does not contain $\top$. If P contains $\top$, the set of relevant atoms of P is $\emptyset$.

Definition 2 (Relevant model extension). *A model extension by a clause* $A_1\sigma \wedge \ldots \wedge A_n\sigma \rightarrow B_1\sigma \vee \ldots \vee B_m\sigma$ *is relevant to the proof if the model extension yields the (sub)proof-tree in the form depicted is Fig. 2 and either* $\forall i(1 \leq i \leq m)B_i\sigma \in Rel(AP_i)$ *or* $\exists i(1 \leq i \leq m)(AP_i$ *contains* $\top)$ *holds.*

We can eliminate irrelevant model extensions. Let AP be an annotated (sub)proof-tree in the form depicted in Fig. 2. If there exists a subproof-tree AP_i $(1 \leq i \leq m)$ such that $B_i\sigma \notin Rel(AP_i)$ and AP_i does not contain $\top$, we can conclude that the model extension forming the root of AP is unnecessary because $B_i\sigma$ does not contribute to AP_i. Therefore, we can delete other subproof-trees $AP_j(1 \leq j \leq m, j \neq i)$ and take NP_i to be a simplified proof-tree of AP where NP_i is a simplified proof-tree of AP_i. When AP contains $\top$, we consider that the model extension forming the root of AP is necessary from a model finding point of view.

Fig. 5 shows a proof simplification procedure which eliminates irrelevant model extensions. The procedure $simplify$ takes an annotated proof-tree and returns a simplified normal proof-tree.

Example 3. Let AP_{S_1} be the annotated proof-tree shown in Fig. 3 (a). Fig. 6 shows the simplification process for AP_{S_1}. Both (a) and (b) show applications of $simplify$ to two leaves in AP_{S_1}, while (c) and (d) show sets of relevant atoms of these leaves. According to these two results, the model extension $p\frown q$ for the bottom left subproof-tree is relevant to the proof. So, the subproof-tree is not simplified as (e) indicates.

On the other hand, (f) says that the set of relevant atoms of the subproof-tree under c does not contain the atom c. This implies that the model extension p—c—d for the outer subproof-tree is irrelevant to the proof. Therefore, it is simplified as (g) indicates. Similarly, we conclude that the model extension a—b is also irrelevant to the proof. Finally, we obtain the simplified proof-tree shown in (h).

On the other hand, the annotated proof-tree APS_2 shown in Fig. 4 (a) contains no irrelevant model extensions, so, APS_2 is not simplified by $simplify$.

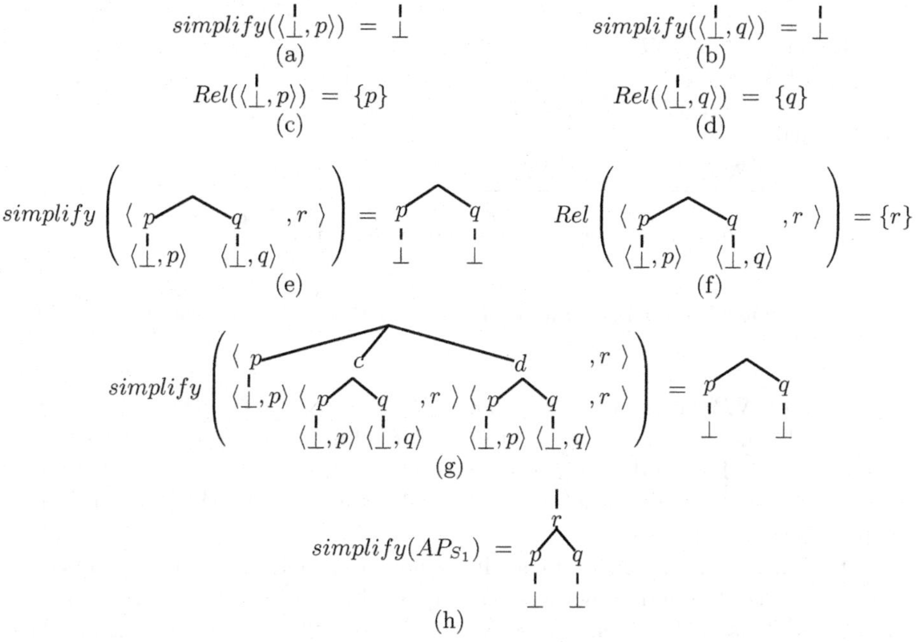

$$simplify(\langle \underset{\perp}{\overset{|}{}},p\rangle) = \underset{\perp}{\overset{|}{}}$$
(a)

$$simplify(\langle \underset{\perp}{\overset{|}{}},q\rangle) = \underset{\perp}{\overset{|}{}}$$
(b)

$$Rel(\langle \underset{\perp}{\overset{|}{}},p\rangle) = \{p\}$$
(c)

$$Rel(\langle \underset{\perp}{\overset{|}{}},q\rangle) = \{q\}$$
(d)

$$simplify\left(\langle \underset{\langle\perp,p\rangle\ \langle\perp,q\rangle}{p \diagup\diagdown q} , r \rangle \right) = \underset{\perp\quad\perp}{p\diagup\diagdown q}$$
(e)

$$Rel\left(\langle \underset{\langle\perp,p\rangle\ \langle\perp,q\rangle}{p \diagup\diagdown q} , r \rangle \right) = \{r\}$$
(f)

$$simplify\left(\langle \underset{\langle\perp,p\rangle\ \langle p\diagup\diagdown q ,r\rangle\ \langle p\diagup\diagdown q ,r\rangle}{p \diagup^{c}\diagdown^{d}} , r \rangle \right) = \underset{\perp\quad\perp}{p\diagup\diagdown q}$$
(g)

$$simplify(APS_1) = \underset{\perp\quad\perp}{\overset{r}{p\diagup\diagdown q}}$$
(h)

Fig. 6. Proof simplification process for APS_1

4   Application to Proof Condensation

Performing proof simplification *during* the proof, instead of *after* the proof has been completed, makes the model generation procedure more efficient. The procedure MG_{simp} in Fig. 7 realizes this idea. MG_{simp} is a combination of the MG in Fig. 1 and Definition 1. MG_{simp} returns a normal proof-tree NP and a set RA of relevant atoms. MG_{simp} builds the proof-tree in a left-first manner.

$B_i\sigma \notin RA_i$ for some $i(1 \leq i \leq m)$ means that the atom $B_i\sigma$ does not contribute to the proof NP_i. That is, the model extension does not contribute

procedure $MG_{simp}(Mc, S) : \langle NP, RA \rangle$;
/* Output(NP): Normal proof-tree, Output(RA): Set of relevant atoms */

1. (Model rejection) If a negative clause $(A_1 \wedge \ldots \wedge A_n \rightarrow false) \in S$ is violated under Mc with a ground substitution σ, **return** $\langle \overset{!}{\perp}, \{A_1\sigma, \ldots, A_n\sigma\} \rangle$;

2. (Model extension) If a positive or mixed clause $(A_1 \wedge \ldots \wedge A_n \rightarrow B_1 \vee \ldots \vee B_m) \in S$ is violated under Mc with a ground substitution σ,
   for $(i = 1; i \leq m; i{+}{+})$ {
   $\quad \langle NP_i, RA_i \rangle = MG_{simp}(Mc \cup \{B_i\sigma\}, S)$;
   $\quad$ **if** $B_i\sigma \notin RA_i$ and NP_i does not contains $\top$ **then** /* irrelevant */
   $\quad\quad$ **return** $\langle NP_i, RA_i \rangle$; /* proof condensation */
   }
   if $\exists i (1 \leq i \leq m)(NP_i$ contains $\top)$ **then** $RA = \emptyset$
   else $RA = \cup_{i=1}^{m}(RA_i \setminus \{B_i\sigma\}) \cup \{A_1\sigma, \ldots, A_n\sigma\}$;

   return $\langle$, $RA \rangle$;

3. (Model finding) If neither 1 nor 2 is applicable, **return** $\langle \overset{!}{\top}, \emptyset \rangle$;

Fig. 7. Model generation procedure with proof condensation

to the proof NP_i. Therefore the proofs $MG'(Mc \cup \{B_j\sigma\}, S)(i < j \leq m)$ can be ignored. Thus $m - i$ among m branches are eliminated after i branches have been explored. The model generation with proof simplification is essentially the same as the proof condensation in the HARP prover [17] and the level cut in the Hyper Tableaux prover [2].

These provers keep a flag for each inner node of proof trees. The flag indicates whether the corresponding node N participates in a subproof below N. Initially, the flag of N is "*off*" which means N does not participate in proof at all. The flag becomes "*on*" when the literal of N resolves away a complement literal in the subproof under N. If the flag remains "*off*" after the subproof below N is completed, we conclude that the extension step yielding N was unnecessary to obtain the subproof. Therefore, we can delete all open (unsolved) sibling nodes of N. Thus, the literal of a node with "*on*" flag is considered as "*relevant*" to the proof in our terminology.

Example 4. Fig. 8 shows model generation with proof condensation for the set S_1 of clauses shown in Example 1. The mark $\times$ indicates a pruned branch. Model extensions are performed in the left-first manner as the figure indicates. The set of the relevant atoms of the inner subproof-tree under c is $\{r\}$ as indicated in Fig. 6 (f). This implies that the model extension $p \overset{c}{\frown} d$ is irrelevant to the proof. Therefore, the proof under d is pruned. In a similar way, the exploration under b is eliminated. Thus, we obtain a proof-tree which has 6 inner nodes and 3 leaves.

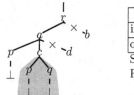

SP	RA
inner	r
outer	r

SP: subproof
RA: relevant atoms

Fig. 8. Eliminating irrelevant model extensions

5   Application to Folding-up

We make use of a set of relevant atoms not only for proof condensation but also to generate a lemma. The following theorem reveals an useful aspect of a set of relevant atoms.

Theorem 1. *Let S be a set of clauses, Mc be a set of ground atoms and $AP = MG(Mc, S)$. If all leaves in AP are labeled with $\perp$, i.e. AP does not contain $\top$, then $S \cup Rel(AP)$ is unsatisfiable.*

Proof. By structural induction on AP.

If $AP = \langle \stackrel{|}{\perp}, A_1\sigma \wedge \ldots \wedge A_n\sigma \rangle$ and $(A_1 \wedge \ldots \wedge A_n \to false)$ $(\in S)$ is the negative clause used for the model rejection, then $Rel(AP) = \{A_1\sigma, \ldots, A_n\sigma\}$. Obviously, $\{A_1 \wedge \ldots \wedge A_n \to false\} \cup Rel(AP)$ is unsatisfiable. Therefore, $S \cup Rel(AP)$ is unsatisfiable because $\{A_1 \wedge \ldots \wedge A_n \to false\} \subset S$.

If AP is in the form depicted in Fig. 2 and $(A_1 \wedge \ldots \wedge A_n \to B_1 \vee \ldots \vee B_m)$ $(\in S)$ is the clause used for the model extension. By the induction hypothesis, $S \cup Rel(AP_i)(i = 1, \ldots, m)$ is unsatisfiable. We have two cases: The model extension is irrelevant or relevant. In the former case, $\exists i(1 \leq i \leq m)(Rel(AP) = Rel(AP_i))$. Therefore $S \cup Rel(AP)$ is unsatisfiable.

In the latter case, we assume $S \cup Rel(AP)$ is satisfiable. Then, there exists a model M of $S \cup Rel(AP)$. It follows from $\forall i(1 \leq i \leq n)(A_i\sigma \in Rel(AP))$ that $\forall i(1 \leq i \leq n)(M \models A_i\sigma)$. Then, $\exists j(1 \leq j \leq m)(M \models B_j\sigma)$. This implies that $S \cup Rel(AP) \cup \{B_{j_0}\sigma\}$ is satisfiable where j_0 is an index satisfying $1 \leq j_0 \leq m$ and $M \models B_{j_0}\sigma$. Here, $Rel(AP) \cup \{B_{j_0}\sigma\} \supset Rel(AP_{j_0})$ and $S \cup Rel(AP_{j_0})$ is unsatisfiable. Then, $S \cup Rel(AP) \cup \{B_{j_0}\sigma\}$ is unsatisfiable. This is a contradiction. Therefore, $S \cup Rel(AP)$ is unsatisfiable.   □

This theorem says that a set of relevant atoms can be considered as a lemma. Consider the model generation procedure shown in Fig. 1. Let Mc be a current model candidate and AP be a subproof-tree which was previously obtained and does not contain $\top$. If $Mc \supset Rel(AP)$ holds, we can reject Mc without further proof because $S \cup Mc$ is unsatisfiable where S is a clause set to be proven. This rejection mechanism, which is a variant of merging [20], can reduce search spaces by orders of magnitude. However, it is expensive to test whether $Mc \supset Rel(AP)$. Therefore, we restrict usage of the rejection mechanism.

Definition 3 (Context unit lemma). *Let S be a set of clauses and AP be a (sub)proof-tree of S in the form depicted in Fig. 2. When $B_i\sigma \in Rel(AP_i)$, $Rel(AP_i) \setminus \{B_i\sigma\} \models_S \neg B_i\sigma$ is called a context unit lemma[2] extracted from AP_i. We call $Rel(AP_i) \setminus \{B_i\sigma\}$ the context of the lemma.*

Note that $B_i\sigma \in Rel(AP_i)$ implies $Rel(AP_i)$ is not empty. Therefore, AP_i does not contain $\top$. Thus, $S \cup Rel(AP_i)$ is unsatisfiable according to Theorem 1. We simply call a context unit lemma with empty context by an *unit lemma*.

The context of the context unit lemma extracted from $AP_i(1 \leq i \leq m)$ is satisfied in model candidates of sibling proofs $AP_j(j \neq i, 1 \leq j \leq m)$, that is, the lemma is available in AP_j. Furthermore, the lemma can be lifted to the nearest ancestor's node which does not satisfied the context (in other words, which is labeled with an atom in the context) and is available in its descendant's proofs.

Fig. 9 shows a model generation procedure which makes use of context unit lemmas. The procedure MG_{fup} takes a model candidate Mc, a set of context unit lemmas UL_I and a set of clauses S to be proven. There is a guarantee that $\forall (\Gamma \models L) \in UL_I(\Gamma \subset Mc)$. That is, every lemma in UL_I is available under Mc. MG_{fup} returns a normal (sub)proof-tree NP of S, a set of relevant atoms RA of NP and a set of context unit lemmas UL_O which can be lifted to an ancestor's node. Each lemma in UL_O is extracted from a subproof-tree of NP according to Definition 3. A leaf labeled with $\star$ indicates that a model candidate rejected by a lemma. This procedure is an implementation of folding-up [12] for model generation. The procedure also accomplishes proof condensation.

Example 5. Fig. 10 shows model generation with folding-up for the set S_2 of clauses shown in Example 2. The set of the relevant atoms of the left inner subproof-tree under q is $\{p, q\}$. So, the context unit lemma $\{p\} \models_{S_2} \neg q$ is extracted from the set. Similarly, we obtain context unit lemma $\{p\} \models_{S_2} \neg s$ and $\emptyset \models_{S_2} \neg p$ from the right inner subproof-tree under s and the outer subproof-tree under p, respectively. The latter lemma is lifted to the root node and the right subproof under p is pruned with it. Thus, a duplicate subproof is eliminated.

There has been a lot of work on refinements for tableaux approaches in order to shorten proofs. Caching [1] is a modified method of lemmas for model elimination calculus. Roughly speaking, cache is a complete database of all unit lemmas produced in past derivations. Merge path [3] is a generalization of folding-up. Merge path allows one to re-use proof trees whether they contain $\top$ or not, while folding-up does not allow one to re-use proof trees which contain $\top$. However, merge path requires an extra mechanism in order to avoid endless derivations which happen even though it apply propositional calculus.

So far little research has been carried out on combining proof condensation and lemma generation within a single framework such as computing relevant atoms. Iwanuma introduced lemma matching [9] which is a systematical treatment of lemma for model elimination calculus. In his work, a set of *levels* works as a set of relevant atoms.

---

[2] $\Gamma \models_S L$ is an abbreviation of $S \cup \Gamma \models L$ where Γ is a set of ground literals, S is a set of clauses, and L is a literal.

---

procedure $MGTP_{fup}(S) : NP$; /* Input(S):Clause set,
                               Output(NP):Normal proof-tree of S */
  $\langle NP, RA, UL\rangle = MG_{fup}(\emptyset, \emptyset, S)$; **return** NP;

procedure $MG_{fup}(Mc, UL_I, S) : \langle NP, RA, UL_O\rangle$;
/* Input(Mc): Model candidate, Input(UL_I): Set of context unit lemmas,
        Output(RA): Set of relevant atoms,
        Output(UL_O): Set of context unit lemmas lifted */

1. (Model rejection) If a negative clause $(A_1 \wedge \ldots \wedge A_n \rightarrow false) \in S$ is violated
   under Mc with a ground substitution σ, **return** $\langle \overset{!}{\perp}, \{A_1\sigma, \ldots, A_n\sigma\}, \emptyset\rangle$;

2. (Model extension) If a positive or mixed clause $(A_1 \wedge \ldots \wedge A_n \rightarrow B_1 \vee \ldots \vee B_m) \in S$
   is violated under Mc with a ground substitution σ,

   for $(i = 1; i \leq m; i++)$ {
       if $\exists(\Gamma \models_S \neg B_i\sigma) \in UL_I \cup \bigcup_{j=1}^{i-1} UL_O^j$ **then** {
           $NP_i = \overset{!}{\star}$; $RA_i = \Gamma$; $UL_O^i = \emptyset$; /* applying a lemma to $B_i\sigma$ */
       } **else** {
           $\langle NP_i, RA_i, UL_O^i\rangle = MG_{fup}(Mc \cup \{B_i\sigma\}, UL_I \cup \bigcup_{j=1}^{i-1} UL_O^j, S)$;
           /* filter out context unit lemmas which cannot to be lifted */
           $UL_O^i = UL_O^i \setminus \{(\Gamma \models_S \neg L) \in UL_O^i \mid B_i\sigma \in \Gamma\}$;
           if $(B_i\sigma \in RA_i)$ **then** /* create a new context unit lemma */
               $UL_O^i = UL_O^i \cup \{(RA_i \setminus \{B_i\sigma\}) \models_S \neg B_i\sigma\}$
           elseif NP_i does not contain $\top$ **then** /* irrelevant model extension */
               return $\langle NP_i, RA_i, \cup_{j=1}^i UL_O^j\rangle$; /* proof condensation */
       }
   }
   if $\exists i(1 \leq i \leq m)(NP_i$ contains $\top)$ **then** $RA = \emptyset$
   else $RA = \cup_{i=1}^m(RA_i \setminus \{B_i\sigma\}) \cup \{A_1\sigma, \ldots, A_n\sigma\}$;

   return $\langle$ $, RA, \cup_{j=1}^m UL_O^j\rangle$;

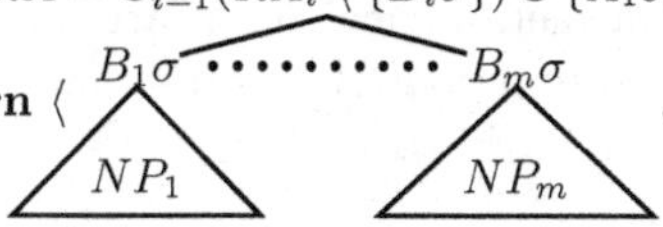

3. (Model finding) If neither 1 nor 2 is applicable, **return** $\langle \overset{!}{\top}, \emptyset, \emptyset\rangle$;

Fig. 9. Model generation procedure with folding-up and proof condensation

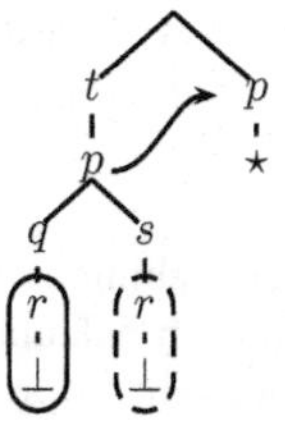

SP	RA	UL
□	p, q	$\{p\} \models_{S2} \neg q$
⊓	p, s	$\{p\} \models_{S2} \neg s$
outer	p	$\emptyset \models_{S2} \neg p$

UL:context unit lemma

Fig. 10. Eliminating a duplicate proof

There are three forms of lemma matching. The first form *unit lemma matching* corresponds to an application of an unit lemma. The second form *identical C-reduction* corresponds to an application of a context unit lemma. The third form *strong contraction* has a similar effect as proof condensation[3].

6   Application to Completeness Proof for Breadth-First Non-Horn Magic Sets

Model generation is forward reasoning in the sense that its proof begins with positive clauses (i.e. facts) and ends with negative clauses (i.e. goals). Non-Horn magic sets (NHM) were designed to enhance forward reasoning provers. The NHM method aims to select only violated clauses that yield relevant model extensions. It is worth noting that a simplified proof obtained by the proof simplification procedure has no irrelevant model extension, that is, every model extension in the proof is relevant.

Considering this, we make a proof-tree transformation procedure that builds a proof-tree of the NHM method from that of model generation. The procedure is obtained by modifying the proof simplification procedure. The transformation procedure gives a syntactic proof of completeness of the NHM method.

There are two versions of the NHM method: the breadth-first NHM and depth-first NHM. In this paper, we consider only the former version, while both versions are considered in [11].

6.1   Breadth-first NHM Transformation Method

Let S be a set of clauses. We introduce a meta-logical predicate $goal/1$ which takes an atom in S as its argument. The literal $goal(A)$ means that an atom A is relevant to the goal and is necessary to be solved.

Definition 4 (Breadth-first NHM Transformation). *The breadth-first NHM transformation is defined as follows. A clause* $A_1 \wedge \ldots \wedge A_n \to B_1 \vee \ldots \vee B_m$ *in* S *is transformed into two clauses:*

$$T_B^1 : goal(B_1) \wedge \ldots \wedge goal(B_m) \to goal(A_1) \wedge \ldots \wedge goal(A_n).$$
$$T_B^2 : goal(B_1) \wedge \ldots \wedge goal(B_m) \wedge A_1 \wedge \ldots \wedge A_n \to B_1 \vee \ldots \vee B_m.$$

Although T_B^1 *has a conjunction of n atoms* $goal(A_1) \wedge \ldots \wedge goal(A_n)$ *in the consequent, we identify* T_B^1 *with n clauses* $goal(B_1) \wedge \ldots \wedge goal(B_m) \to goal(A_i)$ $(1 \le i \le n)$.

---

[3] Note that proof condensation is not applicable to model elimination calculus straightforwardly, because every extension of model elimination calculus is relevant to the proof in our sense. This is because every literal in a chain (a proof tree)becomes *A*-literal, which resolves away its complement literal of an input clause, or participates in reduction operation.

In this transformation, for $n = 0$ (a positive clause), the first transformed clause T_B^1 is omitted. For $m = 0$ (a negative clause), the conjunction of $goal(B_1) \land \ldots \land goal(B_m)$ becomes $true$. For $n \neq 0$, two clauses T_B^1 and T_B^2 are obtained by the transformation. T_B^1 is a clause simulating backward reasoning, and T_B^2 is a clause for forward reasoning that implements relevancy testing.

A set of transformed clauses obtained from S by the breadth-first NHM transformation is denoted by $T_B(S)$. $T_B(S)$ is separated into $T_B^1(S)$ and $T_B^2(S)$. The breadth-first NHM method checks the satisfiability of $T_B(S)$ instead of the original clause set S.

Example 6. Consider the clause set $S1$ described in Example 1. The NHM-transformed clause set $T_B(S1)$ is as follows:

	1.2 $goal(r) \to r$
2.1 $goal(a) \land goal(b) \to goal(r)$	2.2 $goal(a) \land goal(b) \land r \to a \lor b$
3.1 $goal(p) \land goal(c) \land goal(d) \to goal(r)$	
3.2 $goal(p) \land goal(c) \land goal(d) \land r \to p \lor c \lor d$	
4.1 $goal(p) \land goal(q) \to goal(r)$	4.2 $goal(p) \land goal(q) \land r \to p \lor q$
5.1 $true \to goal(p)$	5.2 $p \to false$
6.1 $true \to goal(q)$	6.2 $q \to false$

6.2   Completeness of the breadth-first NHM method

Theorem 2 (Completeness of the breadth-first NHM). *If a set S of clauses is unsatisfiable, then $T_B(S)$ is unsatisfiable.*

The paper [7] gives a semantical proof for the completeness of the NHM method. On the other hand, the present paper gives a syntactical argument. In other words, we give a proof-tree transformation procedure, which maps a proof-tree of a clause set S to that of the breadth-first NHM transformed clause set $T_B(S)$, by modifying the proof simplification procedure. In this proof, we assume the soundness and completeness of the model generation method [13,4].

Theorem 3 (Completeness of Model Generation). *If a set S of clauses is unsatisfiable, then S has a finite proof-tree every leaf of which is labeled with $\bot$.*

Theorem 4 (Soundness of Model Generation). *Let S be a set of clauses. If S has a finite proof-tree every leaf of which is labeled with $\bot$, then S is unsatisfiable.*

Assuming that a set S of clauses is unsatisfiable, by the completeness of model generation, S has a finite proof-tree every leaf of which is labeled with $\bot$. According to the proof-tree transformation procedure described in this section, we can build a finite proof-tree for $T_B(S)$ every leaf of which is labeled with $\bot$. Therefore, we conclude that $T_B(S)$ is unsatisfiable by the soundness of model generation. Thus, the completeness of the breadth-first NHM is proved.

In the following, we assume that a proof-tree to be transformed is finite and all its leaves are labeled with $\bot$.

---

procedure $ToP_{T_B}(AP) : P_{T_B}S$; /* Input(AP): Annotated proof-tree of S,
  Output($P_{T_B}S$): Normal proof-tree of $T_B(S)$ */
  $\langle BP, TP \rangle = ToP_{T_B}0(AP)$; **return** ($\overset{\text{TP}}{\underset{\text{BP}}{\mid}}$);

procedure $ToP_{T_B}0(AP) : \langle BP, TP \rangle$;
  /* Output(BP and TP): Normal subproof-trees of $T_B(S)$ */

1. If $AP = \langle \overset{\mid}{\bot}, A_1\sigma \wedge \ldots \wedge A_n\sigma \rangle$ **then return** $\left(\left\langle \overset{\mid}{\bot} , \begin{array}{c} goal(A_1\sigma) \\ \vdots \\ goal(A_n\sigma) \end{array} \right\rangle \right)$

2. If AP is in the form depicted in Fig. 2 **then**
   if $\forall i(1 \leq i \leq m)B_i\sigma \in Rel(AP_i)$ **then** /* relevant model extension */

   return $\left(\left\langle \begin{array}{c} \overset{\triangle}{\underset{B_1\sigma}{\triangle}} \cdots \cdots \underset{BP_i}{\overset{B_i\sigma}{\triangle}} \cdots \cdots \underset{BP_m}{\overset{B_m\sigma}{\triangle}} \end{array} , \begin{array}{c} TP_1 \\ \vdots \\ TP_m \\ goal(A_1\sigma) \\ \vdots \\ goal(A_n\sigma) \end{array} \right\rangle \right)$

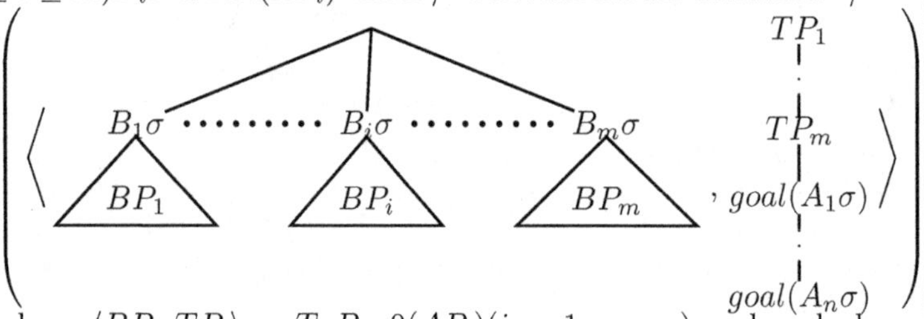

   where $\langle BP_i, TP_i \rangle = ToP_{T_B}0(AP_i)(i = 1, \ldots, m)$ and each duplicate occurrence of $goal(A)$ is avoided by eliminating its lower
   occurrence during constructing the TP part.
   else return $ToP_{T_B}0(AP_{i_0})$ /* irrelevant model extension */
   where i_0 is the minimal index satisfying $1 \leq i_0 \leq m$ and
   $B_{i_0}\sigma \notin Rel(AP_{i_0})$.

Fig. 11. Proof-tree transformation for the breadth-first NHM: ToP_{T_B}

Fig. 11 shows a proof-tree transformation procedure that maps an annotated proof-tree AP of a clause set S to an normal proof-tree of $T_B(S)$. The main part of the procedure is $ToP_{T_B}0$ that takes AP and returns a pair of normal proof-trees: BP is a bottom-up proof part for $T_B(S)$ and TP is a top-down proof part. The concatenation $\overset{\text{TP}}{\underset{\text{BP}}{\mid}}$ forms a proof-tree of $T_B(S)$. Note that TP always has one branch.

The procedure $ToP_{T_B}0$ deals with the proof-tree AP according to the form of AP:

1. When AP is formed by model rejection (Fig. 11 (1)):
   Return a pair of $\langle BP, TP \rangle$. Here, BP represents a subproof-tree built by model rejection with a T_B^2 clause $(A_1\sigma \wedge \ldots \wedge A_n\sigma \rightarrow false)$ and TP represents a subproof-tree built by model extension with a T_B^1 clause $(true \rightarrow goal(A_1\sigma) \wedge \ldots \wedge goal(A_n\sigma))$.
2. When AP is formed by model extension (Fig. 11 (2)):
   If the model extension is relevant to the proof, return a pair of $\langle BP, TP \rangle$. Here, BP represents a subproof-tree built by model extension with a T_B^2

clause $(goal(B_1\sigma) \wedge \ldots \wedge goal(B_m\sigma) \wedge A_1\sigma \wedge \ldots \wedge A_n\sigma \rightarrow B_1\sigma \vee \ldots \vee B_m\sigma)$
and TP represents a subproof-tree built by model extension with a T_B^1 clause
$(goal(B_1\sigma) \wedge \ldots \wedge goal(B_m\sigma) \rightarrow goal(A_1\sigma) \wedge \ldots \wedge goal(A_n\sigma))$.
On the other hand, if the model extension is irrelevant to the proof, return
$\langle TP_{i_0}, BP_{i_0}\rangle$ which is the value of $ToP_{T_B}0(P_{i_0})$ where i_0 satisfies $1 \le i_0 \le m$
and $goal(B_{i_0}\sigma) \notin TP_{i_0}$, so as to remove the model extension from the proof-
tree.

In order to show that the transformed tree $\overset{\downarrow}{\underset{BP}{TP}}(= ToP_{T_B}(AP))$ is a proof-
tree of $T_B(S)$, we prove in the paper [11] that each model extension/rejection in
TP and BP satisfies its applicability conditions.

Example 7. Let AP_{S_1} be the annotated proof-tree shown in Fig. 3 (a). Fig. 12
shows a transformation process for AP_{S_1}. Note that model extensions $a\frown b$ and
$p\frown c\frown d$ are irrelevant, while model extensions $\overset{\uparrow}{r}$ and $p\frown q$ are relevant.
(1) shows a partial value of $ToP_{T_B}0(AP_{S_1})$ where $AP_{S_1}^r$ be a subproof-tree
below r. The model extension $\overset{\uparrow}{r}$ remains after the transformation. (2) and (3)
indicate that model extensions $a\frown b$ and $p\frown c\frown d$ are eliminated by the
transformation.
(4) and (5) show transformations for leaves in AP_{S_1}. Using these two transformations, the transformation for the bottom left subproof-tree is obtained as
shown in (6). From (1), (3) and (6), we obtain $ToP_{T_B}0(AP_{S_1})$ as shown in (7).
Finally, we obtain a normal proof-tree of $T_B(S1)$, which is an concatenation of
the BP part and the TP part of $ToP_{T_B}0(AP_{S_1})$.

7   Experimental Results

We have implemented a model generation theorem prover with proof condensation and folding-up. This prover is written in KLIC [5] version 3.003. KLIC
programs are compiled into C programs.

We select all non-Horn problems (1984 problems) in the TPTP library [19]
version 2.3.0. The problems were run on a SUN Ultra 60 (450MHz, 1GB, So-
laris2.7) workstation with a time limit of 10 minutes and a space limit of 256MB.

Table 1 shows the number of problems solved by model generation without and with proof condensation and folding-up, and SPASS (version 1.0.3)
prover[4] [21]. In the table, "+" indicates that the corresponding method is used,
"-" indicates that it is not used. For example, the column (3) shows the number of problems solved by model generation with folding-up and without proof
condensation.

The number of problems solved has increased by 34 in case of using only proof
condensation (the column (2)), while it has increased by 14 in case of using only
folding-up (the column (3)). According to our experiments, there are many cases
where pruning effect of proof condensation is stronger than that of folding-up.

---

[4] We used the tptp2X utility [19] with options "-q2 -fdfg -t rm_equality:rstfp"
to get SPASS formats from TPTP formats. SPASS ran in automatic mode.

Table 1. Number of problems solved (of 1984 non-Horn problems)

	(1)	(2)	(3)	(4)	S
folding-up	-	-	+	+	PA
proof condensation	-	+	-	+	SS
Number	305	339	319	385	834
(Percentage)	(15.4)	(17.1)	(16.1)	(19.4)	(42.0)

Table 2. Performance comparison

	(1)	(2)	(3)	(4)	S
folding-up	-	-	+	+	PA
proof condensation	-	+	-	+	SS
CIV008-1.002	T.O.	0.11	0.41	0.11	2.50
unsatisfiable		196	1335	195	1784
GEO013-3	T.O.	T.O.	T.O.	76.31	31.15
unsatisfiable				478	5049
GEO033-3	T.O.	1.19	T.O.	1.19	T.O.
unsatisfiable		140		140	
GEO051-3	T.O.	110.95	T.O.	111.02	16.89
unsatisfiable		433		392	2625
KRS013-1	T.O.	0.02	T.O.	0.01	0.03
unsatisfiable		106		77	72
MSC007-2.005	48.49	4.18	35.61	2.63	257.93
unsatisfiable	170875	9827	110093	6180	6927
PRV009-1	379.25	0.71	0.63	0.72	0.01
unsatisfiable	7382541	58	685	56	15
PUZ010-1	T.O.	185.90	T.O.	18.15	211.79
unsatisfiable		925562		69024	453324
PUZ018-2	T.O.	5.08	T.O.	4.14	T.O.
satisfiable		735		371	
SYN437-1	T.O.	T.O.	T.O.	367.42	515.12
satisfiable				6556	1006266
SYN443-1	T.O.	T.O.	27.86	6.89	17.85
unsatisfiable			21523	4246	52659
SYN447-1	T.O.	T.O.	T.O.	214.02	329.47
unsatisfiable				71659	334102
SYN511-1	T.O.	T.O.	385.59	8.62	21.71
unsatisfiable			109990	2961	52513

top: time (sec)
bottom:
    No. of nodes of proof tree ((1)∼(4))
    No. of kept clauses (SPASS)
T.O.: Time out (> 600 sec)

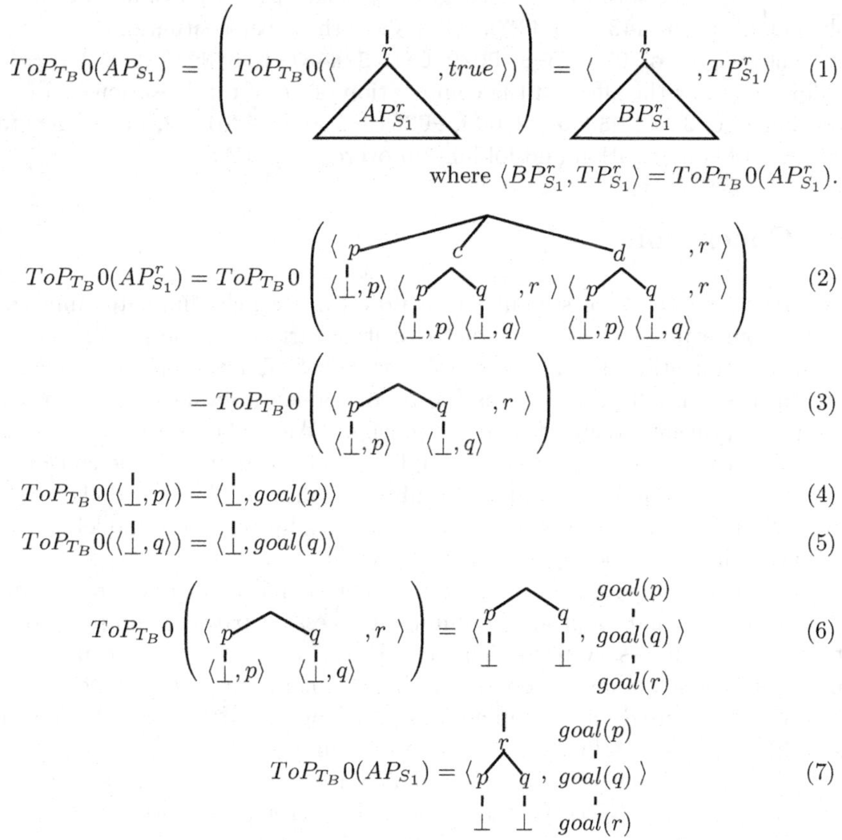

$$ToP_{T_B}0(AP_{S_1}) = \left(ToP_{T_B}0(\langle\ \overset{r}{\underset{AP^r_{S_1}}{\triangle}}\ , true\ \rangle) \right) = \langle\ \overset{r}{\underset{BP^r_{S_1}}{\triangle}}\ , TP^r_{S_1}\rangle \quad (1)$$

where $\langle BP^r_{S_1}, TP^r_{S_1}\rangle = ToP_{T_B}0(AP^r_{S_1})$.

$$ToP_{T_B}0(AP^r_{S_1}) = ToP_{T_B}0\left(\langle\ \overset{\displaystyle p\ \overset{c}{\diagup}\quad\overset{d}{\diagdown}}{\langle\bot,p\rangle\ \langle\ \overset{p\diagup\ \diagdown q}{\langle\bot,p\rangle\ \langle\bot,q\rangle}\ ,r\ \rangle\ \langle\ \overset{p\diagup\ \diagdown q}{\langle\bot,p\rangle\ \langle\bot,q\rangle}\ ,r\ \rangle}\ ,r\ \rangle \right) \quad (2)$$

$$= ToP_{T_B}0\left(\langle\ \overset{p\diagup\ \diagdown q}{\langle\bot,p\rangle\quad\langle\bot,q\rangle}\ ,r\ \rangle \right) \quad (3)$$

$$ToP_{T_B}0(\langle\bot,p\rangle) = \langle\bot,goal(p)\rangle \quad (4)$$

$$ToP_{T_B}0(\langle\bot,q\rangle) = \langle\bot,goal(q)\rangle \quad (5)$$

$$ToP_{T_B}0\left(\langle\ \overset{p\diagup\ \diagdown q}{\langle\bot,p\rangle\quad\langle\bot,q\rangle}\ ,r\ \rangle \right) = \langle\ \overset{p\diagup\ \diagdown q}{\underset{\bot\quad\bot}{}}\ ,\ \begin{array}{l}goal(p)\\ goal(q)\\ goal(r)\end{array}\ \rangle \quad (6)$$

$$ToP_{T_B}0(AP_{S_1}) = \langle\ \overset{r}{\overset{p\diagup\ \diagdown q}{\underset{\bot\quad\bot}{}}}\ ,\ \begin{array}{l}goal(p)\\ goal(q)\\ goal(r)\end{array}\ \rangle \quad (7)$$

Fig. 12. Proof-tree transformation process for AP_{S_1}

This makes it clear that naive model generation performs many irrelevant model extensions. The combination of proof condensation and folding-up has a great effect on pruning search space. The number of problems solved has increased by 80 to 385 (the column (4)), which is 19.4% of the TPTP non-Horn problems.

Compared to SPASS solving 42.0%, this result does not seem to be good. It is considered that this comes from absences of unification[5] and builtin equality treatment in our system.

Table 2 compares the proving performance on several typical problems. The integer of a bottom row is considered as showing search spaces required for proof. All problems exhibit proof condensation or folding-up effects. The entries of GEO033-3, GEO051-3, KRS013-1, MSC007-2.005, PUZ010-1 and PUZ018-2 show

---

[5] Recall that we use ground substitution σ for model extensions. Therefore, unification is not required for model generation.

the pruning effect of proof condensation stronger than that of folding-up, while the entries of SYN443-1 and SYN511-1 show the reverse situation.

The entries of GEO013-3, PUZ010-1, SYN437-1, SYN443-1, SYN447-1 and SYN511-1 show the effect of the combination of proof condensation and folding-up. For GEO033-3, MSC007-2.005, PUZ010-1 and PUZ018-2, model generation with proof condensation and folding-up overcomes SPASS.

8   Conclusion

We have given a proof simplification procedure which eliminates unnecessary parts from a proof so as to extract essential parts from the proof. Performing proof simplification during the proof, instead of after the proof has been completed, is essentially the same as the proof condensation facility which has the ability to prevent irrelevant model extensions. We also have shown that a set of relevant atoms used for the proof simplification procedure can be considered as a lemma with which some duplicate subproofs may be eliminated. This way, we embed folding-up, which eliminates duplicate subproofs, into model generation by modifying the proof simplification procedure.

The proof simplification procedure is considered as a proof transformation procedure that maps a proof containing unnecessary parts to another proof containing no unnecessary parts. This consideration gives a proof transformation procedure, which maps a proof of the model generation procedure to that of the NHM method, used for a completeness proof for the NHM method. Thus, proof simplification is a useful tool of theorem proving.

Experimental results show that orders of magnitude speedup can be achieved for some problems. Nevertheless, state-of-the-art theorem provers such as SPASS and E[6] overcome ours in terms of the number of problems solved. The future work thus includes embedding equality operation into model generation and studying a lifted version of this work.

References

1. O. L. Astrachan and M. E. Stickel. Caching and Lemmaizing in Model Elimination Theorem Provers. In Kapur [10], pages 224–238. 104
2. P. Baumgartner, U. Furbach, and I. Niemelä. Hyper Tableaux. In J. J. Alferes, L. M. Pereira, and E. Orłowska, editors, *Proc. European Workshop: Logics in Artificial Intelligence, JELIA*, volume 1126 of *Lecture Notes in Artificial Intelligence*, pages 1–17. Springer-Verlag, 1996. 96, 102
3. P. Baumgartner, J. D. Horton, and B. Spencer. Merge Path Improvements for Minimal Model Hyper Tableaux. In N. V. Murray, editor, *Automated Reasoning with Analytic Tableaux and Related Methods*, number 1617 in Lecture Notes in Artificial Intelligence, pages 51–65. Springer, June 1999. 104
4. F. Bry and A. Yahya. Minimal Model Generation with Positive Unit Hyper-Resolution Tableaux. In Miglioli et al. [15], pages 143–159. 96, 107

---

[6] E prover [18] solves 32.2% of the TPTP non-Horn problems.

5. T. Chikayama, T. Fujise, and D. Sekita. A Portable and Efficient Implementation of KL1. In *Proceedings International Symposium on Programming Language Implementation and Logic Programming*, number 844 in Lecture Notes in Computer Science, pages 25–39. Springer, 1994. 109

6. H. Ganzinger, editor. *Automated Deduction – CADE-16*, number 1632 in Lecture Notes in Artificial Intelligence. Springer, July 1999. 113, 113

7. R. Hasegawa, K. Inoue, Y. Ohta, and M. Koshimura. Non-Horn Magic Sets to Incorporate Top-down Inference into Bottom-up Theorem Proving. In McCune [14], pages 176–190. 96, 97, 107

8. K. Inoue, M. Koshimura, and R. Hasegawa. Embedding Negation as Failure into a Model Generation Theorem Prover. In Kapur [10], pages 400–415. 96

9. K. Iwanuma. Lemma Matching for a PTTP-based Top-down Theorem Prover. In McCune [14], pages 146–160. 104

10. D. Kapur, editor. *Automated Deduction – CADE-11*, number 607 in Lecture Notes in Artificial Intelligence. Springer, June 1992. 112, 113

11. M. Koshimura and R. Hasegawa. A Proof of Completeness for Non-Horn Magic Sets and Its Application to Proof Condensation. *Automated Reasoning with Analytic Tableaux and Related Methods: POSITION PAPERS*, pages 101–115, June 1999. Technical Report 99-1, University at Albany. 106, 109

12. R. Letz, K. Mayr, and C. Goller. Controlled Integration of the Cut Rule into Connection Tableau Calculi. *Journal of Automated Reasoning*, 13:297–337, 1994. 96, 104

13. R. Manthey and F. Bry. SATCHMO: a theorem prover implemented in Prolog. In E. Lusk and R. Overbeek, editors, *9th International Conference on Automated Deduction*, number 310 in Lecture Notes in Computer Science, pages 415–434. Springer-Verlag, 1988. 96, 107

14. W. McCune, editor. *Automated Deduction – CADE-14*, number 1249 in Lecture Notes in Artificial Intelligence. Springer, July 1997. 113, 113

15. P. Miglioli, U. Moscato, D. Mundici, and M. Ornaghi, editors. *Theorem Proving with Analytic Tableaux and Related Methods*, number 1071 in Lecture Notes in Artificial Intelligence. Springer, May 1996. 112, 113

16. I. Niemelä. A Tableau Calculus for Minimal Model Reasoning. In Miglioli et al. [15], pages 278–294. 96

17. F. Oppacher and E. Suen. HARP: A Tableau-Based Theorem Prover. *Journal of Automated Reasoning*, 4:69–100, 1988. 96, 102

18. S. Schulz. System Abstract: E 0.3. In Ganzinger [6], pages 297–301. 112

19. G. Sutcliffe and C. Suttner. The TPTP Problem Library -CNF Release v1.2.1. *Journal of Automated Reasoning*, 21(2):177–203, October 1998. 109, 109

20. A. Vellino. *The Complexity of Automated Reasoning*. PhD thesis, University of Toronto, 1989. 103

21. C. Weidenbach, B. Afshordel, U. Brahm, C. Cohrs, T. Engel, E. Keen, C. Theobalt, and D. Topić. System Description: SPASS Version 1.0.0. In Ganzinger [6], pages 378–382. 109

Have SPASS with $OCC1N_g^=$

Christian G. Fermüller and Georg Moser[*]

Technische Universität Wien, Austria

Abstract. We prove that a particular superposition based inference operator decides a fragment of clause logic with equality, called $OCC1N_g^=$. We also show that the theorem prover SPASS not only implements the corresponding operator but also generates standard descriptions of unique term models for all satisfiable clause sets in $OCC1N_g^=$.

1   Introduction

Automated model building is a topical research field that extends and complements automated deduction. In a refutational context, its main task can be formulated as follows:

- If a (sound and complete) theorem prover fails to refute a formula F, extract an adequate representation of a model for F from the information produced by the prover.

Obviously, *termination* of the prover is a precondition for model building. Therefore one of the challenges here is to identify non-trivial syntactic classes of formulæ for which the theorem prover in question terminates, i.e., represents a *decision procedure*. Results of this type for equality free clause logic, using resolution methods, are surveyed in [FLTZ93] and [FLHT00]. In presence of (theory free) equality some interesting decidability proofs along this line are provided, e.g., in [BGW93], [FS93], [GD99]. Related results, concerning purely equational classes can be found in [JMW98,Nie96].

Building upon corresponding decidability results, the potential of hyperresolution as a *model builder* is explored in [FL93] and [FL96]. Some of these results are generalized in [FL98] to define an inference based model building procedure for a fragment of clause logic with equality.[1] A related approach to model building, developed by R. Caferra and his collaborators, in particular N. Peltier, consists in augmenting standard calculi for clause logic by additional rules and

---

[*] The research described in this paper was partly supported by Austrian Resarch Fund (FWF) grant No. 14126-MAT

[1] N. Peltier (private communication) recently found a counter example to the claim of [FL98] that the particular inference operator used there terminates on the class $PVD_g^=$. (Fortunately, the decidability of the class in question has also been proved by other means in [Rud00]. The central part of [FL98] — namely the correctness of the suggested "backtracking free" model building procedure — seems not to be affected by the error.)

M. Parigot and A. Voronkov (Eds.): LPAR 2000, LNAI 1955, pp. 114–130, 2000.

constraint handling mechanisms that facilitate the extraction of models during proof search (see, e.g., [CP96,Pel97]).

Here we want to emphasize that the successful and easily accessible theorem prover SPASS [WAB+99,Wei00] can be employed not only as a decision procedure but also as a model builder for large fragments of clause logic. To this aim we investigate a particular class of clause sets $OCC1N_g^=$, that is an extension of an equality free class defined in [FL93] to clause logic with (ground) equalities. The decidability proof for $OCC1N_g^=$ is rather involved. However, the main technical result is that SPASS not only decides $OCC1N_g^=$, but also generates standard descriptions of unique models for all satisfiable inputs from this class. In fact, model building is almost "for free" here, due to the eager use of the splitting rule in SPASS.

It is not our main motivation to establish a decision and model building procedure for yet another fragment of clause logic. We rather prefer to look at the results from a proof theoretic point of view. Giving syntactic criteria (on input formulas) that are sufficient for termination of proof search is a way of characterizing mathematically the "computational strength" of a calculus. In this sense the specific decidability result, together with the remarks on model building, should be seen as mathematically substantiated evidence for the claim that SPASS and its underlying superposition calculus as developed by L. Bachmair and H. Ganzinger are an impressively strong and flexible tool, indeed.

2   Basic Notions

We assume familiarity with clause logic; in particular its semantics. However we need to fix some terminology concerning syntax.

Terms and *atoms* are defined as usual with respect to a given signature Sig of constant, function, and predicate symbols. *Equalities* are atoms involving the binary predicate "$\approx$", interpreted as congruence and denoted in infix notation as usual. A *clause* C is written in form $\Pi \to \Delta$, where Π and Δ are multi-sets of atoms. Π is called the *negative* and Δ the *positive* part of C. $\Pi' \to \Delta'$ is a *sub-clause* of C if $\Pi \subseteq \Pi'$ and $\Delta \subseteq \Delta'$. A *literal* L is an occurrence of an atom, either negative or positive, in a clause. We also write L, Γ for $\{L\} \cup \Gamma$, and Δ, Γ for $\Delta \cup \Gamma$.

It is convenient to view a term as a rooted and ordered tree, where the inner nodes are function symbols and the leaf nodes constants or variables. A *position* p in a term t is a sequence of edges in a path leading from the root to a node n_p of t. By the *depth* of p, denoted by $|p|$, we mean the number of edges it consists of. The sub-term of t that has its root at n_p is denoted as $t|_p$. In writing $t[s]$ we indicate the *occurrence* of a sub-term s in t and use $t[s']$ to denote the term resulting from t by replacing the indicated occurrence of s by s'. These definitions also apply to atoms (i.e. atoms are considered as trees with a predicate symbol as root).

In the following let E be a term or an atom. The set of all variables occurring in E is denoted as vars(E). The *depth* of E is defined as $\tau(E) = \max\{|p| :$

$E|_p$ exists$\}$. The *maximal depth of occurrence* of a sub-term s in E is defined as $\tau_{\max}(s, E) = \max\{|p| : E|_p = s\}$. Similarly, $\tau_{\min}(s, E) = \min\{|p| : E|_p = s\}$. The *maximal depth of variable occurrence* in E is defined as $\tau_v(E) = \max\{|p| : E|_p \in \text{vars}(E)\}$ if $\text{vars}(E) \neq \emptyset$ and $\tau_v(E) = -1$ for ground expressions.

These definitions are generalized to multi-sets of atoms and clauses in the obvious way. E.g., let $C \equiv (\Pi \to \Delta)$ then $\tau(C) = \max\{\tau(E) : E \in \Pi \cup \Delta\}$. $\tau^+(C) = \max\{\tau(E) : E \in \Delta\}$ refers to the positive literals of C only. To avoid undefined cases we define $\tau(C) = -1$ if C is empty and $\tau^+(C) = -1$ if Δ is empty. $\tau_{\max}(t, C)$, $\tau_{\min}(t, C)$, $\tau_v(C)$, and $\tau_v^+(C)$ are defined analogously.

For a set of clauses $\mathcal{S}$, $\tau(\mathcal{S}) = \max\{\tau(C) : C \in \mathcal{S}\}$. $\tau_v(\mathcal{S})$, $\tau^+(\mathcal{S})$ and $\tau_v^+(\mathcal{S})$ are defined analogously. By a *clause set* we mean a finite set of clauses.

An expression (i.e. a term, atom, multi-set of atoms or a clause) is called *ground* if no variable occurs in it. It is called *linear* if each variable occurs at most once in it.

Substitutions are defined as usual. In particular a *most general unifier* of two terms or atoms E_1 and E_2 is denoted by $mgu(E_1, E_2)$. The result of applying a substitution to an expression E is denoted by $E\sigma$.

3   The Class $OCC1N_g^=$

In [FL93] the following class of clause sets was introduced as a non-trivial example of a class for which hyperresolution provides a decision procedure.

Definition 1. $OCC1N^+$ *is the class of all clause sets $\mathcal{S}$, defined over a signature without equality, such that for all $(\Pi \to \Delta) \in \mathcal{S}$:*

(lin) Δ *is linear; i.e. each variable occurs at most once in Δ, and*
(vd) $\tau_{\max}(x, \Delta) \leq \tau_{\min}(x, \Pi)$ *for all $x \in \text{vars}(\Delta) \cap \text{vars}(\Pi)$:*

Let $OCC1N^=$ be defined exactly as $OCC1N^+$, but over a signature *including equality*.

Theorem 1. $OCC1N^=$ *is undecidable.*

Proof. It is known that satisfiability for finite sets $E \equiv \{s_1 \approx t_1, \ldots, s_n \approx t_n\}$ of equalities augmented by a single ground inequality $s_0 \not\approx t_0$ is undecidable, even if the equalities contain only a single variable and only unary function symbols (see, for example, Theorem 4 in [FS93]). We reduce this decision problem to that for $OCC1N^=$.

Let $h^n(v)$ denote $h(h(\ldots(v)\ldots))$ for n iterations of h, and let x be the single variable occurring in $s_i \approx t_i \in E$. We define the following translation $(^\circ)$:

$$(s_i \approx t_i)^\circ =_{\text{def}} (h^n(x) \approx h^n(y) \to s_i\{x \mapsto y\} \approx t_i)$$

where h is a fixed function symbol that does not occur in E, y is a new variable, and $n = \tau(s_i \approx t_i)$. Consider the set of clauses

$$\mathcal{S}_E = \{(s_i \approx t_i)^\circ : s_i \approx t_i \in E\} \cup \{s_0 \approx t_0 \to\}.$$

Clearly, $\mathcal{S}_E \in OCC1N^=$. Since the function symbol h is fresh, any model $\mathcal{M}$ of $E \cup \{s_0 \napprox t_0\}$ can be extended to a model $\mathcal{M}'$ of $\mathcal{S}_E$ by interpreting h as identity function. Conversely, any model of $\mathcal{S}_E$ is also a model of $E \cup \{s_0 \napprox t_0\}$. □

This motivates interest in the class $OCC1N_g^=$, defined exactly as $OCC1N^=$, but requiring all equality literals to be *ground*.

4   The Inference System $\mathcal{G}$

We assume (at least nodding) acquaintance with the *superposition calculus* as defined, e.g., in [BG94].

We use an instance of the inference system called $\mathcal{E}_S$ in [BG94] (equality resolution, ordered factoring, superposition and equality factoring with selection function S). In our version of the calculus, the selection function S selects *all* negative literals. I.e. we employ a *positive superposition-strategy*.

To achieve greater transparency in our decidability proof (Section 5) we separate the cases for equality and non-equality literals, respectively. This is also closer to what is implemented in SPASS (see Section 6). In particular we define an inference rule *positive resolution*[2]. Significant simplifications arise by

- making use of only those order restrictions that are actually needed to decide $OCC1N_g^=$, and
- assuming that all equality literals are ground and that the reduction order is total on ground terms.

The set of inference rules are given in Table 1.

Remark 1. Note that in rules $(of^=)$, (ef), (sd), $(ss^=)$, and $(sr^=)$ no substitution is applied, as in these rules only ground equalities are involved.

Definition 2. *For any set of clauses $\mathcal{S}$, $\mathcal{G}(\mathcal{S})$ denotes the union of $\mathcal{S}$ and all conclusions of an application of one of the above inference rules where the premisses are in $\mathcal{S}$. By $\mathcal{G}^*(\mathcal{S})$ we denote the transitive and reflexive closure of the set operator $\mathcal{G}$.*

5 $\mathcal{G}_c$ Decides $OCC1N_g^=$

We want to prove that (up to renaming of variables) only finitely many different clauses can be derived from any $\mathcal{S} \in OCC1N_g^=$ using the inference system $\mathcal{G}$. In fact, we have to augment $\mathcal{G}$ by a *condensation* mechanism in order to achieve this goal, as we will see below.

---

[2] Positive resolution corresponds to selective superposition combined with removal of the negative equality "$\top \approx \top$", where a non-equality atom A is identified with an equality $A \approx \top$ (like in [BG94]).

Table 1. Inference Rules of System $\mathcal{G}$

Ordered Factoring:
$$\frac{\rightarrow \Delta, L, M}{(\rightarrow \Delta, L)\theta} \ (of)$$
where $\theta = mgu(L, M)$

Superposition Right:
$$\frac{\rightarrow s \approx t, \Delta \qquad \rightarrow \Sigma, L[s']}{(\rightarrow \Delta, \Sigma, L[t])\theta} \ (sr)$$
where $s \succ t$, $s = s'\theta$, $s' \notin \text{vars}(L)$

Equality Factoring:
$$\frac{\rightarrow \Delta, s \approx t, s \approx t'}{t \approx t' \rightarrow \Delta, s \approx t'} \ (ef)$$

Selective Superposition:
$$\frac{\rightarrow s \approx t, \Delta \qquad L[s'], \Gamma \rightarrow \Sigma}{(L[t], \Gamma \rightarrow \Delta, \Sigma)\theta} \ (ss)$$
where $s \succ t$, $s = s'\theta$, $s' \notin \text{vars}(L)$

Positive Resolution:
$$\frac{\rightarrow L, \Delta \qquad M, \Gamma \rightarrow \Sigma}{(\Gamma \rightarrow \Delta, \Sigma)\theta} \ (pr)$$
where $\theta = mgu(L, M)$

Ordered Factoring (of equalities):
$$\frac{\rightarrow \Delta, s \approx t, s \approx t}{\rightarrow \Delta, s \approx t} \ (of^{=})$$

Superposition Right (into equalities):
$$\frac{\rightarrow s \approx t, \Delta \qquad \rightarrow \Sigma, r[s] \approx t'}{\rightarrow \Delta, \Sigma, r[t] \approx t'} \ (sr^{=})$$
where $s \succ t$.

Selective Resolution:
$$\frac{t \approx t, \Gamma \rightarrow \Delta}{\Gamma \rightarrow \Delta} \ (sd)$$

Selective Superposition (into equality literals):
$$\frac{\rightarrow s \approx t, \Delta \qquad r[s] \approx t', \Gamma \rightarrow \Sigma}{(r[t] \approx t', \Gamma \rightarrow \Delta, \Sigma)\theta} \ (ss^{=})$$
where $s \succ t$

Remark 2. The proof is rather involved and therefore broken up into several lemmas, which establish various *invariants* for clause sets in $OCC1N_g^=$ with respect to $\mathcal{G}$. Given the fact that there exists a relatively simple decidability proof for $OCC1N^+$ (see [FLTZ93], Chapter 3) based on *hyperresolution*. the complexity of the proof may appear surprising. However, hyperresolution gets *incomplete* if combined with paramodulation (by simply adding paramodulants of input clauses and hyperresolvents). As a reminder on the subtleties arising by the addition of (ground) equalities to decidable classes of equality-free clause logic consider the following example (due to N. Peltier.) The clause set S consisting of

$$P(f(x)), Q(x) \rightarrow Q(f(x)) \tag{1}$$
$$\rightarrow P(a) \tag{2}$$
$$\rightarrow Q(a) \tag{3}$$

belongs to PVD; a class straightforwardly decidable by hyperresolution (see, e.g., [FLTZ93]). However, adding the ground equality

$$\rightarrow f(a) \approx a$$

to S results in a clause set from which infinitely many different clauses are derivable using $\mathcal{G}$.

We first state a few simple observations about *mgus* involving linear atoms.

Proposition 1. *Let* L *and* M *be variable disjoint atoms and let* $\theta = \{x_1 \mapsto s_1, \ldots, x_n \mapsto s_n\} \cup \{y_1 \mapsto t_1, \ldots, y_m \mapsto t_m\}$ *be the mgu of* L *and* M, *where* $x_i \in \mathrm{vars}(L)$ *and* $y_i \in \mathrm{vars}(M)$. *If* L *is linear then:*

1. *all* t_i *are sub-terms of* L *and therefore linear and pairwise variable disjoint.*
2. *If also* M *is linear, then the following facts hold:*
   (a) $L\theta = M\theta$ *is linear, too,*
   (b) $\tau(L\theta) = \max\{\tau(L), \tau(M)\}$,
   (c) $\tau_{\mathrm{v}}(L\theta) = \max\{\tau_{\mathrm{v}}(L), \tau_{\mathrm{v}}(M)\}$.

Proposition 2. *Property* (vd) *of Definition 1 is stable under substitutions.*

We show that class $OCC1N_g^=$ is closed under our inference operator:

Lemma 1. *If* $\mathcal{S} \in OCC1N_g^=$ *then also* $\mathcal{G}(\mathcal{S}) \in OCC1N_g^=$

Proof. Observe that the two defining conditions for $OCC1N_g^=$, (lin) and (vd), concern variable occurrences only. Also observe that the union of pairwise variable disjoint linear multi-sets of literals is linear, too. Therefore membership in $OCC1N_g^=$ is trivially preserved by the rules $(of^=)$, (ef), $(sr^=)$, (sd), and $(ss^=)$ since only ground terms are manipulated and positive parts of clauses joint. Likewise, (lin) and (vd) remain satisfied if the *mgu* used in the application of a rule is ground. This observation suffices for the cases of selective superposition (sd) and superposition right (sr).

It remains to investigate positive resolution (pr) and ordered factoring (of). Since only positive clauses can be factored, condition (vd) trivially holds for factors. Since positive clauses are linear, Proposition 1 guarantees that also condition (lin) remains satisfied.

For positive resolution consider

$$\frac{C \equiv (\to L, \Delta) \qquad D \equiv (M, \Gamma \to \Sigma)}{E \equiv (\Gamma \to \Delta, \Sigma)\theta} \quad (pr)$$

where $\theta = mgu(L, M)$. Assume that $\{C, D\} \in OCC1N_g^=$. By Proposition 1, the terms $\theta(y_i)$ are linear and pairwise variable disjoint for $y_i \in \mathrm{vars}(M)$. Therefore $\Sigma\theta$ remains linear. Moreover Δ is variable disjoint with L and $\Sigma\theta$. Hence E satisfies condition (lin).

By Proposition 2, condition (vd) not only holds for $\Gamma \to \Sigma$ but also for $\Gamma\theta \to \Sigma\theta$. Since $\Delta\theta(= \Delta)$ is variable disjoint with this sub-clause of E, (vd) also holds for E, which concludes the proof. $\qquad\square$

Observe that the term depth of a derived clause can be strictly greater than the maximal term depth of its parent clauses. E.g., applying selective superposition to the $OCC1N_g^=$-clauses

$$\to f(f(a)) \approx a \quad \text{and} \quad P(g(f(x))) \to Q(g(g(x)))$$

results in the clause

$$P(g(a)) \to Q(g(g(f(a)))).$$

By resolution we can even increase the maximal depth of variable occurrences:
Resolving

$$\to P(f(x)) \quad \text{and} \quad P(y), R(g(y), u) \to Q(y)$$

results in the clause

$$R(g(f(x)), u) \to Q(f(x)).$$

However, we can prove that the maximum of the depth of variable occurrences cannot increase in the *positive* part of a clause. The proof uses ideas from [FLTZ93].

Lemma 2. *For all* $S \in OCC1N_g^= : \tau_v^+(\mathcal{G}(S)) = \tau_v^+(S)$.

Proof. Since all equality literals are ground, the only rules that may increase the depth of occurrences of variables (with respect to the parent clauses) are positive resolution (pr) and ordered factoring (of). Therefore we restrict our investigations to the cases (pr) and (of).

Concerning ordered factoring remember that only positive clauses can be factored. Let $E \equiv (\to \Delta, L)\theta$ be the result of factoring $C \equiv (\to \Delta, L, M)$. Since L and M are linear and variable disjoint Proposition 1 applies and we may conclude that $\tau_v(L\theta) = \max\{\tau_v(L), \tau_v(M)\}$. Moreover, since Δ and L are variable disjoint, we have $\Delta\theta = \Delta$. Hence $\tau_v^+(E) = \tau_v^+(C)$.

For the case of positive resolution consider

$$\frac{C \equiv (\to L, \Delta) \qquad D \equiv (M, \Gamma \to \Sigma)}{E \equiv (\Gamma \to \Delta, \Sigma)\theta} \ (pr)$$

with $\theta = mgu(L, M)$. By condition (**lin**), Δ and L are variable disjoint. Therefore $\Delta\theta = \Delta$ and

$$\tau_v(\Delta\theta) \leq \tau_v(C) = \tau_v^+(C). \tag{4}$$

It remains to establish the appropriate bound with respect to $\Sigma\theta$.

Observe that $\tau_v(\Sigma\theta)$ can only be greater than $\tau_v(\Sigma)$ if there is a variable y occurring in both M and Σ such that

$$\tau_v(\Sigma\theta) = \tau_{\max}(y, \Sigma) + \tau_v(\theta(y)) \tag{5}$$

By Proposition 1, $\theta(y)$ is a sub-term t of L. Moreover, t must occur somewhere in L at the same depth as some occurrence of y in M. Therefore

$$\tau_{\min}(y, M) + \tau_v(\theta(y)) \leq \tau_{\max}(t, L) + \tau_v(t) \leq \tau_v(L) \tag{6}$$

By condition (**vd**) we can connect (5) and (6) to conclude in total that

$$\tau_v(\Sigma\theta) \leq \max\{\tau_v(\Sigma), \tau_v(L)\} \tag{7}$$

Combining (4) and (7) we conclude

$$\tau_v^+(E) = \tau_v^+(\Delta\theta, \Sigma\theta) \leq \max\{\tau_v^+(C), \max\{\tau_v^+(C), \tau_v^+(D)\}\}$$
$$= \max\{\tau_v^+(C), \tau_v^+(D)\}$$

which is *q.e.d.* □

Next, we prove that the maximal *difference* of depths in which a variable can occur within a clause cannot increase. For any clause C let $\mathrm{diffv}(C) = \max\{\tau_{\max}(x, C) - \tau_{\min}(x, C) : x \in \mathrm{vars}(C)\}$ and 0 if C is ground. For sets of clauses S: $\mathrm{diffv}(S) = \max\{\mathrm{diffv}(C) : C \in S\}$.

Lemma 3. *For all $S \in OCC1N_g^= : \mathrm{diffv}(\mathcal{G}(S)) \leq \mathrm{diffv}(S)$.*

Proof. We again distinguish cases corresponding to the different inference rules.

Those rules involving ground substitutions only, or no substitution at all, obviously cannot increase $\mathrm{diffv}(\mathcal{G}(S))$ beyond $\mathrm{diffv}(S)$. By Lemma 1 all positive clauses remain linear. I.e. $\mathrm{diffv}(C) = 0$ for positive $C \in \mathcal{G}(S)$.

It therefore remains to investigate positive resolution:

$$\frac{C \equiv (\to L, \Delta) \qquad D \equiv (M, \Gamma \to \Sigma)}{E \equiv (\Gamma \to \Delta, \Sigma)\theta} \quad (pr)$$

with $\theta = mgu(L, M)$, where $\{C, D\} \in OCC1N_g^=$. By Proposition 1, $\theta = \{x_1 \mapsto s_1, \ldots, x_n \mapsto s_n\} \cup \{y_1 \mapsto t_1, \ldots, y_m \mapsto t_m\}$, where $x_i \in \mathrm{vars}(L)$, $y_i \in \mathrm{vars}(M)$ and all t_i are linear and pairwise variable disjoint. Moreover, the t_i are variable disjoint with Δ and D. Furthermore, by condition (**lin**), $\Delta\theta = \Delta$. It follows that $\mathrm{diffv}(E) \leq \max\{\mathrm{diffv}(C), \mathrm{diffv}(D)\}$.   □

Lemmas 1, 2 and 3 can be combined to prove a global bound for the depth of occurrences of a variable anywhere in a clause.

Lemma 4. *For all $S \in OCC1N_g^= : \tau_v(\mathcal{G}^*(S)) \leq 2\tau_v^+(S) + \tau_v(S)$.*

Proof. We have seen in the proof of Lemma 2 that factorisation does not increase the maximal depth of occurrences of variables and that the only other rule that may affect variable depth is positive resolution. It thus remains to investigate the following case:

$$\frac{C \equiv (\to L, \Delta) \qquad D \equiv (M, \Gamma \to \Sigma)}{E \equiv (\Gamma \to \Delta, \Sigma)\theta} \quad (pr)$$

with $\theta = mgu(L, M)$, where C and D are in $\mathcal{G}^*(S)$. By induction on the number of applications of $\mathcal{G}$ it follows from Lemma 1 that $\{C, D\} \in OCC1N_g^=$. By the definition of E and induction using Lemmas 2 and 1 we conclude that

$$\tau_v(E) \leq \max\{\tau_v(\Gamma\theta), \tau_v^+(E)\} \leq \max\{\tau_v(\Gamma\theta), \tau_v^+(S)\} \tag{8}$$

We now argue in analogy to the proof of Lemma 2, above. $\tau_v(\Gamma\theta)$ can only be greater than $\tau_v(\Gamma)$ if there is an $y \in \mathrm{vars}(M) \cap \mathrm{vars}(\Gamma)$ such that

$$\tau_v(\Gamma\theta) = \tau_{\max}(y, \Gamma) + \tau_v(t)$$

for some sub-term $t = \theta(y)$ of L. Hence

$$\tau_v(\Gamma\theta) \leq \tau_{\min}(y, M) + \mathrm{diffv}(D) + \tau_v(t) \tag{9}$$

Clearly, $\tau_v(t) \le \tau_v(L) \le \tau_v^+(C)$ and moreover $\tau_{\min}(y, M) \le \tau_v(L) \le \tau_v^+(C)$ since t must occurs in L at the same depth as y in M. Thus we have

$$\tau_v(\Gamma\theta) \le 2\tau_v^+(C) + \text{diffv}(D).$$

Now observe that Lemma 3 asserts that $\text{diffv}(D) \le \text{diffv}(S)$ and therefore also $\le \tau_v(S)$. Consequently we obtain

$$\tau_v(\Gamma\theta) \le 2\tau_v^+(C) + \tau_v(S)$$

Combining this with (8) we obtain

$$\tau_v(E) \le \max\{2\tau_v^+(S) + \tau_v(S), \tau_v^+(S)\} = 2\tau_v^+(S) + \tau_v(S). \qquad \Box$$

We have not yet imposed any restriction on the reduction order $\succ$ underlying $\mathcal{G}$. In order to guarantee termination, $\succ$ has to be chosen carefully, as illustrated by the following example, due to R. Niewenhuis:

$$\to f(a) \approx a \tag{1}$$

$$\to a \approx f(b) \tag{2}$$

Let "$\succ$" be a lexicographic path order with the precedence $a \succ_{\text{prec}} f \succ_{\text{prec}} b$. Hence $f(a) \succ a \succ f(b)$. With respect to this order infinitely many new ground equations are derivable from (1) and (2) using $\mathcal{G}$.

To avoid such situations we make use of *Knuth-Bendix orders* ($\succ_{\text{kbo}}$), which are nicely supported in SPASS and turn out to be best suited for our purpose.

We refrain from stating the definition of $\succ_{\text{kbo}}$ in its full generality. Instead we give the definition with respect to the ground term algebra only. This is sufficient to establish Lemma 5.

Definition 3. *If s, t are ground terms then $s \succ_{\text{kbo}} t$ if*

1. weight(s) > weight(t) *or*
2. weight(s) = weight(t) *where $s \equiv f(s_1, \ldots, s_k)$ and $t \equiv g(t_1, \ldots, t_l)$ and*
   (a) $f \succ_{\text{prec}} g$, *or*
   (b) $f \equiv g$ *and* $(s_1, \ldots, s_k) \succ_{\text{kbo}}^{\text{lex}} (t_1, \ldots, t_l)$.

where weight *is a mapping from (ground) terms into non-negative integers; $\succ_{\text{prec}}$ is a strict total order on the signature symbols, and $\succ_{\text{kbo}}^{\text{lex}}$ is the lexicographical extension of $\succ_{\text{kbo}}$ to sequences of (ground) terms.*

For the rest of the paper we assume "$\succ$" to be an extension of $\succ_{\text{kbo}}$, where weight(t) counts the number of constant and function symbols occurring in t. It is proved, e.g.in [BN98] that there exists an order fulfilling the above restrictions that can be extended to a complete simplification order.

Let eqs(S) denote the set of all equality literals occurring in some clause C in S.

Lemma 5. *For all $S \in \text{OCC1N}_g^-$ there is a constant d (depending only on S) such that $\tau(\text{eqs}(\mathcal{G}^*(S))) \le d$.*

Proof. Observe that, since all equality literals are ground, new terms occurring in equality literals can only arise by replacing a ground (sub-)term by another ground term via superposition.

We make use of the fact that $\succ$ respects the size of ground terms. More exactly, let size(t) be the number of function and constant symbols occurring in a ground term t then, by definition of $\succ$:

(s) $s \succ t$ implies size(s) $\geq$ size(t)

for all ground terms s and t.

By the order conditions of the inference rules, any new term t that arises by superimposing one equality into another must be smaller with respect to $\succ$ than some term occurring in a parent clause. Therefore, by **(s)**, it must also be smaller in size. In other words, the size of terms in eqs($\mathcal{G}^*(\mathcal{S})$) is bounded by the maximal size of terms in eqs($\mathcal{G}(\mathcal{S})$). Since there are only finitely many different ground terms of bounded size (over a finite signature) we also obtain a bound on τ(eqs($\mathcal{G}^*(\mathcal{S})$)). $\square$

Lemma 6. *Let* $\mathcal{S} \in OCC1N_g^=$ *and let* d *be the bound on* τ(eqs($\mathcal{G}^*(\mathcal{S})$)) *obtained in Lemma 5, above. Then* $\tau^+(\mathcal{G}^*(\mathcal{S})) \leq \max\{\tau^+(\mathcal{S}), d + \tau_v^+(\mathcal{S})\}$.

Proof. Again, the proof proceeds by induction on the applications of the inference operator $\mathcal{G}$ and case-distinction according to the inference rules.

With respect to ordered factoring and positive resolution the arguments are the same as those in the proof of Lemma 2. (The only difference is that we refer to part *2(b)* instead of *2(c)* of Proposition 1.) I.e. we obtain $\tau^+(E) \leq \tau^+(C)$ for any factor of a (positive) $OCC1N_g^=$-clause C, as well as

$$\tau^+(E) \leq \max\{\tau^+(C), \tau^+(D)\}$$

for any positive resolvent E of $OCC1N_g^=$-clauses C, D.

Consider superposition right:

$$\frac{C \equiv (\rightarrow s{\approx}t, \Delta) \qquad D \equiv (\rightarrow \Sigma, L[s'])}{E \equiv (\rightarrow \Delta, \Sigma, L[t])\theta} \ (sr)$$

$s \succ t$ and $s = s'\theta$, where $C, D \in \mathcal{G}^*(\mathcal{S})$ and therefore, by Lemma 1, $\{C, D\} \in OCC1N_g^=$. Observe that $\theta(y) \leq \tau(s)$ for all y in the domain of θ. Moreover, by condition **(lin)**, $\Delta\theta = \Delta$ and $\Sigma\theta = \Sigma$. By Lemma 5 we thus obtain

$$\tau^+(E) \leq \max\{\tau^+(C), \tau^+(D), d + \tau_v^+(D)\}. \tag{3}$$

By the proof of Lemma 5 superposition right into an equality literal cannot increase the term depth beyond the global bound d.

The only remaining inference rule that can change the maximal term depth of the positive part of a clause is selective superposition:

$$\frac{C =\equiv (\rightarrow s{\approx}t, \Delta) \qquad D \equiv (L[s'], \Gamma \rightarrow \Sigma)}{E \equiv (L[t], \Gamma \rightarrow \Delta, \Sigma)\theta} \ (ss)$$

where $s \succ t$ and $s = s'\theta$. Again, $C, D \in \mathcal{G}^*(\mathcal{S})$ and therefore, by Lemma 1, $\{C, D\} \in OCC1N_g^=$. It suffices to observe that the terms in the range of θ are sub-terms of s to obtain the bound of (3) also for this case.

Summarizing we obtain $\tau^+(\mathcal{G}^*(\mathcal{S})) \leq \max\{\tau^+(\mathcal{S}), d + \tau_v^+(\mathcal{S})\}$. □

The last step in proving a global bound on term depth — going from $\tau^+(\mathcal{G}^*(\mathcal{S}))$ to $\tau(\mathcal{G}^*(\mathcal{S}))$ — is not difficult.

Lemma 7. *For all $\mathcal{S} \in OCC1N_g^= : \tau(\mathcal{G}^*(\mathcal{S})) \leq 2\tau_v^+(\mathcal{S}) + \tau^+(\mathcal{S}) + \tau_v(\mathcal{S}) + d$.*

Proof. From Lemma 4 we obtain the bound

$$2\tau_v^+(\mathcal{S}) + \tau_v(\mathcal{S}) \tag{4}$$

for the maximal depth of occurrence of a variable in $\mathcal{G}^*(\mathcal{S})$. Observe that — by the linearity of the positive part of $OCC1N_g^=$-clauses and the form of our inference rules — only (sub-)terms t occurring in the positive part of one parent clause, replacing a variable from the other parent clause can increase the term depth of a derived clause beyond $\tau_v(\mathcal{G}^*(\mathcal{S}))$. Lemma 6 bounds the depth of t by

$$\max\{\tau_v^+(\mathcal{S}) + d, \tau^+(\mathcal{S})\} \leq \tau^+(\mathcal{S}) + d. \tag{5}$$

Simplifying the sum of (4) and (5) we obtain *q.e.d.* □

The global bound on term depth alone does not yet imply that the inference process converges, i.e., that only finitely many different clauses (up to renaming of variables) can be derived. We also have to bound the length of derived clauses. By the *length* $|C|$ of a clause C we mean the number of literals (i.e. occurrences of atoms) in C. This can be achieved by applying the *condensation rule*. In contrast to the rules of the inference operator $\mathcal{G}$ condensation is a *reduction rule* (compare Section 6). It removes redundant literals from a clause.

Definition 4. *For any clause C we denote by $\mathrm{cond}(C)$ a shortest sub-clause of C' that is also an instance of C. It was proved by Joyner [Joy76] that $\mathrm{cond}(C)$ is unique (up to renaming of variables). We call $\mathrm{cond}(C)$ the condensate of C.*

Definition 5. *By $\mathcal{G}_c(\mathcal{S})$ we denote the set of condensates of all clauses in $\mathcal{G}(\mathcal{S})$. $\mathcal{G}_c^*(\mathcal{S})$ denotes the transitive and reflexive closure of $\mathcal{G}_c$.*

Lemma 8. *If $\mathcal{S} \in OCC1N_g^=$ then $|C| < l_\mathcal{S}$ for all $C \in \mathcal{G}_c^*(\mathcal{S})$ where $l_\mathcal{S}$ is some constant (depending on $\mathcal{S}$ only).*

Proof. The literals of the positive part of any clause in $\mathcal{G}_c^*(\mathcal{S})$ are pairwise variable disjoint. Therefore condensation removes one of the literals L, L' from the clause $(\Gamma \to L, L', \Sigma)$ if $L\nu = L'$, where ν is a renaming of variables not occurring in Γ. From this observation and the bound on $\tau_v^+(\mathcal{G}_c^*(\mathcal{S}))$ (Lemma 2) we obtain a bound on the number of positive literals in a condensed clause, since there are only finitely many (up to variable renaming) different atoms of bounded depth.

Concerning the number of negative literals in a clause, observe that the only rule of $\mathcal{G}$ that can add a negative literal is equality factoring (ef). However, the added equality literals are ground. By Lemma 5 there are only finitely many different equality literals in $\mathcal{G}_c^*(\mathcal{S})$. Therefore we obtain a bound on $|\Gamma|$ for all $(\Gamma \to \Sigma) \in \mathcal{G}_c^*(\mathcal{S})$ from the fact that condensation removes copies of identical literals from any clause.

Summarizing, we have proved that $|C|$ is bounded for $C \in \mathcal{G}_c^*(\mathcal{S})$.   □

Theorem 2. *The inference system* $\mathcal{G}_c$ *provides a decision procedure for class* $OCC1N_g^=$.

Proof. Let $\mathcal{S} \in OCC1N_g^=$. It is easy to check that replacing all clauses by their condensates does not violate the invariants of Lemmas 1-7. Therefore we may combine the bound on the depth of clauses $C \in \mathcal{G}^*(\mathcal{S})$ (see Lemma 7) with the bound on the length of clauses in $\mathcal{G}_c^*(\mathcal{S})$ (see Lemma 8) and conclude that there are only finitely many different clauses in $\mathcal{G}_c^*(\mathcal{S})$.

We can therefore effectively compute $\mathcal{G}_c^*(\mathcal{S})$ (by iteratively applying $\mathcal{G}_c$ until a fixed point is reached) and check whether it contains the empty clause. The decidability of $OCC1N_g^=$ thus follows from the soundness and (refutational) completeness of the inference system $\mathcal{G}_c$.   □

Remark 3. One might want to consider[3] the class $OCC1N_{lg}^=$, arising from $OCC1N_g^=$ by dropping the restriction that equality literals have to be ground for *positive* literals. However, one cannot directly apply the above machinery to this class, since $OCC1N_{lg}^=$ is not closed under applications of *equality factoring*. (Whether another version of the superposition calculus terminates on $OCC1N_{lg}^=$ remains open.)

Similarly, one may investigate the class $OCC1N_{rg}^=$, where only positive equality literals have to be ground. Observe that $OCC1N_{rg}^=$ is not closed under equality resolution (appearing in our version of the superposition calculus as selective resolution).

6   SPASS and $\mathcal{G}_c$

SPASS[4] [WAB+99,Wei00] is a theorem prover for (sorted) first-order logic with equality. Primarily, it is an implementation of the calculi $\mathcal{E}_S$ and $\mathcal{P}_S$[5] presented in [BG94]. SPASS provides a number of options to fine-tune the system that would in principle allow to directly implement $\mathcal{G}_c$ by an appropriate setting of various control flags. However, we claim that one may also take advantage of the following features of SPASS, which are not present in $\mathcal{G}_c$:

---

[3] As suggested by an anonymous referee.

[4] When we refer to specific properties of SPASS, we actually refer to SPASS Version 1.0.X as freely distributed at http://spass.mpi-sb.mpg.de/

[5] The calculus $\mathcal{P}_S$ consists of equality resolution, ordered factoring, superposition and merging paramodulation with selection function S.

- additional *order restrictions* on the inference rules,
- additional *reduction rules*, and
- a form of case analysis called *splitting rule*.

We refer to Section 8 for details on splitting. (We will see that splitting turns SPASS into a model builder for $OCC1N_g^-$.) We do not intend to use the sort constraint handling mechanism of SPASS here.

Additional order restrictions on inference rules obviously cannot spoil the termination of the inference process when $\mathcal{G}_c(\mathcal{S})$ is finite. Concerning reduction, we always assume the *condensation* rule to be used (as it is part of $\mathcal{G}_c$). (Note that condensation is nicely supported by SPASS, compare [Wei00].)

If one wants to employ additional reduction rules of SPASS like *subsumption*, *tautology deletion*, *unit conflict*, *terminator*, *local clause reduction*, *local rewriting* and *unit rewriting* (see [Wei00] for definitions) one has to check that the invariants established for $\mathcal{S} \in OCC1N_g^-$ with respect to $\mathcal{G}$ in Lemmas 1-7 remain valid also with respect to the refined inference operator, which we will call $\mathcal{I}_{spass}^+$ from now on. In all cases this trivially follows form the definitions of the rules in $\mathcal{I}_{spass}^+$ and of $OCC1N_g^-$.

We summarize the observations of this section as a corollary to Theorem 2:

Corollary 1. *An appropriate setting of parameters in* SPASS *results in a sound and complete inference operator, which terminates on inputs* $\mathcal{S} \in OCC1N_g^-$ *and thus decides this class.*

7   Representing Models by Atoms and Ground Equalities

A satisfiable clause set that is closed under a complete inference operator may be considered as a representation of the class of its models. However — when speaking of *model building* — we aim at something more ambitious: we want to extract simple and useful syntactical representations of *single* models from the information generated by standard theorem provers. It is natural to concentrate on *term models* in this context. (We refer to [FL98,FL96,Tam92,Pel97] for a more detailed presentation of this type of model building.)

Various criteria for a representation $\mathcal{R}_\mathcal{M}$ of a (term) model $\mathcal{M}$ to be appropriate (in this context) have been suggested; they include:

- Given $\mathcal{R}_\mathcal{M}$, there should be an efficient algorithm for deciding whether an atom holds in $\mathcal{M}$ or not.
- Evaluation of clauses with respect to $\mathcal{M}$ should be computable.
- Given two representations it should be decidable whether they represent the same model.

In addition, we consider it an advantage if the representations consist in syntactic structures that are already present in the output of a standard theorem prover. (I.e. we want to avoid the use of additional formalisms like, e.g., explicit constraints, tree grammars or term schematizations, if possible.)

Fortunately, a representation format that is suitable for $OCC1N_g^=$ and fulfills all above mentioned criteria — so-called *atomic representations* — has been investigated, e.g., in [FL96,FL98,GP99]. In the presence of equality, *term models* can be viewed as ordinary Herbrand models (i.e. interpretations over the set of all ground terms over the signature[6]) where "$\approx$" is interpreted as a congruence relation. In our context, an atomic representation of a term model with respect to some signature Sig consists in the union of a finite set of *linear* (non-equality) atoms At and a finite set of *ground equalities* Eq (over the Sig). The term model $\mathcal{M}_{At,Eq}^{Sig}$ represented by At and Eq with respect to Sig is defined by

$$A \text{ is } true \text{ in } \mathcal{M}_{At,Eq}^{Sig} \quad \text{iff} \quad At \cup Eq \models A$$

for all ground atoms A over Sig.

Observe that the ground equations in Eq induce a complete and terminating term rewrite system T^{Eq}. To test whether an atom A is true in $\mathcal{M}_{At,Eq}^{Sig}$ it suffices to check whether the normal form of A with respect to T^{Eq} is an instance of some $A' \in At$.

Consider clause evaluation with respect to $\mathcal{M}_{At,Eq}^{Sig}$: Obviously, clauses $C \in S \in OCC1N_g^=$ over Sig can be evaluated by applying the terminating inference operator to $At \cup Eq \cup \{C\}$. If C (not necessarily fulfilling the $OCC1N_g^=$-conditions) does not contain equality literals evaluation procedures given in [FL96] and [FL98] are applicable. We conjecture that evaluating general clauses in which all equality literals are *ground* is possible by similar methods. The fully general case is still open.

For the equivalence test for model representations and the construction of corresponding finite models we refer to [FL96] and [FL98]. The algorithmic complexity of various problems related to atomic representations has been investigated in [Pic98a,Pic98b,GP99].

8   Model Building by Splitting

Splitting may be considered as a special form of "proof by case analysis":

$$\frac{C \equiv (\Gamma_1, \Gamma_2 \rightarrow \Delta_1, \Delta_2)}{C_1 \equiv (\Gamma_1 \rightarrow \Delta_1) \parallel C_2 \equiv (\Gamma_2 \rightarrow \Delta_2)} \; split$$

where $vars(\Gamma_1, \Delta_1) \cap vars(\Gamma_2, \Delta_2) = \emptyset$, and neither C_1 nor C_2 is empty. Obviously we have: $\mathcal{S} \cup \{C_1\} \models \mathcal{S} \cup \{C\}$ and $\mathcal{S} \cup \{C_2\} \models \mathcal{S} \cup \{C\}$

The format of the rule is quite different from the other SPASS-rules and motivates an abstract notion of derivation. A *derivation* from a set of clauses $\mathcal{S}$ is a finitely branching (possible infinite) tree $\mathcal{T}_\mathcal{S}$, the nodes of which are sets of clauses. It is defined inductively according to the different types of rules in SPASS:

---

[6] The signature is augmented by a constant symbol if necessary to prevent the universe from being empty.

- $\mathcal{S}$ is the root of $\mathcal{T}_{\mathcal{S}}$
- A node $\mathcal{S}'$ has either a single successor node $\mathcal{S}'_1$ such that
  - $\mathcal{S}'_1 = \mathcal{S}' \cup \{E\}$ where E is the result of applying an inference rule of $\mathcal{I}^+_{spass}$ to premisses in $\mathcal{S}'$, or
  - $\mathcal{S}'_1$ is the result of applying a reduction rule of $\mathcal{I}^+_{spass}$ to $\mathcal{S}'$

  or it has two successor nodes $\mathcal{S}'_1, \mathcal{S}'_2$ such that
  - $\mathcal{S}'_1 = \mathcal{S}' \cup \{C_1\}$ and $\mathcal{S}'_2 = \mathcal{S}' \cup \{C_2\}$ where C_1 and C_2 represent the cases obtained by the splitting rule applied to some clause C in $\mathcal{S}'$.

A branch $\mathcal{S}, \mathcal{S}_1, \mathcal{S}_2, \ldots$, of $\mathcal{T}_{\mathcal{S}}$ is called *open* if it does not contain the empty clause. Our aim is to show that all open branches of $\mathcal{T}_{\mathcal{S}}$, as generated by SPASS, contain an atomic representation of a model of $\mathcal{S}$ for all satisfiable $\mathcal{S} \in OCC1N_g^=$. For this purpose we briefly recall some concepts from [BG94].

Remember that their superposition calculus is defined with respect to some fixed reduction order $\prec$. This order can be extended to a well-founded order $\prec_C$ on ground clauses in a canonic way.

Let $\mathcal{S}$ be a set of clauses, and C be any ground clause. We call C *redundant* with respect to $\mathcal{S}$, if there exist ground instances $C_1, \ldots, C_k$ of clauses in $\mathcal{S}$ such that $C_1, \ldots, C_k \models C$ and $C \succ_C C_i$ for all $1 \leq i \leq k$. A clause C is called redundant if all its ground instances are redundant.

A set of clauses $\mathcal{S}$ is called *saturated* (with respect to $\mathcal{I}^+_{spass}$) if any conclusion C of an $\mathcal{I}^+_{spass}$-inference from $\mathcal{S}$, not already included in $\mathcal{S}$, is redundant.

For any branch $\mathcal{S}, \mathcal{S}_1, \mathcal{S}_2, \ldots$ in $\mathcal{T}_{\mathcal{S}}$ we call $\mathcal{S}_\infty = \bigcup_j \bigcap_{k \geq j} \mathcal{S}_j$ the *limit* of this branch. A derivation $\mathcal{T}_{\mathcal{S}}$ is *fair*, if for any branch $\mathcal{S}, \mathcal{S}_1, \mathcal{S}_2, \ldots$ in $\mathcal{T}_{\mathcal{S}}$, each clause C that can be derived (using inference rules of $\mathcal{I}^+_{spass}$) from its limit, is contained in some $\mathcal{S}_j$.

The completeness of SPASS can be expressed now as follows:

Proposition 3. *Assume that $\mathcal{T}_{\mathcal{S}}$ is a fair derivation. Every limit $\mathcal{S}_\infty$ of a branch $\mathcal{S}, \mathcal{S}_1, \mathcal{S}_2, \ldots$ in $\mathcal{T}_{\mathcal{S}}$ is saturated and every model of $\mathcal{S}_\infty$ is also a model of $\mathcal{S}$.*

We say that a branch $\mathcal{S}, \mathcal{S}_1, \mathcal{S}_2, \ldots$ of a derivation $\mathcal{T}_{\mathcal{S}}$ *terminates* with a set of clause $\mathcal{S}_k$, occurring on the branch, if $\mathcal{S}_k = \mathcal{S}_\infty$ or contains the empty clause. By Proposition 3, $\mathcal{S}_k$ is saturated if $\mathcal{T}_{\mathcal{S}}$ is fair.

We extend the notion of saturatedness to derivations $\mathcal{T}_{\mathcal{S}}$. $\mathcal{T}_{\mathcal{S}}$ is *saturated* if it is fair and if for every branch $\mathcal{S}, \mathcal{S}_1, \mathcal{S}_2, \ldots$ of $\mathcal{T}_{\mathcal{S}}$, $\mathcal{S}_\infty$ is closed under applications of the splitting rule.

We say that a set of clauses $\mathcal{S}$ *decomposes into* the set of non-positive clauses $\mathcal{N}$, non-equality atoms At and ground equalities Eq if $\mathcal{S} = \mathcal{N} \cup \{(\rightarrow A) : A \in At\} \cup \{(\rightarrow s{\approx}t) : s{\approx}t \in Eq\}$.

Theorem 3. *Let $\mathcal{T}_{\mathcal{S}}$ be a saturated derivation from some satisfiable $\mathcal{S} \in OCC1N_g^=$. Then all open branches of $\mathcal{T}_{\mathcal{S}}$ terminate with some $\mathcal{S}_k$ that decomposes into $\mathcal{N}$, At and Eq. Moreover, $\mathcal{M}^{Sig}_{At,Eq}$ is a term model of $\mathcal{S}$ over the signature Sig of $\mathcal{S}$.*

Proof. Let $\mathcal{S} \in OCC1N_g^=$ be a satisfiable clause set. Let $\mathcal{T}_\mathcal{S}$ be a saturated derivation from $\mathcal{S}$. By our decidability proof, any open branch $\mathcal{S}, \mathcal{S}_1, \mathcal{S}_2, \ldots$ in $\mathcal{T}_\mathcal{S}$ terminates with some $\mathcal{S}_k$. Moreover, we may conclude (from Lemma 1) that $\mathcal{S}_k \in OCC1N_g^=$. Since $\mathcal{T}_\mathcal{S}$ is saturated, $\mathcal{S}_k$ is closed under applications of the splitting rule and therefore decomposes into $\mathcal{N}$, At and Eq.

To simplify notation, we abbreviate $\mathcal{M}_{At,Eq}^{Sig}$ by $\mathcal{M}$. Assume that $\mathcal{M}$ is *not* a model of $\mathcal{S}_k$. Let $False_\mathcal{M}$ be the set of all ground atoms over Sig that are *false* in $\mathcal{M}$. Then, by definition,

$$\mathcal{S}_k \cup \{(B \to) : B \in False_\mathcal{M}\}$$

is unsatisfiable. By compactness there exists a finite, unsatisfiable subset $\mathcal{F}$ of $\{(B \to) : B \in False_\mathcal{M}\}$ such that $\mathcal{S}_k \cup \mathcal{F}$ is unsatisfiable. Therefore the empty clause can be derived from $\mathcal{S}_k \cup \mathcal{F}$ using any refutationally complete inference system. In particular, there exists a finite sequence $\mathcal{Z}_0, \mathcal{Z}_1, \ldots, \mathcal{Z}_l$, where $\mathcal{Z}_0 = \mathcal{S}_k \cup \mathcal{F}$, $\mathcal{Z}_{i+1} = \mathcal{Z}_i \cup \{E_{i+1}\}$ for some clause E_{i+1} that is a conclusion of an $\mathcal{I}_{spass}^+$-inference of clauses in $\mathcal{Z}_i$, and E_l is the empty clause. By the completeness proof for $\mathcal{I}_{spass}^+$ we may assume that none of the E_i is redundant with respect to $\mathcal{S}_k$. Since $\mathcal{S}_k$ is saturated and the clauses in $\mathcal{F}$ are negative (ground unit) clauses and since $\mathcal{I}_{spass}^+$ employs the positive superposition-strategy (see Section 4) this implies that no clauses from $\mathcal{N}$ are involved in this derivation. In other words, the empty clause can already be derived from clauses in $\mathcal{F}$, $\{(\to A) : A \in At\}$ and $\{(\to s \approx t) : s \approx t \in Eq\}$. But, by definition, all those clauses are *true* in $\mathcal{M}$. Since $\mathcal{I}_{spass}^+$ is sound this means that the empty clause cannot be derived from $\mathcal{S}_k \cup \mathcal{F}$. This contradiction concludes the proof. $\qquad\square$

Acknowledgement. We thank Robert Niewenhuis for pointing out an error in a previous version of this paper.

References

BG94.   L. Bachmair and H. Ganzinger. Rewrite-based equational theorem proving with selection and simplification. *Journal of Logic and Computation*, 4(3):217–247, 1994.   117, 117, 117, 125, 128

BGW93.   L. Bachmair, H. Ganzinger, and U. Waldmann. Superposition with simplification as a decision procedure for the monadic class with equality. In *Proc. of KGC 1993*, volume 713 of *LNCS*. Springer, 1993.   114

BN98.   F. Baader and T. Nipkow. *Term Rewriting and All That*. Cambridge University Press, 1998.   122

CP96.   R. Caferra and N. Peltier. Decision procedures using model building techniques. In *Computer Science Logic, Proc. of CSL'95*, volume 1092 of *LNCS*, pages 130–144. Springer, 1996.   115

FL93.   C.G. Fermüller and A. Leitsch. Model building by resolution. In *Computer Science Logic, Proc. of CSL'92*, volume 702 of *LNCS*, pages 134–148. Springer, 1993.   114, 115, 116

FL96.   C.G. Fermüller and A. Leitsch. Model building by resolution. *Journal of Logic and Computation 6*, pages 134–148, 1996.   114, 126, 127, 127, 127

FL98.   C.G. Fermüller and A. Leitsch. Decision procedures and model building in
            equational clause logic. *Logic Journal of the Interest Group in Pure and
            Applied Logics (IGPL)*, 6:17–41, 1998. 114, 114, 114, 126, 127, 127, 127
FLHT00.   C.G. Fermüller, A. Leitsch, U. Hustadt, and T. Tammet. Resolution de-
            cision procedures. In A. Robinson and A. Vorkonov, editors, *Handbook of
            Automated Reasoning*. Elsevier, 2000. To appear. 114
FLTZ93.   C.G. Fermüller, A. Leitsch, T. Tammet, and N. Zamov. *Resolution Methods
            for the Decision Problem*. Springer LNAI 679, 1993. 114, 118, 118, 120
FS93.   C.G. Fermüller and G. Salzer. Ordered paramodulation and resolution as
            decision procedure. In *Proc. of LPAR 1993*, volume 698 of *LNCS*, pages
            122–133. Springer, 1993. 114, 116
GD99.   H. Ganzinger and H. DeNivelle. A superposition decision procedure for the
            guarded fragment with equality. In *Proc. of LICS 1999*, pages 295–305.
            IEEE Computer Society Press, 1999. 114
GP99.   G. Gottlob and R. Pichler. Working with Arms: Complexity results on
            Atomic Representations of Herbrand models. In *Proc. of LICS 1999*, pages
            306–315. IEEE Computer Society Press, 1999. 127, 127
JMW98.   F. Jacquemard, C. Meyer, and C. Weidenbach. Unification in extensions of
            shallow equational theories. In *Proc. of RTA 1998*, pages 76–90. Springer
            LNCS 1379, 1998. 114
Joy76.   W. H. Joyner. Resolution strategies as decision procedures. *Journal of the
            ACM*, 231:398–417, July 1976. 124
Nie96.   Robert Nieuwenhuis. Basic paramodulation and decidable theories. In *Proc.
            of LICS 1996*, pages 437–482. IEEE Computer Society Press, 1996. 114
Pel97.   N. Peltier. Increasing the capabilities of model building by constraint solv-
            ing with terms with integer exponents. *Journal of Symbolic Computation*,
            24:59–101, 1997. 115, 126
Pic98a.   R. Pichler. Algorithms on atomic representations of Herbrand models. In
            Proc. of JELIA 1998, pages 199–215. Springer LNAI 1489, 1998. 127
Pic98b.   R. Pichler. On the complexity of H-subsumption. In *Computer Science
            Logic, Proc. of CSL'98*, volume 1584 of *LNCS*, pages 355–371. Springer,
            1998. 127
Rud00.   T. Rudloff. SHR tableaux — a framework for automated model generation.
            Journal of Symbolic Computation, 2000. To appear. 114
Tam92.   T. Tammet. *Resolution Methods for Decision Problems and Finite Model
            Building*. PhD thesis, Department of Computer Science, Chalmers Univer-
            sity of Technology., Chalmers-Göteborg, 1992. 126
WAB+99.   C. Weidenbach, B. Afshordel, U. Brahm, C. Cohrs, T. Engel, E. Keen,
            C. Theobalt, and D. Topic. System description: SPASS Version 1.0.0. In
            Proc. of CADE-16, volume 1632 of *LNCS*, pages 378–382. Springer, 1999.
            115, 125
Wei00.   C. Weidenbach. SPASS: Combining superposition, sorts and splitting. In
            A. Robinson and A. Voronkov, editors, *Handbook of Automated Reasoning*.
            Elsevier, 2000. To appear.

Compiling and Verifying Security Protocols

Florent Jacquemard[1], Michaël Rusinowitch[1], and Laurent Vigneron[2]

[1] LORIA – INRIA Lorraine
Campus Scientifique, B.P. 239
54506 Vandœuvre-lès-Nancy Cedex, France
Florent.Jacquemard@loria.fr, Michael.Rusinowitch@loria.fr
http://www.loria.fr/~jacquema, http://www.loria.fr/~rusi
[2] LORIA – Université Nancy 2
Campus Scientifique, B.P. 239
54506 Vandœuvre-lès-Nancy Cedex, France
Laurent.Vigneron@loria.fr, http://www.loria.fr/~vigneron

Abstract. We propose a direct and fully automated translation from standard security protocol descriptions to rewrite rules. This compilation defines non-ambiguous operational semantics for protocols and intruder behavior: they are rewrite systems executed by applying a variant of ac-narrowing. The rewrite rules are processed by the theorem-prover daTac. Multiple instances of a protocol can be run simultaneously as well as a model of the intruder (among several possible). The existence of flaws in the protocol is revealed by the derivation of an inconsistency. Our implementation of the compiler CASRUL, together with the prover daTac, permitted us to derive security flaws in many classical cryptographic protocols.

Introduction

Many verification methods have been applied to the analysis of some particular cryptographic protocols [22,5,8,24,34]. Recently, tools have appeared [17,13,9] to automatise the tedious and error-prone process of translating protocol descriptions into low-level languages that can be handled by automated verification systems. In this research stream, we propose a concise algorithm for a direct and fully automated translation of any standard description of a protocol, into rewrite rules. For analysis purposes, the description may include security requirements and malicious agent (intruder) abilities. The asset of our compilation is that it defines non-ambiguous operational semantics for protocols (and intruders): they are rewrite rules executed on initial data by applying a variant of narrowing [15].

In a second part of our work, we have processed the obtained rewrite rules by the theorem-prover daTac [33] based on first order deduction modulo associativity and commutativity axioms (AC). Multiple instances of a protocol can be run simultaneously as well as a model of the intruder (among several possible). The existence of flaws in classical protocols (from [7]) has been revealed by the derivation of an inconsistency with our tool CASRUL.

M. Parigot and A. Voronkov (Eds.): LPAR 2000, LNAI 1955, pp. 131–160, 2000.

In our semantics, the protocol is modelled by a set of transition rules applied on a multiset of objects representing a global state. The global state contains both sent messages and expected ones, as well as every piece of information collected by the intruder. Counters (incremented by narrowing) are used for dynamic generation of nonces (random numbers) and therefore ensure their freshness. The expected messages are automatically generated from the standard protocol description and describes concisely the actions to be taken by an agent when receiving a message. Hence, there is no need to specify manually these actions with special constructs in the protocol description. The verification that a received message corresponds to what was expected is performed by unification between a sent message and an expected one. When there is a unifier, then a transition rule can be fired: the next message in the protocol is composed and sent, and the next expected one is built too. The message to be sent is composed from the previously received ones by simple projections, decryption, encryption and pairing operations. This is made explicit with our formalism. The information available to an intruder is also floating in the messages pool, and used for constructing faked messages, by ac-narrowing too. The intruder-specific rewrite rules are built by the compiler according to abilities of the intruder (for diverting and sending messages) given with the protocol description.

It is possible to specify several systems (in the sense of [17]) running a protocol concurrently. Our compiler generates then a corresponding initial state. Finally, the existence of a security flaw can be detected by the reachability of a specific critical state. One critical state is defined for each security property given in the protocol description by mean of a pattern independent from the protocol.

We believe that a strong advantage of our method is that it is not ad-hoc: the translation is working without user interaction for a wide class of protocols and therefore does not run the risk to be biased towards the detection of a known flaw. To our knowledge, only two systems share this advantage, namely Casper [17] and CAPSL [21]. Therefore, we shall limit our comparison to these works.

Casper is a compiler from protocol specification to process algebra (CSP). The approach is oriented towards finite-state verification by model-checking with FDR [28]. We use almost the same syntax as Casper for protocols description. However, our verification techniques, based on theorem proving methods, will handle infinite states models. This permits to relax many of the strong assumptions for bounding information (to get a finite number of states) in model checking. Especially, our counters technique based on narrowing ensures directly that all randomly generated nonces are pairwise different. This guarantees the freshness of information over sessions. Our approach is based on analysing *infinite* traces by refutational theorem-proving and it captures automatically the traces corresponding to attacks. Note that a recent interesting work by D.Basin [4] proposes a lazy mechanism for the automated analysis of infinite traces.

CAPSL [21] is a specification language for authentication protocols in the flavour of Casper's input. There exists a compiler [9] from CAPSL to an intermediate formalism CIL which may be converted to an input for automated verification tools such as Maude, PVS, NRL [20]. The rewrite rules produced by our compilation is also an intermediate language, which has the advantage to be an idiom understood by many automatic deduction systems. In our case we have a single rule for every protocol message exchange, as opposite to CIL which has two rules. For this reason, we feel that our model is closer to Dolev and Yao original model of protocols [11] than other rewrite models are.

As a back-end system, the advantage of daTac over Maude is that ac-unification is built-in. In [8] it was necessary to program an ad-hoc narrowing algorithm in Maude in order to find flaws in protocols such as Needham-Schroeder Public Key.

We should also mention the works by C. Meadows [19] who was the first to apply narrowing to protocol analysis. Her narrowing rules were however restricted to symbolic encryption equations.

The paper is organised as follows. In Section 1, we describe the syntax for specifying a protocol P to be analysed and to give as input to the translator. Section 2 presents the algorithm implemented in the translator to produce, given P, a set of rewrite rules $R(P)$. This set defines the actions performed by users following the protocol. The intruder won't follow the rules of the protocol, but will rather use various skill to abuse other users. His behaviour is defined by a rewrite system $\mathcal{I}$ given in Section 3. The execution of P in presence of an intruder may be simulated by applying narrowing with the rules of $R(P) \cup \mathcal{I}$ on some initial term. Therefore, this defines an operational semantics for security protocols (Section 4). In Section 5, we show how flaws of P can be detected by pattern matching on execution traces, and Section 6 describes the deduction techniques underlying the theorem prover daTac and some experiments performed with this system. For additional informations the interested reader may refer to http://www.loria.fr/equipes/protheo/SOFTWARES/CASRUL/.

We assume that the reader is familiar with basic notions of cryptography and security protocols (public and symmetric key cryptography, hash functions) [30], and of term rewriting [10].

1   Input Syntax

We present in this section a precise syntax for the description of security protocols. It is very close to the syntax of CAPSL [21] or Casper [17] though it differs on some points – for instance, on those in Casper which concern CSP. The specification of a protocol P comes in seven parts (see Example 1, Figure 1). Three concern the protocol itself and the others describe an instance of the protocol (for a simulation).

1.1   Identifiers Declarations

The identifiers used in the description of a protocol P have to be declared to belong to one of the following types: user (principal name), public_key, `symmetric_key`, `table`, `function`, `number`. The type `number` is an abstraction for any kind of data (numeric, text or record ...) not belonging to one of the other types (`user`, `key` etc). An identifier T of type `table` is a one entry array, which associates public keys to users names ($T[D]$ is a public key of D). Therefore, public keys may be declared alone or by mean of an association table. An identifier F of type `function` is a one-way (hash) function. This means that one cannot retrieve X from the digest $F(X)$.

The unary postfix function symbol $\_^{-1}$ is used to represent the private key associated to some public key. For instance, in Figure 1, $T[D]^{-1}$ is the private key of D.

Among users, we shall distinguish an intruder I (it is not declared). It has been shown by G. Lowe [18] that it is equivalent to consider an arbitrary number of intruders which may communicate and one single intruder.

1.2   Messages

The core of the protocol description is a list of lines specifying the rules for sending messages,

$$(i.\ S_i \to R_i\ :\ M_i)_{1 \le i \le n}$$

For each $i \le n$, the components i (step number), S_i, R_i (`users`, respectively sender and receiver of the message) and M_i (message) are ground terms over a signature $\mathbb{F}$ defined as follows. The declared `identifiers` as well as I are nullary function symbols of $\mathbb{F}$. The symbols of $\mathbb{F}$ with arity greater than 0 are $\_^{-1}$, $\_[\_]$ (for tables access), $\_(\_)$ (for one-way functions access), $\langle\_,\_\rangle$ (pairing), $\{\_\}\_$ (encryption). We assume that multiple arguments in $\langle\_,\dots,\_\rangle$ are right associated. We use the same notation for public key and symmetric key encryption (overloaded operator). Which function is really employed shall be determined unambiguously by the type of the key.

Example 1. Throughout the paper, we illustrate our method on two toy examples of protocols inspired by [36] and presented in Figure 1. These protocols describe messages exchanges in a home cable tv set made of a decoder D and a smartcard C. C is in charge of recording and checking subscription rights to channels of the user. In the first rule of the symmetric key version, the decoder D transmits his name together with an instruction *Ins* to the smartcard C. The instruction *Ins*, summarised in a `number`, may be of the form "(un)subscribe to channel n" or also "check subscription right for channel n". It is encrypted using a symmetric key K known by C and D. The smartcard C executes the instruction *Ins* and if everything is fine (*e.g.* the subscription rights are paid for channel n), he acknowledges to D, with a message containing C, D and the instruction *Ins* encrypted with K. In the public key version, the privates keys of D and C respectively are used for encryption instead of K.

```
protocol TV;  # symmetric key          protocol TV;  # public key
identifiers                            identifiers
C, D : user;                           C, D : user;
Ins  : number;                         Ins  : number;
K    : symmetric_key;                  T    : table;
messages                               messages
```

1. $D \to C : \langle D, \{Ins\}_K \rangle$ 1. $D \to C : \langle D, \{Ins\}_{T[D]^{-1}} \rangle$

2. $C \to D : \langle C, D, \{Ins\}_K \rangle$ 2. $C \to D : \langle C, \{Ins\}_{T[C]^{-1}} \rangle$

```
knowledge                              knowledge
```

$D : C, K;$ $D : C, T, T[D]^{-1};$

$C : K;$ $C : T, T[C]^{-1};$

```
session_instance :                     session_instance :
```

$[D : tv, C : scard, K : key];$ $[D : tv, C : scard, T : key];$

```
intruder : divert, impersonate;        intruder : eaves_dropping;
intruder_knowledge : scard;            intruder_knowledge : key;
```

goal : correspondence_between $scard, tv;$ goal : secrecy_of $Ins;$

Fig. 1. Cable TV toy examples

1.3 Knowledge

At the beginning of a protocol execution, each principal needs some initial knowledge to compose his messages.

The field following `knowledge` associates to each `user` a list of terms of $T(\mathbb{F})$ describing all the data (names, keys, function *etc*) he knows before the protocol starts. We assume that the own name of every user is always implicitly included in his initial knowledge. The intruder's name I may also figure here. In some cases indeed, the intruder's name is known by other (naïve) principals, who shall start to communicate with him because they ignore his bad intentions.

Example 2. In Example 1, D needs the name of the smartcard C to start communication. In the symmetric key version, both C and D know the shared key K. In the public key version, they both know the `table` T. It means that whenever D knows C's name, he can retrieve and use his public key $T[C]$, and conversely. Note that the `number` Ins is not declared in D's knowledge. This value may indeed vary from one protocol execution to one another, because it is created by D at the beginning of a protocol execution. The identifier Ins is therefore called a *fresh* number, or *nonce* (for oNly once), as opposite to persistent identifiers like C, D or K.

Definition 1. *Identifiers which occur in a* `knowledge` *declaration* $U : \ldots$ *(including the user name U) are called* persistent. *Other identifiers are called* fresh.

The subset of $\mathbb{F}$ of fresh identifiers is denoted $\mathbb{F}_{fresh}$. The identifier $ID \in \mathbb{F}_{fresh}$ is said to be *fresh in* M_i, if ID occurs in M_i and does not occur in any M_j for $j < i$. We denote $fresh(M_i)$ the list of identifiers fresh in M_i (occurring in this order). We assume that if there is a public key $K \in fresh(M_i)$ then K^{-1} also occurs in $fresh(M_i)$ (right after K). Fresh identifiers are indeed instantiated by a principal in every protocol session, for use in this session only, and disappear at the end of the session. This is typically the case of nonces. Moreover we assume that the same fresh value cannot be created in two different executions of a protocol. Symmetric keys may either be persistent or fresh.

1.4   Session Instances

This field proposes some possible values to be assigned to the persistent identifiers (*e.g. tv* for D in Figure 1) and thus describes the different systems (in the sense of Casper [17]) for running the protocol. The different sessions can take place concurrently or sequentially an arbitrary number of times.

Example 3. In Figure 1, the field `session_instance` contains only one trivial declaration, where one value is assigned to each identifier. This means that we want a simulation where only one system is running the protocol (*i.e.* the number of concurrent sessions is one, and the number of sequential sessions is unbounded).

1.5   Intruder

The `intruder` field describes which strategies the intruder can use, among passive `eaves_dropping`, `divert`, `impersonate`. These strategies are described in Section 3. A blank line here means that we want a simulation of the protocol without intruder.

1.6   Intruder Knowledge

The `intruder_knowledge` is a set of values introduced in `session_instance`, but not a set of identifiers (like `knowledge` of others principals).

1.7   Goal

This is the kind of flaw we want to detect. There are two families of goals, `correspondence_between` and `secrecy_of` (see Sections 5.4 and 5.3). The secrecy is related to one identifier which must be given in the declaration, and the correspondence is related to two users.

2   Protocol Rules

We shall give a formal description of the possible executions of a given protocol in the formalism of normalised ac-narrowing. More precisely, we give an algorithm which translates a protocol description $\mathcal{P}$ in the above syntax into a set of rewrite rules $R(\mathcal{P})$.

We assume given a protocol $\mathcal{P}$, described by all the fields defined in Section 1, such that

$$R_i = S_{i+1} \text{ for } i = 0 \ldots n - 1$$

This hypothesis is not restrictive since we can add empty messages. For instance, we can replace

$$
\begin{array}{ll}
i.\ A \to B : M & \text{by} \\
i+1.\ C \to D : M' &
\end{array}
\qquad
\begin{array}{l}
i.\ A \to B : M \\
i+1.\ B \to C : \emptyset \\
i+2.\ C \to D : M'
\end{array}
$$

For technical convenience, we let $R_0 = S_1$ and assume that S_0, M_0 are defined and are two arbitrary new constants of $\mathbb{F}$.

As in the model of Dolev and Yao [11] the translation algorithm associates to each step $S_i \to R_i : M_i$ a rewrite rule $l_i \to r_i$. An additional rule $l_{n+1} \to r_{n+1}$ is also created. The left member l_i describes the tests performed by R_{i-1} after receiving the message M_{i-1} – R_{i-1} compares M_{i-1} (by unification) with a *pattern* describing what was expected. The right member r_i describes how $S_i = R_{i-1}$ composes and send the next message M_i, and what is the pattern of the next message expected. This representation makes explicit most of the actions performed during protocol execution (recording information, checking and composing messages), which are generally hidden in protocol description. How to build the message from the pieces has to be carefully (unambiguously) specified. The expected pattern has also to be described precisely.

Example 4. In the symmetric key version of the protocol described in Figure 1, the cipher $\{Ins\}_K$ in last field of message 2 may be composed in two ways: either directly by projection on second field of message 1, or by decryption of this projection (on second field of message 1), and re-encryption of the value Ins obtained, with key K. The first (shortest) case is chosen in our procedure.
The pattern expected by C for message 1 is $\langle C, x_1, \{x_2\}_K \rangle$, because C does not know D's name in advance, nor the number Ins. The pattern expected by D for message 2 is $\langle C, D, \{Ins\}_K \rangle$, because D wants to check that C has sent the right Ins.

2.1   Normalised ac-Narrowing

Our operational semantics for protocols are based on narrowing [15]. To be more precise, each step of an execution of the protocol $\mathcal{P}$ is simulated by a narrowing step using $R(\mathcal{P})$. We recall that narrowing unifies the left-hand side of a rewrite rule with a target term and replaces it with the corresponding right-hand side, unlike standard rewriting which relies on *matching* left-hand sides.

Let $\mathcal{T}(\mathcal{F}, \mathcal{X})$ denote the set of terms constructed from a (finite) set $\mathcal{F}$ of function symbols and a (countable) set $\mathcal{X}$ of variables. The set of ground terms $\mathcal{T}(\mathcal{F}, \emptyset)$ is denoted $\mathcal{T}(\mathcal{F})$. In our notations, every variable starts by the letter x. We use $u[t]_p$ to denote a term that has t as a subterm at position p. We use $u[\cdot]$ to denote the context in which t occurs in the term $u[t]_p$. By $u|_p$, we denote the *subterm* of u rooted at *position* p. A rewrite *rule* over a set of terms is an ordered pair (l, r) of terms and is written $l \to r$. A *rewrite system* $\mathcal{S}$ is a finite set of such rules. The rewrite relation $\to_{\mathcal{S}}$ can be extended to rewrite over congruence classes defined by a set of equations AC, rather than terms. These constitute ac-rewrite systems. In the following the set AC will be $\{x.(y.z) = (x.y).z, x.y = y.x\}$ where $\_.\_$ is a special binary function used for representing multisets of messages. The congruence relation generated by the AC axioms will be denoted $=_{ac}$. For instance $e.h.g =_{ac} g.e.h$. A term s *ac-rewrites by* S to another term t, denoted $s \to_S t$, if $s|_p =_{ac} l\sigma$ and $t = s[r\sigma]_p$, for some rule $l \to r$ in S, position p in s, and substitution σ. When s cannot be rewritten by S in any way we say it is a *normal form* for S. We note $s \downarrow_S t$, or $t = s \downarrow_S$ if there is a finite sequence of rewritings $s \to_S s_1 \to_S \ldots \to_S t$ and t is a *normal form* for S.

In the following we shall consider two rewrite systems $\mathcal{R}$ and $\mathcal{S}$. The role of the system $\mathcal{S}$ is to keep the messages normalised (by rewriting), while $\mathcal{R}$ is used for narrowing. A term s *ac-narrows by* $\mathcal{R}, \mathcal{S}$ to another term t, denoted $s \leadsto_{\mathcal{R},\mathcal{S}} t$, if i) s is a normal form for $\mathcal{S}$, and ii) $s|_p\sigma =_{ac} l\sigma$ and $t = (s[r]_p)\sigma \downarrow_{\mathcal{S}}$, for some rule $l \to r$ in $\mathcal{R}$, position p in s, and substitution σ.

Example 5. Assume $\mathcal{R} = \{a(x).c(x) \to c(x)\}$ and $\mathcal{S} = \{c(x).c(x) \to 0\}$. Then $a(0).b(0).c(x) \leadsto_{\mathcal{R},\mathcal{S}} b(0).c(0)$.

2.2   Messages Algebra

We shall use for the rewrite systems $\mathcal{R}$ and $\mathcal{S}$ a sorted signature $\mathcal{F}$ containing (among other symbols) all the non-nullary symbols of $\mathbb{F}$ of Section 1, and a variable set $\mathcal{X}$ which contains one variable x_t for each term $t \in \mathcal{T}(\mathbb{F})$.

Sorts. The sorts for $\mathcal{F}$ are: user, intruder, iuser = user $\cup$ intruder, public_key, private_key, symmetric_key, table, function, number. Additional sorts are text, a super-sort of all the above sorts, and int, message and list_of.

Signature. All the constants occurring in a declaration session_instance are constant symbols of $\mathcal{F}$ (with the same sort as the identifier in the declaration). The symbol I is the only constant of sort intruder in $\mathcal{F}$. The pairing function $\langle \_, \_ \rangle$ (profile text × text → text) and encryption functions $\{\_\}\_$ (text × public_key → text or text × private_key → text or text × symmetric_key → text) are the same as in $\mathbb{F}$ (see Section 1.2), as well as the unary function $\_^{-1}$ (public_key → private_key or private_key → public_key) for private keys (see Section 1.1), and as the table functions $\_[\_]$ (table × iuser → public_key). We use a unary function symbol $nonce(\_) :$ int → number for the fresh numbers, see Section 2.4. We shall use similar unary functions $K(\_)$ (int → public_key) and $SK(\_)$ (int → symmetric_key) for respectively public and symmetric fresh keys.

At last, the constant 0 (sort int) and unary successor function $s(\_)$ (int → int) will be used for integer (time) encoding. Some other constants $1, \ldots, k$ and $\underline{0}, \underline{1} \ldots$ and some alternative successor functions $s_1(\_), \ldots, s_k(\_)$ are also used. The number k is fixed according to the protocol $\mathcal{P}$ (see page 140).

From now on, x_t, x_{pu}, x_p, x_s, x_{ps}, x_u, x_f are variables of respective sorts table, public_key, public_key $\cup$ private_key, symmetric_key, public_key $\cup$

private_key ∪ symmetric_key, user, and function. K, SK and KA will be arbitrary terms of $\mathcal{T}(\mathbb{F})$ of resp. sorts public_key∪private_key, symmetric_key and public_key ∪ private_key ∪ symmetric_key.

Rewrite system for normalisation. In order to specify the actions performed by the principals, $\mathcal{F}$ contains some destructors. The decryption function applies to a text encrypted with some key, in order to extract its content. It is denoted the same way as the encryption function $\{\_\}\_$. Compound messages can be broken into parts using projections $\pi_1(\_), \pi_2(\_)$. Hence the relations it introduces in the message algebra are:

$$\left\{\{x\}_{x_s}\right\}_{x_s} \to x \tag{1}$$

$$\left\{\{x\}_{x_{pu}}\right\}_{x_{pu}^{-1}} \to x \tag{2}$$

$$\left\{\{x\}_{x_{pu}^{-1}}\right\}_{x_{pu}} \to x \tag{3}$$

$$x^{-1^{-1}} \to x \tag{4}$$

$$\pi_1(\langle x_1, x_2 \rangle) \to x_1 \tag{5}$$

$$\pi_2(\langle x_1, x_2 \rangle) \to x_2 \tag{6}$$

The rule (4) does not correspond to a real implementation of the generation of private key from public key. However, it is just a technical convenience. The terminating rewrite system $(1) - (6)$ is called S_0. It can be easily shown that S_o is convergent [10], hence every message t admits a unique normal form $t \downarrow_{S_0}$ for S_0.

We assume from now on that the protocol $\mathcal{P}$ is *normalised*, in the following sense.

Definition 2. *A protocol $\mathcal{P}$ is called normalised if all the message terms in the field* messages *are in normal form w.r.t. S_0.*

Note that this hypothesis is not restrictive since any protocol $\mathcal{P}$ is equivalent to the normalised protocol $\mathcal{P} \downarrow_{S_0}$.

2.3   Operators on Messages

We define in this section some functions to be called during the construction of the system $\mathcal{R}(\mathcal{P})$ in Section 2.4.

Knowledge decomposition. We denote by $know(U, i)$ the information that a user U has memorised at the end of the step $S_i \to R_i : M_i$ of the protocol $\mathcal{P}$. This information augments incrementally with i:

- if U is the receiver R_i, then he records the received message M_i as well as the sender's (official) name S_i,
- if U is the sender S_i, then he records the fresh elements (nonces...) he has created for composing M_i (and may use latter),

– in any other case, the knowledge of U remains unchanged.

The set $know(U, i)$ contains labelled terms $V : t \in \mathcal{T}(\mathbb{F}) \times \mathcal{T}(\mathcal{F}, \mathcal{X})$. The label t keeps track of the operations to derive V from the knowledge of U at the end of step i, using decryption and projection operators. This term t will be used later for composing new messages.

The informations are not only memorized but also decomposed with the function $CL^{(7-11)}()$ which is the closure of a set of terms using the following four rules:

$$\text{infer } M : \{t\}_{t'} \text{ from } \{M\}_{SK} : t \text{ and } SK : t' \tag{7}$$
$$\text{infer } M : \{t\}_{t'} \text{ from } \{M\}_{K} : t \text{ and } K^{-1} : t' \tag{8}$$
$$\text{infer } M : \{t\}_{t'} \text{ from } \{M\}_{K^{-1}} : t \text{ and } K : t' \tag{9}$$
$$\text{infer } M_1 : \pi_1(t) \tag{10}$$
$$\text{and } M_2 : \pi_2(t) \text{ from } \langle M_1, M_2 \rangle : t \tag{11}$$

The function $know()$ is defined by:

$$know(U, 0) = CL^{(7-11)}(\{T_1 : x_{T_1}, \ldots, T_k : x_{T_k}\})$$
$$\text{where } \mathsf{knowledge}\ U : T_1, \ldots, T_k \text{ is a statement of } \mathcal{P}.$$
$$know(U, i+1) = know(U, i) \quad \text{if } U \neq S_{i+1} \text{ and } U \neq R_{i+1}$$
$$know(R_{i+1}, i+1) = CL^{(7-11)}(know(U, i) \cup \{M_{i+1} : x_{M_{i+1}}, S_{i+1} : x_{S_{i+1}}\})$$
$$know(S_{i+1}, i+1) = CL^{(7-11)}(know(U, i) \cup \{N_1 : x_{N_1}, \ldots, N_k : x_{N_k}\})$$
$$\text{where } N_1, \ldots, N_k = fresh(M_{i+1})$$

Example 6. In the symmetric-key version of the Cable TV example (Figure 1), we have $Ins : \{\pi_2(x_M)\}_K \in know(C, 1)$ where M is the first message and x_M gets instantiated during the execution of a protocol instance.

Message composition. We define now an operator $compose(U, M, i)$ which returns a receipt of $\mathcal{T}(\mathcal{F}, \mathcal{X})$ for the user U for building M from the knowledge gained at the end of step i (hence, U's knowledge at the begining of step $i + 1$). In that way, we formalise the basic operations performed by a sender when he composes the pieces of the message M_{i+1}. In rule (16) below, we assume that M is the kth nonce created in the message M_{i+1}.

$$compose(U, M, i) = t \quad \text{if } M : t \in know(U, i) \tag{12}$$
$$compose(U, \langle M_1, M_2 \rangle, i) = \langle compose(U, M_1, i), compose(U, M_2, i) \rangle \tag{13}$$
$$compose(U, \{M\}_{KA}, i) = \{compose(U, M, i)\}_{compose(U, KA, i)} \tag{14}$$
$$compose(U, T[A], i) = compose(U, T, i)[compose(U, A, i)] \tag{15}$$
$$compose(U, M, i) = nonce(s_k(x_{\text{time}})) \tag{16}$$
$$compose(U, M, i) = \mathbf{Fail} \quad \text{in every other case} \tag{17}$$

The cases of the $compose()$ definition are tried in the given order. Other orders are possible, and more studies are necessary to evaluate their influence on the behaviour of our system.

The construction in case (16) is similar when M is a fresh public key or a fresh symmetric key, with respective terms $K(s_k(x_{\text{time}}))$, and $SK(s_k(x_{\text{time}}))$.

Expected patterns. The term of $T(\mathcal{F}, \mathcal{X})$ returned by the following variant of $compose(U, M, i)$ is a filter used to check received messages by pattern matching. More precisely, the function $expect(U, M, i)$ defined below is called right after the message M_{i+1} has been sent by U (hence with $U = S_{i+1} = R_i$).

$$expect(U, M, i) = t \quad \text{if } M : t \in know(U, i) \tag{18}$$

$$expect(U, \langle M_1, M_2 \rangle, i) = \langle expect(U, M_1, i), expect(U, M_2, i) \rangle \tag{19}$$

$$expect(U, \{M\}_K, i) = \big\{ expect(U, M, i) \big\}_{compose(U, K^{-1}, i)^{-1}} \tag{20}$$

$$expect(U, \{M\}_{K^{-1}}, i) = \big\{ expect(U, M, i) \big\}_{compose(U, K, i)^{-1}} \tag{21}$$

$$expect(U, \{M\}_{SK}, i) = \big\{ expect(U, M, i) \big\}_{compose(U, SK, i)} \tag{22}$$

$$expect(U, T[A], i) = expect(U, T, i)[expect(U, A, i)] \tag{23}$$

$$expect(U, M, i) = x_{U, M, i} \quad \text{in every other case} \tag{24}$$

Note that unless $compose()$, the $expect()$ function cannot fail. If the call to $compose()$ fails in one of the cases (20)–(22), then the case (24) will be applied.

Example 7. The pattern expected by C for message 1 (Figure 1, symmetric key version) is $expect(C, \langle D, \{Ins\}_K \rangle, 1) = \langle x_{C,D,1}, \{x_{C,Ins,1}\}_{x_K} \rangle$ because C does not know D's name in advance, nor the number Ins, but he knows K.

2.4   Narrowing Rules for Standard Messages Exchanges

The global state associated to a step of a protocol instance will be defined as the set of messages $m_1.m_2.\ldots.m_n$ sent and not yet read, union the set of expected messages $w_1.\ldots.w_m$.

A sent message is denoted by $m(i, s', s, r, t, c)$ where i is the protocol step when it is sent, s' is the real sender, s is the official sender, r is the receiver, t is the body of the message and c is a session counter (incremented at the end of each session).

$$m : \texttt{step} \times \texttt{iuser} \times \texttt{iuser} \times \texttt{iuser} \times \texttt{text} \times \texttt{int} \rightarrow \texttt{message}$$

Note that s and s' may differ since messages can be impersonated (the receiver r never knows the identity of the real sender s').

A message expected by a principal is signalled by a term $w(i, s, r, t, \ell)$ with similar meaning for the fields i, s, r, t, and c, and where ℓ is a list containing r's knowledge just before step i.

$$w : \texttt{step} \times \texttt{iuser} \times \texttt{user} \times \texttt{text} \times \texttt{list\_of text} \times \texttt{int} \rightarrow \texttt{message}$$

Nonces and freshness. We describe now a mechanism for the construction of fresh terms, in particular of nonces. This is an important aspect of our method. Indeed, it ensures freshness of the randomly generated nonces or keys over several executions of a protocol. The idea is the following: nonces admit as argument a counter that is incremented at each transition (this argument is therefore the *age* of the nonce). Hence if two nonces are emitted at different steps in an execution trace, their counters do not match. We introduce another term in the global state for representing the counter, with the new unary head symbol h. Each rewrite rule $l \to r$ is extended to $h(s(x_{\text{time}})).l \to h(x_{\text{time}}).r$ in order to update the counter. Note that the variable x_{time} occurs in the argument of $nonce()$ in case (16) of the definition of $compose()$.

Rules. The rules set $R(\mathcal{P})$ generated by our algorithm contains (for $i = 0..n$):

$$
\begin{aligned}
&h(s(x_{\text{time}})). \\
&w\big(i, x_{S_i}, x_{R_i}, x_{M_i}, \ell know(R_i, i), xc\big). \\
&m\big(i, x_r, x_{S_i}, x_{R_i}, x_{M_i}, c\big) \to \\
&\quad h(x_{\text{time}}). \\
&\quad m\big(i+1, x_{R_i}, x_{R_i}, compose(R_i, R_{i+1}, i), compose(R_i, M_{i+1}, i), c\big). \\
&\quad w\big(k_i, compose(R_i, S_{k_i}, i), x_{R_i}, expect(R_i, M_{k_i}, i'), \ell know(R_i, i'), c'\big)
\end{aligned}
$$

where k_i is the next step when R_i expects a message (see definition below), and $\ell know(R_i, i)$, $\ell know(R_i, i')$ are lists of variables described below.

If $i = 0$, the term $m(i, \ldots)$ is missing in left member, and $c = xc$.
If $1 \le i \le n$, then $c = xc'$ (another variable).
If $i = n$, the term $m(i, \ldots)$ is missing in right member.
In every case $(0 \le i \le n)$,
  if $k_i > i$ then $i' = i + 1$ and $c' = xc$,
  if $k_i \le i$ then $i' = 0$ and $c' = s(xc)$.

Note that the calls of $compose()$ may return **Fail**. In this case, the construction of $R(\mathcal{P})$ stops with failure.

After receiving message i (of content x_{M_i}) from x_r (apparently from x_{S_i}), x_{R_i} checks whether he received what he was expecting (by unification of the two instances of x_{M_i}), and then composes and sends message $i + 1$. The term returned by $compose(R_i, M_{i+1}, i)$ contains some variables in the list $\ell know(R_i, i)$. As soon as he is sending the message $i + 1$, x_{R_i} gets into a state where he is waiting for new messages. This will be expressed by deleting the term $w(i, \ldots)$ (previously expected message) and generating the term $w(k_i, \ldots)$ in the right-hand side (next expected message). Hence sending and receiving messages is not synchronous (see e.g. [5]).

The function $\ell know(U, i)$ associates to a user U and a (step) number $i \in \{0..n\}$ a term corresponding to a list of variables, used to refer to the knowledge of U. Below, $\ell :: a$ denotes the appending of the element a at the end of a list ℓ.

$$\ell know(U, 0) = \langle x_U, x_{T_1}, \dots, x_{T_n} \rangle$$
where $\texttt{knowledge } U : T_1, \dots, T_n$ is a statement of $\mathcal{P}$,
$$\ell know(U, i+1) = \ell know(U, i) \text{ if } U \neq R_i$$
$$= \ell know(U, i) :: x_{M_i} :: x_{S_i} :: n_1 :: \dots :: n_k \text{ if } U = R_i$$
where $fresh(M_i) = N_1, \dots, N_k$
and $n_i = x_{N_i}$ if N_i is of sort $\texttt{nonce}$ or $\texttt{symmetric\_key}$,
and $n_i = x_{N_i} :: x_{N_i^{-1}}$ if N_i is of sort $\texttt{public\_key}$,

The algorithm also uses the integer k_i which is the next session step when R_i expects a message. If R_i is not supposed to receive another message in the current session then either he is the session initiator S_1 and k_i is reinitialized to 0, otherwise k_i is the first step in the next session where he should receive a message (and then $k_i < i$). Formally, k_i is defined for $i = 0$ to n as follows:

$$k_i = \min\{j \mid j > i \text{ and } R_j = R_i\} \text{ if this set is not empty;}$$
otherwise $k_i = \min\{j \mid j \leq i \text{ and } R_j = R_i\}$ (recall that $R_0 = S_1$ by hypothesis);

Example 8. In both protocols presented in Figure 1, one has $R_0 = D$, $R_1 = C$, $R_2 = D$, and therefore: $k_0 = 2$, $k_1 = 1$, $k_2 = 0$.

Lemma 1. *k is a bijection from $\{0, \dots, n\}$ to $\{0, \dots, n\}$.*

Example 9. The translator generates the following $R(\mathcal{P})$ for the symmetric key version of the protocol of Figure 1. For sake of readability, in this example and the following ones, the fresh variables are denoted x_i (where i is an integer) instead of the form of the case (24) in the definition of $expect()$.

$$h(s(x_{\text{time}})).w\big(0, x_{S_0}, x_D, x_{M_0}, \langle x_D, x_C, x_K \rangle, xc\big) \to$$
$$h(x_{\text{time}}).m\big(1, x_D, x_D, x_C, \langle x_D, \{nonce(s_1(x_{\text{time}}))\}_{x_K}, xc\big).$$
$$w\big(2, x_C, x_D, \langle x_C, x_D, \{nonce(s_1(x_{\text{time}}))\}_{x_K} \rangle,$$
$$\langle x_D, x_C, x_K, x_{M_0}, x_{S_0}, nonce(s_1(x_{\text{time}})) \rangle, xc\big) \qquad (\text{tvs}_1)$$

$$h(s(x_{\text{time}})).w\big(1, x_D, x_C, x_{M_1}, \langle x_C, x_K \rangle, xc\big).$$
$$m\big(1, x_r, x_D, x_C, x_{M_1}, xc'\big) \to$$
$$h(x_{\text{time}}).m\big(2, x_C, x_C, \pi_1(x_{M_1}), \langle x_C, \pi_1(x_{M_1}), \pi_2(x_{M_1}) \rangle, xc'\big).$$
$$w\big(1, x_D, x_C, \langle x_D, \{x_1\}_{x_K} \rangle, \langle x_C, x_K \rangle, s(xc)\big) \qquad (\text{tvs}_2)$$

$$h(s(x_{\text{time}})).w\big(2, x_C, x_D, x_{M_2}, \langle x_D, x_C, x_K, x_{M_0}, x_{S_0}, x_{Ins} \rangle, xc\big).$$
$$m\big(2, x_r, x_C, x_D, x_{M_2}, xc'\big) \to$$
$$h(x_{\text{time}}).w\big(0, x_{S_0}, x_D, x_{M_0}, \langle x_D, x_C, x_K \rangle, s(xc)\big) \qquad (\text{tvs}_3)$$

3   Intruder Rules

The main difference between the behaviour of a honest principal and the intruder I is that the latter is not forced to follow the protocol, but can send messages arbitrarily. Therefore, there will be no $w()$ terms for I. In order to build messages, the intruder stores some information in the global state with terms of the form $i()$, where i is a new unary function symbol. The rewriting rules corresponding to the various intruder's techniques are detailed below.

The intruder can record the information aimed at him, (25). If divert is selected in the field intruder, the message is removed from the current state (26), but not if eaves_dropping is selected (27).

$$m(x_i, x_u, x_u, I, x, xc) \rightarrow i(x).i(x_u) \tag{25}$$

$$m(x_i, x_u, x_u, x'_u, x, xc) \rightarrow i(x).i(x_u).i(x'_u) \tag{26}$$

$$m(x_i, x_u, x_u, x'_u, x, xc) \rightarrow m(x_i, x_u, x_u, x'_u, x, xc).i(x).i(x_u).i(x'_u) \tag{27}$$

After collecting information, I can decompose it into smaller $i()$ terms. Note that the information which is decomposed (e.g. $\langle x_1, x_2 \rangle$) is not lost during the operation.

$$i\big(\langle x_1, x_2 \rangle\big) \rightarrow i\big(\langle x_1, x_2 \rangle\big).i(x_1).i(x_2) \tag{28}$$

$$i\big(\{x_1\}_{x_p}\big).i\big(x_p^{-1}\big) \rightarrow i\big(\{x_1\}_{x_p}\big).i\big(x_p^{-1}\big).i(x_1) \tag{29}$$

$$i\big(\{x_1\}_{x_s}\big).i(x_s) \rightarrow i\big(\{x_1\}_{x_s}\big).i(x_s).i(x_1) \tag{30}$$

$$i\big(\{x_1\}_{x_p^{-1}}\big).i(x_p) \rightarrow i\big(\{x_1\}_{x_p^{-1}}\big).i(x_p).i(x_1) \tag{31}$$

I is then able to reconstruct terms as he wishes.

$$i(x_1).i(x_2) \rightarrow i(x_1).i(x_2).i\big(\langle x_1, x_2 \rangle\big) \tag{32}$$

$$i(x_1).i(x_{ps}) \rightarrow i(x_1).i(x_{ps}).i\big(\{x_1\}_{x_{ps}}\big) \tag{33}$$

$$i(x_f).i(x) \rightarrow i(x_f).i(x).i\big(x_f(x)\big) \tag{34}$$

$$i(x_t).i(x_u) \rightarrow i(x_t).i(x_u).i\big(x_t[x_u]\big) \tag{35}$$

I can send arbitrary messages in his own name,

$$i(x).i(x_u) \rightarrow i(x).i(x_u).m(j, I, I, x_u, x, \underline{0}) \quad j \le n \tag{36}$$

If moreover impersonate is selected, then I can fake others identity in sent messages.

$$i(x).i(x_u).i(x'_u) \rightarrow i(x).i(x_u).i(x'_u).m(j, I, x_u, x'_u, x, \underline{0}) \quad j \le n \tag{37}$$

Note that the above intruder rules are independent from the protocol $\mathcal{P}$ in consideration. The rewrite system of the intruder (25)–(37) is denoted $\mathcal{I}$.

4   Operational Semantics

4.1   Initial State

After the definition of rules of $R(\mathcal{P})$ and $\mathcal{I}$, the presentation of an operational "state/transition" semantics of protocol executions is completed here by the definition of an initial state $t_{init}(\mathcal{P})$. This state is a term of the form $w(\ldots)$ containing the patterns of the first messages expected by the principals, and their initial knowledge, for every session instance.

We add to the initial state term a set of initial knowledge for the intruder I. More precisely, we let $t_{init}(\mathcal{P}) := t_{init}(\mathcal{P}).i(v_1)\ldots i(v_n)$ if the field `intruder_knowledge:` $v_1, \ldots, v_n$; is declared in $\mathcal{P}$.

Example 10. The initial state for the protocol of Figure 1 (symmetric key version) is: $t_{init}(\mathcal{P}) := h(x_{\text{time}}).w(0, x_1, tv, x_2, \langle tv, scard, key \rangle, \underline{1})$
$$.w(1, x_3, scard, \langle x_3, \{x_4\}_{key}\rangle, \langle scard, key\rangle, \underline{1}).i(scard)$$

4.2   Protocol Executions

Definition 3. *Given a ground term t_0 and rewrite systems R, S the set of executions $EXEC(t_0, R, S)$ is the set of maximal derivations $t_0 \rightsquigarrow_{R,S} t_1 \rightsquigarrow_{R,S} \cdots$*

Maximality is understood w.r.t. the prefix ordering on sequences. The *normal executions* of protocol $\mathcal{P}$ are the elements of the set

$$EXEC_n(\mathcal{P}) := EXEC\big(t_{\text{init}}(\mathcal{P}), R(\mathcal{P}), \mathcal{S}_0\big)$$

Executions in the presence of an intruder are the ones in

$$EXEC_i(\mathcal{P}) := EXEC\big(t_{\text{init}}(\mathcal{P}), R(\mathcal{P}) \cup \mathcal{I}, \mathcal{S}_0\big)$$

4.3   Executability

The following Theorem 1 states that if the construction of $R(\mathcal{P})$ does not fail, then normal executions will not fail (the protocol can always run and restart without deadlock).

Theorem 1. *If $\mathcal{P}$ is normalised, the field `session_instance` of $\mathcal{P}$ contains only one declaration, and the construction of $R(\mathcal{P})$ does not fail on $\mathcal{P}$, then every derivation in $EXEC_n(\mathcal{P})$ is infinite.*

Theorem 1 is not true if the field `session_instance` of $\mathcal{P}$ contains at least two declarations, as explained in the next section. Concurrent executions may interfere and enter a deadlock state.

4.4   Approximations for Intruder Rules

Due to the intruder rules of Section 3 the search space is too large. In particular, the application of rules (32)–(33) is obviously non-terminating. In our experiences, we have used restricted intruder rules for message generation.

Intruder rules guided by expected messages. The first idea is to change rules (36)–(37) so that I sends a faked message $m(i, I, x_u, x'_u, x)$ only if there exists a term of the form $w(i, x_u, x'_u, x, x_\ell, xc)$ in the global state. More precisely, we replace (36), (37) in $\mathcal{I}$ by, respectively,

$$i(x).i(x_u).w(j, I, x_u, x, x_\ell, xc) \rightarrow$$
$$i(x).i(x_u).w(j, I, x_u, x, x_\ell, xc).m(j, I, I, x_u, x, \underline{0}) \qquad \text{where } j \leq n \qquad (36')$$

$$i(x).i(x_u).i(x'_u).w(j, x_u, x'_u, x, x_\ell, xc) \rightarrow$$
$$i(x).i(x_u).i(x'_u).w(j, x_u, x'_u, x, x_\ell, xc).m(j, I, x_u, x'_u, x, \underline{0}) \text{ where } j \leq n \qquad (37')$$

The obtained rewrite system is called $\mathcal{I}_w$.

This approximation is complete: every attack in $EXEC_i(\mathcal{P})$ exists also in the trace generated by the modified system, indeed, the messages in a trace of $EXEC_i(\mathcal{P})$ and not in $EXEC(t_{\text{init}}(\mathcal{P}), R(\mathcal{P}) \cup \mathcal{I}_w, S_0)$ would be rejected by the receiver as non-expected or ill-formed messages. Similar observations are reported independantly in [32]. Therefore, there is no limitation for detecting attacks with this simplification (this strategy prunes only useless branches) but it is still inefficient.

Rules guided approximation. The above strategy is improved by deleting rules (32)–(35) and replacing each rules of (36'), (37') new rules (several for each protocol message), such that a sent message has the form $m(i, I, x_u, x'_u, t, \underline{0})$, where, roughly speaking, t follows the pattern M_i where missing parts are filled with some knowledge of I. Formally, we define a non-deterministic unary operator $* : \mathcal{T}(\mathbb{F}) \rightarrow \mathcal{T}(\mathcal{F}, \mathcal{X})$.

$$\langle M_1, M_2 \rangle^* = \langle M_1^*, M_2^* \rangle \qquad (38)$$

$$\{M\}_K^* = \{M^*\}_{K^*} \quad \big| \quad x_{\{M\}_K} \qquad (39)$$

$$F(M)^* = x_F(M^*) \quad \big| \quad x_{F(M)} \qquad (40)$$

$$T[A]^* = x_T[x_A] \quad \big| \quad x_{T[A]} \qquad (41)$$

$$ID^* = x_{ID} \qquad \text{if } ID \text{ is a nullary function symbol of } \mathbb{F} \qquad (42)$$

Given $T \in \mathcal{T}(\mathbb{F})$ we denote $skel(T)$ the set of possible terms for T^*. Then, we replace (36'), (37') in $\mathcal{I}$ by,
for each $j \in 1..n$, for each $t \in skel(M_j)$, for each distinct identifier A of sort **user**, let $\{x_1, \ldots, x_m\} = Var(t) \cup \{x_A, x_{S_i}, x_{R_i}\}$ (no variable occurrence more than once in the sequence $x_1, \ldots, x_m$):

$$i(x_1). \ldots .i(x_m).w(i, x_{S_i}, x_{R_i}, x, x_\ell, xc) \rightarrow$$
$$i(x_1). \ldots .i(x_m).w(i, x_{S_i}, x_{R_i}, x, x_\ell, xc).m(i, I, I, x_A, t, \underline{0}) \qquad (36'')$$

and, if **impersonate** is selected in the field **intruder** of $\mathcal{P}$, by: for each $i \in 1..n$, for each $t \in skel(M_i)$, for each distinct identifiers A, B of sort **user**, let $\{x_1, \ldots, x_m\} = Var(t) \cup \{x_A, x_B, x_{S_i}, x_{R_i}\}$:

$$i(x_1). \ldots .i(x_m).w(i, x_{S_i}, x_{R_i}, x, x_\ell, xc) \rightarrow$$
$$i(x_1). \ldots .i(x_m).w(i, x_{S_i}, x_{R_i}, x, x_\ell, xc).m(i, I, x_A, x_B, t, \underline{0}) \qquad (37'')$$

Because of deletion of rules (32)-(35), one rule for public key decryption with tables needs to be added:

$$i(\{x_1\}_{x_t[x_u]^{-1}}).i(x_t).i(x_u) \rightarrow i(\{x_1\}_{x_t[x_u]^{-1}}).i(x_p).i(x_u).i(x_1) \qquad (43)$$

The obtained system depends on $\mathcal{P}$. Note that this approximation is not complete. However, it seems to give reasonable results in practice.

5   Flaws

In our state/transition model, a flaw will be detected when the protocol execution reaches some critical state. We define a critical state as a pattern $t_{\text{goal}}(\mathcal{P}) \in \mathcal{T}(\mathcal{F}, \mathcal{X})$, which is constructed automatically from the protocol $\mathcal{P}$. The existence of a flaw is reducible to the following reachability problem, where a can be either i or n:

$$\exists t_0, \ldots, t_{\text{goal}}(\mathcal{P})\sigma \in EXEC_a(\mathcal{P}) \text{ for some substitution } \sigma$$

5.1   Design Flaws

It may happen that the protocol fails to reach its goals even without intruder, *i.e.* only in presence of honest agents following the protocol carefully. In particular, it may be the case that there is an interference between several concurrent runs of the same protocol: confusion between a message $m(i, \ldots)$ from the first run and another $m(i, \ldots)$ from the second one. An example of this situation is given in Appendix A. The critical state in this case is: (recall that xc and xc' correspond to session counters)

$$t_{\text{goal}}(\mathcal{P}) := w(i, x_s, x_r, x_m, x_l, xc).m(i, x_{s'}, x_s, x_r, x_m, xc').[xc \neq xc']$$

where $[xc \neq xc']$ is a constraint that can be checked either by extra rewrite rules or by an internal mechanism as in daTac.

5.2   Attacks, Generalities

Following the classification of Woo and Lam [36], we consider two basic security properties for authentication protocols: secrecy and correspondence. *Secrecy* means that some secret information (*e.g.* a key) exchanged during the protocol is kept secret. *Correspondence* means that every principal was really involved in the protocol execution, *i.e.* that mutual authentication is ensured. The failure of one of these properties in presence of an intruder is called a flaw.

Example 11. The following scenario is a *correspondence attack* for the symmetric key version of the cable tv toy example in Figure 1:

$$1. \quad D \rightarrow I(C) : \langle D, \{Ins\}_K \rangle$$
$$2.\ I(C) \rightarrow \quad D \ : \langle C, D, \{Ins\}_K \rangle$$

Following the traditional notation, the $I(C)$ in step 1 means that I did **divert** the first message of D to C. Note that this ability is selected in Figure 1. It may be performed in real world by interposing a computer between the decoder and the smartcard, with some serial interface and a smartcard reader. The sender $I(C)$ in the second message means that C did **impersonate** C for sending this message. Note that I is able to reconstruct the message of step 2 from the message he diverted at step 1, with a projection π_1 to obtain the name of D and projection π_2 to obtain the cipher $\{Ins\}_K$ and his initial knowledge (the name of the smartcard). Note that the smartcard C did not participate at all to this protocol execution. Such an attack may be used if the intruder wants to watch some channel x which is not registered in his smartcard. See [1] for the description of some real-world hacks on pay TV.

A *secrecy attack* can be performed on the public key version of the protocol in Figure 1. By listening to the message sent by the decoder at step 1, the intruder (with **eaves_dropping** ability) can decode the cipher $\{Ins\}_{T[D]^{-1}}$ since he knows the public key $T[D]$, and thus he will learn the secret instruction Ins. Note that there was no correspondence flaw in this scenario.

5.3   Secrecy Attack

Definition 4. *We say that a principal U of $\mathcal{P}$ shares a (secret) identifier N if there exists j and t such that $N : t \in know(U, j)$.*

In the construction of $R(\mathcal{P})$, we say that the term $t = compose(U, M, j)$ is *bound* to M.

Definition 5. *An execution $t_0, \ldots \in EXEC_i(\mathcal{P})$ satisfies the secrecy property iff for each j, t_j does not contain an instance of $i(t)$ as a subterm, where t is bound to a term N declared in a field* **goal : secrecy of** N *of* $\mathcal{P}$.

To define a critical state corresponding to a secrecy violation in our semantics, we add a binary function symbol $secret(\_, \_)$ to $\mathcal{F}$, which is used to store a term t (nonce or session key) that is bound to some data N declared as secret in $\mathcal{P}$, by **secrecy_of** N. If this term t appears as an argument of $i(\_)$, and I was not supposed to share t, then it means that its secrecy has been corrupted by the intruder I.

We must formalise the condition that "I was not supposed to share t". For this purpose, we add a second argument to $secret(\_, \_)$ which is a term of $\mathcal{T}(\{s, \underline{1}, \ldots, \underline{k}\})$, corresponding to the the value of a session counter, where k is the number of fields **session_instance :** ℓ in $\mathcal{P}$. Let $C = \{\underline{1}, \ldots, \underline{k}\}$. To each field **session_instance** in $\mathcal{P}$ is associated a unique constant in C by the protcedure described in Section 4.1. Let $\mathcal{J} \subseteq C$ be the set of session instances where I has not the role of a principal that shares N.

The critical state $t_{\text{goal}}(\mathcal{P})$ is any of the terms of the set:

$$\{i(x).secret(x, f(\underline{c}))\}_{\underline{c} \in \mathcal{J}}$$

The auxiliary unary function symbol $f(\_)$ scrapes off the $s(\ldots)$ context in the values of session counters, using the following rewrite rule (added to $\mathcal{S}_0$):

$$f(s(x)) \rightarrow f(x) \tag{44}$$

The storage in $secret(\_,\_)$ is performed in the rewrite rule for constructing the message M_{i+1} where N appears for the first time. More precisely, there is a special construction in the rewrite rule for building M_{i+1}. The binding to the secret N is a side effect of the recursive call of the form $compose(U, N, i)$. The ith rule constructed by our algorithm (page 142) will be in this case:

$$\left\{ \begin{array}{l} h(s(x_{\text{time}})). \\ w(i, x_{S_i}, x_{R_i}, x_{M_i}, \ell know(R_i, i), xc). \\ m(i, x_r, x_{S_i}, x_{R_i}, x_{M_i}, xc') \end{array} \right\} \rightarrow$$

$$\left\{ \begin{array}{l} h(x_{\text{time}}). \\ m(i+1, x_{R_i}, x_{R_i}, compose(R_i, R_{i+1}, i), compose(R_i, M_{i+1}, i), xc'). \\ w(k_i, compose(R_i, S_{k_i}, i), x_{R_i}, expect(R_i, M_{k_i}, i'), \ell_i', c'). \\ secret(t, f(xc')) \end{array} \right\}$$

Example 12. The rules generated for the protocol of Figure 1, public key version, are:

$$h(s(x_{\text{time}})).w\big(0, x_{S_0}, x_D, x_{M_0}, \langle x_D, x_C, x_T, x_{T[D]^{-1}} \rangle, xc\big) \rightarrow$$
$$h(x_{\text{time}}).m\big(1, x_D, x_D, x_C, \langle x_D, \{nonce(s_1(x_{\text{time}}))\}_{x_{T[D]^{-1}}} \rangle, xc\big).$$
$$w\big(2, x_C, x_D, \langle x_C, x_D, \{nonce(s_1(x_{\text{time}}))\}_{x_1} \rangle,$$
$$\langle x_D, x_C, x_T, x_{T[D]^{-1}}, x_{M_0}, x_{S_0}, nonce(s_1(x_{\text{time}})) \rangle, xc\big)$$
$$secret(nonce(s_1(x_{\text{time}})), f(xc)) \tag{tvp_1}$$

$$h(s(x_{\text{time}})).w\big(1, x_D, x_C, x_{M_1}, \langle x_C, x_T, x_{T[C]^{-1}} \rangle, xc\big).$$
$$m\big(1, x_r, x_D, x_C, x_{M_1}, xc'\big) \rightarrow$$
$$h(x_{\text{time}}).m\big(2, x_C, x_D, \pi_1(x_{M_1}), \langle x_C, \pi_1(x_{M_1}), \pi_2(x_{M_1}) \rangle, xc'\big).$$
$$w\big(1, x_1, x_{U_1}, \langle x_1, \{x_2\}_{x_K} \rangle, \langle x_C, x_T, x_{T[C]^{-1}} \rangle, s(xc)\big) \tag{tvp_2}$$

$$h(s(x_{\text{time}})).w\big(2, x_C, x_D, x_{M_2}, \langle x_D, x_C, x_T, x_{T[D]^{-1}}, x_{M_0}, x_{S_0}, x_{Ins} \rangle, xc\big).$$
$$m\big(2, x_r, x_C, x_D, x_{M_2}, xc'\big) \rightarrow$$
$$h(x_{\text{time}}).w\big(0, x_{S_0}, x_D, x_{M_0}, \langle x_D, x_C, x_T, x_{T[D]^{-1}} \rangle, s(xc)\big) \tag{tvp_3}$$

Note the term $secret(nonce(s_1(x_{\text{time}})), xc)$ in rule (tvp_1). As described in Example 11, it is easy to see that this protocol has a secrecy flaw. A subterm $secret(nonce(x), \underline{1}).i(nonce(x))$ is obtained in 4 steps, see appendix C.

5.4   Correspondence Attack

The correspondence property between two users U and V means that when U terminates its part of a session c of the protocol (and starts next session $s(c)$), then V must have started his own part, and reciprocally. In Definition 6, we use the notation $first_S(U) = \min\{i \mid S_i = U\}$, assuming $\min(\emptyset) = 0$.

Definition 6. *An execution $t_0, \ldots \in EXEC_i(\mathcal{P})$ satisfies the correspondence property between the (distinct) users U and V iff for each j, t_j does not contain a subterm matching:*

$$w(\text{first}_S(U) - 1, x_s, u, x_t, x_\ell, s(x_c)).w(\text{first}_S(V) - 1, x'_s, x'_r, x'_t, x'_\ell, x_c)$$
$$or \; w(\text{first}_S(V) - 1, x_s, v, x_t, x_\ell, s(x_c)).w(\text{first}_S(U) - 1, x'_s, x'_r, x'_t, x'_\ell, x_c),$$

where $U : u$ and $V : v$ occur in the same line of the field `session_instance`.

The critical state $t_{\text{goal}}(\mathcal{P})$ is therefore any of the two above terms in Definition 6. Again, these terms are independent from $\mathcal{P}$.

Example 13. A critical state for the protocol in Figure 1, symmetric key version, is: $t_{\text{goal}}(\mathcal{P}) := w(0, x_1, tv, x_{M_1}, x_{l_1}, xc).w(1, x_2, scard, x_{M_2}, x_{l_2}, s(xc))$

5.5   Key Compromising Attack

A classical goal of cryptographic protocols is the exchange between two users A and B of new keys – symmetric or public keys. In such a scenario, A may propose to B a new shared symmetric key K or B may ask a trusted server for A's public key K, see Section 5.6 below for this particular second case. In this setting, a technique of attack for the intruder is to introduce a compromised key K': I has built some key K' and he let B think that K' is the key proposed by A or that this is A's public key for instance (see Example 14 for a key compromising attack). The compromising of K may be obtained by exploiting for instance a type flaw as described below. Such an attack is not properly speaking a secrecy attack. However, it can of course be exploited if later on B wants to exchange some secret with A using K (actually the compromised K').

Therefore, a key compromising attack is defined as a secrecy attack for an extended protocol $\mathcal{P}'$ obtained from a protocol $\mathcal{P}$ of the above category as follows:

1. declare a new `identifier` X : `number`;
2. add a rule: $n + 1. \; B \rightarrow A : \{X\}_K$ where n is the number of messages in $\mathcal{P}$ and K is the key to compromise,
3. add the declaration `goal : secrecy_of` X;

5.6   Binding Attack

This is a particular case of key compromising attack, and therefore a particular case of secrecy attack, see Section 5.5. It can occur in protocols where the public keys are distributed by a trusted server (who knows a table K of public keys) because the principals do not know in advance the public keys of others. In some case, the intruder I can do appropriate diverting in order to let some principal learn a fake binding name – public key. For instance, I makes some principal B believe that I's public key $K[I]$ is the public key of a third party A (binding A–$K[I]$). This is what can happen with the protocol SLICE/AS, see [7].

5.7   Type Flaw

This flaw occurs when a principal can accept a term of the wrong type. For instance, he may accept a pair of numbers instead of a new symmetric key, when numbers, pair of numbers and symmetric keys are assumed to have the same type. Therefore, a type flaw refers more to implementation hypotheses than to the protocol itself. Such a flaw may be the cause of one of the above attack, but its detection requires a modification of the sort system of $\mathcal{F}$. The idea it to collapse some sorts, by introducing new sorts equalities. For instance, one may have the equality symmetric_key = text = number. By definition of profiles of {_}_ and ⟨_, _⟩, ciphers and pairs are in this case numbers, and be accepted as symmetric_key.

Example 14. A known key compromising attack on Otway-Rees protocol, see [7], exploits a type flaw of this protocol. We present here the extended version of Otway-Rees, see Section 5.5.

```
protocol Ottway Rees
identifiers
A, B, S      : user;
Kas, Kbs, Kab : symmetric_key;
M, Na, Nb, X : number;
messages
```
1. $A \rightarrow B : \langle M, A, B, \{N_a, M, A, B\}_{K_{as}} \rangle$
2. $B \rightarrow S : \langle M, A, B, \{N_a, M, A, B\}_{K_{as}}, N_b, M, A, B_{K_{bs}} \rangle$
3. $S \rightarrow B : \langle M, \{N_a, K_{ab}\}_{K_{as}}, \{N_b, K_{ab}\}_{K_{bs}} \rangle$
4. $B \rightarrow A : \langle M, \{N_a, K_{ab}\}_{K_{as}} \rangle$
5. $A \rightarrow B : \{X\}_{Kab}$
```
knowledge
A : B, S, Kas;
B : S, Kbs;
S : A, B, Kas, Kbs;
session_instance  [A : a, B : b, S : s, kas : kas, Kbs : kbs];
intruder : divert, impersonate;
intruder_knowledge : ;
goal : secrecy_of X;
```

The symmetric keys K_{as} and K_{bs} are supposed to be only known by A and S, resp. B and S. The identifiers M, N_a, and N_b re nonces. The new symmetric K_{ab} is generated by the trusted server S and transmitted to B and indirectly to A, by mean of the cipher $\{N_a, K_{ab}\}_{K_{as}}$.

If the sorts numbers, text, and symmetric_key are assumed to collapse, then we have the following scenario:

1. $A \rightarrow I(B) : \langle M, A, B, \{N_a, M, A, B\}K_{as} \rangle$
4. $I(B) \rightarrow \quad A \quad : \langle M, \{N_a, M, A, B\}K_{as} \rangle$
5. $A \rightarrow I(B) : \{X\}_{\langle M, A, B\rangle}$

In rule 1, I diverts (and memorises) A's message. In next step 4, I impersonates B and makes him think that the triple $\langle M, A, B \rangle$ is the new shared symmetric key K_{ab}. We recall that $\langle \_, \_ \rangle$ is right associative and thereafter $\langle N_a, M, A, B \rangle$ can be considered as identical to $\langle N_a, \langle M, A, B \rangle \rangle$

6   Verification: Deduction Techniques and Experiments

We have implemented the construction of $R(\mathcal{P})$ in OCaml® and performed experiments using the theorem prover daTac [33] with paramodulation modulo AC. Each rule $l \to r \in R(\mathcal{P})$ is represented as an oriented equation $l = r$, the initial state is represented as a unit positive clause $P(t_{init}(\mathcal{P}))$ and the critical state as a unit negative clause $\neg P(t_{goal}(\mathcal{P}))$.

As for multiset rewriting [8], an ac-operator will take care of concurrency. On the other hand unification will take care of communication in an elegant way. The deduction system combines paramodulation steps with equational rewriting by $\mathcal{S}_0$.

6.1   Deduction Techniques. Generalities

The main deduction technique consists in replacing a term by an equal one in a clause: given a clause $l = r \vee C'$ and clause $C[l']$, the clause $(C' \vee C[r])\sigma$ is deduced, where σ is a unifier of l and l', that is a mapping from variables to terms such that $l\sigma$ is equal to $l'\sigma$.

This deduction rule is called *paramodulation*. It has been introduced by Robinson and Wos [27]. Paramodulation (together with resolution and factoring) was proved refutationally complete by Brand [6] who also shown that applying a replacement in a variable position is useless.

For reducing the number of potential deduction steps, the paramodulation rule has been restricted by an ordering, to guarantee it replaces big terms by smaller ones. This notion of ordered paramodulation has been applied to the Knuth-Bendix completion procedure [16] for avoiding failure in some situations (see [14] and [2]). A lot of work has been devoted to putting more restrictions on paramodulation in order to limit combinatorial explosion [23].

In particular paramodulation is often inefficient with axioms such as associativity and commutativity since these axioms allow for many successful unifications between their subterms and subterms in other clauses. Typically word problems in finitely presented commutative semigroups cannot be decided by standard paramodulation. This gets possible by building the associativity and commutativity in the paramodulation rule using the so-called paramodulation modulo AC and rewriting modulo AC rules.

The integration of associativity and commutativity axioms within theorem-proving systems has been first investigated by Plotkin [26] and Slagle [31]. Rusinowitch and Vigneron [29] have built-in this theory in a way that is compatible with the ordered paramodulation strategy and rewriting and preserves refutational completeness. These techniques are implemented in the daTac system [33].

Another approach has been followed by Wertz [35] and Bachmair and Ganzinger [3], consisting of using the idea of extended clauses developed for the equational case by Peterson and Stickel [25].

In all the approaches, the standard unification calculus has to be replaced by unification modulo associativity and commutativity. This may be very costly since some unification problems have doubly exponentially many minimal solutions [12].

6.2 Deduction Rules for Protocol Verification

We present here the version of paramodulation we have applied for simulating and verifying protocols. States are built with the specific ac-operator ".." for representing the multiset of information components: sent and expected messages, and the knowledge of the intruder.

The definition of our instance of the paramodulation rule is the following.

Definition 7 (Paramodulation). $\dfrac{l = r \qquad P(l')}{P(r.z)\sigma}$ *if σ is an ac-unifier of l.z and l', and z is a new variable.*

This rule is much simpler than the general one in [29]. We only need to apply replacements at the top of the term. In addition the equations are such that the left-hand side is greater than the right-hand side and each clause is unit. So we do not need any strategy for orienting the equations or selecting a literal in a clause.

In the verification of protocols, we encounter only simple unification problems. They reduce to unifying multisets of standard terms, where one of the multisets has no variable as argument of ".". Only one argument of the other multiset is a variable. Hence for handling these problems we have designed a unification algorithm which is more efficient than the standard ac-unification algorithm of daTac.

Let us illustrate this with an example.

Example 15. For performing a paramodulation step from $f(x_1).g(a) = c$ into $P(a.g(x_2).f(b).h(x_3))$, trying to unify $f(x_1).g(a)$ and $a.g(x_2).f(b).h(x_3)$ will not succeed. We have to add a new variable in the left-hand side of the equation for capturing the additional arguments of the ac-operator. The unification problem we have to solve is $f(x_1).g(a).z =_{ac}^? a.g(x_2).f(b).h(x_3)$. Its unique solution σ is $\{x_1 \mapsto b, x_2 \mapsto a, z \mapsto a.h(x_3)\}$. The deduced clause is $P(c.z)\sigma$, that is $P(c.a.h(x_3))$.

The paramodulation rule is used for generating new clauses. We need a rule for detecting a contradiction with the clause representing the goal.

Definition 8 (Contradiction). $\dfrac{P(t) \qquad \neg P(t')}{\square}$ *if σ is an ac-unifier of t and t'.*

In addition to these two deduction rules, we need to simplify clauses by term rewriting, using equations of S_0 (rewrite rules (1)–(6)). For this step we have to compute a match σ of a term l into l', that is a substitution such that $l\sigma = l'$.

Definition 9 (Simplification). $\dfrac{l = r \qquad P(t[l'])}{P(t[r\sigma])}$ _if σ is a match of l into l'._

Applying this rule consists in replacing the initial clause by the simplified one.

6.3   Deduction Strategy

We basically apply a breadth first search strategy. The compilation of the protocol generates four sets of clauses:

(0) the rewrite rules of S_0;
(1) the clauses representing transitions rules (including intruder's rules);
(2) the clause representing the initial state, $P\big(t_{init}(\mathcal{P})\big)$;
(3) the critical state $\big(\neg P\big(t_{\text{goal}}(\mathcal{P})\big)\big)$;

The deduction strategy used by daTac is the following:

Repeat:
    Select a clause C in (2), C contains only a positive literal
    Repeat:
        Select a clause D in (1), D is an equation $l = r$
        Apply Paramodulation from D into C:
            Compute all the most general ac-unifiers
            For each solution σ,
                Generate the resulting clause $C'\sigma$
        Simplify the generated clauses:
            For each generated clause $C'\sigma$,
                Select a rewrite rule $l \to r$ in (0)
                For each subterm s in $C'\sigma$,
                    If s is an instance $l\phi$ of l
                        Then Replace s by $r\phi$ in $C'\sigma$
        Add the simplified generated clauses into (2)
        Try Contradiction between the critical state and each new clause:
            If it applies, Exit with message "_contradiction found_".
    Until no more clause to select in (1)
Until no more clause to select in (2)

Note that any derivation of a contradiction $\square$ with this strategy is a linear derivation from the initial state to the goal and it can be directly interpreted as a scenario for a flaw or an attack.

6.4   Results

The approach has been experimented with several protocols described in [7]. We have been able to find the known flaws with this uniform method in several protocols, in less than 1 minute (including compilation) in every case, see Figure 2.

Protocol	Description	Flaw	Intruder abilities
Encrypted Key Exchange	Key distribution	Correspondence attack	divert impersonate
Needham Shroeder Public Key	Key distribution with authentication	Secrecy attack	divert impersonate
Otway Rees	Key distribution with trusted server	Key compromising = secrecy attack type flaw	divert impersonate
Shamir Rivest Adelman	Transmission of secret information	Secrecy attack	divert impersonate
Tatebayashi Matsuzaki Newman	Key distribution	Key compromising = secrecy attack	eaves_dropping impersonate
Woo and Lam Π	Authentication	Correspondence attack	divert impersonate

Fig. 2. Experiments

See http://www.loria.fr/equipes/protheo/SOFTWARES/CASRUL/ for more details.

7   Conclusion

We have presented a complete, compliant translator from security protocols to rewrite rules and how it is used for the detection of flaws. The advantages of our system are that the automatic translation covers a large class of protocols and that the narrowing execution mechanism permits to handle several aspects like timeliness. A drawback of our approach is that the produced rewrite system can be complex and therefore flaw detection gets time-consuming. However, simplifications should be possible to shorten derivations. For instance, composition and reduction with rules S_0 may be performed in one step.

The translation can be directly extended for handling key systems satisfying algebraic laws such as commutativity (cf. RSA). It can be extended to other kinds of flaws: binding, typing... We plan to analyse E-commerce protocols where our management of freshness should prove to be very useful since fresh data are ubiquitous in electronic forms (order and payment e.g.). We plan to develop a generic daTac proof strategy for reducing the exploration space when searching for flaws. We also conjecture it is possible to modify our approach in order to prove the absence of flaws under some assumptions.

References

1. R. Anderson. Programming Satan's computer. volume 1000 of *Lecture Notes in Computer Science*. Springer-Verlag. 148
2. L. Bachmair, N. Dershowitz, and D. Plaisted. Completion without Failure. In H. Aït-Kaci and M. Nivat, editors, *Resolution of Equations in Algebraic Structures, Volume 2: Rewriting Techniques*, pages 1–30. Academic Press inc., 1989. 152
3. L. Bachmair and H. Ganzinger. Associative-Commutative Superposition. In N. Dershowitz and N. Lindenstrauss, editors, *Proc. 4th CTRS Workshop, Jerusalem (Israel)*, volume 968 of *LNCS*, pages 1–14. Springer-Verlag, 1995. 153
4. D. Basin. Lazy infinite-state analysis of security protocols. In *Secure Networking — CQRE [Secure] '99*, LNCS 1740, pages 30–42. Springer-Verlag, Berlin, 1999. 132
5. D. Bolignano. Towards the formal verification of electronic commerce protocols. In *IEEE Computer Security Foundations Workshop*, pages 133–146. IEEE Computer Society, 1997. 131, 142
6. D. Brand. Proving Theorems with the Modification Method. *SIAM J. of Computing*, 4:412–430, 1975. 152
7. J. Clark and J. Jacob. A survey of authentication protocol literature. http://www.cs.york.ac.uk/~jac/papers/drareviewps.ps, 1997. 131, 150, 151, 155
8. G. Denker, J. Meseguer, and C. Talcott. Protocol specification and analysis in Maude. In *Formal Methods and Security Protocols*, 1998. LICS '98 Workshop. 131, 133, 152
9. G. Denker and J. Millen. Capsl intermediate language. In *Formal Methods and Security Protocols*, 1999. FLOC '99 Workshop. 131, 132
10. N. Dershowitz and J.-P. Jouannaud. *Handbook of Theoretical Computer Science*, volume B, chapter 6: Rewrite Systems, pages 244–320. North-Holland, 1990. 133, 139
11. D. Dolev and A. Yao. On the security of public key protocols. *IEEE Transactions on Information Theory*, IT-29:198–208, 1983. Also STAN-CS-81-854, May 1981, Stanford U. 133, 137
12. E. Domenjoud. A technical note on AC-unification. the number of minimal unifiers of the equation $\alpha x_1 + \cdots + \alpha x_p \doteq_{AC} \beta y_1 + \cdots + \beta y_q$. *JAR*, 8:39–44, 1992. 153
13. R. Focardi and R. Gorrieri. Cvs: A compiler for the analysis of cryptographic protocols. In *12th IEEE Computer Security Foundations Workshop*. IEEE Computer Society, 1999. 131
14. J. Hsiang and M. Rusinowitch. Proving Refutational Completeness of Theorem-Proving Strategies : the Transfinite Semantic Tree Method. *JACM*, 38(3):559–587, July 1991. 152
15. J.-M. Hullot. Canonical forms and unification. In *5th International Conference on Automated Deduction*, volume 87, pages 318–334. Springer-Verlag, LNCS, july 1980. 131, 137
16. D. E. Knuth and P. B. Bendix. Simple Word Problems in Universal Algebras. In J. Leech, editor, *Computational Problems in Abstract Algebra*, pages 263–297. Pergamon Press, Oxford, 1970. 152

17. G. Lowe. Casper: a compiler for the analysis of security protocols. *Journal of Computer Security*, 6(1):53–84, 1998. 131, 132, 132, 133, 136
18. G. Lowe. Towards a completeness result for model checking of security protocols. In *11th IEEE Computer Security Foundations Workshop*, pages 96–105. IEEE Computer Society, 1998. 134
19. C. Meadows. Applying formal methods to the analysis of a key management protocol. *Journal of Computer Security*, 1(1):5–36, 1992. 133
20. C. Meadows. The NRL protocol analyzer: an overview. *Journal of Logic Programming*, 26(2):113–131, 1996. 133
21. J. Millen. CAPSL: Common Authentication Protocol Specification Language. Technical Report MP 97B48, The MITRE Corporation, 1997. 132, 132, 133
22. J. Mitchell, M. Mitchell, and U. Stern. Automated analysis of cryptographic protocols using Murφ. In *IEEE Symposium on Security and Privacy*, pages 141–154. IEEE Computer Society, 1997. 131
23. R. Nieuwenhuis and A. Rubio. Paramodulation-based theorem proving. In J.A. Robinson and A. Voronkov, editors, *Handbook of Automated Reasoning*. Elsevier Science Publishers, 2000. 152
24. L. Paulson. The inductive approach to verifying cryptographic protocols. *Journal of Computer Security*, 6(1):85–128, 1998. 131
25. G. Peterson and M. E. Stickel. Complete sets of reductions for some equational theories. *JACM*, 28:233–264, 1981. 153
26. G. Plotkin. Building-in equational theories. *Machine Intelligence*, 7:73–90, 1972. 152
27. G. A. Robinson and L. T. Wos. Paramodulation and First-Order Theorem Proving. In B. Meltzer and D. Mitchie, editors, *Machine Intelligence 4*, pages 135–150. Edinburgh University Press, 1969. 152
28. A. W. Roscoe. Modelling and verifying key-exchange protocols using CSP and FDR. In *8th IEEE Computer Security Foundations Workshop*, pages 98–107. IEEE Computer Society, 1995. 132
29. M. Rusinowitch and L. Vigneron. Automated Deduction with Associative-Commutative Operators. *Applicable Algebra in Engineering, Communication and Computation*, 6(1):23–56, January 1995. 152, 153
30. B. Schneier. *Applied Cryptography*. John Wiley, 1996. 133
31. J. R. Slagle. Automated Theorem-Proving for theories with Simplifiers, Commutativity and Associativity. *JACM*, 21(4):622–642, 1974. 152
32. P. Syverson, C. Meadows, and I. Cervesato. Dolev-Yao is no better than Machiavelli. In *WITS'00. Workshop on Issues in the Theory of Security*, 2000. 146
33. L. Vigneron. Positive deduction modulo regular theories. In *Proceedings of Computer Science Logic, Paderborn (Germany)*, pages 468–485. LNCS 1092, Springer-Verlag, 1995. 131, 152, 152
34. C. Weidenbach. Towards an automatic analysis of security protocols. In *Proceedings of the 16th International Conference on Automated Deduction*, pages 378–382. LNCS 1632, Springer-Verlag, 1999. 131
35. U. Wertz. First-Order Theorem Proving Modulo Equations. Technical Report MPI-I-92-216, MPI Informatik, April 1992. 153
36. T. Woo and S. Lam. A semantic model for authentication protocols. In *IEEE Symposium on Research in Security and Privacy*, pages 178–194. IEEE Computer Society, 1993. 134, 147

Appendix A: Design Flaws

Example 16.

identifiers	M, B, C : user;
	O, N : number;
	K_b, K_c : public_key;
	$hash$: function;
messages	1. $M \rightarrow C : \{O\}_{K_c}$
	2. $C \rightarrow M : \langle B, \{N\}_{K_b}, hash(N) \rangle$
	3. $M \rightarrow B : \{N\}_{K_b}, hash(O)$
	4. $B \rightarrow M : \left\{ hash\big(hash(N), hash(O)\big) \right\}_{K_b^{-1}}$
knowledge	$C : B, K_b, hash$;
	$M : C, O, K_c, K_b, hash$;
	$B : K_b, K_b^{-1}, hash$;
session_instance	$[M : Merchant, B : Bank, C : Customer,$
	$O : car, K_b : k_b, K_c : k_c]$
	$[M : Merchant, B : Bank, C : Customer,$
	$O : peanut, K_b : k_b, K_c : k_c]$

This is a flawed e-commerce protocol. While browsing an online commerce site, the customer C is offered an object O (together with an order form, price information *etc*) by merchant M. Then, C transmits M a payment form N with his bank account information and the price of O, in order for M to ask directly to C's bank B for the payment. For confidentiality reasons, M must never read the contents of N, and B must not learn O. Therefore, O is encrypted in message 1 with the public key K_c of C. Also, in message 2, N is transmitted by C to M in encrypted form with the bank's public key K_b and in the form of a digest computed with the *hash* one-way function. Then M relays the cipher $\{N\}_{K_b}$ to B together with a digest of O. The bank B makes the verification for the payment and when it is possible, gives his certificate to M in the form of a dual signature.

The problem is that in message 2, there is no occurrence of O, so there may be some interference between two executions of the protocol. Imagine that C is performing simultaneously two transactions with the same merchant M. In the two concurrent execution of the protocol, M sends 1. $M \rightarrow C : \{car\}_{K_c}$ and 1. $M \rightarrow C : \{peanut\}_{K_c}$. C will reply with two distinct corresponding payment forms (the price field will vary) 2. $C \rightarrow M : \langle B, \{N_{car}\}_{K_b}, hash(N_{car}) \rangle$ and 2. $C \rightarrow M : \langle B, \{N_{peanut}\}_{K_b}, hash(N_{peanut}) \rangle$. But after receiving these two messages, M may be confused about which payment form is for which offer (recall that M can not read N_{car} and N_{peanut}), and send the wrong requests to B: 3. $M \rightarrow B : \{N_{car}\}_{K_b}, hash(peanut)$ and 3. $M \rightarrow B : \{N_{peanut}\}_{K_b}, hash(car)$. If the bank refuses the payment of N_{car} but authorises the one of N_{peanut}, it will give a certificate for buying a car and paying peanuts! Fortunately for M, the check of dual signature (by M) will fail and transaction will by aborted, but there is nevertheless a serious interference flaw in this protocol, that can occur even only between two honest agents (without an intruder).

Appendix B: A Correspondence Attack

Trace obtained by daTac of a correspondence attack for the symmetric key TV protocol (Figure 1).

$t_{init}(\mathcal{P}) =$
$h(x_1).w(0, x_2, tv, x_3, \langle tv, scard, key \rangle, \underline{1})$
$\quad .w(1, x_4, scard, \langle x_4, \{x_5\}_{key} \rangle, \langle scard, key \rangle, \underline{1})$
$\quad .i(scard)$

$\leadsto (\text{tvs}_1)$
$h(x_1).m(1, tv, tv, scard, \langle tv, \{nonce(x_1)\}_{key} \rangle, \underline{1})$
$\quad .w(2, scard, tv, \langle scard, tv, \{nonce(x_1)\}_{key} \rangle, \langle tv, scard, key, x_2, nonce(x_1) \rangle, \underline{1})$
$\quad .w(1, x_3, scard, \langle x_3, \{x_4\}_{key} \rangle, \langle scard, key \rangle, \underline{1})$
$\quad .i(scard)$

$\leadsto (26)$
$h(x_1).w(2, scard, tv, \langle scard, tv, \{nonce(x_1)\}_{key} \rangle, \langle tv, scard, key, x_2, nonce(x_1) \rangle, \underline{1})$
$\quad .w(1, x_3, scard, \langle x_3, \{x_4\}_{key} \rangle, \langle scard, key \rangle, \underline{1})$
$\quad .i(tv).i(scard).i(\langle tv, \{nonce(x_1)\}_{key} \rangle)$

$\leadsto (28)$
$h(x_1).w(2, scard, tv, \langle scard, tv, \{nonce(x_1)\}_{key} \rangle, \langle tv, scard, key, x_2, nonce(x_1) \rangle, \underline{1})$
$\quad .w(1, x_3, scard, \langle x_3, \{x_4\}_{key} \rangle, \langle scard, key \rangle, \underline{1})$
$\quad .i(tv).i(scard).i(\{nonce(x_1)\}_{key})$

$\leadsto (37)$
$h(x_1).m(2, I, scard, tv, \langle scard, tv, \{nonce(x_1)\}_{key} \rangle, \underline{0})$
$\quad .w(2, scard, tv, \langle scard, tv, \{nonce(x_1)\}_{key} \rangle, \langle tv, scard, key, x_2, nonce(x_1) \rangle, \underline{1})$
$\quad .w(1, x_3, scard, \langle x_3, \{x_4\}_{key} \rangle, \langle scard, key \rangle, \underline{1})$
$\quad .i(scard).i(tv).i(\{nonce(x_1)\}_{key})$

$\leadsto (\text{tvs}_3)$
$h(x_1).w(0, x_2, tv, x_3, \langle tv, scard, key \rangle, s(\underline{1}))$
$\quad .w(1, x_3, scard, \langle x_3, \{x_4\}_{key} \rangle, \langle scard, key \rangle, \underline{1})$
$\quad .i(scard).i(tv).i(\{nonce(s(x_1))\}_{key})$

One subterm (of the last term) matches the pattern $t_{goal}(\mathcal{P})$.

Appendix C: A Secrecy Attack

Trace obtained by daTac of a secrecy attack for the public key TV protocol (Figure 1).

$t_{init}(\mathcal{P}) =$
$h(x_1).w(0, x_2, tv, x_3, \langle tv, scard, key, key[tv]^{-1} \rangle, \underline{1})$
$\quad .w(1, x_4, scard, \langle x_4, x_5 \rangle, \langle scard, key, key[scard]^{-1} \rangle, \underline{1})$
$\quad .i(key)$

$\leadsto_{(tvp_1)}$
$h(x_1).m(1, tv, tv, scard, \langle tv, \{nonce(x_1)\}_{key[tv]^{-1}} \rangle, \underline{1})$
$\quad .w(2, scard, tv, \langle scard, tv, \{nonce(x_1)\}_{key[tv]^{-1}} \rangle,$
$\quad\quad \langle tv, scard, key, key[tv]^{-1}, x_2, x_3, nonce(x_1) \rangle, \underline{1})$
$\quad .w(1, x_4, scard, \langle x_4, x_5 \rangle, \langle scard, key, key[scard]^{-1} \rangle, \underline{1})$
$\quad .secret(nonce(x_1), f(\underline{1}))$
$\quad .i(key)$

$\leadsto_{(27)}$
$h(x_1).m(1, tv, tv, scard, \langle tv, \{nonce(x_1)\}_{key[tv]^{-1}} \rangle, \underline{1})$
$\quad .w(2, scard, tv, \langle scard, tv, \{nonce(x_1)\}_{key[tv]^{-1}} \rangle,$
$\quad\quad \langle tv, scard, key, key[tv]^{-1}, x_2, x_3, nonce(x_1) \rangle, \underline{1})$
$\quad .w(1, x_4, scard, \langle x_4, x_5 \rangle, \langle scard, key, key[scard]^{-1} \rangle, \underline{1})$
$\quad .secret(nonce(x_1), f(\underline{1}))$
$\quad .i(key).i(tv).i(scard).i(\langle tv, \{nonce(x_1)\}_{key[tv]^{-1}} \rangle)$

$\leadsto_{(28)}$
$h(x_1).m(1, tv, tv, scard, \langle tv, \{nonce(x_1)\}_{key[tv]^{-1}} \rangle, \underline{1})$
$\quad .w(2, scard, tv, \langle scard, tv, \{nonce(x_1)\}_{key[tv]^{-1}} \rangle,$
$\quad\quad \langle tv, scard, key, key[tv]^{-1}, x_2, x_3, nonce(x_1) \rangle, \underline{1})$
$\quad .w(1, x_4, scard, \langle x_4, x_5 \rangle, \langle scard, key, key[scard]^{-1} \rangle, \underline{1})$
$\quad .secret(nonce(x_1), f(\underline{1}))$
$\quad .i(key).i(tv).i(scard).i(\{nonce(x_1)\}_{key[tv]^{-1}})$

$\leadsto_{(35)}$
$h(x_1).m(1, tv, tv, scard, \langle tv, \{nonce(x_1)\}_{key[tv]^{-1}} \rangle, \underline{1})$
$\quad .w(2, scard, tv, \langle scard, tv, \{nonce(x_1)\}_{key[tv]^{-1}} \rangle,$
$\quad\quad \langle tv, scard, key, key[tv]^{-1}, x_2, x_3, nonce(x_1) \rangle, \underline{1})$
$\quad .w(1, x_4, scard, \langle x_4, x_5 \rangle, \langle scard, key, key[scard]^{-1} \rangle, \underline{1})$
$\quad .secret(nonce(x_1), f(\underline{1}))$
$\quad .i(key).i(tv).i(scard).i(\{nonce(x_1)\}_{key[tv]^{-1}}).i(key[tv])$

$\leadsto_{(31)}$
$h(x_1).m(1, tv, tv, scard, \langle tv, \{nonce(x_1)\}_{key[tv]^{-1}} \rangle, \underline{1})$
$\quad .w(2, scard, tv, \langle scard, tv, \{nonce(x_1)\}_{key[tv]^{-1}} \rangle,$
$\quad\quad \langle tv, scard, key, key[tv]^{-1}, x_2, x_3, nonce(x_1) \rangle, \underline{1})$
$\quad .w(1, x_4, scard, \langle x_4, x_5 \rangle, \langle scard, key, key[scard]^{-1} \rangle, \underline{1})$
$\quad .secret(nonce(x_1), f(\underline{1}))$
$\quad .i(key).i(tv).i(scard).i(\{nonce(x_1)\}_{key[tv]^{-1}}).i(key[tv]).i(nonce(x_1))$

The subterm $secret(nonce(x_1), f(\underline{1})).i(nonce(x_1))$ matches the pattern $t_{goal}(\mathcal{P})$.

Equational Binary Decision Diagrams

Jan Friso Groote[1,2] and Jaco van de Pol[1]

[1] CWI, P.O. Box 94079, 1090 GB Amsterdam, The Netherlands
JanFriso.Groote@cwi.nl, Jaco.van.de.Pol@cwi.nl
[2] Department of Computing Science, Eindhoven University of Technology,
P.O. Box 513, 5600 MB Eindhoven, The Netherlands

Abstract. We allow equations in binary decision diagrams (BDD). The resulting objects are called EQ-BDDs. A straightforward notion of reduced ordered EQ-BDDs (EQ-OBDD) is defined, and it is proved that each EQ-BDD is logically equivalent to an EQ-OBDD. Moreover, on EQ-OBDDs satisfiability and tautology checking can be done in constant time.

Several procedures to eliminate equality from BDDs have been reported in the literature. Typical for our approach is that we keep equalities, and as a consequence do not employ the finite domain property. Furthermore, our setting does not strictly require Ackermann's elimination of function symbols. This makes our setting much more amenable to combinations with other techniques in the realm of automatic theorem proving, such as term rewriting.

We introduce an algorithm, which for any propositional formula with equations finds an EQ-OBDD that is equivalent to it. The algorithm has been implemented, and applied to benchmarks known from literature. The performance of a prototype implementation is comparable to existing proposals.

1   Introduction

Binary decision diagrams (BDDs) [5,6,12] are widely used for checking satisfiability and tautology of boolean formulae. Applications include hardware verification and symbolic model checking. Every formula of propositional logic can be efficiently represented as a BDD. BDDs can be reduced and ordered, which in the worst case requires exponential time, but for many interesting applications it can be done in polynomial time. The reduced and ordered BDD (OBDD) is a unique representation for boolean formulae, so satisfiability, tautology and equivalence on OBDDs can be checked in constant time.

Much current research is done on extending the BDD techniques to formulae outside propositional logic. In principle, the boolean variables can be generalized to arbitrary relations. The goal now is to check satisfiability or validity of quantifier free formulae in a certain theory. The main example is the logic of *equality and uninterpreted function symbols* (EUF) [10,7,16]. Another example is the logic of *difference constraints* on integers or reals [13].

M. Parigot and A. Voronkov (Eds.): LPAR 2000, LNAI 1955, pp. 161–178, 2000.
© Springer-Verlag Berlin Heidelberg 2000

EUF formulae have been successfully applied to the verification of pipelined microprocessors [8,7] and of compiler optimizations [16]. In these applications, functions can be viewed as black boxes that are connected in different ways. Hence the concrete functions can be abstracted from, by replacing them by uninterpreted function symbols (i.e., universally quantified function variables). It is clear that if the abstracted formula is valid, then the original formula is. However, the converse is not true, e.g. $x + y = y + x$ is valid, but its abstract version $F(x, y) = F(y, x)$ is not.

Two methods for solving EUF formulae exist. The first method is based on two observations by Ackermann [1]. First, the function variables can be eliminated, essentially by replacing any two subterms of the form $F(x)$ and $F(y)$ by new variables f_1 and f_2, and adding functionality constraints of the form $x = y \to f_1 = f_2$. The second observation is the *finite domain property*, which states that the resulting formula is satisfiable if, and only if, it is satisfiable over a finite domain. Given an upper bound n on this domain, each domain variable can be encoded as a vector of $\lceil \log(n) \rceil$ bits. In this way the original problem is reduced to propositional logic, and can be solved using existing BDD techniques.

The second method extends the BDD data structure, by allowing equations in the nodes of a BDD, instead of boolean variables only. By viewing all atoms as distinct variables, the BDD algorithms can still be used to construct a reduced ordered BDD. Contrary to the propositional case, a path in these OBDDs can be inconsistent, for instance because it violates transitivity constraints. As a consequence, all paths of the resulting OBDD have to be checked in order to conclude satisfiability.

Ultimately, we are interested in the symbolic verification of distributed systems, using high-level descriptions. This involves reasoning about data types (specified algebraically) and control (described by boolean conditions on data). Properties of the system are described using large boolean expressions. We want to use BDD-techniques in order to prove, or at least simplify, boolean expressions containing arbitrary relation and function symbols. In this setting, abstraction doesn't work, as it doesn't preserve logical equivalence. Without abstraction, Ackermann's function elimination cannot be applied, and the finite domain property doesn't hold.

We therefore turn to the second method, allowing equations in the BDD nodes. We will give a new definition of "ordered", such that in ordered BDDs all paths will be consistent. The advantage is that on ordered BDDs with equations, the satisfiability check can be done in constant time. The contribution of this paper is an intermediate step towards the situation where arbitrary relations and function symbols in BDDs are allowed. We restrict to the case of equations, without function symbols.

Technical Contribution. In Section 2 we introduce EQ-BDDs, which are BDDs whose internal nodes may contain equations between variables. We extend the notion of orderedness so that it covers the equality laws for reflexivity, symmetry, transitivity and substitution. The main idea is that in a (reduced) ordered EQ-

BDD (EQ-OBDD) of the form ITE($x = y, P, Q$), y may not occur in P; this can be achieved by substituting occurrences of y by x. By means of term rewriting techniques, we show that every EQ-BDD is equivalent to an EQ-OBDD.

Contrary to OBDDs, EQ-OBDDs are not unique, in the sense that different EQ-OBDDs may still be logically equivalent, so equivalence checking on EQ-OBDDs cannot be done in constant time. However, we show that in an EQ-OBDD, each path from the root to a leaf is consistent. As a corollary, $\mathbf{0}$ is the only contradictory EQ-OBDD, and $\mathbf{1}$ is the only tautological one. Every other EQ-OBDD is satisfiable. So satisfiability and tautology checking on EQ-OBDDs can still be done in constant time.

We present an algorithm for converting propositional formulae with equality into an EQ-OBDD in Section 3. Usually a bottom-up algorithm is used, based on Bryant's APPLY algorithm [5], which implements the logical connectives on OBDDs in polynomial time. In the presence of equalities, APPLY would involve new substitutions, which possibly cause a reordering of the subformulae.

Instead, we use a generalization of the top-down method (cf. [12]). The inefficiency usually attributed to this top-down approach is avoided by using memoization techniques and maximal sharing. We have made a prototype implementation in C, which uses the `ATerm` library [4] to manipulate terms in maximally shared representation. We applied this implementation on the benchmarks used in [16,19]. It appears that our ideas yield a feasible procedure, and that the performance is comparable to the approach in [16].

In EQ-BDDs, interpreted function symbols can be incorporated straightforwardly. A complete term rewrite system for the algebraic data part can be used to reduce the nodes. This always leads to equivalent formulae, but completeness of the method is lost. In future work we plan to investigate under which circumstances completeness can be regained. The fact that equality is incorporated directly, instead of encoded, can give BDD-techniques a much more prominent place in interactive theorem provers like PVS [15]. The fact that the performance of our prototype implementation is comparable with existing proposals indicates that extendibility does not necessarily come with a loss in efficiency.

Related work. After Ackermann [1] proved decidability of quantifier free logic with equality, Shostak [18] and Nelson and Oppen [14] provided practical algorithms for the validity check, based on the congruence closure. Those authors used a transformation to disjunctive normal forms. In [8] this transformation is avoided, by dealing more efficiently with boolean combinations; in particular they incorporate case splitting as in the Davis-Putnam procedure. We next consider papers based on BDDs, that either use the aforementioned method based on the finite domain property, or allow arbitrary atoms in the BDD nodes.

Two recent papers [7,16] refine the method based on finite domains. The main contribution of Bryant et al. [7] is to distinguish between function symbols that occur in *positive equations* only (*p*-symbols) and other function symbols (*g*-symbols). This allows to restrict attention to maximally diverse interpretations, in which *p*-symbols can be interpreted by a fixed value. Also Ackermann's function elimination is improved. Pnueli et al. [16] provide heuristics to obtain lower estimates for the domains. These estimates are also obtained by distinguishing between positive and negative occurrences of equations. Both methods rely on the finite domain property, whereas our solution avoids this.

The other method is closer to our approach. Goel et al. [10] avoid bit vectors for finite domains, by introducing boolean variables e_{ij}, representing the equation $x_i = x_j$. So their method doesn't rely on the finite model property. Similarly, Møller et al. [13] allow difference constraints of the form $x - y \leq c$ in the BDD nodes, with c an integer or real constant. In case the underlying domain consists of integers or reals, $x = y$ can be encoded as $x - y \leq 0 \wedge y - x \leq 0$, leading to two different nodes. For other underlying domains, such as natural numbers or lists, this encoding is not possible, where our approach works for equality in any domain.

Both [10] and [13] first reduce a formula to OBDD, viewing all boolean terms as different variables. Although the nodes on a path are all different after this operation, a path can still be inconsistent, for instance by violating transitivity. Parts of the OBDD are inaccessible, so in general the OBDD is too large. The OBDD can be further reduced in order to check satisfiability (this is called path-reduced in [13]), but this involves the inspection of all paths, of which there can be exponentially many. Indeed, in [10] it is proved that deciding whether an OBDD with e_{ij}-variables has a satisfaction that complies with transitivity is NP-complete. In our case, the paths in the resulting EQ-OBDD are consistent and the test for satisfiability on EQ-OBDDs requires constant time only.

Another approach, mentioned in the full version of [7], considers the addition of transitivity constraints to a formula. Adding all of them usually leads to a blow-up of the BDD. A heuristics is presented to prune the set of needed transitivity constraints. In our approach transitivity constraints are generated on the fly when needed, by performing proper substitutions.

In the implementation, the fundamental data structure is a maximally shared term, partly consisting of boolean connectives, and partly of BDD-nodes. This resembles the Binary Expression Diagrams (BEDs) of [2], for the pure boolean case. We have not thoroughly studied the relationship between our top-down algorithm and their *up-one*. In [17] it is indicated how such a comparison could be made in principle, by using term rewriting theory on strategies. Also a thorough comparison with the algorithm in [8] would be interesting.

2   EQ-BDDs

We now define a syntax for formulae. First assume disjoint sets P and V. Members of P are called proposition (boolean) variables (typically p, q, ...) and V contains domain variables (typically x, y, z, ...).

Definition 1. *Formulae are expressions satisfying the following syntax:*

$$\Phi ::= \mathbf{0} \mid \mathbf{1} \mid P \mid V = V \mid \neg\Phi \mid \Phi \wedge \Phi \mid \mathrm{ITE}(\Phi,\Phi,\Phi)$$

We use $x \neq y$ as an abbreviation of $\neg(x = y)$. In order to avoid confusion, we write $\equiv$ for syntactic equality, so $x \equiv y$ means that x and y are the same variable.

An interpretation consists of a non-empty domain D and interpretation functions $I : V \rightarrow D$ and $J : P \rightarrow \{0, 1\}$. Then the semantics of Φ, denoted by $\Phi_I^J \in \{0, 1\}$, can be defined straightforwardly. In particular, $\text{ITE}(x, y, z)_I^J = y_I^J$ if $x_I^J = 1$, otherwise it equals z_I^J. Equality is interpreted as the identity relation by defining $(x = y)_I^J$ as 1 if $I(x) = I(y)$, 0 otherwise. Now D, I, J forms a model for Φ iff $\Phi_I^J = 1$. Φ is satisfiable iff it has a model and it is tautological (or: universally valid) iff all interpretations are models. Φ and Ψ are logically equivalent iff they have the same models. A theory is a set of formulae. Given a theory S, we write $S \vDash \Phi$ iff all models for S are models of Φ. We rely on the following lemma, which is a theorem of Shostak [18], specialized to the case without function symbols.

Lemma 2. *Let S be a set of equalities and T a set of inequalities. Then $S \cup T$ is satisfiable if and only if for all $x \neq y \in T$, $x = y$ is not in the reflexive, symmetric and transitive closure of S.*

We now turn to the study of EQ-BDDs, which can be seen as a subset of formulae, and consider arbitrary formulae in Section 3. A binary decision diagram (BDD [6,12]) is a DAG, whose internal nodes contain guards, and whose leaves are labeled 0 (low, false) or 1 (high, true). Each node contains two distinguished outgoing edges, called low and high. In ordinary BDDs, the guards solely consist of proposition variables. The only difference between ordinary BDDs and EQ-BDDs is that in the latter, a guard can also consist of equations between domain variables. EQ-BDDs can be depicted as follows (the low/false edges are dashed):

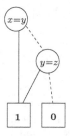

We reason mainly about EQ-BDDs as a restricted subset of formulae, although in implementations we always treat these formulae as maximally shared DAGs. There are constants to represent the nodes 0 or 1. Furthermore, we use the if-then-else function $\text{ITE}(g, t_1, t_2)$ where g is a guard, or label of a node in the BDD, t_1 is the high node and t_2 is the low node. Guards can be proposition variables in P, or equations of the form $x = y$ where x and y are domain variables (V).

Definition 3. *We define the set G of guards and B of EQ-BDDs,*

$$G ::= P \mid V = V$$
$$B ::= 0 \mid 1 \mid \text{ITE}(G, B, B)$$

The EQ-BDD depicted above can be written as: $\text{ITE}(x = y, 1, \text{ITE}(y = z, 1, 0))$.

In order to compute whether an EQ-BDD is tautological or satisfiable, it will first be ordered. In an *ordered* EQ-BDD, the guards on a path may only appear in a fixed order. To this end, we impose a total order on $P \cup V$ (e.g. $x \succ p \succ y \succ z \succ q$). This order is extended lexicographically to guards as follows:

Definition 4 (Order on guards).

$$p \succ q \text{ as given above}$$
$$(x = y) \succ p \text{ if, and only if, } x \succ p$$
$$p \succ (x = y) \text{ if, and only if, } p \succ x$$
$$(x = y) \succ (u = v) \text{ if, and only if, either } x \succ u, \text{ or } x \equiv u \text{ and } y \succ v.$$

Given this order, we can now define what we mean by an ordered EQ-BDD. We use some elementary terminology from term rewrite systems (TRSs), which can for instance be found in [11,3]. In particular, a *normal form* is a term to which no rule can be applied. A system is *terminating* if no infinite rewrite sequence exists.

Definition 5. *An EQ-BDD is ordered if, and only if, it is a normal form w.r.t. the following term rewrite system, called* ORDER. *An EQ-OBDD is an ordered EQ-BDD.:*

1. $\text{ITE}(G, T, T) \to T$.
2. $\text{ITE}(G, \text{ITE}(G, T_1, T_2), T_3) \to \text{ITE}(G, T_1, T_3)$.
3. $\text{ITE}(G, T_1, \text{ITE}(G, T_2, T_3)) \to \text{ITE}(G, T_1, T_3)$.
4. $\text{ITE}(G_1, \text{ITE}(G_2, T_1, T_2), T_3) \to \text{ITE}(G_2, \text{ITE}(G_1, T_1, T_3), \text{ITE}(G_1, T_2, T_3))$,
   provided $G_1 \succ G_2$.
5. $\text{ITE}(G_1, T_1, \text{ITE}(G_2, T_2, T_3)) \to \text{ITE}(G_2, \text{ITE}(G_1, T_1, T_2), \text{ITE}(G_1, T_1, T_3))$,
   provided $G_1 \succ G_2$
6. $\text{ITE}(x = x, T_1, T_2) \to T_1$.
7. $\text{ITE}(y = x, T_1, T_2) \to \text{ITE}(x = y, T_1, T_2)$, provided $x \prec y$
8. $\text{ITE}(x = y, T_1[y], T_2) \to \text{ITE}(x = y, T_1[x], T_2)$, if $x \prec y$ and y occurs in T_1.

Rules 6–8 capture the properties of equality, viz. reflexivity, symmetry, and substitutivity. From these rules, transitivity can be derived, as we demonstrate in Figure 1 (we assume $x \prec y \prec z$). Note that in rule 8 *all* instances of y in T_1 are replaced by x. From a term rewriting perspective this is non-standard, because it is a non-local rule.

In a normal form no rewrite rules are applicable. Hence it is easy to see that in an ordered EQ-BDD, the guards along a path occur in strictly increasing order (otherwise rule 2/3/4/5 would be applicable) and in all guards of the form $x = y$, it must be the case that $x \prec y$ (otherwise rule 6/7 would be applicable). Note that the transformations indicated by the rules are sound, in the sense that they yield logically equivalent EQ-BDDs.

We prove that each EQ-BDD is equivalent to an EQ-OBDD, by showing that the TRS ORDER always terminates. The termination proof uses the powerful *recursive path ordering* (RPO) [9]. For RPO comparisons, we view $\text{ITE}(g, t_1, t_2)$

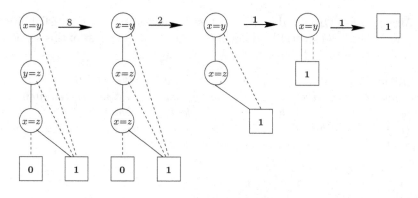

Fig. 1. Derivation of transitivity of equality in EQ-BDDs

as $g(t_1, t_2)$. RPO needs an ordering on the function symbols. For this we just use the total order on guards of Definition 4, extended with $1, 0 \prec g$ for all guards. In this case, RPO specializes to the following relation:

Definition 6. $s \equiv f(s_1, s_2) \succ_{\text{rpo}} t$ *iff* $t \equiv 0$ *or* $t \equiv 1$, *or* $t \equiv g(t_1, t_2)$ *and one of the following holds:*

- *(I)* $s_1 \succeq_{\text{rpo}} t$, *or* $s_2 \succeq_{\text{rpo}} t$;
- *(II)* $f \succ g$ *and* $s \succ_{\text{rpo}} t_1$ *and* $s \succ_{\text{rpo}} t_2$;
- *(III)* $f \equiv g$ *and either* $s_1 \succ_{\text{rpo}} t_1$ *and* $s_2 \succeq_{\text{rpo}} t_2$, *or* $s_2 \succ_{\text{rpo}} t_2$ *and* $s_1 \succeq_{\text{rpo}} t_1$.

Here $x \succeq_{\text{rpo}} y$ means: $x \succ_{\text{rpo}} y$ or $x \equiv y$. Usually in clause (III) the multiset or lexicographic extension is used, but this is not needed for our purposes. From the literature, it is well known that $\succ_{\text{rpo}}$ is an order (in particular the relation is transitive), which is well-founded (because $\succ$ on guards is) and monotone, so it is useful in proving termination.

Lemma 7. *The rewrite system* ORDER *is terminating.*

Proof. It is straightforward to show that rule 1–8 are contained in $\succ_{\text{rpo}}$ (for rule 8 monotonicity of $\succ_{\text{rpo}}$ is used). From this termination follows.   □

Theorem 8. *Every EQ-BDD is equivalent to some EQ-OBDD.*

Traditional OBDDs are unique representations of boolean functions, which makes them useful for checking equivalence between formulae. For EQ-OBDDs, however, this uniqueness property fails, as the following example shows.

Example 9. Let $x \prec y \prec z$. Consider the EQ-BDDs $\text{ITE}(x = y, 1, \text{ITE}(y = z, 0, 1))$ and $\text{ITE}(x = z, 1, \text{ITE}(y = z, 0, 1))$. These represent the predicates $y = z \to x = y$ and $y = z \to x = z$, which are logically equivalent. Both are ordered, because no rewrite rule is applicable. But they are not identical.   □

Although EQ-OBDDs do not have the uniqueness property, satisfiability or tautology checking can still be done in constant time. The rest of this section is devoted to the proof of this statement.

Definition 10. Path*s are sequences of* 0*'s and* 1*'s. We let letters* α*,* β *and* γ *range over paths, and write* ε *for the empty sequence,* $\alpha.\beta$ *for the concatenation, and* $\alpha \sqsubseteq \beta$ *if* α *is a prefix of* β*. With* $seq(T)$ *we denote the sequences that correspond to a path in EQ-BDD* T*. For a path* $\alpha \in seq(T)$ *we write* $T|_\alpha$ *for the* guard *at the end of path* α*, inductively defined by:*

- $\text{ITE}(G, T, U)|_\varepsilon = G$.
- $\text{ITE}(G, T, U)|_{1.\alpha} = T|_\alpha$ *(the high branch).*
- $\text{ITE}(G, T, U)|_{0.\alpha} = U|_\alpha$ *(the low branch).*

We also define the theory up to the node corresponding to path $\alpha \in seq(T)$*, notation* $Th(T, \alpha)$*, inductively on an EQ-BDD* T*:*

- $Th(T, \varepsilon) = \emptyset$.
- $Th(T, \alpha.1) = Th(T, \alpha) \cup \{T|_\alpha\}$.
- $Th(T, \alpha.0) = Th(T, \alpha) \cup \{\neg T|_\alpha\}$.

Finally, $\alpha \in seq(T)$ *is called* consistent *iff* $Th(T, \alpha)$ *is satisfiable.*

Example 11. Let $T \equiv \text{ITE}(x = y, \mathbf{1}, \text{ITE}(y = z, \text{ITE}(x = z, \mathbf{1}, \mathbf{0}), \mathbf{1}))$. Then the guard at path 0.1 is: $T|_{0.1} \equiv x = z$. The theory at that point is: $Th(T, 0.1) = \{x \neq y, y = z\}$ which is satisfiable, so 0.1 is consistent.   □

The analysis of EQ-OBDDs depends on the following rather syntactic lemma. The first states that in EQ-OBDDs y does not occur below the high branch of $x = y$; the second states that y does not occur positively above $x = y$.

Lemma 12. *Let* T *be an EQ-OBDD, and* $\alpha, \beta \in seq(T)$ *be consistent paths.*

1. *If* $T|_\alpha \equiv x = y$ *and* $\alpha.1 \sqsubseteq \beta$*, then* $T|_\beta \not\equiv z = y$ *and* $T|_\beta \not\equiv y = z$.
2. *If* $T|_\alpha \equiv x = y$ *and* $\beta.1 \sqsubseteq \alpha$*, then* $T|_\beta \not\equiv z = y$ *and* $T|_\beta \not\equiv y = z$.
3. *If* $Th(T, \alpha) \vDash x = z$ *and* $x \prec z$*, then for some* y*,* $y = z \in Th(T, \alpha)$.

Proof. (1) If $T|_\beta$ contains y, rewrite step 8 would be applicable, which contradicts orderedness.

(2) If $T|_\beta \equiv z = y$ rewrite step (8) is applicable, contradicting orderedness. Assume $T|_\beta \equiv y = z$. Note that $x \prec y$, as $x = y$ appears in the EQ-OBDD, so $x = y \prec y = z$. Hence, on the path between the nodes labeled with $y = z$ and $x = y$, at least one of the steps (4,5) would be applicable. This contradicts orderedness of T.

(3) Let $Th(T, \alpha) \vDash x = z$. Note that $Th(T, \alpha)$ is satisfiable, but $Th(T, \alpha) \cup \{x \neq z\}$ is not. Hence by two applications of Lemma 2, $x = z$ is in the reflexive, symmetric, transitive closure of the positive equations in $Th(T, \alpha)$. I.e. there exist n and x_i $(0 \leq i \leq n)$, such that $x_0 \equiv x$, $x_n \equiv z$ and for all i $(0 \leq i < n)$, $x_i = x_{i+1} \in Th(T, \alpha)$ or $x_{i+1} = x_i \in Th(T, \alpha)$. Because $x \prec z$, we have $n \geq 1$.

Consider the last equation in this sequence, which is either $x_{n-1} = z \in Th(T, \alpha)$, in which case we are done, or it is $z = x_{n-1} \in Th(T, \alpha)$. In this case, x_{n-1} doesn't occur in any other equation (it cannot occur positively above $z = x_{n-1}$ in T by (2), nor can it occur below it by (1)). Hence $n = 1$ and $z = x \in Th(T, \alpha)$. This contradicts orderedness of T, because $x \prec z$. $\square$

We can now prove that each guard in an EQ-OBDD is logically independent from those occurring above it.

Lemma 13. *Let T be an EQ-OBDD and let $\alpha \in seq(T)$ be consistent. Then*

1. $Th(T, \alpha) \not\models T|_\alpha$ *and*
2. $Th(T, \alpha) \not\models \neg T|_\alpha$.

Proof. If $T|_\alpha \equiv p$ $(p \in P)$, then by orderedness, p does not occur in $Th(T, \alpha)$, so the lemma follows (this is similar to the traditional BDD-case). Now let $T|_\alpha \equiv x = z$. Hence, $x \prec z$.

(1) Assume $Th(T, \alpha) \models x = z$. By Lemma 12.3, for some y, $y = z \in Th(T, \alpha)$. Then rewrite step 8 is applicable, which contradicts orderedness.

(2) Assume $Th(T, \alpha) \models x \neq z$. Using Lemma 2 it can be proved that for some y and v, $Th(T, \alpha) \models \{x = y, v = z\}$ and either $y \neq v \in Th(T, \alpha)$ or $v \neq y \in Th(T, \alpha)$. By Lemma 12.2, no positive equations containing z occur in $Th(T, \alpha)$, so $z \equiv v$. Now if $z \neq y \in Th(T, \alpha)$, $z = y$ occurs above $x = z$ in the ordered EQ-BDD T, so $z \prec x$, contradicting $x \prec z$. Hence, $y \neq z \in Th(T, \alpha)$. Note that as T is ordered and $y = z$ occurs above $x = z$, $y \prec x$. Now by Lemma 12.3, for some w, $w = x \in Th(T, \alpha)$. But then rewrite step 8 would be applicable, which contradicts orderedness. $\square$

Theorem 14. *Satisfiability and tautology on EQ-OBDDs can be checked in constant time.*

Proof. Using Lemma 13 it can be proved that each path to a leaf in an EQ-OBDD is consistent, so all leaves are reachable by some interpretation. Hence if the EQ-OBDD is a tautology, all leaves must be syntactically equal to **1**, and by rule (1) of ORDER, the EQ-OBDD must be the node **1**. In a similar way, the only contradictory EQ-OBDD is **0**. Hence an EQ-OBDD is satisfiable if, and only if, it is syntactically different from **0**. $\square$

3 Algorithm for Checking Tautology and Satisfiability

We are now interested in constructing EQ-BDDs out of formulae. In traditional BDDs, a formula is transformed into an OBDD in a bottom-up fashion. Given two ordered BDDs, the logical operations (conjunction, disjunction, etc.) can be performed in polynomial time by Bryant's APPLY algorithm. If two EQ-OBDDs are combined in this way, new substitutions must be done in both of them, which destroy the ordering. We can of course re-order them by using the rewrite system ORDER, but the advantage of having a polynomial APPLY has been lost.

As an alternative, we use a top-down approach, which in the context of OBDDs has for instance been described in [12]. This approach is based on the Shannon expansion. For propositional logic, this reads: $\Phi \Longleftrightarrow \text{ITE}(p, \Phi|_p, \Phi|_{\neg p})$, where in $\Phi|_p$ all occurrences of p are replaced by $\mathbf{1}$, and in $\Phi|_{\neg p}$ by $\mathbf{0}$. Taking for p the smallest propositional variable in the ordering, this Shannon expansion can be used to create a root node for p, and recursively continuing with two subformulae that do not contain p. The number of variables in the formula decreases. So, this process terminates. Because at each step the smallest variable is taken, the resulting BDD is ordered.

When p is an equation, say $x = y$, the Shannon expansion still holds. In the formula $\Phi|_{x=y}$, we assume that $x = y$, so we are allowed to substitute y for x. This leads to the following variant of the Shannon expansion:

$$\Phi \Longleftrightarrow \text{ITE}(x = y, \Phi[x := y], \Phi[(x = y) := \mathbf{0}])$$

This is recursively applied, with $x = y$ the smallest equation in Φ, oriented in such a way that $x \prec y$ in the variable order. Due to the substitutions it is not guaranteed that the resulting EQ-BDD is ordered. However, we will show that repeatedly applying the Shannon expansion does lead to an EQ-OBDD.

3.1   A Topdown Algorithm

We now describe the algorithm precisely. We introduce a term rewrite system SIMPLIFY, which removes superfluous occurrences of $\mathbf{0}$ and $\mathbf{1}$ and orients all guards. It is clearly terminating and confluent.

Definition 15. *The TRS* SIMPLIFY *consists of the following rules:*

$$
\begin{array}{llll}
\mathbf{0} \wedge T \to \mathbf{0} & \neg \mathbf{1} & \to \mathbf{0} & \\
T \wedge \mathbf{0} \to \mathbf{0} & \neg \mathbf{0} & \to \mathbf{1} & x = x \to \mathbf{1} \\
\mathbf{1} \wedge T \to T & \text{ITE}(\mathbf{1}, T, U) \to T & & y = x \to x = y \quad \text{if } x \prec y \\
T \wedge \mathbf{1} \to T & \text{ITE}(\mathbf{0}, T, U) \to U & &
\end{array}
$$

We write $\Phi\!\downarrow$ for the normal form of Φ obtained by this rewrite system. Φ is called simplified, if $\Phi \equiv \Phi\!\downarrow$.

Note that every closed formula rewrites to $\mathbf{0}$ or $\mathbf{1}$. Furthermore, on EQ-BDDs only the last four rules are applicable. Finally, note that ordered EQ-BDDs are simplified. We introduce an auxiliary operation $\Phi|_s$, where Φ is a formula and s a guard or the negation of a guard. We assume that Φ is simplified.

Definition 16. *We define $\Phi|_s$, where s is p, $\neg p$, $x = y$ or $x \neq y$ as follows: If $s \equiv p$, then $\Phi|_s$ consists of replacing all occurrences of p by $\mathbf{1}$; in $\Phi|_{\neg s}$ all occurrences of p are replaced by $\mathbf{0}$. In case $s \equiv x = y$, we obtain $\Phi|_s$ by replacing all occurrences of y by x, and $\Phi|_{\neg s}$ by replacing $x = y$ by $\mathbf{0}$ everywhere.*

Example 17. Let $\Phi \equiv x = z \wedge y = z$ and $g \equiv x = z$ and assume $x \prec y \prec z$. Then $\Phi|_g \equiv x = x \wedge y = x$ and $\Phi|_{\neg g} \equiv \mathbf{0} \wedge y = z$. After simplification, we get: $\Phi|_g\!\downarrow \equiv x = y$ and $\Phi|_{\neg g}\!\downarrow \equiv \mathbf{0}$. □

We are now ready to define the basic top-down transformation algorithm:

Definition 18. *Assume that Φ be a simplified formula. We define the algorithm* TOPDOWN *on input Φ as follows:*

- TOPDOWN($\mathbf{1}$) $\equiv \mathbf{1}$
- TOPDOWN($\mathbf{0}$) $\equiv \mathbf{0}$
- *Otherwise, let g be the smallest guard occurring in Φ. Then*

$$\text{TOPDOWN}(\Phi) \equiv \overline{\text{ITE}}(g, \text{TOPDOWN}(\Phi|_g\downarrow), \text{TOPDOWN}(\Phi|_{\neg g}\downarrow))$$

where
$$\overline{\text{ITE}}(g, T, U) \equiv \begin{cases} T & \text{if } T \equiv U \\ \text{ITE}(g, T, U) & \text{otherwise.} \end{cases}$$

Note that a closed formula simplifies to $\mathbf{1}$ or $\mathbf{0}$, so in the other case it must contain a guard. Note that due to substitutions, new equalities can be introduced on the fly. We now prove termination and soundness of the algorithm TOPDOWN. With $\#(\Phi)$ we denote the number of guard occurrences in the completely unfolded tree of Φ. Note that none of the rules from SIMPLIFY increases the number of guards, so we have the following:

Lemma 19. *For any formula Φ, we have $\#(\Phi) \geq \#(\Phi\downarrow)$.*

Lemma 20. *Let Φ be a simplified formula, and let g be a simplified guard.*

(1) $\#(\Phi) \geq \#(\Phi|_g)$ (3) *if g occurs in Φ, then $\#(\Phi) > \#(\Phi|_g)$*

(2) $\#(\Phi) \geq \#(\Phi|_{\neg g})$ (4) *if g occurs in Φ, then $\#(\Phi) > \#(\Phi|_{\neg g})$*

Proof. Simultaneous formula induction on Φ. This boils down to checking that in Definition 16, each guard is replaced by at most one other guard. □

Theorem 21. *The algorithm* TOPDOWN(Φ) *always terminates.*

Proof. With each recursive call, $\#(\Phi)$ strictly decreases. □

Theorem 22 (soundness). *For any formula Φ, we have:* $\Phi \Longleftrightarrow$ TOPDOWN(Φ)

Proof. Induction over the number of calls to TOPDOWN. The induction step uses that $\Phi \Longleftrightarrow \Phi\downarrow$ and $g \Rightarrow (\Phi \Longleftrightarrow \Phi|_g)$ and similar for $\neg g$. □

3.2   Iteration of TOPDOWN

Unfortunately, it is *not* the case that TOPDOWN(Φ) is always ordered, as the following example shows.

Example 23. Assume $x \prec y \prec z$. Then TOPDOWN($x \neq y \wedge (x = z \wedge y = z)$) $\equiv$ ITE($x = y, \mathbf{0}, $ITE($x = z, $ITE($x = y, \mathbf{1}, \mathbf{0}), \mathbf{0}$)). See Figure 2, where the formulae in square brackets denote the arguments to TOPDOWN, and the dashed nodes occur in the call graph, but are suppressed in the resulting EQ-BDD. In the low branch, $x = y$ is replaced by $\mathbf{0}$, but due to substitutions in the recursive call, new occurrences of $x = y$ are generated. Note that this is dangerous, as after one application of TOPDOWN it still contains unsatisfiable paths, which erroneously could lead one to believe that the EQ-BDD represents a satisfiable formula. □

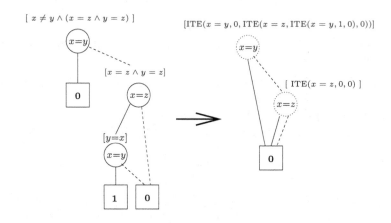

Fig. 2. Two call-graphs to TOPDOWN.

Note that in the previous example, an EQ-OBDD is found by another application of TOPDOWN. We propose to apply TOPDOWN repeatedly to a formula Φ, until a fixed point is reached. In the benchmarks presented in Section 3.3 at most two iterations of TOPDOWN were required to obtain an EQ-OBDD. In the rest of this section we prove that the fixed point can be reached in a finite number of steps, and that it is an ordered EQ-BDD.

Lemma 24. *Let Φ be a simplified EQ-BDD and g be a simplified guard. Then*

(1) $\Phi \succeq_{\mathrm{rpo}} \Phi|_g\!\downarrow$ (3) *if g occurs in Φ, then $\Phi \succ_{\mathrm{rpo}} \Phi|_g\!\downarrow$*
(2) $\Phi \succeq_{\mathrm{rpo}} \Phi|_{\neg g}\!\downarrow$ (4) *if g occurs in Φ, then $\Phi \succ_{\mathrm{rpo}} \Phi|_{\neg g}\!\downarrow$*

Proof. We apply simultaneous induction on the structure of Φ. We only present two interesting fragments of the proof of case (1) and (3), where $\Phi \equiv \mathrm{ITE}(u = v, T, U)$ and $g \equiv x = y$. Note that $x \prec y$ and $u \prec v$, because Φ and g are simplified.

First consider case (1). By definition $\Phi|_g\!\downarrow \equiv \mathrm{ITE}((u = v)|_g\!\downarrow, T|_g\!\downarrow, U|_g\!\downarrow)\!\downarrow$. Observe that $(u = v)|_g\!\downarrow$ either equals $\mathbf{1}$, $x = v$ (if $u \equiv y$), $u = x$ (if $v \equiv y$ and $u \prec x$), $x = u$ (if $v \equiv y$ and $x \prec u$) or $u = v$. The case $v = x$ does not occur, for we would have $v \prec x \prec y \equiv u \prec v$.

In the first case $\Phi|_g\!\downarrow \equiv T|_g\!\downarrow$. Using the induction hypothesis, $T \succeq_{\mathrm{rpo}} T|_g\!\downarrow$. By property (I) of recursive path orderings it follows that $\Phi \succ_{\mathrm{rpo}} T$ and hence $\Phi \succ_{\mathrm{rpo}} \Phi|_g\!\downarrow$. In the next three cases, it is obvious that $x = v \prec u = v$ and $u = x \prec u = v$ and $x = u \prec u = v$, respectively. Now using a similar argument as above, we can show that $\Phi \succ_{\mathrm{rpo}} T|_g\!\downarrow$ and $\Phi \succ_{\mathrm{rpo}} U|_g\!\downarrow$. So, by property (II) of RPO it follows that $\Phi \succ_{\mathrm{rpo}} \Phi|_g\!\downarrow$. In the last case, where $(u = v)|_g\!\downarrow \equiv u = v$, we find by the induction hypothesis $T \succeq_{\mathrm{rpo}} T|_g\!\downarrow$ and $U \succeq_{\mathrm{rpo}} U|_g\!\downarrow$. By property (III) of RPO it follows that $\Phi \succeq_{\mathrm{rpo}} \Phi|_g\!\downarrow$.

Now consider case (3). Note that in case (1) we proved that $\Phi \succ_{\mathrm{rpo}} \Phi|_g\!\downarrow$ in all but the case where $(u = v)|_g\!\downarrow \equiv u = v$. So, we only need to consider this case. As g occurs in Φ, it must occur in T or in U. As the cases are symmetric, we can without loss of generality assume that g occurs in T. Via the induction hypothesis it follows that $T \succ_{\mathrm{rpo}} T|_g\downarrow$. Furthermore, by case (1) $U \succeq_{\mathrm{rpo}} U|_g\downarrow$. So, by property (III) of RPO we can conclude that

$$\Phi \equiv \mathrm{ITE}(u = v, T, U) \succ_{\mathrm{rpo}} \mathrm{ITE}(u = v, T|_g\downarrow, U|_g\downarrow) \equiv \Phi|_g\downarrow. \qquad \square$$

Lemma 25. *Let Φ be a simplified EQ-BDD.*

1. $\Phi \succeq_{\mathrm{rpo}} \mathrm{TOPDOWN}(\Phi)$.
2. Φ is ordered iff $\Phi \equiv \mathrm{TOPDOWN}(\Phi)$.

Proof. Part 1 is proved by induction on $\#(\Phi)$. Note that if Φ does not contain a guard then it is equal to $\mathbf{1}$ or $\mathbf{0}$, and this theorem is trivial. So, assume Φ contains at least one guard and let g be the smallest guard occurring in Φ. Recall from Lemma 19, 20 that $\#(\Phi) > \#(\Phi|_g\downarrow)$ and similar for $\neg g$. Then $\mathrm{TOPDOWN}(\Phi) = \overline{\mathrm{ITE}}(g, \mathrm{TOPDOWN}(\Phi|_g\downarrow), \mathrm{TOPDOWN}(\Phi|_{\neg g}\downarrow))$. By induction hypothesis and Lemma 24, we have:

$$(*) \qquad \begin{array}{l} \Phi \succ_{\mathrm{rpo}} \Phi|_g\downarrow \succeq_{\mathrm{rpo}} \mathrm{TOPDOWN}(\Phi|_g\downarrow) \\ \Phi \succ_{\mathrm{rpo}} \Phi|_{\neg g}\downarrow \succeq_{\mathrm{rpo}} \mathrm{TOPDOWN}(\Phi|_{\neg g}\downarrow) \end{array}$$

First, assume $\mathrm{TOPDOWN}(\Phi|_g\downarrow) \equiv \mathrm{TOPDOWN}(\Phi|_{\neg g}\downarrow)$. Then $\mathrm{TOPDOWN}(\Phi) \equiv \mathrm{TOPDOWN}(\Phi|_g\downarrow)$ and we are done by (*). Now assume $\mathrm{TOPDOWN}(\Phi|_g\downarrow) \not\equiv \mathrm{TOPDOWN}(\Phi|_{\neg g}\downarrow)$, and assume that $\Phi \equiv \mathrm{ITE}(h, T, U)$. Then $\mathrm{TOPDOWN}(\Phi) \equiv \mathrm{ITE}(g, \mathrm{TOPDOWN}(\Phi|_g\downarrow), \mathrm{TOPDOWN}(\Phi|_{\neg g}\downarrow))$. As g is the smallest guard, one of the following two cases must hold.

- $g \equiv h$. In this case $\Phi|_g\downarrow \equiv T|_g\downarrow$. Using Lemma 24 and the induction hypothesis, we can conclude $T \succeq_{\mathrm{rpo}} T|_g\downarrow \equiv \Phi|_g\downarrow \succeq_{\mathrm{rpo}} \mathrm{TOPDOWN}(\Phi|_g\downarrow)$. Similarly, $U \succeq_{\mathrm{rpo}} \mathrm{TOPDOWN}(\Phi|_{\neg g}\downarrow)$. By case (III) of RPO it follows that $\Phi \succeq_{\mathrm{rpo}} \mathrm{TOPDOWN}(\Phi)$.
- $h \succ g$. Using (*) we can immediately apply case (II) of RPO and conclude that $\Phi \succ_{\mathrm{rpo}} \mathrm{TOPDOWN}(\Phi)$.

Part 2. Both directions are proved by structural induction on Φ. $\Longrightarrow$: We must show that if Φ is ordered, then $\Phi \equiv \mathrm{TOPDOWN}(\Phi)$. The case where Φ equals $\mathbf{0}$ or $\mathbf{1}$ is trivial. So, consider the case where $\Phi \equiv \mathrm{ITE}(g, T, U)$. As Φ is ordered, g must be the smallest guard of Φ and cannot occur in T or U. Also, if $g \equiv x = y$, y does not occur in T. Moreover, T and U are ordered, hence also simplified. So, $\Phi|_g\downarrow \equiv T$ and $\Phi|_{\neg g}\downarrow \equiv U$. Note that $T \not\equiv U$.

$$\begin{array}{ll} \mathrm{TOPDOWN}(\Phi) \equiv & \\ \overline{\mathrm{ITE}}(g, \mathrm{TOPDOWN}(\Phi|_g\downarrow), \mathrm{TOPDOWN}(\Phi|_{\neg g}\downarrow)) \equiv & \\ \overline{\mathrm{ITE}}(g, \mathrm{TOPDOWN}(T), \mathrm{TOPDOWN}(U)) \equiv & \text{(Induction hypothesis)} \\ \mathrm{ITE}(g, T, U) \equiv & \\ \Phi & \end{array}$$

$\Longleftarrow$: Assume $\Phi = \mathrm{TOPDOWN}(\Phi)$. If Φ is $\mathbf{1}$ or $\mathbf{0}$ then it is trivially ordered. So assume $\Phi = \mathrm{ITE}(g, \Phi_1, \Phi_2)$. Then $\mathrm{TOPDOWN}(\Phi) = \overline{\mathrm{ITE}}(h, \Psi_1, \Psi_2)$, where h is the smallest guard in Φ, $\Psi_1 \equiv \text{TOPDOWN}(\Phi|_h\downarrow)$ and $\Psi_2 \equiv \text{TOPDOWN}(\Phi|_{\neg h}\downarrow)$. If $\Psi_1 \equiv \Psi_2$, then $\Phi \equiv \Psi_1$ and using Lemma 24.3 and 25.1 we get the following contradiction: $\Phi \succ_{\text{rpo}} \Phi|_h\downarrow \succeq_{\text{rpo}} \Psi_1 \equiv \Phi$.

Hence $\Psi_1 \not\equiv \Psi_2$. Then it must be the case that $g \equiv h$, $\Phi_1 \equiv \Psi_1$ and $\Phi_2 \equiv \Psi_2$. Note that then $\Phi|_h\downarrow \equiv \Phi_1|_h\downarrow$. Now, as $\Phi_1 \succeq_{\text{rpo}} \Phi_1|_h\downarrow \equiv \Phi|_h\downarrow \succeq_{\text{rpo}} \Psi_1$ it must be the case that $\Phi_1 \equiv \Phi_1|_h\downarrow$, hence $\Phi_1 \equiv \text{TOPDOWN}(\Phi_1)$. Similarly, $\Phi_2 \equiv \text{TOPDOWN}(\Phi_2)$.

We must show that Φ is ordered. By induction hypothesis, Φ_1 and Φ_2 are ordered, so no rule of the TRS ORDER is applicable to a strict subterm of Φ. We now show that no rule (1–8) is applicable to the root of Φ:

If rule 1 is applicable, then $\Psi_1 \equiv \Psi_2$, which we excluded already. In case of rule 2, $\Phi_1 \equiv \text{ITE}(g, T, U)$, and we obtain the following contradiction: $\Phi_1 \succ_{\text{rpo}} T \succeq_{\text{rpo}} T|_g\downarrow \equiv \Phi_1|_g\downarrow \equiv \Phi_1$. Rule 3 is excluded similarly. Rule 4 and 5 are not applicable because $g \equiv h$, which is the smallest guard in Φ. Rule 6 and 7 are not applicable because Φ is simplified. Finally, if rule 8 were applicable, $g \equiv x = y$ and y occurs in Φ_1. Then, using monotonicity of $\succ_{\text{rpo}}$, we have the following contradiction: $\Phi_1 \succ_{\text{rpo}} \Phi_1[y := x] \equiv \Phi_1|_g \succeq_{\text{rpo}} \Phi_1|_g\downarrow \equiv \Phi_1$. The last inequality uses the fact that the applicable rules of SIMPLIFY are contained in $\succ_{\text{rpo}}$. □

Theorem 26. *Let Φ be a simplified formula. Iterated application of* TOPDOWN *to Φ leads in a finite number of steps to an EQ-OBDD equivalent to Φ.*

Proof. After one application of TOPDOWN, Φ is transformed into a simplified EQ-BDD. So, iterated application of TOPDOWN leads to a sequence $\Phi, \Phi_1, \Phi_2, \ldots$ of which each Φ_i $(i \geq 1)$ is a simplified EQ-BDD. By Lemma 25.1 the sequence $\Phi_1, \Phi_2, \ldots$ is decreasing in a well-founded way. Hence, at a certain point in the sequence we find that $\Phi_i \equiv \Phi_{i+1}$. By Lemma 25.2 Φ_i is the required EQ-OBDD. Note that by Lemma 25.2 Φ_i is the first ordered EQ-BDD in the sequence. □

We conclude with the complete algorithm to transform an arbitrary formula Φ to EQ-OBDD, which is just a repeated application of TOPDOWN until a fixed point is reached:

$$\text{EQ-OBDD}(\Phi) = \texttt{fixedpoint}(\text{TOPDOWN})(\Phi\downarrow)$$

We stress that in the benchmarks we never needed more than 2 iterations. This is not generally the case:

Example 27. Given $a \prec b \prec c \prec d \prec e \prec f$, the following EQ-BDD needs 4 iterations: $\text{ITE}(a = f, \text{ITE}(a = e, d = e, c = d), b = c)$. The intermediate EQ-BDDs have size 9, 13, 23 and 21, respectively. This can be checked with our implementation. □

3.3   Implementation and Benchmarks

In order to study the performance of TOPDOWN, we made an implementation and used it to try the benchmarks reported in [16,19]. The authors report to have comparable performance as in [10]. Unfortunately, we could not obtain the benchmarks used in [7]. We first describe the implementation, including some variable orderings we used and then present the results.

Prototype implementation. We have made a prototype implementation of the TOPDOWN algorithm. As programming language we used C, including the `ATerm`-library [4]. The basic data types in this library are `ATerms` and `ATermTables`. `ATerms` are terms, which are internally represented as maximally shared DAGs. As a consequence, syntactical equality of terms can be tested in constant time. The basic operations are term formation and decomposition, which are also performed in constant time. `ATermTables` implement hash tables of dynamic size, with the usual operations. The `ATerm`-library also provides memory management functionality, by automatically garbage collecting unreferenced terms. By representing formulae and BDDs as `ATerms`, we are sure that they are always a maximally shared DAG.

Care has to be taken in order to avoid that during some computation, shared subterms are processed more than once. Therefore all recursive procedures, like "find the smallest variable", "simplify" and $\Phi|_s$ are implemented using a hash table to implement memoization. In this way, syntactically equal terms are processed only once, and the time complexity for computing these functions is linear in the number of nodes in the DAG, which is the number of *different* subterms in the formulae. Also the TOPDOWN-function itself uses a hash table for memoization. This contributes to its efficiency: Consider a formula Ψ which is symmetric in p and q (for instance: $(p \wedge q) \vee \Phi$, or $(p \wedge \Phi) \vee (q \wedge \Phi)$). Then $(\Psi|_p\downarrow)|_{\neg q}\downarrow \equiv (\Psi|_{\neg p}\downarrow)|_q\downarrow$. Thanks to memoization, only one of them will actually be computed. Still, the TOPDOWN function has worst case exponential behavior, which is unavoidable, because in the propositional case (i.e. excluding equations) it builds an OBDD from a propositional formula in one iteration. Due to memoization of TOPDOWN's arguments, the memory demands are rather high.

Results. Benchmark formulae can be obtained from [19] and most of them could be solved with the methods described in [16]. Each formula is known to be a tautology. They originate from compiler optimization; each formula expresses that the source and target code of a compilation step are equivalent. We used the versions where Ackermann's function elimination has been applied [1], but domain minimization [16] has not yet been applied. In fact, our method does not rely on the finiteness of domains at all. The benchmark formulae extend the formulae of Definition 1 in various ways, but these extensions could be dealt with easily.

It is well known that the variable ordering has an important effect on the performance. We therefore tried a number of orderings: With 't' we denote the textual order of the variables as given in [19]. With 'r' we denote the reverse of this textual order. Finally, 'bt' ('br') denotes the textual (reverse) order, except that boolean variables always precede domain variables.

We can now present the results. They can be found in Figure 3. The first column contains the number of the files, as given in [19]. The next three columns give an indication of the size of the formula: #b is the number of boolean variables, #d the number of domain variables, and #n is the number of nodes in a maximally shared representation of the formula. The fifth column contains the

Nr. file	#d	#b	#n	[16,19]	t	bt	r	br
022	59	49	993	:0.16	:13	:16	17:01	7:50
025	45	55	285	:0.2	:0.3	:0.3	:0.1	:0.1
027	21	60	569	:1.7	12:37	10:55	—	—
032	16	48	525	:0.1	:3.2	:3.2	5:02	4:12
037	12	26	942	:0.15	2:17	:2.3	7:28	:12
038	6	14	844	:0.18	:17	:0.4	:6.8	:0.3
043	158	72	1717	—	—	—	—	—
044	39	14	383	:0.1	:3.7	:2.0	0:28	:1.6
046	68	35	667	:0.13	—	—	—	—
049	163	75	1717	—	—	—	:0.3	:0.1

Fig. 3. Timing results for the benchmarks

times reported in [19], obtained by the method of [16]. The other columns show our results, using various variable orderings. Each entry is in minutes, i.e. $a : b.c$ means a minutes, and $b.c$ seconds. With — we denote that a particular instance could not be solved, due to lack of memory. The times are including the time to start the executable, I/O and transforming the benchmarks to the ATerm format. We used an IRIX machine with 300 MHz and where the processes could use up to 1.5 GB internal memory.

The table shows that we can solve 8 out of 10 formulae. In this respect our method is comparable to [16]. The exact times are not relevant, because we have made a prototype implementation, without incorporating all well-known optimizations applied in BDD-packages, whereas [19] used an existing BDD-package.

It is also clear that the variable ordering is rather important. In most cases, it is a good idea to split on boolean variables first, before splitting on equalities. The reason probably is that splitting on an equality introduces new guards, which can be rather costly. We also counted the number of iterations of TOPDOWN that were needed in order to reach an EQ-OBDD. Remarkably, the maximum number of iterations was 2 and nearly all time was spent in the first iteration. Most benchmarks even reached a fixed point in the first iteration.

We conclude that the algorithm TOPDOWN is feasible. This is quite remarkable, as the top-down method is usually regarded as inefficient. We attribute this to the use of maximal sharing and memoization. In the next standard example, it is even more effective than using APPLY.

Example 28. Consider the formula $X \equiv p \wedge (\Phi \wedge \neg p)$. In case p is the smallest variable, TOPDOWN terminates in one call, because $X|_p\downarrow \equiv \mathbf{0}$ and $X|_{\neg p}\downarrow \equiv \mathbf{0}$ and a contradiction is detected.   □

The usual APPLY algorithm will completely build the tree for Φ, potentially resulting in an exponential blow-up. Many heuristics for providing a variable ordering will make p minimal, so this is a realistic scenario. In [2] an adaptation to the original APPLY algorithm is described, which also solves this formula in constant time.

4   Future Work

Our motivation originates from investigations in the computer-aided analysis of distributed systems and protocols, where data is usually specified by algebraic data types, and automated reasoning is generally based on term rewriting. For this reason, function symbols cannot be eliminated, and the domains are generally structured and often infinite. For instance, as soon as we introduce the successor function on natural numbers, all interesting models are infinite.

Our approach forms an extendible basis. We may allow function symbols in EQ-BDDs. In the algorithm, the rewrite rules of the data domain can be added to the TRS SIMPLIFY. In this way, one is able to prove for instance that $x \leq y \vee x \neq y$ is a tautology. Obviously this is not true when the interpretation of functions is free (e.g. interpret $\leq$ as $<$). However, consider the following definition of $\leq$ in terms of rewrite rules, where S denotes the successor function:

$$x < 0 \to \mathbf{0} \qquad x < S(y) \to x \leq y \qquad x \leq y \to x < y \vee x = y$$

An EQ-OBDD proof with auxiliary rewrite rules of $x \leq y \vee x \neq y$ looks as follows:

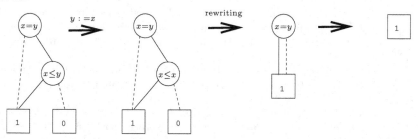

Also, $x \leq 0 \wedge y = 0 \to x = y$ can be proved in this way. Note that this doesn't hold on the integers or reals, so the logic of difference constraints [13] cannot be used here.

As future work we plan to investigate under which conditions such extensions are complete. For instance, in the example above we at least additionally need the following rules:

$$0 = S(x) \to \mathbf{0} \qquad S(x) = S(y) \to x = y \qquad x = S^n(x) \to \mathbf{0}.$$

We also plan to improve and extend our algorithm in the presence of function symbols. One of the main issues here is how to extend the ordering on the new nodes.

Acknowledgments. We like to thank Ofer Shtrichman for making his benchmarks publicly available and discussing them. We are also indebted to the anonymous referees for improving some of the proofs.

References

1. W. Ackermann. *Solvable Cases of the Decision Problem.* Studies in Logic and the Foundations of Mathematics. North-Holland, Amsterdam, 1954. 162, 163, 175
2. H. R. Andersen and H. Hulgaard. Boolean expression diagrams. In *Twelfth Annual IEEE Symposium on Logic in Computer Science,* pages 88–98, Warsaw, Poland, 1997. IEEE Computer Society. 164, 176
3. F. Baader and T. Nipkow. *Term Rewriting and All That.* Cambridge University Press, 1998. 166
4. M.G.J. van den Brand, H.A. de Jong, P. Klint, and P.A. Olivier. Efficient Annotated Terms. *Software – Practice & Experience,* 30:259–291, 2000. See also http://www.cwi.nl/projects/MetaEnv/aterm/. 163, 175
5. R.E. Bryant. Graph-based algorithms for boolean function manipulation. *IEEE Transactions on Computers,* C-35(8):677–691, 1986. 161, 163
6. R.E. Bryant. Symbolic boolean manipulation with ordered binary-decision diagrams. *ACM Computing Surveys,* 24(3):293–318, 1992. 161, 165
7. R.E. Bryant, S. German, and M.N. Velev. Exploiting positive equality in a logic of equality with uninterpreted functions. In N. Halbwachs and D Peled, editors, *Proc. of Computer Aided Verification, CAV'99,* LNCS 1633. Springer-Verlag, 1999. 161, 162, 163, 163, 164, 174
8. Burch, J.R. and D.L. Dill. Automatic verification of pipelined microprocessors control. In D. L. Dill, editor, *Proceedings of Computer Aided Verification, CAV'94,* LNCS 818, pages 68–80. Springer-Verlag, June 1994. 162, 163, 164
9. N. Dershowitz. Termination of rewriting. *Journal of Symbolic Computation,* 3(1–2):69–115, 1987. 166
10. A. Goel, K. Sajid, H. Zhou, and A. Aziz. BDD based procedures for a theory of equality with uninterpreted functions. In *Proceedings of Computer Aided Verification, CAV'98,* LNCS 1427, pages 244–255. Springer-Verlag, 1998. 161, 164, 164, 164, 174
11. J.W. Klop. Term rewriting systems. In D. Gabbay S. Abramski and T. Maibaum, editors, *Handbook of Logic in Computer Science,* volume 1. Oxford University Press, 1991. 166
12. C. Meinel and T. Theobald. *Algorithms and Data Structures in VLSI Design: OBDD-Foundations and Applications.* Springer-Verlag, 1998. 161, 163, 165, 170
13. J. Møller, J. Lichtenberg, H. R. Andersen, and H. Hulgaard. Difference decision diagrams. In J. Flum and M. Rodriguez-Artalejo, editors, *Computer Science Logic (CSL'99),* LNCS 1683. Springer-Verlag, 1999. 161, 164, 164, 164, 177
14. G. Nelson and D.C. Oppen. Fast decision procedures based on congruence closure. *Journal of the Association for Computing Machinery,* 27(2):356–364, 1980. 163
15. S. Owre, S. Rajan, J.M. Rushby, N. Shankar, and M.K. Srivas. PVS: Combining specification, proof checking, and model checking. In R. Alur and T.A. Henzinger, editors, *Proceedings of Computer Aided Verification CAV'96,* LNCS 1102, pages 411–414. Springer-Verlag, 1996. 163
16. A. Pnueli, Y. Rodeh, O. Shtrichman, and M. Siegel. Deciding equality formulas by small domains instantiations. In N. Halbwachs and D Peled, editors, *Proceedings of Computer Aided Verification, CAV'99,* LNCS 1633. Springer-Verlag, 1999. 161, 162, 163, 163, 163, 174, 175, 175, 176, 176, 176
17. J.C. van de Pol and H. Zantema. Binary decision diagrams by shared rewriting. In M. Nielsen and B. Rovan, editors, *Proceedings of Mathematical Foundations of Computer Science, MFCS'00,* LNCS 1893. Springer-Verlag, 2000. 164
18. R.E. Shostak. An algorithm for reasoning about equality. *Communications of the ACM,* 21(7):583–585, 1978. 163, 165
19. O. Shtrichman. Benchmarks for satisfiability checking of equality formulas. See http://www.wisdom.weizmann.ac.il/~ofers/sat/bench.htm, 1999. 163, 174, 175, 175, 175, 176, 176, 176

A PVS Proof Obligation Generator for Lustre Programs*

Cécile Canovas-Dumas and Paul Caspi
{cdumas,caspi}@imag.fr

Laboratoire Verimag (CNRS, UJF, INPG)
http://www-verimag.imag.fr

Abstract. This paper presents a tool for proving safety properties of Lustre programs in PVS, based on continuous induction. The tool applies off-line a repeated induction strategy and generates proof obligations left to PVS. We show on examples how it avoids some drawbacks of co-induction which needs to consider "absent elements" in the case of clocked streams.

1 Introduction

Co-induction has been advocated as providing a good theoretical framework for proving stream programs and several experiments and tools [5,16,13,9,14,7,2] have been recently designed in this setting, mostly based on Coq [6] and PVS [15]. Two main proof principles have been used, the "Bisimulation Proof Principle" originated from Park's work and the "Infinite Proof Principle" due to Coquand. However, when we tried to apply these principles to the proof of Lustre [8] programs, we found that they were less efficient than the one which arises from the old, semantic based, Kahn's theory of data-flow[11].

This observation mainly arises when considering multi-clock systems where sub-streams are extracted from streams by some filtering procedure. A way of dealing with such sub-streams within the co-inductive framework consists of introducing an "absent" element which takes the place of erased elements in the filtering process.[1] The point is that, as we show in section 5, bisimulation proofs rely on the observations (destructors of the co-inductive type) we can draw from a running process, and an "absent" element is sometimes a quite poor observation of what is actually taking place in the process. This may oblige us to add the observation of some internal state values or, equivalently, to strengthen the property we want to prove by extending it to these state values. As for infinite proofs, "absent" elements require more complex case analysis.

An alternate solution arises from the remark that stream programming, as found in Lustre, describes programs as functions from streams to streams. The

---

* This work has been partially supported by Esprit project Syrf and Inria action Presysa.
[1] This absent element is the analog of "silent" elements introduced by Milner in synchronizing trees [12].

M. Parigot and A. Voronkov (Eds.): LPAR 2000, LNAI 1955, pp. 179–188, 2000.

co-inductive point of view comes from the fact that the co-domain type of our functions (streams) is a co-inductive type. Now another possible proof principle can be derived from the recursive nature of the domain type (also streams). Induction allows us to prove properties of functions over inductive types, but these, in general only yield finite objects. This is where continuity is helpful, by bridging the gap between such finite objects and the infinite ones (streams) we deal with. When taking this point of view, we shall see that we don't anymore need "absent" elements. Thus every observation we draw from a process is meaningful and proofs get simpler and shorter.

In order to exemplify this observation, we shall first say some words of stream programming based on Lustre (section 2). Section 3 shows how to prove safety properties of Lustre programs by continuous induction and section 4 briefly describes the tool we designed for automatically generating PVS proof obligations from Lustre programs and safety properties. In section 5 we provide a tentative comparison with co-induction and finally discuss related works.

2   Lustre and Stream Programming

2.1   Kahn Semantic

The main idea of Kahn network semantic is to consider :

- The complete partial order (CPO) $(D^\infty, \leq, \epsilon)$ where $D^\infty = D^* + D^\omega$ is the set of finite and infinite sequences of some set D, with respect to the prefix order of sequences $x \leq y \Rightarrow \exists z : y = x@z$ and the empty sequence $x = \epsilon@x = x@\epsilon$, where @ is the concatenation of sequences.
  Completeness, here, amounts to the fact that every chain $C = \{x_0 \leq x_1 \ldots \leq x_n \ldots\}$ has a least upper bound $lub\ C$.
- The CPO of functions over D^∞ tuples and higher order extensions.
- Least fixed points of continuous functions, where continuity refers to lub preservation. In case of higher order functions, these fixed points are known to be continuous functions.

In this setting, continuity implies monotony which, in turn, can be interpreted as causality: if x is a prefix of y, then $(f\ x)$ is a prefix of $(f\ y)$.

2.2   Lustre Primitives

Table 1 displays the inductive definitions of Lustre primitives. In this table:

- . is the usual sequence constructor;
- any constant is lifted to infinite sequences and similarly, every operator is lifted to operate point-wise on sequences;
- ->pre is a unit delay;
- when is a filtering operator;
- current is a converse hold operator.

It is easy to show that every primitive is continuous and every composition of primitives is continuous.

$$
\begin{array}{ll}
xs + \epsilon & = \epsilon \\
\epsilon + ys & = \epsilon \\
x.xs + y.ys & = (x + y).(xs + ys) \\[6pt]
\epsilon \text{ ->pre } ys & = \epsilon \\
x.xs \text{ ->pre } \epsilon & = x.\epsilon \\
x.xs \text{ ->pre } y.ys & = x.(y.ys \text{ ->pre } ys) \\[6pt]
\epsilon \text{ when } cs & = \epsilon \\
xs \text{ when } \epsilon & = \epsilon \\
x.xs \text{ when } true.cs & = x.(xs \text{ when } cs) \\
x.xs \text{ when } false.cs & = xs \text{ when } cs \\[6pt]
\text{current } v \; \epsilon \; xs & = \epsilon \\
\text{current } v \; false.cs \; xs & = v.(\text{current } v \; cs \; xs) \\
\text{current } v \; true.cs \; \epsilon & = \epsilon \\
\text{current } v \; true.cs \; x.xs & = x.(\text{current } x \; cs \; xs)
\end{array}
$$

Table 1. Inductive definition of Lustre primitives

2.3 Lustre Programs

Then Lustre programs are sets of (mutually recursive) definitions built on these primitives. For instance, fib defines the sequence of Fibonacci numbers:

```
fib = 1 -> pre(fib +(0 -> pre fib));
```

Functions (nodes) allow definitions to be encapsulated:

```
node sum(delta, x : real) returns (y : real);
      let
            y = x*delta + (0 -> pre y);
      tel
```

in such a way that $z = \texttt{sum(2.0*delta, x when half)}$ represents an integrator operating half rate. This functional style makes Lustre a useful language for programming control and hardware systems.

2.4 Synchrony

Though not essential here, let us say a word on synchrony: it consists of rejecting, thank to some statics analysis referred to as "clock calculus", expressions like x + (x when half) whose execution requires an unbounded memory.

3 Induction and Continuity

The idea is to look at safety properties (*All* modality) as inductively defined predicates over input sequences. Yet, infinite sequences are not an algebraic type and cannot be considered as an initial algebra but, as we already noticed, they can be considered as *lubs* of finite sequence chains. If our predicates are continuous and if we prove that a property holds for every finite input sequence, it will hold by continuity on the *lubs*.

Table 2 displays the proof rules for such safety properties:

- the rules assume x, xs and xss are not free in H;
- (IND) is the ordinary induction on lists;
- (REC) is the usual fix-point rule, which also extends by continuity.

$$(EMPTY)\frac{}{H \vdash All\ \epsilon} \qquad (CONC)\frac{H \vdash x \quad H \vdash All\ xs}{H \vdash All\ x.xs}$$

$$(HYP1)\frac{}{H, All\ x.xs \vdash x} \qquad (HYP2)\frac{}{H, All\ x.xs \vdash All\ xs}$$

$$(IND)\frac{H \vdash (h\ \epsilon) \quad H,(h\ xs) \vdash (h\ x.xs)}{H \vdash.(h\ xss)} \qquad (REC)\frac{H \vdash (h\ \epsilon) \quad H,(h\ x) \vdash (h\ (e\ x))}{H \vdash h\ (rec\ x : (e\ x))}$$

Table 2. Proof rules

Example 1: We want to prove: $\forall xs, cs : All\ xs \Rightarrow All\ (\text{current}\ true\ cs\ xs)$

First step: we can choose here to induct on cs. This gives three sub-goals:

$$\frac{All\ xs}{All\ (\text{current}\ true\ \epsilon\ xs)},$$ which holds by expanding **current**, and

$$\frac{All\ xs \quad \forall xs : All\ xs \Rightarrow All\ (\text{current}\ true\ cs\ xs)}{All\ (\text{current}\ true\ false.cs\ xs)},$$ which holds by expanding **current** and using the induction hypothesis, and

$$\frac{All\ xs \quad \forall xs : All\ xs \Rightarrow All\ (\text{current}\ true\ cs\ xs)}{All\ (\text{current}\ true\ true.cs\ xs)},$$ which yields, by expanding **current** and remarking that the $xs = \epsilon$ case holds directly:

$$\frac{All\ x.xs \quad \forall xs : All\ xs \Rightarrow All\ (\text{current}\ true\ cs\ xs)}{All\ (\text{current}\ x\ cs\ xs)}$$

General step: at this point, we have got rid of initialization and deal with the general case which can be restated as
$\forall x, xs : All\ x.xs \Rightarrow All\ (\text{current}\ x\ cs\ xs)$

By induction on cs, we get three new goals; the ϵ and $false.cs$ work as previously, and the last one leads to:

All x.xs

$$\frac{\forall x, xs : All\ x.xs \Rightarrow All\ (\text{current}\ x\ cs\ xs)}{All\ (\text{current}\ x\ true.cs\ xs)}.$$ The $xs = \epsilon$ case holds directly and

the $xs = x'.xs'$ case holds by instantiating the induction hypothesis.

We can see here that we have exactly examined four cases at each step, the initial step and the general case step and this is clearly the minimum number of cases needed to prove the property. Furthermore, the cases involving ϵ are obvious and yield direct proofs.

4   A Proof Obligation Generator for Lustre Programs

Initially, we defined Kahn semantic in PVS[15], but trying to directly use this theory through PVS strategies is rather inefficient.[2] So we decided to unfold the strategies off-line and to only leave to PVS the remaining proof obligations.

For instance a proof goal like $\dfrac{H}{All\ x.xs}$ will generate the proof obligation $\dfrac{H}{x}$ left to PVS, while the remaining goal $\dfrac{H}{All\ xs}$ will be analyzed off-line. Furthermore ϵ-like goals will be in general discharged within the tool and do not generate proof obligation.

A heuristic strategy has been tried, which seems to provide sensible results; it consists of:

- First, apply the continuity rule until there is no more recursive definitions.
- Then, eliminate initial values. This corresponds to the first step of example 1.
- Finally, deal with the general case.

It should be noted that the initial value elimination step may require unfolding several times the property, depending on the number of initial values to eliminate, until the induction hypothesis can be used. This allows us to tune the number of unfoldings to the needs of the property to be proved. This is, for instance, the case when trying to prove the property:

Example 2 All(fib ≥ 0)

Here, our initial value elimination strategy provides us with the right number of unfoldings needed to prove the property.

A translation tool, written in Caml, performs the unfolding of the strategy, starting from a Lustre program and a given *All* property, and generates PVS proof obligations. Table 3 displays the resulting PVS proof obligations of example 2. It is easy to see from this example that the proof obligations are obvious and can be discharged with a straightforward PVS strategy.

On the contrary, the strategy fails to provide proof obligations when dealing with non synchronous problems like:

---

[2] Just think of writing a strategy for automatically obtaining the unfolding of example 1.

```
example2_ex2_node : THEORY
 BEGIN
 IMPORTING streams
fib : VAR stream[int]

ex2_prop1_0 : THEOREM
(    (hd(fib)   >= 0)
) => (1 >= 0)

ex2_prop1_00 : THEOREM
(    (hd(tl(fib)))   >= 0)
AND (hd(fib)   >= 0)
) => ((hd(fib)   + 0) >= 0)

ex2_prop1_000 : THEOREM
(    (hd(tl(tl(fib)))   >= 0)
AND (hd(tl(fib))   >= 0)
AND (hd(fib)   >= 0)
) => ((hd(tl(fib))   + hd(fib) ) >= 0)

    END example2_ex2_node
```

Table 3. PVS proof obligations for example 2

Example 3: $All(x \geq 0) \Rightarrow All(x + (x \ when \ c) \geq 0)$

The problem, here, is that the induction hypotheses never allow the remaining goals to be discharged because there is always an extra tl which appears in the goal and not in the hypothesis. In this sense, our strategy looks very much like Wadler's deforestation [19] in the particular context of proof generation.

5   Co-induction

5.1   Processes and Co-inductive Definitions

Processes (co-algebras), (X, D, f, g) are defined by:

- a set of states X and a set of values D;
- a transition function f from X to X;
- an output function g from X to D;

Then, it can be shown [10] that (D^ω, D, hd, tl) is a final co-algebra in the sense that there exists an unique (behavior) function h from X to D^ω such that, for all x in X, $hd(h \ x) = g \ x$ and $tl(h \ x) = h(f \ x)$ where hd, tl are the usual sequence destructors. Equivalently, $h \ x = (g \ x).h(f \ x)$ (Here the uniqueness of h comes simply from the fact that, for any x and n, $h \ x \ n = g(f^n \ x)$).

Such a definition is said co-inductive. Uniqueness gives sense to recursively defined functions, such as $rec \ h : \lambda \ x : (g \ x).h(f \ x)$ provided their co-domains lie within some final co-algebra.

Bisimulation Proofs are based on the observation of the property truth value, i.e., $D = \{false, true\}$. Then an always true property is such that the observation yields the sequence $true^\omega$. If we want to prove that the property holds for some x_0 in X, we must find some property P such that $P(x)$ implies $g(x)$ and $P(f(x))$ and then prove that $P(x_0)$ holds. A "natural" choice for P can be to choose $g^{-1}(true)$.

Infinite Proofs correspond to the `cofix` tactic in Coq and avoid the drawback of having to choose a property. They are based on quite the same rules as induction rules of table 2 but for the ϵ cases which can be discarded. In this sense, co-induction can be seen as simpler than induction.

5.2   The Clocked Stream Case

The idea here, like in [4,14], is to replace filtered values by an absent value a. Table 4 displays the expression of Lustre primitives in this setting.[3]

$$
\begin{array}{ll}
a.xs + a.ys & = a.(xs + ys) \\
x.xs + a.ys & = a.(x.xs + ys) \\
a.xs + y.ys & = a.(xs + y.ys) \\
x.xs + y.ys & = (x + y).(xs + ys) \\
\\
a.xs \text{ ->pre } ys & = a.(xs \text{ ->pre } ys) \\
x.xs \text{ ->pre } a.ys & = x.(a.ys \text{ ->pre } ys) \\
x.xs \text{ ->pre } y.ys & = x.(y.ys \text{ ->pre } ys) \\
\\
a.xs \text{ when } a.cs & = a.(xs \text{ when } cs) \\
a.xs \text{ when } c.cs & = a.(xs \text{ when } c.cs) \\
x.xs \text{ when } a.cs & = a.(x.xs \text{ when } cs) \\
x.xs \text{ when } true.cs & = x.(xs \text{ when } cs) \\
x.xs \text{ when } false.cs & = a.(xs \text{ when } cs) \\
\\
\text{current } v\ a.cs\ a.xs & = a.(\text{current } v\ cs\ xs) \\
\text{current } v\ a.cs\ x.xs & = a.(\text{current } v\ cs\ x.xs) \\
\text{current } v\ false.cs\ xs & = v.(\text{current } v\ cs\ xs) \\
\text{current } v\ true.cs\ a.xs & = a.(\text{current } v\ true.cs\ xs) \\
\text{current } v\ true.cs\ x.xs & = x.(\text{current } x\ cs\ xs)
\end{array}
$$

Table 4. Co-inductive definition of Lustre primitives

Observing properties of such processes leads now to consider D as a three valued set: $\{a, false, true\}$, and we have to decide what is a good observation;

---

[3] This table could have been made simpler by merging cases. However the given expressions easily distinguish synchronous and non synchronous cases.

we can decide to say that a property is always true if the observation belongs to $\{a, true\}^{\omega}$. This consists of saying that a property holds as long as its truth value is not *false*. Then the previous proof rules remain valid.

Example 1: We want to prove that the function *current true c x* is always true as soon as x is always true, i.e. belongs to $\{a, true\}^{\omega}$.

Here, the bisimulation rule does not work:

- the initial proof is obvious by case analysis because, initially, v and x evaluate to *true*;
- but the induction step does not hold; the observation of the absent value a does not tell us anything on the stored value v that can be observed at the next step and v can be *false*. Notice that cases 2 and 4 could have been eliminated by some clock analysis (they are not synchronous), but the first one is synchronous and cannot be eliminated.

A solution in this case consists of strengthening the property by adding an observation of the stored internal state value v. However, intuitively, we can see that this would not have been necessary if we had used the definitions of section 2.2: here, the observed value always tells us which is the stored value.

Moreover, we can see that the definitions of table 1 are always simpler than the ones of table 4. Thus, infinite proofs are more complex than inductive ones.

5.3   Comparison with Induction

We can show that continuous induction is as powerful a proof principle as co-induction.

Theorem 1. *Continuous induction generates the same proof obligations as bisimulation.*

We have seen that the bisimulation proof of

$$All \ (rec \ h : \lambda \ x : (g \ x).h(f \ x))x_0$$

yields the proof obligations $\dfrac{}{(g \ x_0)}$ and $\dfrac{(g \ x)}{(g \ (f \ x))}$. By coding this property as:

$$All \ (\mathtt{map}(g) \ (rec \ y : x_0.(\mathtt{map}(f) \ y)))$$

and by applying the unfolding strategy of the paper, we easily obtain the same proof obligations.

The comparison with infinite proofs is even easier as both are base on almost the same rules.

6   Related Works

Continuous induction has been used for long in proving programs [17]. Yet, it seems that the first attempt to use it for proving properties on streams goes back to Ashroft and Wadge [3], and our work can be seen as simply an update of this one. However, it is devoted to the Lucid context which was somewhat different from the Kahn-Lustre one. Furthermore the proof of when and current based programs was not addressed while it constitutes a major motivation here. More recently, Pavlović [18] describes an approach which seems quite similar to ours, but much more generally presented in a categorical framework that makes it quite difficult to understand. Furthermore it does not seem to draw the practical consequences we have drawn from the comparison between continuous induction and co-induction.

7   Conclusion

The problem of dealing with clocks when proving stream programs leads either to consider "absent" elements and apply usual co-induction or to consider both finite and infinite sequences.[4] In the later case, continuous induction provides us with simple and efficient proof schemes and seems a better choice.

Yet, formalizing it within the PVS framework appeared quite inefficient. This is why we chose to run these strategies off-line, in the translation tool itself, and to only generate PVS proof obligations, that is to say, what remains to prove, once the unfolding of inference rules has been performed.

Then, this tool appears very efficient: it allows us to handle multiple clocks in a functional style, it is fast and it often finds the right number of unfoldings which allows a property to be proved.

Furthermore, the framework would offer the possibility of dealing with asynchronous and possibly unbounded Kahn networks, and thus the possibility to model and prove mixed synchronous-asynchronous designs. However, in this case, proofs are more difficult to obtain because the generalization step may not converge. Then, this generalization should be provided manually, in the same way as when some invariant strengthening is needed. But this has still to be developed.

Several other points remain to be studied:

- Experiments have only dealt with safety properties. Liveness ones have to be addressed.
- Only flat programs have been studied. This is likely to only apply to small programs and raises the question of addressing large, structured ones.
- One way of answering this last question could be to study some refinement-based design method à la B[1], adapted to the synchronous programming context. In this sense, a proof obligation generator like the one described here could be a good starting point.

---

[4] Co-induction on finite and infinite sequences also requires absent elements.

References

1. J.-R. Abrial. *The B-Book*. Cambridge University Press, 1995. 187
2. R. Amadio and S.Coupet-Grimal. Analysis of a guard condition in type theory. In M.Nivat, editor, *Foundations of Software Science and Computation Structures*, volume 1378 of *Lecture Notes in Computer Science*. Springer Verlag, 1998. 179
3. E.A. Ashcroft and W.W. Wadge. Lucid, a formal system for writing and proving programs. *SIAM j. Comp.*, 3:336–354, 1976. 187
4. P. Caspi and M. Pouzet. A co-iterative characterization of synchronous stream functions. In *Proceedings of the Workshop on Coalgebraic Methods in Computer Science, Lisbon*, volume 11 of *Electronic Notes in Theoretical Computer Science*. Elsevier, 1998. 185
5. Th. Coquand. Infinite objects in type theory. In *Types for Proofs and Programs*, volume 806 of *Lecture Notes in Computer Science*. Springer Verlag, 1993. 179
6. Th. Coquand and G. Huet. The calculus of construction. *Information and Computation*, 76(2), 1988. 179
7. E. Gimenez. Codifying guarded definitions with recursive schemes. In *Types for Proofs and Programs, TYPES'94*, volume 996 of *Lecture Notes in Computer Science*. Springer Verlag, 1995. 179
8. N. Halbwachs, P. Caspi, P. Raymond, and D. Pilaud. The synchronous dataflow programming language LUSTRE. *Proceedings of the IEEE*, 79(9):1305–1320, September 1991. 179
9. U. Hensel and B. Jacobs. Coalgebraic theories of sequences in PVS. Technical Report CSI-R9708, Computer Science Institute, University of Nijmegen, 1997. 179
10. B. Jacobs and J. Rutten. A tutorial on (co)algebras and (co)induction. *Bulletin of EATCS*, 62:229–259, 1997. 184
11. G. Kahn. The semantics of a simple language for parallel programming. In *IFIP 74*. North Holland, 1974. 179
12. R. Milner. *A Calculus of Communicating Systems*, volume 92 of *Lecture Notes in Computer Science*. Springer Verlag, 1980. 179
13. P.S. Miner and S.D. Johnson. Verification of an optimized fault-tolerant clock synchronization circuit. In *Designing Correct Circuits*, Electronic Workshops in Computing, Bastad, Sweden, 1996. Springer-Verlag. 179
14. D. Nowak, J.R. Beauvais, and J.P. Talpin. Co-inductive axiomatization of a synchronous language. In *Theorem Proving in Higher Order Logics*, volume 1479 of *Lecture Notes in Computer Science*, pages 387–399. Springer Verlag, 1998. 179, 185
15. S. Owre, J. Rushby, and N. Shankar. PVS: a prototype verification system. In *11th Conf. on Automated Deduction*, volume 607 of *Lecture Notes in Computer Science*, pages 748–752. Springer Verlag, 1992. 179, 183
16. C. Paulin-Mohring. Circuits as streams in Coq, verification of a sequential multiplier. Research Report 95-16, Laboratoire de l'Informatique du Parallélisme, September 1995. 179
17. L. Paulson. *Logic and Computation, Interactive Proof with Cambridge LCF*. Cambridge University Press, 1987. 187
18. D. Pavlović. Guarded induction on final coalgebras. In *Proceedings of the Workshop on Coalgebraic Methods in Computer Science, Lisbon*, volume 11 of *Electronic Notes in Theoretical Computer Science*, 1998. 187
19. P. Wadler. Deforestation: transforming programs to eliminate trees. *Theoretical Computer Science*, 73:231–248, 1990. 184

Efficient Structural Information Analysis
for Real CLP Languages[*]

Roberto Bagnara[1], Patricia M. Hill[2], and Enea Zaffanella[1]

[1] Department of Mathematics, University of Parma, Italy.
{bagnara,zaffanella}@cs.unipr.it
[2] School of Computing, University of Leeds, U. K.
hill@comp.leeds.ac.uk

Abstract. We present the rational construction of a generic domain for structural information analysis of CLP languages called Pattern($\mathcal{D}^\sharp$), where the parameter $\mathcal{D}^\sharp$ is an abstract domain satisfying certain properties. Our domain builds on the parameterized domain for the analysis of logic programs Pat($\Re$), which is due to Cortesi et al. However, the formalization of our CLP abstract domain is independent from specific implementation techniques: Pat($\Re$) (suitably extended in order to deal with CLP systems omitting the occur-check) is one of the possible implementations. Reasoning at a higher level of abstraction we are able to appeal to familiar notions of unification theory. This higher level of abstraction also gives considerable more latitude for the implementer. Indeed, as demonstrated by the results summarized here, an analyzer that incorporates structural information analysis based on our approach can be highly competitive both from the precision *and*, contrary to popular belief, from the efficiency point of view.

1 Introduction

Most interesting CLP languages [16] offer a constraint domain that is an amalgamation of a domain of syntactic trees — like the classical domain of finite trees (also called the *Herbrand* domain) or the domain of rational trees [9] — with a set of "non-syntactic" domains, like finite domains, the domain of rational numbers and so forth. The inclusion of uninterpreted functors is essential for preserving Prolog programming techniques. Moreover, the availability of syntactic constraints greatly contributes to the expressive power of the overall language. When syntactic structures can be used to build aggregates of interpreted terms one can express, for instance, "records" or "unbounded containers" of numerical quantities.

From the experience gained with the first prototype version of the CHINA data-flow analyzer [1] it was clear that, in order to attain a significant precision

---

[*] This work has been partly supported by MURST project "Certificazione automatica di programmi mediante interpretazione astratta." Some of this work was done during a visit of the first and third authors to Leeds, funded by EPSRC under grant M05645.

M. Parigot and A. Voronkov (Eds.): LPAR 2000, LNAI 1955, pp. 189–208, 2000.

in the analysis of numerical constraints in CLP languages, one must keep at least part of the uninterpreted terms in concrete form. Note that almost any analysis is more precise when this kind of structural information is retained to some extent: in the case mentioned here the precision loss was just particularly acute. Of course, structural information is very valuable in itself. When exploited for optimized compilation it allows for enhanced clause indexing and simplified unification. Moreover, several program verification techniques are highly dependent on this kind of information.

Cortesi et al. [10,11], after the work of Musumbu [21], put forward a very nice proposal for dealing with structural information in the analysis of logic programs. Using their terminology, they defined a generic abstract domain $\text{Pat}(\Re)$ that automatically upgrades a domain $\Re$ (which must support a certain set of elementary operations) with structural information.

As far as the overall approach is concerned, we extend the work described in [11] by allowing for the analysis of any CLP language [16]. Most importantly, we do *not* assume that the analyzed language performs the *occur-check* in the unification procedure. This is an important contribution, since the vast majority of *real* (i.e., implemented) CLP languages (in particular, almost all Prolog systems) do omit the occur-check, either as a mere efficiency measure or because they are based upon a theory of extended rational trees [9]. We describe a generic construction for structural analysis of CLP languages. Given an abstract domain $\mathcal{D}^\sharp$ satisfying a small set of very reasonable and weak properties, the structural abstract domain $\text{Pattern}(\mathcal{D}^\sharp)$ is obtained automatically by means of this construction. In contrast to [11], where the authors define a specific implementation of the generic structural domain (e.g., of the representation of term-tuples), the formalization of $\text{Pattern}(\cdot)$ is implementation-independent: $\text{Pat}(\Re)$ (suitably extended in order to deal with CLP languages and with the occur-check problem) is a possible base for the implementation. Reasoning at a higher level of abstraction we are able to appeal to familiar notions of unification theory [18]. One advantage is that we can identify an important parameter (a common anti-instance function) that gives some control over the precision and computational cost of the resulting structural domain. In addition, we believe our implementation-independent treatment can be more easily adapted to different analysis frameworks/systems.

One of the merits of $\text{Pat}(\Re)$ is to define a generic implementation that works on any domain $\Re$ that provides a certain set of elementary, fine-grained operations. Because of the simplicity of these operations it is particularly easy to extend an existing domain in order to accommodate them. However, this simplicity has a high cost in terms of efficiency: the execution of many isolated small operations over the underlying domain is much more expensive than performing few macro-operations where global effects can be taken into account. The operations that the underlying domain must provide are thus more complicated in our approach. However, this extra complication and the higher level of abstraction give considerable more latitude for the implementer. Indeed, as demonstrated by the results summarized here, an analyzer that incorporates structural information analysis based on our approach can be highly competitive both from the precision *and* the efficiency point of view. One of the contributions of this paper is that it disproves the common belief (now reinforced by [8]) whereby abstract domains enhanced with structural information are inherently inefficient.

The paper is structured as follows: Section 2 introduces some basic concepts and the notation that will be used in the paper; Section 3 presents the main ideas behind the tracking of explicit structural information for the analysis of CLP languages; Section 4 introduces the $\mathcal{D}^\sharp$ and Pattern($\mathcal{D}^\sharp$) domains and explains how an abstract semantics based on $\mathcal{D}^\sharp$ can systematically be upgraded to one on Pattern($\mathcal{D}^\sharp$); Section 5 summarizes the extensive experimental evaluation that has been conducted to validate the ideas presented in this paper; Section 6 presents a brief discussion of related work and, finally, Section 7 concludes with some final remarks.

2 Preliminaries

Let U be a set. The cardinality of U is denoted by $|U|$. We will denote by U^n the set of n-tuples of elements drawn from U, whereas U^* denotes $\bigcup_{n\in\mathbb{N}} U^n$. Elements of U^* will be referred to as *tuples* or as *sequences*. The *empty sequence*, i.e., the only element of U^0, is denoted by ε. Throughout the paper all variables denoting sequences will be written with a "bar accent" like in $\bar{s}$. For $\bar{s} \in U^*$, the *length* of $\bar{s}$ will be denoted by $|\bar{s}|$. The concatenation of the sequences $\bar{s}_1, \bar{s}_2 \in U^*$ is denoted by $\bar{s}_1 :: \bar{s}_2$. For each $\bar{s} \in U^*$ and each set $X \in \wp_{\mathrm{f}}(U)$, the sequence $\bar{s} \setminus X$ is obtained by removing from $\bar{s}$ all the elements that appear in X. The *projection mappings* $\pi_i : U^n \to U$ are defined, for $i = 1, \ldots, n$, by $\pi_i\big((e_1, \ldots, e_n)\big) = e_i$. We will also use the liftings $\pi_i : \wp(U^n) \to \wp(U)$ given by $\pi_i(S) = \big\{\, \pi_i(\bar{s}) \mid \bar{s} \in S \,\big\}$. If a sequence $\bar{s}$ is such that $|\bar{s}| \geq i$, we let $\mathrm{prefix}_i(\bar{s})$ denote the sequence of the first i elements of $\bar{s}$.

Let *Vars* denote a denumerable and totally ordered set of variable symbols. We assume that *Vars* contains (among others) two infinite, disjoint subsets: $\mathbf{z}$ and $\mathbf{z}'$. Since *Vars* is totally ordered, $\mathbf{z}$ and $\mathbf{z}'$ are as well. Thus we assume $\mathbf{z} = (Z_1, Z_2, Z_3, \ldots$ and $\mathbf{z}' = (Z_1', Z_2', Z_3', \ldots$. If $W \subseteq Vars$ we will denote by T_W the set of terms with variables in W. For any term or a tuple of terms t we will denote the set of variables occurring in t by $vars(t)$. We will also denote by $vseq(t)$ the sequence of first occurrences of variables that are found on a depth-first, left-to-right traversal of t. For instance, $vseq\big((f(g(X), Y), h(X))\big) = (X, Y)$.

We implement the "renaming apart" mechanism by making use of two strong normal forms for tuples of terms. Specifically, the set of n-*tuples in* $\mathbf{z}$-*form* is given by $\mathbf{T}_{\mathbf{z}}^n = \big\{\, \bar{t} \in T_{Vars}^n \mid vseq(\bar{t}) = \big(Z_1, Z_2, \ldots, Z_{|vars(\bar{t})|}\big) \,\big\}$. The set of all the tuples in $\mathbf{z}$-form is denoted by $\mathbf{T}_{\mathbf{z}}^*$. The definitions for $\mathbf{T}_{\mathbf{z}'}^n$ and $\mathbf{T}_{\mathbf{z}'}^*$ are obtained in a similar way, by replacing $\mathbf{z}$ with $\mathbf{z}'$. There is a useful device for toggling between $\mathbf{z}$- and $\mathbf{z}'$-forms. Let $\bar{t} \in \mathbf{T}_{\mathbf{z}}^n \cup \mathbf{T}_{\mathbf{z}'}^n$ and $\big|vars(\bar{t})\big| = m$. Then $\bar{t}' = \bar{t}[Z_1'/Z_1, \ldots, Z_m'/Z_m]$, if $\bar{t} \in \mathbf{T}_{\mathbf{z}}^n$, and $\bar{t}[Z_1/Z_1', \ldots, Z_m/Z_m']$, if $\bar{t} \in \mathbf{T}_{\mathbf{z}'}^n$. Notice that $\bar{t}'' = \big(\bar{t}'\big)' = \bar{t}$.

When $\bar{V} \in \mathit{Vars}^m$ and $\bar{t} \in \mathcal{T}_{\mathit{Vars}}^m$ we use $[\bar{t}/\bar{V}]$ as a shorthand for the substitution $[\pi_1(\bar{t})/\pi_1(\bar{V}), \ldots, \pi_m(\bar{t})/\pi_m(\bar{V})]$, if $m > 0$, and to denote the empty substitution if $m = 0$. If $\mathit{vars}(\bar{t}) \cap \bar{V} = \varnothing$, then $[\bar{t}/\bar{V}]$ is *idempotent*. Suppose that $\bar{s} = (s_1, \ldots, s_m) \in \mathcal{T}_{\mathit{Vars}}^m$ and $\bar{t} = (t_1, \ldots, t_m) \in \mathcal{T}_{\mathit{Vars}}^m$, then, $\bar{s} = \bar{t}$ denotes $(s_1 = t_1, \ldots, s_m = t_m)$. It is also useful to sometimes regard a substitution $[\bar{t}/\bar{V}]$ as the finite set of equations $\bar{V} = \bar{t}$. A couple of observations are useful for what follows. If $\bar{s} \in \mathbf{T}_{\mathbf{z}}^*$ and $\bar{u} \in \mathbf{T}_{\mathbf{z}}^{|\mathit{vars}(\bar{s})|}$, then $\bar{s}'[\bar{u}/\mathit{vseq}(\bar{s}')] \in \mathbf{T}_{\mathbf{z}}^*$. Moreover $\mathit{vseq}(\bar{s}'[\bar{u}/\mathit{vseq}(\bar{s}')]) = \mathit{vseq}(\bar{u})$.

The logical theory underlying a CLP constraint system [16] is denoted by $\mathfrak{T}$. To simplify the notation, we drop the outermost universal quantifiers from (closed) formulas so that if F is a formula with free variables $\bar{Z}$, then we write $\mathfrak{T} \models F$ to denote the expression $\mathfrak{T} \models \forall \bar{Z} : F$.

The notation $f \colon A \rightarrowtail B$ signifies that f is a *partial* function from A to B.

3 Making the Herbrand Information Explicit

A quite general picture for the analysis of a CLP language is as follows. We want to describe a (possibly infinite) set of constraint stores over a tuple of *variables of interest* $\bar{V} = (V_1, \ldots, V_k)$. Each constraint store can be represented, at some level of abstraction, by a formula of the kind $\exists_\Delta \ . \ ((\bar{V} = \bar{t}) \land C)$, where $(\bar{V} = \bar{t})$, with $\bar{t} \in \mathcal{T}_{\mathit{Vars}}^k$, is a system of Herbrand equations in solved form, $C \in \mathcal{C}^\flat$ is a constraint on the concrete constraint domain $\mathcal{C}^\flat$, and the set $\Delta = \mathit{vars}(C) \cup \mathit{vars}(\bar{t})$ is such that $\Delta \cap \bar{V} = \varnothing$. Roughly speaking, C limits the values that the quantified variables occurring in $\bar{t}$ can take. Notice that this treatment does not exclude the possibility of dealing with domains of rational trees: the non-Herbrand constraints will simply live in the constraint component. For example, the constraint store resulting from execution of the SICStus goal '?- X = f(a, X)' may be captured by $\exists X \ . \ (\{V_1 = X\} \land X = f(a, X))$ but also by $\exists X \ . \ (\{V_1 = f(a, X)\} \land X = f(a, X))$.

Once variables $\bar{V}$ have been fixed, the Herbrand part of the constraint store can be represented as a k-tuple of terms. We are thus assuming a concrete domain where the Herbrand information is explicit and other kinds of information are captured by some given constraint domain $\mathcal{C}^\flat$. For instance, if the target language of the analysis is $\mathrm{CLP}(\mathcal{R})$ [17], $\mathcal{C}^\flat$ may encode conjunctions of equations and inequations over arithmetic expressions, the mechanisms for delaying non-linear constraints, and other peculiarities of the arithmetic part of the language. We assume constraints are modeled by logical formulas, so that it makes sense to talk about the *free variables* of $C^\flat \in \mathcal{C}^\flat$, denoted by $FV(C^\flat)$. These are the variables that the constraint solver makes visible to the Herbrand engine, all the other variables being restricted in scope to the solver itself. Since we want to characterize any set of constraint stores, our concrete domain is

$$\mathcal{D}^\flat = \bigcup_{n \in \mathbb{N}} \wp \Big(\big\{ \, (\bar{s}, C^\flat) \mid \bar{s} \in \mathbf{T}_{\mathbf{z}}^n, C^\flat \in \mathcal{C}^\flat, FV(C^\flat) \subseteq \mathit{vars}(\bar{s}) \, \big\} \Big)$$

partially ordered by subset inclusion.

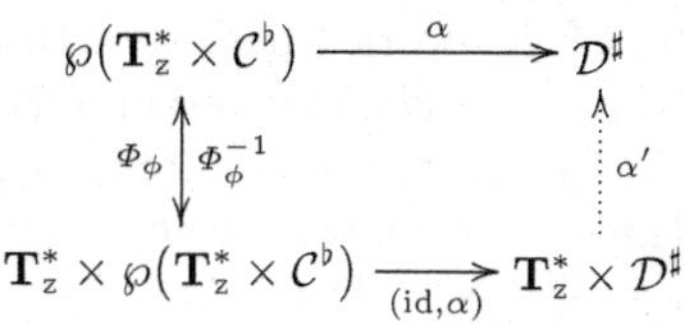

$$\begin{array}{ccc} \wp\!\left(\mathbf{T}^*_z \times \mathcal{C}^b\right) & \xrightarrow{\ \ \alpha\ \ } & \mathcal{D}^\sharp \\[2pt] {\scriptstyle \Phi_\phi}\Big\uparrow{\scriptstyle \Phi_\phi^{-1}} & & \Big\uparrow{\scriptstyle \alpha'} \\[2pt] \mathbf{T}^*_z \times \wp\!\left(\mathbf{T}^*_z \times \mathcal{C}^b\right) & \xrightarrow[(\mathrm{id},\alpha)]{} & \mathbf{T}^*_z \times \mathcal{D}^\sharp \end{array}$$

Fig. 1. Upgrading a domain with structural information.

An abstract interpretation [12] of $\mathcal{D}^b$ can be specified by choosing an abstract domain $\mathcal{D}^\sharp$ and a suitable abstraction function $\alpha\colon \mathcal{D}^b \to \mathcal{D}^\sharp$. If $\mathcal{D}^\sharp$ is not able to encode enough structural information from $\mathcal{C}^b$ so as to achieve the desired precision, it is possible to improve the situation by keeping some Herbrand information explicit. One way of doing that is to perform a change of representation for $\mathcal{D}^b$ and use the new representation as the basis for abstraction. The new representation is obtained by factoring out some common Herbrand information. The meaning of 'some' is encoded by a function.

Definition 1. (Common anti-instance function.) *For each $n \in \mathbb{N}$, a function $\phi\colon \wp(\mathbf{T}^n_{\mathbf{z}}) \to \mathbf{T}^n_{\mathbf{z}'}$ is called a* common anti-instance function *if and only if the following holds: whenever $T \in \wp(\mathbf{T}^n_{\mathbf{z}})$, if $\phi(T) = \bar{r}'$ and $\big|vars(\bar{r})\big| = m$ with $m \geq 0$, then $\forall \bar{t} \in T\,:\, \exists \bar{u} \in \mathbf{T}^m_{\mathbf{z}}\,.\, \bar{r}'\big[\bar{u}/vseq(\bar{r}')\big] = \bar{t}$. In words, $\phi(T)$ is an anti-instance [18], in $\mathbf{z}'$-form, of each $\bar{t} \in T$.*

Any choice of ϕ induces a function $\Phi_\phi\colon \mathcal{D}^b \to \mathbf{T}^*_{\mathbf{z}} \times \mathcal{D}^b$, which is given, for each $E^b \in \mathcal{D}^b$, by $\Phi_\phi(E^b) = \big(\bar{s},\, \{\, (\bar{u}, G^b) \mid (\bar{t}, G^b) \in E^b,\, \bar{s}'\big[\bar{u}/vseq(\bar{s}')\big] = \bar{t}\,\}\big)$, where $\bar{s}' = \phi\big(\pi_1(E^b)\big)$. The corestriction to the image of Φ_ϕ, that is the function $\Phi_\phi\colon \mathcal{D}^b \to \Phi_\phi(\mathcal{D}^b)$, is an isomorphism, the inverse being given, for each $F^b \in \mathcal{D}^b$, by $\Phi_\phi^{-1}\big((\bar{s}, F^b)\big) = \{\, \big(\bar{s}'\big[\bar{u}/vseq(\bar{s}')\big], G^b\big) \mid (\bar{u}, G^b) \in F^b\,\}$.

So far, we have just chosen a different representation for $\mathcal{D}^b$, that is $\Phi_\phi(\mathcal{D}^b)$. The idea behind structural information analysis is to leave the first component of the new representation (the *pattern component*) untouched, while abstracting the second component by means of α, as illustrated in Figure 1. The dotted arrow indicates a *residual abstraction function* α'. As we will see in Section 4.2, such a function is implicitly required in order to define an important operation over the new abstract domain $\mathbf{T}^*_{\mathbf{z}} \times \mathcal{D}^\sharp$. Notice that, in general, α' does not make the diagram of Figure 1 commute.

This approach has several advantages. First, factoring out common structural information improves the analysis precision, since part of the approximated k-tuples of terms is recorded, *in concrete form*, into the first component of $\mathbf{T}^*_{\mathbf{z}} \times \mathcal{D}^\sharp$. Secondly, the above construction is adjustable by means of the parameter ϕ. The most precise choice consists in taking ϕ to be a *least common anti-instance* (lca) function. For example, the set $E^b = \{\langle(s(0), Z_1), C_1\rangle, \langle(s(s(0)), Z_1), C_2\rangle\}$, is mapped onto $\Phi_{\mathrm{lca}}(E^b) = \big((s(Z_1), Z_2), \{\langle(0, Z_1), C_1\rangle, \langle(s(0), Z_1), C_2\rangle\}\big)$, where $C_1, C_2 \in \mathcal{C}^b$. At the other end of the spectrum is the possibility of choosing ϕ so that it returns a k-tuple of distinct variables for each set of k-tuples of terms. This corresponds to a framework where structural information is simply discarded. With this choice, $E^\flat$ would be mapped onto $\left((Z_1, Z_2), E^\flat\right)$. In-between these two extremes there are a number of possibilities that help to manage the complexity/precision tradeoff. The tuples returned by ϕ can be limited in *depth*, for instance. Another possibility is to limit them in *size*, that is, limiting the number of occurrences of symbols or the number of variables. This flexibility enables the analysis' domains to be designed without considering the structural information: the problem for the domain designers is to approximate the elements of $\wp\left(\mathbf{T}_\mathbf{z}^k \times \mathcal{C}^\flat\right)$ with respect to the property of interest. It does not really matter whether k is fixed by the arity of a predicate or k is the number of variables occurring in a pattern.

4   Parametric Structural Information Analysis

In this section we describe how a complete abstract semantics — which includes an abstract domain plus all the operations needed to approximate the concrete semantics — can be turned into one keeping track of structural information.

We first need some assumptions on the domain $\mathcal{C}^\flat$, which represents the non-Herbrand part of constraint stores. Following [14], it is not at all restrictive to assume that, in order to define the concrete semantics of programs, four operations over $\mathcal{C}^\flat$ need to be characterized. These model the constraint accumulation process, parameter passing, projection, and renaming apart (see also [1,2] on this subject).

Constraint accumulation is modeled by the binary operator '$\otimes$': $\mathcal{C}^\flat \times \mathcal{C}^\flat \to \mathcal{C}^\flat$ and the unsatisfiability condition in the constraint solver is modeled by the special value $\perp^\flat \in \mathcal{C}^\flat$. Notice that, while '$\otimes$' may be reasonably expected to satisfy certain properties, such as $\forall C^\flat \in \mathcal{C}^\flat : \perp^\flat \otimes C^\flat = \perp^\flat$, these are not really required for what follows. The same applies to all the other operators we will introduce: only properties that are actually used will be singled out.

Parameter passing requires, roughly speaking, the ability of adding equality constraints to a constraint store. Notice that we assume $\mathcal{C}^\flat$ and its operations encode both the proper *constraint solver* and the so called *interface* between the *Herbrand engine* and the solver [16]. In particular, the interface is responsible for *type-checking* of the equations it receives. For example in CLP($\mathcal{R}$) the interface is responsible for the fact that $X = a$ cannot be consistently added to a constraint store where X was previously classified as numeric.

Another ingredient for defining the concrete semantics of any CLP system is the projection of a satisfiable constraint store onto a set of variables. This is modeled by the family of operators $\left\{ \P_\Delta^\flat : \mathcal{C}^\flat \to \mathcal{C}^\flat \mid \Delta \in \wp_\mathrm{f}(Vars) \right\}$. If Δ is a finite set of variables and $C^\flat \in \mathcal{C}^\flat$ represents a satisfiable constraint store (i.e., $C^\flat \neq \perp^\flat$), then $\P_\Delta^\flat C^\flat$ represents the projection of $C^\flat$ onto the variables in Δ.

For each $\bar{s}, \bar{t} \in \mathbf{T}_\mathbf{z}^*$, we write $\varrho_{\bar{s}}(\bar{t})$ (read "rename $\bar{t}$ away from $\bar{s}$") to denote $\bar{t}[Z_{n+1}/Z_1, \ldots, Z_{n+m}/Z_m]$, where $n = \left|vars(\bar{s})\right|$ and $m = \left|vars(\bar{t})\right|$. The ϱ operator is useful for concatenating normalized term-tuples, still obtaining a normalized term-tuple, since we have $\bar{s} :: \varrho_{\bar{s}}(\bar{t}) \in \mathbf{T}_\mathbf{z}^*$. The renaming apart has to be extended to elements of $\mathcal{D}^\flat$. Let $C^\flat \in \mathcal{C}^\flat$ such that $FV(C^\flat) \subseteq vars(\bar{t})$. Then $\varrho_{\bar{s}}((\bar{t}, C^\flat))$ denotes the pair $(\varrho_{\bar{s}}(\bar{t}), C_1^\flat)$, where $C_1^\flat \in \mathcal{C}^\flat$ is obtained from $C^\flat$ by applying the same renaming applied to $\bar{t}$ in order to obtain $\varrho_{\bar{s}}(\bar{t})$.

Term tuples are normalized by a *normalization function* $\eta\colon \mathcal{T}^*_{Vars} \to \mathbf{T}^*_{\mathbf{z}}$ such that, for each $\bar{u} \in \mathcal{T}^*_{Vars}$, the resulting tuple $\eta(\bar{u}) \in \mathbf{T}^*_{\mathbf{z}}$ is a variant of $\bar{u}$. As for ϱ, the normalization function has to be extended to elements of $\mathcal{D}^\flat$. Suppose that $G^\flat \in \mathcal{C}^\flat$ where $FV(G^\flat) \subseteq vars(\bar{u})$. then $\eta((\bar{u}, G^\flat))$ denotes $(\eta(\bar{u}), G_1^\flat) \in \mathcal{D}^\flat$ where it is assumed that $G_1^\flat$ can be obtained from $G^\flat$ by applying the same renaming applied to $\bar{u}$ in order to obtain $\eta(\bar{u})$.

We will now show how any abstract domain can be upgraded so as to capture structural information by means of the Pattern$(\cdot)$ construction. Then we will focus our attention on the abstract semantic operators.

4.1   From $\mathcal{D}^\sharp$ to Pattern$(\mathcal{D}^\sharp)$

Since one of the driving aims of this work is maximum generality, we refer to a very weak abstract interpretation framework [12]. To start with, we assume very little on abstract domains.

Definition 2. (Abstract domain for $\mathcal{D}^\flat$.) *An abstract domain for $\mathcal{D}^\flat$ is a set $\mathcal{P}^\sharp$ equipped with a preorder relation '$\preceq$'$\subseteq \mathcal{P}^\sharp \times \mathcal{P}^\sharp$, an order preserving function $\gamma\colon \mathcal{P}^\sharp \to \mathcal{D}^\flat$, and a least element $\perp^\sharp$ such that $\gamma(\perp^\sharp) = \varnothing$. Moreover, γ is such that if $(\bar{p}_1, C^\flat) \in \gamma(E^\sharp)$, and $\mathfrak{T} \models C^\flat \to \bar{p}_1 = \bar{p}_2$, then $\eta((\bar{p}_2, C^\flat)) \in \gamma(E^\sharp)$.*

Informally, $\mathcal{P}^\sharp$ is a set of abstract properties on which the notion of "relative precision" is captured by the preorder '$\preceq$'. Moreover, $\mathcal{P}^\sharp$ is related to the concrete domain $\mathcal{D}^\flat$ by means of a *concretization function* γ that specifies the soundness correspondence between $\mathcal{D}^\flat$ and $\mathcal{P}^\sharp$. The distinguished element $\perp^\sharp$ models an impossible state of affairs. In this framework, $d^\sharp \in \mathcal{P}^\sharp$ is a safe approximation of $d^\flat \in \mathcal{D}^\flat$ if and only if $d^\flat \subseteq \gamma(d^\sharp)$.

Suppose we are given an abstract domain complying with Definition 2. Here is how it can be upgraded with explicit structural information.

Definition 3. (The Pattern$(\cdot)$ construction.) *Let $\mathcal{D}^\sharp$ be an abstract domain for $\mathcal{D}^\flat$ and let γ be its concretization function. Then*

$$\text{Pattern}(\mathcal{D}^\sharp) = \{\perp^\sharp_p\} \cup \left\{ (\bar{s}, E^\sharp) \in \mathbf{T}^*_{\mathbf{z}} \times \mathcal{D}^\sharp \;\middle|\; \gamma(E^\sharp) \subseteq \mathbf{T}^{|vars(\bar{s})|}_{\mathbf{z}} \times \mathcal{C}^\flat \right\}.$$

The meaning of each element $(\bar{s}, E^\sharp) \in \text{Pattern}(\mathcal{D}^\sharp)$ is given by the concretization function $\gamma_p\colon \text{Pattern}(\mathcal{D}^\sharp) \to \mathcal{D}^\flat$ such that $\gamma_p(\perp^\sharp_p) = \varnothing$ and

$$\gamma_p((\bar{s}, E^\sharp)) = \left\{ \eta((\bar{r},\, C^\flat)) \;\middle|\; \begin{array}{l} (\bar{u}, C^\flat) \in \gamma(E^\sharp) \\ \mathfrak{T} \models C^\flat \to \bar{r} = \bar{s}\,[\bar{u}/vseq(\bar{s}')] \end{array} \right\}.$$

We also define the binary relation '$\preceq_p$'$\subseteq \text{Pattern}(\mathcal{D}^\sharp) \times \text{Pattern}(\mathcal{D}^\sharp)$ given, for each $d_1^\sharp, d_2^\sharp \in \text{Pattern}(\mathcal{D}^\sharp)$, by $d_1^\sharp \preceq_p d_2^\sharp \iff \gamma_p(d_1^\sharp) \subseteq \gamma_p(d_2^\sharp)$.

It can be seen that Pattern($\mathcal{D}^\sharp$) is an abstract domain in the sense of Definition 2 provided $\mathcal{D}^\sharp$ is. Thus Pattern($\mathcal{D}^\sharp$) can constitute the basis for designing an abstract semantics for CLP. This will usually require selecting an abstract semantic function on Pattern($\mathcal{D}^\sharp$), an effective convergence criterion for the abstract iteration sequence (notice that the '$\preceq$' and '$\preceq_p$' relations are not required to be computable), and perhaps a convergence acceleration method ensuring rapid termination of the abstract interpreter [12]. The last ingredient to complete the recipe is a computable way to associate an abstract description $d^\sharp \in$ Pattern($\mathcal{D}^\sharp$) to each concrete property $d^\flat \in \mathcal{D}^\flat$. For this purpose, the existence of a computable function $\alpha_p \colon \mathcal{D}^\flat \to$ Pattern($\mathcal{D}^\sharp$) such that, for each $d^\flat \in \mathcal{D}^\flat$, $d^\flat \subseteq \gamma_p\big(\alpha_p(d^\flat)\big)$ is assumed.

While one option is to design an abstract semantics based on Pattern($\mathcal{D}^\sharp$) from scratch, it is more interesting to start with an abstract semantics centered around $\mathcal{D}^\sharp$. In this case, it is possible to systematically lift the semantic construction to Pattern($\mathcal{D}^\sharp$).

4.2   Operations over $\mathcal{D}^\sharp$ and Pattern($\mathcal{D}^\sharp$)

We now present the abstract operations we assume on $\mathcal{D}^\sharp$ and the derived operations over Pattern($\mathcal{D}^\sharp$). Each operator on $\mathcal{D}^\sharp$ is introduced by means of safety conditions that ensure the safety of the derived operators over Pattern($\mathcal{D}^\sharp$).

Given the abstract domain, there are still many degrees of freedom for the design of a constructive abstract semantics. Thus, choices have to be made in order to give a precise characterization. In what follows we continue to strive for maximum generality. Where this is not possible we detail the design choices we have made in the development of the CHINA analyzer [1]. While some things may need adjustments for other analysis frameworks, the general principles should be clear enough for anyone to make the necessary changes.

Meet with Renaming Apart   We call *meet with renaming apart* (denoted by '$\rhd$') the operation of taking two descriptions in $\mathcal{D}^\sharp$ and, roughly speaking, juxtaposing them. This is needed when "solving" a clause body with respect to the current interpretation and corresponds, at the concrete level, to a renaming followed by an application of the '$\otimes$' operator. Its counterpart on Pattern($\mathcal{D}^\sharp$) is denoted by 'rmeet' and defined as follows.

Definition 4. ('$\rhd$' and 'rmeet') *Let* '$\rhd$'$\colon \mathcal{D}^\sharp \times \mathcal{D}^\sharp \to \mathcal{D}^\sharp$ *be such that, for each* $E_1^\sharp, E_2^\sharp \in \mathcal{D}^\sharp$,

$$\gamma(E_1^\sharp \rhd E_2^\sharp) = \left\{ \eta\big((\bar{r}, C_1^\flat \otimes G_2^\flat)\big) \left| \begin{array}{l} (\bar{r}_1, C_1^\flat) \in \gamma(E_1^\sharp) \\ (\bar{r}_2, C_2^\flat) \in \gamma(E_2^\sharp) \\ (\bar{w}_2, G_2^\flat) = \varrho_{\bar{r}_1}\big((\bar{r}_2, C_2^\flat)\big) \\ \mathfrak{T} \models (C_1^\flat \otimes G_2^\flat) \to \bar{r} = \bar{r}_1 :: \bar{w}_2 \end{array} \right. \right\}.$$

Then, we define $\mathrm{rmeet}\big((\bar{s}_1, E_1^\sharp), (\bar{s}_2, E_2^\sharp)\big) = \big(\bar{s}_1 :: \varrho_{\bar{s}_1}(\bar{s}_2), E_1^\sharp \rhd E_2^\sharp\big)$, *for each* $(\bar{s}_1, E_1^\sharp), (\bar{s}_2, E_2^\sharp) \in$ Pattern($\mathcal{D}^\sharp$).

A consequence of this definition is that there is no precision loss in 'rmeet' [3].

Parameter Passing Concrete parameter passing is realized by an extended unification procedure. Unification is *extended* because it must involve the constraint solver(s). Remember that our notion of "constraint solver" includes also the interface between the Herbrand engine and the proper solver [16]. The interface needs to be notified about all the bindings performed by the Herbrand engine in order to maintain consistency between the solver and the Herbrand part. We also assume that CLP programs are normalized in such a way that interpreted function symbols only occur in explicit constraints (note that this is either required by the language syntax itself, as in the case of the clp(Q, R) libraries of SICStus Prolog, or is performed automatically by the CLP system).

At the abstract level we do not prescribe the use of any particular algorithm. This is to keep our approach as general as possible. For instance, an implementor is not forced to use any particular representation for term-tuples (as in [11]). Similarly, one can choose any sound unification procedure that works well with the selected representation. Of particular interest is the possibility of choosing a representation and procedure that closely match the ones employed in the concrete language being analyzed. In this case, all the easy steps typical of any unification procedure (functor name/arity checks, peeling, and so on) will be handled, at the abstract level, exactly as they are at the concrete level. The only crucial operation in abstract parameter passing over $\text{Pattern}(\mathcal{D}^\sharp)$ is the binding of an abstract variable to an abstract term. This is performed by first applying a non-cyclic approximation of the binding to the pattern component and then notifying the original (possibly cyclic) binding to the abstract constraint component. The correctness of this approach can be proved [3] by assuming the existence of a bind operator on the underlying abstract constraint system satisfying the following condition.

Definition 5. (bind) *Let $E^\sharp \in \mathcal{D}^\sharp$ be a description such that $\gamma(E^\sharp) \subseteq \mathbf{T}_{\bar{\mathbf{z}}}^m \times \mathcal{C}^b$. Let $\bar{Z} = (Z_1, \ldots, Z_m)$, $u \in \mathcal{T}_{\bar{Z}}$, $vseq(u) = (Z_{j_1}, \ldots, Z_{j_l})$ and let $1 \leq h \leq m$. Then, define $(k_1, \ldots, k_{m_1}) = \big((1, \ldots, h-1) :: ((j_1, \ldots, j_l) \setminus \{1, \ldots, h-1\}) :: ((h+1, \ldots, m) \setminus \{j_1, \ldots, j_l\})\big)$. If $E_1^\sharp = \text{bind}(E^\sharp, u, Z_h)$, then,*

$$
\gamma(E_1^\sharp) \supseteq \left\{ \eta\big((\bar{q}\theta, C_1^b)\big) \; \middle| \;
\begin{array}{l}
(\bar{p}, C^b) \in \gamma(E^\sharp) \\
\bar{p} = (p_1, \ldots, p_m) \\
\bar{q} = (p_{k_1}, \ldots, p_{k_{m_1}}) \\
\theta \text{ is an idempotent substitution} \\
FV(C_1^b) \subseteq vars(\bar{q}\theta) \\
\mathfrak{T} \models \theta \leftarrow (Z_h' = u')[\bar{p}/\bar{Z}'] \\
\mathfrak{T} \models C_1^b \leftrightarrow \big((Z_h' = u')[\bar{p}/\bar{Z}'] \wedge C^b\big)\theta
\end{array}
\right\}.
$$

Note that $m_1 = m - 1$ if $Z_h \notin vars(u)$, and $m_1 = m$, otherwise.

To motivate and explain the above condition on $E_1^\sharp = \text{bind}(E^\sharp, u, Z_h)$, suppose that $\bar{p}$ is the pattern component and C^b the constraint component of an element in the concretization of $E^\sharp$. Now, the pattern components of elements of the abstract domain Pattern($\mathcal{D}^\sharp$) are always in normal form and thus, after applying the binding $[u'/Z'_h]$ to an element of $E^\sharp$, we must apply the normalization function so that the result is also in Pattern($\mathcal{D}^\sharp$). This will first remove the h-th term Z_h in the case that Z_h does not occur in u and then permute the remaining elements of $\bar{Z}$. A corresponding operation is applied to the pattern $\bar{p}$. That is, $\bar{q}$ is constructed from $\bar{p}$ first by removing the h-th term p_h in the case that Z_h does not occur in u and then by applying the same permutation as before on the remaining elements of $\bar{p}$. As a most general solution ϕ to $(Z'_h = u')[\bar{p}/\bar{Z}']$ may be cyclic, only an approximation of ϕ, the idempotent substitution θ, is applied to $\bar{q}$. The actual solution ϕ together with $C^\flat \theta$ is captured by the constraint $C^\flat_1$. Finally, note that the new pattern component $\bar{q}\theta$ may not be in normal form, so that in the condition for bind it is the normalized variant of $(\bar{q}\theta, C^\flat_1)$ that must be in the concretization of $E^\sharp_1$.

We refer the reader to [3] for a description of how any correct unification algorithm can be transformed into a correct (abstract) unification algorithm for Pattern($\mathcal{D}^\sharp$) using the bind operator and the normalization function η.

Projection When all the goals in a clause body have been solved, projection is used to restrict the abstract description to the tuple of arguments of the clause's head. The projection operations on $\mathcal{D}^\flat$ consist simply in dropping a suffix of the term-tuple component, with the consequent projection on the underlying constraint domain.

Definition 6. ('project$^\flat_k$') $\{ \text{project}^\flat_k \colon \mathcal{D}^\flat \to \mathcal{D}^\flat \mid k \in \mathbb{N} \}$ *is a family of operations such that, for each $k \in \mathbb{N}$ and each $(\bar{u}, C^\flat) \in \mathcal{D}^\flat$ with $|\bar{u}| \geq k$, if we define $\Delta = vars\big(\text{prefix}_k(\bar{u})\big)$, then $\text{project}^\flat_k\big((\bar{u}, C^\flat)\big) = \big(\text{prefix}_k(\bar{u}), \P^\flat_\Delta C^\flat\big)$.*

We now introduce the corresponding projection operations on Pattern($\mathcal{D}^\sharp$) and, in order to establish their correctness, we impose a safety condition on the projection operations of $\mathcal{D}^\sharp$.

Definition 7. ($\P^\sharp_k$ *and* project$^\sharp_k$) *Assume we are given a family of operations $\{ \P^\sharp_k \colon \mathcal{D}^\sharp \to \mathcal{D}^\sharp \mid k \in \mathbb{N} \}$ such that, for each $E^\sharp \in \mathcal{D}^\sharp$ with $\gamma(E^\sharp) \subseteq \mathbf{T}^m_\mathbf{z} \times C^\flat$ and each $k \leq m$, $\gamma(\P^\sharp_k E^\sharp) \supseteq \{ \text{project}^\flat_k\big((\bar{u}, C^\flat)\big) \mid (\bar{u}, C^\flat) \in \gamma(E^\sharp) \}$. Then, for each $(\bar{s}, E^\sharp) \in$ Pattern($\mathcal{D}^\sharp$) such that $\bar{s} \in \mathbf{T}^m_\mathbf{z}$ and each $k \leq m$, we define $\text{project}^\sharp_k\big((\bar{s}, E^\sharp)\big) = \big(\text{prefix}_k(\bar{s}), \P^\sharp_j E^\sharp\big)$, where $j = \big| vars\big(\text{prefix}_k(\bar{s})\big) \big|$.*

With these definitions 'project$^\sharp_k$' is correct with respect to 'project$^\flat_k$' [3].

Remapping The operation of *remapping* is used to adapt a description in Pattern($\mathcal{D}^\sharp$) to a different, less precise, pattern component. Remapping is essential to the definition of various *join* and *widening* operators. Consider a description $(\bar{s}, E^\sharp_{\bar{s}}) \in$ Pattern($\mathcal{D}^\sharp$) and a pattern $\bar{r}' \in \mathbf{T}^*_{\mathbf{z}'}$ such that $\bar{r}'$ is an anti-instance of $\bar{s}$. We want to obtain $E^\sharp_{\bar{r}} \in \mathcal{D}^\sharp$ such that $\gamma_p\big((\bar{r}, E^\sharp_{\bar{r}})\big) \supseteq \gamma_p\big((\bar{s}, E^\sharp_{\bar{s}})\big)$. This is what we call *remapping* $(\bar{s}, E^\sharp_{\bar{s}})$ to $\bar{r}'$.

Definition 8. ('remap') *Let $(\bar{s}, E_{\bar{s}}^{\sharp}) \in \text{Pattern}(\mathcal{D}^{\sharp})$ be a description with $\bar{s} \in \mathbf{T_z^k}$ and let $\bar{r}' \in \mathbf{T_{z'}^k}$ be an anti-instance of $\bar{s}$. Assume also $|vars(\bar{r})| = m$ and let $\bar{u} \in \mathbf{T_z^m}$ be the unique tuple such that $\bar{r}'[\bar{u}/vseq(\bar{r}')] = \bar{s}$. Then the operation $\text{remap}(\bar{s}, E_{\bar{s}}^{\sharp}, \bar{r}')$ yields $E_{\bar{r}}^{\sharp}$ such that $\gamma(E_{\bar{r}}^{\sharp}) \supseteq \gamma_p((\bar{u}, E_{\bar{s}}^{\sharp}))$.*

Observe that the remap function is closely related to the residual abstraction function α' of Figure 1.[1] With this definition, the specification of 'remap' meets our original requirement [3].

Upper Bound Operators A concrete (collecting) semantics for CLP will typically use set union to gather results coming from different computation paths. We assume that our base domain $\mathcal{D}^{\sharp}$ captures this operation by means of an upper bound operator '$\oplus$'. Namely, for each $E_1^{\sharp}, E_2^{\sharp} \in \mathcal{D}^{\sharp}$ and each $i = 1, 2$, we have that $E_i^{\sharp} \preceq E_1^{\sharp} \oplus E_2^{\sharp}$. This is used to merge descriptions arising from the different computation paths explored during the analysis.

The operation of merging two descriptions in $\text{Pattern}(\mathcal{D}^{\sharp})$ is defined in terms of 'remap'. Let $(\bar{s}_1, E_1^{\sharp})$ and $(\bar{s}_2, E_2^{\sharp})$ be two descriptions with $\bar{s}_1, \bar{s}_2 \in \mathbf{T_z^k}$. The resulting description is $(\bar{r}, E_1^{\sharp} \oplus E_2^{\sharp})$, where $\bar{r} \in \mathbf{T_{z'}^k}$ is an anti-instance of both $\bar{s}_1$ and $\bar{s}_2$, and $E_i^{\sharp} = \text{remap}(\bar{s}_i, E_i^{\sharp}, \bar{r}')$, for $i = 1, 2$. We note again that $\bar{r}'$ might be the least common anti-instance of $\bar{s}_1$ and $\bar{s}_2$, or it can be a further approximation of $lca(\bar{s}_1, \bar{s}_2)$: this is one of the degrees of freedom of the framework. Thus, the family of operations we are about to present is parameterized with respect to a common anti-instance function and the analyzer may dynamically choose which anti-instance function is used at each step.

Definition 9. ('join$_\phi$') *Let ϕ be any common anti-instance function. The operation (partial function) $\text{join}_\phi \colon \wp_f(\text{Pattern}(\mathcal{D}^{\sharp})) \rightarrowtail \text{Pattern}(\mathcal{D}^{\sharp})$ is defined as follows. For each $k \in \mathbb{N}$ and each finite family $F = \{(\bar{s}_i, E_i^{\sharp}) \mid i \in I\}$ of elements of $\text{Pattern}(\mathcal{D}^{\sharp})$ such that $\bar{s}_i \in \mathbf{T_z^k}$ for each $i \in I$, $\text{join}_\phi(F) = (\bar{r}, E^{\sharp})$, where $\bar{r}' = \phi(\{\bar{s}_i \mid i \in I\})$ and $E^{\sharp} = \bigoplus_{i \in I} \text{remap}(\bar{s}_i, E_i^{\sharp}, \bar{r}')$.*

If ϕ is any common anti-instance function then 'join$_\phi$' is an upper bound operator [3].

Widenings It is possible to devise a (completely unnatural) abstract domain $\mathcal{D}^{\sharp}$ that enjoys the *ascending chain condition*[2] still preventing $\text{Pattern}(\mathcal{D}^{\sharp})$ from possessing the same property. This despite the fact that any element of $\mathbf{T_z^n}$ has a finite number of distinct anti-instances in $\mathbf{T_{z'}^n}$. However, this problem is of no practical interest if the analysis applies 'join$_\phi$' at each step of the iteration sequence. In this case, if we denote by $(\bar{s}_j, E_j^{\sharp}) \in \text{Pattern}(\mathcal{D}^{\sharp})$ the description at step $j \in \mathbb{N}$, we have $(\bar{s}_{i+1}, E_{i+1}^{\sharp}) = \text{join}_\phi(\{(\bar{s}_i, E_i^{\sharp}), \ldots\})$, assuming no widening

---

[1] Indeed, one can define $\alpha' = \lambda(\bar{s}, E^{\sharp}) \in \mathbf{T_z^k} \times \mathcal{D}^{\sharp} . \text{remap}(\bar{s}, E^{\sharp}, (Z_1', \ldots, Z_k'))$.

[2] Namely, each strictly increasing chain is finite.

is employed. This implies that $\bar{s}'_{i+1}$ is an anti-instance of $\bar{s}_i$. As any ascending chain in $\mathbf{T_z^n}$ is finite, the iteration sequence will eventually stabilize if $\mathcal{D}^\sharp$ enjoys the ascending chain condition.

In some cases, however, rapid termination of the analysis on $\mathcal{D}^\sharp$ can only be ensured by using one or more widening operators $\nabla\colon \mathcal{D}^\sharp \times \mathcal{D}^\sharp \to \mathcal{D}^\sharp$ [13]. These can be lifted to work on $\mathrm{Pattern}(\mathcal{D}^\sharp)$. As an example, we show the default lifting used by the CHINA analyzer:

$$\mathrm{widen}\big((\bar{s}_1, E_1^\sharp), (\bar{s}_2, E_2^\sharp)\big) = \begin{cases} (\bar{s}_2, E_2^\sharp), & \text{if } \bar{s}_1 \neq \bar{s}_2; \\ (\bar{s}_2, E_1^\sharp \nabla E_2^\sharp), & \text{if } \bar{s}_1 = \bar{s}_2. \end{cases} \quad (1)$$

This operator refrains from widening unless the pattern component has stabilized. A more drastic choice for a widening is given by

$$\mathrm{Widen}\big((\bar{s}_1, E_1^\sharp), (\bar{s}_2, E_2^\sharp)\big) = \big(\bar{s}_2, \mathrm{remap}(\bar{s}_1, E_1^\sharp, \bar{s}'_2) \nabla E_2^\sharp\big). \quad (2)$$

Widening operators only need to be evaluated over $(\bar{s}_1, E_1^\sharp)$ and $(\bar{s}_2, E_2^\sharp)$ when $\bar{s}'_2$ is an anti-instance of $\bar{s}_1$. Thus, as $\mathbf{T_z^n}$ satisfies the ascending chain condition, 'widen' and 'Widen' are well-defined widening operators on $\mathrm{Pattern}(\mathcal{D}^\sharp)$ [3].

Besides ensuring termination, widening operators are also used to accelerate convergence of the analysis. It is therefore important to be able to define widening operators on $\mathrm{Pattern}(\mathcal{D}^\sharp)$ without relying on the existence of corresponding widenings on $\mathcal{D}^\sharp$. There are many possibilities in this direction and some of them are currently under experimental evaluation. Just note that any upper bound operator 'join$_\phi$' can be regarded as a widening as soon as the common anti-instance function ϕ is different from the lca. In order to ensure the convergence of the abstract computation, we will only consider widening operators on $\mathrm{Pattern}(\mathcal{D}^\sharp)$ satisfying the following (very reasonable) condition: if $(\bar{s}, E^\sharp)$ is the result of the widening applied to $(\bar{s}_1, E_1^\sharp)$ and $(\bar{s}_2, E_2^\sharp)$, where $\bar{s}'_2$ is an anti-instance of $\bar{s}_1$, then $\bar{s}'$ is an anti-instance of $\bar{s}_2$. Both widen and Widen comply with this restriction.

Comparing Descriptions The *comparison* operation on $\mathrm{Pattern}(\mathcal{D}^\sharp)$ is used by the analyzer in order to check whether a local fixpoint has been reached.

Definition 10. ('compare') *Let '$\precsim$' $\subseteq \mathcal{D}^\sharp \times \mathcal{D}^\sharp$ be a computable preorder that correctly approximates '$\preceq$', that is, for each $E_1^\sharp, E_2^\sharp \in \mathcal{D}^\sharp$, we have $E_1^\sharp \preceq E_2^\sharp$ whenever $E_1^\sharp \precsim E_2^\sharp$. The approximated ordering relation over $\mathrm{Pattern}(\mathcal{D}^\sharp)$, denoted by 'compare' $\subseteq \mathrm{Pattern}(\mathcal{D}^\sharp) \times \mathrm{Pattern}(\mathcal{D}^\sharp)$, is defined, for each $(\bar{s}_1, E_1^\sharp), (\bar{s}_2, E_2^\sharp) \in \mathrm{Pattern}(\mathcal{D}^\sharp)$, by $\mathrm{compare}\big((\bar{s}_1, E_1^\sharp), (\bar{s}_2, E_2^\sharp)\big) \Longleftrightarrow \big(\bar{s}_1 = \bar{s}_2 \wedge E_1^\sharp \precsim E_2^\sharp\big)$.*

It must be stressed that the above ordering is "approximate" since it does not take into account the peculiarities of $\mathcal{D}^\sharp$. More refined orderings can be obtained in a domain-dependent way, namely, when $\mathcal{D}^\sharp$ has been fixed. It is easy to show that compare is a preorder over $\mathrm{Pattern}(\mathcal{D}^\sharp)$ that correctly approximates the approximation ordering '$\preceq_p$' [3]. The ability of comparing descriptions only when they have the same pattern is not restrictive in our setting. Indeed, the definition of join_ϕ and the condition we imposed on widenings ensure that any two descriptions arising from consecutive steps of the iteration sequence are ordered by the anti-instance relation. When combined with the ascending chain condition of the pattern component, this allows to inherit termination from the underlying domain $\mathcal{D}^\sharp$.

5   Experimental Evaluation

We have conducted an extensive experimentation on the analysis using the Pattern($\cdot$) construction: this allowed us to tune the implementation and gain insight on the implications of keeping track of explicit structural information. To put ourselves in a realistic situation, we assessed the impact of the Pattern($\cdot$) construction on *Modes*, a very precise and complex domain for mode analysis. This captures information on simple types, groundness, boundedness, pair-sharing, freeness, and linearity. It is a combination of, among other things, two copies of the GER representation for *Pos* [5] — one for groundness and one for boundedness — and the non-redundant pair-sharing domain *PSD* [4] with widening as described in [22]. Each of these domains has been suitably extended to ensure correctness and precision of the analysis even for systems that omit the occur-check [1,15]. Some details on how the domains are combined can be found in [6].

The benchmark suite used for the development and tuning of the CHINA analyzer is probably the largest one ever employed for this purpose. The suite comprises all the programs we have access to (i.e., everything we could find by systematically dredging the Internet): 300 programs, 16 MB of code, 500 K lines, the largest program containing 10063 clauses in 45658 lines of code.

The comparison between *Modes* and Pattern(*Modes*) involves the two usual things: *precision* and *efficiency*. However, how are we going to compare the precision of the domain with explicit structural information with one without it? That is something that should be established in advance. Let us consider a simple but not trivial Prolog program: `mastermind.pl`.[3] Consider also the only direct query for which it has been written, '?- play.', and focus the attention on the procedure `extend_code/1`. A standard goal-dependent analysis of the program with the *Modes* domain is only able to tell something like

```
extend_code(A) :- list(A).
```

This means: "during any execution of the program, whenever `extend_code/1` succeeds it will have its argument bound to a *list cell* (i.e., a term whose principal functor is either '.'/2 or []/0)". Not much indeed. Especially because this can be established instantly by visual inspection: `extend_code/1` is *always* called with a list argument and this completes the proof. If we perform the analysis with Pattern(*Modes*) the situation changes radically. Here is what such a domain allows CHINA to derive:[4]

---

[3] Available at `http://www.cs.unipr.it/China/Benchmarks/Prolog/mastermind.pl`.

[4] Some extra groundness information obtained by the analysis has been omitted.

$x = \%$inc.	indep		ground		linear		free		bound	
	GI	GD	GI	GD	GI	GD	GI	GD	GI	GD
$x < 0$	0	1	0	0	0	1	0	0	0	0
$x = 0$	222	211	228	223	213	205	244	245	230	220
$0 < x \leq 2$	36	35	24	26	46	44	26	21	49	45
$2 < x \leq 5$	22	27	17	17	17	18	11	13	9	15
$5 < x \leq 10$	7	8	10	11	9	11	10	8	9	8
$x \geq 10$	13	18	21	23	15	21	9	13	3	12

Table 1. A summary of the *Modes* precision gained using structural information.

```
extend_code([([A|B],C,D)|E]) :- list(B), list(E),
  (functor(C,_,1);integer(C)), (functor(D,_,1);integer(D)),
  ground([C,D]), may_share([[A,B,E]]).
```

Under the circumstances mentioned above, this means: "the argument of procedure extend_code/1 will be bound to a term of the form [([A|B],C,D)|E], where B and E are bound to list cells; C is either bound to a functor of arity 1 or to an integer, and likewise for D; both C and D are ground, and (consequently) pair-sharing may only occur between A, B, and E".

It is clear that the analysis with Pattern(*Modes*) yields much more information. However, it is not clear at all how to define a fair measure for this precision gain. The approach we have chosen is simple though unsatisfactory: throw away all the structural information at the end of the analysis and compare the usual numbers (i.e., number of ground variables, number of free variables and so on). With reference to the above example, this metric pretends that explicit structural information gives no precision improvements on the analysis of extend_code/1 in mastermind.pl. In fact, once all the structural information has been discarded, the analysis with Pattern(*Modes*) only specifies that, upon success, the argument of extend_code/1 will be a list cell. In other words, we are measuring how the explicit structural information present in Pattern(*Modes*) improves the precision on *Modes* itself, which is only a tiny part of the real gain in accuracy. The value of this extra precision can only be measured from the point of view of the target application of the analysis.

It is important to note that the experimental results we are about to report have been obtained without using any widening on the pattern component. The widening operations are only propagated to the underlying *Modes* domain by means of the 'widen' operator given in Eq. (1). Moreover, the merge operation employed is always 'join$_{lca}$'. For space limitations, here we can only summarize the results of the experimentation. The interested reader can find all the details at http://www.cs.unipr.it/China. As far as precision is concerned, we measure five different quantities: the total number of independent argument pairs (*indep*); the total number of *ground* argument positions; the total number of *linear* argument positions; the total number of *free* argument positions; and the total number of *bound* (or *nonvar*) argument positions.

time difference in seconds	# prog.		% prog.	
	GI	GD	GI	GD
degradation ≥ 1	9	20	100.0	100.0
$0.5 \leq$ degradation < 1	2	4	97.0	93.3
$0.2 \leq$ degradation < 0.5	15	18	96.3	92.0
degradation < 0.2	105	106	91.3	86.0
same time	90	77	56.3	50.7
improvement < 0.2	34	31	26.3	25.0
$0.2 \leq$ improvement < 0.5	11	11	15.0	14.7
$0.5 \leq$ improvement < 1	9	5	11.3	11.0
improvement ≥ 1	25	28	8.3	9.3

Table 2. A summary on efficiency: the distribution of analysis time differences.

Since we are completely disregarding the precision gains coming from structural information in itself, our results give a (very pessimistic) lower bound on the overall precision improvement. The results are summarized by partitioning the benchmark suite into six classes of programs, identified by the percentage increase in precision due to the Pattern($\cdot$) construction. Table 1 gives the cardinalities of these classes for both goal-independent (GI) and goal-dependent (GD) analyses. A precision increase, on at least one of the measured quantities, is observed on more than one third of the benchmarks. The only precision decrease is due to the interaction between the Pattern($\cdot$) construction and the widenings used in the *Modes* domain. It is also worth observing that, on average, goal-dependent analysis is more likely to benefit from the addition of structural information.

In order to evaluate the impact on efficiency of the Pattern($\cdot$) transformation we computed the fixpoint evaluation time for all the programs, both with the *Modes* and with the Pattern(*Modes*) domains. Results are summarized by partitioning the benchmark suite into a number of classes and giving the cardinality of each class. As a first parameter, we considered the absolute time difference observed for each program.[5] Table 2 gives the cardinality of 9 classes, distinguishing between GI and GD analyses. The numbers show that the full range of possible behaviors is indeed observable. Quite surprisingly, it is not uncommon, although inherently more precise and complex, for the case with the Pattern($\cdot$) construction to result in significant time improvements. The reason for this is only partly due to the enhanced ability of the Pattern component to be able to detect and hence prune failed computation paths. Most importantly, the description of a set of tuples of terms in Pattern(*Modes*) is often much more efficient

---

[5] As the benchmark suite comprises several real programs of very respectable size, we believe that absolute time comparison is what really matters to assess the feasibility of the Pattern($\cdot$) construction with respect to the underlying domain. A time difference less than one second is an approximation of "the user will not notice." The experiments were conducted on a PC equipped with an AMD Athlon clocked at 700 MHz, 256 MB of RAM, and running Linux 2.2.16.

T = time in secs.	GI			GD		
	w/o SI	w SI	diff.	w/o SI	w SI	diff.
$T \geq 10$	15	11	-4	22	17	-5
$5 \leq T < 10$	9	5	-4	10	11	$+1$
$1 \leq T < 5$	32	35	$+3$	35	43	$+8$
$0.5 \leq T < 1$	11	21	$+10$	20	23	$+3$
$0.2 \leq T < 0.5$	27	38	$+11$	37	46	$+9$
$0.1 \leq T < 0.2$	164	158	-6	155	146	-9
$T < 0.1$	42	32	-10	21	14	-7

Table 3. A summary on efficiency: the distribution of analysis times.

than the corresponding description in *Modes*. Percentages in the columns on the right show how many programs are at least as good as the corresponding class. For instance, more than 85% of the benchmarks either reduce the analysis time or increase it by at most 0.2 secs. Since the occasional bad-behaving cases can be dealt with by defining a suitable widening operator on the pattern component, these results disprove the common belief that structural information has a heavy impact on the efficiency of the analysis.

As a second criterion, Table 3 partitions the benchmark suite into 7 classes based on their total fixpoint computation time, again distinguishing between GI and GD analysis. The columns labeled 'diff.' show how each class grows or shrinks because of the addition of structural information. It can be seen that the Pattern($\cdot$) construction causes only a minor change to the distribution, decreasing the number of benchmarks in both the fastest and the slowest classes.

6   Related Work

The use of explicit structural information has also been studied in [7], where abstract equation systems are integrated into an analysis domain tracking set-sharing, freeness, linearity and compoundness. While allowing for an implementation independent definition, this proposal still assumes the occur-check, therefore resulting in an unsound analysis for implemented CLP languages. An experimental evaluation on a small benchmark suite (19 programs) was reported by Mulkers et al. in [19,20]. Here the investigation mainly focused on the comparison between different instances of the underlying domain, showing the positive impact of freeness and linearity information on both the precision and performance of the classical set-sharing analysis. The experiments on the integration of structural information, by means of a depth-k abstraction (replacing all subterms occurring at a depth greater or equal to k with fresh abstract variables) for values of k between 0 and 3, showed that the domain they employed was not suitable to the analysis of real programs and, in fact, even the analysis of a modest-sized program like 'ann' could only be carried out with depth-0 abstraction (i.e., with no structural information at all).

In [8], an alternative technique is proposed for augmenting a data-flow analysis with structural information. Instead of upgrading the analysis domain, this technique relies on program transformations. In this approach, called *untupling*, the data-flow analysis of a given program would be performed in four distinct phases. This new analysis technique is advocated for its simplicity and efficiency. Comparing their limited experimental evaluation to the one conducted in [11], the authors of [8] claim that the untupling approach is inherently more efficient than abstract domain enhancement. Our new performance results suggest that this conclusion may need reconsidering. On the other hand, the proposal in [8] may be simpler to implement despite the four phases required, especially if one has to start *from scratch*. However, the Pattern(·) construction, besides being more precise and particularly efficient, is already implemented and has been thoroughly tested on a large number of benchmarks using the very expressive abstract domain *Modes*. Furthermore, as the implementation is in the form of a C++ template, only a very limited effort is required to upgrade any other abstract domain with structural information.

7 Conclusion

We have presented the rational construction of a generic domain for structural analysis of real CLP languages: $\text{Pattern}(\mathcal{D}^\sharp)$, where the parameter $\mathcal{D}^\sharp$ is an abstract domain satisfying certain properties. We build on the parameterized $\text{Pat}(\Re)$ domain of Cortesi et al. [10,11], which is restricted to logic programs and requires the occur-check to be performed. However, while $\text{Pat}(\Re)$ is presented as a *specific implementation* of a generic structural domain, our formalization is implementation-independent. Reasoning at a higher level of abstraction we are able to appeal to familiar notions of unification theory, while leaving considerable more latitude for the implementer. Indeed our results show that, contrary to popular belief, an analyzer incorporating structural information analysis based on our approach can be highly competitive even from the efficiency point of view.

References

1. R. Bagnara. *Data-Flow Analysis for Constraint Logic-Based Languages*. PhD thesis, Dipartimento di Informatica, Università di Pisa, Italy, March 1997. 190, 195, 197, 202
2. R. Bagnara. A hierarchy of constraint systems for data-flow analysis of constraint logic-based languages. *Science of Computer Programming*, 30(1-2):119-155, 1998. 195
3. R. Bagnara, P. M. Hill, and E. Zaffanella. Efficient structural information analysis for real CLP languages. Quaderno 229, Dipartimento di Matematica, Università di Parma, 2000. Available at http://www.cs.unipr.it/~bagnara. 198, 198, 199, 199, 200, 200, 201, 201
4. R. Bagnara, P. M. Hill, and E. Zaffanella. Set-sharing is redundant for pair-sharing. *Theoretical Computer Science*, 2000. To appear. 202
5. R. Bagnara and P. Schachte. Factorizing equivalent variable pairs in ROBDD-based implementations of *Pos*. In A. M. Haeberer, editor, *Proc. of the "7th Int'l Conf. on Algebraic Methodology and Software Technology"*, vol. 1548 of *Lecture Notes in Computer Science*, pages 471-485, Amazonia, Brazil, 1999. Springer-Verlag, Berlin. 202

6. R. Bagnara, E. Zaffanella, and P. M. Hill. Enhancing Sharing for precision. In M. C. Meo and M. Vilares Ferro, editors, *Proc. of the "AGP'99 Joint Conf. on Declarative Programming"*, pages 213–227, L'Aquila, Italy, 1999. 202

7. M. Bruynooghe, M. Codish, and A. Mulkers. Abstract unification for a composite domain deriving sharing and freeness properties of program variables. In F. S. de Boer and M. Gabbrielli, editors, *Verification and Analysis of Logic Languages, Proc. of the W2 Post-Conference Workshop, Int'l Conf. on Logic Programming*, pages 213–230, Santa Margherita Ligure, Italy, 1994. 205

8. M. Codish, K. Marriott, and C. Taboch. Improving program analyses by structure untupling. *Journal of Logic Programming*, 43(3):251–263, 2000. 192, 206, 206, 206

9. A. Colmerauer. Prolog and infinite trees. In K. L. Clark and S. Å. Tärnlund, editors, *Logic Programming, APIC Studies in Data Processing*, vol. 16, pages 231–251. Academic Press, New York, 1982. 190, 191

10. A. Cortesi, B. Le Charlier, and P. Van Hentenryck. Conceptual and software support for abstract domain design: Generic structural domain and open product. Technical Report CS-93-13, Brown University, Providence, RI, 1993. 191, 206

11. A. Cortesi, B. Le Charlier, and P. Van Hentenryck. Combinations of abstract domains for logic programming: Open product and generic pattern construction. *Science of Computer Programming*, 38(1–3), 2000. 191, 191, 191, 198, 206, 206

12. P. Cousot and R. Cousot. Abstract interpretation frameworks. *Journal of Logic and Computation*, 2(4):511–547, 1992. 194, 196, 197

13. P. Cousot and R. Cousot. Comparing the Galois connection and widening/narrowing approaches to abstract interpretation. In M. Bruynooghe and M. Wirsing, editors, *Proc. of the 4th Int'l Symp. on Programming Language Implementation and Logic Programming*, vol. 631 of *Lecture Notes in Computer Science*, pages 269–295, Leuven, Belgium, 1992. Springer-Verlag, Berlin. 201

14. R. Giacobazzi, S. K. Debray, and G. Levi. Generalized semantics and abstract interpretation for constraint logic programs. *Journal of Logic Programming*, 25(3):191–247, 1995. 195

15. P. M. Hill, R. Bagnara, and E. Zaffanella. The correctness of set-sharing. In G. Levi, editor, *Static Analysis: Proc. of the 5th Int'l Symp.*, vol. 1503 of *Lecture Notes in Computer Science*, pages 99–114, Pisa, Italy, 1998. Springer-Verlag, Berlin. 202

16. J. Jaffar and M. Maher. Constraint logic programming: A survey. *Journal of Logic Programming*, 19&20:503–582, 1994. 190, 191, 193, 195, 198

17. J. Jaffar, S. Michaylov, P. Stuckey, and R. Yap. The CLP($\mathcal{R}$) language and system. *ACM Transactions on Programming Languages and Systems*, 14(3):339–395, 1992. 193

18. J.-L. Lassez, M. J. Maher, and K. Marriott. Unification revisited. In J. Minker, editor, *Foundations of Deductive Databases and Logic Programming*, pages 587–625. Morgan Kaufmann, Los Altos, Ca., 1988. 191, 194

19. A. Mulkers, W. Simoens, G. Janssens, and M. Bruynooghe. On the practicality of abstract equation systems. Report CW 198, Department of Computer Science, K. U. Leuven, Belgium, 1994. 205

20. A. Mulkers, W. Simoens, G. Janssens, and M. Bruynooghe. On the practicality of abstract equation systems. In L. Sterling, editor, *Logic Programming: Proc. of the 12th Int'l Conf. on Logic Programming*, MIT Press Series in Logic Programming, pages 781–795, Kanagawa, Japan, 1995. The MIT Press. 205

21. K. Musumbu. *Interprétation Abstraite des Programmes Prolog*. PhD thesis, Institut d'Informatique, Facultés Univ. Notre-Dame de la Paix, Namur, Belgium, 1990. 191

22. E. Zaffanella, R. Bagnara, and P. M. Hill. Widening Sharing. In G. Nadathur, editor, *Principles and Practice of Declarative Programming*, vol. 1702 of *Lecture Notes in Computer Science*, pages 414–431, Paris, 1999. Springer-Verlag, Berlin. 202

Playing Logic Programs with the Alpha-Beta Algorithm

Jean-Vincent Loddo and Roberto Di Cosmo

Laboratoire Preuves Programmes et Systèmes (PPS)
Université Paris 7 - France
{loddo,dicosmo}@pps.jussieu.fr

Abstract. Alpha-Beta is a well known optimized algorithm used to compute the values of classical combinatorial games, like chess and checkers . The known proofs of correctness of Alpha-Beta do rely on very specific properties of the values used in the classical context (integers or reals), and on the finiteness of the game tree. In this paper we prove that Alpha-Beta correctly computes the value of a game tree even when these values are chosen in a much wider set of partially ordered domains, which can be pretty far apart from integer and reals, like in the case of the lattice of idempotent substitutions or ex-equations used in logic programming. We do so in a more general setting that allows us to deal with infinite games, and we actually prove that for potentially infinite games Alpha-Beta correctly computes the value of the game *whenever it terminates*. This correctness proofs allows us to apply Alpha-Beta to new domains, like constraint logic programming.

1 Introduction

Game theory has found various applications in the research field of programming languages semantics, so that game theory is a very active research subject in computer science. After the preliminary works of Lamarche [11], Blass [2] and Joyal [9] in the early 90s, the works of Abramsky, Malacaria and Jagadeesan [1] lead to the first fully abstract semantics for functional (PCF) or imperative (Idealized Algol) languages. Then, more recently, specialists of Linear Logics got interested in links between games and the geometry of interaction [13], whereas Curien and Herbelin showed that certain classical abstract machines could be interpreted in terms of games [4].

But these relevant works use more the *vocabulary* of games (player, move, game, strategy) than the results and the techniques of traditional Game Theory: typically, nobody is interested to know, in those games, if there is a winner, and what he wins; the focus there is on the dynamic aspect of player *interaction*, and *game composition*, not on the only interesting notion of classical game theory, *the gain*. This should not be taken as a criticism, but as proof of the richness of Game Theory, which can be useful even when one only takes its vocabulary: the generality of the concepts it manipulates (arenas, multiple and independent agents, strategies of cooperation or of non-cooperation, quantification of

M. Parigot and A. Voronkov (Eds.): LPAR 2000, LNAI 1955, pp. 207–224, 2000.

the remuneration after each game) and their intuitive nature, already provide a powerful metalanguage that allows us to tackle many of the aspects of modern programming languages.

In a paper written with S. Nicolet [5], the authors showed for the first time that classical notions like payoff, propagation functions and evaluation of a game tree are not sterile in the semantics of programming languages, by introducing a two player combinatorial game whose *value*, defined by means of von Neumann's MiniMax theorem [16], is the result of the execution of a logic program. In that work, we had to introduce an ad hoc framework to deal with substitutions as game values, and to prove the correctness of this game semantics with respect to the traditional semantics of logic programs [3].

In this paper, we present a general framework for classical two player games, but relaxing many traditional restrictions: infinite plays are allowed, values are no longer required to be totally ordered, and propagation functions can be chosen from a wide set of candidates. In this setting, the formal definition of the value of a game is given.

Then, we focus on the problem of *efficiently computing* the value of the game, by using the Alpha-Beta *algorithm*, not just on the existence of the value as formally defined.

Surprisingly enough, we can show that the Alpha-Beta algorithm computes the correct value of the game under a few general assumptions on the domain of values, thus greatly broadening its applicability: it can be, for example, used as a computational engine for logic programming.

But we do not content ourselves with proving correctness of Alpha-Beta on games whose value domain is more abstract than the usual integers and reals; we go much further by introducing the notions necessary to deal with potentially infinite games, and we prove that Alpha-Beta is (partially) correct on potentially infinite games: whenever it terminates, it computes the value of the game. This step takes us out of the usual domain of combinatorial game theory, where game trees can be huge but not infinite, and this is, to our best knowledge, the first correctness proof in this setting.

Then, we present an application to logic programming, and exhibit an example to show how Alpha-Beta can give us a significant gain in performance, or even terminate where other engines may loop. Finally, we conclude with a selection of future directions for research and application.

2   An Abstract Theory of Two Players Combinatorics Games

In this section, we introduce our formal framework for two player games, together with some fundamental notions, like that of a game-tree and the value of a finite game [14,15,6]. We then present the notion of an *approximation* of a game value, as is found in the theory of combinatorial games like chess or go, which are too big to be fully developed, and use it as a key notion to extend the framework to deal with infinite games too.

2.1   The Rules of the Game

A two player game is the simplest game in combinatorial Game Theory: we find two players, Player and Opponent, each opposed to the other, that play in turn one after the other and that must behave *rationally* (i.e. their move are deterministic, and they both seek to win). The game is assumed to be finite, and once one knows the terminal position in a play, the gain (or loss) of each player is known also; besides, what one looses, the other wins, so there always is precisely one winner.

A game is given once one knows its "syntax", that is the set of all possible plays, and its "semantics", that is the *value* of each of these plays (what Player gains or looses towards Opponent). We will formalize each of these aspects in turn, starting here from the "syntax".

2.2   Basic Definitions: Syntax

There are two ways of knowing all possible plays, either by giving them extensively as a set, an approach quite inadequate to handle real-world games where this set can be enormous, even if finite; or by giving a set of "positions" and "rules" that allow to produce all possible plays (like in chess or go) [12].
We will take in what follows this second approach: the following definitions are essentially the traditional ones.

Definition 1 (Syntax of a game). *The syntax of a game is formally defined as a tuple*

$$G = (WPOS, BPOS, IPOS, \to_{\mathcal{R}})$$

Here $WPOS$ is the set of Player position, while $BPOS$ is the set of opponent positions and we write POS for the disjoint sum $WPOS \oplus BPOS$, and π for a generic position in POS. The third component, $IPOS \subseteq POS$ is the set of initial positions in the game. Finally, $\to_{\mathcal{R}} \subseteq POS \times POS$ is a *locally finite* (i.e. only a finite number of pairs in the relation can share their first component) transition relation that represents all the possible moves in a play as transitions between positions.
We require that the moves in the game are *alternating*, that is to say that whenever $(\pi, \pi') \in \to_{\mathcal{R}}$ we have that if π is a player position, then π' is an opponent position and vice-versa.
Defining $\to_{\mathcal{R}}$ is equivalent to defining a function $moves : POS \to 2^{POS}$, whose value is $moves(\pi) = \{\pi' \mid \pi \to_{\mathcal{R}} \pi'\}$ and that explicitly gives the possible moves out of a given position.

Once the syntax of a game is known, we have all the necessary information to determine when a game is finished (we have reached a terminal position, or not).

Definition 2 (Terminal, non terminal positions). *Given a game G, we identify the following derived notions*

terminal positions $TPOS = \{\pi \in POS \mid not \; \exists \pi' \; such \; that \; \pi \to_{\mathcal{R}} \pi'\}$
player terminal positions $WTPOS = TPOS \cap WPOS$
opponent terminal positions $BTPOS = TPOS \cap BPOS$
non terminal positions $NTPOS = \{\pi \in POS \mid \exists \pi' \; such \; that \; \pi \to_{\mathcal{R}} \pi'\}$
player non terminal positions $WNTPOS = NTPOS \cap WPOS$
opponent non terminal positions $BNTPOS = NTPOS \cap BPOS$

Finally, all the possible plays starting from a given initial position can be represented as a tree, known as a *game tree*.

Definition 3 (Game tree [14]). *A game tree Γ_π, for an initial position π, is a tree having positions as nodes, and representing all possible plays starting at π in the traditional way. Since a game tree can be infinite, it should be formally defined as a fix-point of a suitable monotone function, but we do not enter into the details here. Since the $\to_{\mathcal{R}}$ relation is locally finite, the game tree is finitely branching. A game is* finite *if its game tree is,* infinite *otherwise.*

2.3   Basic Definitions: Semantics

It is now time to turn to the essential aspect of a game in classical game theory: its *value*. To each terminal position, which is reached when a play is complete, is associated a *gain* for Player, taken out of some domain D (traditionally, the integers), given by an evaluation function h. Of course, this gain for Player is actually a *loss* for opponent, and given a set of possible moves, Player chooses the move that maximizes its gain, while Opponent chooses the move that minimizes its loss; this rational choice can be abstracted by two functions $\uparrow$ and $\downarrow$ on the domain of values. These functions usually operate on the *set* of possible choices to provide a value, but we prefer to see the choices presented orderly as a tuple, not a set, and we require that the result of the choice function is independent of the order. All these elements give us the semantics part of the game.

Definition 4 (Evaluation structure). *An evaluation structure $\mathcal{E}$ is a tuple*

$$(D\;,\;\uparrow\;,\downarrow\;,\;h)$$

where D is the domain of values, $\uparrow: D^n \to D$ and $\downarrow: D^n \to D$ are functions of all finite arities $n \geq 1$ materializing rational choices of Player and Opponent, and $h : TPOS \to D$ is the evaluation function giving the value (gain) of each terminal position. We require the $\uparrow$ and $\downarrow$ functions to be order-independent, that is, for any permutation $\sigma : n \to n$, $\uparrow (a_1, \ldots, a_n) = \uparrow (a_{\sigma 1}, \ldots, a_{\sigma n})$ (and similarly for $\downarrow$).

Remark 1. The requirement that the choice-functions be order-independent is not necessary for the proofs, that go through seamlessly without it. But the traditional framework of combinatorial games always enforces this condition, and we keep it here to make the presentation of the result more intuitive.

Given an evaluation structure, it is possible to compute the *value* of any finite game (this definition mirrors the traditional one).

Definition 5 (Value of a finite game (Minimax propagation)). *Given a game G and an* evaluation structure, $\mathcal{E}$ *the function* $Val : POS \to D$ *defined as follows associates to each position in G its value:*

$$Val\ \pi = \quad h(\pi) \qquad\qquad if\ \pi \in TPOS$$

$$Val\ \pi = \quad \underset{\pi' \in moves(\pi)}{\uparrow}\ Val\ \pi' \qquad if\ \pi \in WNTPOS$$

$$Val\ \pi = \quad \underset{\pi' \in moves(\pi)}{\downarrow}\ Val\ \pi' \qquad if\ \pi \in BNTPOS$$

2.4   From Huge to Infinite Games

Many finite games, like chess and go and unlike tic-tac-toe, are so huge that computing their value is not feasible. This is why one needs sometimes to try to compute an approximation of the value of a game from a given position. For that, we simply stop exploring the huge tree at some internal nodes, whose value is arbitrarily provided by some heuristic function. We will see that, while heuristics are simply useful to approximate the value of huge finite games, they are essential to *define* the value of an infinite game.

But to compute approximations, one needs to be able to compare values, that is, in what follows we assume that D is actually equipped with a partial order relation $\leq$. Also, for the approximations to be useful, it is necessary that the choice functions $\uparrow$ and $\downarrow$ be monotone w.r.t. the product partial order induced on D^n by $\leq$.

In what follows, we will always assume this monotonicity, which is always satisfied by the max and min functions used for traditional combinatorial games.

Heuristics and approximations A heuristic function is just a function of type $NTPOS \to D$ assigning arbitrary values to nonterminal positions in a game, but only some heuristics are interesting, and one usually distinguish between optimistic and pessimistic heuristics according to the ability of the heuristic to provide an approximation greater of, or inferior to, the actual value.

Definition 6 (Admissible heuristics). *A heuristic function* $\varphi : NTPOS \to D$ *is an* admissible pessimistic heuristic *(resp.* admissible optimistic heuristics *) for a finite game iff* $\forall \pi \in NTPOS.\ \varphi(\pi) \leq Val\ \pi \qquad (resp.\ \geq).$

Unfortunately, determining if a heuristic is admissible can be quite hard, and another definition can be more useful in practice: we say a heuristic is monotone if the approximation it provides are consistent among themselves.

Definition 7 (Monotone heuristics). *A heuristic function* $\varphi : NTPOS \to D$ *is* monotone pessimistic *(resp.* monotone optimistic*) iff :*

$$\forall \pi \in NTPOS. \quad \begin{cases} \varphi(\pi) \leq h(\pi') \ \ \forall \pi' \ s.t. \ \pi \to_{\mathcal{R}} \pi' \ and \ \pi' \in TPOS \quad (resp. \geq) \\ \varphi(\pi) \leq \varphi(\pi') \ \forall \pi' \ s.t. \ \pi \to_{\mathcal{R}} \pi' \ and \ \pi' \in NTPOS \ (resp. \geq) \end{cases}$$

Once we have a heuristic, we can build out of our evaluation structure a structure useful to compute approximations.

Definition 8 (Approximation structure). *A* pessimistic approximation structure *(resp.* optimistic*) is a tuple*

$$(D \, , \, \leq \, , \, \uparrow \, , \, \downarrow \, , \, h \, , \, \varphi)$$

where $(D \, , \, \uparrow \, , \, \downarrow \, , \, h)$ *is an evaluation structure, and such that*

- $\uparrow$ *and* $\downarrow$ *are monotone w.r.t. the product order induced on* D^n *by* $\leq$
- *the function* $\varphi' : POS \to D$ *defined as* $\varphi' = \varphi \oplus h$ *(i.e. the disjunctive union of relations* $\varphi : NTPOS \to D$ *and* $h : TPOS \to D$*) satisfies:*

$$\varphi'(\pi) \leq \underset{\pi' \in moves(\pi)}{\uparrow} \varphi'(\pi') \qquad if \ \pi \in WNTPOS \qquad (resp. \geq)$$

$$\varphi'(\pi) \leq \underset{\pi' \in moves(\pi)}{\downarrow} \varphi'(\pi') \qquad if \ \pi \in BNTPOS \qquad (resp. \geq)$$

If D contains a minimal element $\perp$, the heuristic $\lambda\pi.\perp$ will give us a *canonical* pessimistic approximation structure. Similarly, if D contains a maximal element $\top$, the heuristic $\lambda\pi.\top$ will give us a *canonical* optimistic approximation structure.

The monotonicity property allows us to easily prove the following result.

Proposition 1 (Monotonicity and approximations).
    If $(D, \leq)$ *is a* distributive lattice, *and we take* $\uparrow = \vee$ *(the sup) and* $\downarrow = \wedge$ *(the inf) on* D*, then any* monotone pessimistic *heuristics (resp.* optimistic*) gives raise to a* pessimistic approximation structure *(resp.* optimistic*).*

Once we have an approximation structure at hand, we can compute an approximation of the value of a game, by cutting the tree branches at some internal nodes, obtaining another (smaller) tree, and computing the value of this cut tree. Of course, the approximation thus computed depends on where the cut actually take place, so an approximation structure really gives rise to a whole set of approximations. This can be put more formally as follows:

Definition 9 (Set of approximations of a game ($SetOfVal_\varphi$)**).**
    Given a game G *and an evaluation structure* $(D \, , \, \uparrow \, , \downarrow \, , \, h)$*, and heuristic* $\varphi : NTPOS \to D$*, we define the* approximation function for the game G, *relative to* φ*, written* $SetOfVal_\varphi : POS \to 2^D$*, as the smallest function* $S : POS \to 2^D$ *(w.r.t. to the point-wise partial order on* $POS \to 2^D$ *derived from the partial order* $\subseteq$ *in* 2^D *) such that:*

1. $S(\pi)$ *contains* $h(\pi)$ *if* $\pi \in TPOS$
2. $S(\pi)$ *contains the element* $\varphi(\pi)$ *if* $\pi \in NTPOS$
3. $S(\pi)$ *contains the element* $\overset{n}{\underset{i=1}{\uparrow}} v_i$ *if* $\pi \in WNTPOS$, *where* $moves(\pi) =$
   $\{\pi_1, \pi_2, \ldots, \pi_n\}$ *and* $v_i \in S(\pi_i)$
4. $S(\pi)$ *contains the element* $\overset{n}{\underset{i=1}{\downarrow}} v_i$ *if* $\pi \in BNTPOS$, *where* $moves(\pi) =$
   $\{\pi_1, \pi_2, \ldots, \pi_n\}$ *and* $v_i \in S(\pi_i)$

We remark here that $SetOfVal_\varphi(\pi)$ is well defined for all positions π as its value is the least fixed point of a monotone function over the complete lattice 2^D, which always exists due to Knaster-Tarski's fix-point theorem.

Actually, this really defines a function $SetOfVal : (NTPOS \rightarrow D) \rightarrow POS \rightarrow 2^D$, which we will apply to monotone heuristic functions in order to obtain a set of approximations having a reasonable algebraic structure.

Proposition 2 (Algebraic structure of the monotone approximations).
If $(D , \leq , \uparrow , \downarrow , h , \varphi)$ *is a* pessimistic approximation structure *(resp.* optimistic), *then for all position* $\pi \in POS$ *the set* $SetOfVal_\varphi(\pi) \in 2^D$ *is an* upper-semi-lattice *(i.e.* $\forall x, y \in S.x \vee y \in S$) *(resp.* lower-semi-lattice).
Moreover, if the game is finite, *then* $SetOfVal_\varphi(\pi)$ *contains its upper (resp lower) limit* $Val\,\pi$.

To put it in other terms, for finite games, given any monotone pessimistic heuristic h_{pess} and any monotone optimistic approximation h_{opt}, we have that

$$SetOfVal_{h_{pess}}(\pi) \cap SetOfVal_{h_{opt}}(\pi) = Val\,\pi$$

Example 1. In the following picture, we give the game tree of a finite game (on the left), and three smaller game trees obtained by cutting the full tree at some positions. The evaluation domain D is $Nat \times Nat$ with $\leq$ the product order and we use the trivial heuristic $h_{pess} = \lambda\pi.(0,0)$. Each tree is labelled with its valuation (the values of the trees on the right approximations in $SetOfVal_{h_{pess}}(\pi)$): notice that the set of approximations is not totally ordered. Terminal nodes are grey, no terminal are white; player nodes are circles, opponent nodes are squares, and cut nodes are crossed.

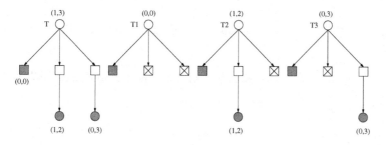

2.5 Infinite Games

The framework set up to approximate the value of huge games can be used as is to handle infinite games too. Indeed the *SetOfVal* function, defined as we did, is well defined both on finite *and* infinite games. This allows us to formally define two canonical approximations of the value of an infinite game, as partial functions assigning to each position the limit of its approximations, if it exists.

Definition 10 (Limit values of a game, value). *If* $(D, \leq, \uparrow, \downarrow, h_{pess}, \varphi)$ *is a* pessimistic *approximation structure, then*

$$Val_{h_{pess}} \pi \doteq \sup SetOfVal_{h_{pess}}(\pi)$$

If $(D, \leq, \uparrow, \downarrow, h_{opt}, \varphi)$ *is an* optimistic *approximation structure, then*

$$Val_{h_{opt}} \pi \doteq \inf SetOfVal_{h_{opt}}(\pi)$$

where we write $x \doteq y$ *for "x is equal to y if y defined".*
If they coincide, that defines the value $Val \, \pi$ *of the infinite game on* π.

Notice that $Val_{h_{pess}}$ is defined everywhere if the domain is a complete partial order (CPO), and $Val_{h_{opt}}$ is defined everywhere if if the domain is a co-complete partial order (co-CPO). Also, on finite games both approximations coincide with the value of the finite game.

Proposition 3 (Relating approximations). *Given a pessimistic approximation structure and an optimistic approximation structure sharing the same evaluation structure (hence different only for the heuristic functions, h_{pess} and h_{opt}), such that $h_{pess}(\pi) \leq h_{opt}(\pi)$ for all non terminal positions π, then we have that*

$$h_{pess}(\pi) \leq y \quad \forall y \in SetOfVal_{h_{opt}}(\pi) \quad ; \quad h_{opt}(\pi) \geq x \quad \forall x \in SetOfVal_{h_{pess}}(\pi)$$

for all non terminal positions π.

Proof. We give the proof of the first inequality by induction on the definition of $y \in SetOfVal_{h_{opt}}$. The second inequality is proved similarly.

- Base case: $y = h_{opt}(\pi)$ and $h_{pess}(\pi) \leq y$ by hypothesis
- Inductive step: suppose $\pi \in WNTPOS$ (the case $\pi \in BNTPOS$ is proved similarly, using the monotonicity of $\downarrow$); we have that

$$h_{pess}(\pi) \leq \overset{n}{\underset{i=1}{\uparrow}} h_{pess}(\pi_i) \quad \text{because } (\uparrow, \downarrow, h_{pess}) \text{ is an approximation structure}$$

$$\leq \overset{n}{\underset{i=1}{\uparrow}} y_i \qquad \text{by induction hypothesis and monotonicity of } \uparrow$$

$$= \quad y$$

Corollary 1 (Finite computations for infinite games). *Under the hypothesis of the previous proposition, we only have two possible cases*

1. $SetOfVal_{h_{pess}}(\pi) \cap SetOfVal_{h_{opt}}(\pi) = \{v\}$ *where* $v = Val\,\pi$
2. $SetOfVal_{h_{pess}}(\pi) \cap SetOfVal_{h_{opt}}(\pi) = \emptyset$

The first case gives us a sufficient condition to stop the computation of the approximations on an infinite game tree: as soon as the intersection is non empty, we know we have the value of the game, without needing to fully compute the set of approximations.

We will use these properties in our analysis of the Alpha-Beta algorithm on infinite games.

3 Correctness of the Alpha-Beta Algorithm

We turn now to the central result of the paper: the correctness proof of the Alpha-Beta algorithm on approximation structures that are distributive lattices.
Let us first recall the definition of the Alpha-Beta algorithm, extended with heuristic functions (see for example [14] for an excellent introduction to Alpha-Beta), where the α and β parameters provide a lowest and an upper bound on the value that we want the algorithm to provide us with.

Definition 11. *Sequential Alpha-Beta algorithm The Alpha-Beta algorithm is defined, in pseudo-language, as follows*

```
function  AlphaBeta ( π : Pos  ;  α ,β : D ): D
begin
if π ∈ TPOS  then return(h(π));
if π ∈ WNTPOS  and  moves(π) = {π₁,...,πₙ}  then begin
        v := α ↑ h_pess(π);
        i := 1;
        while (not v ≥ β) and (i ≤ n) do begin
                v := v ↑ AlphaBeta(πᵢ,v,β);
                i := i + 1;
                end;
        end;
if  π ∈ BNTPOS  and  moves(π) = {π₁,...,πₙ}  then begin
        v := β ↓ h_opt(π);
        i := 1;
        while (not v ≤ α) and (i ≤ n) do begin
                v := v ↓ AlphaBeta(πᵢ,α ,v) ;
                i := i + 1 ;
                end;
        end;
return( v );
end;
```

We will prove in the following a key result of *relative* correctness of Alpha-Beta: if the algorithm is called with an α and a β parameters, it will never return anything outside the interval α, β, so we can only expect to prove that the value it computes is correct *up to* the interval we gave. But this result is a sort of induction loading needed in the proof: the interesting point is the corollary: when we call it with $\alpha = \bot$ and $\beta = \top$, it gives *the* correct result whenever it terminates.

Let us first establish a simple property of distributive lattices

Lemma 1 (Insertion). *Let* $(D, \leq)$ *be a distributive lattice and* $\uparrow = \vee$ *(the sup)* $\downarrow = \wedge$ *(the inf) on* D. *Then for all* α, β, x *we have that* $\alpha \uparrow [\beta \downarrow x] = \alpha \uparrow [\beta \downarrow (\alpha \uparrow x)]$ *and* $\beta \downarrow [\alpha \uparrow x] = \beta \downarrow [\alpha \uparrow (\beta \downarrow x)]$

Proof.

$$
\begin{aligned}
\alpha \uparrow [\beta \downarrow (\alpha \uparrow x)] &= \alpha \uparrow [(\beta \downarrow \alpha) \uparrow (\beta \downarrow x)] \quad \text{by distributivity} \\
&= \quad \alpha \uparrow (\beta \downarrow x) \qquad \text{because } (\beta \downarrow \alpha) \leq \alpha
\end{aligned}
$$

The second equation is proved similarly.

Definition 12 (Equality modulo $\alpha\beta$)). *We will write* $AlphaBeta(\pi, \alpha, \beta) = _{\alpha\beta} z$, *where* $z \in D$, *as an abbreviation of one of the following equations over* D, *according to the type of the position* π:

$$
\begin{aligned}
\alpha \uparrow [\beta \downarrow AlphaBeta(\pi, \alpha, \beta)] &= \alpha \uparrow [\beta \downarrow z] \qquad if \, \pi \in WPOS \\
\beta \downarrow [\alpha \uparrow AlphaBeta(\pi, \alpha, \beta)] &= \beta \downarrow [\alpha \uparrow z] \qquad if \, \pi \in BPOS
\end{aligned}
$$

Theorem 1 (Relative correctness of the *AlphaBeta* method)). *Let* $(D, \leq)$ *be a distributive lattice and suppose to apply the method AlphaBeta with* $\uparrow = \vee$ *(the sup) and* $\downarrow = \wedge$ *(the inf) of* D, *and with a couple of monotone heuristics* h_{pess} *and* h_{opt}, *respectively pessimistic and optimistic and such that* $h_{pess}(\pi) \leq h_{opt}(\pi)$ *for all position* $\pi \in NTPOS$. *Then, for all position* $\pi \in POS$, *and for all* $\alpha, \beta \in D$, *if the function* $AlphaBeta(\pi, \alpha, \beta)$ *terminates, then there exist* $x \in SetOfVal_{pess}(\pi)$ *and* $y \in SetOfVal_{opt}(\pi)$ *such that:*

$$
\begin{cases}
AlphaBeta(\pi, \alpha, \beta) =_{\alpha\beta} x \\
AlphaBeta(\pi, \alpha, \beta) =_{\alpha\beta} y
\end{cases}
$$

Proof. This is a long induction on the structure of the finite part of the tree visited by the algorithm whenever it terminates. Full details are given in the appendix.

Corollary 2 (Correctness of the *AlphaBeta* method)). *In the hypothesis of the theorem 1, suppose also that* D *contains a minimal element* $\bot$ *and a maximal element* $\top$. *Then, if the method* $AlphaBeta(\pi, \alpha, \beta)$ *terminates returning the value* v *then*

$$
v = Val_{h_{pess}}(\pi) = Val_{h_{opt}}(\pi) = Val \, \pi
$$

4   An Application to Logic Programming

We believe that, while the core result of the paper is really the proof of correctness we provided, it would not be satisfactory to conclude it without providing a significant example of application outside the scope of traditional game-playing like chess. So we turn now to our favorite example, that was also the starting point of [5]: the world of logic programming. In what follows, we radically improve w.r.t. [5] by actually treating *constraint* logic programming (that includes traditional logic programming via the special case of Herbrand constraints).

We suppose that the constraints C come equipped with an intersection operator $\wedge$ (for Herbrand terms, this is the usual most general unifier for substitutions), and a disjunction operator $\vee$. We also suppose that the relation of logical implication of constraints $\Rightarrow$ is a partial order on C, and we assume the empty constraint *true*, and the never satisfied constraint *false*.

We now give the syntax and semantics of the game of a constraint logic program, but we need to assume, due to lack of space, familiarity with constraint logic programming (CLP) (see [8] for an introduction).

4.1   Syntax of the Game

Definition 13 (Positions). *The set of positions $\pi \in POS$ is defined as follows*

$$WPOS \ni \pi ::= (A, c)$$
$$BPOS \ni \pi ::= [G, c]$$

where G is a positive (conjunctive) goal, A is a positive atom and c is a (conjunctive) constraint in C.

We follow the traditional notation in game theory that uses circles (here, parentheses) for player positions, and squares (here square parentheses) for opponent positions.

Definition 14 (Initial Positions). *The game starts from an opponent position having an empty constraint:*

$$IPOS \ni \pi ::= [G, true]$$

We now define the transition relation by giving explicitly the *moves* function.

Definition 15 (Rules of a CLP game, $\rightarrow_{CLP}$). *The possible moves, that define the transition relation $\rightarrow_{CLP}$ for the CLP game for a given CLP program P, are as follows:*

player moves *in a player position (A, c), player can choose any (renaming of) a rule $A \leftarrow c' \mid A_1, A_2, ..., A_n$ of P, s.t. $c \wedge c'$ is satisfiable in the constraint algebra of C, and reach an opponent position $[(A_1, A_2, ..., A_n), c \wedge c']$*

opponent moves *in an opponent position $[(A_1, A_2, ..., A_n), c]$, opponent can choose any atom A_i and reach a player position (A_i, c).*

4.2   Semantics of a CLP(C) Game

The game of a CLP program can easily be infinite, so we need an approximation structure and some heuristic functions to obtain an approximation of the game value. The values we have to handle are "disjunctions" of constraints, represented as sets of constraints, hence subsets of the powerset of C. Such disjunctions denotes a solution space in any model of C, and then we say that a set of constraint is greater than another when its denotation is bigger (covers) the denotation of the other. More formally

Definition 16 (The partially ordered domain of values). *The evaluation domain is $D = 2^C$, the powerset of the constraints. We equip D with a partial order $\leq$ which is the covering extension [7] of the partial order induced on the constraint by the logical implication of constraints:*

$$d_1 \leq d_2 \text{ iff } \forall c_1 \in d_1 \exists e_1 \ldots e_n \in d_2 \text{ t.q. } c_1 \Rightarrow \bigvee_{i=1,\ldots,n} e_i$$

Definition 17 (Evaluation of terminal positions). *The evaluation function on terminal nodes is defined as*

$$\begin{cases} h\,(A,c) = \emptyset \\ h\,[\Diamond,c] = c \end{cases}$$

where $\Diamond$ is the empty goal.

Player takes the set theoretic union of values, while opponent takes the intersection of the denotations of the constraints, more formally

Definition 18 (Player and opponent choice functions). *The functions $\uparrow$,$\downarrow$: $(2^C)^n \to 2^C$, are defined as $\uparrow = \cup$ and $\downarrow = \wedge$ where $\wedge$ is defined as $d_1 \wedge \ldots \wedge d_n = \{c_1 \wedge \ldots \wedge c_n \mid c_1 \in d_1, \ldots, c_n \in d_n\}$. Both functions are monotone w.r.t. the order on D.*

We define now two heuristics as follows

Definition 19 (Heuristics). *For any nonterminal position $\pi = (A,c)$ or $\pi = [G,c]$*

$$\begin{cases} h_{pess}(\pi) = \emptyset \\ h_{opt}(\pi) = \{c\} \end{cases}$$

In this game, if $\pi \to_{\mathcal{CLP}} \pi'$, then the constraint component c' of π' is either the constraint component c of π, or $c \wedge c''$ for some of the constraint c''. In both case, $c' \Rightarrow c$, hence $h_{opt}(\pi) \geq h_{opt}(\pi')$. Also, trivially $h_{pess}(\pi) \leq h_{pess}(\pi')$, so both heuristics are monotone, and we have at hand both an optimistic and a pessimistic approximation structure for the game of $CLP(C)$.

This general construction yields an interesting object when the evaluation domain 2^C satisfies the condition for applying Alpha-Beta; this depends in general on C, but there are many examples that satisfy the requirement, like the case of *ex-equations* Eqn (see [10]), for which it is easy to verify that $\uparrow = \cup$ and $\downarrow = $ *most general unifier* $(for\ ex - equations)$ are the sup and inf on the domain 2^{Eqn} and that they satisfy the hypothesis of the Alpha-Beta theorem, that can then be used to compute the answers.

Let us see how Alpha-Beta works on an example

Example 2. Consider the (classical) logic program

```
1. p(f(Y)).
2. p(X)  :- q(X),r(X).
3. q(f(a)).
4. q(f(f(a))).
5. r(X)  :- ...
```

Here, if we consider the goal $p(X)$, once we use rule 1, and we get the answer $\exists Y.X = f(Y)$, it is useless to try rule 2, as rule 2 finds answers for $q(X)$ of the shape $\{X = f(a), X = f(f(a))\}$. Since the most general unifier gives always a result which is less than its arguments, we cannot find anything better than $\{X = f(a), X = f(f(a))\}$. Without even knowing the answer for $r(X)$, we can give up the search right there. This is what Alpha-Beta does, by cutting the subtree rooted at $r(X)$ and exiting with the answer $\{\exists Y.X = f(Y)\}$:

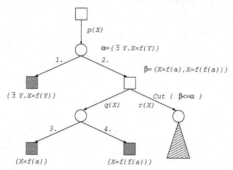

We provide a second example where the cut takes place at a player node.

Example 3. Consider the program

```
1. p(X,Y,Z)  :- q(X,Y),r(X,Z).
2. q(a,b).
3. r(a,Z).
4. r(X,Z)  :- ...
```

Starting with a goal $p(X, Y, Z)$, we use rule 1, then we evaluate the sub-goal $q(X, Y)$, that forces $X = a$, and we use rule 3 to solve $r(X, Z)$. After that, we have a solution where $X = a$ and we force no constraint on Z. That means that there is no interest in looking for other solutions to $r(X, Z)$, as $X = a$ is forced by $q(X, Y)$. Again, this is what AlphaBeta really does, with a cut condition $\alpha \geq \beta$:

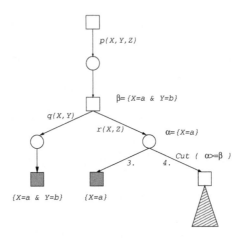

5   Conclusions and Future Work

We have introduced a formal framework to extend the traditional notion of value of a combinatorial game in order to deal with games whose value can now range on arbitrary partially ordered domains. Also, by means of the notion of approximation structure (pessimistic and optimistic), we can now give a precise meaning to the value of a potentially infinite game. In this general framework, we have generalized the well-known Alpha-Beta algorithm to compute values in an arbitrary distributive lattice, and formally proved that this algorithm is partially correct in general on infinite games, and correct on finite ones. This is, in our opinion, a significant achievement, as the algorithm can now be confidently applied to many different fields, quite far apart from the traditional fields of application of combinatorial game theory. As an example, we have provided an instantiation of the game framework for (constraint) logic programming, that includes the minimax characterization of the semantics of logic programming formally introduced by the authors in [5], and we have shown how the Alpha-Beta method can be used to compute (efficiently) the set of answers for a given goal.

Another very promising direction of research is in the field of abstract interpretation, where the very same algorithm could be used to compute the abstract value of a program, by just changing the value domain, thus providing a more efficient means of performing static analysis.

Finally, we believe that this approach can be interesting also for the study of logic programming in the presence of negative atoms in the clauses.

References

1. S. Abramsky and R. Jagadeesan. Games and full completeness for multiplicative linear logic. *The Journal of Symbolic Logic*, 59(2):543–574, 1994. 207
2. A. Blass. A game semantics for linear logic. *Annals of Pure and Applied Logic*, 56:pages 183–220, 1992. 207

3. A. Bossi, M. Gabbrielli, G. Levi, and M. Martelli. The s-semantics approach: Theory and applications. *Journal of Logic Programming*, 19-20:149–197, 1994. 208

4. P. Curien and H. Herbelin. Computing with abstract bohm trees. 1996. 207

5. R. Di Cosmo, J.-V. Loddo, and S. Nicolet. A game semantics foundation for logic programming. In C. Palamidessi, H. Glaser, and K. Meinke, editors, *PLILP'98*, volume 1490 of *Lecture Notes in Computer Science*, pages 355–373, 1998. 208, 217, 217, 220

6. G. Diderich. A survey on minimax trees and associated algorithms. *Minimax and its Applications, Kluwer Academic Publishers*, 1995. 208

7. F. Fages. Constructive negation by pruning. *Journal of Logic Proigramming*, 2(32):85–118, 1997. 218

8. J. Jaffar and M. J. Maher. Constraint logic programming: A survey. *Journal of Logic Programming*, 19/20:503–581, 1994. 217

9. A. Joyal. Free lattices, communication and money games. *Proceedings of the 10th International Congress of Logic, Methodology, and Philosophy of Science*, 1995. 207

10. N. J. Kim Marriot, Harald Søndergaard. Denotational abstract interpretation of logic programs. *ACM Transaction on Programming Languages and Systems*, 16(3):607–648, 1994. 219

11. F. Lamarche. Game semantics for full propositional linear logic. *Proceedings of the 10th Annual IEEE Symposium on Logic in Computer Science*, pages 464–473, 1995. 207

12. G. Owen. *Game Theory*. W.B. Saunders, 1968. 209

13. V. D. P. Baillot and T. Ehrhard. Believe it or not, AJM's games model is a model of classical linear logic. *Proceedings of the 12th Annual IEEE Symposium on Logic in Computer Science*, pages pages 68–75, 1997. 207

14. J. Pearl. *Heuristics. Intelligent Search Strategies for Computer Problem Solving*. Addison Wesley, 1984. 208, 210, 215

15. A. Plaat. *Research Re: search & Re-search*. PhD thesis, Tinbergen Institute and Department of Computer Science, Erasmus University Rotterdam, 1996. 208

16. J. von Neumann. Zur Theorie der Gesellschaftsspiele. *Mathaematische Annalen*, (100):195–320, 1928. 208

A   Proof of the Relative Correctness Theorem

Here is the proof of Theorem 1

Proof. If the method terminates this means that it computes the value visiting only a finite part of the game tree (possibly infinite). So we can proceed by induction on the structure of this finite part. We give the proof of the case $\pi \in WPOS$, i.e. the player positions. For opponent positions the proof is similar.

Base case. The method visits a unique node and returns the value $v = AlphaBeta(\pi, \alpha, \beta) \in D$. There are only two possibilities:

– The node is terminal: $\pi \in TPOS$

Implies $v = h(\pi)$ so $x = h(\pi) \in SetOfVal_{pess}(\pi)$ and $y = h(\pi) \in SetOfVal_{opt}(\pi)$ verify trivially the equality modulo $\alpha\beta$ with v.

– The node was immediately cut because $v \geq \beta$

Let $x = h_{pess}(\pi) \in SetOfVal_{pess}(\pi)$ and $y = h_{opt}(\pi) \in SetOfVal_{opt}(\pi)$. Then

$$
\begin{aligned}
\alpha \uparrow [\beta \downarrow v] &= \alpha \uparrow [\beta \downarrow (\alpha \uparrow h_{pess}(\pi))] && because\, v = \alpha \uparrow h_{pess}(\pi) \\
&= \alpha \uparrow [\beta \downarrow h_{pess}(\pi)] && by\, lemma\, 1 \\
&= \alpha \uparrow [\beta \downarrow x]
\end{aligned}
$$

So $v =_{\alpha\beta} x$. Moreover, since $\alpha \uparrow h_{pess}(\pi) \geq \beta$ we have also $\alpha \uparrow [\beta \downarrow v] = \alpha \uparrow \beta$. Using the hypothesis $h_{pess}(\pi) \leq h_{opt}(\pi)$, we obtain, by monotonicity of the operator $\uparrow$, that also $\alpha \uparrow h_{opt}(\pi) \geq \beta$ so $\alpha \uparrow [\beta \downarrow y] = \alpha \uparrow [\beta \downarrow (\alpha \uparrow h_{opt}(\pi))] = \alpha \uparrow \beta$, that means $v =_{\alpha\beta} y$.

Structural induction. In the position π, we suppose that the algorithm has analyzed n moves of the player among the m possibles moves (with $m \geq n$) before returning the result. We now proceed by a sub-induction on n.

– Case $n = 1$

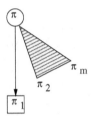

Implies $v = AlphaBeta(\pi, \alpha, \beta) = \alpha \uparrow h_{pess}(\pi) \uparrow AlphaBeta(\pi_1, v_0, \beta)$ $(*)$ where $v_0 = \alpha \uparrow h_{pess}(\pi)$. By inductive hypothesis on structure of tree we have that exist $x_1 \in SetOfVal_{pess}(\pi_1)$ and $y_1 \in SetOfVal_{opt}(\pi_1)$ such that:

$$
\begin{cases}
AlphaBeta(\pi_1, v_0, \beta) =_{v_0\beta} x_1 \\
AlphaBeta(\pi_1, v_0, \beta) =_{v_0\beta} y_1
\end{cases}
$$

Since $\pi_1 \in BPOS$ is an opponent position, that means:

$$
\beta \downarrow [v_0 \uparrow AlphaBeta(\pi_1, v_0, \beta)] = \beta \downarrow [v_0 \uparrow x_1] = \beta \downarrow [v_0 \uparrow y_1] \qquad (**)
$$

We distinguish now between two reasons of termination of the algorithm: the case $n = m = 1$ (all the moves have been evaluated) and the case $n = 1 < m$ (some moves have been cut).

– Sub-case $n = m = 1$

We define the approximations: $x =\uparrow x_1 = x_1$ and $y =\uparrow y_1 = y_1$. By definition $x \in SetOfVal_{pess}(\pi)$ et $y \in SetOfVal_{opt}(\pi)$, so:

$$
\begin{aligned}
\alpha \uparrow [\beta \downarrow v] &= \alpha \uparrow [\beta \downarrow (v_0 \uparrow AlphaBeta(\pi_1, v_0, \beta))] && by\, (*) \\
&= \alpha \uparrow [\beta \downarrow (\alpha \uparrow h_{pess}(\pi) \uparrow x_1)] && by\, (**) \\
&= \alpha \uparrow [\beta \downarrow (h_{pess}(\pi) \uparrow x)] && by\, Lemma\, 1\, and\, x = x_1 \\
&= \alpha \uparrow [\beta \downarrow x] && because\, h_{pess}(\pi) \leq x
\end{aligned}
$$

The last fact in proof $h_{pess}(\pi) \leq x$ is insured because h_{pess} is a monotone pessimistic. With the same logical steps we prove $\alpha \uparrow [\beta \downarrow v] = \alpha \uparrow [\beta \downarrow y]$ where the necessary condition $h_{pess}(\pi) \leq y$ is insured by prop. 3 using the hypothesis $h_{pess}(\pi) \leq h_{opt}(\pi)$.

– Sub-case $n = 1 < m$

If the algorithm has cut the $m-1$ remaining moves, this means that the condition $v_0 \uparrow v_1 \geq \beta$, where $v_1 = AlphaBeta(\pi_1, v_0, \beta)$, was true. This implies $\alpha \uparrow [\beta \downarrow v] = \alpha \uparrow [\beta \downarrow (v_0 \uparrow v_1)] = \alpha \uparrow \beta$. We can define the approximations $x = x_1 \uparrow x_1'$ where $x_1' = h_{pess}(\pi_2) \uparrow \ldots \uparrow h_{pess}(\pi_m)$ and $y = y_1 \uparrow y_1'$ where $y_1' = h_{opt}(\pi_2) \uparrow \ldots \uparrow h_{opt}(\pi_m)$. By definition $x \in SetOfVal_{pess}(\pi)$ and $y \in SetOfVal_{opt}(\pi)$. Then

$$
\begin{aligned}
\alpha \uparrow [\beta \downarrow x] = \quad & \alpha \uparrow [\beta \downarrow (\alpha \uparrow x)] && by\ lemma\ 1 \\
= \ & \alpha \uparrow [\beta \downarrow (\alpha \uparrow h_{pess}(\pi) \uparrow x)] && because\ h_{pess}(\pi) \leq x \\
= \ & \alpha \uparrow [\beta \downarrow (v_0 \uparrow x_1 \uparrow x_1')] && by\ def.\ of\ v_0\ and\ x \\
= \ & \alpha \uparrow [\beta \downarrow (v_0 \uparrow x_1)] \uparrow [\beta \downarrow x_1'] && by\ distributive\ law \\
= \ & \alpha \uparrow [\beta \downarrow (v_0 \uparrow v_1)] \uparrow [\beta \downarrow x_1'] && by\ (**) \\
= \ & \alpha \uparrow \beta \uparrow [\beta \downarrow x_1'] && because\ v_0 \uparrow v_1 \geq \beta \\
= \ & \alpha \uparrow \beta && because\ \beta \downarrow x_1' \leq \beta
\end{aligned}
$$

Using the prop. 3 we can infer $h_{pess}(\pi) \leq y$ and use it to prove $\alpha \uparrow [\beta \downarrow y] = \alpha \uparrow \beta$ with these same logical steps.

– Case $n > 1$

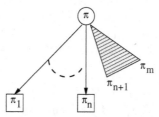

By definition of the method, if we consider a fake position π' with only the first $n-1$ moves of π, and another fake position π'' having only the nth move, we can write $v = AlphaBeta(\pi, \alpha, \beta) = AlphaBeta(\pi', \alpha, \beta) \uparrow AlphaBeta(\pi'', v_1, \beta)$ where $v_1 = AlphaBeta(\pi', \alpha, \beta)$, considering $h_{pess}(\pi') = h_{pess}(\pi'') = h_{pess}(\pi)$. Then, by inductive hypothesis on n there exists $x' \in SetOfVal_{pess}(\pi')$ and $y' \in SetOfVal_{opt}(\pi')$ such that

$$
(\#) \begin{cases} AlphaBeta(\pi', \alpha, \beta) =_{\alpha\beta} x' \\ AlphaBeta(\pi', \alpha, \beta) =_{\alpha\beta} y' \end{cases}
$$

and exist $x'' \in SetOfVal_{pess}(\pi'')$ and $y'' \in SetOfVal_{opt}(\pi'')$ such that

$$
(\#\#) \begin{cases} AlphaBeta(\pi'', v_1, \beta) =_{v_1\beta} x'' \\ AlphaBeta(\pi'', v_1, \beta) =_{v_1\beta} y'' \end{cases}
$$

Then

$$
\begin{aligned}
\alpha \uparrow [\beta \downarrow v] =\ & \alpha \uparrow [\beta \downarrow (v_1 \uparrow AlphaBeta(\pi'', v_1, \beta))] && by\, def.\, of\, v\, and\, v_1 \\
=\ & \alpha \uparrow [\beta \downarrow (v_1 \uparrow (\beta \downarrow AlphaBeta(\pi'', v_1, \beta)))] && by\, lemma\, 1 \\
=\ & \alpha \uparrow [\beta \downarrow (v_1 \uparrow (\beta \downarrow x''))] && by\, (\#\#) \\
=\ & \alpha \uparrow [\beta \downarrow (v_1 \uparrow x'')] && by\, lemma\, 1 \\
=\ & \alpha \uparrow [\beta \downarrow v_1] \uparrow [\beta \downarrow x''] && by\, distributive\, law \\
=\ & \alpha \uparrow [\beta \downarrow x'] \uparrow [\beta \downarrow x''] && by\, (\#) \\
=\ & \alpha \uparrow [\beta \downarrow (x' \uparrow x'')] && by\, distributive\, law
\end{aligned}
$$

We have proved $\alpha \uparrow [\beta \downarrow v] = \alpha \uparrow [\beta \downarrow (x' \uparrow x'')]$ $(*)$ and, with the same arguments, we can prove $\alpha \uparrow [\beta \downarrow v] = \alpha \uparrow [\beta \downarrow (y' \uparrow y'')]$ $(**)$. We distinguish now between two reasons of termination of the algorithm: the case $n = m$ (all the moves have been evaluated) and the case $n < m$ (some moves have been cut).

– Sub-case $1 < n = m$

The moves of π are exactly the set of moves of π' together with the moves of π''. Then, defining the approximations: $x = x' \uparrow x''$ and $y = y' \uparrow y''$ we have $x \in SetOfVal_{pess}(\pi)$ and $y \in SetOfVal_{opt}(\pi)$. Hence conditions $(*)$ and $(**)$ give our thesis.

– Sub-case $1 < n < m$

If the algorithm has cut the $m - n$ remaining moves, this means that condition $v_0 \uparrow v_1 \geq \beta$ where $v_0 = AlphaBeta(\pi', \alpha, \beta)$ was true. Then $\alpha \uparrow [\beta \downarrow v] = \alpha \uparrow [\beta \downarrow (v_0 \uparrow v_1)] = \alpha \uparrow \beta$ $(***)$. We define the approximations: $x = x' \uparrow x'' \uparrow x'''$ where $x''' = h_{pess}(\pi_{n+1}) \uparrow \ldots \uparrow h_{pess}(\pi_m)$ and $y = y' \uparrow y'' \uparrow y'''$ where $y''' = h_{opt}(\pi_{n+1}) \uparrow \ldots \uparrow h_{opt}(\pi_m)$. By definition $x \in SetOfVal_{pess}(\pi)$ and $y \in SetOfVal_{opt}(\pi)$. Then

$$
\begin{aligned}
\alpha \uparrow [\beta \downarrow x] =\ & \alpha \uparrow [\beta \downarrow (x' \uparrow x'' \uparrow x''')] && by\, def.\, of\, x \\
=\ & \alpha \uparrow [\beta \downarrow (x' \uparrow x'')] \uparrow [\beta \downarrow x'''] && by\, associative\, and\, distributive\, laws \\
=\ & \alpha \uparrow [\beta \downarrow v] \uparrow [\beta \downarrow x'''] && by\, (*) \\
=\ & \alpha \uparrow \beta \uparrow [\beta \downarrow x'''] && by\, (***) \\
=\ & \alpha \uparrow \beta && because\, \beta \downarrow x''' \leq \beta
\end{aligned}
$$

With the same logical steps, simply using $(**)$ instead of $(*)$, we obtain $\alpha \uparrow [\beta \downarrow y] = \alpha \uparrow \beta$.

Logic Programming Approaches for Representing and Solving Constraint Satisfaction Problems: A Comparison

Nikolay Pelov, Emmanuel De Mot, and Marc Denecker

Department of Computer Science, K.U.Leuven
Celestijnenlaan 200A, B-3001 Heverlee, Belgium
E-mail: {pelov,emmanuel,marcd}@cs.kuleuven.ac.be

Abstract. Many logic programming based approaches can be used to describe and solve combinatorial search problems. On the one hand there is constraint logic programming which computes a solution as an answer substitution to a query containing the variables of the constraint satisfaction problem. On the other hand there are systems based on stable model semantics, abductive systems, and first order logic model generators which compute solutions as models of some theory. This paper compares these different approaches from the point of view of knowledge representation (how declarative are the programs) and from the point of view of performance (how good are they at solving typical problems).

1 Introduction

Consistency techniques are widely used for solving finite domain constraint satisfaction problems (CSP) [19]. These techniques have been integrated in logic programming, resulting in finite domain constraint logic programming (CLP) [20]. In this paradigm, a program typically creates a data structure holding the variables of the CSP to be solved, sets up the constraints and uses a labelling technique to assign values to the variables. The constraint solver uses consistency techniques to prune the search. This leads to a rather procedural programming style. Moreover, the problem description is not very declarative because the mapping between domain variables and their value has an indirect representation in a term structure.

In this paper, we compare CLP and three computational paradigms allowing problem solving based on more declarative representations. A common feature of these approaches is that the relation between the CSP variables and their values is encoded as a predicate or function relating identifiers of the CSP variables with their value. E.g. in the graph coloring problem, the predicate relates node numbers with colors. This representation allows for a more natural declarative representation of the problem.

One approach is specification in first order logic. As pointed out in [12], one can represent a CSP as a first order logic theory such that (part of) its models

M. Parigot and A. Voronkov (Eds.): LPAR 2000, LNAI 1955, pp. 225–239, 2000.

correspond to the solutions of the CSP. Hence first order model generators such as SEM [24] can be used to solve such problems.

The two other approaches use extensions of logic programming. Recently, a logic programming paradigm based on stable model semantics [6] has emerged. Niemelä [14] proposes it as a constraint programming paradigm, Marek and Truszczyński [13] introduce Stable Logic Programming and Lifschitz [11] proposes Answer Set Programming. As described in [13], the methodology of these approaches is to encode a computational problem by a logic program such that its stable models represent the solutions. A number of efficient systems for computing stable models have been developed. Of these, Niemelä's SMODELS [15, 14] is considered one of the most performant systems.

Abduction [8] uses a similar predicate representation for the relation between the identifiers of CSP variables and their value. This predicate is declared to be open or abducible. Constraining this relation to be a solution, an abductive system will return models of the abducible which are solutions of the CSP.

We use some typical CSP problems to compare the merits of the various approaches. One experiment is in graph coloring. We have compared the representation and the performance of CLP with the three other approaches in a sequence of experiments where the size of the graph increases and the number of colors remains constant. Another experiment is the n-queens problem where both the domain size and the number of constraints increases with increasing problem size. We also report on experiments using CLP, stable logic programming and abduction for solving a complex real world scheduling problem. For each different system, we have tried to use any special features provided by it.

In Section 2 we review in more detail the various approaches and systems, focusing mainly on the knowledge representation aspects. Section 3 reports on the experiments and we conclude in Section 4.

We are not aware of any previous work which compares this wide range of logic based systems for their suitability in solving CSP problems. Mackworth [12] explores the space of possible CSP formalizations but assesses neither the quality from point of view of knowledge representation nor the performance of actual systems. Also, approaches based on stable model semantics and abduction are not included in his work. This paper is an extension and revision of [17] which focuses more on the formal relations between the declarative specifications of the problems on the different systems.

One more problem which uses aggregate functions is included in the present paper. So is an additional experiment for finding all solutions of the n-queens problem. Finally, some comments from the authors of the different systems were taken into account.

2   Formalisms and Systems

A *constraint satisfaction problem* (CSP) is usually defined as a finite set of *constraint variables* $\mathcal{X} = \{X_1, \ldots, X_n\}$ (the variables of the CSP), a finite domain D_i of possible values for each variable X_i, and a finite set of *constraint relations*

$\mathcal{R}$ where each $r \in \mathcal{R}$ is a constraint between a subset of the set $\mathcal{X}$ of variables. A *solution* is an instantiation of the variables of $\mathcal{X}$ which satisfies all the constraints in $\mathcal{R}$.

2.1   Constraint Logic Programming

Constraint logic programming (CLP) [7] is an extension of logic programming where some of the predicate and function symbols have a fixed interpretation over some subdomain (e.g. finite trees or real numbers). Special purpose constraint solvers are integrated with a logic programming system for efficient reasoning on these symbols. This results in a very expressive language which can efficiently solve problems in many domains.

Van Hentenryck [20] pioneered the work on finite domain constraint logic programming, CLP(FD), by introducing domain declarations for the logic variables and integrating consistency techniques as part of the SLD proof procedure. A CLP(FD) system supports standard arithmetic relations $(=, \neq, <)$ and functions $(+, -, *)$ on the natural numbers. A typical formulation of the n-queens problem is as follows:

$$queens(N, L) \leftarrow$$
$$length(L, N),$$
$$domain(L, 1, N),$$
$$constrain\_all(L),$$
$$labeling(L).$$
$$constrain\_all([]).$$
$$constrain\_all([X|Xs]) \leftarrow$$
$$constrain\_between(X, Xs, 1)$$
$$constrain\_all(Xs).$$
$$constrain\_between(X, [], N).$$
$$constrain\_between(X, [Y|Ys], N) \leftarrow$$
$$safe(X, Y, N),$$
$$N_1 \ is \ N + 1,$$
$$constrain\_between(X, Ys, N_1).$$
$$safe(X_1, X_2, D) \leftarrow$$
$$X_1 \neq X_2, abs(X_1 - X_2) \neq D.$$

Executing the query $queens(n, L)$ first creates a list L with n variables where the ith variable gives the column position of the queen on row i. Then the constraints expressed with the $safe/3$ predicate are added by using two nested recursive predicates. Such procedural code for setting up constraints and the encoding of the solution in a large data structure results in a rather procedural style which is typical for the CLP approach.

2.2   First Order Logic: Model Generation

The most elegant solution for the n-queens problem is using many sorted first order logic and first order model generation. Systems like FINDER and SEM

[24] are examples. One can introduce functions (with the sorts of their domain and range) and predicates (with the sorts of their domains and the sort *bool* as range). In addition, functions can be restricted to be injective, bijective, ... This allows to express the n-queens problem very concisely as:

$$D = \{1..n\}$$

$$pos : D \to D \quad (bijection)$$

$$abs(pos(X_1) - pos(X_2)) \neq X_2 - X_1 \leftarrow X_1 < X_2.$$

The first line declares D as a sort with interpretation consisting of the set of integers 1 to n. The following line introduces the function $pos/1$ as a bijection from D to D. Hence, the range of the function is a permutation of its domain. This function represents the column positions of the queens. The only remaining constraint is that queens have to be on different diagonals. This is expressed by the formula on the third line using the predefined functions $abs/1$ and $-/2$. Due to symmetry, one need only to verify the constraint for pairs of queens X_1, X_2 such that $X_1 < X_2$.

Solutions are given by the interpretation of the $pos/1$ function in the models of this theory. In principle, this approach is applicable on any CSP problem by representing the CSP variables by logical constants. However, in most cases, CSP variables are just an encoding of some attribute of a set of first order objects, such as the position of a queen or the color of a node in a graph. In such cases, there is no need to introduce the CSP variable. The attribute can be represented directly as a function or predicate on these objects (e.g. *pos*).

As the domains of all sorts are finite, SEM first computes the grounding of the theory and then uses backtracking combined with various inference and simplification rules to guide the search for models [24].

2.3 Stable Logic Programming

In [14], Niemelä proposes logic programming with the stable model semantics [6] as a constraint logic programming paradigm. The underlying idea is to represent a problem as a set of rules, each rule being the declarative expression of a piece of knowledge about the problem domain and such that the stable models of the whole program are constrained to be solutions of the problem.

The SMODELS system [15] is an efficient implementation of the stable model semantics. It works with propositional rules and a special pre-processing program is used for grounding strongly range restricted logic programs. The implementation combines bottom-up inference with backtracking search and employs powerful pruning methods. A recent extension of the system [16] introduces choice rules:

$$l \{l_1, l_2, \ldots l_n\} u \leftarrow B.$$

where $l_1, l_2, \ldots l_n$ are literals. The semantics of such a rule is that if the body B is true then at least l and at most u literals among l_i should be true in a stable model of the program.

Following [14] and [16], the program for the n-queens problems can be formulated as:

$d(1..n)$.

$1 \{pos(X, Y) : d(Y)\} 1 \leftarrow d(X)$.
$1 \{pos(X, Y) : d(X)\} 1 \leftarrow d(Y)$.

$\leftarrow d(X_1), d(Y_1), d(X_2), d(Y_2), pos(X_1, Y_1), pos(X_2, Y_2),$
$\quad X_1 < X_2, X_2 - X_1 = abs(Y_1 - Y_2)$.

Solutions are given by the $pos(i, j)$ atoms in the stable models of the program. The first line defines that $d/1$ is a domain with elements $1..n$ with n the size of the board. The first choice rule is used to define the solution space of the problem by stating that for each X in the domain $d(X)$, there exists exactly one Y such that $pos(X, Y)$ is true. The colon notation denotes an expansion of $pos(X, Y)$ for every value of Y. Similarly, the second choice rule expresses that there is exactly one queen on each column. The last rule defines the final constraint of the problem: no two queens on the same diagonal. Again, the "<" constraints in these rules eliminate instances which are redundant due to symmetry. The main difference with the first order logic specification is that the mapping between queens and their position is now represented by a predicate. Declaring that this predicate represents a bijective function is succinctly expressed by the two choice rules.

2.4 Abduction

Abductive logic programming [8] extends the logic programming paradigm with abductive reasoning. An abductive logic program has three components: (1) a logic program P, (2) a set of predicates A called abducibles or open predicates, and (3) a set of integrity constraints I. The abducibles are predicates not defined in the program. The task of an abductive system is to find a set Δ of ground abducible atoms such that the integrity constraints are true in the logic program consisting of $P \cup \Delta$; formally: $P \cup \Delta \models I$.

Kakas and Michael proposed an integration of CLP and an abductive logic programming system [9]. Originally, it was defined only for definite programs and integrity constraints and in [10] it was extended to deal with negation as failure through abduction in a similar way as in [5]. One restriction of ACLP is that integrity constraints need to be of the form $\leftarrow a(\bar{X}), B$, where a is an abducible. As we will see, this forces sometimes to reformulate some constraints by an additional recursion. Such restrictions are not present in SLDNFAC [3], a more recent integration of an abductive system with CLP that is based on the more general abductive procedure SLDNFA [2].

The SLDNFAC system uses ID-Logic [1] as specification language which is transformed into an abductive logic program by using a Lloyd-Topor transformation. The specification of the n-queens problem is:

$d(1..n).$

$open\_function(pos(d, d)).$

$$Y_1 \neq Y_2 \wedge X_2 - X_1 \neq Y_2 - Y_1 \wedge X_2 - X_1 \neq Y_1 - Y_2$$
$$\Leftarrow pos(X_1, Y_1) \wedge pos(X_2, Y_2) \wedge X_1 < X_2.$$

The first line of the program defines $d/1$ as a domain predicate with the integers $1..n$ as elements (defining rows and columns). The next line states that the predicate $pos/2$ represents an open function in the defined domain. It is used to represent the column position of a queen in a row. Finally there is a constraint saying that two queens can not be on the same column and diagonal. This representation is almost identical to the FOL specification of section 2.2. The main difference is that the open function is represented by a predicate.

As mentioned, ACLP does not allow function declarations. Consequently, the fact that pos predicate represents a function must be expressed by explicit constraints. A standard way to axiomatize that the abductive predicate $pos(X, Y)$ should be true for each X in the domain $d(X)$ is by using the following rule and integrity constraints:

$has\_pos(X) \leftarrow d(Y), pos(X, Y).$
$\leftarrow d(X), not\ has\_pos(X).$

Unfortunately, the integrity constraint does not satisfy the ACLP's restriction that at least one positive abductive atom should occur in it. Hence, these axioms have to be reformulated using a recursive program which generates a position for each queen. The specification for the ACLP system is:

$A = \{pos/2\}$

$problem(N) \leftarrow nqueens(N, N).$
$nqueens(0, N).$
$nqueens(X, N) \leftarrow X > 0, Y\ in\ 1..N,\ pos(X, Y),$
$\quad\quad X_{next}\ is\ X - 1,\ nqueens(X_{next}, N).$

$attack(X_1, Y_1, X_2, Y_2) \leftarrow Y_1 = Y_2.$
$attack(X_1, Y_1, X_2, Y_2) \leftarrow Y_1 + X_1 = Y_2 + X_2.$
$attack(X_1, Y_1, X_2, Y_2) \leftarrow Y_1 - X_1 = Y_2 - X_2.$

$\leftarrow pos(X_1, Y_1),\ pos(X_2, Y_2),\ X_1 < X_2,\ attack(X_1, Y_1, X_2, Y_2).$

The n-queens problem is solved by solutions of the abductive query $\leftarrow problem(n)$. The ACLP representation is in the middle of the declarative FOL representation and the more procedural CLP representation.

3 Experiments

3.1 The Systems

The finite domain CLP package is the one provided with ECL^iPS^e version 4.2.

Both abductive systems, ACLP [10] and SLDNFAC, [3] are meta interpreters written in Prolog, running on ECL^iPS^e version 4.2 and making use of its finite domain library. For all these systems, a search strategy which first selects variables with the smallest domain which participate in the largest number of constraints was used.

The model generator SEM version 1.7 is a fine tuned package written in C. SMODELS version 2.25, the system for computing stable models, is implemented in C++ and the associated program used for grounding is LPARSE version 0.99.54. All experiments have been done on the same hardware, namely Pentium II.

3.2 Graph Coloring

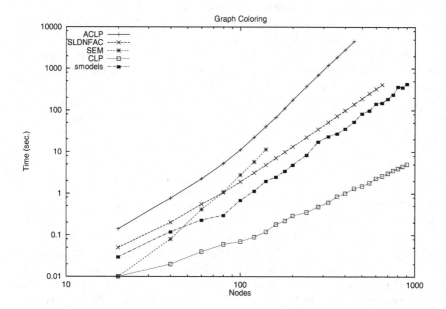

Fig. 1. Graph coloring

Our first experiment is done with 4-colorable graphs. We used a graph generator[1] program which is available from address http://web.cs.ualberta.ca/

---

[1] The graphs have been generated with the following parameters: 0, 13, 6, n, 4, 0.2, 1, 0 where n is the number of vertices. Graph-coloring problems generated with these parameters are difficult.

`~joe/Coloring/Generators/generate.html`. We applied the systems in a sequence of experiments with graphs of increasing size and constant number of colors. We have modified only one parameter of the problem namely the number of vertices. Figure 1 gives the results of solving the problem with the different systems. Both axes are plotted in a logarithmic scale. On the x-axis we have put the number of vertices. Not surprisingly, CLP is the fastest system. The times for SMODELS is second best on this problem. We assume it is in part because of the very concise formulation. Using the so called technique of rules with exceptions [14], the two rules needed to describe the space of candidate solutions also encode the constraint that the color is a function of the vertex. Hence there is only one other rule, namely the constraint that two adjacent vertices must have a different color. The difference with CLP is almost two orders of magnitude for the largest problems. The times reported for SMODELS do not include the time for grounding the problem, these times only consist of a small part of the total time. Grounding the problem for 650 nodes takes only 10 seconds, whereas solving the problem takes over 100 seconds. SLDNFAC is slightly worse than SMODELS. Although meta-interpretation overhead tends to increase with problems size, the difference with SMODELS grows very slowly. The model generator SEM deteriorates much faster and runs out of memory for the larger problems. The fact that it grounds the whole theory is a likely explanation. The difference with SMODELS supports the claim that SMODELS has better techniques for grounding. ACLP performs substantially worse than SLDNFAC and also deteriorates faster. The difference is likely due to the function-specification available in SLDNFAC. Contrary to ACLP, SLDNFAC exploits the knowledge that the abducible encodes a function to reduce the number of explicitly stored integrity constraints.

3.3   N-Queens

Figure 2 gives the running times for the different systems for finding a first solution. Both axes are plotted on a linear scale. The time consumed while grounding is again not included in the graph (for 18 queens, half a second). Again, CLP gives the best results. SLDNFAC is second best and, although meta-interpretation overhead increases with problem size, deteriorates very slowly. ACLP is third[2], with a small difference, probably due to the lack of the function-specification mentioned in the section above. The next one is SEM. It runs out of memory for large problems (it needs about 120MB for 27 queens). SMODELS performs very poorly on this problem, in particular when compared with its performance on the graph coloring problem. It is well-known that to obtain good results for computing the first solution for the n-queens problem, a good search heuristic is needed, like the first fail principle used by the systems based on CLP. We believe that the bad performance of SMODELS is explained by the absence of

---

[2] The results with ACLP are substantially better than those in the previous paper [17]. This is due to the removal of a redundant and time consuming complete consistency check after the processing of each new CLP constraint.

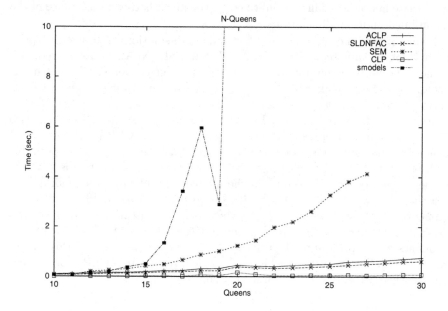

Fig. 2. N-queens: one solution

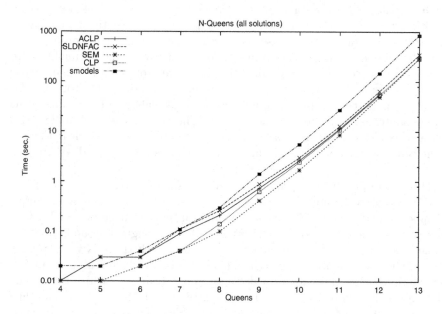

Fig. 3. N-queens: all solutions

appropriate heuristics. This is confirmed by the much better performance of the system in computing all solutions.

Figure 3 gives the running times for finding all solutions. The y-axis is plotted on a logarithmic scale. The CLP, ACLP and SLDNFAC systems are based on the same finite domain constraint solver, so their convergence is not unexpected. Indeed, the abductive system generates a constraint problem which is equivalent to the problem generated by the CLP program and no backtracking occurs in the abductive system. Hence, its overhead becomes ignorable. Also the SEM system converges to the same performance as CLP (but runs out of memory for big problems). In this experiment, the SMODELS system performs much better but is still the slowest system. A likely reason for this is that the number of propositional variables in the n-queens problem grows quadratically with the problem size, in contrast with the graph coloring problem where the number of variables grows only linearly (because of a constant number of colors). Consequently, the grounding grows faster for this problem. The CLP consistency techniques seem to be much less sensitive to the domain size, and this carries over to the abductive systems which reduce the problem to a CLP problem and then use the CLP solver to search for the solution.

3.4   A Real World Problem

A Belgian electricity company has a number of power plants divided in geographic areas. Each power plant has a number of power generating units, each of which must receive a given number (usually 1 or 2) of preventive maintenances with a fixed duration in the course of one year. The computational problem is to schedule these maintenances according to some constraints and optimality criteria. Some of the constraints are: some time slots are prohibited for maintenance for some units; for each power plant, there is an upper limit on the total number of units in maintenance per week for reasons of availability of personnel; some of the maintenances are fixed in advance, ... The objective of the problem is to find a schedule that maximizes the minimal weekly reserve, which is the sum of the capacity of all units not in maintenance minus the expected weekly peak load.

This is a rather difficult problem in several aspects. Firstly, the specification uses aggregate expressions like cardinality and sum (e.g. for each area, there is an upper limit to the total capacity for units in maintenance per week). Only CLP, SMODELS and SLDNFAC support some form of aggregates and only these systems were used in our experiment. Also, the search space is very large, as there are 56 maintenances to be scheduled in 52 weeks which makes about 56^{52} combinations[3]. The company provided a set of constraints for which the optimal solution was known to have a minimal week reserve of 2100 (100%). The three systems found correct schedules but none was able to find this optimal solution.

---

[3] The maintenances with duration of more than one week cannot be scheduled in week 52, hence this number is only an upper approximation.

This application was first considered in a context of a master's thesis [18] and then reported in [4], where a first attempt was done for integrating the SLDNFA proof procedure with the CLP system ROPE [23, 22]. This early system needed 24 hours to reduce the problem to a constraint store. Later on, in [22] several different direct encodings in CLP of the problem were presented and compared. Recently, [21] discussed an extension of the SLDNFAC system with aggregate functions and this problem was used as a benchmark.

The first version of the SMODELS system did not support aggregate expressions. A more recent version of the system added a limited support for rules with a body consisting of a single cardinality or sum constraint [16] and allowed us to specify the problem. However, these aggregate constraints cannot be used for computing the sum or the cardinality of a set of atoms and we were not able to express the optimization function. By setting increasing lower bounds on the reserve capacity, branch and bound can be simulated manually. It should be noted that, because of the very large size of the problem, the specification of the problem in the SMODELS system had to be redesigned with special care in order to produce a ground program not exceeding the limits of the system.

Table 1 summarizes the results of executing the problem with the different systems. The first row "Setup" gives the time used for pre-processing the problem specification. For the abductive systems, this is the time for reducing the high-level specification to a set of constraints. For the SMODELS system this is the time for grounding the program. The rest of the rows give the times used by the constraint solver to find a solution with the given quality. The results for CLP are taken from [22] for a standard encoding of the problem[4] and the program was run under SICStus Prolog.

Reserve	CLP	SLDNFAC	SMODELS
Setup		45	36.4
1900		63.2	8.07
2000	7.71	62.9	>8h
2010	25.85	63.8	
2020	43.73	62.9	
2030	57.28	63.0	
2040	71.63	261.1	
2050	26843.50	871.3	

Table 1. Power plant scheduling

In the case of SLDNFAC, it can be seen in Table 1 that substantial progress was made. Rather than the 24h needed in the earlier version [4], the current SLDNFAC procedure only needs 45 seconds for reducing the problem and about 15 minutes for finding a solution of level 2050 (97.6%). A solution with reserve

---

[4] Without using global constraints, like *cumulative*.

capacity of 2030 (96.5%) was found in less than two minutes. Note that the timings for a solution with a reserve capacity of 1900 up to 2030 are similar. This is explained by the fact that in the five cases the same solution with reserve capacity of 2030 was computed. The small differences in timings are due to noise in the measurements. Strange enough, CLP deteriorates when it reaches a solution for a reserve capacity of 2050 whereas the SLDNFAC solution does not. This must be due to the fact that the constraint store built by the CLP solution differs from the one built by the SLDNFAC solution. This is accidental: in general, constraint stores constructed by a hand made CLP program are more efficient than the ones computed by SLDNFAC. The SMODELS system needed 40 seconds for grounding and the best solution we were able to find was 1900 (90.5%) in 8 seconds. We did not find better solutions in reasonable time.

4   Conclusion

Finite domain CLP is widely accepted as an excellent tool for CSP solving. However CLP programs have drawbacks from the point of view of knowledge representation. As explained in Section 2.1, the variables of the CSP have to be organized in a data structure and "procedural" code is required to create this data structure and to set up the constraints. This level of indirection increases the conceptual distance between the program and the problem and makes programs less declarative. Recently, several attempts have been made to introduce formalisms allowing more declarative formalizations. They are based on stable model semantics [11, 13, 14] and on abduction [9, 10, 3]. Although these systems have an expressivity beyond what is needed to describe a CSP (they address non-monotonic reasoning while CSP solving requires only negation of primitive constraints), it is worthwhile to compare these systems with CLP which is state of the art for CSP solving. Because both stable models and abduction express solutions to problems as models of their theory, we have also included first order model generators in our study [24]. As argued in Section 2, these three approaches are better than CLP from knowledge representation point of view, the formalizations are more natural, more readable, conceptually closer to the problem, in short they are more declarative than CLP programs. Which one of the three discussed mechanisms is the most declarative is likely a matter of taste and familiarity.

 ·   Inevitably there is a price to be paid for these higher level descriptions. None of the "declarative" systems experimented with comes close to the performance level of CLP. This result holds although the CLP system is not favored by the problem choice. Indeed, in both graph coloring and n-queens problem, all constraints are disequality constraints which are known to give little propagation.

Our experiments show that first order model generators do not scale well and run out of memory for large problem instances even though the size of the ground program is smaller compared to SMODELS. We think that this is not an inherent limitation of the approach but rather that such systems were written with the goal of fast performance and this is visible in our experiemnts. In contrast, SMODELS runs in linear space wrt the size of the grounding [15] and was able to solve all problem sizes. Of the two abductive systems, SLDNFAC supports a substantially richer formalism and is performing slightly better than ACLP. As the two systems follow more or less the same strategy of top-down reduction of integrity constraints and of forwarding the reduced ones to the CLP solver and as both are implemented as a Prolog meta-interpreter, the difference seems to be mainly due to the support of function specifications. The fact that the SLDNFAC meta-interpreter outperforms SEM (a fine tuned C implementation) on both problems and compares very well with the C++ implementation of SMODELS (it is much better on the n-queens problem while it reaches almost the same performance on the graph coloring problem) suggest that its overall strategy is the best one of the three systems for CSP solving. Also the experiments with the large scheduling problem suggest this: the setup time is acceptable and differences in search time seem to be due to differences in the order of traversing the search space. While the difference with CLP is substantial, a low level implementation or compilation should be able to come close to the performance levels of CLP, offering the best of both worlds: declarative problem formulations and efficient execution. However, SLDNFA, the procedure underlying SLDNFAC, is complex, hence building a direct implementation is a hard task. We believe the development of such a system is a worthwhile topic for future research.

Acknowledgements

Nikolay Pelov, Emmanuel De Mot and Marc Denecker are supported by the GOA project LP+. We want to thank Maurice Bruynooghe for his contribution on the topic and anonymous referees for their useful comments.

References

[1] M. Denecker. Extending classical logic with inductive definitions. In J. Lloyd et al., editors, *First International Conference on Computational Logic*, volume 1861 of *Lecture Notes in Artificial Intelligence*, pages 703–717, London, U.K., July 2000. Springer. 230

[2] M. Denecker and D. De Schreye. SLDNFA: an abductive procedure for abductive logic programs. *Journal of Logic Programming*, 34(2):201–226, Feb. 1998. 229

[3] M. Denecker and B. Van Nuffelen. Experiments for integration CLP and abduction. In K. Apt, A. Kakas, E. Monfroy, and F. Rossi, editors, *Proceedings of the 1999 ERCIM/COMPULOG Workshop on Constraints*, Paphos, Cyprus, Oct. 1999. University of Cyprus. 229, 231, 236

[4] M. Denecker, H. Vandecasteele, D. De Schreye, G. Seghers, and T. Bayens. Scheduling by "abductive execution" of a classical logic specification. In *ERCIM/COMPULOG Workshop on Constraints*, Schloss Hagenberg, Austria, Oct.27–28 1997. 235, 235

[5] K. Eshghi and R. Kowalski. Abduction compared with negation by failure. In G. Levi and M. Martelli, editors, *Proceedings of the Sixth International Conference on Logic Programming*, pages 234–254. Lisbon, Portugal, MIT Press, June 1989. 229

[6] M. Gelfond and V. Lifschitz. The stable model semantics for logic programming. In R. A. Kowalski and K. A. Bowen, editors, *Logic Programming, Proceedings of the Fifth International Conference and Symposium*, pages 1070–1080, Seattle, Washington, Aug. 1988. MIT Press. 226, 228

[7] J. Jaffar and M. Maher. Constraint logic programming: A survey. *Journal of Logic Programming*, 19/20:503–581, 1994. 227

[8] A. C. Kakas, R. Kowalski, and F. Toni. Abductive logic programming. *Journal of Logic and Computation*, 2(6):719–770, Dec. 1992. 226, 229

[9] A. C. Kakas and A. Michael. Integrating abductive and constraint logic programming. In L. Sterling, editor, *Proceedings of the 12th International Conference on Logic Programming*, pages 399–413. Tokyo, Japan, MIT Press, 1995. 229, 236

[10] A. C. Kakas, A. Michael, and C. Mourlas. ACLP: Abductive constraint logic programming. *Journal of Logic Programming*, 44(1–3):129–177, 2000. 229, 231, 236

[11] V. Lifschitz. Answer set planning. In D. De Schreye, editor, *Proceedings of the 16th International Conference on Logic Programming*, pages 23–37. MIT Press, Dec. 1999. 226, 236

[12] A. K. Mackworth. The logic of constraint satisfaction. *Journal of Artificial Intelligence*, 58(1–3):3–20, Dec. 1992. 225, 226

[13] V. W. Marek and M. Truszczyński. Stable models and an alternative logic programming paradigm. In K. R. Apt, V. W. Marek, M. Truszczyński, and D. S. Warren, editors, *The Logic Programming Paradigm: A 25-Year Perespective*, pages 375–398. Springer, 1999. 226, 226, 236

[14] I. Niemelä. Logic programs with stable model semantics as a constraint programming paradigm. *Annals of Mathematics and Artificial Intelligence*, 25(3,4):241–273, 1999. 226, 226, 228, 229, 232, 236

[15] I. Niemelä and P. Simons. Efficient implementation of the well-founded and stable model semantics. In M. Maher, editor, *Logic Programming, Proceedings of the 1996 Joint International Conference and Syposium*, pages 289–303, Bonn, Germany, Sept. 1996. MIT Press. 226, 228, 237

[16] I. Niemelä, P. Simons, and T. Soininen. Stable model semantics of weight constraint rules. In M. Gelfond, N. Leone, and G. Pfeifer, editors, *Proceedings of the Fifth International Conference on Logic Programming and Nonmonotonic Reasoning*, volume 1730 of *Lecture Nortes in Computer Science*, pages 317–331, El Paso, Texas, USA, Dec. 1999. Springer-Verlag. 228, 229, 235

[17] N. Pelov, E. De Mot, and M. Bruynooghe. A comparison of logic programming approaches for representation and solving of constraint satisfaction problems. In M. Denecker, A. Kakas, and F. Toni, editors, *8th International Workshop on Non-Monotonic Reasoning, Special Session on Abduction*, Breckenridge, Colorado, USA, Apr. 2000. 226, 232

[18] G. Seghers and T. Bayens. Solving a combinatorial maintenance problem for electrical power plants, developed in OLP-FOL and CLP. Master's thesis, Department of Computer Science, K.U.Leuven, 1996. 235

[19] E. Tsang. *Foundations of Constraint Satisfaction*. Computation in Cognitive Science. Academic Press, 1993. 225

[20] P. Van Hentenryck. *Constraint Satisfaction in Logic Programming*. MIT Press, 1989. 225, 227

[21] B. Van Nuffelen and M. Denecker. Problem solving in ID-logic with aggregates. In A. K. Mark Denecker, editor, *Eight International Workshop on Nonmonotonic Reasoning, special track on Abductive Reasoning*, Breckenridge, Colorado, USA, 2000. Workshop associated with KR'2000. 235

[22] H. Vandecasteele. *Constraint Logic Programming: Applications and Implementation*. PhD thesis, K.U.Leuven, 1999. 235, 235, 235

[23] H. Vandecasteele and D. De Schreye. Implementing a finite-domain CLP-language on top of Prolog: a transformational approach. In F. Pfenning, editor, *Proceedings of Logic Programming and Automated Reasoning*, volume 822 of *Lecture Notes in Artificial Intelligence*, pages 84–98, Kiev, Ukraine, 1994. Springer-Verlag. 235

[24] J. Zhang and H. Zhang. Constraint propagation in model generation. In U. Montanari and F. Rossi, editors, *Proc. of 1st International Conference on Principles and Practice of Constraint Programming*, volume 976 of *Lecture Notes in Computer Science*, pages 398–414, France, Sept. 1995. Springer. 226, 228, 228, 236

Quantified Propositional Gödel Logics*

Matthias Baaz[1], Agata Ciabattoni[1,**], and Richard Zach[2]

[1] Institut für Algebra und Computermathematik E118.2,
Technische Universität Wien, A–1040 Vienna, Austria
[baaz, agata]@logic.at
[2] Institut für Computersprachen E185.2,
Technische Universität Wien, A–1040 Vienna, Austria
zach@logic.at

Abstract. It is shown that $\mathbf{G}_\uparrow^{\mathrm{qp}}$, the quantified propositional Gödel logic based on the truth-value set $V_\uparrow = \{1 - 1/n : n \geq 1\} \cup \{1\}$, is decidable. This result is obtained by reduction to Büchi's theory S1S. An alternative proof based on elimination of quantifiers is also given, which yields both an axiomatization and a characterization of $\mathbf{G}_\uparrow^{\mathrm{qp}}$ as the intersection of all finite-valued quantified propositional Gödel logics.

1 Introduction

In 1932, Gödel [10] introduced a family of finite-valued propositional logics to show that intuitionistic logic does not have a characteristic finite matrix. Dummett [7] later generalized these to an infinite set of truth-values, and showed that the set of its tautologies **LC** is axiomatized by intuitionistic logic extended by the linearity axiom $(A \supset B) \vee (B \supset A)$. Gödel-Dummett logic naturally turns up in a number of different areas of logic and computer science. For instance, Dunn and Meyer [8] pointed out its relation to relevance logic; Visser [15] employed it in investigations of the provability logic of Heyting arithmetic; Pearce used it to analyze inference in extended logic programming [13]; and eventually it was recognized as one of the most important formalizations of fuzzy logic [11].

The propositional Gödel logics are well understood: Any infinite set of truth-values characterizes the same set of tautologies. **LC** is also characterized as the intersection of the sets of tautologies of all finite-valued Gödel logics $\mathbf{G}_k$ [7], and as the logic determined either by linearly ordered Kripke frames or linearly ordered Heyting algebras [12].

When Gödel logic is extended beyond pure propositional logic, however, the situation is more complex. For the cases of propositional entailment and extension to first-order validity, infinite truth-value sets with different order types determine different logics with different properties. There are infinitely many sets of truth values which give rise to distinct logics. As an example, consider the truth-value sets

2000 Mathematics Subject Classification: Primary 03B50; Secondary 03B55.
* Research supported by the Austrian Science Fund under grant P–12652 MAT
** Research supported by EC Marie Curie fellowship HPMF–CT–19 99–00301

M. Parigot and A. Voronkov (Eds.): LPAR 2000, LNAI 1955, pp. 240–256, 2000.
© Springer-Verlag Berlin Heidelberg 2000

$$V_\infty = [0,1]$$
$$V_\downarrow = \{0\} \cup \{1/n : n \geq 1\}$$
$$V_\uparrow = \{1\} \cup \{1 - 1/n : n \geq 1\}$$
$$V_k = \{1\} \cup \{1 - 1/n : n = 1, \ldots, k-1\}$$

Propositional entailment with respect to V_∞ is compact, but not with respect to $V_\downarrow$ or $V_\uparrow$. If a formula A is entailed by a set Γ with respect to V_k for every k, then it is also entailed with respect to $V_\uparrow$, but not necessarily with respect to V_∞ or $V_\downarrow$ [5]. Similarly, the first-order logic based on V_∞ is axiomatizable (this is Takeuti and Titani's intuitionistic fuzzy logic [14]), while those based on $V_\uparrow$ and $V_\downarrow$ are not [2]. The first-order Gödel logic based on $V_\uparrow$ is the intersection of all finite-valued first-order Gödel logics.

Another interesting generalization of propositional logic is obtained by adding quantifiers over propositional variables. In classical logic, propositional quantification does not increase expressive power per se. It does, however, allow expressing complicated properties more naturally and succinctly, e.g., satisfiability and validity of formulas are easily expressible within the logic once such quantifiers are available. This fact can be used to provide efficient proof search methods for several non-monotonic reasoning formalisms [9].

For Gödel logic the increase in expressive power is witnessed by the fact that statements about the topological structure of the set of truth-values (taken as infinite subsets of the real interval $[0,1]$) can be expressed using propositional quantifiers [4]. In [4] it is also shown that there is an uncountable number of different quantified propositional infinite-valued Gödel logics. The same paper investigates the quantified propositional Gödel logic $\mathbf{G}_\infty^{qp}$ based on the set of truth-values $[0,1]$, which was shown to be decidable. It is of some interest to characterize the intersection of all finite-valued quantified propositional Gödel logics. As was pointed out in [4], $\mathbf{G}_\infty^{qp}$ does not provide such a characterization.

In this paper we study the quantified propositional Gödel logic $\mathbf{G}_\uparrow^{qp}$ based on the truth-value set $V_\uparrow$. We show that $\mathbf{G}_\uparrow^{qp}$ is decidable. In general, it is not obvious that a quantified propositional logic is decidable or even axiomatizable. For instance, neither the closely related quantified propositional intuitionistic logic, nor the set of valid first-order formulas on the truth-value set $V_\uparrow$ are r.e. Although our result can be obtained by reduction to Büchi's monadic second order theory of one successor S1S [6], we also give a more informative proof based on elimination of propositional quantifiers. This proof allows us to characterize $\mathbf{G}_\uparrow^{qp}$ as the intersection of all finite-valued quantified propositional Gödel logics, and moreover yields an axiomatization of $\mathbf{G}_\uparrow^{qp}$.

A remark is in order about the relationship between the approach taken here using truth-value semantics and Kripke semantics. As was pointed out above, **LC** is often defined as the propositional logic of linearly ordered Kripke frames. In Kripke semantics, quantified propositional **LC** would then result by adding quantifiers over propositions (subsets of the set of worlds closed under accessibility). Here different classes of linear Kripke structures which all define **LC** in the pure propositional case in general do not define the same quantified propositional logic. In particular, the logic obtained by just taking Kripke models of order type ω is not the same as that defined by the class of all finite linear orders. It follows from the results of this paper that the logic of all finite linear Kripke structures coincides with $\mathbf{G}_\uparrow^{\mathrm{qp}}$.

2   Gödel Logics

Syntax. We work in the language of propositional logic containing a countably infinite set $Var = \{p, q, \ldots\}$ of (propositional) variables, the constants $\bot, \top$, as well as the connectives $\wedge, \vee$, and $\supset$. Propositional variables and constants are considered atomic formulas. Uppercase letters will serve as meta-variables for formulas. If $A(p)$ is a formula containing the variable p free, then $A(X)$ denotes the formula with all occurrences of the variable p replaced by the formula X. $Var(A)$ is the set of variables occurring in the formula A. We use the abbreviations $\neg A$ for $A \supset \bot$ and $A \leftrightarrow B$ for $(A \supset B) \wedge (B \supset A)$.

Semantics. The most important form of Gödel logic is defined over the real unit interval $V_\infty = [0, 1]$; in a more general framework, the truth-values are taken from a set V such that $\{0, 1\} \subseteq V \subseteq [0, 1]$. In the case of k-valued Gödel logic $\mathbf{G}_k$, we take $V_k = \{1 - 1/i : i = 1, \ldots, k - 1\} \cup \{1\}$. The logic we will be most interested in is based on the set $V_\uparrow = \{1 - 1/i : i \geq 1\} \cup \{1\}$.

A *valuation* $v\colon Var \to V$ is an assignment of values in V to the propositional variables. It can be extended to formulas using the following truth functions introduced by Gödel [10]:

$$v(\bot) = 0 \qquad\qquad v(A \vee B) = \max(v(A), v(B))$$
$$v(\top) = 1$$
$$v(A \wedge B) = \min(v(A), v(B)) \qquad v(A \supset B) = \begin{cases} 1 & \text{if } v(A) \leq v(B) \\ v(B) & \text{otherwise} \end{cases}$$

A formula A is a *tautology* over a truth-value set $V \subseteq [0, 1]$ if for all valuations $v\colon Var \to V$, $v(A) = 1$. The *propositional logics* $\mathbf{LC}$, $\mathbf{G}_\uparrow$ and $\mathbf{G}_k$ are the sets of tautologies over the corresponding truth value sets, e.g., $\mathbf{LC} = \mathbf{G}_\infty = \{A : A \text{ a tautology over } V_\infty\}$. We also write $\mathbf{G} \models A$ for $A \in \mathbf{G}$ ($\mathbf{G} \in \{\mathbf{LC}, \mathbf{G}_\uparrow, \mathbf{G}_k\}$).

It is easily seen that $\mathbf{LC} \supseteq \mathbf{G}_\uparrow \supseteq \mathbf{G}_k$. Dummett [7] showed that $\mathbf{LC} = \mathbf{G}_\uparrow$ and that $\mathbf{LC} = \bigcap_{k \geq 2} \mathbf{G}_k$.

The abbreviation $A \prec B$ for $(A \supset B) \wedge ((B \supset A) \supset A)$ will be used extensively below. It expresses strict linear order in the sense that

$$v(A \prec B) = \begin{cases} 1 & \text{if } v(A) < v(B) \text{ or } v(B) = 1 \\ \min(v(A), v(B)) & \text{otherwise} \end{cases}$$

Propositional Quantification. In *classical* propositional logic we define $(\exists p)A(p)$ by $A(\bot) \vee A(\top)$ and $(\forall p)A(p)$ by $A(\bot) \wedge A(\top)$. In other words, propositional quantification is semantically defined by the supremum and infimum, respectively, of truth functions (with respect to the usual ordering "$0 < 1$" over the classical truth-values $\{0,1\}$). This can be extended to Gödel logic by using *fuzzy quantifiers*. Syntactically, this means that we allow formulas $(\forall p)A$ and $(\exists p)A$ in the language. Free and bound occurrences of variables are defined in the usual way. Given a valuation v and $w \in V$, define $v[w/p]$ by $v[w/p](p) = w$ and $v[w/p](q) = v(q)$ for $q \not\equiv p$. The semantics of fuzzy quantifiers is then defined as follows:

$$v((\exists p)A) = \sup\{v[w/p](A) : w \in V\} \qquad v((\forall p)A) = \inf\{v[w/p](A) : w \in V\}$$

When we consider quantifiers, V has to be closed under infima and suprema, since otherwise truth values for quantified formulas are not defined.

We also add the additional unary connective $\circ$ to the language. The truth function for $\circ$ is given by $v(\circ A) = v((\forall p)((p \supset A) \vee p))$. In $\mathbf{G}_{\uparrow}^{\mathrm{qp}}$, this makes

$$v(\circ A) = \begin{cases} 1 & \text{if } v(A) = 1 \\ 1 - \frac{1}{n+1} & \text{if } v(A) = 1 - \frac{1}{n} \end{cases}$$

We abbreviate $\circ \ldots \circ A$ (n occurrences of $\circ$) by $\circ^n A$.

Using the above definitions, it is straightforward to extend the notion of tautologyhood to the new language. We write $\mathbf{G}_{\uparrow}^{\mathrm{qp}}$ ($\mathbf{G}_{\infty}^{\mathrm{qp}}$, $\mathbf{G}_{k}^{\mathrm{qp}}$) for the set of tautologies in the extended language over $V_{\uparrow}$ (V_{∞}, V_k).

We will show below that every quantified propositional formula is equivalent in $\mathbf{G}_{\uparrow}^{\mathrm{qp}}$ to a quantifier-free formula, which in general can contain $\circ$. $\circ A$ itself (or the equivalent formula $(\forall p)((p \supset A) \vee p)$), however, is not in general equivalent to a quantifier-free formula not containing $\circ$. Inspection of the truth tables shows that a quantifier-free formula containing only the variable q takes one of 0, $v(q)$, or 1 as its value under a given valuation v, and thus no such formula can define $\circ q$.

3   Hilbert-Style Calculi

All the calculi we consider are based on the following set of axioms:

I1	$A \supset (B \supset A)$	I7	$(A \wedge \neg A) \supset B$
I2	$(A \wedge B) \supset A$	I8	$(A \supset \neg A) \supset \neg A$
I3	$(A \wedge B) \supset B$	I9	$\bot \supset A$
I4	$A \supset (B \supset (A \wedge B))$	I10	$A \supset \top$
I5	$A \supset (A \vee B)$	I11	$(A \supset (B \supset C)) \supset ((A \supset B) \supset (A \supset C))$
I6	$B \supset (A \vee B)$	I12	$((A \supset C) \wedge (B \supset C)) \supset ((A \vee B) \supset C)$

These axioms, together with the rule of modus ponens, define the system IPC that is sound and complete for intuitionistic propositional logic. The system LC is obtained by adding to IPC the linearity axiom

$$\text{LC} \quad (A \supset B) \vee (B \supset A).$$

It is well known [7] that IPC and LC are sound for all propositional Gödel logics, and that LC is complete for all infinite-valued propositional Gödel logics. We will make frequent use of this fact below, and omit derivations of formulas which are (instances of) quantifier- and $\circ$-free tautologies in $\mathbf{G}_\uparrow$. These omissions are indicated by pointing out that the formula follows already in LC or IPC. In particular, familiar inference patterns such as the chain rule or case distinction are derivable in LC and its extensions.

When we turn to quantified propositional logics, a natural system $\mathsf{IPC}^{\mathrm{qp}}$ to start with is obtained by adding to IPC the following two axioms:

$$\supset\exists \quad A(C) \supset (\exists p)A(p) \qquad\qquad \supset\forall \quad (\forall p)A(p) \supset A(C)$$

and the rules:

$$\frac{A(p) \supset B^{(p)}}{(\exists p)A(p) \supset B^{(p)}}\mathrm{R}\exists \qquad\qquad \frac{B^{(p)} \supset A(p)}{B^{(p)} \supset (\forall p)A(p)}\mathrm{R}\forall$$

where for any formula C, the notation $C^{(p)}$ indicates that p does not occur free in C, i.e., p is a (propositional) *eigenvariable*.

Let $\mathsf{QG}_\uparrow^{\mathrm{qp}}$ be the system obtained by adding to $\mathsf{IPC}^{\mathrm{qp}}$ the axioms (LC),

$$\forall\vee \quad (\forall p)[A \vee B(p))] \supset [A \vee (\forall p)B(p)]$$

where $p \notin A$, and the following:

G1 $\circ(A \supset B) \leftrightarrow (\circ A \supset \circ B)$ G4 $(A \supset \circ B) \supset ((A \supset C) \vee (C \supset B))$

G2 $A \prec \circ A$ G5 $(A \leftrightarrow \bot) \vee (\exists p)(A \leftrightarrow \circ p)$

G3 $(\circ A \supset \circ B) \supset ((A \supset B) \vee \circ B)$ G6 $(A \prec B) \supset (\circ A \supset B)$

Proposition 1. *The system $\mathsf{QG}_\uparrow^{\mathrm{qp}}$ is sound for $\mathbf{G}_k^{\mathrm{qp}}$ and $\mathbf{G}_\uparrow^{\mathrm{qp}}$.*

Proof. It is easily seen that the rules of inference preserve validity. For instance, if $B \supset A(p)$ is valid, then, for any valuation v, $v[w/p](B) \leq v[w/p](A(p))$ where $w \in V$. If p does not occur in B, then $v(B) = v[w/p](B)$ and we have $v(B) \leq \inf\{v[w/p](A(p)) : w \in V\}$. That LC is sound for arbitrary Gödel logics was shown in [7]. The tedious but straightforward verification that the remaining axioms ($\vee\forall$) and (G1)–(G6) are valid is left to the reader.

Remark 2. In [4] it was shown that a system sound and complete for $\mathbf{G}_\infty^{\mathrm{qp}}$, the quantified propositional Gödel logic based on the truth-value set $[0,1]$, is obtained by extending $\mathsf{IPC}^{\mathrm{qp}}$ with (LC), ($\vee\forall$) and the axiom

$$(\forall p)[(A^{(p)} \supset p) \vee (p \supset B^{(p)})] \supset (A^{(p)} \supset B^{(p)}).$$

This schema is not valid in $\mathbf{G}_\uparrow^{\mathrm{qp}}$ (it comes out $= 0$ under any v with $v(A) = 1/2$ and $v(B) = 0$). On the other hand, it is easy to see that $v(\circ A) = v(A)$ in V_∞, and hence axiom (G2) is not valid in $\mathbf{G}_\infty^{\mathrm{qp}}$. Thus neither of $\mathbf{G}_\infty^{\mathrm{qp}}$ and $\mathbf{G}_\uparrow^{\mathrm{qp}}$ is included in the other. This is in contrast to the situation in propositional entailment and first-order logic, where V_∞ defines the smallest Gödel logic and is included in all others.

4 Decidability

In this section we prove that $\mathbf{G}_\uparrow^{\mathrm{qp}}$ is decidable. This is done by defining a reduction of tautologyhood in $\mathbf{G}_\uparrow^{\mathrm{qp}}$ to S1S, the monadic theory of one successor, which was shown to be decidable by Büchi [6].

S1S is the set of second-order formulas in the language with second-order quantification restricted to monadic set variables X, Y, ... with one unary function $'$ (successor) which are true in the model $\langle \omega, ' \rangle$. For the purposes of this section we consider $\bigcirc A$ to be an abbreviation of $(\forall p)((p \supset A) \vee p)$.

Suppose A is a quantified propositional formula, and B is a formula in the language of S1S with only x free. Let $TV(B(x))$ abbreviate $(\forall z)(B(z') \supset B(z))$. We define A^x by:

$$p^x = X_p(x)$$
$$\bot^x = X_\bot(x)$$
$$\top^x = (\forall z)(z = z)$$
$$(B \wedge C)^x = B^x \wedge C^x$$
$$(B \vee C)^x = B^x \vee C^x$$
$$(B \supset C)^x = (\forall y)(B^y \supset C^y) \vee (\exists y)(B^y \wedge \neg C^y) \wedge C^x$$
$$(\forall p)B^x = (\forall X_p)(TV(X_p(x)) \supset B^x)$$
$$(\exists p)B^x = (\exists X_p)(TV(X_p(x)) \wedge B^x)$$

Consider the following reduction:

$$\Phi(A) = (\forall X_\bot)((\forall x)\neg X_\bot(x) \supset (\forall x)A^x)$$

The idea behind this is to correlate truth-values in $V_\uparrow$ with subsets of ω which are closed under predecessor, i.e., predicates in

$$TV = \{P \subseteq \omega : \text{if } n \in P \text{ then } m \in P \text{ for all } m \leq n\}.$$

Under this correlation, 1 corresponds to ω, and $1 - 1/n$ corresponds to $\{1, \ldots, n\}$.

Let s be an interpretation of the language of S1S, mapping variables to elements or subsets of ω. We denote by $s[n/x]$ the interpretation which is just like s except that it assigns n to x. Then $TV(A(x))$ obviously expresses the condition that the predicate $A(x)[s] = \{n : S1S \models A(x)[s[n/x]]\}$ defined by $A(x)$ in s is closed under predecessor. If a monadic predicate P is closed under predecessor, we define its truth value by

$$tv(P) = \sup\{1 - \frac{1}{n} : 1^n \in P\}.$$

Conversely, every truth-value $v \in V_\uparrow$ corresponds to a monadic predicate

$$mp(v) = \begin{cases} \{k : k \leq n\} & \text{if } v = 1 - 1/n \\ \omega & \text{if } v = 1. \end{cases}$$

Note that for $P, Q \in TV$, $P \subseteq Q$ iff $tv(P) \leq tv(Q)$, and conversely, for $v, w \in V_\uparrow$, $v \leq w$ iff $mp(v) \subseteq mp(w)$.

Lemma 3. *Let v be a valuation and s be the interpretation defined by $s(X_p) = mp(v(p))$ and $s(X_\perp) = \emptyset$. Then we have $tv(A^x[s]) = v(A)$.*

Proof. By induction on the complexity of A. The claim is obvious for atomic formulas, conjunction and disjunction. If $A \equiv B \supset C$ we have to distinguish two cases. Suppose first that $v(B) \leq v(C)$. By induction hypothesis, $B^x[s] = mp(v(B)) \subseteq mp(v(C)) = C^x[s]$, and hence the first disjunct in the definition of $(B \supset C)^x$ is true. Thus $(B \supset C)^x$ defines ω and $tv((B \supset C)^x[s]) = 1$. Now suppose that $v(B) > v(C)$. Then $tv(B^x[s]) \supsetneq tv(C^x[s])$, $S1S \not\models (\forall y)(B^y \supset C^y)$ $[s]$ and $S1S \models (\exists y)(B^y \wedge \neg C^y)$ $[s]$, and thus $(B \supset C)^x[s] = C^x[s]$.

If $A \equiv (\exists p)B$, let $v[w/p]$ be the valuation which is just like v except that $v[w/p](p) = w$, and let $s[mp(w)/X_p]$ be the corresponding interpretation which is like s except that it assigns $mp(w)$ to X_p.

By induction hypothesis, $tv(B^x[s[mp(w)/X_p]]) = v[w/p](B)$. We again have two cases. Suppose first that $\sup\{v[w/p](B) : w \in V_\uparrow\} = 1 - 1/n$. For all $m > n$, $S1S \not\models B^x[m/x, mp(w)/X_p]$, since $v[w/p](B^x) < 1 - 1/m$ by induction hypothesis. On the other hand, $S1S \models TV(P_p) \supset B^x$ $[s[k/x, mp(1 - 1/n)/P_p]]$ for all $k \leq n$, and so $tv((\exists p)B^x[s]) = 1 - 1/n$. Now consider the case where $\sup\{v[w/p](B) : w \in V_\uparrow\} = 1$. Here there is no bound n on the the members of sets defined by $B^x[s[mp(w)/X_p]]$ where $w \in V_\uparrow$. Hence, $mp((\exists p)B)^x[s]) = \omega$ and $tv((\exists p)B^x[s]) = 1$.

The case $A \equiv (\forall p)B$ is similar. □

Lemma 4. *Let s be an interpretation with $s(X_\perp) = \emptyset$ and $s(X_p) \in TV$. Let v be defined by $v(p) = tv(s(X_p))$. Then $A^x[s] \in TV$, and $v(A) = tv(A^x[s])$.*

Proof. By induction on the complexity of A. The claim is again trivial for atomic formulas, conjunctions or disjunctions. If $A \equiv B \supset C$, two cases occur. If $S1S \models (\forall y)(B^y \supset C^y)$, then $B^y[s] \subseteq C^y[s]$. By induction hypothesis, $v(B) \leq v(C)$, and hence $v(B \supset C) = 1 = tv((B \supset C)^x[s])$. Otherwise, for some n we have $n \in B^y[s]$ but $n \notin C^y[s]$. So $(\exists y)(B^y \wedge \neg C^y)$ must be true and the predicate defined is the same as $C^y[s]$.

Now for the case $A \equiv (\exists p)B$: If $S1S \models (\exists X_p)(TV(X_p) \supset B^x)[s[n/x]]$, then there is a prefix closed witness P so that $S1S \models B^x[s[n/x, P/X_p]]$. By induction hypothesis, $B^x[s[P/X_p]] \in TV$, and hence $S1S \models TV(X_p) \supset B^x$ $[s[m/x, P/X_p]]$ for all $m \leq n$, and thus $((\exists p)B)^x[s] \in TV$ as well.

Consider $N = ((\exists p)B)^x[s]$. First, suppose that $\sup N = k$. That means that for some $P \in TV$, $1^k \in B^x[s[P/X_p]]$, and for no $Q \in TV$ and no $j > k$, $j \in B^x[s[Q/X_p]]$. By induction hypothesis, $v[tv(P)/p](B) = 1 - 1/k$ and for all $w \in V_\uparrow$, $v[w/p](B) \leq 1 - 1/k$. Hence $v((\exists p)B) = 1 - 1/k$.

If $\sup N$ does not exist, for each k there is a witness $Q_k \in TV$ with $k \in B^x[s[Q_k/X_p]]$. By induction hypothesis, for each k we have $v[tv(Q_k)/p](B) \geq 1 - 1/k$, and so $v((\exists p)B) = 1$.

The case $A \equiv (\forall p)B$ is similar. □

Theorem 5. $G_\uparrow^{qp}$ *is decidable.*

Proof. If there is a valuation v such that $v(A) < 1$, then by Lemma 3 there is an s with $s(P_\perp) = \emptyset$ and n so that $n \notin A^x[s]$, and hence $S1S \nvDash \Phi(A)$.

Conversely, suppose $S1S \nvDash \Phi(A)$. We may assume, without loss of generality, that all propositional variables in A are bound. Then there is an interpretation s with $X_\perp(x)[s] = \emptyset$ so that some $n \notin A^x[s]$. By Lemma 4, $A^x[s] \in TV$. Hence, if $n \notin A^x[s]$, then $k \notin A^x[s]$ for all $k \geq n$, and, also by Lemma 4, $v(A) = tv(A^x[s]) < 1$.

Thus a formula A is a tautology in $\mathbf{G}_\uparrow^{\mathrm{qp}}$ iff $S1S \models \Phi(A)$. The claim follows by the decidability of $S1S$. $\qquad\square$

5   Properties and Normal Forms

In this section we introduce suitable normal forms for formulas of $\mathsf{QG}_\uparrow^{\mathrm{qp}}$ and prove some useful properties of $\mathsf{QG}_\uparrow^{\mathrm{qp}}$. These results will be crucial in the proof of the elimination of quantifiers.

Proposition 6.   *1.* $\mathsf{QG}_\uparrow^{\mathrm{qp}} \vdash (A \supset B) \supset (\circ A \supset \circ B)$

2. $\mathsf{QG}_\uparrow^{\mathrm{qp}} \vdash \circ(A \wedge B) \leftrightarrow (\circ A \wedge \circ B)$

3. $\mathsf{QG}_\uparrow^{\mathrm{qp}} \vdash \circ(A \vee B) \leftrightarrow (\circ A \vee \circ B)$

Proof. (1) From (G2) we have $(A \supset B) \supset \circ(A \supset B)$, which, together with the left-to-right direction of (G1) yields the result.

(2) The left-to-right implication immediately follows from axioms (I2) and (I3) together with Prop. 6(1). For the converse, replace B by $B \supset (A \wedge B)$ in Prop. 6(1) and use (I4) to derive $\circ A \supset \circ(B \supset (A \wedge B))$. Then, using (G1), one has $\circ A \supset (\circ B \supset \circ(A \wedge B))$. The claim follows by IPC.

(3) In LC, we have $(A \vee B) \leftrightarrow (A \supset B) \supset B) \wedge (B \supset A) \supset A)$. Replacing A by $\circ A$ and B by $\circ B$, we have $(\circ A \vee \circ B) \leftrightarrow (\circ A \supset \circ B) \supset \circ B) \wedge (\circ B \supset \circ A) \supset \circ A)$. The result follows using (G1) and IPC. $\qquad\square$

Proposition 7.   *1. If p does not occur bound in $C(p)$, then*

$$\mathsf{QG}_\uparrow^{\mathrm{qp}} \vdash (\forall \bar{q})(A \leftrightarrow B) \supset (C(A) \supset C(B))$$

where $\bar{q}$ are the propositional variables occurring free in A and B.

2. If $C(p)$ is quantifier-free, we also have

$$\mathsf{QG}_\uparrow^{\mathrm{qp}} \vdash (A \leftrightarrow B) \supset (C(A) \supset C(B))$$

Proof. By induction on the complexity of C. Cases for $\wedge$, $\vee$, and $\supset$ are easy. If $C(p) \equiv \circ D(p)$, we use the induction hypothesis and Prop. 6(1). If $C(p) \equiv$

$(\exists r)D(p,r)$, we argue:

$$
\begin{array}{lll}
(1) & (\forall \bar{q})(A \leftrightarrow B) \supset (D(A,r) \supset D(B,r)) & \text{by IH} \\
(2) & ((\forall \bar{q})(A \leftrightarrow B) \wedge D(A,r)) \supset D(B,r)) & (1), \text{IPC} \\
(3) & D(B,r) \supset (\exists r)D(B,r) & \supset\exists \\
(4) & (\forall \bar{q})(A \leftrightarrow B) \wedge D(A,r)) \supset (\exists r)D(B,r) & (2), (3) \\
(5) & D(A,r) \supset ((\forall \bar{q})(A \leftrightarrow B) \supset (\exists r)D(B,r)) & (4), \text{IPC} \\
(6) & (\exists r)(D(A,r) \supset ((\forall \bar{q})(A \leftrightarrow B) \supset (\exists r)D(B,r))) & (5), \text{R}\exists \\
(7) & (\forall \bar{q})(A \leftrightarrow B) \supset ((\exists r)D(A,r) \supset (\exists r)D(B,r)) & (6), \text{IPC}
\end{array}
$$

The case of $C \equiv (\forall r)D(p,r)$ is handled similarly. $\qquad\qquad\square$

Definition 8. A formula A of $\mathsf{QG}_\uparrow^{\mathrm{qp}}$ is in $\bigcirc$-*normal form* if it is quantifier-free and for all subformulas $\bigcirc B$ of A, $B \in \{\bot, \top\} \cup \mathit{Var}$ or $B \equiv \bigcirc B'$.

Proposition 9. *Let A be a quantifier-free formula of $\mathsf{QG}_\uparrow^{\mathrm{qp}}$. Then there exists a formula A' of $\mathsf{QG}_\uparrow^{\mathrm{qp}}$ in $\bigcirc$-normal form such that $\mathsf{QG}_\uparrow^{\mathrm{qp}} \vdash A \leftrightarrow A'$.*

Proof. Follows from axiom (G1), Prop. 6(2) and (3) using Prop. 7(2). $\qquad\square$

Proposition 10. *For every $n \geq 0$, $\mathsf{QG}_\uparrow^{\mathrm{qp}} \vdash \bigcirc^n \top \leftrightarrow \top$.*

Proof. $\bigcirc^n \top \supset \top$ is already derivable intuitionistically. For $\top \supset \bigcirc^n \top$, use (G2), Prop. 6(1), and induction on n. $\qquad\qquad\square$

For propositional Gödel logic, a normal form similar to the disjunctive normal form of classical logic has been introduced in [1] (see also [3, 4]). This so-called *chain normal form* is based on the fact that, in a sense, the truth value of a formula only depends on the ordering of the variables occurring in the formula induced by the valuation under consideration. The chain normal form can then be constructed by enumerating all such orderings (using $\prec$ and $\leftrightarrow$ to encode the ordering) in a way similar to how one constructs a disjunctive normal form by enumerating all possible truth value assignments. We extend the notion of chain normal form and the results of [3] in order to deal with the $\bigcirc$ connective. This is possible, since by Prop. 9 we can always push the $\bigcirc$ in front of atomic subformulas, so we only need to consider orderings of subformulas of the form $\bigcirc^j B$ with B atomic. Let Γ be a finite subset of $\{\bigcirc^j p, \bigcirc^j \bot : p \in \mathit{Var}, j \in \omega\} \cup \{\top\}$ and $\top, \bot \in \Gamma$.

Definition 11. A $\bigcirc$-*chain* over Γ is an expression of the form

$$
(S_1 \star_1 S_2) \wedge \cdots \wedge (S_{n-1} \star_{n-1} S_n)
$$

such that $\Gamma = \{S_1, \ldots, S_n\}$, $S_1 \equiv \bot$, $S_n \equiv \top$, and $\star_i \in \{\leftrightarrow, \prec\}$, for all $i = 1, \ldots, n$.

Every O-chain C uniquely determines a partition $\Pi_1^C, \ldots, \Pi_k^C$ of Γ so that $\Pi_i^C = \{S_{j_i}, \ldots, S_{j_{i+1}-1}\}$ where $j_1 = 1$, $j_{k+1} = n + 1$, $j_i < j_{i+1}$, $\star_{j_i} = \cdots = \star_{j_{i+1}-2} = \,\leftrightarrow$, and $\star_{j_{i+1}-1} = \,\prec$. Conversely, every such partition determines a O-chain up to provable equivalences. It is easily seen that if C is such a chain, then $\mathsf{QG}_\uparrow^{\mathrm{qp}} \vdash C \supset (S_i \leftrightarrow S_j)$ if $S_i, S_j \in \Pi_l^C$ for some l, and $\mathsf{QG}_\uparrow^{\mathrm{qp}} \vdash C \supset (S_i \prec S_{i'})$ if $S_i \in \Pi_j^C$, $S_{i'} \in \Pi_{j'}^C$ and $j < j'$. Thus C also uniquely corresponds to an ordering of Γ which we denote $<_C$, defined by $S_i <_C S_{i'}$ iff $S_i \in \Pi_j^C$, $S_{i'} \in \Pi_{j'}^C$ and $j < j'$. This order is total, the Π_i^C are maximal anti-chains, $\bot$ is minimal, and $\top$ is maximal.

Suppose now that A is in O-normal form, and that Γ contains all the subformulas of A of the form $O^j p$ or $O^j \bot$, as well as $\top$; that C is an O-chain on Γ; and that the valuation v agrees with $<_C$, i.e., $S_i <_C S_j$ iff $v(S_i) < v(S_j)$. Using the same idea as in the proof of Lemma 3 in [3], one can find $A^C \in \Gamma$, the "value" of A under C, so that $v(A^C) = v(A)$, and the choice of A^C depends only on $<_C$, not on v itself. Specifically, A^C can be constructed as follows: (1) If $A \in \Gamma$, then $A^C \equiv A$. (2) If $A \equiv D \wedge E$, then $A^C \equiv D^C$ if $D^C <_C E^C$ and $\equiv E^C$ otherwise. (3) If $A \equiv D \vee E$, then $A^C \equiv D^C$ if $E^C <_C D^C$, and $\equiv E^C$ otherwise. (4) If $A \equiv D \supset E$, then $A^C \equiv E^C$ if $E^C <_C D^C$, and $\equiv \top$ otherwise. This "evaluation" of A is provable in the sense that $\mathsf{QG}_\uparrow^{\mathrm{qp}} \vdash C \supset (A \leftrightarrow A^C)$. This follows easily using the following theorems of **LC**:

$$
\begin{array}{ll}
(D \prec E) \supset (D \wedge E \leftrightarrow D) & (E \prec D) \supset (D \wedge E \leftrightarrow E) \\
(D \leftrightarrow E) \supset (D \wedge E \leftrightarrow D) & (D \prec E) \supset (D \vee E \leftrightarrow E) \\
(E \prec D) \supset (D \vee E \leftrightarrow D) & (E \leftrightarrow D) \supset (D \vee E \leftrightarrow E) \\
(D \prec E) \supset (D \supset E \leftrightarrow \top) & (E \prec D) \supset (D \supset E \leftrightarrow E) \\
(E \leftrightarrow D) \supset (D \supset E \leftrightarrow \top)
\end{array}
$$

Definition 12. Let A be a quantifier free formula in O-normal form, Γ_A be the set of all subformulas of A of the form $O^j p, O^k \bot, \top$, $\Gamma \supseteq \Gamma_A$, and $C(\Gamma)$ the set of all possible O-chains over Γ. Then

$$
\bigvee_{C \in C(\Gamma)} C \wedge A^C
$$

is the O-*chain normal form* for A over Γ.

Theorem 13. *Let A and Γ be as above, and A' be the O-chain normal form for A over Γ. Then $\mathsf{QG}_\uparrow^{\mathrm{qp}} \vdash A \leftrightarrow A'$.*

Proof. (See also Thm. 4 of [3].) First note that $\bigvee_{C \in C(\Gamma)} C$ is a tautology and provable in LC. Since for each $C \in C(\Gamma)$ we have $\mathsf{QG}_\uparrow^{\mathrm{qp}} \vdash (C \wedge A^C) \supset A$, the right-to-left implication $A' \supset A$ follows by case distinction.

For the left-to-right implication, consider $A \supset (A \wedge \bigvee_{C \in C(\Gamma)} C)$. This is provable, since $\bigvee_{C \in C(\Gamma)} C$ is provable. By distributivity of $\wedge$ over $\vee$, we have $A \supset \bigvee_{C \in C(\Gamma)}(A \wedge C)$. We also have $(A \wedge C) \supset (C \wedge A^C)$ for each $C \in C(\Gamma)$ from $\mathsf{QG}_\uparrow^{\mathrm{qp}} \vdash C \supset (A \leftrightarrow A^C)$. Together we get $A \supset \bigvee_{C \in C(\Gamma)}(C \wedge A^C)$. $\square$

We now strengthen the $\circ$-normal form result so that only $\circ$-chains that are intuitively "possible" need to be considered. For this, we have to verify that we can exclude chains C which result in orders which, e.g., have $\circ S <_C S$.

Definition 14. A formula A is in *minimal normal form* over Γ if it is of the form $\bigvee_{C \in \mathcal{C} \subseteq C(\Gamma)} C$, where each C is a $\circ$-chain over Γ, and so that the corresponding ordered partition $\Pi_1^C, \ldots, \Pi_k^C$ satisfies

1. for no $i < j$ and $S \in \Gamma$ do we have $\circ^{r+s} S \in \Pi_i^C$ and $\circ^r S \in \Pi_j^C$ with $s > 0$;
2. for all $S \in \Gamma$, if $\circ^s S \in \Pi_i^C$ ($i < k$), then $\circ^r S \notin \Pi_i^C$ if $r \neq s$; and
3. for no j, j' and $S \in \Gamma$ do we have both $\circ^i S \in \Pi_j^C$ and $\circ^{i+1} S \in \Pi_{j'}^C$ with $j' > j + 1$.

Theorem 15. *Let A be in $\circ$-normal form. There exists a formula A^{nf} in minimal normal form such that $\mathsf{QG}_\uparrow^{\mathrm{qp}} \vdash A \leftrightarrow A^{\mathrm{nf}}$.*

Proof. By Thm. 13, $\mathsf{QG}_\uparrow^{\mathrm{qp}} \vdash A \leftrightarrow A'$ where A' is a $\circ$-chain normal form over Γ. Consider a disjunct of A' of the form $C \wedge A^C$, where $\Pi_1^C, \ldots, \Pi_k^C$ is the ordered partition of Γ corresponding to C. If $A^C \in \Pi_k^C$, then $\mathsf{QG}_\uparrow^{\mathrm{qp}} \vdash (C \wedge A^C) \leftrightarrow C$, since $\mathsf{QG}_\uparrow^{\mathrm{qp}} \vdash A^C \leftrightarrow (A^C \leftrightarrow \top)$. Otherwise, $A^C \in \Pi_i^C$ with $i < k$. Then the sequence $\Pi_i^C, \ldots, \Pi_k^C$ corresponds to a conjunction

$$C' \equiv (A^C \star_1 S_1') \wedge \ldots \wedge (S_{j-1}' \star_j \top)$$

where for at least one $l \leq j$, $\star_j = \prec$, and $\mathsf{QG}_\uparrow^{\mathrm{qp}} \vdash C \leftrightarrow C'' \wedge C'$, where C'' is the part of C corresponding to $\Pi_1^C, \ldots, \Pi_{i-1}^C$. Since $\mathsf{QG}_\uparrow^{\mathrm{qp}} \vdash A^C \leftrightarrow (A^C \leftrightarrow \top)$, we have

$$\mathsf{QG}_\uparrow^{\mathrm{qp}} \vdash (C' \wedge A^C) \leftrightarrow (C' \wedge (\top \leftrightarrow A^C)) \tag{1}$$

As is easily seen, the right-hand side of (1) is provably equivalent to

$$C''' \equiv (A^C \leftrightarrow S_1') \wedge \ldots \wedge (S_{j-1}' \leftrightarrow \top)$$

In sum, $\mathsf{QG}_\uparrow^{\mathrm{qp}} \vdash (C \wedge A^C) \leftrightarrow (C'' \wedge C''')$, and $C'' \wedge C'''$ is a $\circ$-chain.

By induction on the number of disjuncts in A' one shows that there is A'' which is a disjunction of $\circ$-chains such that $\mathsf{QG}_\uparrow^{\mathrm{qp}} \vdash A \leftrightarrow A''$. Now we have to prove that there exists a disjunction of $\circ$-chains A^{nf} satisfying 1–3 of Def. 14 so that $\mathsf{QG}_\uparrow^{\mathrm{qp}} \vdash A'' \leftrightarrow A^{\mathrm{nf}}$.

Suppose that for some disjunct C in A'' we have $\circ^{r+s} S \in \Pi_i^C$ and $\circ^r S \in \Pi_j^C$ where $s > 0$ and $i < j$. Then, since $\mathsf{QG}_\uparrow^{\mathrm{qp}} \vdash (\circ^{r+s} A \prec \circ^r A) \leftrightarrow \circ^r A$ we have $\mathsf{QG}_\uparrow^{\mathrm{qp}} \vdash C \leftrightarrow C'$ where C' is the $\circ$-chain corresponding to $\Pi_1^C, \ldots, \Pi_{i-1}^C, \Pi_i^C \cup \ldots \cup \Pi_k^C$.

Consider a disjunct C of A'' where for some $i < k$, both $\circ^r S \in \Pi_i^C$ and $\circ^s S \in \Pi_i^C$ where $r < s$. Then $\mathsf{QG}_\uparrow^{\mathrm{qp}} \vdash C \supset (\circ^s S \leftrightarrow \top)$. To see this, recall that $\mathsf{QG}_\uparrow^{\mathrm{qp}} \vdash \circ^r v \prec \circ^s S$ if $r < s$. By definition of $\prec$, that means that

$$\mathsf{QG}_\uparrow^{\mathrm{qp}} \vdash ((\circ^s S \supset \circ^r S) \supset \circ^r S) \wedge (\circ^r S \supset \circ^s S). \tag{2}$$

Since $\mathsf{QG}_\uparrow^{\mathrm{qp}} \vdash C \supset (\circ^s S \leftrightarrow \circ^r S)$, we have $\mathsf{QG}_\uparrow^{\mathrm{qp}} \vdash C \supset (\circ^s S \supset \circ^r S)$ which together with the left conjunct of (2) gives $\mathsf{QG}_\uparrow^{\mathrm{qp}} \vdash C \supset \circ^r S$. Thus, as before, C is provably equivalent to the $\circ$-chain corresponding to $\Pi_1^C, \ldots, \Pi_i^C \cup \ldots \cup \Pi_k^C$.

Lastly, suppose that for a disjunct C of A'' we have both $\circ^i S \in \Pi_j^C$ and $\circ^{i+1} S \in \Pi_{j'}^C$ for some j, j' such that $j' > j + 1$. Then by axiom (G6) together with transitivity we get $C \supset (\circ^{i+1} S \prec \circ^{i+1} S)$, and since $\mathsf{QG}_\uparrow^{\mathrm{qp}} \vdash (B \prec B) \leftrightarrow B$ we have $\mathsf{QG}_\uparrow^{\mathrm{qp}} \vdash C \leftrightarrow C'$ where C' is the $\circ$-chain corresponding to $\Pi_1^C, \ldots, \Pi_{j-1}^C, \Pi_j^C \cup \ldots \cup \Pi_{j'}^C \ldots \cup \Pi_k^C$.

By induction on the number of disjuncts in A'' we obtain the desired A^{nf}. □

6   Quantifier Elimination

In this section we prove quantifier elimination for $\mathsf{QG}_\uparrow^{\mathrm{qp}}$. As a corollary of this result we show that the system $\mathsf{QG}_\uparrow^{\mathrm{qp}}$ is sound and complete for $\mathbf{G}_\uparrow^{\mathrm{qp}}$ and that the latter is the intersection of all finite-valued quantified propositional Gödel logics $\mathbf{G}_k^{\mathrm{qp}}$.

Proposition 16.   *1.* $\mathsf{QG}_\uparrow^{\mathrm{qp}} \vdash (\forall p) A(p) \leftrightarrow (A(\bot) \wedge (\forall p) A(\circ p))$
2. $\mathsf{QG}_\uparrow^{\mathrm{qp}} \vdash (\exists p) A(p) \leftrightarrow (A(\bot) \vee (\exists p) A(\circ p))$.

Proof. (1) The left-to-right implication follows easily from the two instances of $(\supset \forall)$

$$(\forall p) A(p) \supset A(\bot) \qquad \text{and} \qquad (\forall p) A(p) \supset A(\circ p).$$

For right-to-left, consider

$$(q \leftrightarrow \bot) \supset (A(\bot) \wedge (\forall p) A(\circ p)) \supset A(q) \tag{3}$$

$$(q \leftrightarrow \circ p) \supset (A(\bot) \wedge (\forall p) A(\circ p)) \supset A(q) \tag{4}$$

which are derived easily from Prop. 7(2) using $\mathsf{IPC}^{\mathrm{qp}}$. Use (R∃) to introduce the existential quantifier in the antecedent of (4), and then (I12) to obtain

$$[(q \leftrightarrow \bot) \vee (\exists p)(q \leftrightarrow \circ p)] \supset (A(\bot) \wedge (\forall p) A(\circ p)) \supset A(q) \tag{5}$$

The antecedent of (5) is an instance of (G5), and so

$$\mathsf{QG}_\uparrow^{\mathrm{qp}} \vdash (A(\bot) \wedge (\forall p) A(\circ p)) \supset A(q)$$

from which the right-to-left direction of (1) follows by (R∀).
(2) The argument is analogous to the derivation of (1).   □

Definition 17. For $\Gamma \subseteq Var \cup \{\bot, \top\}$, let $OP_\Gamma(A)$ be the set of formulas inductively defined as follows:

$$OP_\Gamma(A * B) = OP_\Gamma(A) \cup OP_\Gamma(B), \quad \text{where } * \in \{\vee, \wedge, \supset\}$$
$$OP_\Gamma((Qp) A) = OP_\Gamma(A), \quad \text{where } Q \in \{\forall, \exists\}$$
$$OP_\Gamma(\circ^k v) = \begin{cases} \{\circ^k v\} & \text{if } v \in \Gamma \\ \emptyset & \text{otherwise} \end{cases}$$

Then $\exp_\Gamma(A) = \{k : \circ^k q \in OP_\Gamma(A)\}$

Definition 18. The *quantifier depth* qd(A) of a formula is defined by:

$$qd(p) = qd(\bot) = 0 \qquad qd((\forall p)B) = qd((\exists p)B) = qd(B) + 1$$
$$qd(B * C) = \max(qd(B), qd(C)) \text{ for } * \in \{\wedge, \vee, \supset\}$$

Lemma 19. *Let A be a closed formula such that (a) every quantifier free sub-formula of A is in $\circ$-normal form and (b) no two quantifier occurrences bind the same variable. Let $\Delta = \{p_1, \ldots, p_j\}$ be the set of variables belonging to the innermost quantifiers in A, and $\Gamma = Var(A) \setminus \Delta$. Then there is a formula $A^\sharp$ so that*

1. *$QG_\uparrow^{qp} \vdash A \leftrightarrow A^\sharp$,*
2. *$\max \exp_\Delta(A^\sharp) \leq \min \exp_\Gamma(A^\sharp)$,*
3. *$\max \exp_{Var(A^\sharp)}(A^\sharp) \leq 2 \cdot \max \exp_{Var(A)}(A)$,*
4. *$qd(A^\sharp) \leq qd(A)$.*

Proof. Suppose $\Gamma = \{q_1, \ldots, q_l\}$. Let $A_0 = A$, $m = \max \exp_\Delta(A)$. At stage i, pick the non-innermost quantified subformula $(\forall q_i)B_i(q_i)$ or $(\exists q_i)B_i(q_i)$ of A_i corresponding to q_i and replace

$$(\forall q_i)B_i(q_i) \text{ by } B_i(\bot) \wedge \ldots \wedge B_i(\circ^{m-1}\bot) \wedge (\forall p)B_i(\circ^m q_i)$$
$$(\exists q_i)B_i(p) \text{ by } B_i(\bot) \vee \ldots \vee B_i(\circ^{m-1}\bot) \vee (\exists q_i)B_i(\circ^m q_i)$$

to obtain A_{i+1}. The procedure terminates with $A_l = A^\sharp$.

At each stage $QG_\uparrow^{qp} \vdash A_i \leftrightarrow A_{i+1}$ follows by induction on m from Prop. 16. The lower bounds are obvious from the construction of $A^\sharp$. □

Lemma 20. *Suppose $A(p)$ is in $\circ$-normal form and*

$$\max \exp_{\{p\}} A \leq \min \exp_{Var(A) \setminus \{p\}} A.$$

There is a formula $A^\exists$, with $Var(A^\exists) \subseteq Var(A) \setminus \{p\}$ so that

$$QG_\uparrow^{qp} \vdash (\exists p)A \leftrightarrow A^\exists$$

and $\max \exp_{Var(A^\exists) \cup \{\bot\}} A^\exists \leq \max \exp_{Var(A) \cup \{\bot\}} A + 1$.

Proof. Let $m = \max \exp_{Var(A) \cup \{\bot\}} A$ be the maximal exponent of a subformula $\circ^j S$ and let $\Gamma = \{\circ^i S : S \in Var \cup \{\bot\}, i \leq m\}$.

Theorem 15 provides us with A^{nf} in minimal normal form over Γ so that $QG_\uparrow^{qp} \vdash (\exists p)A \leftrightarrow (\exists p)A^{nf}$. Since $\exists$ distributes over $\vee$, we only have to consider formulas of the form $(\exists p)C$ where C is a $\circ$-chain and satisfies the conditions of Thm. 15. C corresponds to an ordered partition $\Pi_1, \ldots, \Pi_k$ over Γ. We prove that $QG_\uparrow^{qp} \vdash (\exists p)C \leftrightarrow C'$ for some quantifier-free C' by induction on k.

If $k = 2$, then either $p \in \Pi_1$ or $p \in \Pi_k$. In the first case, $QG_\uparrow^{qp} \vdash (\exists p)C(p) \leftrightarrow C(\bot)$, in the second one, $QG_\uparrow^{qp} \vdash (\exists p)C(p) \leftrightarrow C(\top)$.

Now suppose $k > 2$. Three cases arise, according to how the equivalence classes containing p are distributed.

(1) The partition corresponding to C is of the form

$$\Pi_1, \ldots, \Pi_i, \{p\}, \{\bigcirc p\}, \ldots, \{\bigcirc^j p\} \cup \Pi_k$$

Then $C(p)$ is of the form

$$B \wedge \underbrace{(v \prec p) \wedge (p \prec \bigcirc p) \wedge \ldots \wedge (\bigcirc^j p \leftrightarrow \top)}_{D(p)} \wedge E$$

Since $D(\top)$ is provable, $\mathsf{QG}^{\mathrm{qp}}_\uparrow \vdash (\exists p)C \leftrightarrow B \wedge v \prec \top \wedge E$.

(2) The partition corresponding to C is of the form

$$\Pi_1, \ldots, \Pi_i, \{p\}, \{\bigcirc p\}, \ldots, \{\bigcirc^j p\}, \Pi_{i'}, \ldots, \Pi_k$$

and $\bigcirc^j p \notin \Pi_{i'}$. Then $C(p)$ is of the form

$$B \wedge \underbrace{(S \prec p) \wedge (p \prec \bigcirc p) \wedge \ldots \wedge (\bigcirc^j p \prec S')}_{D(p)} \wedge E$$

We first show that $\mathsf{QG}^{\mathrm{qp}}_\uparrow \vdash (\exists p)D(p) \leftrightarrow (\bigcirc^{j+1} S \prec S')$. For the right-to-left direction, observe that

$$\mathsf{QG}^{\mathrm{qp}}_\uparrow \vdash (\bigcirc^{j+1} S \prec S') \supset [(S \prec \bigcirc S) \wedge \ldots \wedge (\bigcirc^j S \prec \bigcirc^{j+1} S) \wedge (\bigcirc^{j+1} S \prec S'),$$

from which the claim follows by (R∃). The left-to-right direction is proved by induction on j, using axiom (G6). In sum, we have

$$\mathsf{QG}^{\mathrm{qp}}_\uparrow \vdash (\exists p)C(p) \leftrightarrow (B \wedge (\bigcirc^{j+1} S \prec S') \wedge E)$$

(3) The partition corresponding to C is of the form

$$\Pi_1, \ldots, \Pi_i, \{p\}, \{\bigcirc p\}, \ldots, \{\bigcirc^j p\} \cup \Pi, \Pi_{i'}, \ldots, \Pi_k \quad \cdot$$

with $S \in \Pi$, $S \neq \bigcirc^j p$. Because of the condition on $\max \exp_{\{p\}} A$ we can assume that $S \equiv \bigcirc^n q$ with $n \geq j$.

We proceed by induction on j. If $j = 0$, then we have a conjunct $p \leftrightarrow S$, and $(\exists p)C \equiv C(S)$. Otherwise, we have a conjunct $\bigcirc^j p \leftrightarrow \bigcirc^n q$ with $n \geq j$. Using (G3), this conjunct is provably equivalent to $(\bigcirc^{j-1} p \leftrightarrow \bigcirc^{n-1} q) \vee (\bigcirc^j p \wedge \bigcirc^n q)$. Hence, C is equivalent to the disjunction of two $\bigcirc$-chains corresponding to

$$\Pi_1, \ldots, \Pi_i, \{p\}, \{\bigcirc p\}, \ldots, \ \{\bigcirc^{j-1} p, \bigcirc^{n-1} q\}, \Pi, \Pi_{i'}, \ldots, \Pi_k$$
$$\Pi_1, \ldots, \Pi_i, \{p\}, \{\bigcirc p\}, \ldots, \ \{\bigcirc^j p\} \cup \Pi \cup \Pi_{i'} \cup \ldots \cup \Pi_k$$

For the first $\bigcirc$-chain, the maximum exponent of p is smaller and hence the induction hypothesis of the present subcase applies. The second $\bigcirc$-chain is shorter overall, and hence the induction hypothesis based on number of equivalence classes applies. □

Lemma 21. *Let $A(p)$ be in $\circ$-normal form, and so that*

$$\max \exp_{\{p\}} A \leq \min \exp_{Var(A)\setminus\{p\}} A.$$

There is a formula $A^{\forall}$, with $Var(A^{\forall}) \subseteq Var(A) \setminus \{p\}$ so that

$$QG_{\uparrow}^{qp} \vdash (\forall p)A \leftrightarrow A^{\forall}$$

and $\max \exp_{Var(A^{\forall}) \cup \{\perp\}} A^{\forall} \leq \max \exp_{Var(A) \cup \{\perp\}} A + 1.$

Proof. Let A^{nf} be the minimal normal form of A. It is provably equivalent to the formula obtained from A^{nf} by replacing each element of a chain $S \prec S'$ by $\circ S \supset S'$. By distributivity then, $A \leftrightarrow A'$ where A' is a conjunction of disjunctions of implications of the form $\circ^i S \supset \circ^j S'$. Any such disjunct of the form $\circ^i p \supset \circ^j p$ is provably equivalent to $\top$ if $i \leq j$ (in which case the entire disjunction can be deleted), or to $\top \supset \circ^j p$ if $i > j$. The part of a disjunction in A' containing p thus can be assumed to be of the form

$$\bigvee_i (D_i \supset \circ^{n_i} p) \vee \bigvee_j (\circ^{m_j} p \supset E_j)$$

where $p \notin D_i, E_i$. This, in turn, is equivalent to a conjunction of disjunctions of the form

$$\bigvee_i (D \supset \circ^{n_i} p) \vee \bigvee_j (\circ^{m_j} p \supset E)$$

This can again be simplified by taking $n = \max\{n_i\}$ and $m = \min\{m_j\}$, since $QG_{\uparrow}^{qp} \vdash (A \supset B) \vee (A \supset C) \leftrightarrow (A \supset C)$ if $QG_{\uparrow}^{qp} \vdash B \supset C$.

Since $QG_{\uparrow}^{qp} \vdash (\forall p)(A \wedge B) \leftrightarrow (\forall p)A \wedge (\forall p)B$ and $QG_{\uparrow}^{qp} \vdash (\forall p)(A(p) \vee B) \leftrightarrow (\forall p)A(p) \vee B$ if $p \notin B$, it suffices to show that a formula of the form

$$F \equiv (\forall p)(D \supset \circ^n p) \vee (\circ^m p \supset E))$$

is equivalent to a quantifier free formula. We distinguish three cases:

(1) $E \equiv \circ^k \top$, $k \geq 0$. Then $QG_{\uparrow}^{qp} \vdash (\circ^m p \supset E)$ and hence $QG_{\uparrow}^{qp} \vdash F \leftrightarrow \top$.

(2) $E \equiv \circ^k \perp$, $k < m$. Then $QG_{\uparrow}^{qp} \vdash (\circ^m p \supset E) \leftrightarrow E$, and hence $QG_{\uparrow}^{qp} \vdash F \leftrightarrow (A \supset \circ^n \perp) \vee E$.

(3) Since $\max \exp_{\{p\}} A \leq \min \exp_{Var(A)\setminus\{p\}} A$ by assumption, this leaves only the case $E \equiv \circ^m S$. Then $QG_{\uparrow}^{qp} \vdash F \leftrightarrow (A \supset \circ^{n+1} S) \vee \circ^m S$. The left-to-right implication is obvious by $(\supset \forall)$, instantiating p by $\circ S$. For the right-to-left implication two cases arise:

(a) $n \leq m$. By (G4), we have $QG_{\uparrow}^{qp} \vdash (A \supset \circ^{n+1} S) \supset [(A \supset \circ^n p) \vee (\circ^n p \supset \circ^n S)]$. Furthermore, $QG_{\uparrow}^{qp} \vdash (\circ^n p \supset \circ^n S) \supset (\circ^m p \supset \circ^m S)$. In sum, we have

$$[(A \supset \circ^{n+1} S) \vee \circ^m S] \supset [(A \supset \circ^n p) \vee (\circ^m p \supset \circ^m S) \vee \circ^m S]$$

Since $QG_{\uparrow}^{qp} \vdash \circ^m S \supset (\circ^m p \vee \circ^m S)$, we have $QG_{\uparrow}^{qp} \vdash [(A \supset \circ^{n+1} S) \vee \circ^m S] \supset F$.

(b) $n > m$. By (G2), $\mathrm{QG}_\uparrow^{\mathrm{qp}} \vdash \bigcirc^m S \supset \bigcirc^{n+1} S$, and so $\mathrm{QG}_\uparrow^{\mathrm{qp}} \vdash [(A \supset \bigcirc^{n+1} S) \vee \bigcirc^m S] \supset (A \supset \bigcirc^{n+1} S]$. Using induction and (G4), it is easy to show that

$$\mathrm{QG}_\uparrow^{\mathrm{qp}} \vdash (A \supset \bigcirc^{n+1} S) \supset [(A \supset \bigcirc^n p) \vee \underbrace{\bigvee_{i=m}^{n-1} (\bigcirc^{i+1} p \supset \bigcirc^i p)}_{D} \vee (\bigcirc^m p \supset \bigcirc^m S].$$

Each of the disjuncts $\bigcirc^{i+1} p \supset \bigcirc^i p$ implies $\bigcirc^i p$, which in turn implies $A \supset \bigcirc^n p$, so $\mathrm{QG}_\uparrow^{\mathrm{qp}} \vdash D \supset (A \supset \bigcirc^n p)$. In sum, we have again $\mathrm{QG}_\uparrow^{\mathrm{qp}} \vdash [(A \supset \bigcirc^{n+1} S) \vee \bigcirc^m S] \supset F$.

The bound on $\max \exp_{Var(A^\vee) \cup \{\perp\}} A$ follows by inspection. $\square$

Theorem 22. *For every closed formula A of $\mathrm{QG}_\uparrow^{\mathrm{qp}}$ there exists a variable-free formula A^{qf} such that $\mathrm{QG}_\uparrow^{\mathrm{qp}} \vdash A \leftrightarrow A^{\mathrm{qf}}$, and $\max \exp_{\{\perp\}} A^{\mathrm{qf}} \leq 2^{\mathrm{qd}(A)+l}$ where $l = \max \exp_{Var(A) \cup \{\perp\}}$.*

Proof. We may assume, renaming variables if necessary, that each variable in A is bound by only one quantifier occurrence. By induction on $\mathrm{qd}(A)$. If $\mathrm{qd}(A) = 0$, there is nothing to prove. If $\mathrm{qd}(A) > 0$, let $A^\sharp$ be as in Lemma 19. Replace each innermost quantified formula $(\exists p)B$, $(\forall p)B$ by $B^\exists$ or $B^\forall$, respectively. The resulting formula A' satisfies $\mathrm{qd}(A') \leq \mathrm{qd}(A) - 1$ and $\max \exp_{Var(A) \cup \{\perp\}} A' \leq 2 \max \exp_{Var(A) \cup \{\perp\}} A + 1$. $\square$

Proposition 23. *Let A be variable-free, and in $\bigcirc$-normal form. Then either $\mathrm{QG}_\uparrow^{\mathrm{qp}} \vdash A \leftrightarrow \top$ or $\mathrm{QG}_\uparrow^{\mathrm{qp}} \vdash A \leftrightarrow \bigcirc^k(\perp)$ where $k \leq \max \exp_{\{\perp\}} A = n$.*

Proof. Consider the minimal normal form A^{nf} of A over $\{\bigcirc^k(\perp) : k \leq n\}$. Each chain in A^{nf} is of one of two forms

$$C = (\perp \prec \bigcirc(\perp)) \wedge (\bigcirc(\perp) \prec \bigcirc\bigcirc(\perp)) \wedge \ldots \wedge (\bigcirc^{n-1}\perp \prec \bigcirc^n(\perp))$$

$$C_m = (\perp \prec \bigcirc(\perp)) \wedge (\bigcirc(\perp) \prec \bigcirc\bigcirc(\perp)) \wedge \ldots \wedge (\bigcirc^{m-1}\perp \prec \bigcirc^m(\perp)) \wedge \bigwedge_{k=m}^{n} \bigcirc^k(\perp)$$

C is provable, so $\mathrm{QG}_\uparrow^{\mathrm{qp}} \vdash C \leftrightarrow \top$, and $\mathrm{QG}_\uparrow^{\mathrm{qp}} \vdash C_m \leftrightarrow \bigcirc^m(\perp)$. So if A^{nf} contains C, then $\mathrm{QG}_\uparrow^{\mathrm{qp}} \vdash A \leftrightarrow \top$, otherwise $\mathrm{QG}_\uparrow^{\mathrm{qp}} \vdash A \leftrightarrow \bigcirc^k(\perp)$, where k is the maximum of C_i occurring in A^{nf}. $\square$

Corollary 24. *Let A be closed and not containing $\bigcirc$. Then either $\mathrm{QG}_\uparrow^{\mathrm{qp}} \vdash A$ or $\mathrm{QG}_\uparrow^{\mathrm{qp}} \vdash A \leftrightarrow \bigcirc^k(\perp)$, where $k \leq 2^{\mathrm{qd}(A)}$.*

Corollary 25. *The calculus $\mathrm{QG}_\uparrow^{\mathrm{qp}}$ is complete for $\mathbf{G}_\uparrow^{\mathrm{qp}}$.*

Proof. If $\mathrm{QG}_\uparrow^{\mathrm{qp}} \not\vdash A$, then $\mathrm{QG}_\uparrow^{\mathrm{qp}} \vdash A \leftrightarrow \bigcirc^k \perp$ for some k. Since $\mathbf{G}_\uparrow^{\mathrm{qp}} \not\models \bigcirc^k \perp$ for all k, $\mathbf{G}_\uparrow^{\mathrm{qp}} \not\models A$.

Theorem 26. $\mathbf{G}_\uparrow^{\mathrm{qp}}$ *is the intersection of all finite-valued quantified propositional Gödel logics.*

Proof. $\mathsf{QG}_\uparrow^{\mathrm{qp}}$ is sound for each finite-valued Gödel logic, so $\mathbf{G}_\uparrow^{\mathrm{qp}} \subseteq \mathbf{G}_k^{\mathrm{qp}}$ for each k. Conversely, if $\mathbf{G}_\uparrow^{\mathrm{qp}} \not\models A$, then $\mathsf{QG}_\uparrow^{\mathrm{qp}} \vdash A \leftrightarrow \mathsf{O}^k(\bot)$ for some k. Since $\mathsf{QG}_\uparrow^{\mathrm{qp}}$ is sound for $\mathbf{G}_{k+2}$, we have $\mathbf{G}_{k+2} \not\models A$ as obviously $\mathbf{G}_{k+2} \not\models \mathsf{O}^k \bot$.

References

[1] Baaz, M.: Infinite-valued Gödel logics with 0-1-projections and relativizations. In *Gödel 96. Kurt Gödel's Legacy.* Proceedings. LNL 6, Springer, 23–33. 248

[2] Baaz, M., Leitsch, A., Zach, R.: Incompleteness of an infinite-valued first-order Gödel logic and of some temporal logics of programs. In *Computer Science Logic. Selected Papers from CSL'95.* Springer, 1996, 1–15. 241

[3] Baaz, M., Veith, H.: Interpolation in fuzzy logic. *Arch. Math. Logic,* 38 (1999), 461–489. 248, 248, 249, 249

[4] Baaz, M., Veith, H.: An axiomatization of quantified propositional Gödel logic using the Takeuti-Titani rule. In *Logic Colloquium 1998.* Proceedings. LNL 13, Association for Symbolic Logic, 91–104. 241, 241, 241, 244, 248

[5] Baaz, M., Zach R.: Compact propositional Gödel logics. In *28th International Symposium on Multiple Valued Logic.* Proceedings. IEEE Press, 1998, 108–113. 241

[6] Büchi, J. R.: On a decision method in restricted second order arithmetic. In *Logic, Methodology, and Philosophy of Science,* Proceedings of the 1960 Congress, Stanford University Press, 1–11. 241, 245

[7] Dummett, M.: A propositional calculus with denumerable matrix. *J. Symbolic Logic,* 24(1959), 97–106. 240, 240, 242, 244, 244

[8] Dunn, J. M., Meyer, R. K.: Algebraic completeness results for Dummett's *LC* and its extensions. *Z. Math. Logik Grundlagen Math.,* 17 (1971), 225–230. 240

[9] Egly, U., Eiter, T., Tompits, H., Woltran, S.: Solving advanced reasoning tasks using quantified boolean formulas, In *AAAI-2000.* Proceedings. to appear. 241

[10] Gödel, K.: Zum intuitionistischen Aussagenkalkül. *Anz. Akad. Wiss. Wien,* 69 (1932), 65–66. 240, 242

[11] Hájek, P.: *Metamathematics of Fuzzy Logic.* Kluwer, 1998. 240

[12] Horn, A.: Logic with truth values in a linearly ordered Heyting algebra. *J. Symbolic Logic,* 27(1962), 159–170. 240

[13] Pearce, D.: Stable inference as intuitionistic validity. *J. Logic Programming,* 38 (1999), 79–91. 240

[14] Takeuti, G., Titani, S.: Intuitionistic fuzzy logic and intuitionistic fuzzy set theory. *J. Symbolic Logic,* 49 (1984), 851–866. 241

[15] Visser, A.: On the completeness principle: a study of provability in Heyting's Arithmetic. *Annals Math. Logic,* 22 (1982), 263–295. 240

Proof-Search in Implicative Linear Logic as a Matching Problem

Philippe de Groote

LORIA UMR n° 7503 – INRIA
Campus Scientifique, B.P. 239
54506 Vandœuvre lès Nancy Cedex – France
e-mail: degroote@loria.fr

Abstract. We reduce the provability of fragments of multiplicative linear logic to matching problems consisting in finding a one-one-correspondence between two sets of first-order terms together with a unifier that equates the corresponding terms. According to the kind of structure to which these first-order terms belong our matching problem corresponds to provability in the implicative fragment of multiplicative linear logic, in the Lambek calculus, or in the non-associative Lambek calculus.

1 Introduction

Four decades ago, Lambek introduced a non-commutative logical calculus, known as **L**, intended to give a mathematical account of the structure of natural languages [13]. This calculus, which serves as a basis for modern categorial grammars [17,18,24], appears a posteriori to be the intuitionistic non-commutative fragment of Girard's multiplicative linear logic [7].

In a categorial grammar, sentence parsing amounts to automatic deduction in the underlying logical calculus. This gives a practical interest to proof-search algorithms for **L**. Nevertheless, the complexity of **L** provability is still an open problem, even in the case of its implicative fragment. It is known, however, that the calculus obtained by allowing **L** to be commutative (i.e., the intuitionistic fragment of multiplicative linear logic) is NP-complete. This result, due to Kanovitch, remains valid in the purely implicative case [9]. On the other hand, **NL** (the non-associative variant of **L** that Lambek introduced in [14]) is known to be polynomial. This has been established by Aarts and Trautwein for the implicative fragment of **NL** [1], and by ourself for the full system [6].

In this paper, we try to get some new insight into the complexity of **L**. To this end, we reduce provability in the implicative fragment of **L** to a matching problem consisting in finding a one-one-correspondence between two sets of first-order terms ranging over the free monoid, together with a unifier that equates the corresponding terms. Interestingly enough, when the terms range over the free groupoid or over the free commutative monoid, our matching problem corresponds to provability in the implicative fragments of **NL** or multiplicative linear logic, respectively. This sheds light on the role played by associativity, and commutativity.

M. Parigot and A. Voronkov (Eds.): LPAR 2000, LNAI 1955, pp. 257–274, 2000.

Our reduction, which is inspired by the language models of **L** [20], is not entirely new. Indeed, in his thesis [22], Roorda shows how to associate to any formula **L** a matching problem akin to ours. With respect to this, our contribution is twofold:

- we show that a formula is provable if and only if the associated matching problem admits a solution (Roorda only proves the easy part of this statement, i.e., the necessity of the condition);
- we define a notion of PN-matching that characterises exactly the matching problems that are associated to formulas; consequently, our reduction works in both direction.

We also define a general proof-search procedure that works for the implicative fragments of **NL**, **L**, and multiplicative linear logic. This procedure, which is based on our notion of PN-matching, is specified by a non-deterministic transition system. Here, our main contribution is to show that each transition is history independent, which allows dynamic programming techniques to be used.

2   Intuitionistic Implicative Linear Logic

In this section, we present three variants of intuitionistic implicative linear logic: the implicative fragment of Girard's multiplicative linear logic [7], the implicative fragment of Lambek's calculus of syntactic types (also known as *the Lambek calculus*) [13], and the implicative fragment of the so-called non associative Lambek calculus [14]. These three calculi will be called **IMLL**, **IL**, and **INL**, respectively. As we will see, **IMLL** may be seen as the commutative extension of **IL** which may be seen as the associative extension of **INL**:

We start with a presentation of the weakest system. The formulas of **INL** are built up from a set of atomic formulas $\mathcal{A}$ and the connectives $\multimap$ and $\mathbin{\rotatebox[origin=c]{180}{$\multimap$}}$ according to the following grammar:

$$\mathcal{F} ::= \mathcal{A} \mid (\mathcal{F} \multimap \mathcal{F}) \mid (\mathcal{F} \mathbin{\rotatebox[origin=c]{180}{$\multimap$}} \mathcal{F})$$

The consequence relation of **INL** is specified by the following Gentzen-like sequent calculus. The sequents of this calculus have the form $\Gamma \vdash A$ where Γ is a (possibly empty)[1] binary tree of formulas, i.e., a fully bracketed structure. We take for granted the notion of context, i.e., a binary tree with a hole. If $\Gamma[\,]$ is such a context, $\Gamma[\Delta]$ denotes the binary tree obtained by filling the hole in $\Gamma[\,]$ with the binary tree Δ.

$$A \vdash A \quad \text{(Id)}$$

$$\frac{\Gamma \vdash A \quad \Delta[B] \vdash C}{\Delta[\,(\Gamma, (A \multimap B))\,] \vdash C} \;(\multimap\text{-L}) \qquad \frac{(A, \Gamma) \vdash B}{\Gamma \vdash (A \multimap B)} \;(\multimap\text{-R})$$

---

[1] This is a slight departure from the original (non associative) Lambek calculus that requires sequents whose antecedents are non empty.

$$\frac{\Gamma \vdash A \quad \Delta[B] \vdash C}{\Delta[((B \multimapinv A), \Gamma)] \vdash C} \quad (\multimapinv\text{-L}) \qquad\qquad \frac{(\Gamma, A) \vdash B}{\Gamma \vdash (B \multimapinv A)} \quad (\multimapinv\text{-R})$$

The above system does not include any structural rule. As a consequence **INL**, seen as logical system, is quite weak (without being trivial). For instance, the two connectives "$\multimap$" and "$\multimapinv$" do not satisfy the following transitivity rules:

$$(A \multimap B, B \multimap C) \vdash A \multimap C \qquad\qquad (C \multimapinv B, B \multimapinv A) \vdash C \multimapinv A$$

Indeed, these two rules suppose the associativity of the binary operation whose residuals are "$\multimap$" and "$\multimapinv$".

Now, by extending **INL** with the following structural rules, which allow for associativity, we obtain the Lambek calculus **IL**:

$$\frac{\Gamma[(\Delta, (\Theta, \Lambda))] \vdash A}{\Gamma[((\Delta, \Theta), \Lambda)] \vdash A} \quad (\text{assoc}_1) \qquad\qquad \frac{\Gamma[((\Delta, \Theta), \Lambda)] \vdash A}{\Gamma[(\Delta, (\Theta, \Lambda))] \vdash A} \quad (\text{assoc}_2)$$

In fact, the usual presentation of **IL** leaves Rules (assoc_1) and (assoc_2) implicit by defining the antecedents of the sequents to be sequences of formulas rather than binary trees.

Finally, by extending **IL** with the following exchange rule:

$$\frac{\Gamma[(\Delta, \Theta)] \vdash A}{\Gamma[(\Theta, \Delta)] \vdash A} \quad (\text{exchange})$$

one obtains the implicative fragment of Girard's multiplicative linear logic. In this case, there is no longer any need for distinguishing between two kinds of implications because the formulas $A \multimap B$ and $B \multimapinv A$ are provably equivalent.

It is well-known that **IMLL**, **IL**, and **INL** are such that any sequent $(A, \Gamma) \vdash B$ is provable if and only if $\Gamma \vdash A \multimap B$ is provable. This allows the provability problem for sequents to be reduced to the provability problems for sequents made of only one formula. In the sequel of this paper, for the sake of simplicity, we will only consider such one-formula sequents.

3   Intuitionistic Proof-Nets

In Girard's multiplicative linear logic, implication is not taken as a primitive. The formulas are built upon a set of literals—i.e., atomic formulas $(A, B, C, \ldots)$ or negated atomic formulas $(A^{\perp}, B^{\perp}, C^{\perp}, \ldots)$—by means of two connectives ($\otimes$ and $\invamp$) that correspond to multiplicative conjunction and disjunction, respectively. Then, implication is defined according to de Morgan's laws. This gives rise to the following translation of the implicative formulas introduced in the previous section:

$$[\![A]\!]^{+} = A \qquad\qquad\qquad [\![A]\!]^{-} = A^{\perp}$$
$$[\![\alpha \multimap \beta]\!]^{+} = [\![\alpha]\!]^{-} \invamp [\![\beta]\!]^{+} \qquad [\![\alpha \multimap \beta]\!]^{-} = [\![\beta]\!]^{-} \otimes [\![\alpha]\!]^{+}$$
$$[\![\alpha \multimapinv \beta]\!]^{+} = [\![\alpha]\!]^{+} \invamp [\![\beta]\!]^{-} \qquad [\![\alpha \multimapinv \beta]\!]^{-} = [\![\beta]\!]^{+} \otimes [\![\alpha]\!]^{-}$$

Example 1. Let $F = (C \multimap (B \multimap A)) \multimap (C \multimap (B \multimap ((E \multimap E) \multimap ((D \multimap D) \multimap A))))$. Then, we have

$$[\![F]\!]^+ = ((A \otimes B^\perp) \otimes C) \,\mathscr{P}\, (C^\perp \,\mathscr{P}\, (B \,\mathscr{P}\, ((A^\perp \otimes (D^\perp \,\mathscr{P}\, D)) \otimes (E \,\mathscr{P}\, E^\perp))))$$

This translation allows proof-nets to be defined for **IMLL**, **IL**, and **INL**. Proof-nets are a graph-theoretic representation of proofs. Their definition comes in two rounds. One first defines a notion of proof-structure, which corresponds to a class of graphs intended to represent proofs. Then one gives a correctness criterion that allows one to distinguish the proof-structures that correspond to actual proofs from the other ones.

There exist several correctness criteria in the literature [3,4,7,8,10] (including criteria adapted to the non-commutative case [12,19,21,22]), among which the most well known are Girard's long trip condition [7], and the Danos-Regnier criterion [4]. These criteria might be used in the present intuitionistic setting because (contrarily to classical logic) multiplicative linear logic is a conservative extension of its intuitionistic fragment. Nevertheless, it is possible to define criteria that are intrinsically intuitionistic. This is the case of the criterion we give here, which is taken from [5]. This criterion has also the advantage of being easily adaptable to the non-commutative and the non-associative cases.

Proof-nets and proof-structures being simple graphs (whose vertices are decorated with literals and connectives), we use freely elementary graph-theoretic concepts that can be found in any textbook. In particular, we adopt the terminology of [2], and we will write $P = \langle V, E \rangle$ for a proof-structure (or a proof-net) P whose set of vertices is V, and set of edges is E. We also take for granted the notion of parse tree of a multiplicative formula. The leaves of such a parse tree are decorated with literals, and its nodes are decorated either with the connective $\otimes$ or the connective $\mathscr{P}$.

We first introduce a notion of *proof-frame*. Then we define the notions of *proof-structure* and *proof-net*.

Definition 1. *Let A be an implicative formula. The* proof-frame *of A is defined to be the parse tree of the multiplicative formula $[\![A]\!]^+$.* ∎

The translation $[\![\ ]\!]^+$ implicitly assigns polarities (positive or negative) to all the sub-formulas of a given implicative formula. This assignment is reflected on the proof-frames as follows:

- each leaf that is decorated with a positive literal $(A, B, C, \ldots)$ is assigned the positive polarity;
- each leaf that is decorated with a negative literal $(A^\perp, B^\perp, C^\perp, \ldots)$ is assigned the negative polarity;
- each node that is decorated with $\mathscr{P}$ is assigned the positive polarity;
- each node that is decorated with $\otimes$ is assigned the negative polarity.

In a proof-frame, a subgraph made of one node together with its two daughters is called a link. The left and right daughters of a link are respectively called its left and right premises. The mother is called the conclusion of the link. Note that the two premises of any link are assigned opposite polarities by the translation $[\![\ ]\!]^+$. Consequently, one distinguishes between four sorts of links according to the polarities that are assigned to their vertices. The *npp-links* are defined to be the links whose left premise is negative, whose conclusion is positive, and whose right premise is positive. The *ppn-links*, *nnp-links*, and *pnn-links* are defined accordingly. The *npp-* and *ppn-*links are also called $\mathcal{V}$-links, according to the connective that decorates their conclusions. Similarly, the *nnp-* and *pnn-*links are called ⊗-links.

Definition 2. *Let A be an implicative formula. A proof-structure of A (if any) is a simple decorated graph made of:*

(a) *the proof-frame of A,*
(b) *a perfect matching on the leaves of this proof-frame that relates any leaf decorated with a positive literal A to some leaf decorated with the negative literal $A^{\perp}$.*

The edges defining the perfect matching on the leaves of the proof-frame are called the axiom links of the proof-structure.   ∎

In a proof-structure, the two leaves of the underlying proof-frame that are related by a given axiom link are called the conclusions of this axiom link. We also define the *principal inputs* of a proof-structure (or a proof-frame) to be the negative premises of its $\mathcal{V}$-links. Similarly, we define its *principal outputs* to be the positive premises of its ⊗-links (this notion will be only needed in Section 4).

Let Σ be a countably infinite set, whose elements will be called the *constants*. We write $\mathcal{T}(\Sigma)$ for the carrier set of the groupoid[2] $\langle \mathcal{T}(\Sigma), \cdot, \epsilon \rangle$ freely generated by Σ. We also write Σ^* (respectively, $\mathbb{N}^{\Sigma}$) for the carrier sets of the monoid $\langle \Sigma^*, \cdot, \epsilon \rangle$ (respectively, the commutative monoid[3] $\langle \mathbb{N}^{\Sigma}, \cdot, \epsilon \rangle$) freely generated by Σ.

Definition 3. *Let A be an implicative formula, An **INL** (respectively, **IL**, **IMLL**) proof-net of A (if any) is a proof-structure of A, $P = \langle V, E \rangle$, together with an application $\rho : V \to \mathcal{T}(\Sigma)$ (respectively, $\rho : V \to \Sigma^*$, $\rho : V \to \mathbb{N}^{\Sigma}$) such that:*

(a) *the value assigned by ρ to the root of the underlying proof-frame is ϵ;*
(b) *the values assigned by ρ to the principal inputs of P are constants that are pairwise different;*
(c) *the values assigned by ρ to the two conclusions of an axiom-link are equal;*

---

[2] I.e, an algebraic structure with a (non necessarily associative) binary operation "·" that admits an identity element ϵ.

[3] Remark that the set of functions from a set Σ to $\mathbb{N}$, together with the pointwise addition, corresponds indeed to the commutative monoid freely generated by Σ. Therefore, in this case, it woud be more natural to write "+" for the binary operation of the structure. Nevertheless, for the sake of uniformity, we will stick to the product notation.

(d) *the values assigned by ρ obey the constraints given in Figure 1, i.e.:*
  (d1) *the value assigned to the positive premise of a npp-link must be equal to the product of the value assigned to its conclusion with the value assigned to its negative premise,*
  (d2) *the value assigned to the positive premise of a ppn-link must be equal to the product of the value assigned to its negative premise with the value assigned to its conclusion;*
  (d3) *the value assigned to the negative premise of a nnp-link must be equal to the product of the value assigned to its conclusion with the value assigned to its positive premise;*
  (d4) *the value assigned to the negative premise of a pnn-link must be equal to the product of the value assigned to its positive premise with the value assigned to its conclusion.*   ∎

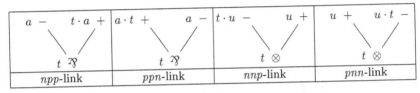

Fig. 1. Constraints on the links of a proof-net

In [5], the notion of *dynamic graph underlying a proof-net* is introduced. Using this notion, it is easy to prove that, for any given proof-net, the valuation ρ is unique up to the renaming of the atoms assigned to the principal inputs.

Example 2. Figure 2 gives a proof-net for the formula of Example 1.

In his thesis [22], Roorda noted that it is possible to assign labels that obey the contraints of Definition 3 to the vertices of any correct proof-structure (i.e., a proof-structure that corresponds to some sequent derivation). On the other hand, he did not prove that the existence of such an assignement is sufficient to ensure correctness. He stated it as an open problem. We solve the question in [5] where, indeed, we proved that the condition is sufficient.[4] Consequently, we have the following proposition.

Proposition 1. *Let A be an implicative formula. A is **INL**(respectively, **IL**, **IMLL**) provable if and only if there exists an **INL**(respectively, **IL**, **IMLL**) proof-net for it.*   □

4   Proof-Search as a Matching Problem

Definition 3 suggests almost immediately a proof-search procedure, which may be roughly described as follows:

---

[4] In fact, Roorda conjectured that the condition was not sufficient.

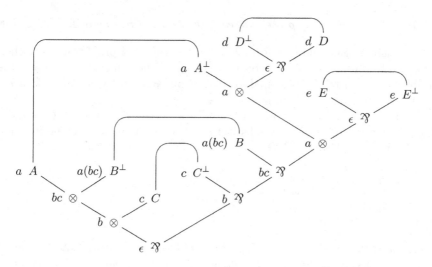

Fig. 2. A proof-net

- given a formula, assign to the vertices of its proof-frame values that obey the constraints of Figure 1;
- try to find a set of axiom links such that the values assigned to the conclusions of any axiom link are equal.

Now, the problem in trying to assign values to a proof-frame is that the constants assigned to its principal inputs are not sufficient to determine the values assigned to its other vertices. The way out is to assign variables to some of the vertices, and then to search for a set of axiom links such that the terms assigned to the conclusions of any axiom link are unifiable. To make this idea precise, we introduce the notion of *valuated proof-frame*.

Let $\mathcal{X}$ be a countably infinite set disjoint from Σ, whose elements will be called the *variables*. We write $\mathcal{T}(\Sigma, \mathcal{X})$ for the set of terms generated by Σ and $\mathcal{X}$ (including the identity element ϵ). We have $\mathcal{T}(\Sigma) \subset \mathcal{T}(\Sigma, \mathcal{X})$ and, in this setting, the elements of $\mathcal{T}(\Sigma)$ are called the *ground terms*. When t is a term, we write $\mathrm{var}(t)$ (respectively, $\mathrm{cst}(t)$) to denote the set of variables (respectively, constants) occurring in t. We extend these notations to sets of terms in the obvious way.

Definition 4. *Let A be an implicative formula. A valuated proof-frame of A consists of the proof-frame of A, $P = \langle V, E \rangle$, together with an application $\rho : V \to \mathcal{T}(\Sigma, \mathcal{X})$ such that:*

(a) *the value assigned by ρ to the root of P is ϵ;*
(b) *the values assigned by ρ to the principal inputs of P are elements of Σ that are pairwise different;*
(c) *the values assigned by ρ to the principal outputs of P are elements of $\mathcal{X}$ that are pairwise different;*

(d) *the values assigned by ρ obey the constraints given in Figure 1.*   ∎

One easily shows that any formula admits a valuated proof-frame, which is unique up to a renaming of the constants assigned the principal inputs and the variables assigned to the principal outputs. Consequently, we will speak of *the* valuated proof-frame of a formula.

A substitution skeleton σ is defined to be a partial function $\sigma : \mathcal{X} \to \mathcal{T}(\Sigma)$ whose domain is finite. Such a substitution skeleton induces a unique function $\hat{\sigma} : \mathcal{T}(\Sigma, \mathcal{X}) \to \mathcal{T}(\Sigma, \mathcal{X})$ such that:

(a) $\hat{\sigma}(\epsilon) = \epsilon$,
(b) $\hat{\sigma}(x) = \sigma(x)$, for $x \in \mathrm{dom}(\sigma)$,
(c) $\hat{\sigma}(a) = a$, for $a \in (\Sigma \cup \mathcal{X}) \setminus \mathrm{dom}(\sigma)$,
(d) $\hat{\sigma}(\alpha \cdot \beta) = \hat{\sigma}(\alpha) \cdot \hat{\sigma}(\beta)$.

A function such as $\hat{\sigma}$ is called a substitution, and we write $\mathrm{Subst}(\mathcal{X}, \Sigma)$ for the set of substitutions. By a slight abuse of language, we will speak of the domain of a substitution to mean the domain of its skeleton. If σ and τ are two substitutions whose domains are disjoint, we have that $\sigma \circ \tau = \tau \circ \sigma$, and the skeleton of the substitution is the union of the two skeletons. In such a case, again by abuse of language, we will write $\sigma \cup \tau$ for $\sigma \circ \tau$.

The next lemma is almost immediate.

Lemma 1. *Let A be an implicative formula, and let $P = \langle \langle V, E \rangle, \rho \rangle$ be its valuated proof-frame. Then, A is **INL**(respectively, **IL**, **IMLL**) provable if and only if there exists a one-one correspondence R between the positive leaves and the negative leaves of P, together with a substitution $\sigma \in \mathrm{Subst}(\mathcal{X}, \Sigma)$ such that for any positive leave p and any negative leave n, $p\,R\,n$ implies that:*

(a) *if p is decorated with A then n is decorated with $A^{\perp}$;*
(b) *$\sigma(\rho(p)) = \sigma(\rho(n))$ (respectively, modulo associativity, modulo associativity and commutativity).*

Proof. Imagine there exist such a correspondence R and such a substitution σ. It is easy to check that $\langle \langle V, E \cup R \rangle, \sigma \circ \rho \rangle$ is a proof-net. Consequently, by Proposition 1, A is provable.

Conversely, suppose that A is provable, and consequently, that there exists a proof-net $\langle \langle V, E' \rangle, \rho' \rangle$. Take $R = E' \setminus E$ (the axiom links of the proof-nets). We have that $p\,R\,n$ implies $\rho'(p) = \rho'(n)$. It is then easy to show, by induction on the proof-frame of A, that there exists a substitution such that $\rho' = \sigma \circ \rho$.   □

From the above lemma, we have that to any implicative formula A correspond two sets of terms P and N such that A is provable if and only if there exists a one-one-correspondence between P and N together with a substitution that unifies the corresponding terms. In general, the converse is not true: one cannot associate an implicative formula to any pair of sets of terms. The main goal of this section is to characterize the pairs of sets (P, N) which correspond to implicative formulas.

We define the sets of positive $(\mathcal{P})$ and negative $(\mathcal{N})$ terms as follows:

$$\mathcal{P} ::= \mathcal{X} \mid (\mathcal{P} \cdot \Sigma) \mid (\Sigma \cdot \mathcal{P})$$
$$\mathcal{N} ::= \Sigma \mid (\mathcal{N} \cdot \mathcal{X}) \mid (\mathcal{X} \cdot \mathcal{N})$$

The unique variable occurring in a positive term is called the head of the term. Similarly, the head of a negative term is the unique constant occurring in it. We also define the set of positive ground terms $(\mathcal{G})$. These are positive terms whose head as been instantiated by a constant:

$$\mathcal{G} ::= \Sigma \mid (\mathcal{G} \cdot \Sigma) \mid (\Sigma \cdot \mathcal{G})$$

We define the accessibility relation $\prec$ on $\mathcal{G} \cup \mathcal{P} \cup \mathcal{N}$ as follows. Let $t, u \in \mathcal{G} \cup \mathcal{P} \cup \mathcal{N}$. Then, $t \prec u$ if and only if:

- either $t \in \mathcal{G} \cup \mathcal{P}$, $u \in \mathcal{N}$, and the head of u occurs in t;
- or $t \in \mathcal{N}$, $u \in \mathcal{P}$, and the head of u occurs in t.

We now define the central notion of this paper

Definition 5. *A* PN-*matching problem consists of two finite sets of terms P and N such that:*

(a) *P contains one positive ground term, called the root of the problem, and all its other elements are positive terms;*
(b) *all the elements of N are negative terms;*
(c) *all the heads of the positive (respectively, negative) terms are different;*
(d) *the head of each positive (respectively, negative) term occurs in exactly one negative (respectively, positive or positive ground) term;*
(e) *each constant (respectively, variable) that occurs in a positive or positive ground (respectively, negative) term is the head of a negative (respectively, positive) term;*
(f) *$(\forall t \in P \cup N)\ r \prec^* t$, where r is the root of the problem, and $\prec^*$ is the transitive reflexive closure of the accessibility relation.*

Such a PN-*matching problem admits a free-solution (respectively,* A-*solution,* AC-*solution) if and only if there exists a one-one-correspondence R between P and N together with a substitution $\sigma \in \mathrm{Subst}(\mathcal{X}, \Sigma)$ such that $(\forall p \in P)(\forall n \in N)\ p\,R\,n$ implies $\sigma(p) = \sigma(n)$ (respectively, modulo associativity, modulo associativity and commutativity).* ∎

As we will see, the above notion of PN-matching corresponds to provability of one-literal formulas. It is not difficult to get rid of this restriction. It suffices to add to the problem (P, N) a set of constraints $C \in P \times N$ such that

$$(\forall p_1, p_2 \in P)(\forall n_1, n_2 \in N)\ (p_1, n_1), (p_1, n_2), (p_2, n_2) \in C \Rightarrow (p_2, n_1) \in C,$$

and require that any solution (R, σ) is such that $R \subset C$. Nevertheless, we prefer not to consider such a set of constraints C in order to keep the notion of PN-matching as simple as possible. One of our goals is to gain some insight into the complexity of the Lambek calculus. With respect to this aim, there is no harm in considering only one-literal formulas. Indeed, it is a direct consequence of [15] that one-literal multiplicative formulas are not easier to prove than many-literal multiplicative formulas.

We will speak of free PN-matching, associative PN-matching, or associative commutative PN-matching according to the kind of solutions we consider (free-solution, A-solution, or AC-solution, respectively). On the other hand, when stating properties that are common to the three kinds of problems, we will simply say PN-matching.

It is not difficult to prove that the definitional properties of a PN-matching problem (P, N) imply that the accessibility relation on $P \cup N$ is a tree. This property will be useful in the sequel.

Clearly, a necessary condition for a PN-matching problem to admit a solution is that P and N have the same cardinality. We will come back to this in Section 5.

We say that a substitution σ is relative to a set of terms T if and only if $\mathrm{dom}(\sigma) \subset \mathrm{var}(T)$. It is easy to show that, whenever a PN-matching problem (P, N) admits a solution (R, σ), it admits a solution (R, σ') where σ' is relative to P. In the sequel of this paper, we will only consider such solutions.

We end this section by proving that free, associative, and associative commutative PN-matching are equivalent to provability in **INL**, **IL**, and **IMLL**.

Proposition 2. *For any one-literal implicative formula A, there exists a* PN-*matching problem (P, N) such that A is* **INL** *(respectively, **IL**, **IMLL**) provable if and only if (P, N) admits a free solution (respectively, A-solution, AC-solution).*

Proof. Take P to be the set of terms assigned to the positive leaves of the valuated proof-frame of A. Similarly, take N to be the set of terms assigned to its negative leaves. One may easily show, by induction on the proof-frame of A, that (P, N) is a PN-matching problem. Then, by Lemma 1, this problem admits a solution if and only if A is provable.   □

As a corollary of this proposition, we have that associative commutative PN-matching is NP-complete since provability in **IMLL** is known to be NP-complete [9].

To show the converse of proposition 2, we associate a valuated partial parse tree $\mathcal{T}(t)$ to each term $t \in \mathcal{P}$ as follows:

(a) $\mathcal{T}(X)$ consists of a simple positive node A, which is assigned X;
(b) $\mathcal{T}(a \cdot t)$ is obtained as follows: replace in $\mathcal{T}(t)$ the positive leaf which is assigned t by a *ppn*-link whose positive premise A is assigned $a \cdot t$ and whose negative premise $A^{\perp}$ is assigned a;
(c) $\mathcal{T}(t \cdot a)$ is obtained as follows: replace in $\mathcal{T}(t)$ the positive leave which is assigned t by a *npp*-link whose negative premise $A^{\perp}$ is assigned a and whose positive premise A is assigned $t \cdot a$.

Similarly, one defines $\mathcal{T}(t)$, for $t \in \mathcal{N}$:

(a) $\mathcal{T}(a)$ consists of a simple negative node $A^{\perp}$, which is assigned a;

(b) $\mathcal{T}(X \cdot t)$ is obtained as follows: replace in $\mathcal{T}(t)$ the negative leave which is assigned t by a pnn-link whose positive premise A is assigned X and whose negative premise $A^{\perp}$ is assigned $X \cdot t$;

(c) $\mathcal{T}(t \cdot X)$ is obtained as follows: replace in $\mathcal{T}(t)$ the negative leave which is assigned t by a nnp-link whose negative premise $A^{\perp}$ is assigned $t \cdot X$ and whose positive premise A is assigned X.

Finally, one defines $\mathcal{T}(t)$, for $t \in \mathcal{G}$:

(a) $\mathcal{T}(a)$ consists of a npp-link whose both premises are assigned a and whose conclusion is assigned ϵ;

(b) $\mathcal{T}(a \cdot t)$, where t is not atomic, is obtained as follows: replace in $\mathcal{T}(t)$ the positive leave which is assigned t by a ppn-link whose positive premise A is assigned $a \cdot t$ and whose negative premise $A^{\perp}$ is assigned a;

(c) $\mathcal{T}(t \cdot a)$ is obtained as follows: replace in $\mathcal{T}(t)$ the positive leave which is assigned t by a npp-link whose negative premise $A^{\perp}$ is assigned a and whose positive premise A is assigned $t \cdot a$.

Proposition 3. *For any* PN-*matching problem* (P, N), *there exists a one-literal implicative formula* A *such that* A *is* **INL***(respectively,* **IL***,* **IMLL***) provable if and only if* (P, N) *admits a free solution (respectively,* A-*solution,* AC-*solution).*

Proof. Let P and N be two sets of terms that satisfiy Conditions (b), (c), (d), and (f) of Definition 5—but that does not necessarily satisfy Condition (e). We construct a valuated proof-frame F by induction on the accessibility relation. If N is empty, and consequently, $P = \{r\}$ where r is the root of the problem, we take $F = \mathcal{T}(r)$. Otherwise, let $t \in P \cup N$ be a term that is maximal with respect of $\prec$. Let F' be the valuated proof-frame associated, by induction hypothesis, to the problem obtained by removing t from (P, N). F is then constructed by grafting $\mathcal{T}(t)$ in place of the unique leave of F' that is assigned the head of t and that has the same polarity as t. It is not difficult to check that F is indeed the valuated proof-frame of a one-literal formula A, and that the positive and negative leaves of F are respectively assigned the elements of P and N. Hence, the proof of the proposition follows by Lemma 1. □

As a corollary of this proposition, we have that free PN-matching is polynomial since provability in **INL** is known to be polynomial [1]. Finally, as a corollary of both Proposition 2 and 3, we have that provability of one-literal formulas in **IL** is NP-complete or polynomial if and only if associative PN-matching is NP-complete or polynomial, respectively. Consequently, the complexity problem of the Lambek calculus may be studied through our notion of PN-matching

5   A pn-Matching Algorithm

In this section, we give a general PN-matching algorithm. We first specify it by means of a non deterministic transition system. Then we explain how this algorithm may be implemented in a more efficient way.

In what follows, we assume that the terms defining a PN-matching problem are assigned different integers, and if t is such an indexed term $\#t$ denotes the integer assigned to t. We also assume that a substitution applied on an indexed term does not affect the integer, i.e., $\#\sigma(t) = \#t$. This is only needed to keep a trace of the original terms when applying a substitution and to allow correspondences between terms to be represented as relations on integers.

Let P and N be two sets of indexed terms, $R \subset \mathbb{N} \times \mathbb{N}$, and $\sigma \in \mathrm{Subst}(\mathcal{X}, \Sigma)$. Consider the following transition:

$$\langle P, N, R, \sigma \rangle \longrightarrow \langle \tau(P \setminus \{t\}), \tau(N \setminus \{u\}), R \cup \{(\#t, \#u)\}, \tau \circ \sigma \rangle \quad (1)$$

where $t \in P$ is a positive ground term, $u \in N$, and τ is a substitution whose domain is $\mathrm{var}(u)$ and such that $t = \tau(u)$.

We will prove that a PN-matching problem (P, N) admits a solution (R, σ) if and only if there exists a sequence of transitions such that:

$$\langle P, N, \varnothing, id \rangle \longrightarrow^* \langle \varnothing, \varnothing, R, \sigma \rangle \quad (2)$$

Clearly there cannot be infinite sequences such as (2). Moreover, the branching due to the non-determinism of Transition (1) is finite. Consequently, Transition (1) specifies indeed a non deterministic algorithm. Proving the correctness of this algorithm (i.e., the if-part of the above statement) is straightforward.

Proposition 4. *Let (P, N) be a* PN-*matching problem such that*

$$\langle P, N, \varnothing, id \rangle \longrightarrow^* \langle \varnothing, \varnothing, R, \sigma \rangle$$

Then (R, σ) is a solution to (P, N).

Proof. A straightforward induction on the sequence of transitions. $\quad\square$

In order to prove the completeness of the algorithm, we first establish a lemma.

Lemma 2. *Let (P_1, N_1) and (P_2, N_2) be two* PN-*matching problems such that* $\mathrm{var}(P_1) \cap \mathrm{var}(P_2) = \varnothing$. *If there exist sequences of transitions such that*

$$\langle P_1, N_1, \varnothing, id \rangle \longrightarrow^* \langle \varnothing, \varnothing, R_1, \sigma_1 \rangle \quad and \quad \langle P_2, N_2, \varnothing, id \rangle \longrightarrow^* \langle \varnothing, \varnothing, R_2, \sigma_2 \rangle$$

then there exist a sequence of transition such that

$$\langle P_1 \cup P_2, N_1 \cup N_2, R_0, \sigma_0 \rangle \longrightarrow^* \langle \varnothing, \varnothing, R_2 \cup R_1 \cup R_0, \sigma_2 \cup \sigma_1 \cup \sigma_0 \rangle$$

where R_0 is any relation, and σ_0 is a substitution whose domain is disjoint from $\mathrm{var}(P_1)$ *and* $\mathrm{var}(P_2)$.

Proof. Since $\mathrm{var}(P_1) \cap \mathrm{var}(P_2) = \varnothing$, we have that $\sigma_1(P_2) = P_2$ and $\sigma_1(N_2) = N_2$. It is then straightforward to prove that:

$$\langle P_1 \cup P_2, N_1 \cup N_2, R_0, \sigma_0 \rangle \longrightarrow^* \langle P_2, N_2, R_1 \cup R_0, \sigma_1 \circ \sigma_0 \rangle$$
$$\longrightarrow^* \langle \varnothing, \varnothing, R_2 \cup R_1 \cup R_0, \sigma_2 \circ \sigma_1 \circ \sigma_0 \rangle$$

Moreover, we have that the domain of σ_0, σ_1, and σ_2 are pairwise disjoint. Hence, $\sigma_2 \circ \sigma_1 \circ \sigma_0 = \sigma_2 \cup \sigma_1 \cup \sigma_0$. $\quad\square$

We now prove the completeness of the algorithm.

Proposition 5. *Let* (P, N) *be a* PN-*matching problem that admits a solution* (R, σ). *Then there exists a sequence of transitions such that:*

$$\langle P, N, \varnothing, id \rangle \longrightarrow^* \langle \varnothing, \varnothing, R, \sigma \rangle$$

Proof. Let $r \in P$ be the root of the problem, and let $u \in N$ be such that $(\#r, \#u) \in R$. Then, let σ_u be the substitution σ restricted to $\mathrm{var}(u)$, and let $(t_i)_{i \in n}$ be the positive terms such that $u \prec t_i$. Define the following sets, relations, and substitutions:

$$P_i = \{t \in P \mid \sigma_u(t_i) \prec^* \sigma_u(t)\}$$
$$N_i = \{t \in N \mid \sigma_u(t_i) \prec^* \sigma_u(t)\}$$
$$R_i = R \cap (\#P_i \times \#N_i)$$
$$\sigma_i \text{ is the substitution } \sigma \text{ restricted to } \mathrm{var}(P_i)$$

It is easy to show, from the definitional properties of a PN-matching problem that:

- $(\forall i, j \in n)\ i \neq j$ implies $P_i \cap P_j = \varnothing$ and $N_i \cap N_j = \varnothing$,
- $\bigcup_{i \in n} P_i = P \setminus \{r\}$ and $\bigcup_{i \in n} N_i = N \setminus \{u\}$,
- $\bigcup_{i \in n} R_i = R \setminus \{(\#r, \#u)\}$ and $(\bigcup_{i \in n} \sigma_i) \cup \sigma_u = \sigma$,
- $((\sigma_u(P_i), \sigma_u(N_i)))_{i \in n}$ is a family of PN-matching problems, with $(t_i)_{i \in n}$ as roots, that admits the family of solutions $((R_i, \sigma_i))_{i \in n}$.

Then, by induction hypothesis, there exist sequences of transitions such that:

$$\langle \sigma_u(P_i), \sigma_u(N_i), \varnothing, id \rangle \longrightarrow^* \langle \varnothing, \varnothing, R_i, \sigma_i \rangle,$$

and, by iterating Lemma 2,

$$\langle \bigcup_{i \in n} \sigma_u(P_i), \bigcup_{i \in n} \sigma_u(N_i), \{(\#r, \#u)\}, \sigma_u \rangle \longrightarrow^* \langle \varnothing, \varnothing, R, \sigma \rangle,$$

which allows us to conclude since

$$\langle P, N, \varnothing, id \rangle \longrightarrow \langle \bigcup_{i \in n} \sigma_u(P_i), \bigcup_{i \in n} \sigma_u(N_i), \{(\#r, \#u)\}, \sigma_u \rangle$$

$\square$

There are different sources of non-determinism in our PN-matching algorithm:

(a) the choice of the positive ground term t according to which the transition is done,
(b) the choice of the negative term u to be matched with t,
(c) the choice of the substitution τ such that $t = \tau(u)$.

One cannot avoid (b) and (c). On the other hand, the non-determinism due to (a) may be circumvented as we will explain by transforming our algorithm.

We first show that there is no need for updating N in Transition (1). Consider the following sequence of transitions:

$$\langle P, N, \varnothing, id \rangle \longrightarrow^* \langle P', N', R, \sigma \rangle$$
$$\longrightarrow \langle \tau(P' \setminus \{t\}), \tau(N' \setminus \{u\}), R \cup \{(\#t, \#u)\}, \tau \circ \sigma \rangle$$

where (P, N) is a PN-matching problem. It is a direct consequence of Properties (c), (d) and (e) of Definition 5 that the substitution τ does not affect $N' \setminus \{u\}$, i.e., $\tau(N' \setminus \{u\}) = N' \setminus \{u\}$. Moreover, because of these same properties, there cannot be any positive ground term $t' \in \tau(P' \setminus \{t\})$ such that $t' = \tau'(u)$ for some substitution τ'. Hence, updating N is only needed in order to ensure that P and N have the same number of elements. But this may be checked once and for all before starting any sequence of transition. Therefore, one may assume that N is an invariant datum that is global to all the possible sequences of transitions.

Now consider the set $P' \setminus \{t\}$ that appears in the above transition. This set may be partitioned into two set P_1 and P_2 as follows:

$$P_1 = \{t \in P' \mid u \prec t\} \quad \text{and} \quad P_2 = P' \setminus (P_1 \cup \{t\})$$

Again by the definitional properties of a PN-matching problem, one may prove that all the terms in $\tau(P_1)$ are positive ground terms and that $\tau(P_2) = P_2$. Moreover, using Property (f) of Definition 5, one proves that, whenever $\tau(P' \setminus \{t\})$ does not contain any positive ground term, we have $P' \setminus \{t\} = \varnothing$.

These observations lead us to the definition of a new transition:

$$\langle G, R, \sigma \rangle \xrightarrow{P, N} \langle (G \setminus \{t\}) \cup \tau(Q), R \cup \{(\#t, \#u)\}, \tau \circ \sigma \rangle \tag{3}$$

where:

- P, G, and N are sets of positive, positive ground, and negative terms respectively;
- $R \subset \mathbb{N} \times \mathbb{N}$ and $\sigma \in \text{Subst}(\mathcal{X}, \Sigma)$;
- $t \in G$, $u \in N$, and τ is a substitution whose domain is $\text{var}(u)$ and such that $t = \tau(u)$;
- $Q = \{t \in P \mid u \prec t\}$.

It follows from the above discussion that a PN-matching problem (P, N) with root r admits a solution (R, σ) if and only if there exists a sequence of transitions such that:

$$\langle \{r\}, \varnothing, id \rangle \xrightarrow{P, N}^* \langle \varnothing, R, \sigma \rangle \tag{4}$$

provided that P and N have the same number of elements.

Finally, let G, R, and σ be such that

$$\langle \{r\}, \varnothing, id \rangle \xrightarrow{P, N}^* \langle G, R, \sigma \rangle \tag{5}$$

where (P, N) is a PN-matching problem whose root is r. Assume that there exists two different transitions:

$$\langle G, R, \sigma \rangle \xrightarrow{P,N} \langle (G \setminus \{t_1\}) \cup \tau_1(Q_1), R \cup \{(\#t_1, \#u_1)\}, \tau_1 \circ \sigma \rangle$$

$$\langle G, R, \sigma \rangle \xrightarrow{P,N} \langle (G \setminus \{t_2\}) \cup \tau_2(Q_2), R \cup \{(\#t_2, \#u_2)\}, \tau_2 \circ \sigma \rangle$$

It is easy to prove, by induction on the length of Sequence (5) that $\mathrm{cst}(t_1) \cap \mathrm{cst}(t_2) = \varnothing$. Consequently, we have $u_1 \neq u_2$, $\mathrm{var}(u_1) \cap \mathrm{var}(u_2) = \varnothing$, and $Q_1 \cap Q_2 = \varnothing$. This implies that there exist two transitions such that:

$$\langle (G \setminus \{t_1\}) \cup \tau_1(Q_1), R \cup \{(\#t_1, \#u_1)\}, \tau_1 \circ \sigma \rangle \xrightarrow{N,P}$$
$$\langle (G \setminus \{t_1, t_2\}) \cup \tau_1(Q_1) \cup \tau_2(Q_2), R \cup \{(\#t_1, \#u_1), (\#t_2, \#u_2)\}, (\tau_1 \cup \tau_2) \circ \sigma \rangle$$

$$\langle (G \setminus \{t_2\}) \cup \tau_2(Q_2), R \cup \{(\#t_2, \#u_2)\}, \tau_2 \circ \sigma \rangle \xrightarrow{N,P}$$
$$\langle (G \setminus \{t_1, t_2\}) \cup \tau_1(Q_1) \cup \tau_2(Q_2), R \cup \{(\#t_1, \#u_1), (\#t_2, \#u_2)\}, (\tau_1 \cup \tau_2) \circ \sigma \rangle$$

Consequently, there is no source of non-determinism in the choice of the positive ground term according to which the transition is done. This means that the search for a successful sequence of transitions may be organised as an and/or-tree.

Figure 3 gives such an and/or-tree is given for the following associative PN-matching problem:

$$P = \{1 : abc, 2 : Z, 3 : Y, 4 : eX, 5 : Wd\} \quad \text{and}$$
$$N = \{1' : c, 2' : YbZ, 3' : e, 4' : d, 5' : aXW\}$$

The main nodes, in this tree, are labelled with ground terms and the edges growing from these correspond to the different negative terms that match with the ground terms labelling the main nodes. Each such edge is labelled with the index of the corresponding negative term. Then there is a possible or-node with leaving edges corresponding to the possible different unifiers. Finally, each possible unifier gives rise to a and-node whose leaving edges reach the new positive ground terms resulting from a transition.

In such an and/or-tree, the subtree growing out of a main node is history independent: it is completely determined by the ground term labelling the main node. Consequently, the proof-search space may be organised as a DAG rather than as a tree (by using memoization or dynamic programming techniques, for instance).

6   Conclusions and Future Work

We have reduced **INL**, **IL**, and **IMLL** provability to matching problems that emphasise the part played by associativity in the case of **IL**, and associativity and commutativity in the case of **IMLL**. As we said in the introduction, we hope that this reduction will give an insight into the complexity of the Lambek calculus.

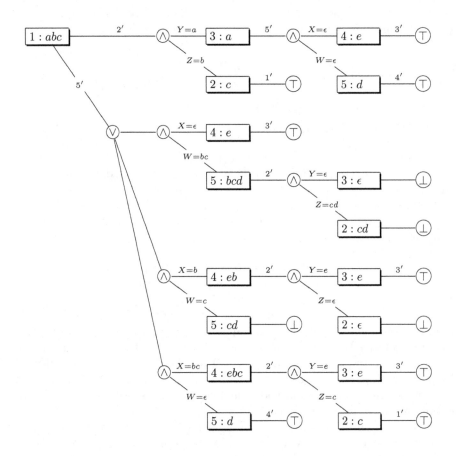

Fig. 3. Proof-search space organized as an and/or tree

An important question, in this context, is to know whether our PN-matching algorithm runs in polynomial time in the non-associative case. In fact, even in this simple case, it is not difficult to construct families of formulas for which the proof-search space, when organised as a tree, has an exponential number of nodes. Consequently, the only hope of obtaining a polynomial algorithm is to organise some sharing as we suggested at the end of the previous section. If we do so, our PN-matching algorithm runs in polynomial time provided that the number of different positive ground terms involved in the search is polynomial. Several experimental results suggest that this is the case. Unfortunately, we do not know how to prove it in general. Therefore, the next step of this work will be to solve this question.

The first experiments we have conducted seem to indicate that our PN-matching algorithm has a good behaviour in practice. Hence, it would be interesting to see how our approach compete with other methods [16,23]. In this practical setting, working only in the implicative fragment of mutiplicative linear logic is a limitation. This raised the question of extending our procedure to fragments including additives and exponentials. In this respect, [11] might be a source of inspiration. Indeed, in the purely implicative case, there is a strong connection between our proof-search algorithm and Lamarche's games.

References

1. E. Aarts and K. Trautwein. Non-associative Lambek categorial grammar in polynomial time. *Mathematical Logic Quaterly*, 41:476–484, 1995. 257, 267
2. C. Berge. *Graphs*. North-Holland, second revised edition edition, 1985. 260
3. V. Danos. *Une application de la logique linéaire à l'étude des processus de normalisation et principalement du lambda calcul*. Thèse de doctorat, Université de Paris VII, 1990. 260
4. V. Danos and L. Regnier. The structure of multiplicatives. *Archive for Mathematical Logic*, 28:181–203, 1989. 260, 260
5. Ph. de Groote. An algebraic correctness criterion for intuitionistic multiplicative proofnets. *Theoretical Computer Science*, 224:115–134, 1999. 260, 262, 262
6. Ph. de Groote. The non-associative lambek calculus with product in polynomial time. In *International Conference on Theorem Proving with Analytic Tableaux and Related Methods*, volume 1617 of *Lecture Notes in Artificial Intelligence*, pages 128–139. Springer Verlag, 1999. 257
7. J.-Y. Girard. Linear logic. *Theoretical Computer Science*, 50:1–102, 1987. 257, 258, 260, 260
8. S. Guerrini. Correctness of multiplicative proof-nets is linear. In *Proceedings of the fourteenth annual IEEE symposium on logic in computer science*, pages 454–463, 1999. 260
9. M. Kanovich. Horn programming in linear logic is np-complete. In *7-th annual IEEE Symposium on Logic in Computer Science*, pages 200–210. IEEE Computer Society Press, 1992. 257, 266
10. Yves Lafont. From proof nets to interaction nets. In J.-Y. Girard, Y. Lafont, and L. Regnier, editors, *Advances in Linear Logic*, pages 225–247. Cambridge University Press, 1995. 260

11. F. Lamarche. Games semantics for full propositional linear logic. In *Ninth Annual IEEE Symposium on Logic in Computer Science*. IEEE Press, 1995. 273

12. F. Lamarche and C. Retoré. Proof nets for the Lambek calculus. In M. Abrusci and C. Casadio, editors, *Proofs and Linguistic Categories, Proceedings 1996 Roma Workshop*. Cooperativa Libraria Universitaria Editrice Bologna, 1996. 260

13. J. Lambek. The mathematics of sentence structure. *Amer. Math. Monthly*, 65:154–170, 1958. 257, 258

14. J. Lambek. On the calculus of syntactic types. In *Studies of Language and its Mathematical Aspects*, pages 166–178, Providence, 1961. Proc. of the 12th Symp. Appl. Math.. 257, 258

15. P. Lincoln and T. Winkler. Constant-only multiplicative linear logic is NP-complete. *Theoretical Computer Science*, 135:155–169, 1994. 266

16. H. Mantel and J. Otten. linTAP: A tableau pover for linear logic. In *International Conference on Theorem Proving with Analytic Tableaux and Related Methods*, volume 1617 of *Lecture Notes in Artificial Intelligence*, pages 217–231. Springer Verlag, 1999. 273

17. M. Moortgat. Categorial type logic. In J. van Benthem and A. ter Meulen, editors, *Handbook of Logic and Language*, chapter 2. Elsevier, 1997. 257

18. G. Morrill. *Type Logical Grammar: Categorial Logic of Signs*. Kluwer Academic Publishers, Dordrecht, 1994. 257

19. M. Okada. A graph-theoretic characterization theorem for multiplicative fragment of non-commutative linear logic. *Electronic Notes on Theoretical Computer Science*, 3, 1996. 260

20. M. Pentus. Language completeness of the Lambek calculus. In *Proceedings of the ninth annual IEEE symposium on logic in computer science*, pages 487–496, 1994. 258

21. C. Retoré. Calcul de Lambek et logique linéaire. *Traitement Automatique des Langues*, 37(2):39–70, 1997. 260

22. D. Roorda. *Resource Logics: proof-theoretical investigations*. PhD thesis, University of Amsterdam, 1991. 258, 260, 262

23. T. Tammet. Proof strategies in linear logic. *Journal of Automated Reasoning*, 12:273–304, 1994. 273

24. J. van Benthem. *Language in action: Categories, Lambdas and Dynamic Logic*, volume 130 of *Sudies in Logic and the foundation of mathematics*. North-Holland, Amsterdam, 1991. 257

A New Model Construction for the Polymorphic Lambda Calculus

Dieter Spreen

Fachbereich Mathematik, Theoretische Informatik
Universität Siegen, 57068 Siegen, Germany
spreen@informatik.uni-siegen.de

Abstract. Various models for the Girard-Reynolds second-order lambda calculus have been presented in the literature. Except for the term model they are either realizability or domain models. In this paper a further model construction is introduced. Types are interpreted as inverse limits of ω-cochains of finite sets. The corresponding morphisms are sequences of maps acting locally on the finite sets in the ω-cochains. The model can easily be turned into an effectively given one. Moreover, it can be arranged in such a way that the universally quantified type $\forall t.t$ representing absurdity in the higher-order logic defined by the type structure is interpreted by the empty set, which means that it is also a model of this logic.

1 Introduction

Type systems originally introduced in logic and the foundations of mathematics have been proved quite useful in computer science. By the Curry-Howard isomorphism typed expressions can be interpreted in various ways. If types are considered as formulae of a logical calculus, the expressions of a certain type are proofs of the corresponding formula. In case we think of a type as a data structure, an expression is a program which evaluates to a value in this data structure. But we can also consider types as formulae of a specification language. Then the statement that an expression is of a certain type means that this program results in a value which meets the specification given by the type.

Various type systems of different computational and expressive power have been considered in the literature. In this paper we will be mainly concerned with the polymorphic lambda calculus.

The polymorphic lambda calculus, introduced independently by Girard [13,14] and Reynolds [21], is an extension of the usual typed lambda calculus that allows a form of parametric polymorphism. Types include universally quantified types which are types of polymorphic terms, thought of as describing those functions which are defined in a uniform manner at all types. Terms can be applied to types and in this sense can be parameterised by types.

In order to achieve this, type variables are introduced into the typed lambda calculus. So, for instance, $\lambda x : \sigma.\ x$ should be thought of as the identity function on the type denoted by σ. The polymorphic identity function, the term

M. Parigot and A. Voronkov (Eds.): LPAR 2000, LNAI 1955, pp. 275–292, 2000.

which stands for the identity function on any type, is given by the expression $\Lambda t.\ \lambda x\colon t.\ x$. It has a universally quantified type denoted by $\forall t.\ t \to t$. Given a type σ_1, a term $\Lambda t.\ M$ of universally quantified type $\forall t.\ \sigma_2$ can be instantiated to a term $\{\sigma_1/t\}M$ which then has type $\{\sigma_1/t\}\sigma_2$, and so, for instance, the polymorphic identity above instantiates at type σ to the identity $\lambda x\colon \sigma.\ x$ of type $\sigma \to \sigma$.

While the pioneering work of Girard contains most of the results on the syntax of the calculus, an understanding of its models has developed more slowly. There is a trivial model obtained by interpreting types as either the empty or the one-point set. But this is obviously inadequate as a model of polymorphism and the many useful data structures definable in the calculus. The difficulty of providing nontrivial models arises essentially from the impredicative nature of the calculus: in the abstraction of a universally quantified type $\forall t.\ \sigma$ the type variable t is understood to range over all types including the universally quantified type itself. Reynolds [23] showed that no model of the polymorphic lambda calculus in which the function-space constructor behaves set theoretically is possible, classically. But, by a result of Pitts [20], such a model can be constructed in constructive set theory.

Nontrivial models, term and realizability models, were already presented by Girard [14] and Troelstra [27]. McCracken [18], building on ideas from Scott [25] produced the first correct domain-theoretic model. It was constructed from Scott's universal domain $\mathcal{P}\omega$, using closures (a special kind of retracts) to represent types. Following a suggestion of Scott [26], McCracken [19] has as well shown that finitary retracts over certain finitary complete partial orders can be used to represent types. Amadio, Bruce, and Longo [1], again using ideas appearing in several papers by Scott, have also constructed a model using finitary projections over complete partial orders. All these domain models are models for stronger calculi with a type of all types, a fact which is used in giving meaning to universally quantified types. But by a result of Girard [14], such systems are inconsistent, when they are considered as logical calculi.

In his paper [15], Girard produced an interesting new model in which types of the polymorphic lambda calculus are represented as certain kinds of objects called qualitative domains. In this construction types with free type variables, called "variable types" by Girard, are interpreted as nicely behaving functors on a category of these domains. The central observation was that the behaviour of such functors is already determined by what they do on finite qualitative domains. So he got rid of the circularity in the construction of universally quantified types. Building upon these ideas, Coquand, Gunter, and Winskel presented a domain model for the polymorphic lambda calculus in which types are interpreted as dI-domains [7] and Scott domains [8], respectively. In the last model a universally quantified type is interpreted as a domain (considered as a category) of continuous sections of the Grothendieck fibration of a continuous functor.

By using an observation of the present author, Gruchalski [17] showed that the construction of the Scott domain model can be simplified so that only domain-theoretic methods are used. Girard's qualitative domains can be represented by certain graphs such that the usual domain constructions can be carried out directly on the graphs [16]. Using similar structures Berardi and Berline [3] managed to construct a large family of concrete models in a noncategorical way.

A major drawback of the domain models is that the type $\forall t.\ t$, which represents absurdity in the logical calculus given by the type system, is interpreted by a nonempty set, which means that this semantics is not adequate when one wants to give meaning to the logic, known to be consistent. Moreover, even when one is only interested in the functional language, then one should be able to introduce some notion of computability in the model, which seems to be impossible at least in the case of the dI- and Scott domain model of Coquand, Gunter and Winskel.

In the model we present in this paper types are interpreted as sets which are approximated by sequences of finite sets, more exactly, as inverse limits of ω-cochains of finite sets. The points in these limits are certain sequences of elements of the approximating finite sets. As morphisms in the new category **SFS** (**S**equences of **F**inite **S**ets), which we are considering, we take those maps that can be represented as sequences of mappings which act locally on the approximating finite sets and commute with the connecting projections of the chain. For any two sequences of elements of the approximating finite sets they preserve the longest initial segment in which these coincide. In the context of rank-ordered sets Bruce and Mitchell [5] called such maps rank-preserving. As it turns out, the constructors for products and exponents are itself rank-preserving. Note here that the objects in **SFS** are also sequences. Modulo some coding, a universally quantified type $\forall t.\ \sigma$ is then interpreted as the set of all rank-preserving sections with respect to the fibration of the rank-preserving constructor obtained from the interpretation of σ.

The model can easily be turned into an effectively given one. Moreover, without any restrictions on the collection of finite sets used for approximation the empty set is an object of our category, which implies that the universally quantified type representing absurdity is interpreted by the empty set. Thus, the model not only gives meaning to the polymorphic lambda calculus viewed as a functional language but also when considered as a logical calculus.

The construction of the model allows some variations. If, e.g., one requires the approximating finite sets not to be empty, the interpretation of $\forall t.\ t$ is nonempty, too. In case all the finite sets are T_0-spaces and the connecting projections are continuous, one obtains a model in which every type is interpreted by a directed-complete partial order (not necessarily with a smallest element). Note that in this case the morphisms have to be continuous as well. However, requiring the topology on the finite sets to satisfy stronger separation conditions will produce no further models, since a topology on a finite set which satisfies the T_1-axiom is already discrete. In general, one can consider any property of sets that is closed under the inverse limit construction. With respect to the canonical metric defined on sets of infinite sequences the spaces in **SFS** are complete ultrametric spaces.

Beside the polymorphic lambda calculus the category **SFS** can also be used to give meaning to other type systems. Here, we consider only the slight extension of the second-order lambda calculus studied by Bruce, Meyer and Mitchell [6].

Extensions of the polymorphic lambda calculus concern the kind structure built upon the type expressions of the calculus. Essentially, kinds are the "types" that appear in type expressions. The object set of the category **SFS** is a complete projection space. Projection spaces have been studied by Ehrig et al. [9,10,11,12] as a generalization of the projective model of process algebra of Bergstra and Klop [4]. They are a nonempty sets with a family of commuting projection functions. The projections assign a canonical sequence of approximations to each element of the set. With rank-preserving maps as morphisms the category of complete projection spaces is Cartesian closed.

The paper is organised as follows. In Sect. 2 the syntax of the polymorphic lambda calculus is recalled and in Sect. 3 a modification of Bruce, Meyer and Mitchell's notion of a second-order environment model is given which includes the case that types can be empty. In both cases we follow the presentation in [6].

The category **SFS** is considered in Sect. 4 and in Sect. 5 complete projection spaces are introduced. It is shown that the collection of all **SFS** objects is a complete projection space. Section 6 deals with representations in **SFS** of products of rank-preserving maps over the object set of **SFS**. In Sect. 7 the new model for the polymorphic lambda calculus is defined. Concluding remarks appear in Sect. 8.

2   Syntax of the Polymorphic Lambda Calculus

2.1   Constructors and Kinds

Every term of the calculus we are going to consider has a type and every subexpression of a type expression has a kind. The subexpressions of type expressions, which may be type expressions or operators like $\rightarrow$ and $\forall$, will be called constructors. We will define the sets of kinds and constructor expressions before introducing the syntax and type checking rules for terms.

We use the constant T to denote the kind consisting of all types. The set of kind expressions is given by

$$\kappa ::= T \mid \kappa_1 \Rightarrow \kappa_2 \ .$$

Let $\mathcal{V}_{\mathrm{cst}}$ be a set of variables v^κ, each with a specified kind. We assume that we have infinitely many variables for each kind. Moreover, let $\mathcal{C}_{\mathrm{cst}}$ be the set containing the function-type constructor constant $\rightarrow$ and the polymorphic-type constructor constant $\forall$. As usual, we write $\rightarrow$ as an infix operator and write $\forall t.\sigma$ for $\forall(\lambda t.\sigma)$. The constructor expressions over $\mathcal{C}_{\mathrm{cst}}$ and $\mathcal{V}_{\mathrm{cst}}$, and their kinds, are defined by the derivation system

$$\rightarrow: T \Rightarrow (T \Rightarrow T), \; \forall : (T \Rightarrow T) \Rightarrow T, \; v^\kappa : \kappa$$

$$\frac{\mu : \kappa_1 \Rightarrow \kappa_2, \; \nu : \kappa_1}{\mu\nu : \kappa_2}, \qquad \frac{\mu : \kappa_2}{\lambda v^{\kappa_1}.\mu : \kappa_1 \Rightarrow \kappa_2} \; .$$

A subset of the constructor expressions are the type expressions, the constructor expressions of kind T. We use the following metavariables: $r, s, t, \ldots$ stand for arbitrary type variables and $\rho, \sigma, \tau, \ldots$ stand for arbitrary type expressions. As in the definition above, we will generally use μ and ν for constructor expressions.

Since we have a "kinded" lambda calculus, there are many nontrivial equations between types and constructors, which follow from the familiar axioms and inference rules of the ordinary simply typed lambda calculus. If $\mu = \nu$ is provable from the axioms and rules for constructors, we write $\vdash_c \mu = \nu$. The constructor axiom system will be used to assign types to terms, since equal types will be associated with the same set of terms.

2.2 Terms and Their Types

We write free variables without type labels. However, we always assign types to free variables using a technical device called context.

Let $\mathcal{V}_{\text{term}}$ be an infinite collection of variables. We will use the notation $x, y, z, \ldots$ for these variables. The set of *pre-terms* over variables from $\mathcal{V}_{\text{cst}}$ and $\mathcal{V}_{\text{term}}$ is defined by

$$M ::= x \mid \lambda x : \sigma.M \mid MN \mid \Lambda t.M \mid M\sigma \; ,$$

where $x \in \mathcal{V}_{\text{term}}$, t is a type variable, and σ is a type expression over $\mathcal{C}_{\text{cst}}$ and $\mathcal{V}_{\text{cst}}$. We will define the well-typed terms below.

The type of a second-order lambda term will depend on the context in which it occurs. We must know the types of all free variables before assigning a type. A *context* Γ is a finite set $\Gamma = \{x_1 : \sigma_1, \ldots, x_k : \sigma_k\}$ of associations of types to variables, with no variable appearing twice in Γ. If x does not occur in a context Γ, then we write $\Gamma, x : \sigma$ for the context $\Gamma, x : \sigma = \Gamma \cup \{x : \sigma\}$.

The typing relation is a three-place relation between contexts, pre-terms and type expressions. Let Γ be a context, M be a pre-term, and $\sigma : T$ a type expression. We define $\Gamma \vdash M : \sigma$, which is read "M has type σ with respect to Γ," by the derivation system below. The axiom about the typing relation is

$$x : \sigma \vdash x : \sigma \; .$$

The type derivation rules are

$$(\rightarrow E) \; \frac{\Gamma \vdash M : \sigma \rightarrow \tau, \; \Gamma \vdash N : \sigma}{\Gamma \vdash MN : \tau}, \qquad (\rightarrow I) \; \frac{\Gamma, x : \sigma \vdash M : \tau}{\Gamma \vdash \lambda x : \sigma.M : \sigma \rightarrow \tau}$$

$$(\forall E) \; \frac{\Gamma \vdash M : \forall \mu}{\Gamma \vdash M\tau : \mu\tau}, \qquad (\forall I) \; \frac{\Gamma \vdash M : \tau}{\Gamma \vdash \Lambda t.M : \forall t.\tau} \; t \text{ not free in } \Gamma$$

and two rules that apply to terms of any form

$$\text{(add hyp)} \quad \frac{\Gamma \vdash M : \tau}{\Gamma, x : \sigma \vdash M : \tau} \quad x \text{ not in } \Gamma, \qquad \text{(type eq)} \quad \frac{\Gamma \vdash M : \sigma \quad \vdash_c \sigma = \tau}{\Gamma \vdash M : \tau} \ .$$

We say that M is a *term* if $\Gamma \vdash M : \sigma$ for some Γ and σ. In writing $\Gamma \vdash M : \sigma$, we will mean that the typing $\Gamma \vdash M : \sigma$ is derivable.

2.3   Equations between Terms

Since we write terms with type assignments, it is natural to include type assignments in equations as well. By *equation*, we will mean an expression

$$\Gamma \vdash M = N : \sigma \ ,$$

where $\Gamma \vdash M : \sigma$ and $\Gamma \vdash N : \sigma$. Intuitively, an equation $\{x_1 : \sigma_1, \ldots, x_k : \sigma_k\} \vdash M = N : \sigma$ means, "if the variables $x_1, \ldots, x_k$ have types $\sigma_1, \ldots, \sigma_k$, respectively, then terms M and N denote the same element of type σ.

The axioms and inference rules for equations between second-order lambda terms are similar to the axioms and rules of the ordinary typed lambda calculus. The main difference is that we tend to have two versions of each axiom or rule, one for ordinary function abstraction or application, and another for type abstraction or application. For lack of space we do not list the axioms and rules and refer the reader to [6] instead.

3   Second-Order Environment Models

Models for second-order lambda calculus have several parts: "kind frames" are used to interpret kinds and constructors and additional sets indexed by types to interpret terms. All these parts are collected together in what is called a frame. Models are defined as frames which satisfy an additional condition involving the meaning of terms.

3.1   Semantics of Constructor Expressions

Constructor expressions are interpreted using kind frames, which are essentially frames for the simply typed lambda calculus. A *kind frame*, *Kind* for a set $\mathcal{C}_{\mathrm{cst}}$ of constructor constants is a tuple

$$Kind = (\{\, \mathrm{Kind}^\kappa \mid \kappa \text{ a kind}\,\}, \{\, \varPhi_{\kappa_1, \kappa_2} \mid \kappa_1, \kappa_2 \text{ kinds}\,\}, \mathcal{I}) \ ,$$

where

$$\varPhi_{\kappa_1, \kappa_2} : \mathrm{Kind}^{\kappa_1 \Rightarrow \kappa_2} \to [\mathrm{Kind}^{\kappa_1} \to \mathrm{Kind}^{\kappa_2}]$$

is a bijection between $\mathrm{Kind}^{\kappa_1 \Rightarrow \kappa_2}$ and some set $[\mathrm{Kind}^{\kappa_1} \to \mathrm{Kind}^{\kappa_2}]$ of maps from Kind^{κ_1} to Kind^{κ_2}, and

$$\mathcal{I} : \mathcal{C}_{\mathrm{cst}} \to \bigcup_\kappa \mathrm{Kind}^\kappa$$

preserves kinds, that is, $\mathcal{I}(c^\kappa) \in \text{Kind}^\kappa$. Since constructor expressions include all typed lambda expressions one is interested in kind frames which are models of the simply typed lambda calculus.

Let η be an environment mapping constructor variables to $\bigcup_\kappa \text{Kind}^\kappa$ such that for each v^κ, one has $\eta(v^\kappa) \in \text{Kind}^\kappa$. The meaning $[\![\mu]\!]_\eta$ of a constructor expression μ in environment η is defined as follows:

$$[\![v^\kappa]\!]_\eta = \eta(v^\kappa),$$
$$[\![c^\kappa]\!]_\eta = \mathcal{I}(c^\kappa),$$
$$[\![\mu\nu]\!]_\eta = \Phi_{\kappa_1,\kappa_2}([\![\mu]\!]_\eta)([\![\nu]\!]_\eta),$$
$$[\![\lambda v^\kappa.\mu]\!]_\eta = \Phi_{\kappa_1,\kappa_2}^{-1}(f), \qquad \text{where } f(a) = [\![\mu]\!]_{\eta[a/v^\kappa]} \text{ for all } a \in \text{Kind}^\kappa.$$

Here, $\eta[a/v^\kappa]$ is the environment that maps v^κ onto a and every other variable $u^{\kappa'}$ onto $\eta(u^{\kappa'})$.

Note that the above conditions do not entail that the map f is in the range of Φ_{κ_1,κ_2}. Therefore, the meaning may not be defined for all constructor expressions. *Kind* is said to be a *kind environment model* for $\mathcal{C}_{\text{cst}}$ if every constructor expression over $\mathcal{C}_{\text{cst}}$ has a meaning in every environment for *Kind*.

3.2   Frames and Environment Models

As in the definition of a kind environment model, first a structure, called frame, is defined and then models are defined by distinguishing frames which interpret all terms from those that do not. Second-order frames include versions of the maps $\Phi_{\cdot,\cdot}$, now indexed by types, plus an additional collection of such maps for polymorphic types. Intuitively, a polymorphic term $\Lambda t.M$ denotes a map from the set of types to elements of types. More precisely, the meaning of $\Lambda t.M$ is regarded as an element of the Cartesian product $\Pi_{a \in \text{Kind}^T} \text{Dom}^{f(a)}$ for some map $f \colon \text{Kind}^{T \Rightarrow T}$ determined from the typing of M. Therefore, for every map $f \in \text{Kind}^{T \Rightarrow T}$, a second-order model has a map Φ_f mapping $\text{Dom}^{\forall(f)}$ to some subset $[\Pi_{a \in \text{Kind}^T} \text{Dom}^{f(a)}]$ of $\Pi_{a \in \text{Kind}^T} \text{Dom}^{f(a)}$. Here $\forall(f)$ denotes the element $\Phi_{T \Rightarrow T,T}(\mathcal{I}(\forall))(f)$ of Kind^T. In the same way we write $a \to b$ to mean $\Phi_{T,T}(\Phi_{T,T \Rightarrow T}(\mathcal{I}(\to))(a))(b)$ in the next definition.

A *second-order frame* $\mathcal{F}$ for terms over constants from $\mathcal{C}_{\text{cst}}$ is a tuple

$$\mathcal{F} = (\mathit{Kind}, \mathit{Dom}, \{\, \Psi_{a,b} \mid a,b \in \text{Kind}^T \,\}, \{\, \Psi_f \mid f \in \text{Kind}^{T \Rightarrow T} \,\})$$

satisfying conditions (1) through (4):

1. $\mathit{Kind} = (\{\text{Kind}^\kappa\}, \{\Phi_{\kappa_1,\kappa_2}\}, \mathcal{I})$ is a kind frame for $\mathcal{C}_{\text{cst}}$.
2. $\mathit{Dom} = \{\, \text{Dom}^a \mid a \in \text{Kind}^T \,\}$ is a family of sets Dom^a indexed by elements $a \in \text{Kind}^T$.
3. For each $a,b \in \text{Kind}^T$, there is a set $[\text{Dom}^a \to \text{Dom}^b]$ of maps from Dom^a to Dom^b with bijection $\Psi_{a,b} \colon \text{Dom}^{a \to b} \to [\text{Dom}^a \to \text{Dom}^b]$.
4. For every $f \in \text{Kind}^{[T \Rightarrow T]}$, there is a subset $[\Pi_{a \in \text{Kind}^T} \text{Dom}^{f(a)}]$ of $\Pi_{a \in \text{Kind}^T} \text{Dom}^{f(a)}$ with bijection $\Psi_f \colon \text{Dom}^{\forall(f)} \to [\Pi_{a \in \text{Kind}^T} \text{Dom}^{f(a)}]$.

Essentially, condition (3) states that $\mathrm{Dom}^{a\to b}$ must "represent" some set $[\mathrm{Dom}^a \to \mathrm{Dom}^b]$ of maps from Dom^a to Dom^b. Similarly, condition (4) specifies that $\mathrm{Dom}^{\forall(f)}$ must represent some subset $[\Pi_{a\in\mathrm{Kind}^T}\mathrm{Dom}^{f(a)}]$ of the product $\Pi_{a\in\mathrm{Kind}^T}\mathrm{Dom}^{f(a)}$.

Note that since in the model we are going to construct types are allowed to be empty, in what follows environments are partial maps mapping variables in $\mathcal{V}_{\mathrm{cst}} \cup \mathcal{V}_{\mathrm{term}}$ to certain values which are defined for all elements of $\mathcal{V}_{\mathrm{cst}}$, but need not be defined for all ordinary variables.

Terms are interpreted using Ψ's for application and Ψ^{-1}'s for abstraction. Since different Ψ and Ψ^{-1} maps are used, depending on the types of terms, the type of a term will be used to define its meaning. If Γ is a context and η an environment mapping $\mathcal{V}_{\mathrm{cst}}$ to elements of the appropriate kinds and $\mathcal{V}_{\mathrm{term}}$ to elements of $\bigcup\{\,\mathrm{Dom}^a \mid a \in \mathrm{Kind}^T\,\}$, one says that η *satisfies* Γ, written $\eta \models \Gamma$, if for every $x\colon \sigma \in \Gamma$, $\eta(x)$ is defined with

$$\eta(x) \in \mathrm{Dom}^{[\![\sigma]\!]_\eta} \ ,$$

in case $\mathrm{Dom}^{[\![\sigma]\!]_\eta}$ is not empty, and $\eta(x)$ is undefined, otherwise.

Let $\mathcal{F}$ be a second-order frame. For any well-typed term $\Gamma \vdash M\colon \sigma$ and environment $\eta \models \Gamma$ the meaning $[\![\Gamma \vdash M\colon \sigma]\!]_\eta$ is defined by induction on typing derivations. The inductive clauses of the meaning function are given in the same order as the typing rules in Sect. 2.2, with rules $(\to E)$, $(\to I)$, $(\forall E)$, and $(\forall I)$ preceding rules (add hyp) and (type eq) which do not rely on the form of terms:

$[\![\Gamma \vdash x\colon \sigma]\!]_\eta = \eta(x)$,

$[\![\Gamma \vdash MN\colon \tau]\!]_\eta = \Psi_{a,b}([\![\Gamma \vdash M\colon \sigma \to \tau]\!]_\eta)([\![\Gamma \vdash N\colon \sigma]\!]_\eta)$,

  where $a = [\![\sigma]\!]_\eta$ and $b = [\![\tau]\!]_\eta$,

$[\![\Gamma \vdash \lambda x\colon \sigma.M\colon \sigma \to \tau]\!]_\eta = \Psi^{-1}_{a,b}(g)$, where $a = [\![\sigma]\!]_\eta$, $b = [\![\tau]\!]_\eta$ and

  $g(d) = [\![\Gamma, x\colon \sigma \vdash M\colon \tau]\!]_{\eta[d/x]}$ for all $d \in \mathrm{Dom}^a$,

$[\![\Gamma \vdash M\tau\colon \mu\tau]\!]_\eta = \Psi_f([\![\Gamma \vdash M\colon \forall\mu]\!]_\eta)([\![\tau]\!]_\eta)$, where $f = [\![\mu]\!]_\eta$,

$[\![\Gamma \vdash \Lambda t.M\colon \forall t.\sigma]\!]_\eta = \Psi^{-1}_f(g)$, where $f \in \mathrm{Kind}^{T\Rightarrow T}$ is the map $[\![\Lambda t.\sigma]\!]_\eta$ and

  $g(a) = [\![\Gamma \vdash M\colon \sigma]\!]_{\eta[a/t]}$ for all $a \in \mathrm{Kind}^T$,

$[\![\Gamma, x\colon \sigma \vdash M\colon \tau]\!]_\eta = [\![\Gamma \vdash M\colon \tau]\!]_\eta$, where the left-hand typing follows from

  the rule (add hyp),

$[\![\Gamma \vdash M\colon \tau]\!]_\eta = [\![\Gamma \vdash M\colon \sigma]\!]_\eta$, where the left-hand typing follows by rule

  (type eq).

It is relatively easy to see that the environments on the right-hand sides of these clauses all satisfy the appropriate contexts (cf. [6]).

In the above definition of meaning, there is no guarantee that the map g in the $\lambda x\colon \sigma.M$ case is in the domain of $\Psi^{-1}_{a,b}$, and similarly for the map g in the $\Lambda t.M$ case. A second-order frame

$$\mathcal{F} = (Kind, Dom, \{\,\Psi_{a,b} \mid a,b \in \mathrm{Kind}^T\,\}, \{\,\Psi_f \mid f \in \mathrm{Kind}^{T\Rightarrow T}\,\})$$

is an *environment model* if $Kind$ is a kind environment model and for every term $\Gamma \vdash M : \sigma$ and every environment $\eta \models \Gamma$, the maps g in the $\lambda x : \sigma.M$ case and g in the $\Lambda t.M$ case, respectively, of the definition of the meaning $[\![\Gamma \vdash M : \sigma]\!]_\eta$ are in the domains of $\Psi_{a,b}^{-1}$ and Ψ_f^{-1}.

It is easy to check that the meanings of the terms have the appropriate semantic types:

Lemma 1. *Let η be an environment for a model* $(Kind, Dom, \{\Psi_{a,b}\}, \{\Psi_f\})$ *and* $\Gamma \vdash M : \sigma$ *be a term. If $\eta \models \Gamma$ and $Dom^{[\![\tau]\!]\eta}$ is not empty, for every $x : \tau \in \Gamma$ such that x occurs free in M, then $[\![\Gamma \vdash M : \sigma]\!]_\eta \in Dom^{[\![\sigma]\!]\eta}$.*

The only nontrivial case is abstraction by rule $(\to I)$. Since we assume that $\Gamma \vdash \lambda x : \sigma.M : \sigma \to \tau$ follows from $\Gamma, x : \sigma \vdash M : \tau$, we have to distinguish the cases whether $Dom^{[\![\sigma]\!]\eta}$ is empty or not. The second case is obvious and in the first case we have that the map g is the empty map, which is the only map from $Dom^{[\![\sigma]\!]\eta}$ to $Dom^{[\![\tau]\!]\eta}$.

As in [6] it moreover follows that the meaning of a well-typed term $\Gamma \vdash M : \sigma$ does not depend on the derivation of the typing.

An environment $\eta \models \Gamma$ for model $\mathcal{F}$ *satisfies* an equation $\Gamma \vdash M = N : \sigma$, if $[\![\Gamma \vdash M : \sigma]\!]_\eta = [\![\Gamma \vdash N : \sigma]\!]_\eta$. A model $\mathcal{F}$ *satisfies* an equation $\Gamma \vdash M = N : \sigma$, if $\mathcal{F}$ and η satisfy $\Gamma \vdash M = N : \sigma$ for all $\eta \models \Gamma$. In the same way as in [6] one obtains that the axioms and inference rules presented in Sect. 2.3 are sound for environment models.

Proposition 1 (Soundness). *Let $\Gamma \vdash M = N : \sigma$ be a provable equation. Then $\Gamma \vdash M = N : \sigma$ is satisfied by every environment model.*

4 Sequences of Finite Sets

4.1 Basic Definitions

The objects by which we will interpret types in the model we are going to present are inverse limits T of ω-cochains $T_0 \xleftarrow{p_0} T_1 \xleftarrow{p_1} T_2 \xleftarrow{p_2} \cdots$ of finite sets T_i of natural numbers such that for every $i > 0$ either T_i is empty and p_{i-1} is the empty map, or T_i is not empty and p_{i-1} is surjective. The elements of such a limit are sequences $(y_i)_{i \in \omega}$ with $y_i \in T_i$ such that $y_i = p_i(y_{i+1})$. If $x \in T$ then we denote the ith element of the sequence by x_i.

It should be observed that if for some index i the finite set T_i in the cochain is empty, then for all $j > i$ the sets T_j and hence the limit T must be empty as well.

At first sight, the restriction to subsets of ω seems to be an unnecessary complication of the construction. The reason for it will become clear later. As a consequence of this restriction an encoding of the objects obtained is necessary after most of the construction steps. But in order to make the construction more transparent, we only indicate in the following how the encoding has to be done and then suppress it as much as possible. So, e.g., we always identify a finite object with its code. Moreover, we write $\{\ldots\}^c$ to indicate that not the set $\{\ldots\}$ but the set of codes of its elements is meant.

Let $\langle\,,\ldots,\,\rangle\colon \omega^n \to \omega$ be a computable n-tuple encoding with decoding functions π_i^n $(1 \le i \le n)$ which is monotone in each argument. Moreover, let D be a canonical coding of finite sets of natural numbers (cf. [24]). If A, A', B and B' are finite sets of natural numbers and $p\colon A \to B$ and $p'\colon A' \to B'$ are finite functions, then

$$A \times^c A' = \{\,\langle a, a'\rangle \mid a \in A \wedge a' \in A'\,\}$$

and $p \times^c p'\colon A \times^c A' \to B \times^c B'$ is the function defined by

$$p \times^c p'(\langle a, a'\rangle) = \langle p(a), p'(a')\rangle \ .$$

A finite function $g\colon A \to B$ is coded by the number $\langle n, a, b\rangle$, where n, a, and b are such that $D_a = A$, $D_b = B$, and $D_n = \{\,\langle d, e\rangle \mid g(d) = e\,\}$. Obviously, the value of the function for an argument can easily be obtained from the code.

The morphisms we will consider are such that the degree of coincidence between any two sequences in their domain is preserved.

Definition 1. *Let T and T' be inverse limits of ω-cochains. A map $f\colon T \to T'$ is said to be* rank-preserving *if for all x, $y \in T$ and all $i \in \omega$ the following condition holds:*

$$x_i = y_i \Rightarrow f(x)_i = f(y)_i \ .$$

As follows from the definition a rank-preserving map is determined by its behaviour on the finite approximation of its domain space, i.e. by its own finite approximations. Note that in the context of rank-ordered sets the given definition is equivalent to the one given by Bruce and Mitchell [5]. In the case of projection spaces maps with this property are called projection compatible (cf. [9]).

Let **SFS** be the category which has as

- *objects*   inverse limits T of ω-cochains $T_0 \xleftarrow{p_0} T_1 \xleftarrow{p_1} T_2 \xleftarrow{p_2} \cdots$ of finite sets T_i of natural numbers such that either T_{i+1} is empty and p_i is the empty map, or T_{i+1} is not empty and p_i is surjective, and as
- *morphisms*   rank-preserving maps.

We denote its object set by SFS and for any two objects T and T' the set of all morphisms from T to T' by **SFS**$[T, T']$. The empty set as inverse limit of the cochain $\emptyset \xleftarrow{\emptyset} \emptyset \xleftarrow{\emptyset} \cdots$ is obviously initial in this category and the inverse limit of the cochain $\{i\} \leftarrow \{i\} \leftarrow \cdots$ is terminal, for any $i \in \omega$.

4.2   Product and Exponentials

The product is defined componentwise. Let T, T', T'', and T''' be **SFS** objects such that T and T', respectively, are the inverse limits of the cochains $T_0 \xleftarrow{p_0}$

$T_1 \xleftarrow{p_1} T_2 \xleftarrow{p_2} \cdots$ and $T'_0 \xleftarrow{p'_0} T'_1 \xleftarrow{p'_1} T'_2 \xleftarrow{p'_2} \cdots$. Construct a new cochain $T_0^\times \xleftarrow{p_0^\times} T_1^\times \xleftarrow{p_1^\times} T_2^\times \xleftarrow{p_2^\times} \cdots$ by setting

$$T_i^\times = T_i \times^c T'_i \quad \text{and} \quad p_i^\times = p_i \times^c p'_i$$

and define $\boldsymbol{T} \otimes \boldsymbol{T}'$ to be its inverse limit. We denote elements of $\boldsymbol{T} \otimes \boldsymbol{T}'$ by $\ll\boldsymbol{x}, \boldsymbol{x}'\gg$ with $\boldsymbol{x} \in \boldsymbol{T}$ and $\boldsymbol{x}' \in \boldsymbol{T}'$. Let $\mathrm{pr}(\ll\boldsymbol{x}, \boldsymbol{x}'\gg) = \boldsymbol{x}$ and $\mathrm{pr}'(\ll\boldsymbol{x}, \boldsymbol{x}'\gg) = \boldsymbol{x}'$. Moreover, for two morphisms $f \in \mathbf{SFS}[\boldsymbol{T}, \boldsymbol{T}'']$ and $g \in \mathbf{SFS}[\boldsymbol{T}', \boldsymbol{T}''']$ let $(f \otimes g)(\ll\boldsymbol{x}, \boldsymbol{x}'\gg) = \ll f(\boldsymbol{x}), g(\boldsymbol{x}')\gg$. Then $\mathrm{pr}, \mathrm{pr}'$ and $f \otimes g$ are rank-preserving. Obviously, one obtains a product in $\mathbf{SFS}$ by this means. So, the category is Cartesian.

Next, we want to show that $\mathbf{SFS}$ is also Cartesian closed. To this end we first show that the function space can itself be represented as an inverse limit of an ω-cochain of finite sets of natural numbers. Let to this end $p_{ij} = p_j \circ \cdots \circ p_{i-1}$, for $j < i$. The idea is to represent a rank-preserving map from $\boldsymbol{T}$ to $\boldsymbol{T}'$ by a sequence of locally acting functions from T_i to T'_i. Set

$$[T_i \to T'_i] =$$
$$\{\, h \colon T_i \to T'_i \mid (\forall y, z \in T_i)(\forall j < i)[p_{ij}(y) = p_{ij}(z) \Rightarrow p'_{ij}(h(y)) = p'_{ij}(h(z))] \,\}^c$$

and define $q_i \colon [T_{i+1} \to T'_{i+1}] \to [T_i \to T'_i]$ by

$$q_i(h)(y) = p'_i(h(z)) \ ,$$

for some $z \in p_i^{-1}(\{y\})$. Obviously, the value of $q_i(h)(y)$ is independent of the choice of z. Moreover, q_i is surjective.

$[T_i \to T'_i]$ is a set of codes of finite functions. Note that in the definition of q_i we omitted the corresponding coding and decoding functions and dealt with the finite functions directly, according to what has been said earlier that we identify finite objects and their codes.

Let $[\boldsymbol{T} \Rightarrow \boldsymbol{T}']$ be the inverse limit of the cochain $([T_i \to T'_i], q_i)_{i \in \omega}$. There is a one-to-one correspondence between $\mathbf{SFS}[\boldsymbol{T}, \boldsymbol{T}']$ and $[\boldsymbol{T} \Rightarrow \boldsymbol{T}']$. As it is easily seen, both sets are not empty exactly if $\boldsymbol{T}'$ is not empty or both $\boldsymbol{T}$ and $\boldsymbol{T}'$ are empty. In this case define $\Theta_{\boldsymbol{T}, \boldsymbol{T}'} \colon \mathbf{SFS}[\boldsymbol{T}, \boldsymbol{T}'] \to [\boldsymbol{T} \Rightarrow \boldsymbol{T}']$ by letting $\Theta_{\boldsymbol{T}, \boldsymbol{T}'}(f)_i$ be the (code of the) finite function that maps $y \in T_i$ to $f(z)_i$, for some $z \in \boldsymbol{T}$ with $y = z_i$. Since f is rank-preserving, this definition is independent of the choice of $\boldsymbol{z}$ and the condition in the definition of $[T_i \to T'_i]$ is satisfied. Moreover $q_i(\Theta_{\boldsymbol{T}, \boldsymbol{T}'}(f)_{i+1}) = \Theta_{\boldsymbol{T}, \boldsymbol{T}'}(f)_i$. Thus $\Theta_{\boldsymbol{T}, \boldsymbol{T}'}(f) \in [\boldsymbol{T} \Rightarrow \boldsymbol{T}']$. Conversely, define $\Psi_{\boldsymbol{T}, \boldsymbol{T}'} \colon [\boldsymbol{T} \Rightarrow \boldsymbol{T}'] \to \mathbf{SFS}[\boldsymbol{T}, \boldsymbol{T}']$ by

$$\Psi_{\boldsymbol{T}, \boldsymbol{T}'}(\boldsymbol{g})(\boldsymbol{x})_i = g_i(x_i) \ .$$

Since

$$p'_i(g_{i+1}(x_{i+1})) = q_i(g_{i+1})(p_i(x_{i+1})) = g_i(x_i) \ ,$$

we have that $\Psi_{\boldsymbol{T}, \boldsymbol{T}'}(\boldsymbol{g})(\boldsymbol{x}) \in \boldsymbol{T}'$. Moreover, $\Psi_{\boldsymbol{T}, \boldsymbol{T}'}(\boldsymbol{g})$ is rank-preserving. As is readily verified, both maps $\Theta_{\boldsymbol{T}, \boldsymbol{T}'}$ and $\Psi_{\boldsymbol{T}, \boldsymbol{T}'}$ are inverse to each other.

Proposition 2. *For all* $T, T' \in$ SFS, *the map* $\Psi_{T,T'}$ *is a bijection from* $[T \Rightarrow T']$ *onto* $\mathbf{SFS}[T, T']$.

We will now show that $[T \Rightarrow T']$ is the exponent of T and T' in **SFS**. Define eval $\in \mathbf{SFS}[[T \Rightarrow T'] \otimes T, T']$ by

$$\mathrm{eval}(\ll g, x \gg) = \Psi_{T,T'}(g)(x) \ ,$$

where $g \in [T \Rightarrow T']$ and $x \in T$, and curry: $\mathbf{SFS}[T \otimes T', T''] \to \mathbf{SFS}[T, [T' \Rightarrow T'']]$ by

$$\mathrm{curry}(f)(x)_i(y) = \Theta_{T \otimes T', T''}(f)_i(\langle x_i, y \rangle) \ ,$$

for $f \in \mathbf{SFS}[T \otimes T', T'']$, $x \in T$, and $y \in T_i$. Then one has for $h \in \mathbf{SFS}[T \otimes T', T']$, $k \in \mathbf{SFS}[T, [T' \Rightarrow T'']]$, $x \in T$, and $z \in T'$ that

$$
\begin{aligned}
(\mathrm{eval} \circ (\mathrm{curry}(h) \otimes \mathrm{id}_{T'}))(\ll x, z \gg)_i &= \mathrm{eval}(\ll \mathrm{curry}(h)(x), z \gg)_i \\
&= \Psi_{T', T''}(\mathrm{curry}(h)(x))(z)_i \\
&= \mathrm{curry}(h)(x)_i(z_i) \\
&= \Theta_{T \otimes T', T''}(h)_i(\langle x_i, z_i \rangle) \\
&= h(\ll x, z \gg)_i
\end{aligned}
$$

and

$$
\begin{aligned}
\mathrm{curry}(\mathrm{eval} \circ (k \otimes \mathrm{id}_{T'}))(x)_i(z_i) &= \Theta_{T \otimes T', T''}(\mathrm{eval} \circ (k \otimes \mathrm{id}_{T'}))_i(\langle x_i, z_i \rangle) \\
&= \mathrm{eval}(k \otimes \mathrm{id}_{T'}(\ll x, z \gg))_i \\
&= \mathrm{eval}(\ll k(x), z \gg)_i \\
&= \Psi_{T', T''}(k(x))(z)_i \\
&= k(x)_i(z_i) \ .
\end{aligned}
$$

Summing up what we have shown so far we obtain the following result.

Theorem 1. *The category* **SFS** *of inverse limits of* ω-*cochains of finite sets of natural numbers and rank-preserving maps is Cartesian closed.*

Thus, the category **SFS** gives rise to a model of the simply typed lambda calculus, even to a model of an extension of this calculus by explicit pairs (cf. [2]).

5   Complete Projection Spaces

5.1   Basic Definitions

In this section we study the structure of SFS. As will turn out, SFS is a complete projection space. We have already seen that the usual constructions on SFS are rank-preserving. With rank-preserving maps as morphisms the category of complete projection spaces is Cartesian closed.

Definition 2. *Let P be a nonempty set and $([\cdot]_i)_{i \in \omega}$ be a family of maps from P into P.*

1. $(P, ([\cdot]_i)_{i \in \omega})$ *is a* projection space, *if for all $i, j \in \omega$*

$$[\cdot]_i \circ [\cdot]_j = [\cdot]_{\min\{i,j\}} \ .$$

2. $(P, ([\cdot]_i)_{i \in \omega})$ *is* complete *if for any sequence $(x_i)_{i \in \omega}$ from P with $x_i = [x_{i+1}]_i$, for $i \in \omega$, there is a unique element $x \in P$ such that for all i, $x_i = [x]_i$.*

For $i \in \omega$ set $P_i = \{ x \in P \mid x = [x]_i \}$.

Projection spaces have been studied by Ehrig et al. [9,10,11,12] as a generalization of the projective model of process algebra by Bergstra and Klop [4]. Bruce and Mitchell [5] considers a subclass of complete projection spaces, called rank-ordered sets, which are such that the map $[\cdot]_0$ projects the whole space onto a distinguished element $\bot$.

The mapping $[\cdot]_i$ can be thought of as a map that takes an element x to its ith approximation. In the case of an inverse limit $T \in \text{SFS}$ the ith approximation $[T]_i^{\text{SFS}}$ is the inverse limit of the cochain

$$T_0 \xleftarrow{p_0} T_1 \xleftarrow{p_1} T_2 \xleftarrow{p_2} \cdots \xleftarrow{p_{i-1}} T_i \xleftarrow{p_i} T_i \xleftarrow{p_{i+1}} \cdots \ ,$$

with $p_j = \text{id}_{T_i}$, for $j \geq i$, which is one-to-one correspondence with T_i.

Proposition 3. $(\text{SFS}, ([\cdot]_i^{\text{SFS}})_{i \in \omega})$ *is a complete projection space.*

Note that each projection space P is the inverse limit of the cochain $(P_i, [\cdot]_i \restriction P_{i+1})_{i \in \omega}$, but the sets P_i need not be finite. As is readily verified, a map $F \colon P \to Q$ between projection spaces P and Q is rank-preserving exactly if for all $x \in P$ and all $i, j \in \omega$ with $j \geq i$, $[F([x]_j^P)]_i^Q = [F(x)]_i^Q$.

Let **CP** be the category which has as

− *objects*   complete projection spaces and as
− *morphisms*   rank-preserving maps.

We denote the object set by CP and for any two objects P and Q the set of all morphisms from P to Q by **CP**$[P, Q]$.

5.2   Products and Exponentials

Products in the category of complete projections spaces are formed as Cartesian products $P \times Q$ with projections $[\cdot]_i^\times$ given by

$$[(x,y)]_i^\times = ([x]_i^P, [y]_i^Q) \ .$$

Exponentials are the sets **CP**$[P, Q]$ of all rank-preserving maps $F \colon P \to Q$. Define projections $[\cdot]_i^\rightarrow$ by

$$[F]_i^\rightarrow (x) = [F(x)]_i^Q \ .$$

Then $\mathbf{CP}[P,Q]$ is a complete projection space again. Moreover, let Eval and Curry be given in the usual way, that is, for $F \in \mathbf{CP}[P,Q]$, $G \in \mathbf{CP}[P \times Q, R]$, $x \in P$ and $y \in Q$ let

$$\text{Eval}(F, x) = F(x) \ ,$$
$$\text{Curry}(G)(x)(y) = G(x, y) \ .$$

Then Eval $\in \mathbf{CP}[\mathbf{CP}[P,Q] \times P, Q]$ and Curry $\in \mathbf{CP}[\mathbf{CP}[P \times Q, R], \mathbf{CP}[P, \mathbf{CP}[Q, R]]]$.

Obviously, $(P \times Q)_i = P_i \times Q_i$ and $\mathbf{CP}[P,Q]_i = \{ F \colon K \to L \mid (\forall x \in P)F(x) = [F([x]_i^P)]_i^Q \}$.

Theorem 2. *The category* $\mathbf{CP}$ *of complete projection spaces and rank-preserving maps is Cartesian closed.*

6   The Product Type Construction

Let $F \colon \text{SFS} \to \text{SFS}$ be rank-preserving. We call F a *parameterisation*. A map $f \colon \text{SFS} \to \bigcup \{ F(\boldsymbol{T}) \mid \boldsymbol{T} \in \text{SFS} \}$ is a *section* of F if $f(\boldsymbol{T}) \in F(\boldsymbol{T})$, for all $\boldsymbol{T} \in \text{SFS}$.

Definition 3. *Let* F *be a parameterisation. A section* f *of* F *is* rank-preserving *if for all* $\boldsymbol{T}, \boldsymbol{T}' \in \text{SFS}$ *and all* $i \in \omega$ *the following condition holds:*

$$[\boldsymbol{T}]_i^{\text{SFS}} = [\boldsymbol{T}']_i^{\text{SFS}} \Rightarrow f(\boldsymbol{T})_i = f(\boldsymbol{T}')_i \ .$$

We would like to interpret $\forall t.\sigma$ by the set of all rank-preserving sections of the parameterisation associated with σ. This set can be represented as an inverse limit of sets, in the same way as the morphism set $\mathbf{SFS}[\boldsymbol{T}, \boldsymbol{T}']$, but in general the approximating sets are not finite, as the sets SFS_i are not finite. Therefore, we construct a set which *is* an inverse limit of an ω-chain of finite sets and is in a one-to-one correspondence with the set of rank-preserving sections of this parameterisation.

Let F be a parameterisation. Then the idea is the following: In order to know $F(\boldsymbol{T})$ we have to know its approximations $F(\boldsymbol{T})_i$, which depend only on the approximation T_j for $j \leq i$, since F is rank-preserving. Thus, instead of taking a product over all $\boldsymbol{T} \in \text{SFS}$ it suffices in the ith approximation to consider the product over all finite cochains

$$T_0 \xleftarrow{p_0} T_1 \xleftarrow{p_1} \cdots \xleftarrow{p_{i-1}} T_i \ .$$

Each such cochain is uniquely determined by the sequence $(p_0, \ldots, p_{i-1})$ of projections. Let

$$\text{Seq} = \{ \sigma \mid \sigma = (p_0, \ldots, p_n) \wedge (\forall j < n) \text{dom}(p_j) = \text{range}(p_{j+1}) \}$$

and for $\sigma \in \text{Seq}$ define $\text{Ext}(\sigma)$ to be the inverse limit of the cochain

$$T_0 \xleftarrow{q_0} T_1 \xleftarrow{q_1} T_2 \xleftarrow{q_2} \cdots \xleftarrow{q_{i-1}} T_i \xleftarrow{q_i} T_{i+1} \xleftarrow{q_{i+1}} \cdots \ ,$$

with $T_i = \text{range}(p_i)$, for $i \leq n$, and $T_i = \text{dom}(p_n)$, otherwise, and $q_i = p_i$, for $i \leq n$, and $q_i = \text{id}_{\text{dom}(p_n)}$, otherwise. Then $\text{Ext}(\sigma)$ is in bijective correspondence with $\text{dom}(p_n)$.

One can easily construct an onto and one-to-one enumeration seq: $\omega \to \text{Seq}$ of Seq such that $\text{lth}(\text{seq}(m)) \leq m + 1$, where $\text{lth}(\text{seq}(m))$ is the length of the sequence $\text{seq}(m)$. This can e.g. be achieved by enumerating the sequences $(p_0, \ldots, p_i)$ such that both i and all elements in the domains and ranges of the p_j $(j \leq i)$ are bounded by n ahead of those sequences $(p_0, \ldots, p_i)$ such that i and the elements in the domains and ranges of the p_j $(j \leq i)$ are bounded by $n+1$. Define $\Pi_{\leq n}F$ to be the set of all $\langle a_0, \ldots, a_n \rangle$ with $a_m \in F(\text{Ext}(\text{seq}(m)))_{\text{lth}(\text{seq}(m))-1}$, for all $m \leq n$, such that the following consistency condition holds: for all $m_1, m_2 \leq n$ with $m_1 \neq m_2$, if the sequence $\text{seq}(m_2)$ extends the sequence $\text{seq}(m_1)$ then

$$a_{m_1} = p_{\text{lth}(\text{seq}(m_1))-1}^{F(\text{Ext}(\text{seq}(m_2)))} (\cdots (p_{\text{lth}(\text{seq}(m_2))-2}^{F(\text{Ext}(\text{seq}(m_2)))} (a_{m_2})) \cdots) \ .$$

Moreover, let $p_n^{\Pi F} \colon \Pi_{\leq n+1}F \to \Pi_{\leq n}F$ be the projection onto the (coded) first $n + 1$ components and define ΠF to be the inverse limit of the cochain $(\Pi_{\leq n}F, p_n^{\Pi F})_{n \geq 0}$.

In the finite approximations of ΠF the information about the finite approximations of the values under F is collected. Every $t \in \Pi F$ contains the information about the behaviour of F on all finite approximating cochains. In order to obtain a value in $F(\boldsymbol{T})$ for some object $\boldsymbol{T}$, the type application function Apply has thus to single out the necessary information from t. To make this idea precise, let $\text{init}(\boldsymbol{T}, i)$ be the uniquely determined number m with $\text{seq}(m) = (p_0^{\boldsymbol{T}}, \ldots, p_i^{\boldsymbol{T}})$. Then set

$$\text{Apply}(t, \boldsymbol{T})_i = \pi_{\text{init}(\boldsymbol{T}, i)}^{\text{init}(\boldsymbol{T}, i)} (t_{\text{init}(\boldsymbol{T}, i)}) \ .$$

In order to obtain the ith approximation of $\text{Apply}(t, \boldsymbol{T})$, we need to compute $t_{\text{init}(\boldsymbol{T}, i))} \in \Pi_{\leq \text{init}(\boldsymbol{T}, i)}F$. Then $t_{\text{init}(\boldsymbol{T}, i))}$ is an $(\text{init}(\boldsymbol{T}, i) + 1)$-tuple and

$$\pi_{\text{init}(\boldsymbol{T}, i)}^{\text{init}(\boldsymbol{T}, i)} (t_{\text{init}(\boldsymbol{T}, i)}) \in F(\text{Ext}(\text{seq}(\text{init}(\boldsymbol{T}, i))))_{\text{lth}(\text{seq}(\text{init}(\boldsymbol{T}, i)))-1} \ .$$

By the definition of $\text{init}(\boldsymbol{T}, i)$, $\text{lth}(\text{seq}(\text{init}(\boldsymbol{T}, i))) = i+1$. Moreover, we have that $\text{Ext}(\text{seq}(\text{init}(\boldsymbol{T}, i)))$ is the ω-cochain $T_0 \leftarrow T_1 \leftarrow \cdots \leftarrow T_i \leftarrow T_{i+1} \leftarrow T_{i+1} \leftarrow \cdots$. Since F is rank-preserving, it thus follows that

$$F(\text{Ext}(\text{seq}(\text{init}(\boldsymbol{T}, i))))_{\text{lth}(\text{seq}(\text{init}(\boldsymbol{T}, i)))-1} = F(\boldsymbol{T})_i \ .$$

Hence, $\text{Apply}(t, \boldsymbol{T}) \in F(\boldsymbol{T})$.

Proposition 4. *Let F be a parameterisation. Then the following two statements hold:*

1. *For every $t \in \Pi F$, the map $\underline{\lambda} \boldsymbol{T}.\text{Apply}(t, \boldsymbol{T})$ is a rank-preserving section of F.*
2. *The map $\Psi_F \colon t \mapsto \underline{\lambda} \boldsymbol{T}.\text{Apply}(t, \boldsymbol{T})$ is bijection from ΠF onto the set of all rank-preserving sections of F.*

Statement (1) follows as the foregoing remark. For the proof of (2) let f be a rank-preserving section of F and define

$$\Theta_F(f)_n = \langle f(\text{Ext}(\text{seq}(0)))_{\text{lth}(\text{seq}(0))-1}, \ldots, f(\text{Ext}(\text{seq}(n)))_{\text{lth}(\text{seq}(n))-1} \rangle \ .$$

Then a straightforward calculation shows that $\Theta_F(f) \in \Pi F$. In addition, Θ_F and Ψ_F are inverse to each other.

Because of the property that always $\text{lth}(\text{seq}(m)) \leq m + 1$ we moreover have that ΠF depends on $F \in \mathbf{CP}[\text{SFS}, \text{SFS}]$ in a rank-preserving way.

Proposition 5. $\Pi \in \mathbf{CP}[\mathbf{CP}[\text{SFS}, \text{SFS}], \text{SFS}]$.

7   Semantics of the Polymorphic Lambda Calculus

Set $\text{Kind}^T = \text{SFS}$ and for kind expressions κ_1, κ_2 define $\text{Kind}^{\kappa_1 \Rightarrow \kappa_2} = \mathbf{CP}[\text{Kind}^{\kappa_1}, \text{Kind}^{\kappa_2}]$. Moreover, let $\Phi_{\kappa_1, \kappa_2}$ be the identity on this set. Finally, for $T, T' \in \text{SFS}$ and $F \in \mathbf{CP}[\text{SFS}, \text{SFS}]$ set

$$\mathcal{I}(\rightarrow)(T)(T') = [T \Rightarrow T'] \qquad \text{and} \qquad \mathcal{I}(\forall)(F) = \Pi F \ .$$

Then $Kind = (\{\, \text{Kind}^\kappa \mid \kappa \text{ a kind} \,\}, \{\, \Phi_{\kappa_1, \kappa_2} \mid \kappa_1, \kappa_2 \text{ kinds} \,\}, \mathcal{I})$ is a kind environment model.

Lemma 2. *For any constructor expression μ, any constructor variable v^κ occurring free in μ and any environment η mapping constructor variables to elements of the appropriate kinds the map $\lambda a \in \text{Kind}^\kappa.[\![\mu]\!]_{\eta[a/v^\kappa]}$ is rank-preserving.*

The proof proceeds by structural induction. Note hereto that for projection spaces P and Q and rank-preserving maps $F: P \rightarrow Q$, $[F(a)]_i^Q = [F]_i^{\rightarrow}([a]_i^P)$.

Next, for $T, T' \in \text{SFS}$ and $F \in \mathbf{CP}[\text{SFS}, \text{SFS}]$, set $\text{Dom}^T = T$ and let the maps $\Psi_{T, T'}$ and Ψ_F, respectively, be as in Sects. 4.2 and 6. Then it follows with Propositions 2 and 4(2) that

$$\mathcal{R} = (Kind, \text{SFS}, \{\, \Psi_{T, T'} \mid T, T' \in \text{SFS} \,\}, \{\, \Psi_F \mid F \in \mathbf{CP}[\text{SFS}, \text{SFS}] \,\})$$

is a second-order frame.

Theorem 3. $\mathcal{R}$ *is an environment model.*

One has to verify that the maps g in clauses (3) and (5), respectively, of the extension of an environment to terms are in the domains of $\Psi_{T, T'}$ and Ψ_F.

8   Final Remarks

In this paper a new model for the Girard-Reynolds second-order lambda calculus is presented, which is not based on domains. This is in accordance with Reynolds, who argued in [22] that "types are not limited to computation" and that "they should be explicable without invoking constructs, such as Scott domains, that are peculiar to the theory of computation."

The model allows empty types. As a consequence of this, the type $\forall t. \, t$ is interpreted by the empty set, which entails that it is also a model of the logic associated with the type structure of the calculus. This is not the case with domain models.

The model is simple and constructive.

Acknowledgement

The author is grateful to Furio Honsell for useful hints. Moreover, he wants to thank the referees for their detailed reports.

References

1. Amadio, R., Bruce, K., Longo, G.: The Finitary Projection Model for Second Order Lambda Calculus and Solutions to Higher Domain Equations. In: Proceedings of the IEEE Symposium on Logic in Computer Science. IEEE Computer Society Press, Los Alamitos CA (1986) 122–130 276
2. Asperti, A., Longo, G.: Categories, Types, and Structures. MIT Press, Cambridge MA (1991) 286
3. Berardi, S., Berline, C.: Building Continuous Webbed Models for System F. In: Spreen, D. (ed.): Workshop on Domains IV. Electronic Notes in Theoretical Computer Science, Vol. 35. Elsevier, Amsterdam (2000) 277
4. Bergstra, J.A., Klop, J.W.: A Convergence Theorem in Process Algebra. Technical Report CS-R8733. CWI, Amsterdam (1987) 278, 287
5. Bruce, K., Mitchell, J.C.: PER Models of Subtyping, Recursive Types and Higher-Order Polymorphism. In: Proceedings of the ACM Symposium on Principles of Programming Languages. Ass. for Comp. Machinery, New York (1992) 316–327 277, 284, 287
6. Bruce, K., Meyer, A.R., Mitchell, J.C.: The Semantics of Second-Order Lambda Calculus. Inform. and Computation 85 (1990) 76–134 278, 278, 280, 282, 283, 283
7. Coquand, T., Gunter, C., Winskel, G.: DI-domains as a Model of Polymorphism. In: Main, M. et al. (eds.): Proceedings of the Third Workshop on the Mathematical Foundations of Programming Language Semantics. Lecture Notes in Computer Science, Vol. 298. Springer-Verlag, Berlin (1987) 344–363 276
8. Coquand, T., Gunter, C., Winskel, G.: Domain Theoretic Models of Polymorphism, Inform. and Computation 81 (1989) 123–167 276
9. Ehrig, H., Parisi Presicce, F., Boehm, P., Rieckhoff, C., Dimitrovici, C., Große-Rhode, M.: Algebraic Data Types and Process Specification Based on Projection Spaces. In: Sannella, D., Tarlecki, A. (eds.): Recent Trends in Data Type Specifications. Lecture Notes in Computer Science, Vol. 332. Springer-Verlag, Berlin (1988) 23–43 278, 284, 287
10. Große-Rhode, M.: Parameterized Data Type and Process Specification Using Projection Algebras. In: Ehrig, H. et al. (eds.): Categorical Methods in Computer Science with Aspects from Topology. Lecture Notes in Computer Science, Vol. 393. Springer-Verlag, Berlin (1989) 185–197 278, 287

11. Große-Rhode, M., Ehrig, H.: Transformation of Combined Data Type and Process Specification Using Projection Algebras. In: de Bakker, J.W. et al. (eds.): Stepwise Refinement of Distributed Systems. Lecture Notes in Computer Science, Vol. 430. Springer-Verlag, Berlin (1989) 301–339   278, 287
12. Herrlich, H., Ehrig, H.: The Construct PRO of Projection Spaces: Its Internal Structure. In: Ehrig, H. et al. (eds.): Categorical Methods in Computer Science with Aspects from Topology. Lecture Notes in Computer Science, Vol. 393, Springer-Verlag, Berlin (1989) 286–293   278, 287
13. Girard, J.-Y.: Une extension de l'interprétation fonctionelle de Gödel à l'analyse et son application à l'élimination des coupures dans l'analyse et la théorie des types. In: Fenstad, J.F. (ed.): Proceedings of the Second Scandinavian Logic Symposium. North-Holland, Amsterdam (1971) 63–92   275
14. Girard, J.-Y.: Interprétation fonctionelle et élimination des coupures de l'arithmétique d'ordre supérieure. These D'Etat. Université Paris VII (1972)   275, 276, 276
15. Girard, J.-Y.: The System F of Variable Types, Fifteen Years Later. Theoretical Computer Science 45 (1986) 159–192   276
16. Girard, J.-Y., Lafont, Y., Taylor, P.: Proofs and Types. Cambridge University Press, Cambridge (1989)   277
17. Gruchalski, A.: Constructive Domain Models of Typed Lambda Calculi. Ph.D. Thesis. University of Siegen (1995)   276
18. McCracken, N.: An Investigation of a Programming Language with a Polymorphic Type Structure. Ph.D. Thesis. Syracuse University (1979)   276
19. McCracken, N.: A Finitary Retract Model for the Polymorphic Lambda-Calculus. Manuscript (1982)   276
20. Pitts, A.M.: Polymorphism is Set-Theoretic, Constructively. In: Pitt, D. et al. (eds.): Proceedings of the Summer Conference on Category Theory and Computer Science. Lecture Notes in Computer Science, Vol. 283. Springer-Verlag, Berlin (1987) 12–39   276
21. Reynolds, J.C.: Towards a Theory of Type Structures. In: Robinet, B. (ed.): Colloque sur la programmation. Lecture Notes in Computer Science, Vol. 19, Springer-Verlag, Berlin (1974) 408–425   275
22. Reynolds, J.C.: Types, Abstraction, and Parametric Polymorphism. In: Mason, R.E.A. (ed.): Information Processing 83. Elsevier, Amsterdam (1983) 513–523   290
23. Reynolds, J.C.: Polymorphism is Not Set-Theoretic. In: Kahn, G. et al. (eds.): Proceedings of the International Symposium on Semantics of Data Types. Lecture Notes in Computer Science, Vol. 173, Springer-Verlag, Berlin (1984) 145–156   276
24. Rogers, H., Jr.: Theory of Recursive Functions and Effective Computability. McGraw-Hill, New York (1967)   284
25. Scott, D.: Data Types as Lattices. Siam J. Computing 5 (1976) 522–587   276
26. Scott, D.: A Space of Retracts. Manuscript. Merton College, Oxford (1980)   276
27. Troelstra, A.S.: Notes on Intuitionistic Second Order Arithmetic. In: Mathias, A.R.D., Rogers, H. (eds.): Cambridge Summer School in Mathematical Logic. Lecture Notes in Mathematics, Vol. 337. Springer-Verlag, Berlin (1973) 171–205   276

Church's Lambda Delta Calculus

Rick Statman*

Department of Mathematical Sciences
Carnegie Mellon University
Pittsburgh, PA 15213
statman@cs.cmu.edu

1 Introduction

In 1941 Church [2] introduced the lambda-delta calculus in an untyped context. The purpose of this note is to investigate Church's calculus in a simply typed setting and to establish the fundamental properties of this calculus. Toward this end we add to classical type theory a conditional d (definition by cases functional) at all finite types $A \to (A \to (B \to (B \to B)))$. This functional (IF- THEN-ELSE) is defined by the non-equational condition

$$\bigwedge xyuv. \ (x = y \Rightarrow dxyuv = u \ \& \sim x = y \Rightarrow dxyuv = v).$$

2 Preliminaries

Simple types are built up from 0 by $\to$. For each pair A, B of simple types we introduce a new constant

$$d : A \to (A \to (B \to (B \to B)))$$

satisfying the defining condition

$$\bigwedge xyuv. \ (x = y \Rightarrow dxyuv = u \ \& \sim x = y \Rightarrow dxyuv = v)$$

We shall fix a formulation of type theory in the language with $\bigwedge, \Rightarrow, =$ and λ (lambda). In particular, $K = \lambda xy.x$ and $K^* = \lambda xy.y$. We define FALSE $:= K = K^*$ and $\sim A := A \Rightarrow$ FALSE. For terms we adopt the simply typed lambda calculus with beta-eta conversion;

$$(\lambda x.X)Y = [Y/x]X \qquad\qquad (beta)$$
$$\lambda x.(Xx) = X (x \text{ not free in } X) \qquad\qquad (eta)$$

$$\frac{X = U \qquad\qquad Y = V}{XY = UV} \qquad\qquad \text{(application)}$$

---

* Research supported by the National Science Foundation Grant CCR 920 1893

M. Parigot and A. Voronkov (Eds.): LPAR 2000, LNAI 1955, pp. 293–307, 2000.

$$\frac{X = Y}{\lambda x.X = \lambda x.Y \ (x \text{ not free in any assumption}}$$ (abstraction)

$$X = X$$ (identity)

$$\frac{X = Y \qquad Y = Z}{X = Z}$$ (transitivity)

$$\frac{X = Y}{Y = X}$$ (symmetry)

and for logical rules we adopt the natural deduction rules $\Rightarrow I$, $\Rightarrow E$, $\bigwedge I$, $\bigwedge E$ and the classical rule

$[\sim X = Y]$

$$\frac{\text{FALSE}}{X = Y}$$ ($\sim$ rule)

viz

$[A]$

$$\frac{B}{A \Rightarrow B}$$ ($\Rightarrow I$)

$$\frac{A \Rightarrow B \qquad A}{B}$$ ($\Rightarrow E$)

$$\frac{A}{\bigwedge x.A \ (x \text{ not free in any assumption)}}$$ ($\bigwedge I$)

$$\frac{\bigwedge x.A}{[X/x]\,A}$$ ($\bigwedge E$)

The following identities are easily proved using the above together with the defining condition for d; if $x, y : A$ and $u, v : B$

$\bigwedge xyuv \ dxxuv = u$	(identity)
$\bigwedge xyu \ dxyuu = u$	(reflexivity)
$\bigwedge xyuv \ dxyuv = dyxuv$	(symmetry)
$\bigwedge xy \ dxyxy = y$	(hypothesis)

and in addition if $x, y : A$ $u, v : ((B \to (B \to B)) \to B$ and $w : B \to C$ then

$\bigwedge xyuv\ dxy(u(dxy))(v(dxy)) = dxy(uK)(vK^*)$ (transitivity)
$\bigwedge xyuvw\ w(dxyuv) = dxy(wu)(wv)$ (monotonicity)

The reader should compare this to [1]. If $B = B1 \rightarrow (\ldots(Bb \rightarrow 0)\ldots)$ then for $x, y : A$, $u, v : B$, and $zi : Bi$ for $i = 1, \ldots, b$ we have in type theory

$$d = \lambda xyuv\lambda z1 \ldots zb.\ dxy(uz1 \ldots zb)(vz1 \ldots zb).$$

Thus, it suffices to have only d with $B = 0$ since the others are definable. We shall assume that this is true below. We can also make some simplifications in deductions.

In particular, we can always assume that in the $\sim$ rule the equation $X = Y$ is between terms of type 0 since higher type equations can be replaced as follows

$$
\begin{array}{ccc}
\begin{array}{c}
[\sim X = Y] \\
D \\
\underline{L = R} \\
X = Y
\end{array}
&
\longrightarrow
&
\end{array}
$$

$$
\begin{array}{cc}
& \dfrac{X = Y}{} \qquad \dfrac{x = x}{} \\
\sim Xx = Yx & Xx = Yx \\
\end{array}
$$

$$
\dfrac{L = R}{\begin{array}{c}[\sim X = Y] \\ D \\ \underline{L = R} \\ Xx = Yx\end{array}}
$$

$$
\dfrac{\lambda x.Xx = x}{X = \lambda x.Xx} \qquad \dfrac{\lambda x.Xx = \lambda x.Yx \qquad \lambda x.Yx = Y}{\lambda x.Xx = Y}
$$

$$X = Y.$$

3 Systems

DELTA   = classical type theory + beta-eta conversion +
            the defining equation for d for all types A

DELTA(n) = classical type theory + beta-eta conversion +
            the defining equation for d for types A with rank $(A) < n + 1$

Delta   = beta-eta conversion + the axioms of identity,
            reflexivity, symmetry, hypothesis, transitivity
            and monotonicity for d for all types A

Delta(n) = beta-eta conversion + the axioms of identity,
            reflexivity, symmetry, hypothesis, transitivity,
            and monotonicity for d for all types A with rank$(A) < n + 1$

$\vdash$ refers to provability in classical type theory.
$\mapsto$ refers to equational provability.

When using the axioms of DELTA and Delta it is convenient to take all instances of the axioms instead of the universally quantified axioms. The universally quantified axioms can be inferred by $\bigwedge I$ since axioms are not assumptions and the variable restriction is therefore met.

4   Conservation of DELTA over Delta

We begin by showing that Delta gives a finite (schematic) equational axiomatization of the equational consequences of DELTA.

Recall that a natural deduction is normal if it has no $\bigwedge I$ immediately followed by a $\bigwedge E$ and no $\Rightarrow I$ immediately followed by an $\Rightarrow E$. Define the relation of reduction between deductions by

$$\frac{\dfrac{\dfrac{D}{A}}{\bigwedge xA}}{[X/x]A} \qquad \text{red.} \qquad \frac{[X/x]\,(\,D\,)}{[X/x]A}$$

$$\frac{\dfrac{[\,A\,]}{\dfrac{D1}{B}}}{\dfrac{A \to B \quad A}{B}} \qquad D2 \qquad \text{red.} \qquad \dfrac{\dfrac{D2}{[\,A\,]}}{\dfrac{D1}{B}}$$

Proposition 1 (Prawitz) : Every reduction sequence of deductions terminates in a unique normal deduction

Proposition 2:

$$\begin{aligned}
&\text{Delta, } dxyuv = v, && duvwz = z & \mapsto\ & dxywz = z \\
&\text{Delta, } dxyuv = v, && dxyvw = w & \mapsto\ & dxyuw = w \\
&\text{Delta, } dxyuv = u, && dxyvw = v & \mapsto\ & dxyuw = u \\
&\text{Delta, } dxyuv = v, && & \mapsto\ & dxyvu = u \\
&\text{Delta, } dxyuv = v, && dxyab = b & \mapsto\ & dxy(ua)(vb) = vb \\
&\text{Delta, } dxyuv + u, && dxyab = a & \mapsto\ & dxy(ua)(vb) = ua \\
&\text{Delta, } dxyKK^* = K && & \mapsto\ & x = y \\
&\text{Delta, } d(dxyKK^*)(K^*)(K)(K^*) = (K^*) && & \mapsto\ & x = y
\end{aligned}$$

For sets S of equations and negations of equations. Define $x\#y \Leftrightarrow dxyKK^* = K^*$ and S^* be S with each negation $\sim X = Y$ replaced by $X\#Y$. If D is a natural deduction define depth(D) and length(D) as follows. If D is an axiom of assumption then depth$(D) = \text{length}(D) = 0$. If

$$D = \frac{D(1) \ldots D(n)}{\qquad\qquad}R$$

$$A$$

for R some rule of inference then

$$\text{length}(D) = 1 + \text{length}(D(1)) + = \ldots + \text{length}(D(n)) \text{and}$$
$$\text{depth}(D) = 1 + \max\{\text{depth}(D(1)), \ldots, \text{depth}(D(n))\} \text{ if } R \text{ is a}$$
$$\text{one of the logical rules} \Rightarrow I, \Rightarrow, \lambda, \lambda E, \sim \text{ and}$$
$$= \max\{\text{depth}(D(1)), \ldots, \text{depth}(D(n))\} \text{ otherwise}$$

Lemma 1:

(1) If there is a deduction of $X = Y$ from DELTA, Delta, $S \cup (U = V)$ which is normal and depth $< n + 1$ then there is a deduction of $dUVXY = Y$ from DELTA, Delta, S which is normal and depth $< n + 1$.
(2) If there is a deduction of $X = Y$ from DELTA, Delta, $S \cup (\sim U = V)$ which is normal and depth $< n + 1$ then there is a deduction of $dUVXY = X$ from DELTA, Delta, S which is normal and depth $< n + 1$.

PROOF: The proof is on the ordinal number $\text{ord}(D) = \text{omega} * \text{depth}(D) + \text{length}(D)$ of the normal deduction D.
Basis: $\text{ord}(D) = 0$.

(1) $X = Y$ is an axiom, a member of Delta or a member of $S \cup (U = V)$. In case $X = Y$ is an axiom, a member of Delta, or a member of S then we have the following deduction:

$$\frac{dUVY = dUVY}{dUVXY = DUVYY} \qquad X = Y \qquad \frac{}{dUVYY = Y}$$
$$\frac{}{dUVXY = Y.}$$

In case $X = Y$ is $U = V$ we have that $dUVUV = V$ is in Delta.
(2) $X = Y$ is an axiom, a member of Delta, or a member of S. We have the following deduction:

$$\frac{dUVX = dUVX}{dUVXY = dUVXX} \qquad \frac{X = Y}{Y = X} \qquad dUVXX = X$$
$$\frac{}{dUVXY = X.}$$

Induction step; $\text{ord}(D) > 0$.

Case 1: D ends in $\sim$

$$D = \quad \begin{array}{c} [\sim X = Y] \\ D\hat{} \\ \frac{L = R}{X = Y} \end{array}$$

(1) By induction hypothesis there is a deduction of $dXYLR = L$ from DELTA, Delta, $S \cup (U = V)$ which is normal and depth $< n$. Thus, there is a deduction of $dXYXY = X$ from DELTA, Delta, $S \cup (U = V)$ which is normal and depth $< n$. Thus, there is a deduction of $X = Y$ from DELTA, Delta, $S \cup (U = V)$ which is normal and depth $< n$. Thus, by induction hypothesis there is a deduction of $dUVXY = Y$ from DELTA, Delta, S which is normal and depth $< n < n + 1$.

(2) Similar to (1).

Case 2: D ends in $\Rightarrow E$. The first possibility is $D =$

$$\frac{\sim Z(1) = Z(2) \Rightarrow dZ(1)Z(2)Z(3)Z(4) = Z(4) \quad \overset{D^{\char94}}{\sim Z(1) = Z(2)}}{dZ(1)Z(2)Z(3)Z(4) = Z(4)}$$

with $X = dZ(1)Z(2)Z(3)Z(4)$ and $Y = Z(4)$. We distinguish two subcases.

Subcase 1: $D^{\char94}$ ends in $\Rightarrow I$. Say $D^{\char94} =$

$$\frac{\begin{array}{c}[Z(1) = Z(2)]\\ D^{\char94\char94}\\ L = R\end{array}}{\sim Z(1) = Z(2).}$$

(1) By induction hypothesis there is a deduction of $dZ(1)Z(2)LR = R$ from DELTA, Delta, $S \cup \{U = V\}$ which is normal and depth $< n - 1$. Thus, there is a deduction of $dZ(1)Z(2)Z(3)Z(4) = Z(4)$ from DELTA, Delta, $S \cup \{U = V\}$ which is normal and depth $< n - 1$. We have the following: Thus, by induction hypothesis there is a deduction of $dWZLR = R$ from DELTA, Delta, S which is normal and depth < 2. Thus, by proposition 2 there is a deduction of $dUVXY = Y$ from DELTA, Delta, S which is normal and depth $< n + 1$.

(2) Similar to (1)

Subcase 2; $D^{\char94}$ is empty. In case $\sim Z(1) = Z(2)$ belongs to S we have
(1) the deduction

$$\frac{\dfrac{dUVX = dUVX \qquad \overset{D}{X = Y}}{dUVXX = dUVXY}}{\dfrac{dUVXY = dUVXX \qquad dUVXX = X}{dUVXY = Y}}$$

(2) Similar to (1).

In case $\sim Z(1) = Z(2)$ is $\sim U = V$ and we are in (2). We have the following member of Delta (instance of transitivity).

$$dZ(1)Z(\overset{.}{2})(dZ(1)Z(2)Z(3)Z(4))Z(4) = dZ(1)Z(2)Z(3)Z(4).$$

The only other possibility is $D =$

$$
\begin{array}{cc}
 & \hat{D} \\
\sim W = Z & W = Z
\end{array}
$$
$$\rule{6cm}{0.4pt}$$
$$L = R.$$

(1) By induction hypothesis there is a deduction of $dUVWZ = Z$ from DELTA, Delta, S which is normal and depth $< n$. In addition, there is a deduction of $dWZLR = R$ from DELTA, Delta, S which is normal and depth < 2. Thus, by Proposition 2, there is a deduction of $dUVXY = Y$ from DELTA, Delta, S which is normal and depth $< n + 1$.

(2) We distinguish two subcases.

Subcase 1: $\sim U = V$ is $\sim W = Z$.

By induction hypothesis there is a deduction of $dUVUV = U$ from DELTA, Delta, S which is normal and depth $< n$. Thus, there is a deduction of $U = V$ from DELTA, Delta, S which is normal and depth $< n$. Thus, there is a deduction of $dUVLR = L$ from DELTA, Delta, S which is normal and depth $< n + 1$.

Subcase 2: Otherwise.

By induction hypothesis there is a deduction of $dUVWZ = W$ from DELTA, Delta, S which is normal and depth $< n$. In addition, there is a deduction of $dWZLR = R$ from DELTA, Delta, S which is normal and depth < 2. Thus, by proposition 2 there is a deduction of $dUVLR = R$ from DELTA, Delta, S which is normal and depth $< n + 1$.

Case 3: D ends in one of the rules of equality or beta-eta conversion. These cases follow from Proposition 2.

End of Proof.

Proposition 3:

DELTA, $S \cup \{U = V\} \vdash X = Y \Rightarrow$ Delta, $S* \mapsto dUVXY = X$

DELTA, $S \cup \{\sim U = V\} \vdash X = Y \Longrightarrow$ Delta, $S* \mapsto dUVXY = X$

From Proposition 3 we obtain

Theorem 1 : DELTA $\vdash X = Y \Rightarrow$

Delta $\mapsto X = Y.$

5 Reduction of DELTA to Delta

Next we combine III with a well known reduction of type theory to equations to give a reduction of typed logical reasoning to typed equational reasoning.

If A is an equation $X = Y$ we put $A[L] = X$ and $A[R] = Y$. We translate statements of type theory into equations by the operation $+$ as follows

$(X = Y)+ := X = Y$

$(A \to B)+ := d(A + [L])(A + [R])(B + [L])(B + [R]) = B + [R]$

$(\bigwedge xA)+ := \lambda x.A + [L] = \lambda x.A + [R].$

Proposition 4: DELTA $\vdash A \Leftrightarrow A+$.

We obtain the following:

Corollary: DELTA $\vdash A \Leftrightarrow$ Delta $\mapsto A+$.

Proposition can be improved given certain consequences of the axiom of choice; in particular, the existence of extensionality functionals.

For each pair C, D of simple types we introduce a new constant $e : (C \to D) \to ((C \to D) \to C)$ satisfying the "defining" condition

$$\bigwedge xy. \quad x(exy) = y(exy_{\Rightarrow} x = y.$$

Given the "defining" condition for e and $A = C \to D$ we can define $d : A \to (A \to (0 \to (0 \to 0)))$ from $d : C \to (C \to (0 \to (0 \to 0)))$ by the term

$$\lambda xy. \, d(x(exy)(y(exy))$$

and derive the defining condition for $d : A \to (A \to (0 \to (0 \to 0)))$. Moreover, the "defining" condition for e follows from Delta and the equation

$$\lambda xy. \, d(x(exy)(y(exy))xy = K.$$

Let the set of all these equations be Ext. We conclude

Proposition 5: DELTA(0), Delta, Ext $\vdash A \Leftrightarrow A+$.

VI. Delta(1)

IV can be sharpened to DELTA(n) and Delta (n), and, in the case of $n < 2$, DELTA is conservative over Delta (n). Delta(1) corresponds to 1st order logic and is thus undecidable.

Proposition 6: DELTA$(n) \mapsto X = Y \Leftrightarrow$ Delta$(n) \mapsto X = Y$.

Let OMEGA be the full type structure over a countable ground domain as in [3].

Lemma 2: If M is any countable model of DELTA(1) then there exists a partial surjective homomorphism $h :$ OMEGA $\to M$ such that $h([[d]]) = [[d]]$.

Theorem 2: If X and Y contain only d with A of rank < 2 then DELTA $\vdash X = Y \Leftrightarrow$ Delta(1) $\mapsto X = Y$.

The sentence A of type theory is said to be first-order if each equation in A is between d free terms of type with rank < 2 and each quantifier in A is of type $= 0$. If A is first-order then $A+$ contains at most d of corresponding rank < 2. Thus

Corollary: If A is first-order then

$$\text{DELTA } \vdash A \Leftrightarrow \text{ Delta}(1) \mapsto A + .$$

Gödel's famous observation [4] that the consistency of nth order arithmetic can be proved in $n + 1$th order shows that this corollary does not extend to larger values of n.

Corollary: The problem of determining whether

$$\text{Delta}(1) \vdash M = N$$
$$\text{is recursively unsolvable.}$$

6   Delta(0)

We extend the language of type theory by adding infinitely many distinct type 0 constant $c(0), c(1), \ldots, c(n), \ldots$. In addition, we supplement beta-eta reduction by the following form of d reduction

(d) $\qquad\qquad\qquad$ $dc(i)c(j)XY \to \begin{pmatrix} X \text{ if } & i = j \\ Y \text{ if } \sim i = j \end{pmatrix}$

The resulting notion of reduction $\twoheadrightarrow$ (beta-eta-delta) is obviously terminating Church-Rosser. We shall also consider the language with additional constants $F : A = A(1) \to (\ldots (A(a) \to 0)$ and reduction rules

(Eval) $\qquad\qquad\qquad$ $FM(1) \ldots M(a) \to c$

where each $M(i)$ is a closed term of type $A(i)$. It will be convenient to refer to the rules (d) and (Eval) together by the notation '$\mapsto$'. In particular, we consider the following conditions

(1) The number of (Eval) rules if finite.
(2) Each (Eval) rule has a left hand side which is in long beta-eta normal form and is $\mapsto$ normal.
(3) Each closed term of type 0 appears at most once as the left hand side of an $\mapsto$ rule.

(4) Whenever

$$FM(1)\ldots M(a) \rightarrowtail c(i)$$

$$FN(1)\ldots N(a) \rightarrowtail c(j)$$

and $\sim i = j$ there exists $0 < k < a+1$ with $A(k) = B(1) \rightarrow (\ldots (B(b) \rightarrow 0)\ldots)$,
and there exists $F(1) : B(1)\, ,\ldots,\, F(b) : B(b),\, c(r) : 0,\, c(s) : 0$ with $\sim r = s$,
and we have
$M(k) = \lambda x(1)\; \ldots\; x(b)\; X$
$N(k) = \lambda y(1)\; \ldots\; y(b)\; Y$

with $X, Y : 0$
such that

$$[F(1)/x(1),\; \ldots,\; F(b)/x(b)]\, Y \rightarrowtail c(r)$$

$$[F(1)/y(1),\; \ldots,\; F(b)/y(b)]\, Y \rightarrowtail c(s)$$

Proposition 7: A notion of reduction $\rightarrowtail$ satisfying (2) and (3) is terminating Church-Rosser.

Let M be a closed term in long beta-eta normal form with Bohm tree $\mathrm{BOHM}(M)$. We define the decorated Bohm tree of M, $\mathrm{BOHM}+(M)$ as follows. The nodes of $\mathrm{BOHM}+(M)$ are the same as the nodes of $\mathrm{BOHM}(M)$ except certain nodes are labelled with matrices of constants as follows. The node v whose Bohm tree label has prefix $\mathrm{PREFIX}(v) =$

$$\lambda x(1) : A(1)\; \ldots\; x(t) : A(t)$$

is labelled also with the matrix $\mathrm{MATRIX}(v) =$

$$F(1,1),\; \ldots\; ,\; F(1,t)$$
$$\vdots$$
$$F(s,1),\; \ldots\; ,F(s,t)$$

where $s > 0$ depends on v and $F(i,j) : A(j)$ is a new constant. $s = s(v)$ is computed as follows. First, the tree ordering of nodes is extended to the Kleene-Brouwer linear ordering and then reversed. We shall refer to this as the one co-K-B ordering. s is computed recursively over the co-K-B order. For given v enumerate all the nodes below v in the co-K-B order of the same prefix type. For each such node w consider all the simultaneous selections of rows from the matrices already labelling all nodes below v. For each such pair of selections include a row in $\mathrm{MATRIX}(v)$. Evidently, $s(v)$ is bounded by an elementary function in the rank, in the sense of ordered sets, of v. Moreover, each row of

MATRIX(v) can be associated with a node w with similar prefix below v and a pair of selections of rows from matrices labelling nodes below w and below v.

With each node v in BOHM+(M) and each selection of rows from the matrices labelling nodes w below v we associate a substitution SUB(v, selection) defined as follows. If

PREFIX(w) $= \lambda x(1) : B(1) \ \ldots \ x(r) : B(r)$,
MATRIX(w) $= G(1,1), \ \ldots \ , \ G(1,r)$

$$
\begin{array}{ccc}
. & & . \\
. & & . \\
. & & .
\end{array}
$$

$$G(s,1), \ \ldots \ , \ G(1,r)$$

and row j is selected then SUB(v,selection) includes

$$[G(j,1)/x(1), \ \ldots \ , \ G(j,r)/x(r)] \, .$$

Now the term implicit at v with this selection is defined to be IMPLICIT(v, selection) $=$ SUB(v, selection) BOHM(M)(v).

Proposition 7: Given a notion of reduction $\rightarrowtail$ satisfying (1), (2), (3), and (4) there exists a substitution \$ for the constants that appear in the (Eval) rules such that for each (Eval) rule as above we have (\$$F$) (\$$M(1)$) $\ldots$ (\$$M(a)$) $\twoheadrightarrow c$.
PROOF: Suppose that $F : A$. SUB(F) is defined by recursion on A. Suppose that $A = A(1) \rightarrow (\ldots (A(m) \rightarrow 0) \ldots)$, $A(k) = A(k,1) \rightarrow (\ldots (A(k,m(k)) \rightarrow 0) \ldots)$, and
$FM(1,1) \ \ldots \ M(1,m) \rightarrowtail c(1)$

.

.

.

$FM(n,1) \ \ldots \ M(n,m) \rightarrowtail c(n)$
are all the EVAL rules which begin with F. For each pair $i < j$ there exists a $k = k(i,j)$ such that if

$M(i,k) = \lambda x(k,1) : A(k,1) \ \ldots \ x(k,m(k)) : A(k,m(k)) \ X,$
$M(j,k) = \lambda x(k,1) : A(k,1) \ \ldots \ x(k,m(k)) : A(k,m(k)))Y,$
with
$X, Y : 0$

there are constants $F(i,j,1) : A(k,1), \ \ldots \ , F(i,j,m(k)) : A(k,m(k)), \ c(p(i,j)) : 0 \ c(q(i,j)) : 0$ such that $p(i,j)$ is distinct from $q(i,j)$,
$[F(i,j,1)/x(k,1), \ \ldots \ , F(i,j,m(k))/x(k,m(k))] \, X \rightarrowtail c(p(i,j))$, and
$[F(i,j,1)/x(k,1), \ \ldots \ , F(i,j,m(k))/x(k,m(k))] \, Y \rightarrowtail c(q(i,j))$.
Let $x(i) : A(i)$. For $i = 1,\ldots,n$ and $j = 1,\ldots,i-1,i+1,\ldots n$ define terms $Q(i,j) : A$ and $P(i) : 0 \rightarrow A$
$$Q(i,j) = \lambda x(1) \ldots x(m) \ x(k(i,j))(\mathrm{SUB}(F(i,j,1)))\ldots(\mathrm{SUB}(F(i,j,m(k))))$$
$$P(i) = \lambda x(1)\ldots \ x(m)\lambda z : 0 \ d(Q(i,1)x(1) \ \ldots \ x(m))(c(q(i,q))))$$
$$(\ldots \ (d(Q(i,n)x(1) \ \ldots \ x(m))c(q(i,n)))c(i)z) \ \ldots \ )z.$$

Put

$$\mathrm{SUB}(F) = \lambda x(1) \ldots x(m).\, P(n)x(1) \ldots x(m)(\ldots (P(1)x(1) \ldots x(m)c(0)) \ldots)).$$

We claim that $\mathrm{SUB}(F)$ has the desired properties. The proof is by induction on the length of the left hand side $FM(1) \ldots M(m)$. We suppose that $FM(1) \ldots M(m) \rightarrowtail c$ is the tth Eval rule for t some value between 1 and n so $M(i) = M(t, i)$ and $c = c(t)$

We have

$$\mathrm{SUB}(F)(\mathrm{SUB}(M(1))) \ldots (\mathrm{SUB}(M(m))) \twoheadrightarrow$$
$$P(n)(\mathrm{SUB}(M(1))) \ldots (\mathrm{SUB}(M(m)))(\ldots$$
$$(P(1)(\mathrm{SUB}(M(1))) \ldots (\mathrm{SUB}(M(m)))c(0)) \ldots).$$

Let

$$U(i) = P(i)(\mathrm{SUB}(M(1))) \ldots (\mathrm{SUB}(M(m)))(\ldots$$
$$(P(1)(\mathrm{SUB}(M(1))) \ldots (\mathrm{SUB}(M(m)))c(0)) \ldots).$$

Then

$$U(i) \twoheadrightarrow (d((Q(i,1))(\mathrm{SUB}(M(1))) \ldots (\mathrm{SUB}(M(m))))(c(q(i,1))$$
$$(\ldots (d(Q(i,n)(\mathrm{SUB}(M(1))) \ldots$$
$$(\mathrm{SUB}(M(m)))c(q(i,n)))c(1)U(i-1)) \ldots)U(i-1).$$

In addition,
$(Q(i,j))(\mathrm{SUB}(M(1))) \ldots (\mathrm{SUB}(M(m))) \twoheadrightarrow$
$\mathrm{SUB}(M(k(i,j)))(\mathrm{SUB}(F(i,j,1))) \ldots (\mathrm{SUB}(F(i,j,m(k(i,j))))) =$
$\mathrm{SUB}(M(k(i,j)))F(i,j,1) \ldots F(i,j,m(k(i,j))))$.
Now $M(k(i,j))F(i,j,1) \ldots F(i,j,m(k(i,j))) \rightarrowtail c(p(i,j)) = / = c(q(i,j))$
when $t = j$ and this reduction uses only EVAL rules of smaller left hand side. Thus, by induction hypothesis

$$\mathrm{SUB}(M(k(i,j)))(\mathrm{SUB}(F(i,j,1))) \ldots (\mathrm{SUB}(F(i,j,m(k(i,j)))))$$

beta-eta-delta converts to $c(p(i,j))$. Thus, for $i = / = t$ we have $U(i)$ beta-eta-delta converts to $U(i-1)$. Now when $i = t$ for all j
$M(k(i,j))F(i,j,1) \ldots F(i,j,m)m(k(i,j))) \rightarrowtail c(q(i,j))$ and this reduction uses only EVAL rules of smaller left hand side. Thus, by induction hypothesis for all j
$\mathrm{SUB}(M(k(i,j)))(\mathrm{SUB}(F(i,j,1)))\ldots(\mathrm{SUB}(F(i,j,m(k(i,j)))))$ beta-eta-delta converts to $c(q(i,j))$ and $U(i)$ beta-eta-delta converts to $c(i)$. This completes the proof.

Proposition 8: Suppose that $M = \lambda x(1) \ldots x(a).\, X : A = A(1) \rightarrow (\ldots A(a) \rightarrow 0) \ldots)$ and $N = \lambda y(1) \ldots \lambda y(a).\, Y : A$, and $\sim$ Delta $(0) \vdash M = N$. Then there exist $F(1) : A(1), \ldots, F(a) : A(a)$, $c(i) : 0$, $c(j) : 0$ with $\sim i = j$, and a notion of reduction $\rightarrowtail$ satisfying conditions (1), (2), (3), (4) such that
$[F(1)/x(1), \ldots, F(a)/x(a)] X \rightarrowtail c(i)$
$[F(1)/y(1), \ldots, F(a)/y(z)] Y \rightarrowtail c(j)$

PROOF: We have that in P_ω $[[M]]$ and $[[N]]$ are distinct so there are $f(1), \ldots, f(t)$, $c(0)$, and $c(1)$ such that $[[M]] f(1) \ldots f(t) = c(0)$ and $[[N]] f(1) \ldots f(t) = c(1)$. We shall now interpret the constants in the matrix labels in the decorated Bohm tree of $P =$

$$\lambda x(1) : A(1) \ \ldots \ x(t) : A(t) \ \lambda x : 0 \ y : 0. \ dXYxy.$$

If the root node constants are $F(1), \ldots, F(t), c(0), c(1)$. Suppose that all the constants at nodes below v have been interpreted and suppose that row i is associated with node w and pair of selections of constants. If $[[\mathrm{SUB}(w, \mathrm{selection})\mathrm{BOHM}(P)(w)]]$ and $[[\mathrm{SUB}(v, \mathrm{selection})\mathrm{BOHM}(P)(v)]]$ are distinct then there exist $f(1), \ldots, f(t)$ in P_ω such that $[[\mathrm{SUB}(w, \mathrm{selection})\mathrm{BOHM}(P)(w)]] f(1) \ \ldots \ f(t) \ = / =$ $[[\mathrm{SUB}(v, \mathrm{selection})\mathrm{BOHM}(P)(v)]] f(1) \ \ldots \ f(t)$ and we set

$$[[F(i,1)]] = f(1) , \ \ldots \ , \ [[F(i,t)]] = f(t).$$

Otherwise we interpret the constants of the ith row arbitrarily.

Now we define a notion $\rightarrowtail$ of reduction. For each closed type 0

$$FM(1) \ldots M(m)$$

subterm of a term implicit at some node in $\mathrm{BOHM}(P)$ with some selection we associate the reduction

$$FM(1) \ \ldots \ M(m) \rightarrowtail c(i)$$

for $c(i) = [[FM(1) \ \ldots \ M(m)]]$. These reductions together with the (d) delta reductions are not a notion of reduction as we have defined it because the left hand sides of some of the EVAL reductions are not $\rightarrowtail$ normal. However, since the reductions are all length decreasing, the Knuth-Bendix completion of this set of reductions is generated by a set of rules which satisfies
(2) each EVAL rule has a left hand side which is in long beta-eta-$\rightarrowtail$ normal form
(3) each left hand side is the left hand side of at most one $\rightarrowtail$ rule.
(1) the set of $\rightarrowtail$ rules is finite
Finally we must show that this set of rules satisfies (4). Clearly, it suffices to show that the original set satisfies (4). Suppose that we are given
$FM(1) \ \ldots \ M(m) \rightarrowtail c(i)$
$FN(1) \ \ldots \ N(m) \rightarrowtail c(j)$
with i different from j. Then $[[F]] [[M(1)]] \ldots [[M(m)]] = [[FM(1) \ \ldots \ M(m)]] =$ $/ = [[FN(1) \ \ldots \ N(m)]] = [[F]] [[N(1)]] \ \ldots \ [[N(m)]]$ so for some k $[[M(k)]]$ is different from $[[N(k)]]$. Let
$M(k) = \lambda x(1) \ \ldots \ x(t) \ X$
$N(k) = \lambda x(1) \ \ldots \ x(t) \ Y$
$X, Y : 0$
Now $M(k)$ and $N(k)$ are instances of subterms of P say W and V respectively which appear at nodes w and v respectively in $\mathrm{BOHM}(P)$.W.L.O.G. assume

that v is above w in the co-K-B ordering. Now there are selections selection(1) and selection(2) such that

$\text{SUB}(w, \text{selection}(1))\text{BOHM}(w) = \text{SUB}(w, \text{selection}(1))W = M(k)$
$\text{SUB}(v, \text{selection}(2))\text{BOHM}(v) = \text{SUB}(v, \text{selection}(2))V = N(k).$

Since these are distinct, the row

$$F(1) \ \ldots \ F(t)$$

of $\text{MATRIX}(v)$ associated with w and these selections is interpreted by members of $P_\omega f(1), \ \ldots, f(t)$ such that $[[M(k)]] f(1) \ \ldots \ f(t) = c(p)$ is distinct from $[[N(k)]] f(1) \ \ldots \ f(t) = c(q)$. It follows easily that
$[F(1)/x(1), \ \ldots, F(t)/x(t)] X \rightarrowtail c(p)$
and
$[F(1)/x(1), \ \ldots, F(t)/x(t)] Y \rightarrowtail c(q)$
This completes the proof.

Theorem 3: If M and N are closed terms : $A = A(1) \to (\ \ldots \ (A(a) \to 0) \ \ldots)$ and $\sim \text{Delta}(0) \mapsto M = N$ then there exists closed terms
$T(1) : A(1) , \ \ldots , \ T(a) : A(a),$
$c(i) : 0, \ c(j) : 0$ with $\sim i = j$
such that
$MT(1) \ \ldots \ T(a) \twoheadrightarrow c(i)$
$NT(1) \ \ldots \ T(a) \twoheadrightarrow c(j).$

Corollaries:

(1) The problem of determining whether $\text{Delta}(0) \vdash M = N$ is decidable.
(2) The set of hereditarily finite full type structures is complete for $\text{Delta}(0)$. Kreisel's hereditarily continuous functionals is complete for $\text{Delta}(0)$. Indeed, any single infinite model is complete.
(3) The problem of determining whether
    $\text{Delta}(0), \ M(1) = N(1), \ \ldots , \ M(m) = N(m) \vdash M = N$ is decidable. Indeed, consistency is decidable and $M = N$ is consistent $\Leftrightarrow M = N$ is true in the two element model.

We now consider the $\text{Delta}(0)$ unification problem. Given closed terms M, N : $A \to B$ we wish to decide whether there exists a closed term P such that $\text{Delta}(0) \mapsto MP = NP$. Such a P is called a unifier or a solution as in [5]. A special case of this is the "matching problem" when $N = KQ$ or equivalently to decide whether there exists P such that $\text{Delta}(0) \mapsto MP = Q$. In the presence of d of lowest type the general unification problem is reducible to matching as follows. *
If $B = B(1) \to (\ \ldots \ (B(b) \to 0) \ \ldots)$ then M and N are unifiable

$$\Leftrightarrow$$

$\lambda x \ \lambda x(1) \ \dots \ x(b). \ d(Mxx(1) \ \dots x(b) \ )(Nxx(1) \ \dots \ x(b))K \ K* $ and $\lambda(x) \ \dots \ x(b)$. K can be matched. The following is well known from unification theory.

Proposition 9: If the set of Church numerals is the set of solutions to a unification problem then unification and, therefore, matching is recursively unsolvable.

Lemma 2: The set of Church numerals is the set of unifiers of the terms

$$\lambda x. \ \lambda uv. \ u(xuv) \ \text{and} \lambda x. \ \lambda uv.xu(uv).$$

Corollary: Unification and matching in Delta(0) are undecidable.

References

1. Bloom and Tindell, "Varieties of the if-then-else", *SIAM Journal of computing*, **12**, (1983).
2. Church, "The Calculi of Lambda Conversion", Princeton University Press, 1941.
3. Friedman, "Equality between Functionals", *SLNM*, **453**, 1975.
4. Gödel, "Uber formal unentscheidbare Satze der Principia Mathematica und verwandter Systeme, I", *Monatsh. Math. Phys.*, **38**, 1931.
5. Huet, "Unification in typed lambda calculus", *SLNCS*, **37**, 1975.
6. Kreisel, "Interpretation of analysis by means of constructive functionals of finite type in Constructivity in Mathematics", A. Heyting, ed., North Holland, 1959.
7. Prawitz, "Natural Deduction", Almquist and Wiksell, 1965.

Querying Inconsistent Databases[*]

Sergio Greco[1] and Ester Zumpano[1]

DEIS, Univ. della Calabria, 87030 Rende, Italy
email: {greco,zumpano}@si.deis.unical.it

Abstract. In this paper we consider the problem of answering queries consistently in the presence of inconsistent data, i.e. data violating integrity constraints. We propose a technique based on the rewriting of integrity constraints into disjunctive rules with two different forms of negation (negation as failure and classical negation). The disjunctive program can be used i) to generate 'repairs' for the database and ii) to produce consistent answers, i.e. maximal set of atoms which do not violate the constraints. We show that our technique is sound, complete and more general than techniques previously proposed.

1 Introduction

Integrity constraints represent an important source of information about the real world. They are usually used to define constraints on data (functional dependencies, inclusion dependencies, etc.). An integrity constraint can be considered as a query which must always be true after a modification of the database. Integrity constraints have nowadays a wide applicability in several context such as semantic query optimization, cooperative query answering, database integration, view update and others. Since the satisfaction of integrity constraints cannot be, generally, guaranteed, in the evaluation of queries, we must compute answers which are consistent with the integrity constraints.

The presence of inconsistencies might arise, for instance, when the database is obtained from the integration of different information sources. The integration of knowledge from multiple sources is an important aspect in several areas such as data warehousing, database integration, automated reasoning systems, active reactive databases. The following example shows a typical case of inconsistency.

Example 1. Consider the following database schema consisting of the single binary relation *Teaches(Course, Professor)* where the attribute *Course* is a key for the relation. Assume there are two different instances for the relations *Teaches* as reported in the following figure.

---

[*] Work partially supported by a MURST grant under the project "Data-X" and by a EC grant under the project "Contact". The first author is also supported by ISI-CNR.

M. Parigot and A. Voronkov (Eds.): LPAR 2000, LNAI 1955, pp. 308–325, 2000.

Course	Professor
c_1	p_1
c_2	p_2

Course	Professor
c_1	p_1
c_2	p_3

The two instances satisfy the constraint that *Course* is a key but, from the union of the two databases, we derive a relation which does not satisfy the constraint since there are two distinct tuples with the same value for the attribute *Course*.

In the integration of two conflicting databases simple solutions could be based on the definition of preference criteria such as a partial order on the source information or majority criteria [19]. However, these solution are not satisfactory in the general case and more interesting solutions are those based on 1) the computation of 'repairs' for the database, 2) the computation of consistent answers [5].

The computation of repairs is based on the insertion and deletion of tuples so that the resulting database satisfies all constraints. The computation of consistent answers is based on the identification of tuples satisfying integrity constraints and the selection of tuples matching the goal. For instance, for the integrated database of Example 1, we have two alternative repairs consisting in the deletion of one of the tuples (c_2, p_2) and (c_2, p_3). The consistent answer to a query over the relation *Teaches* contains the unique tuple (c_1, p_1) so that we don't know which professor teaches course c_2. The following example presents another case of inconsistent database.

Example 2. [22,5] Consider a database D consisting of the following two relations

Supplier	Department	Item
c_1	d_1	i_1
c_2	d_2	i_2

Supply

Item	Type
i_1	t
i_2	t

Class

with the integrity constraint, defined by the following first order formula

$$(\forall X \,\forall Y \,\forall Z) \; [\; Supply(X, Y, Z) \wedge Class(Z, t) \supset X = c_1 \;]$$

stating that only supplier c_1 can supply items of type t. The database $D = \{Supply(c_1, d_1, i_1),\ Supply(c_2, d_2, i_2),\ Class(i_1, t),\ Class(i_2, t) \}$ is inconsistent because the integrity constraint is not satisfied (supplier c_2 also supplies an item of type t). From the integrity constraint we can derive two alternative repaired databases: $D_1 = \{Supply(c_1, d_1, i_1),\ Class(i_1, t),\ Class(i_2, t) \}$ and $D_2 = \{Supply(c_1, d_1, i_1),\ Supply(c_2, d_2, i_2),\ Class(i_1, t) \}$ derived by deleting either the atom $Supply(c_2, d_2, i_2)$ or the atom $Class(i_2, t)$.

Moreover, while we are not able to answer a query $Supply(c_2, X, Y)$, asking for the department and item supplied by c_2, we are able to answer a query $Supply(c_1, X, Y)$ asking for the department and item supplied by c_1. Further, a query $Class(Z, t)$, asking for the items of type "t" can be also answered (with

$Z = i_1$, i.e. i_1 is the only item of type t) whereas a query $not\,Class(i_2, t)$, asking if item i_2 is not of type t, cannot be answered (unknown fact).

Therefore, it is very important, in the presence of inconsistent data, to compute the set of consistent answers, but also to know which facts are unknown and if there are possible repairs for the database. In our approach it can be possible to compute the tuples which are consistent with the integrity constraint and answer queries by considering as true facts those contained in every repaired database, false facts those not contained in all repaired databases and unknown the remaining facts.

We point out that, recently, there have been several proposals considering the integration of databases as well as the computation of queries over inconsistent databases [1,2,5,6,17,18,19,20]. All these techniques work for restricted cases and the most general technique so far introduced is complete only for universal quantified binary constraints [5]. Techniques for the integration of knowledge bases, expressed by means of first order formulas, have been proposed as well [3,4,23,14].

The main contribution of the paper is the introduction of a technique which maximizes the correct answers derivable from an inconsistent database. Our technique is based on the rewriting of integrity constraints into disjunctive rules with two different forms of negation (negation as failure and classical negation). The disjunctive program can be used i) to generate 'repairs' for the database and ii) to produce consistent answers, i.e. maximal set of atoms which do not violate the constraints. We show that our technique is sound, complete and more general than techniques previously proposed.

2 Disjunctive Deductive Databases

A *(disjunctive Datalog) rule* r is a clause of the form

$$A_1 \vee \cdots \vee A_k \leftarrow B_1, \cdots, B_m, not\,C_1, \cdots, not\,C_n, \qquad k + m + n > 0.$$

$A_1, \cdots, A_k, B_1, \cdots, B_m, C_1, \cdots, C_n$ are atoms of the form $p(t_1, ..., t_h)$, where p is a *predicate* of arity h and the terms $t_1, ..., t_h$ are constants or variables. The disjunction $A_1 \vee \cdots \vee A_k$ is the *head* of r, while the conjunction $B_1, \cdots, B_m, not\,C_1, \cdots, not\,C_n$ is the *body* of r. We also assume the existence of the binary built-in predicate symbols (comparison operators) which can be used only in the body of rules. A *(disjunctive Datalog) program* is a finite set of rules. A *not*-free (resp. $\vee$-free) program is called *positive* (resp. *normal*).

As usual, a literal is an atom A or a negated atom $not\,A$; in the former case, it is positive, and in the latter negative. Two literals are *complementary*, if they are of the form A and $not\,A$, for some atom A. For a set S of literals, $not\,S = \{not\,L \mid L \in S\}$.

The *Herbrand Universe* $U_\mathcal{P}$ of a program $\mathcal{P}$ is the set of all constants appearing in $\mathcal{P}$, and its *Herbrand Base* $B_\mathcal{P}$ is the set of all ground atoms constructed from the predicates appearing in $\mathcal{P}$ and the constants from $U_{\mathcal{P}}$. A term, (resp. an atom, a literal, a rule or a program) is *ground* if no variables occur in it. A rule r' is a *ground instance* of a rule r, if r' is obtained from r by replacing every variable in r with some constant in $U_{\mathcal{P}}$. We denote by $ground(\mathcal{P})$ the set of all ground instances of the rules in $\mathcal{P}$.

An interpretation of $\mathcal{P}$ is any subset of $B_{\mathcal{P}}$. The value of a ground atom L w.r.t. an interpretation I, $value_I(L)$, is *true* if $L \in I$ and *false* otherwise. The value of a ground negated literal $not\ L$ is $not\ value_I(L)$. The truth value of a conjunction of ground literals $C = L_1, \ldots, L_n$ is the minimum over the values of the L_i, i.e., $value_I(C) = min(\{value_I(L_i) \mid 1 \leq i \leq n\})$, while the value $value_I(D)$ of a disjunction $D = L_1 \vee \ldots \vee L_n$ is their maximum, i.e., $value_I(D) = max(\{value_I(L_i) \mid 1 \leq i \leq n\})$; if $n = 0$, then $value_I(C) = true$ and $value_I(D) = false$.

A ground rule r is *satisfied* by I if $value_I(Head(r)) \geq value_I(Body(r))$. Thus, a rule r with empty body is satisfied by I if $value_I(Head(r)) = true$ whereas a rule r' with empty head is satisfied by I if $value_I(Body(r)) = false$. In the following we also assume the existence of rules with empty head which defines denials[1], i.e. rules which are satisfied only if the body is false. An interpretation M for $\mathcal{P}$ is a model of $\mathcal{P}$ if M satisfies each rule in $ground(\mathcal{P})$.

The (model-theoretic) semantics for positive $\mathcal{P}$ assigns to $\mathcal{P}$ the set of its *minimal models* $MM(\mathcal{P})$, where a model M for $\mathcal{P}$ is minimal, if no proper subset of M is a model for $\mathcal{P}$ [21]. Accordingly, the program $\mathcal{P} = \{a \vee b \leftarrow\}$ has the two minimal models $\{a\}$ and $\{b\}$, i.e. $MM(\mathcal{P}) = \{\ \{a\},\ \{b\}\ \}$. The more general *disjunctive stable model semantics* also applies to programs with (unstratified) negation [10]. Disjunctive stable model semantics generalizes stable model semantics, previously defined for normal programs [9].

For any interpretation I, denote with $\mathcal{P}^I$ the ground positive program derived from $ground(\mathcal{P})$ 1) by removing all rules that contain a negative literal $not\ a$ in the body and $a \in I$, and 2) by removing all negative literals from the remaining rules. An interpretation M is a (disjunctive) stable model of $\mathcal{P}$ if and only if $M \in MM(\mathcal{P}^M)$.

For general $\mathcal{P}$, the stable model semantics assigns to $\mathcal{P}$ the set $SM(\mathcal{P})$ of its *stable models*. It is well known that stable models are minimal models (i.e. $SM(\mathcal{P}) \subseteq MM(\mathcal{P})$) and that for negation free programs minimal and stable model semantics coincide (i.e. $SM(\mathcal{P}) = MM(\mathcal{P})$). Observe that stable models are minimal models which are "supported", i.e. their atoms can be derived from the program. For instance, the program consisting of the rule $a \vee b \leftarrow not\ c$ has three minimal models $M_1 = \{a\}$, $M_1 = \{b\}$ and $M_3 = \{c\}$. However, only M_1 and M_2 are stable.

---

[1] Under total semantics

2.1 Classical Negation

Traditional declarative semantics of Datalog and logic programming uses the closed world assumption and each ground atom which does not follow from the database is assumed to be false. *Extended Datalog* programs extend standard Datalog programs with a different form of negation, known as *classical* or *strong negation*, which can also appear in the head of rules. Thus, while standard programs provide negative information implicitly, extended programs provide negative information explicitly and we can distinguish queries which fail in the sense that they do not succeed and queries which fail in the stronger sense that negation succeeds [10,16,11].

An extended atom is either an atom, say A or its negation $\neg A$. An extended Datalog program is a set of rules of the form

$$A_0 \lor \ldots \lor A_k \leftarrow B_1, \ldots, B_m, not\ B_{m+1}, \ldots, not\ B_n \qquad k + n > 0$$

where $A_0, \ldots, A_k, B_1, \ldots, B_n$ are extended atoms. A (2-valued) interpretation I for an extended program $\mathcal{P}$ is a pair $\langle T, F \rangle$ where T and F define a partition of $B_{\mathcal{P}} \cup \neg B_{\mathcal{P}}$ and $\neg B_{\mathcal{P}} = \{\neg A | A \in B_{\mathcal{P}}\}$. The truth value of an extended atom $L \in B_{\mathcal{P}} \cup \neg B_{\mathcal{P}}$ w.r.t. an interpretation I is equal to (i) *true* if $L \in T$ and, (ii) *false* if $A \in F$. Moreover, we say that an interpretation $I = \langle T, F \rangle$ is *consistent* if there is no atom A such that $A \in T$ and $\neg A \in T$. The semantics of an extended program $\mathcal{P}$ is defined by considering each negated predicate symbol, say $\neg p$, as a new symbol syntactically different from p and by adding to the program, for each predicate symbol p with arity n the constraint $\leftarrow p(X_1, \ldots, X_n), \neg p(X_1, \ldots, X_n)$.

The existence of a (2-valued) model for an extended program is not guaranteed, also in the case of negation (as-failure) free programs. For instance, the program consisting of the two facts a and ¬a does not admit any (2-valued) model.

In the following, for the sake of simplicity, we shall also use rules whose bodies may contain disjunctions. Such rules, called generalized disjunctive rules, are used as shorthands for multiple standard disjunctive rules. More specifically, a generalized disjunctive rule of the form

$$A_1 \lor \ldots \lor A_k \leftarrow (B_{1,1} \lor \ldots \lor B_{1,m_1}), \ldots, (B_{n,1} \lor \ldots \lor B_{n,m_n})$$

denotes the set of standard rules

$$A_1 \lor \ldots \lor A_k \leftarrow B_{1,i_1}, \ldots, B_{n,i_n} \qquad \forall j, i : 1 \leq j \leq n \ and \ 1 \leq i_j \leq m_j$$

Given a generalized disjunctive program $\mathcal{P}$, $st(\mathcal{P})$ denotes the standard disjunctive programs derived from $\mathcal{P}$ by rewriting body disjunctions.

2.2 Disjunctive Queries

Predicate symbols are partitioned into two distinct sets: *base predicates* (also called EDB predicates) and *derived predicates* (also called IDB predicates). Base predicates correspond to database relations defined over a given domain and they do not appear in the head of any rule whereas derived predicates are defined by means of rules.

Given a database D, a predicate symbol r and a program $\mathcal{P}$, $D(r)$ denotes the set of r-tuples in D whereas $\mathcal{P}_D$ denotes the program derived from the union of $\mathcal{P}$ with the tuples in D, i.e., $\mathcal{P}_D = \mathcal{P} \cup \{r(t) \leftarrow \mid t \in D(r)\}$. In the following a tuple t of a relation r will be also denoted as a fact $r(t)$.

The semantics of $\mathcal{P}_D$ is given by the set of its stable models by considering either their union (*possible semantics* or *brave reasoning*) or their intersection (*certain semantics* or *cautious reasoning*).

A *query* Q is a pair $(g, \mathcal{P})$ where g is a predicate symbol, called the *query goal*, and $\mathcal{P}$ is a program. The answer of a query $Q = (g, \mathcal{P})$ over a database D, under the possible (resp. certain) semantics is given by $D'(g)$ where $D' = \bigcup_{M \in SM(\mathcal{P}_D)} M$ (resp. $D' = \bigcap_{M \in SM(\mathcal{P}_D)} M$).

3 Databases with Constraints

Databases contain, other than data, intentional knowledge expressed by means of integrity constraints. Database schemata contain the knowledge on the structure of data, i.e. they give constraints on the form the data must have. The relationships among data are usually defined by constraints such as functional dependencies, inclusion dependencies, etc. Integrity constraints and relation schemata are introduced to prevent the insertion or deletion of data which could produce incorrect states. Generally, databases contain explicit representation of intentional knowledge.

A database D has associated a schema $\mathcal{DS} = (Rs, \mathcal{IC})$ which defines the intentional properties of D. In particular, Rs denotes the structure of the relations whereas $\mathcal{IC}$ contains the set of integrity constraints. Integrity constraints are used to define properties which are supposed to be satisfied by all instances of a database schema. Early works have considered the case of general integrity constraints expressed by arbitrary sentences from first-order logic. However, feasibility considerations have led to the study of more restricted classes of constraints, usually called *dependencies* such as functional dependencies, inclusion dependencies, join dependencies and others.

Definition 1. *An integrity constraint (or embedded dependency) is a formula of the first order predicate calculus of the form:*

$$(\forall x_1 ... \forall x_n) [\ \Phi(x_1, ..., x_n) \supset (\exists z_1 ... \exists z_k) \Psi(y_1, ..., y_m) \ ]$$

where $\Phi(x_1, ..., x_n)$ and $\Psi(y_1, ..., y_m)$ are two conjunctions of literals such that $x_1, ..., x_n$ and $y_1, ..., y_m$ are the distinct variables appearing in Φ and Ψ respectively, $\{z_1, ..., z_k\} = \{y_1, ..., y_m\} - \{x_1, ..., x_n\}$ is the set of variables existentially quantified.

In the definition above, conjunction Φ is called the *body* and conjunction Ψ the *head* of the integrity constraint. Without loss of generality, it is possible to assume that the equality symbol only occurs in the head Ψ and only between variables that also appear in the body Φ.

There are six common restrictions on embedded dependencies that give us six classes of dependencies:

1. The *full* (or *universal*) are those not containing existential quantified variables.
2. The *unirelational* are those with one relation symbol only; dependencies with more than one relation symbols are called *multirelational*.
3. The *single-head* are those with a single atom in the head; dependencies with more than one atom in the head are called *multi-head*.
4. The *tuple-generating* are those without the equality symbol.
5. The *equality-generating* are full, single-head, with an equality atom in the head.
6. The *typed* are those whose variables are assigned to fixed positions of base atoms and every equality atom involves a pair of variables assigned to the same position of the same base atom; dependencies which are not typed will be called *untyped*.

Moreover, an embedded dependency is said to be *positive* if no negated literal occur in it[2]. Most of the dependencies developed in database theory are restricted cases of some of the above classes. For instance, functional dependencies are positive, full, single-head, unirelational, equality-generating constraints.

Without loss of generality, it is possible to assume that the equality symbol only occurs in the head Ψ and only between variables that also appear in the body Φ. In the rest of this section we concentrate on universal, single-head dependencies. Therefore, an integrity constraint is a formula of the form

$$\forall X \; [\; B_1 \wedge ... \wedge B_k \wedge not \; B_{k+1} \wedge ... \wedge not \; B_n \wedge \phi \supset B_0 \;]$$

where $B_1, ... B_n$ are base literals, B_0 can be either a base literal or a built-in literal, X denotes the list of all variables appearing in $B_0, ..., B_n$ and ψ is a conjunction of built-in literals. Observe that a multi-head constraint of the form

$$\forall X \; [\; B_1 \wedge ... \wedge B_k \wedge not \; B_{k+1} \wedge ... \wedge not \; B_n \wedge \phi \supset A_1 \wedge ... \wedge A_m \;]$$

with $m > 1$, can be rewritten into m single head constraints

$$\forall X \; [\; B_1 \wedge ... \wedge B_k \wedge not \; B_{k+1} \wedge ... \wedge not \; B_n \wedge \phi \supset A_i \;] \qquad 1 \leq i \leq m$$

Definition 2. *Given a database schema $\mathcal{DS} = \langle Rs, \mathcal{IC} \rangle$ and a database instance D over $\mathcal{DS}$, we say that D is* consistent *if $D \models \mathcal{IC}$, i.e. if all integrity constraints in $\mathcal{IC}$ are satisfied by D, otherwise it is* inconsistent.

Example 3. The database of Example 1, derived from the union of the two source databases, is inconsistent since there is an instance of the constraint which is not satisfied, namely $Teaches(c_2, p_2) \wedge Teaches(c_2, p_3) \supset p_2 = p_3$.

---

[2] Classical definitions of embedded dependencies only consider positive constraints.

Definition 3. *Given a database schema* $\mathcal{DS} = \langle Rs, \mathcal{IC} \rangle$ *and a database D over* $\mathcal{DS}$. *A* repair *for D is a pair of sets of atoms* (R^+, R^-) *such that 1)* $R^+ \cap R^- = \emptyset$, *2)* $D \cup R^+ - R^- \models \mathcal{IC}$ *and 3) there is no pair* $(S^+, S^-) \neq (R^+, R^-)$ *such that* $S^+ \subseteq R^+$, $S^- \subseteq R^-$ *and* $D \cup S^+ - S^- \models \mathcal{IC}$. *The database* $D \cup R^+ - R^-$ *will be called the* repaired database.

Thus, repaired databases are consistent databases which are derived from the source database by means of a minimal[3] sets of insertion and deletion of tuples.

Example 4. Assume we are given a database whose schema contains two unary relations p and q with the *inclusion dependency* $\quad \forall(X) \ [\ p(X) \supset q(X) \ ]$ and the database instance consisting of the following set $D = \{p(a), p(b), q(a), q(c)\}$. D is inconsistent since $p(b) \supset q(b)$ is not satisfied. The repairs for D are $R_1 = (\{q(b)\}, \emptyset)$ and $R_2 = (\emptyset, \{p(b)\})$ which produce, respectively, the repaired databases $D_1 = \{p(a), p(b), q(a), q(c), q(b)\}$ and $D_2 = \{p(a), q(a), q(c)\}$.

A (relational) query over a database defines a function from the database to a relation. It can be expressed by means of alternative equivalent languages such as relational algebra, 'safe' relational calculus or 'safe' non recursive Datalog [24]. In the following we shall use Datalog. Thus, a query is a pair $(g, \mathcal{P})$ where $\mathcal{P}$ is a safe non-recursive Datalog program and g is a predicate symbol specifying the output (derived) relation.

Observe that relational queries define a restricted case of disjunctive queries. The reason to consider relational and disjunctive queries is that, as we shall show in the next section, relational queries over databases with constraints can be rewritten into extended disjunctive queries over databases without constraints.

Definition 4. *Given a database schema* $\mathcal{DS} = \langle Rs, \mathcal{IC} \rangle$ *and a database D over* $\mathcal{DS}$. *An atom A is true (resp. false) with respect to* $(D, \mathcal{IC})$ *if A belongs to all repaired databases (resp. there is no repaired database containing A). The set of atoms which are neither true nor false are* undefined.

Thus, true atoms appear in all repaired databases whereas undefined atoms appear in a proper subset of repaired databases. Given a database D and a set of integrity constraints $\mathcal{IC}$, the application of $\mathcal{IC}$ to D, denoted by $\mathcal{IC}(D)$, defines three distinct sets of atoms: the set of true atoms $\mathcal{IC}(D)^+$, the set of undefined atoms $\mathcal{IC}(D)^u$ and the set of false atoms $\mathcal{IC}(D)^-$.

Definition 5. *Given a database schema* $\mathcal{DS} = \langle Rs, \mathcal{IC} \rangle$, *a database D over* $\mathcal{DS}$ *and a query* $Q = (g, \mathcal{P})$. *The* consistent answer *of the query Q on the database D, denoted as* $Q(D, \mathcal{IC})$, *is the set of g-tuples contained in all repaired databases.*

Fact 1 *Given a database schema* $\mathcal{DS} = \langle Rs, \mathcal{IC} \rangle$, *a database D over* $\mathcal{DS}$ *and a positive relational query Q, then* $Q(D, \mathcal{IC}) = Q(\mathcal{IC}(D)^+)$.

---

[3] Minimal w.r.t. set inclusion.

Example 5. Consider again the integrated database D of Example 4 and the query $Q = (p, \emptyset)$ The answer of the query $Q(D, \mathcal{IC})$ contains the facts $p(a)$. The answer of the query $(q, \emptyset)(D, \mathcal{IC})$ gives the facts $q(a)$ and $q(c)$.

4   Querying and Repairing Inconsistent Databases

In this section we present a technique which permits us to compute consistent answers and repairs for possible inconsistent databases. The technique is based on the generation of an extended disjunctive program $\mathcal{LP}$ derived from the set of integrity constraints. The repairs for the database can be generated from the stable models of $\mathcal{LP}$ whereas the computation of the consistent answers of a query $(g, \mathcal{P})$ can be derived by considering the stable models of the program $\mathcal{P} \cup \mathcal{LP}$ over the database D.

Definition 6. *Let c be a universally quantified constraint of the form*

$$\forall X (B_1 \wedge ... \wedge B_k \wedge not\ B_{k+1} \wedge ... \wedge not\ B_n \wedge \phi \supset B_0)$$

then, $dj(c)$ denotes the extended disjunctive rule

$$\neg B'_1 \vee ... \vee \neg B'_k \vee B'_{k+1} \vee ... \vee B'_0 \leftarrow (B_1 \vee B'_1), ..., (B_k \vee B'_k), \phi,$$
$$(not\ B_{k+1} \vee \neg B_{k+1}), ..., (not\ B_{n+1} \vee \neg B_0)$$

where B'_i denotes the atom derived from B_i, by replacing of the predicate symbol p with the new symbol p_d if B_i is a base atom otherwise is equal to false.

Let $\mathcal{IC}$ be a set of universally quantified integrity constraints, then $\mathcal{DP}(\mathcal{IC})$ $= \{\ dj(c)\ |\ c \in \mathcal{IC}\ \}$ and $\mathcal{LP}(\mathcal{IC}) = st(\mathcal{DP}(\mathcal{IC}))$.

Thus, $\mathcal{DP}(\mathcal{IC})$ denotes the set of generalized disjunctive rules derived from the rewriting of $\mathcal{IC}$ whereas $\mathcal{LP}(\mathcal{IC})$ denotes the set of standard disjunctive rules derived from $\mathcal{DP}(\mathcal{IC})$. Clearly, given a database D and a set of constraints $\mathcal{IC}$, $\mathcal{LP}(\mathcal{IC})_D$ denotes the program derived from the union of the rules in $\mathcal{LP}(\mathcal{IC})$ with the facts in D whereas $SM(\mathcal{LP}(\mathcal{IC})_D)$ denotes the set of stable models of $\mathcal{LP}(\mathcal{IC})_D$. Observe that every stable model is consistent, according to the definition of consistent set given in Section 2, since it cannot contain two atoms of the form A and $\neg A$.

Example 6. Consider the following integrity constraints:

1. $\forall X\ [\ p(X) \wedge not\ s(X) \supset q(X)\ ]$
2. $\forall X\ [\ q(X) \supset r(X)\ ]$

and the database D containing the facts $p(a), p(b), s(a)$ and $q(a)$. The derived generalized extended disjunctive program is defined as follows:

$$\neg p_d(X) \vee s_d(X) \vee q_d(X) \leftarrow (p(X) \vee p_d(X)),\ (not\ s(X) \vee \neg s_d(X)),$$
$$(not\ q(X) \vee \neg q_d(X)).$$
$$\neg q_d(X) \vee r_d(X) \qquad \leftarrow (q(X) \vee q_d(X)),\ (not\ r(X) \vee \neg r_d(X)).$$

The above rules can be now rewritten in standard form by eliminating body disjunctions. Let P be the corresponding extended disjunctive Datalog program, the computation of the program $\mathcal{P}_D$, derived from the union of $\mathcal{P}$ with the facts in D, gives the following stable models:

$$M_1 = D \cup \{\neg p_d(b), \neg q_d(a)\}, \qquad M_4 = D \cup \{r_d(a), s_d(b)\},$$
$$M_2 = D \cup \{\neg p_d(b), r_d(a)\}, \qquad M_5 = D \cup \{q_d(b), \neg q_d(a), r_d(b)\},$$
$$M_3 = D \cup \{\neg q_d(a), s_d(b)\}, \qquad M_6 = D \cup \{q_d(b), r_d(a), r_d(b)\}.$$

Observe that a (generalized) extended disjunctive Datalog program can be simplified by eliminating from the body rules all literals whose predicate symbols are derived and do not appear in the head of any rule (these literals cannot be true). For instance, the generalized rules of the above example can be rewritten as

$$\neg p_d(X) \vee s_d(X) \vee q_d(X) \leftarrow p(X),\ not\ s(X),\ (not\ q(X) \vee \neg q_d(X)).$$
$$\neg q_d(X) \vee r_d(X) \qquad\quad \leftarrow (q(X) \vee q_d(X)),\ not\ r(X).$$

because the predicate symbols p_d, $\neg r_d$ and $\neg s_d$ do not appear in the head of any rule. As mentioned in the Introduction, in the presence of inconsistencies, generally, there are two possible alternative solutions: i) compute repairs making the database consistent through the insertion and deletion of tuples, or ii) compute consistent answers but leaving the database inconsistent. The rewriting of constraints into disjunctive rules is useful for both solutions.

4.1   Computing Database Repairs

Every stable model can be used to define a possible repair for the database by interpreting new derived atoms (denoted by the subscript "d") as insertions and deletions of tuples. Thus, if a stable model M contains two atoms $\neg p_d(t)$ (derived atom) and $p(t)$ (base atom) we deduce that the atom $p(t)$ violates some constraint and, therefore, it must be deleted. Analogously, if M contains the derived atoms $p_d(t)$ and do not contain $p(t)$ (i.e. $p(t)$ is not in the database) we deduce that the atom $p(t)$ should be inserted in the database. We now formalize the definition of repaired database.

Definition 7. *Given a database schema* $\mathcal{DS} = \langle Rs, \mathcal{IC} \rangle$ *and a database D over* $\mathcal{DS}$. *Let M be a stable model of* $\mathcal{LP}(\mathcal{IC})_D$. *Then,* $R(M) = (\ \{p(t)\ |\ p_d(t) \in M\ \wedge\ p(t) \notin D\},\ \{p(t)\ |\ \neg p_d(t) \in M\ \wedge\ p(t) \in D\}\ )$.

Theorem 2. *Given a database schema* $\mathcal{DS} = \langle Rs, \mathcal{IC} \rangle$ *and a database D over* $\mathcal{DS}$. *Then*

1. *(Soundness) for every stable model M of* $\mathcal{LP}(\mathcal{IC})_D$, $R(M)$ *is a repair for D;*
2. *(Completeness) for every database repair S for D there exists a stable model M for* $\mathcal{LP}(\mathcal{IC})_D$ *such that* $S = R(M)$.

Proof (Sketch) Soundness derives from the fact that every stable model M is consistent and is minimal. Completeness derives from the fact that stable models are the only minimal models which are 'supported'.

Example 7. Consider the database of Example 4. Algorithm 1, applied to the integrity constraint $\forall X\ [\ p(X) \supset q(X)\ ]$, produces the disjunctive rules

$$r : \neg p_d(X) \vee q_d(X) \leftarrow (p(X) \vee p_d(X)), (not\ q(X) \vee \neg q(X)).$$

which can be rewritten into the simpler form

$$r' : \neg p_d(X) \vee q_d(X) \leftarrow p(X), not\ q(X).$$

The program $\mathcal{P}_D$, where $\mathcal{P}$ is the program consisting of the disjunctive rule r', has two stable models $M_1 = D \cup \{\ \neg p_d(b)\}$ and $M_2 = D \cup \{\ q_d(b)\}$. The derived repairs are $R(M_1) = (\{q(b)\}, \emptyset)$ and $R(M_2) = (\emptyset, \{p(b)\})$ corresponding, respectively, to the insertion of $q(b)$ and the deletion of $p(b)$.

4.2   Computing Consistent Answers

We consider now the problem of computing a consistent answer without modifying the (possibly inconsistent) database. We assume that tuples contained in the database or implied by the constraints may be either *true* or *false* or *undefined*.

From the results of Section 4.1 we derive

1. $\mathcal{IC}(D)^+ = \{\ p(t) \in D \mid \nexists M \in \mathrm{SM}(\mathcal{LP}(\mathcal{IC})_D)\ s.t.\ \neg p_d(t) \in M\ \} \cup$
   $\{\ p(t) \notin D \mid \forall M \in \mathrm{SM}(\mathcal{LP}(\mathcal{IC})_D)\ s.t.\ p_d(t) \in M\ \}$,
2. $\mathcal{IC}(D)^- = \{\ p(t) \notin D \mid \nexists M \in \mathrm{SM}(\mathcal{LP}(\mathcal{IC})_D)\ s.t.\ p_d(t) \in M\ \} \cup$
   $\{\ p(t) \in D \mid \forall M \in \mathrm{SM}(\mathcal{LP}(\mathcal{IC})_D)\ s.t.\ \neg p_d(t) \in M\ \}$,
3. $\mathcal{IC}(D)^u = \{\ p(t) \mid \exists M_1, M_2 \in \mathrm{SM}(\mathcal{LP}(\mathcal{IC})_D)\ s.t.\ p_d(t) \in M_1\ and$
   $\neg p_d(t) \in M_2\ \}$.

Observe that the sets $\mathcal{IC}(D)^+$, $\mathcal{IC}(D)^-$ and $\mathcal{IC}(D)^u$ are disjoint and that $\mathcal{IC}(D)^+ \cup \mathcal{IC}(D)^-$ defines a set of consistent atoms. Thus, the set of undefined atoms $\mathcal{IC}(D)^u$ contains the tuples which cannot be assumed neither true nor false:

$$\mathcal{IC}(D)^u = \{\ p(t) \mid p(t) \in D \vee \exists M \in \mathrm{SM}(\mathcal{LP}(\mathcal{IC})_D)$$
$$s.t.\ p_d(t) \in M\ or\ \neg p_d(t) \in M\ \} - (\ \mathcal{IC}(D)^+ \cup \mathcal{IC}(D)^-\ ).$$

We are now in the position to introduce the definition of consistent answer.

The consistent answer for the query $Q = (g, \mathcal{P})$ over the database D under constraints $\mathcal{IC}$ is as follows:

$$Q(D, \mathcal{IC})^+ = \{\ g(t) \in D \mid \nexists M \in \mathrm{SM}((\mathcal{P} \cup \mathcal{LP}(\mathcal{IC}))_D)\ s.t.\ \neg g_d(t) \in M\ \} \cup$$
$$\{\ g(t) \notin D \mid \forall M \in \mathrm{SM}((\mathcal{P} \cup \mathcal{LP}(\mathcal{IC}))_D)\ s.t.\ g_d(t) \in M\ \},$$
$$Q(D, \mathcal{IC})^u = \{\ g(t) \mid \exists M_1, M_2 \in \mathrm{SM}((\mathcal{P} \cup \mathcal{LP}(\mathcal{IC}))_D)$$
$$s.t.\ g_d(t) \in M_1\ and\ \neg g_d(t) \in M_2\ \}$$

whereas the set of atoms which are neither true nor undefined can be assumed to be false.

For instance, in Example 7, the set of true tuples are those belonging to the intersection of the two models, that is $p(a), q(a)$ and $q(c)$, whereas the set of undefined tuples are those belonging to the union of the two models and not belonging to their intersection.

Example 8. Consider the database of Example 2. To answer a query it is necessary to define, first, the atoms which are true, undefined and false:

1. $\mathcal{IC}(D)^+ = \{Supply(c_1, d_1, i_1), Class(i_1, t)\}$, the set of true atoms,
2. $\mathcal{IC}(D)^u = \{Supply(c_2, d_2, i_2), Class(i_2, t)\}$, the set of undefined atoms,
3. The atoms not belonging to $\mathcal{IC}(D)^+$ and $\mathcal{IC}(D)^u$ are false.

The answer to the query $(Class, \emptyset)$ gives the tuple (i_1, t).

Observe that for every database D over a given schema $\mathcal{DS} = \langle Rs, \mathcal{IC} \rangle$, for every query $Q = (g, \mathcal{P})$ and for every repaired database D'

1. each atom in $A \in Q(D, \mathcal{IC})^+$ belongs to the stable model of $\mathcal{P}_{D'}$ (soundness)
2. each atom in $A \in Q(D, \mathcal{IC})^-$ does not belongs to the stable model of $\mathcal{P}_{D'}$ (completeness).

Example 9.   Consider   the   integrated   database $D = \{ Teaches(c_1, p_1),$ $Teaches(c_2, p_2),$ $Teaches(c_2, p_3) \}$ of Example 1. The functional dependency defined by the key of relation $Teaches$ can be defined as

$$\forall(X, Y) \, [\, Teaches(X, Y) \wedge Teaches(X, Z) \supset Y = Z \,]$$

The corresponding disjunctive program P consists of the rule

$$\neg Teaches_d(X, Y) \vee \neg Teaches_d(X, Z) \leftarrow Teaches(X, Y), Teaches(X, Z), X \neq Z$$

The program P_D has two stable models: $M_1 = D \cup \{\neg Teaches_d(c_2, p_2)\}$ and $M_2 = D \cup \{\neg Teaches_d(c_2, p_3)\}$. Therefore, the set of facts which we can assume to be true contains the single fact $Teaches(c_1, p_1)$.

4.3   Complexity

The technique proposed is general but expensive. In this section we define an upper bound for the general case and present results for special cases.

Theorem 3. *Let D be a database, $Q = (g, \mathcal{P})$ a query and $\mathcal{IC}$ be a set of (full) single-head integrity constraints. Then*

1. *A repair for D exists.*
2. *Checking if there is exists a repair for D such that the answer of Q is not empty is in Σ_2^P.*

3. Checking if a fact belongs to the consistent answer of Q is in Π_2^P.

Proof (Sketch) We have seen that the answer of a query $Q = (g, \mathcal{P})$ over a database D with integrity constraint $\mathcal{IC}$, is computed by considering the stable models of the disjunctive program $P' = (\mathcal{P} \cup \mathcal{LP}(\mathcal{IC}))_D$. The results derive from the fact that

1. P' is stratified w.r.t. negation by default (i.e. *not*) and therefore it has always stable models,
2. Checking if there is exists a stable model for P' containing a given tuple is in Σ_2^P-complete.
3. Checking if all stable models for P' contain a given tuple is in Σ_2^P-complete.

Moreover, for restricted cases of integrity constraints, answers can be computed very efficiently.

Theorem 4. *Let D be a database, $\mathcal{FD}$ a set of functional dependencies over D, $\mathcal{RC}$ a set of (full) referential dependencies over D and Q a query. Then,*

1. *$Q(D, \mathcal{FD})$ can be computed in polynomial time,*
2. *$Q(D, \mathcal{RC})$ can be computed in polynomial time.*

5   Generalizing Constraints and Answers

In the previous section we have considered universal, single-head constraints. Here we extend our framework by considering more general constraints and partially defined answers. First of all observe that constraints with disjunctive heads can be rewritten into disjunctive rules and that for universally quantified integrity constraints it is possible to move literals from the head to the body and vice versa. This is not true for disjunctive Datalog rules under stable model semantic (for instance, the rules $a \vee b \leftarrow$ and $a \leftarrow not\ b$ have the same minimal models but different stable models).

5.1   Existential Quantified Constraints

In the presence of existential quantified variables we modify Definition 1 as follows.

Definition 8. *Let r be a constraint rule and let $r' = dj(r)$ be the corresponding generalized disjunctive rule. For each predicate $q(X, Y)$ in r, where X is the list of universal quantified variable and Y is the list of existential quantified variables*

– *Add to the disjunctive program the rules*

$$q'(X) \leftarrow q(X, Y)$$
$$q'_d(X) \leftarrow q_d(X, Y)$$

where q' and q'_d are new predicate symbols storing the projection of q and q_d on the universal quantified variables (specified by X).

- *Replace $q_d(X,Y)$ appearing in the head of r' with $q_d(X,\perp)$, where $\perp$ denotes a list of unknown values.*
- *Replace every $q(X,Y)$ and $q_d(X,Y)$ appearing in the body of r' with $q'(X)$ and $q'_d(X)$, respectively.*

The computation of repairs and consistent answer can be done by considering the disjunctive program modified as above.

Example 10. Consider the referential constraint $\forall X\ [emp(X) \supset \exists Y\, ss\#(X,Y)]$ stating that every employee must have a social security number and the database $D = \{emp(a),\, emp(b),\, ss\#(a,1)\}$. From the rewriting of the integrity constraint, after the elimination of redundant literals, we get the rule

$$r_2 :\ \neg emp_d(X) \vee ss\#_d(X,\perp) \leftarrow emp(X),\, not\ ss\#1(X).$$

where $ss\#1$ is defined by the rule

$$r_1 :\ ss\#1(X) \leftarrow ss\#(X,Y)$$

The output program P consists of the rules r_1 and r_2. The program P_D has two stable models $M_1 = D \cup \{ss\#1(a), \neg emp_d(b)\}$ and $M_2 = D \cup \{ss\#1(a), ss\#_d(b,\perp)\}$.

From the two models we derive the two repairs $R(M_1) = (\emptyset, \{emp(b)\})$ and $R(M_2) = (\{ss\#(b,\perp)\}, \emptyset)$ which produce, respectively, the two repaired database $D_1 = \{emp(a), ss\#(a,1)\}$ and $D_2 = \{emp(a), emp(b), ss\#(a,1),\ ss\#(b,\perp)\}$.

The answer to a query $(emp, \emptyset)$ contains the only fact $emp(a)$ whereas the answer to the query $(ss\#, \emptyset)$ contains the only fact $ss\#(a,1)$. The atoms $emp(b)$ and $ss\#(b,\perp)$ are undefined.

5.2   Partially Defined Answers

In the framework introduced in the previous section we have assigned to every ground atom a truth value. Here we extend our framework and also consider partially defined atoms, i.e. atoms with 'unknown' attributes.

Given two ground atoms $A = p(t_1, ..., t_n)$ and $B = p(u_1, ..., u_n)$ we say that A *subsumes* B (written $A \preceq B$) if $\forall i$ either $t_i = u_i$ or $t_i = \perp$. Given two set of ground atoms S_1 and S_2, S_1 *subsumes* S_2 (written $S_1 \preceq S_2$) if $\forall B \in S_2\ \exists A \in S_1$ s.t. $A \preceq B$ and $\forall A \in S_1\ \exists B \in S_2$ s.t. $A \preceq B$.

Given a set of sets of ground atoms S and a set of ground atoms T we say that T *approximates* S (written $T \sqsubseteq S$) if for each $S_i \in S$ is $T \preceq S_i$. Moreover, we say that T is the *maximal approximation* of S if $T \sqsubseteq S$ and there is no set U such that $S \preceq U \sqsubseteq T$. For instance, the maximal approximation of $\{p(a,b)\}$ and $\{p(a,d), p(c,b)\}$ is $\{p(a,\perp), p(\perp,b)\}$.

Let D be a database, $\mathcal{IC}$ a set of constraints and S the set of stable models of $\mathcal{LP}(\mathcal{IC})_D$. Let N be the maximal approximation of S (clearly $\bigcap_{M \in S} M \subseteq N$). Then, the set of partially defined atoms which are true, false and undefined are:

1. $\mathcal{IC}(D)^+ = \{ p(t) \in D \mid \not\exists M \in \mathrm{SM}(\mathcal{LP}(\mathcal{IC})_D) \ s.t. \ \neg p_d(t) \in M \ \} \cup$
   $\{ p(t) \notin D \mid p_d(t) \in N \ \}$,
2. $\mathcal{IC}(D)^- = \{ p(t) \notin D \mid \not\exists M \in \mathrm{SM}(\mathcal{LP}(\mathcal{IC})_D) \ s.t. \ p_d(t) \in M \ \} \cup$
   $\{ p(t) \in D \mid \forall M \in \mathrm{SM}(\mathcal{LP}(\mathcal{IC})_D) \ s.t. \ \neg p_d(t) \in M \ \}$,
3. $\mathcal{IC}(D)^u = \{ p(t) \mid \exists M_1, M_2 \in \mathrm{SM}(\mathcal{LP}(\mathcal{IC})_D) \ s.t. \ p_d(t) \in M_1 \ and$
   $\neg p_d(t) \in M_2 \ \}$.

Definition 9. *Given a database D, a set of integrity constraints $\mathcal{IC}$ and a query $Q = (g, \mathcal{P})$. The generalized consistent answer for the query $Q = (g, \mathcal{P})$ over the database D is*

$$Q(D, \mathcal{IC})^+ = \{ g(t) \in D \mid \not\exists M \in \mathrm{SM}(P \cup \mathcal{LP}(\mathcal{IC})_D) \ s.t \ \neg g_d(t) \in M \ \} \cup$$
$$\{ g(t) \notin D \mid g_d(t) \in N \ (the \ max. \ approx. \ of \ SM((P \cup \mathcal{LP}(\mathcal{IC}))_D)) \ \}.$$
$$Q(D, \mathcal{IC})^- = \{ g(t) \notin D \mid \not\exists M \in \mathrm{SM}(P \cup \mathcal{LP}(\mathcal{IC})_D) \ s.t \ \neg g_d(t) \in M \ \} \cup$$
$$\{ g(t) \in D \mid g_d(t) \in N \ (the \ max. \ approx. \ of \ SM((P \cup \mathcal{LP}(\mathcal{IC}))_D)) \ \}.$$

Example 11. Consider the database $D = \{p(c_1, p_1), p(c_2, p_2), p(c_2, p_3)\}$ of Example 1 with the integrity constraint $\forall X \ [\ p(X, Y), p(X, Z) \supset Y = Z \ ]$. The disjunctive program derived from the integrity constraint has two stable models $M_1 = D \cup \{\neg p(c_2, p_2)\}$ and $M_2 = D \cup \{\neg p(c_2, p_3)\}$. The best approximation of M_1 and M_2 is $D \cup \{\neg p(c_2, \bot)\}$. The set of positive atoms contains $p(c_1, p_1)$ and $p(c_2, \bot)$. Thus, we know that c_2 is a course but the professor teaching that course is unknown.

Observe that partially defined tuples generalizes knowledge of undefined atoms. For the database of Example 1, we have derived that the facts $p(c_2, p_2)$ and $p(c_2, p_3)$ are undefined. By using partially defined atoms, we add the knowledge that c_2 is a course.

6   Related Work

The problem of managing inconsistent databases has been deeply investigated in the last few years, mainly in the areas of databases and artificial intelligence.

Agarwal et al. proposed an extention of relational algebra, called *flexible algebra*, to deal with inconsistent data [1]. The flexible algebra extends relational algebra through the introduction of *flexible relations*, i.e. non 1NF relations that contain sets of non-key attributes. Their technique only considers constraints defining functional dependencies. However, flexible algebra is sound only for the class of databases having only dependencies determined by the primary key. An extention of flexible algebra for other keys functional dependencies, called *integrated relational calculus*, was proposed by Dung [6].

An alternative approach, taking the disjunction of the maximal consistent subsets of the union of the databases, has been proposed in [3]. For instance, assuming that there are two sources containing respectively $p(a, c)$ and $p(a, b)$ where the first argument is a key, the solution is to store $p(a, b) \vee p(a, c)$. A refinement of this technique has been presented in [20] were it was proposed to take into account the majority view of the knowledge bases. A different framework, based on annotated logic programming, was introduced in [23].

An interesting technique has been recently proposed in [5]. This technique is based on the computation of an equivalent query $T_w(Q)$ derived from the source query Q. The definition of $T_w(Q)$ is based on the notion of residue developed in the context of semantic query optimization. A classical example is showed below.

Example 12. [5] For the constraint of Example 2 with the query goal $Class(Z, t)$, the technique proposed in [5] generates, first, the association:

$$Class(Z, W) \longrightarrow Class(Z, W)\{\forall(X, Y)(\neg Supply(X, Y, Z) \lor W \neq t \lor X = c_1).$$

so that if a query $Q = Class(Z, t)$ is submitted, it is generated a new query $T_w(Q) = T_w(Class(Z, t))$ equal to:

$$\{Class(Z, t), Class(Z, t) \land \forall(X, Y)(\neg Supply(X, Y, Z) \lor X = c_1)\}.$$

The evaluation of $T_w(Q)$ produces the unique consistent answer $Z = i_1$.

This technique has been showed to be complete for universal binary integrity constraints and universal quantified queries. A binary integrity constraints is of the form: $\forall X(B_1 \lor B_2 \lor \phi)$ where B_1, B_2 are literals and ϕ is a conjunctive formula. The following example shows a case where the technique proposed in [5] is not complete.

Example 13. Consider the integrity constraint $\forall(X, Y, Z)\,[\,p(X, Y) \land p(X, Z) \supset Y = X\,]$ and the database $D = \{p(a, b), p(a, c)\}$ and the query $Q = \exists U p(a, U)$ (we are using here the formalism used in [5]). The technique proposed in [5] generate the new query $T_w(Q) = \exists U[p(a, U) \land \forall Z(\neg p(a, Z) \lor Z = U)]$. which $T_w(Q)$ is not satisfied, thus contradicting the expected answer which is true.

In our framework the query Q can be expressed as $(p1, P)$ where P consists of the rule

$$r_1: \quad p1(X) \leftarrow p(X, U), X = a.$$

whereas the constraint produces the following generalized disjunctive rule

$$\neg p_d(X, Y) \lor p_d(X, Z) \leftarrow (p(X, Y) \lor p_d(X, Y)), (p(X, Z) \lor p_d(X, Z)), Y \neq Z.$$

which can rewritten into the simpler form

$$r_2: \quad \neg p_d(X, Y) \lor p_d(X, Z) \leftarrow p(X, Y), p(X, Z), Y \neq Z.$$

The program P'_D, where $P = \{r_1, r_2\}$, has two stable models $M_1 = D \cup \{\neg p_d(a, c), p1(a)\}$ and $M_2 = D \cup \{\neg p_d(a, b), p1(a)\}$. The set of true facts is $\{p1(a)\}$ and, therefore, the answer to the query consists of the fact $p1(a)$.

Thus, our technique is more general of the technique proposed in [5]. We conclude by mentioning that a simple prototype has been implemented on the top of the system dlv [15].

An interesting technique has been recently proposed in [5]. This technique is based on the computation of an equivalent query $T_w(Q)$ derived from the source query Q. The definition of $T_w(Q)$ is based on the notion of residue developed in the context of semantic query optimization.

This technique has been showed to be complete for universal binary integrity constraints and universal quantified queries. A binary integrity constraints is of the form $\forall X(B_1 \vee B_2 \vee \phi)$, where B_1, B_2 are literals and ϕ is a conjunctive formula. Our technique for universally quantified dependencies is complete and, therefore, more general. These results can be easily extended to more general classes of constraints. We conclude by mentioning that a simple prototype has been implemented on the top of the system dlv [15].

7   Conclusions

The paper has proposed a technique for querying inconsistent databases. The technique is based on the use of disjunctive programs and stable model semantics. Our technique is sound and complete and more general than previous proposed techniques. We are currently investigating the identification of classes of constraints which permit an efficient computation (this is also one of most interesting topics currently investigated in disjunctive databases and nonmonotonic reasoning areas). Another interesting issue is the extension of the framework to allow users to specify preferences on the insertion or deletion of atoms and on the source information (in the case of inconsistencies derived from the integration of multiple information sources).

References

1. S.Argaval, A.M. Keller, G.Wiederhold, and K. Saraswat. Flexible Relation: an Approach for Integrating Data from Multiple, Possibly Inconsistent Databases. In IEEE Int. Conf. on Data Engineering,1995.   310, 322
2. Bry, F., Query Answering in Information System with Integrity Constraints, In IFIP WG 11.5 Working Conf. on Integrity and Control in Inform. System, 1997.   310
3. C. Baral, S. Kraus, J. Minker, Combining Multiple Knowledge Bases. IEEE-Trans. on Knowledge and Data Engineering, 3(2): 208-220 (1991)   310, 322
4. C. Baral, S. Kraus, F. Minker, V. S. Subrahmanian, Combining Knowledge Bases Consisting of First Order Theories. ISMIS 1991: 92-101.   310
5. M. Arenas, L. Bertossi, J. Chomicki   Consistent Query Answers in Inconsistent Databases. Proc. PODS 1999, pp. 68–79, 1999.   309, 309, 310, 310, 323, 323, 323, 323, 323, 323, 323, 324
6. P. M. Dung, Integrating Data from Possibly Inconsistent Databases. Proc. Int. Conf. on Cooperative Information Systems, 1996: 58-65   310, 322
7. T. Eiter, G. Gottlob and H. Mannila, Disjunctive Datalog, ACM Transactions on Database Systems, 22(3):364–418, 1997
8. Fernandez, J. A., and Minker, J. Computing perfect models of disjunctive stratified databases. In Proc. ILPS'91 W. on Disj. Logic Progr., pp. 110-117, 1991.

9. Gelfond, M., Lifschitz, V. The Stable Model Semantics for Logic Programming, *in Proc. of Fifth Conf. on Logic Programming,* pp. 1070–1080, 1988. 311

10. Gelfond, M. and Lifschitz, V. (1991), Classical Negation in Logic Programs and Disjunctive Databases, *New Generation Computing, 9,* 365–385. 311, 312

11. S. Greco, D. Saccà, Negative Logic Programs. in *North American Conference on Logic Programming,* pages 480-497, 1990. 312

12. Greco, S., Binding Propagation in Disjunctive Databases, *Proc. Int. Conf. on Very Large Data Bases,* 1997.

13. Greco, S., Minimal founded semantics for disjunctive logic programming, *Int. Conf. on Logic Programming and Nonmonotonic Reasoning,* 1999.

14. J. Grant, V. S. Subrahmanian: Reasoning in Inconsistent Knowledge Bases. *IEEE-Trans. on Knowledge and Data Eng.,* 7(1): 177-189 (1995) 310

15. Eiter T., N. Leone, C. Mateis, G. Pfeifer and F. Scarcello. A Deductive System for Non-monotonic Reasoning. *Proc. LPNMR Conf.,* 1997. 363-374. 323, 324

16. R. A. Kowalski, F. Sadri, Logic Programs with Exceptions. *New Generation Computing,* Vol. 9, No. 3/4, pages 387-400, 1991. 312

17. J. Lin, A Semantics for Reasoning Consistently in the Presence of Inconsistency. *Artificial Intelligence* 86(1): 75-95 (1996). 310

18. J. Lin, Integration of Weighted Knowledge Bases. *Artificial Intelligence* 83(2): 363-378 (1996) 310

19. J. Lin, A. O. Mendelzon, Merging Databases Under Constraints. *Int. Journal of Cooperative Information Systems* 7(1): 55-76 (1998) 309, 310

20. J. Lin, A. O. Mendelzon, Knowledge Base Merging by Majority, in R. Pareschi and B. Fronhoefer (eds.), *Dynamic Worlds,* Kluwer, 1999. 310, 322

21. Minker, J. On Indefinite Data Bases and the Closed World Assumption, *Proc. 6-th Conf. on Automated Deduction,* pp. 292–308, 1982. 311

22. J.M. Nicolas, Logic for Improving Integrity Checking in Relational Data Bases. *Acta Informatica,* No. 18, pages 227-253, 1982. 309

23. V.S. Subrahmanian, Amalgamating Knowledge Bases. TODS 19(2): 291-331 (1994) 310, 323

24. J.K. Ullman, *Principles of Database and Knowledge-Base Systems,* Vol. 1, Computer Science Press, Rockville, Md., 1988. 315

How to Decide Query Containment under Constraints Using a Description Logic

Ian Horrocks[1], Ulrike Sattler[2], Sergio Tessaris[1], and Stephan Tobies[2]

[1] Department of Computer Science, University of Manchester, UK
[2] LuFg Theoretical Computer Science, RWTH Aachen, Germany

Abstract. We present a procedure for deciding (database) query containment under constraints. The technique is to extend the logic $\mathcal{DLR}$ with an ABox, and to transform query subsumption problems into $\mathcal{DLR}$ ABox satisfiability problems. Such problems can then be decided, via a reification transformation, using a highly optimised reasoner for the $\mathcal{SHIQ}$ description logic. We use a simple example to support our hypothesis that this procedure will work well with realistic problems.

1 Introduction

Query containment under constraints is the problem of determining whether the result of one query is contained in the result of another query for every database satisfying a given set of constraints (derived, for example, from a schema). This problem is of particular importance in information integration (see [10]) and data warehousing where, in addition to the constraints derived from the source schemas and the global schema, inter-schema constraints can be used to specify relationships between objects in different schemas (see [6]).

In [12], query containment without constraints was shown to be NP-complete, and a subsequent analysis identified cycles in queries as the main source of complexity [13]. Query containment under different forms of constraints have, e.g., been studied in [23] (containment w.r.t. functional and inclusion dependencies) and [11, 24] (containment w.r.t. *is-a* hierarchies).

Calvanese et al. [4] have established a theoretical framework using the logic $\mathcal{DLR}$,[1] presented several (un)decidability results, and described a method for solving the decidable cases using an embedding in the propositional dynamic logic $CPDL_g$ [17, 15]. The importance of this framework is due to the high expressive power of $\mathcal{DLR}$, which allows Extended Entity-Relationship (EER) schemas and inter-schema constraints to be captured. However, the embedding technique does not lead directly to a practical decision procedure as there is no (known) implementation of a $CPDL_g$ reasoner. Moreover, even if such an implementation were to exist, similar embedding techniques [14] have resulted in severe tractability problems when used, for example, to embed the $\mathcal{SHIF}$ description logic in $\mathcal{SHF}$ by eliminating inverse roles [18].

---

[1] Set semantics is assumed in this framework.

M. Parigot and A. Voronkov (Eds.): LPAR 2000, LNAI 1955, pp. 326–343, 2000.
© Springer-Verlag Berlin Heidelberg 2000

In this paper we present a practical decision procedure for the case where neither the queries nor the constraints contain regular expressions. This represents a restriction with respect to the framework described in Calvanese et al., where it was shown that the problem is still decidable if regular expressions are allowed in the schema and the (possibly) containing query, but this seems to be acceptable when modelling classical relational information systems, where regular expressions are seldom used [7, 6]. When excluding regular expressions, constraints imposed by EER schemas can still be captured, so the restriction (to contain no regular expressions) is only relevant to inter-schema constraints. Hence, the use of $\mathcal{DLR}$ in both schema and queries still allows for relatively expressive queries, and by staying within a strictly first order setting we are able to use a decision procedure that has demonstrated good empirical tractability.

The procedure is based on the method described by Calvanese et al., but extends $\mathcal{DLR}$ by defining an *ABox*, a set of axioms that assert facts about named *individuals* and tuples of named individuals (see [5]). This leads to a much more natural encoding of queries (there is a direct correspondence between variables and individuals), and allows the problem to be reduced to that of determining the satisfiability of a $\mathcal{DLR}$ *knowledge base* (KB), i.e., a combined schema and ABox. This problem can in turn be reduced to a KB satisfiability problem in the $\mathcal{SHIQ}$ description logic, with n-ary relations reduced to binary ones by reification. In [24], a similar approach is presented. However, the underlying description logic ($\mathcal{ALCNR}$) is less expressive than $\mathcal{DLR}$ and $\mathcal{SHIQ}$ (for example, it is not able to capture Entity-Relationship schemas).

We have good reasons to believe that this approach represents a practical solution. In the FaCT system [18], we already have an (optimised) implementation of the decision procedure for $\mathcal{SHIQ}$ schema satisfiability described in [21], and using FaCT we have been able to reason very efficiently with a realistic schema derived from the integration of several Extended Entity-Relationship schemas using $\mathcal{DLR}$ inter-schema constraints (the schemas and constraints were taken from a case study undertaken as part of the Esprit DWQ project [7, 6]). In Section 4, we use the FaCT system to demonstrate the empirical tractability of a simple query containment problem with respect to the integrated DWQ schema. FaCT's schema satisfiability algorithm can be straightforwardly extended to deal with ABox axioms (and thus arbitrary query containment problems) [22], and as the number of individuals generated by the encoding of realistic query containment problems will be relatively small, this extension should not compromise empirical tractability.

Most proofs are either omitted or given only as outlines in this paper. For full details, please refer to [20] .

2   Preliminaries

In this section we will (briefly) define the key components of our framework, namely the logic $\mathcal{DLR}$, (conjunctive) queries, and the logic $\mathcal{SHIQ}$.

2.1 The Logic $\mathcal{DLR}$

We will begin with $\mathcal{DLR}$ as it is used in the definition of both schemas and queries. $\mathcal{DLR}$ is a description logic (DL) extended with the ability to describe relations of any arity. It was first introduced in [9].

Definition 1. *Given a set of atomic concept names* NC *and a set of atomic relation names* NR, *every* $C \in$ NC *is a concept and every* $\boldsymbol{R} \in$ NR *is a relation, with every* $\boldsymbol{R}$ *having an associated arity. If* C, D *are concepts,* $\boldsymbol{R}, \boldsymbol{S}$ *are relations of arity* n, i *is an integer* $1 \leqslant i \leqslant n$, *and* k *is a non-negative integer, then*

$$\top, \neg C, C \sqcap D, \exists[\$i]\boldsymbol{R}, (\leq k[\$i]\boldsymbol{R}) \text{ are } \mathcal{DLR} \text{ concepts, and}$$
$$\top_n, \neg \boldsymbol{R}, \boldsymbol{R} \sqcap \boldsymbol{S}, (\$i/n : C) \qquad \text{are } \mathcal{DLR} \text{ relations with arity } n.$$

Relation expressions must be well-typed in the sense that only relations with the same arity can be conjoined, and in constructs like $\exists[\$i]\boldsymbol{R}$ *the value of* i *must be less than or equal to the arity of* $\boldsymbol{R}$.

The semantics of $\mathcal{DLR}$ *is given in terms of interpretations* $\mathcal{I} = (\Delta^{\mathcal{I}}, \cdot^{\mathcal{I}})$, *where* $\Delta^{\mathcal{I}}$ *is the domain (a non-empty set), and* $\cdot^{\mathcal{I}}$ *is an interpretation function that maps every concept to a subset of* $\Delta^{\mathcal{I}}$ *and every n-ary relation to a subset of* $(\Delta^{\mathcal{I}})^n$ *such that the following equations are satisfied ("$\sharp$" denotes set cardinality).*

$$\top^{\mathcal{I}} = \Delta^{\mathcal{I}} \qquad (C \sqcap D)^{\mathcal{I}} = C^{\mathcal{I}} \cap D^{\mathcal{I}}$$
$$\neg C^{\mathcal{I}} = \Delta^{\mathcal{I}} \setminus C^{\mathcal{I}} \qquad (\exists[\$i]\boldsymbol{R})^{\mathcal{I}} = \{d \in \Delta^{\mathcal{I}} \mid \exists(d_1,\ldots,d_n) \in \boldsymbol{R}^{\mathcal{I}}.d_i = d\}$$
$$(\leq k[\$i]\boldsymbol{R})^{\mathcal{I}} = \{d \in \Delta^{\mathcal{I}} \mid \sharp\{(d_1,\ldots,d_n) \in \boldsymbol{R}^{\mathcal{I}} : d_i = d\} \leq k\}$$
$$\top_n{}^{\mathcal{I}} \subseteq (\Delta^{\mathcal{I}})^n \qquad \boldsymbol{R}^{\mathcal{I}} \subseteq \top_n{}^{\mathcal{I}}$$
$$(\neg \boldsymbol{R})^{\mathcal{I}} = \top_n{}^{\mathcal{I}} \setminus \boldsymbol{R}^{\mathcal{I}} \qquad (\boldsymbol{R} \sqcap \boldsymbol{S})^{\mathcal{I}} = \boldsymbol{R}^{\mathcal{I}} \cap \boldsymbol{S}^{\mathcal{I}}$$
$$(\$i/n : C)^{\mathcal{I}} = \{(d_1,\ldots,d_n) \in \top_n{}^{\mathcal{I}} \mid d_i \in C^{\mathcal{I}}\}$$

Note that $\top_n$ does not need to be interpreted as the set of all tuples of arity n, but only as a subset of them, and that the negation of a relation $\boldsymbol{R}$ with arity n is relative to $\top_n$.

In our framework, a schema consists of a set of logical inclusion axioms expressed in $\mathcal{DLR}$. These axioms could be derived from the translation into $\mathcal{DLR}$ of schemas expressed in some other data modelling formalism (such as Entity-Relationship modelling [3, 8]), or could directly stem from the use of $\mathcal{DLR}$ to express, for example, inter-schema constraints to be used in data warehousing, (see [6]).

Definition 2. *A* $\mathcal{DLR}$ *schema* $\mathcal{S}$ *is a set of* axioms *of the form* $C \sqsubseteq D$ *and* $\boldsymbol{R} \sqsubseteq \boldsymbol{S}$, *where* C, D *are* $\mathcal{DLR}$ *concepts and* $\boldsymbol{R}, \boldsymbol{S}$ *are* $\mathcal{DLR}$ *relations of the same arity; an interpretation* $\mathcal{I}$ *satisfies* $C \sqsubseteq D$ *iff* $C^{\mathcal{I}} \subseteq D^{\mathcal{I}}$, *and it satisfies* $\boldsymbol{R} \sqsubseteq \boldsymbol{S}$ *iff* $\boldsymbol{R}^{\mathcal{I}} \subseteq \boldsymbol{S}^{\mathcal{I}}$. *An interpretation* $\mathcal{I}$ *satisfies a schema* $\mathcal{S}$ *iff* $\mathcal{I}$ *satisfies every axiom in* $\mathcal{S}$.

Crucially, we extend $\mathcal{DLR}$ to assert properties of *individuals*, names representing single elements of the domain. An *ABox* is a set of axioms asserting facts about individuals and tuples of individuals.

Definition 3. *Given a set of individuals* NI, *a* $\mathcal{DLR}$ *ABox* $\mathcal{A}$ *is a set of* axioms *of the form* $w{:}C$ *and* $\boldsymbol{w}{:}\boldsymbol{R}$, *where* C *is a concept,* $\boldsymbol{R}$ *is a relation of arity* n, w *is an individual and* $\boldsymbol{w}$ *is an* n-*tuple* $\langle w_1, \ldots, w_n \rangle$ *such that* $w_1, \ldots, w_n$ *are individuals. We will often write* w_i *to refer to the ith element of an* n-*tuple* $\boldsymbol{w}$, *where* $1 \leqslant i \leqslant n$.

Additionally, the interpretation function $\cdot^{\mathcal{I}}$ *maps every individual to an element of* $\Delta^{\mathcal{I}}$ *and thus also tuples of individuals to tuples of elements of* $\Delta^{\mathcal{I}}$. *An interpretation* $\mathcal{I}$ satisfies *an axiom* $w{:}C$ *iff* $w^{\mathcal{I}} \in C^{\mathcal{I}}$, *and it* satisfies *an axiom* $\boldsymbol{w}{:}\boldsymbol{R}$ *iff* $\boldsymbol{w}^{\mathcal{I}} \in \boldsymbol{R}^{\mathcal{I}}$. *An interpretation* $\mathcal{I}$ satisfies *an ABox* $\mathcal{A}$ *iff* $\mathcal{I}$ *satisfies every axiom in* $\mathcal{A}$.

A knowledge base (KB) $\mathcal{K}$ *is a pair* $\langle \mathcal{S}, \mathcal{A} \rangle$, *where* $\mathcal{S}$ *is a schema and* $\mathcal{A}$ *is an ABox. An interpretation* $\mathcal{I}$ *satisfies a KB* $\mathcal{K}$ *iff it satisfies both* $\mathcal{S}$ *and* $\mathcal{A}$.

If an interpretation $\mathcal{I}$ *satisfies a concept, axiom, schema, or ABox* X, *then we say that* $\mathcal{I}$ *is a* model *of* X, *call* X satisfiable, *and write* $\mathcal{I} \models X$.

Note that it is not assumed that individuals with different names are mapped to different elements in the domain (the so-called unique name assumption).

Definition 4. *If* $\mathcal{K}$ *is a KB,* $\mathcal{I}$ *is a model of* $\mathcal{K}$, *and* $\mathcal{A}$ *is an ABox, then* $\mathcal{I}'$ *is called an* extension *of* $\mathcal{I}$ *to* $\mathcal{A}$ *iff* $\mathcal{I}'$ *satisfies* $\mathcal{A}$, $\Delta^{\mathcal{I}} = \Delta^{\mathcal{I}'}$, *and all concepts, relations, and individuals occuring in* $\mathcal{K}$ *are interpreted identically by* $\mathcal{I}$ *and* $\mathcal{I}'$.

Given two ABoxes $\mathcal{A}, \mathcal{A}'$ *and a schema* $\mathcal{S}$, $\mathcal{A}$ *is* included *in* $\mathcal{A}'$ *w.r.t.* $\mathcal{S}$ *(written* $\langle \mathcal{S}, \mathcal{A} \rangle \approx \mathcal{A}'$*) iff every model* $\mathcal{I}$ *of* $\langle \mathcal{S}, \mathcal{A} \rangle$ *can be extended to* $\mathcal{A}'$.

2.2 Queries

In this paper we will focus on conjunctive queries (see [1, chap. 4]), and describe only briefly (in Section 5) how the technique can be extended to deal with disjunctions of conjunctive queries (for full details please refer to [20]). A *conjunctive query* q is an expression

$$q(\boldsymbol{x}) \leftarrow term_1(\boldsymbol{x}, \boldsymbol{y}, \boldsymbol{c}) \wedge \ldots \wedge term_n(\boldsymbol{x}, \boldsymbol{y}, \boldsymbol{c})$$

where $\boldsymbol{x}$, $\boldsymbol{y}$, and $\boldsymbol{c}$ are tuples of *distinguished* variables, variables, and constants, respectively (distinguished variables appear in the answer, "ordinary" variables are used only in the query expression, and constants are fixed values). Each term $term_i(\boldsymbol{x}, \boldsymbol{y}, \boldsymbol{c})$ is called an atom in q and is in one of the forms $C(w)$ or $\boldsymbol{R}(\boldsymbol{w})$, where w (resp. $\boldsymbol{w}$) is a variable or constant (resp. tuple of variables and constants) in $\boldsymbol{x}$, $\boldsymbol{y}$ or $\boldsymbol{c}$, C is a $\mathcal{DLR}$ concept, and $\boldsymbol{R}$ is a $\mathcal{DLR}$ relation.[2]

For example, a query designed to return the bus number of the city buses travelling in both directions between two stops is:

$$\text{BUS}(nr) \leftarrow \text{bus\_route}(nr, stop_1, stop_2) \wedge \text{bus\_route}(nr, stop_2, stop_1) \wedge \text{city\_bus}(nr)$$

where nr is a distinguished variable (it appears in the answer), $stop_1$ and $stop_2$ are non-distinguished variables, city_bus is a $\mathcal{DLR}$ concept and bus_route is a $\mathcal{DLR}$ relation.

---

[2] The fact that these concepts and relations can also appear in the schema is one of the distinguishing features of this approach.

In this framework, the *evaluation* $q(\mathcal{I})$ of a query q with n distinguished variables w.r.t. a $\mathcal{DLR}$ interpretation $\mathcal{I}$ (here perceived as standard FO interpretation) is the set of n-tuples $\boldsymbol{d} \in (\Delta^{\mathcal{I}})^n$ such that

$$\mathcal{I} \models \exists \boldsymbol{y}.term_1(\boldsymbol{d}, \boldsymbol{y}, \boldsymbol{c}) \wedge \ldots \wedge term_n(\boldsymbol{d}, \boldsymbol{y}, \boldsymbol{c}).$$

As usual, we require unique interpretation of constants, i.e., in the following we will only consider those intepretations $\mathcal{I}$ with $c^{\mathcal{I}} \neq d^{\mathcal{I}}$ for any two constants $c \neq d$. A query $q(\boldsymbol{x})$ is called *satisfiable* w.r.t a schema $\mathcal{S}$ iff there is an interpretation $\mathcal{I}$ with $\mathcal{I} \models \mathcal{S}$ and $q(\mathcal{I}) \neq \emptyset$. A query $q_1(\boldsymbol{x})$ is *contained* in a query $q_2(\boldsymbol{x})$ w.r.t. a schema $\mathcal{S}$ (written $\mathcal{S} \models q_1 \sqsubseteq q_2$), iff, for every model $\mathcal{I}$ of $\mathcal{S}$, $q_1(\mathcal{I}) \subseteq q_2(\mathcal{I})$. Two queries q_1, q_2 are called *equivalent* w.r.t. $\mathcal{S}$ iff $\mathcal{S} \models q_1 \sqsubseteq q_2$ and $\mathcal{S} \models q_2 \sqsubseteq q_1$.

For example, the schema containing the axioms

$$(\text{bus\_route} \sqcap (\$1/3 : \text{city\_bus})) \sqsubseteq \text{city\_bus\_route}$$

$$\text{city\_bus\_route} \sqsubseteq (\text{bus\_route} \sqcap (\$1/3 : \text{city\_bus})),$$

states that the relation city_bus_route contains exactly the bus_route information that concerns city buses. It is easy to see that the following CITY_BUS query

$$\text{CITY\_BUS}(nr) \leftarrow \text{city\_bus\_route}(nr, stop_1, stop_2) \wedge \text{city\_bus\_route}(nr, stop_2, stop_1)$$

is equivalent to the previous BUS query w.r.t. the given schema. In an information integration scenario, for example, this could be exploited by reformulating the BUS query as a CITY_BUS query ranging over a smaller database without any loss of information.

2.3   The Logic $\mathcal{SHIQ}$

$\mathcal{SHIQ}$ is a standard DL, in the sense that it deals with concepts and (only) binary relations (called *roles*), but it is unusually expressive in that it supports reasoning with inverse roles, qualifying number restrictions on roles, transitive roles, and role inclusion axioms.

Definition 5. *Given a set of atomic concept names* NC *and a set of atomic role names* NR *with transitive role names* $\text{NR}_+ \subseteq \text{NR}$, *every* $C \in \text{NC}$ *is a concept, every* $R \in \text{NR}$ *is a role, and every* $R \in \text{NR}_+$ *is a transitive role. If R is a role, then* R^- *is also a role (and if* $R \in \text{NR}_+$ *then* R^- *is also a transitive role). If S is a (possibly inverse) role, C, D are concepts, and k is a non-negative integer, then*

$$\top, \neg C, C \sqcap D, \exists S.C, \leqslant kS.C \text{ are also } \mathcal{SHIQ} \text{ concepts.}$$

The semantics of $\mathcal{SHIQ}$ is given in terms of interpretations $\mathcal{I} = (\Delta^{\mathcal{I}}, \cdot^{\mathcal{I}})$, *where* $\Delta^{\mathcal{I}}$ *is the domain (a non-empty set), and* $\cdot^{\mathcal{I}}$ *is an interpretation function that maps every concept to a subset of $\Delta^{\mathcal{I}}$ and every role to a subset of $(\Delta^{\mathcal{I}})^2$ such that the following equations are satisfied.*

$$\top^{\mathcal{I}} = \Delta^{\mathcal{I}} \qquad (\exists S.C)^{\mathcal{I}} = \{d \mid \exists d'.(d, d') \in S^{\mathcal{I}} \text{ and } d' \in C^{\mathcal{I}}\}$$

$$\neg C^{\mathcal{I}} = \Delta^{\mathcal{I}} \setminus C^{\mathcal{I}} \quad (\leqslant kS.C)^{\mathcal{I}} = \{d \mid \sharp\{d' : (d, d') \in S^{\mathcal{I}} \text{ and } d' \in C^{\mathcal{I}}\} \leqslant k\}$$

$$(C \sqcap D)^{\mathcal{I}} = C^{\mathcal{I}} \cap D^{\mathcal{I}} \qquad R^{\mathcal{I}} = (R^{\mathcal{I}})^+ \text{ for all } R \in \text{NR}_+$$

$$(R^-)^{\mathcal{I}} = \{(d', d) \mid (d, d') \in R^{\mathcal{I}}\}$$

$\mathcal{SHIQ}$ schemas, ABoxes, and KBs are defined similarly to those for $\mathcal{DLR}$: if C, D are concepts, R, S are roles, and v, w are individuals, then a schema $\mathcal{S}$ consists of axioms of the form $C \sqsubseteq D$ and $R \sqsubseteq S$, and an ABox $\mathcal{A}$ consists of axioms of the form $w{:}C$ and $\langle v, w \rangle{:}R$. Again, a KB $\mathcal{K}$ is a pair $\langle \mathcal{S}, \mathcal{A} \rangle$, where $\mathcal{S}$ is a schema and $\mathcal{A}$ is an ABox.

The definitions of interpretations, satisfiability, and models also parallel those for $\mathcal{DLR}$, and there is again no unique name assumption.

Note that, in order to maintain decidability, the roles that can appear in number restrictions are restricted [21]: if a role S occurs in a number restriction $\leqslant kS.C$, then neither S nor any of its sub roles may be transitive (i.e., if the schema contains a $\sqsubseteq$-path from S' to S, then S' is not transitive).

3 Determining Query Containment

In this section we will describe how the problem of deciding whether one query is contained in another one w.r.t. a $\mathcal{DLR}$ schema can be reduced to the problem of deciding KB satisfiability in the $\mathcal{SHIQ}$ description logic. There are three steps to this reduction. Firstly, the queries are transformed into $\mathcal{DLR}$ ABoxes $\mathcal{A}_1$ and $\mathcal{A}_2$ such that $\mathcal{S} \models q_1 \sqsubseteq q_2$ iff $\langle \mathcal{S}, \mathcal{A}_1 \rangle \not\approx \mathcal{A}_2$ (see Definition 4). Secondly, the ABox inclusion problem is transformed into one or more KB satisfiability problems. Finally, we show how a $\mathcal{DLR}$ KB can be transformed into an equisatisfiable $\mathcal{SHIQ}$ KB.

3.1 Transforming Query Containment into ABox Inclusion

We will first show how a query can be transformed into a *canonical $\mathcal{DLR}$ ABox*. Such an ABox represents a generic pattern that must be matched by all tuples in the evaluation of the query, similar to the tableau queries one encounters in the treatment of simple query containment for conjunctive queries [1].

Definition 6. *Let q be a conjunctive query. The* canonical ABox *for q is defined by*

$$\mathcal{A}_q = \{w{:}\boldsymbol{R} \mid \boldsymbol{R}(\boldsymbol{w}) \text{ is an atom in } q\} \cup \{w{:}C \mid C(w) \text{ is an atom in } q\}.$$

We introduce a new atomic concept P_w for every individual w in $\mathcal{A}$ and define the completed canonical ABox *for q by*

$$\widehat{\mathcal{A}}_q = \mathcal{A}_q \cup \{w{:}P_w \mid w \text{ occurs in } \mathcal{A}_q\} \cup \{w_i{:}\neg P_{w_j} \mid w_i, w_j \text{ constants in } q \text{ and } i \neq j\}.$$

The axioms $w{:}P_w$ in $\widehat{\mathcal{A}}_q$ introduce representative concepts *for each individual w in $\mathcal{A}_q$. They are used (in the axioms $w_i{:}\neg P_{w_j}$) to ensure that individuals corresponding to different constants in q cannot have the same interpretation, and will also be useful in the transformation to KB satisfiability.*

By abuse of notation, we will say that an interpretation $\mathcal{I}$ and an assignment ρ of distinguished variables, non-distinguished variables and constants to elements in the domain of $\mathcal{I}$ such that $\mathcal{I} \models \rho(q)$ define a model for $\mathcal{A}_q$ with the interpretation of the individuals corresponding with ρ and the interpretation $P_w^{\mathcal{I}} = \{w^{\mathcal{I}}\}$.

We can use this definition to transform the query containment problem into a (very similar) problem involving $\mathcal{DLR}$ ABoxes. We can assume that the names of the non-distinguished variables in q_2 differ from those in q_1 (arbitrary names can be chosen without affecting the evaluation of the query), and that the names of distinguished variables and constants appear in both queries (if a name is missing in one of the queries, it can be simply added using a term like $\top(v)$).

The following Theorem shows that a canonical ABox really captures the structure of a query, allowing the query containment problem to be restated as an ABox inclusion problem.

Theorem 1 *Given a schema $\mathcal{S}$ and queries q_1 and q_2, $\mathcal{S} \models q_1 \sqsubseteq q_2$ iff $\langle \mathcal{S}, \widehat{\mathcal{A}}_{q_1} \rangle \not\approx \mathcal{A}_{q_2}$.*

Before we prove Theorem 1, note that, in general, this theorem no longer holds if we replace $\mathcal{A}_{q_2}$ by $\widehat{\mathcal{A}}_{q_2}$. Let $\mathcal{S}$ be a schema and q_1, q_2 be two queries such that q_1 is satisfiable w.r.t. $\mathcal{S}$ and q_2 contains at least one non-distinguished variable z. Then the completion $\widehat{\mathcal{A}}_{q_2}$ contains the assertion $z{:}P_z$ where P_z is a new atomic concept. Since q_1 is satisfiable w.r.t. $\mathcal{S}$ and P_z does not occur in $\mathcal{S}$ or q_1, $\langle \mathcal{S}, \widehat{\mathcal{A}}_{q_1} \rangle$ has a model $\mathcal{I}$ with $P_z^{\mathcal{I}} = \emptyset$. Such a model $\mathcal{I}$ cannot be extended to a model $\mathcal{I}'$ of $\widehat{\mathcal{A}}_{q_2}$ because there is no possible interpretation for z that would satisfy $z^{\mathcal{I}'} \in P_z^{\mathcal{I}'}$. Hence, $\langle \mathcal{S}, \widehat{\mathcal{A}}_{q_1} \rangle \not\approx \widehat{\mathcal{A}}_{q_2}$ regardless of whether $\mathcal{S} \models q_1 \sqsubseteq q_2$ holds or not. In the next section we will see how to deal with the non-distinguished individuals in $\mathcal{A}_{q_2}$ without the introduction of new representative concepts.

PROOF OF THEOREM 1: For the if direction, assume $\mathcal{S} \not\models q_1 \sqsubseteq q_2$. Then there exists a model $\mathcal{I}$ of $\mathcal{S}$ and a tuple $(d_1, \ldots, d_n) \in (\Delta^{\mathcal{I}})^n$ such that $(d_1, \ldots, d_n) \in q_1(\mathcal{I})$ and $(d_1, \ldots, d_n) \notin q_2(\mathcal{I})$. $\mathcal{I}$ and the assignment of variables leading to $(d_1, \ldots, d_n)$ define a model for $\widehat{\mathcal{A}}_{q_1}$. If $\cdot^{\mathcal{I}}$ could be extended to satisfy $\mathcal{A}_{q_2}$, then the extension would correspond to an assignment of the non-distinguished variables in q_2 such that $(d_1, \ldots, d_n) \in q_2(\mathcal{I})$, thus contradicting the assumption.

For the only if direction, assume there is a model $\mathcal{I}$ of both $\mathcal{S}$ and $\widehat{\mathcal{A}}_{q_1}$ that cannot be extended to a model of $\mathcal{A}_{q_2}$. Hence there is a tuple $(d_1, \ldots, d_n) \in q_1(\mathcal{I})$ and a corresponding assignment of variables that define $\mathcal{I}$. If there is an assignment of the non-distinguished variables in q_2 such that $(d_1, \ldots, d_n) \in q_2(\mathcal{I})$, then this assignment would define the extension of $\mathcal{I}$ such that $\mathcal{A}_{q_2}$ is also satisfied. $\qquad\square$

3.2   Transforming ABox Inclusion into ABox Satisfiability

Next, we will show how to transform the ABox inclusion problem into one or more KB satisfiability problems. In order to do this, there are two main difficulties that must be overcome. The first is that, in order to transform inclusion into satisfiability, we would like to be able to "negate" axioms. This is easy for axioms of the form $w{:}C$, because an interpretation satisfies $w{:}\neg C$ iff it does not satisfy $w{:}C$. However, we cannot deal with axioms of the form $w{:}\mathbf{R}$ in this way, because $\mathcal{DLR}$ only has a weak form of negation for relations relative to $\top_n$. Our solution is to transform all axioms in $\mathcal{A}_{q_2}$ into the form $w{:}C$.

The second difficulty is that $\mathcal{A}_{q_2}$ may contain individuals corresponding to non-distinguished variables in q_2 (given the symmetry between queries and ABoxes, we will refer to them from now on as non-distinguished individuals). These individuals introduce an extra level of quantification that we cannot deal with using our standard reasoning procedures: $\langle \mathcal{S}, \widehat{\mathcal{A}}_{q_1} \rangle \approx \mathcal{A}_{q_2}$ iff for all models $\mathcal{I}$ of $\langle \mathcal{S}, \widehat{\mathcal{A}}_{q_1} \rangle$ *there exists* some extension of $\mathcal{I}$ to $\mathcal{A}_{q_2}$. We deal with this problem by eliminating the non-distinguished individuals from $\mathcal{A}_{q_2}$.

We will begin by exploiting some general properties of ABoxes that allow us to *compact* $\mathcal{A}_{q_2}$ so that it contains only one axiom $\boldsymbol{w}$:**R** for each tuple $\boldsymbol{w}$, and one axiom $\boldsymbol{w}$:C for each individual w that is not an element in any tuple. It is obvious from the semantics that we can combine all ABox axioms relating to the same individual or tuple: $\mathcal{I} \models \{\boldsymbol{w}{:}C, \boldsymbol{w}{:}D\}$ (resp. $\{\boldsymbol{w}{:}\mathbf{R}, \boldsymbol{w}{:}\mathbf{S}\}$) iff $\mathcal{I} \models \{\boldsymbol{w}{:}(C \sqcap D)\}$ (resp. $\{\boldsymbol{w}{:}(\mathbf{R} \sqcap \mathbf{S})\}$). The following lemma shows that we can also absorb $w_i{:}C$ into $\boldsymbol{w}$:**R** when w_i is an element of $\boldsymbol{w}$.

Lemma 1 *Let $\mathcal{A}$ be a $\mathcal{DLR}$ ABox with $\{w_i{:}C, \boldsymbol{w}{:}\mathbf{R}\} \subseteq \mathcal{A}$, where w_i is the ith element in $\boldsymbol{w}$. Then $\mathcal{I} \models \mathcal{A}$ iff $\mathcal{I} \models \{\boldsymbol{w}{:}(\mathbf{R} \sqcap \$i : C)\} \cup \mathcal{A} \setminus \{w_i{:}C, \boldsymbol{w}{:}\mathbf{R}\}$.*

PROOF: From the semantics, if $\boldsymbol{w}^{\mathcal{I}} \in (\mathbf{R} \sqcap \$i : C)^{\mathcal{I}}$, then $\boldsymbol{w}^{\mathcal{I}} \in \mathbf{R}^{\mathcal{I}}$ and $w_i^{\mathcal{I}} \in C^{\mathcal{I}}$, and if $w_i^{\mathcal{I}} \in C^{\mathcal{I}}$ and $\boldsymbol{w}^{\mathcal{I}} \in \mathbf{R}^{\mathcal{I}}$, then $\boldsymbol{w}^{\mathcal{I}} \in (\mathbf{R} \sqcap \$i : C)^{\mathcal{I}}$.   □

The ABox resulting from exhaustive application of Lemma 1 can be represented as a graph, with a node for each tuple, a node for each individual, and edges connecting tuples with the individuals that compose them. The graph will consist of one or more connected components, where each component is either a single individual (representing an axiom $w{:}C$, where w is not an element in any tuple) or a set of tuples linked by common elements (representing axioms of the form $\boldsymbol{w}$:**R**). As the connected components do not have any individuals in common, we can deal independently with the inclusion problem for each connected set of axioms: $\langle \mathcal{S}, \mathcal{A} \rangle \approx \mathcal{A}'$ iff $\langle \mathcal{S}, \mathcal{A} \rangle \approx \mathcal{G}$ for every connected set of axioms $\mathcal{G} \subseteq \mathcal{A}'$. As an example, Figure 1 shows the graph that corresponds to the ABox $\mathcal{A}_{q_2}$ from Example 1.

Returning to our original problem, we will now show how we can *collapse* a connected component $\mathcal{G}$ by a graph traversal into a single axiom of the form $w{:}C$, where w is an element of a tuple occurring in $\mathcal{G}$ (an arbitrarily chosen "root" individual), and C is a concept that describes $\mathcal{G}$ from the point of view of w. An example for this process will be given later in this section.

This would be easy if we were able to refer to individuals in C (i.e., if our logic included *nominals* [25]), which is not the case. However, as we will see, it is sufficient to refer to the distinguished individuals w_i in $\mathcal{G}$ (which also occur in $\widehat{\mathcal{A}}_{q_1}$) by their representative concepts P_{w_i}. Moreover, we can refer to non-distinguished individuals z_i by using $\top$ as their representative concept (this is only valid for z_i that are encountered only once during the traversal of $\mathcal{G}$, but we will see later that we can, without loss of generality, restrict our attention to this case). Informally, the use of $\top$ as the representative concept for such z_i can be justified by the fact that when an interpretation $\mathcal{I}$ is extended to $\mathcal{G}$, z_i can be interpreted as any element in $\Delta^{\mathcal{I}}$ ($= \top^{\mathcal{I}}$).[3]

---

[3] For full details, the reader is again referred to [20].

The following lemma shows how we can use the representative concepts to transform an axiom of the form $w{:}R$ into an axiom of the form $w_i{:}C$.

Lemma 2 *If S is a schema, $\widehat{A}$ is a completed canonical ABox and A' is an ABox with $w{:}R \in A'$, then $\langle S, \widehat{A} \rangle \approx A'$ iff $\langle S, \widehat{A} \rangle \approx (\{w_i{:}C\} \cup A' \setminus \{w{:}R\})$, where $w = \langle w_1, \ldots, w_n \rangle$, w_i is the ith element in w, C is the concept*

$$\exists[\$i](R \sqcap \bigsqcap_{1 \leqslant j \leqslant n . j \neq i} (\$j/n : P_j)),$$

and P_j is the appropriate representative concept for w_j ($\top$ if w_j is a non-distinguished individual, P_{w_j} otherwise).

PROOF (sketch): For the only if direction, it is easy to see that, if $\mathcal{I} \models \langle S, \widehat{A}_{q_1} \rangle$, and $\mathcal{I}'$ is an extension of $\mathcal{I}$ that satisfies $w{:}R$, then $\mathcal{I}'$ also satisfies $w_i{:}C$.

The converse direction is more complicated, and exploits the fact that, for every model $\mathcal{I}$ of $\langle S, \widehat{A}_{q_1} \rangle$, there is a similar model $\mathcal{I}'$ in which every representative concept P_{w_i} is interpreted as $\{w_i^{\mathcal{I}'}\}$. If $\mathcal{I}$ cannot be extended to satisfy $w{:}R$, then neither can $\mathcal{I}'$, and, given the interpretations of the P_{w_i}, it is possible to show that $\mathcal{I}'$ cannot be extended to satisfy $w_i{:}C$ either. $\square$

All that now remains is to choose the order in which we apply the transformations from Lemma 1 and 2 to the axioms in G, so that, whenever we use Lemma 2 to transform $w{:}R$ into $w_i{:}C$, we can then use Lemma 1 to absorb $w_i{:}C$ into another axiom $v{:}R$, where w_i is an element of v. We can do this using a recursive traversal of the graphical representation of G (a similar technique is used in [4] to transform queries into concepts). A traversal starts at an individual node w (the "root") and proceeds as follows.

- At an individual node w_i, the node is first marked as visited. Then, while there remains an unmarked tuple node connected to w_i, one of these, w, is selected, visited, and the axiom $w{:}R$ transformed into an axiom $w_i{:}C$. Finally, any axioms $w_i{:}C_1, \ldots, w_i{:}C_n$ resulting from these transformations are merged into a single axiom $w_i{:}(C_1 \sqcap \ldots \sqcap C_n)$.
- At a tuple node w, the node is first marked as visited. Then, while there remains an unmarked individual node connected to w, one of these, w_i, is selected, visited, and any axiom $w_i{:}C$ that results from the visit is merged into the axiom $w{:}R$ using Lemma 1.

Note that the correctness of the collapsing procedure does not depend on the traversal (whose purpose is simply to choose a suitable ordering), but only on the individual transformations.

Having collapsed a component G, we finally have a problem that we can decide using KB satisfiability:

Lemma 3 *If S is a schema and $\widehat{A}$ is a completed canonical ABox, then $\langle S, \widehat{A} \rangle \approx \{w{:}C\}$ iff w is an individual in $\widehat{A}$ and $\langle S, (\widehat{A} \cup \{w{:}\neg C\}) \rangle$ is not satisfiable, or w is not an individual in $\widehat{A}$ and $\langle (S \cup \{\top \sqsubseteq \neg C\}), \widehat{A} \rangle$ is not satisfiable.*

PROOF (sketch): If w is an individual in $\widehat{A}$, $\langle S, \widehat{A}\rangle \models \{w{:}C\}$ implies that every model $\mathcal{I}$ of $\langle S, \widehat{A}\rangle$ must also satisfy $w{:}C$, and this is true iff $\mathcal{I}$ does not satisfy $w{:}\neg C$. In the case where w is not an individual in $\widehat{A}$, a model $\mathcal{I}$ of $\langle S, \widehat{A}\rangle$ can be extended to $\{w{:}C\}$ iff $C^{\mathcal{I}} \neq \emptyset$, which is true iff $\Delta^{\mathcal{I}} \not\subseteq (\neg C)^{\mathcal{I}}$. □

If a non-distinguished individual z_i is encountered more than once during a traversal, then it is enforcing a co-reference that closes a cycle in the query. In this case we cannot simply use $\top$ to refer to it, as this would fail to capture the fact that z_i must be interpreted as the same element of $\Delta^{\mathcal{I}}$ on each occasion.

In [4] this problem is dealt with by replacing the non-distinguished variables occurring in a cycle in q_2 with variables or constants from q_1, and forming a disjunction of the concepts resulting from each possible replacement. This is justified by the fact that cycles cannot be expressed in the $\mathcal{DLR}$ schema and so must be present in q_1. However, this fails to take into account the fact that identifying two or more of the non-distinguished variables in q_2 could eliminate the cycle.

We overcome this problem by introducing an additional layer of disjunction in which non-distinguished individuals occurring in cycles are identified (in every possible way) with other individuals occurring in the same cycle. We then continue as in [4], but only replacing those individuals that actually enforce a co-reference, i.e., that would be encountered more than once during the graph traversal.[4]

Example 1 To illustrate the inclusion to satisfiability transformation, we will refer to the example given in Section 2.2. The containment of BUS in CITY_BUS w.r.t. the schema is demonstrated by the inclusion $\langle S, \widehat{A}_1\rangle \models A_2$, where S, $\widehat{A}_1$ and A_2 are the schema and two canonical ABoxes (completed in the case of $\widehat{A}_1$) corresponding to the given queries:

$$S = \left\{ \begin{array}{l} (\text{bus\_route} \sqcap (\$1/3 : \text{city\_bus})) \sqsubseteq \text{city\_bus\_route}, \\ \text{city\_bus\_route} \sqsubseteq (\text{bus\_route} \sqcap (\$1/3 : \text{city\_bus})) \end{array} \right\}$$

$$\widehat{A}_1 = \left\{ \langle n, y_1, y_2\rangle{:}\text{bus\_route}, \langle n, y_2, y_1\rangle{:}\text{bus\_route}, n{:}\text{city\_bus}, n{:}P_n, y_1{:}P_{y_1}, y_2{:}P_{y_2} \right\}$$

$$A_2 = \left\{ \langle n, z_1, z_2\rangle{:}\text{city\_bus\_route}, \langle n, z_2, z_1\rangle{:}\text{city\_bus\_route} \right\}$$

The two axioms in A_2 are connected, and can be collapsed into a single axiom using the described procedure. Figure 1 shows a traversal of the graph $\mathcal{G}$ corresponding to A_{q_2} that starts at z_1 and traverses the edges in the indicated sequence.[5] The resulting axiom (describing A_2 from the point of view of z_1) is $z_1{:}C$, where C is the concept

$$\underset{1}{\exists[\$2]}(\text{city\_bus\_route} \sqcap (\underset{2}{\$3} : (P_{z_2} \sqcap \underset{3}{\exists[\$2]}(\text{city\_bus\_route} \sqcap \underset{4}{\$1} : P_n \sqcap \underset{5}{\$3} : P_{z_1}))) \sqcap \underset{6}{\$1} : P_n)$$

P_{z_1}, P_{z_2} are "place-holders" for z_1, z_2[6] and the numbers below the $\mathcal{DLR}$ operators denote the edges which correspond to the respective subconcept of C. As z_2 is encountered only once in the traversal, P_{z_2} can be replaced with $\top$, but as z_1 is encountered

---

[4] Note that the graph traversal must always start from the same root.

[5] We will ignore the first non-deterministic step as no individual identifications are required in order to prove the containment.

[6] In practice, we use such "place-holders" during the collapsing procedure and then make appropriate (possibly non-deterministic) substitutions.

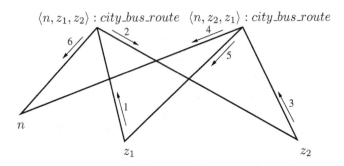

$\langle n, z_1, z_2 \rangle : city\_bus\_route$ $\langle n, z_2, z_1 \rangle : city\_bus\_route$

Fig. 1. A traversal of the graph corresponding to $\mathcal{A}_{q_2}$

twice (as the root and as P_{z_1}), it must be replaced (non-deterministically) with an individual i occurring in $\widehat{\mathcal{A}}_1$ (we will refer to the resulting concepts as $C_{[z_1/i]}$), and thus $\langle \mathcal{S}, \widehat{\mathcal{A}}_1 \rangle \not\approx \mathcal{A}_2$ iff $\langle \mathcal{S}, \widehat{\mathcal{A}}_1 \rangle \not\approx \{i : C_{[z_1/i]}\}$. Taking $i = y_1$ we have $\langle \mathcal{S}, \widehat{\mathcal{A}}_1 \rangle \not\approx \{y_1 : C_{[z_1/y_1]}\}$ because $\langle \mathcal{S}, (\widehat{\mathcal{A}}_1 \cup \{y_1 : \neg C_{[z_1/y_1]}\}) \rangle$ is not satisfiable.

Summing up, we thus have:

Theorem 2 *For a $\mathcal{DLR}$ KB $\mathcal{K} = \langle \mathcal{S}, \mathcal{A} \rangle$ and a $\mathcal{DLR}$ ABox $\mathcal{A}'$, the problem of deciding whether $\mathcal{A}$ is included in $\mathcal{A}'$ w.r.t. $\mathcal{S}$ can be reduced to (possibly several) $\mathcal{DLR}$ ABox satisfiability problems.*

Concerning the practicability of this reduction, it is easy to see that, for any fixed choice of substitutions for the non-distinguished individuals in $\mathcal{G}$, the reduction from Theorem 2 can be computed in polynomial time. More problematically, it is necessary to consider each possible identification of non-distinguished individuals occuring in cycles in $\mathcal{G}$, and for each of these *all* possible mappings from the set Z of non-distinguished individuals that occur more than once in the collapsed $\mathcal{G}$ to to the set W of individuals that occur in $\widehat{\mathcal{A}}_1$ (of which there are $|W|^{|Z|}$ many). However, both Z and W will typically be quite small, especially Z which will consist only of those non-distinguished individuals that occur in a cycle in $\mathcal{G}$ and are actually used to enforce a co-reference (i.e., to "close" the cycle). This represents a useful refinement over the procedure described in [4], where all z_i that occur in cycles are non-deterministically replaced with some w_i, regardless of whether or not they are used to enforce a co-reference. Moreover, it is easy to show that most individual identifications cannot contribute to the solution, and can thus be ignored. Therefore, we do not believe that this additional non-determinism compromises the feasibility of our approach.

Interestingly, also in [13], cycles in queries are identified as a main cause for complexity. There it is shown that query containment without constraints is decidable in polynomial time for acyclic queries whereas the problem for possibly cyclic queries is NP-complete [12].

3.3 Transforming $\mathcal{DLR}$ Satisfiability into $\mathcal{SHIQ}$ Satisfiability

We decide satisfiability of $\mathcal{DLR}$ knowledge bases by means of a satisfiability-preserving translation $\sigma(\cdot)$ from $\mathcal{DLR}$ KBs to $\mathcal{SHIQ}$ KBs. This translation must deal with the fact that $\mathcal{DLR}$ allows for arbitrary n-ary relations while $\mathcal{SHIQ}$ only allows for unary predicates and binary relations; this is achieved by a process called *reification* (see, for example [16]). The main idea behind this is easily described: each n-ary tuple in a $\mathcal{DLR}$-interpretation is represented by an individual in a $\mathcal{SHIQ}$-interpretation that is linked via the dedicated functional relations $f_1, \ldots, f_n$ to the elements of the tuple.

For $\mathcal{DLR}$ without regular expressions, the mapping $\sigma(\cdot)$ (given by [4])

$$\sigma(\top_n) = \top_n \qquad\qquad \sigma(\top) = \top_1$$
$$\sigma(\mathbf{P}) = \mathbf{P} \qquad\qquad \sigma(A) = A$$
$$\sigma(\$i/n : C) = \top_n \sqcap \exists f_i.\sigma(C) \qquad \sigma(\neg C) = \neg\sigma(C)$$
$$\sigma(\neg\mathbf{R}) = \top_n \sqcap \neg\sigma(\mathbf{R}) \qquad \sigma(C_1 \sqcap C_2) = \sigma(C_1) \sqcap \sigma(C_2)$$
$$\sigma(\mathbf{R}_1 \sqcap \mathbf{R}_2) = \sigma(\mathbf{R}_1) \sqcap \sigma(\mathbf{R}_2) \qquad \sigma(\exists[\$i]\mathbf{R}) = \exists f_i^-.\sigma(\mathbf{R})$$
$$\sigma(\le k[\$i]\mathbf{R}) = (\le k\ f_i^-\ \sigma(\mathbf{R}))$$

reifies $\mathcal{DLR}$ expressions into $\mathcal{SHIQ}$-concepts. This mapping can be extended to a knowledge base (KB) as follows.

Definition 7. *Let* $\mathcal{K} = (\mathcal{S}, \mathcal{A})$ *be a* $\mathcal{DLR}$ *KB. The reification of* $\mathcal{S}$ *is given by*

$$\{(\sigma(\mathbf{R}_1) \sqsubseteq \sigma(\mathbf{R}_2)) \mid (\mathbf{R}_1 \sqsubseteq \mathbf{R}_2) \in \mathcal{S}\} \cup \{(\sigma(C_1) \sqsubseteq \sigma(C_2)) \mid (C_1 \sqsubseteq C_2) \in \mathcal{S}\}.$$

To reify the ABox $\mathcal{A}$*, we have to reify all tuples appearing in the axioms. For each distinct tuple* $\mathbf{w} = \langle w_1, \ldots, w_n \rangle$ *occurring in* $\mathcal{A}$*, we chose a distinct individual* $t_{\mathbf{w}}$ *(called the "reification of* $\mathbf{w}$*") and define:*

$$\sigma(\mathbf{w}{:}\mathbf{R}) = \{t_{\mathbf{w}}{:}\sigma(\mathbf{R})\} \cup \{\langle t_{\mathbf{w}}, w_i \rangle{:}f_i \mid 1 \le i \le n\} \quad \text{and}$$
$$\sigma(\mathcal{A}) = \bigcup \{\sigma(\mathbf{w}{:}\mathbf{R}) \mid \mathbf{w}{:}\mathbf{R} \in \mathcal{A}\} \cup \{w{:}\sigma(C) \mid w{:}C \in \mathcal{A}\}.$$

We need a few additional inclusion and ABox axioms to guarantee that any model of $(\sigma(\mathcal{S}), \sigma(\mathcal{A}))$ *can be "un-reified" into a model of* $(\mathcal{S}, \mathcal{A})$*. Let* n_{max} *denote the maximum arity of the* $\mathcal{DLR}$ *relations appearing in* $\mathcal{K}$*. We define* $f(\mathcal{S})$ *to consist of the following axioms (where* $x \equiv y$ *is an abbreviation for* $x \sqsubseteq y$ *and* $y \sqsubseteq x$*):*

$$\top \equiv \top_1 \sqcup \cdots \sqcup \top_{n_{max}}$$
$$\top \sqsubseteq (\le 1\ f_1) \sqcap \cdots \sqcap (\le 1\ f_{n_{max}})$$
$$\forall f_i.\bot \sqsubseteq \forall f_{i+1}.\bot \qquad\qquad\qquad \text{for } 2 \le i < n_{max}$$
$$\top_i \equiv \exists f_1.\top_1 \sqcap \cdots \sqcap \exists f_i.\top_1 \sqcap \forall f_{i+1}.\bot \quad \text{for } 2 \le i \le n_{max}$$
$$\mathbf{P} \sqsubseteq \top_n \qquad\qquad\qquad\qquad\qquad \text{for each atomic relation } \mathbf{P} \text{ of arity } n$$
$$A \sqsubseteq \top_1 \qquad\qquad\qquad\qquad\qquad \text{for each atomic concept } A$$

These are standard reification axioms, and can already be found in [4].

We introduce a new atomic concept Q_w *for every individual* w *in* $\mathcal{A}$ *and define* $f(\mathcal{A})$ *to consist of the following axioms:*

$$f(\mathcal{A}) = \{w{:}Q_w \mid w \text{ occurs in } \mathcal{A}\} \cup$$
$$\{w_1{:}\le 1\ f_1^-.(\top_n \sqcap \exists f_2.Q_{w_2} \sqcap \ldots \sqcap \exists f_n.Q_{w_n}) \mid \langle w_1, \ldots, w_n \rangle \text{ occurs in } \mathcal{A}\}$$

These axioms are crucial when dealing with the problem of tuple-admissibility (see below) in the presence of ABoxes.

Finally, we define $\sigma(\mathcal{K}) = \langle(\sigma(\mathcal{S}) \cup f(\mathcal{S})), (\sigma(\mathcal{A}) \cup f(\mathcal{A}))\rangle$.

Theorem 3 *Let $\mathcal{K} = \langle\mathcal{S}, \mathcal{A}\rangle$ be a $\mathcal{DLR}$ knowledge-base. $\mathcal{K}$ is satisfiable iff the $\mathcal{SHIQ}$-KB $\sigma(\mathcal{K})$ is satisfiable.*

PROOF (sketch): The same techniques that were used in [2] can be adapted to the DL $\mathcal{SHIQ}$, and extended to deal with ABox axioms. The only-if direction is straightforward. A model $\mathcal{I}$ of $\mathcal{K}$ can be transformed into a model of $\sigma(\mathcal{K})$ by introducing, for every arity n with $2 \leq n \leq n_{\max}$ and every n-tuple of elements $\boldsymbol{d} \in (\Delta^{\mathcal{I}})^n$, a new element $t_{\boldsymbol{d}}$ that is linked to the elements of $\boldsymbol{d}$ by the functional relations $f_1, \ldots, f_n$. If we interpret $\top_1$ by $\Delta^{\mathcal{I}}$, $\top_n$ by the reifications of all elements in $\top_n^{\mathcal{I}}$, and, for every w that occurs in $\mathcal{A}$, Q_w by $w^{\mathcal{I}}$, then it is easy to show that we have constructed a model of $\sigma(\mathcal{K})$.

The converse direction is more complicated since a model of $\sigma(\mathcal{K})$ is not necessarily *tuple-admissible*, i.e., in general there may be distinct elements t, t' that are reifications of the same tuple $\boldsymbol{d}$. In the "un-reification" of such a model, $\boldsymbol{d}$ would only appear once which may conflict with assertions in the $\mathcal{DLR}$ KB about the number of tuples in certain relations. However, it can be shown that every satisfiable KB $\sigma(\mathcal{K})$ also has a tuple-admissible model. It is easy to show that such a model, by "un-reification", induces a model for the original KB $\mathcal{K}$.   □

We now have the machinery to transform a query containment problem into one or more $\mathcal{SHIQ}$ schema and ABox satisfiability problems. In the FaCT system we already have a decision procedure for $\mathcal{SHIQ}$ schema satisfiability, and this can be straightforwardly extended to deal with ABox axioms [22].

We have already argued why we believe our approach to be feasible. It should also be mentioned, that our approach matches the known worst-case complexity of the problem, which was determined as EXPTIME-complete in [4]. Satisfiability of a $\mathcal{SHIQ}$-KB can be determined in EXPTIME.[7] All reduction steps can be computed in deterministic polynomial time, with the exception of the reduction used in Theorem 2, which requires consideration of exponentially many mappings. Yet, for every fixed mapping, the reduction is polynomial, which yields that our approach decides query containment in EXPTIME.

4   The FaCT System

It is claimed in Section 1 that one of the main benefits of our approach is that it leads to a practical solution to the query containment problem. In this section we will substantiate this claim by presenting the results of a simple experiment in which the FaCT

---

[7] This does not follow from the algorithm presented in [22], which focuses on feasibility rather than worst-case complexity. It can be shown using a precompletion strategy similar to the one used in [26] together with the EXPTIME-completeness of $\mathcal{CIQ}$ [15].

system is used to decide a query containment problem with respect to the DWQ schema mentioned in Section 1.

The FaCT system includes an optimised implementation of a schema satisfiability testing algorithm for the DL $\mathcal{SHIQ}$. As the extension of FaCT to include the ABox satisfiability testing algorithm described in [22] has not yet been completed, FaCT is currently only able to test the satisfiability of a KB $\langle S, \mathcal{A} \rangle$ in the case where the $\mathcal{A}$ contains a single axiom of the form $w{:}C$ (this is equivalent to testing the satisfiability of the concept C w.r.t. the schema S). We have therefore chosen a query containment problem that can be reduced to a $\mathcal{SHIQ}$ KB satisfiability problem of this form using the methodology described in Section 3.

The DWQ schema is derived from the integration of several Extended Entity-Relationship (EER) schemas using $\mathcal{DLR}$ axioms to define inter-schema constraints [7]. One of the schemas, called the *enterprise* schema, represents the global concepts and relationships that are of interest in the Data Warehouse; a fragment of the enterprise schema that will be relevant to the query containment example is shown in Figure 2. A total of 5 source schemas representing (portions of) actual data sources are integrated with the enterprise schema using $\mathcal{DLR}$ axioms to establish the relationship between entities and relations in the source and enterprise schemas (the resulting integrated schema contains 48 entities, 29 relations and 49 $\mathcal{DLR}$ axioms). For example, one of the $\mathcal{DLR}$ axioms defining the relationship between the enterprise schema and the entity "Business-Customer" in the source schema describing business contracts is

$$\text{Business-Customer} \sqsubseteq (\text{Company} \sqcap \exists[\$1](\text{agreement} \sqcap$$
$$(\$2/3 : (\text{Contract} \sqcap \exists[\$1](\text{contract-company} \sqcap$$
$$(\$2/2 : \text{Telecom-company}))))))).$$

This axiom states, roughly speaking, that a Business-Customer is a kind of Company that has an agreement where the contract is with a Telecom-company.

As a result of this axiom, it is relatively easy to see that the query

$$q_1(x) \leftarrow \text{Business-Customer}(x)$$

is contained in the query

$$q_2(x) \leftarrow \text{agreement}(x, y_1, y_2) \wedge \text{Contract}(y_1) \wedge \text{Service}(y_2) \wedge$$
$$\text{contract-company}(y_1, y_3) \wedge \text{Telecom-company}(y_3)$$

with respect to the DWQ schema S, written $S \models q_1 \sqsubseteq q_2$.

The two queries can be transformed into the following (completed) canonical $\mathcal{DLR}$ ABoxes

$$\widehat{\mathcal{A}}_{q_1} = \{x{:}\text{Business-Customer}, x{:}P_x\}$$
$$\mathcal{A}_{q_2} = \{\langle x, y_1, y_2 \rangle{:}\text{agreement}, y_1{:}\text{Contract}, y_2{:}\text{Service},$$
$$\langle y_1, y_3 \rangle{:}\text{contract-company}, y_3{:}\text{Telecom-company}\},$$

where P_x is the representative concept for x. We can now compact and collapse $\mathcal{A}_{q_2}$ to give an ABox $\{x{:}C_{q_2}\}$, where

$$C_{q_2} = \exists[\$1](\text{agreement} \sqcap (\$2/3 : P_{y_1}) \sqcap (\$3/3 : P_{y_2}) \sqcap (\$2/3 : \text{Contract}) \sqcap$$
$$(\$3/3 : \text{Service}) \sqcap (\$2/3 : (\exists[\$1] \text{contract-company} \sqcap (\$2/2 : P_{y_3}) \sqcap$$
$$(\$2/2 : \text{Telecom-company})))).$$

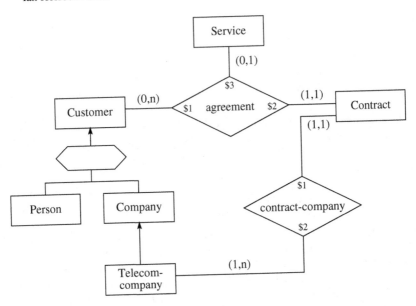

Fig. 2. A fragment of the DWQ enterprise schema

As each of the place-holders P_{y_1}, P_{y_2} and P_{y_3} occurs only once in the ABox, they can be replaced with $\top$, and C_{q_2} can be simplified to give

$$C'_{q_2} = \exists[\$1](\text{agreement} \sqcap (\$2/3 : \text{Contract}) \sqcap (\$3/3 : \text{Service}) \sqcap$$
$$(\$2/3 : (\exists[\$1]\text{contract-company} \sqcap (\$2/2 : \text{Telecom-company})))).$$

We can now determine if the query containment $\mathcal{S} \models q_1 \sqsubseteq q_2$ holds by testing the satisfiability of the KB $\langle \mathcal{S}, \mathcal{A} \rangle$, where $\mathcal{A} = \{x{:}\text{Business-Customer}, x{:}P_x, x{:}\neg C'_{q_2}\}$. Moreover, $\mathcal{A}$ can be compacted to give $\{x{:}C\}$, where $C = \text{Business-Customer} \sqcap P_x \sqcap \neg C'_{q_2}$, and the KB satisfiability problem can be decided by using FaCT to test the satisfiability of the concept $\sigma(C)$ w.r.t. the schema $\sigma(\mathcal{S})$. Thus we have $\mathcal{S} \models q_1 \sqsubseteq q_2$ iff $\sigma(C)$ is not satisfiable w.r.t. $\sigma(\mathcal{S})$.

The FaCT system is implemented in Common Lisp, and the tests were performed using Allegro CL Enterprise Edition 5.0 running under Red Hat Linux on a 450MHz Pentium III with 128Mb of RAM. Excluding the time taken to load the schema from disk (60ms), FaCT takes only 60ms to determine that $\sigma(C)$ is not satisfiable w.r.t. $\sigma(\mathcal{S})$. Moreover, if $\sigma(\mathcal{S})$ is first *classified* (i.e., the subsumption partial ordering of all named concepts in $\sigma(\mathcal{S})$ is computed and cached), the time taken to determine the unsatisfiability is reduced to only 20ms. The classification procedure itself takes 3.5s (312 satisfiability tests are performed at an average of ≈11ms per satisfiability test), but this only needs to be done once for a given schema.

Although the above example is relatively trivial, it still requires FaCT to perform quite complex reasoning, the result of which depends on the presence of $\mathcal{DLR}$ inter-schema constraint axioms; in the absence of such axioms (e.g., in the case of a single

EER schema), reasoning should be even more efficient. Of course deciding arbitrary query containment problems would, in general, require full ABox reasoning. However, the above tests still give a useful indication of the kind of performance that could be expected: the algorithm for deciding $\mathcal{SHIQ}$ ABox satisfiability presented [22] is similar to the algorithm implemented in FaCT, and as the number of individuals generated by the encoding of realistic query containment problems will be relatively small, extending FaCT to deal with such problems should not compromise the demonstrated empirical tractability. Moreover, given the kind of performance exhibited by FaCT, the limited amount of additional non-determinism that might be introduced as a result of cycles in the containing query would easily be manageable.

The results presented here are also substantiate our claim that transforming $\mathcal{DLR}$ satisfiability problems into $\mathcal{SHIQ}$ leads to greatly improved empirical tractability with respect to the embedding technique described in Calvanese et al. [4]. During the DWQ project, attempts were made to classify the DWQ schema using a similar embedding in the less expressive $\mathcal{SHIF}$ logic [19] implemented in an earlier version of the FaCT system. These attempts were abandoned after several days of CPU time had been spent in an unsuccessful effort to solve a single satisfiability problem. This is in contrast to the 3.5s taken by the new $\mathcal{SHIQ}$ reasoner to perform the 312 satisfiability tests required to classify the whole schema.

5 Discussion

In this paper we have shown how the problem of query containment under constraints can be decided using a KB (schema plus ABox) satisfiability tester for the $\mathcal{SHIQ}$ description logic, and we have indicated how a $\mathcal{SHIQ}$ schema satisfiability testing algorithm can be extended to deal with an ABox. We have only talked about conjunctive queries, but extending the procedure to deal with disjunctions of conjunctive queries is straightforward. The procedure for verifying containment between disjunctions of conjunctive queries is not very different from the one described for simple conjunctive queries. The main difference is that, although each conjunctive part becomes an ABox (as described in Section 3.1), the object representing the whole disjunctive query is a set of alternative ABoxes. This results in one more non-deterministic step, whose complexity is determined by the number of disjuncts appearing in both queries. Full details can be found in [20].

Although there is some loss of expressive power with respect to the framework presented in [4] this seems to be acceptable when modelling classical relational information systems, where regular expressions are seldom used.

As we have shown in Section 4, the FaCT implementation of the $\mathcal{SHIQ}$ schema satisfiability algorithm works well with realistic problems, and given that the number of individuals generated by query containment problems will be relatively small, there is good reason to believe that a combination of the ABox encoding and the extended algorithm will lead to a practical decision procedure for query containment problems. Work is underway to test this hypothesis by extending the FaCT system to deal with $\mathcal{SHIQ}$ ABoxes.

Bibliography

[1] S. Abiteboul, R. Hull, and V. Vianu. *Foundations of databases*. Addison-Wesley, 1995. 329, 331

[2] D. Calvanese. *Unrestricted and Finite Model Reasoning in Class-Based Representation Formalisms*. PhD thesis, Dipartimento di Informatica e Sistemistica, Università di Roma "La Sapienza", 1996. 338

[3] D. Calvanese, G. De Giacomo, and M. Lenzerini. Structured objects: modeling and reasoning. In *Proc. of DOOD'95*, number 1013 in LNCS, pages 229–246. Springer-Verlag, 1995. 328

[4] D. Calvanese, G. De Giacomo, and M. Lenzerini. On the decidability of query containment under constraints. In *Proc. of PODS'98*, 1998. 326, 327, 327, 334, 335, 335, 336, 337, 337, 338, 341, 341

[5] D. Calvanese, G. De Giacomo, and M. Lenzerini. Answering queries using views in description logics. In *Proc. of KRDB'96*, pages 6–10. CEUR, 1999. 327

[6] D. Calvanese, G. De Giacomo, M. Lenzerini, D. Nardi, and R. Rosati. Source integration in data warehousing. In *Proc. of DEXA-98*, pages 192–197. IEEE Computer Society Press, 1998. 326, 327, 327, 328

[7] D. Calvanese, G. De Giacomo, M. Lenzerini, D. Nardi, and R. Rosati. Use of the data reconciliation tool at telecom italia. DWQ deliverable D.4.3, Foundations of Data Warehouse Quality (DWQ), 1999. 327, 327, 339

[8] D. Calvanese, M. Lenzerini, and D. Nardi. Description logics for conceptual data modeling. In Jan Chomicki and Günter Saake, editors, *Logics for Databases and Information Systems*, pages 229–263. Kluwer Academic Publisher, 1998. 328

[9] Diego Calvanese, Giuseppe De Giacomo, Maurizio Lenzerini, Daniele Nardi, and Riccardo Rosati. Description logic framework for information integration. In *Proc. of KR-98*, pages 2–13, San Francisco, 1998. Morgan Kaufmann. 328

[10] T. Catarci and M. Lenzerini. Representing and using interschema knowledge in cooperative information systems. *Journal for Intelligent and Cooperative Information Systems*, 2(4):375–399, 1993. 326

[11] Edward P. F. Chan. Containment and minimization of positive conjunctive queries in OODB's. In *Proc. of PODS'92*, pages 202–211. ACM Press, 1992. 326

[12] A. K. Chandra and P. M. Merlin. Optimal implementation of conjunctive queries in relational databases. In *Proc. of 9th Annual ACM Symposium on the Theory of Computing*, pages 77–90. Assoc. for Computing Machinery, 1977. 326, 336

[13] C. Chekuri and A. Rajaraman. Conjunctive query containment revisited. In *Proc. of ICDT-97*, number 1186 in LNCS, pages 56–70. Springer-Verlag, 1997. 326, 336

[14] G. De Giacomo. *Decidability of Class-Based Knowledge Representation Formalisms*. PhD thesis, Dipartimento di Informatica e Sistemistica, Università di Roma "La Sapienza", 1995. 326

[15] G. De Giacomo and M. Lenzerini. TBox and ABox reasoning in expressive description logics. In *Proc. of KR'96*, pages 316–327. Morgan Kaufmann, 1996. 326, 338

[16] Giuseppe De Giacomo and Maurizio Lenzerini. What's in an aggregate: foundations for description logics with tuples and sets. In *Proc. of IJCAI'95*, pages 801–807. Morgan Kaufmann, 1995. 337

[17] M. J. Fischer and R. E. Ladner. Propositional dynamic logic of regular programs. *Journal of Computer and System Sciences*, 18:194–211, 1979. 326

[18] I. Horrocks. FaCT and iFaCT. In *Proc. of DL'99*, pages 133–135. CEUR, 1999. 326, 327

[19] I. Horrocks and U. Sattler. A description logic with transitive and inverse roles and role hierarchies. *Journal of Logic and Computation*, 9(3):385–410, 1999. 341

[20] I. Horrocks, U. Sattler, S. Tessaris, and S. Tobies. Query containment using a DLR ABox. LTCS-Report 99-15, LuFG Theoretical Computer Science, RWTH Aachen, Germany, 1999. See http://www-lti.informatik.rwth-aachen.de/Forschung/Reports.html. 327, 329, 333, 341

[21] I. Horrocks, U. Sattler, and S. Tobies. Practical reasoning for expressive description logics. In *Proc. of LPAR'99*, pages 161–180, 1999. 327, 331

[22] I. Horrocks, U. Sattler, and S. Tobies. Reasoning with individuals for the description logic shiq. In *Proc. of CADE-17*, number 1831 in LNCS, pages 482–496. Springer-Verlag, 2000. 327, 338, 338, 339, 341

[23] D. S. Johnson and A. Klug. Testing containment of conjunctive queries under functional and inclusion dependencies. *Journal of Computer and System Sciences*, 28(1):167–189, 1984. 326

[24] A. Y. Levy and M.-C. Rousset. CARIN: A representation language combining horn rules and description logics. In *Proc. of ECAI'96*, pages 323–327. John Wiley, 1996. 326, 327

[25] A. Schaerf. Reasoning with individuals in concept languages. *Data and Knowledge Engineering*, 13(2):141–176, 1994. 333

[26] S. Tessaris and G. Gough. Abox reasoning with transitive roles and axioms. In *Proc. of DL'99*. CEUR, 1999. 338

Static Reduction Analysis
for Imperative Object Oriented Languages

Gilles Barthe and Bernard Paul Serpette

INRIA Sophia-Antipolis, France
{Gilles.Barthe,Bernard.Serpette}@inria.fr

Abstract. We define a generic control-flow sensitive static analysis, *Static Reduction Analysis* (SRA), for an untyped object-oriented language featuring side-effects and exceptions. While its aims and range of applications closely relate to *Control Flow Analysis* (CFA), SRA exhibits a distinguishing feature: it only deals with abstract syntax tree (AST) nodes and does not involve approximations of environments nor stores.

1 Introduction

Static analysis [9,11] is a collection of compile-time techniques to predict program properties and a prerequisite for many program transformations/compiler optimisations (e.g. dead code elimination, partial evaluation or parallelisation) and program verifications (e.g. array-bound checking, pointer analysis or model-checking). In a nutshell, a static analysis computes, given a program P and a program point p, a finite set $\mathcal{S}(p)$ of *approximations* of p such that if P has value v at p, then $v \in \mathcal{S}(p)$ must hold. Knowledge about $\mathcal{S}(p)$ for each program point p of P is then used to perform transformations/verifications on P.

The main purpose of this paper is to report on a generic static analyser for an untyped, imperative object-oriented language featuring inheritance, overloading and exceptions. Our analyser presents four distinguishing features:

- context-sensitive: most analyses are control-flow insensitive and assume that statements and calls may be executed in an arbritrary order. Such analyses lead to unecessarily coarse approximations. For example, a control-flow insensitive analysis for the program

$$\texttt{x=1 ; x=2;}$$

simply merges all the possible values of x in the approximation and returns both values 1 and 2. We see this loss of precision as a weakness of existing analyses:

  - in a language like Java, variables are initialised by default [4, Section 4.5.4]. These default values will always be included in approximations, yielding imprecise results. Indeed, the example Java program

M. Parigot and A. Voronkov (Eds.): LPAR 2000, LNAI 1955, pp. 344–361, 2000.

```
class Point {
        int x;
static   void main (){
        o = new Point();
        o.x=10;
        o.x;
        }
}
```
is approximated to {0,10};

- the loss of precision incurred by control-flow insensitivity is not acceptable for some applications, such as information flow analysis [19]. Indeed, consider a slightly modified version of our first example:

x=secret; x=public;

It is approximated by the set {public, secret} so that if x is deemed public, a context-flow insensitive analysis will reveal a possible information flow leak (a public variable is approximated by a secret variable) whereas one knows that the value of x is public. Further examples of the phenomenon may be found in [8] and include programs such as

x=secret; x=6;

x=y; y=x;

where x is deemed public and y is deemed secret.[1]

- generic: in order to maintain a high-level of abstraction (and generality), method lookup is left unspecified and is relegated to an external lookup function *Lookup*, which determines the method that is activated in a method call. Specific approaches to method lookup may then be recovered by choosing a suitable instantiation of the *Lookup* function.

- parameterised: the quality of approximations, i.e. the size of $S(p)$, is determined by an approximation function, and is dictated by applications. For example, static analyses used in compiler optimisations favour efficiency against precision, whereas static analyses used in program verification tend to give up efficiency for increased precision. Following [16], we account for the different uses of static analyses by parameterising our analysis over an approximation function $\mathcal{F}$. Specific analyses may then be recovered by choosing a suitable instantiation of the $\mathcal{F}$ function.

- AST based: the salient feature of our analysis is to deal only with abstract syntax objects (AST nodes). The basic idea is to derive the analysis from an evaluator that handles bindings through a global store rather than through environments. Being AST based, Static Reduction Analysis does not require approximating new run-time values, such as environments; this is in strong contrast with k-CFA and related analyses, see e.g. [7,10,16] for functional languages and [5,13] for object-oriented languages, which introduce complex notions such as contours. It makes SRA simpler than these analyses without compromising generality (we can recover the k-CFA analyses by taking suitable instantiations of the approximation function) nor efficiency (the only

---

[1] Currently our analysis does not accept all the examples from [8] because we only compute arithmetic expressions when they appear in a comparison.

overheads of SRA, as compared to CFA, are those incurred by control-flow sensitivity). Again, this is in strong contrast with Set-Based Analysis, see e.g. [3,6], which trades generality for simplicity.

The main purpose of this paper is to define Static Reduction Analysis for an object-oriented language. Its applications to program transformation, including partial evaluation, and program verification, including security, will be reported elsewhere.

The remaining of the paper is organised as follows. In Section 2, we briefly review Static Reduction Analysis for λ-calculus. In Section 3, we introduce the language $\mathcal{L}_{ioe}$ and give the associated SRA rule. Section 4 introduces a family of approximations functions and compare their relative merits. Finally, we conclude in Section 5 with a brief description of a prototype implementation of SRA and directions for future work.

2   Static Reduction Analysis for the λ-Calculus

The purpose of this section is to define SRA for a call-by-value λ-calculus; most of the ideas originate from [14], where SRA is first introduced. For the clarity of the presentation, we first recall the traditional definition of evaluators based on environments and closures:

$$
\begin{aligned}
&[\![\lambda x.B]\!]_\varepsilon = <\lambda x.B, \varepsilon> && \text{(Abstraction)} \\
&[\![x]\!]_\varepsilon = \varepsilon(x) && \text{(Variable)} \\
&[\![F@A]\!]_\varepsilon = [\![B]\!]_{\varepsilon'[x \to [\![A]\!]_\varepsilon]} && \text{(Call in the extended environment)} \\
&\quad \textbf{where } [\![F]\!]_\varepsilon = <\lambda x.B, \varepsilon'> && \text{(Getting a closure)}
\end{aligned}
$$

Such evaluators allow to link values to expressions but introduce runtime objects, namely environments and closures, which need to be approximated [16]. An alternative for describing evaluators without introducing new runtime objects is to switch to a global store for handling bindings:

$$
\begin{aligned}
&[\![\lambda x.B]\!] = \lambda x.B && \text{(Abstraction)} \\
&[\![F@A]\!] = [\![B']\!] && \text{(Call in global environment)} \\
&\quad \textbf{where } [\![F]\!] = \lambda x.B && \text{(Function value)} \\
&\qquad g = gensym() && \text{(A fresh variable)} \\
&\qquad B' = B\{x \leftarrow g\} && \text{(α-conversion)} \\
&\qquad [\![g]\!] = [\![A]\!] && \text{(Variable definition)}
\end{aligned}
$$

The last clause generates a new equation fixing the value of g in the global store. This procedure lets us emulate shallow binding and maintains a tight relationship between source and generated terms as the evaluator only performs an α-conversion between $\lambda x.B$ and $\lambda g.B'$ (we say that $\lambda g.B'$ is a *copy* of $\lambda x.B$). While the latter evaluator may look rather unusual, it is a better starting point for static analyses. First, only AST objects need to be approximated. Second, approximations are simply obtained by merging multiple copies into a single one. These two facts largely contribute to the simplicity of the analysis.

In [14], the decision to merge or not multiple copies of a function is relegated to an external function $\mathcal{F}$, called the *approximation function*. This yields a family of analyses including the k-CFA analyses. Technically, $\mathcal{F}$ takes as argument an abstraction and a program point at which the abstraction is applied and returns another abstraction equal to an α-conversion of its second argument—we do not identify terms up to α-conversion and rather see bound variables as memory locations. The set $\mathcal{S}(e)$ of a program point e is then defined as the least solution to the system of constraints generated by the rules:

$$
\begin{array}{ll}
\mathcal{S}(\lambda x.B) = \{\lambda x.B\} & \text{(Values of abstraction)} \\
\mathcal{S}(F@A) = \bigcup_i \mathcal{S}(B'_i) & \text{(Values of call)} \\
\quad \textbf{where } \lambda x_i.B_i \in \mathcal{S}(F) & \text{(Each possible function)} \\
\quad\quad \lambda x'_i.B'_i = \mathcal{F}(F@A, \lambda x_i.B_i) & \text{(Approximated functions)} \\
\quad\quad \mathcal{S}(x'_i) \supseteq \mathcal{S}(A) & \text{(Values of variables)}
\end{array}
$$

By taking appropriate instantiations for $\mathcal{F}$, one obtains analyses of different precision: e.g. one may recover the evaluator by forcing $\mathcal{F}$ to make a fresh copy always, and one may recover the well-known 0-CFA analysis [16] by forcing $\mathcal{F}$ to return its second argument always. These results shall scale up to the $\mathcal{L}_{ioe}$-calculus, which we define below.

3   Static Reduction Analysis for the $\mathcal{L}_{ioe}$-Calculus

3.1   The Imperative Object-Oriented Language $\mathcal{L}_{ioe}$

The $\mathcal{L}_{ioe}$ language is an untyped object-oriented language featuring method calls, assignments, control structures (conditionals and exceptions), pointer equality, arithmetic operations and comparison operators. Programs are defined as a list of classes, which are themselves described by their lists of fields and methods. Inheritance between classes is left abstract and given by a partial order $\leq$ on *Class*. The full syntax of the language is described in Figure 1. Note that:

- in order to simplify the presentation, methods are assumed to have at least one argument. To handle static methods without arguments, a dummy argument must be added;
- *New* expressions allocate a new object with explicit initial values. In an expression $New(c.a^*)$, the argument a_i provides the initial value for the field f_i of the resulting object;
- *Vector* expressions are declared with the vector length and an initial value that is shared by all cells;
- there are no global variables (static variables in Java). However, static variables can be emulated by introducing a specific class Global having a unique instance created at beginning of the program. This instance holds all the global variables and, to be accessed, must be given as first argument for all methods. All other Java's modifiers are handled by the lookup function;
- the **this** variable must appear explicitly as the first parameter of non-static methods (implicitly added in our examples);

– like in Java, Try expressions include a list of guards and introduce methods which may include free variables. These methods have a specific status during the copying process, see the end of Subsection 3.2, and may be used to emulate sequences, local variables and while loops.

Evaluation is call-by-value and arguments are evaluated from left to right. The evaluator is derived as an instance of the analysis by taking as approximation function the function $\mathcal{F}_\infty$ which always makes a fresh copy of method calls. In

$Program$	$= Class^*$	
$Class$	$= Field^* \times Method^*$	
$Method$	$= Var^* \times Expr$	
$Guard$	$= Class \times Method$	
$Expr$	$=$ Union of types listed below	
$Data$	$= \mathbb{Z}$	(integer)
New	$= Class \times Expr^*$	(object allocation with initial values)
$Vector$	$= Expr \times Expr$	(vector allocation with shared initial value)
Op	$= Prim \times Expr \times Expr$	(arithmetic operation)
$Call$	$= Method \times Expr^*$	(method invocation)
$GetVar$	$= Var$	(variable)
$SetVar$	$= Var \times Expr$	(assignment)
$GetField$	$= Expr \times Field$	(read inside object)
$SetField$	$= Expr \times Field \times Expr$	(write inside object)
If	$= Expr \times Expr \times Expr$	(conditional)
$Comp$	$= Prim \times Expr \times Expr$	(arithmetic comparison)
Eq	$= Expr \times Expr$	(pointer equality)
$GetVector$	$= Expr \times Expr$	(read inside vector)
$SetVector$	$= Expr \times Expr \times Expr$	(write inside vector)
$Length$	$= Expr$	(length of vector)
$Throw$	$= Expr$	(throwing an exception)
Try	$= Expr \times Guard^*$	(catching an exception)

Fig. 1. THE LANGUAGE

the sequel, we adopt the following useful conventions:

– We let $Cons(a.b)$ denote an element of a product type $Cons = A \times B$ and a^* denote an element of a vector type A^*. Moreover, we let $\|a^*\|$ denote the length of a^*. For $1 \leq i \leq n$, we let a_i denote the ith element of a^* and, if $\|a^*\| \neq 0$ we let $Tail(a^*)$ denote the tail of a^*.
– The cardinal of a set A is denoted by $\|A\|$ and its powerset is denoted by $\mathcal{P}(A)$. Moreover, relations or graphs on A are specified as functions $f : A \to \mathcal{P}(A)$ and the transitive closure of f, $f^+ : A \to \mathcal{P}(A)$, is defined by the clauses:

$$y \in f(x) \ \Rightarrow \ y \in f^+(x)$$
$$(y \in f^+(x) \ \wedge \ z \in f(y)) \ \Rightarrow \ z \in f^+(x)$$

The inverse $f^{-1} : A \to \mathcal{P}(A)$ of $f : A \to \mathcal{P}(A)$ is defined by:

$$f^{-1}(y) = \{x \in A \mid y \in f(x)\}$$

- If $c = Class(f^*.m^*)$ then $Index(g, c)$ is the unique (if it exists) i such that $f_i = g$.

3.2 Static Reduction Rules for the $\mathcal{L}_{ioe}$-Calculus

Static Reduction Analysis assigns to each expression of a program a set of possible values. Values are themselves expressions and can be numerals, arithmetic operations (viewed as $\mathbb{Z}$ constructors so as to ensure termination of the analysis), vectors or New objects. In other words, Static Reduction Analysis may be viewed as a function $\mathcal{S} : Expr \cup Var \to \mathcal{P}(Value)$ with

$$Value = Data \cup Op \cup Vector \cup New$$

Below we assume given:

- a lookup function $Lookup : Method, Value^* \to Method$ for method resolution;
- an approximation function $\mathcal{F} : Call, Method, Value^* \to Method$ for deciding about method copying.

Moreover, we assume given two specific boolean values $true$ and $false$—e.g. they can be the unique instances of two classes $True$ and $False$ and thus $true, false$ are included in New.

Auxiliary functions The analysis relies on several auxiliary functions, which we describe below:

- a function down $\mathcal{D}$, which returns the left leaf of an expression;
- a function next $\mathcal{N}$, which approximates evaluation flow;
- a function $\mathcal{U}$, which approximates dynamic scope and the call graph.

The down function $\mathcal{D} : Expr \to Expr$ returns the left leaf of an expression.

$$
\begin{array}{lll}
\text{DATA}_\mathcal{D}: & e \in Data(n) & \Rightarrow \mathcal{D}(e) = e \\
\text{GETVAR}_\mathcal{D}: & e = GetVar(v) & \Rightarrow \mathcal{D}(e) = e \\
\text{NEW}_\mathcal{D}: & e = New(c.a^*) & \Rightarrow \mathcal{D}(e) = \textbf{if } \|a^*\| = 0 \textbf{ then } e \textbf{ else } \mathcal{D}(a_1) \\
\text{OP}_\mathcal{D}: & e = Op(f.a.b) & \Rightarrow \mathcal{D}(e) = \mathcal{D}(a)
\end{array}
$$

The definition of $\mathcal{D}$ on other expressions is straightforward and similar to $\text{OP}_\mathcal{D}$.

The next function $\mathcal{N} : (Expr \cup Method) \to \mathcal{P}(Expr \cup Method)$ approximates evaluation flow. Intuitively, $e_2 \in \mathcal{N}^+(e_1)$ if the evaluator would have computed e_1 before computing e_2. Note that the definition of $\mathcal{N}$ relies on the function $\mathcal{A}$ defined in Subsection 3.2, which approximates the arguments and corresponding methods activated on a call site.

$$
\begin{array}{lll}
\text{METHOD}_\mathcal{N}\colon & m = Method(v^*.b) & \Rightarrow \mathcal{N}(m) \ni \mathcal{D}(b) \\
\text{NEW}_\mathcal{N}\colon & e = New(c.a^*) \wedge 1 \le i < \|a^*\| & \Rightarrow \mathcal{N}(a_i) \ni \mathcal{D}(a_{i+1}) \\
& e = New(c.a^*) \wedge \|a^*\| > 0 & \Rightarrow \mathcal{N}(a_{\|a^*\|}) \ni e \\
\text{IF}_\mathcal{N}\colon & e = If(p.a.b) \wedge true \in \mathcal{S}(p) & \Rightarrow \mathcal{N}(p) \ni \mathcal{D}(a) \wedge \mathcal{N}(a) \ni e \\
& e = If(p.a.b) \wedge false \in \mathcal{S}(p) & \Rightarrow \mathcal{N}(p) \ni \mathcal{D}(b) \wedge \mathcal{N}(b) \ni e \\
\text{CALL}_\mathcal{N}\colon & e = Call(m.a^*) \wedge 1 \le i < \|a^*\| & \Rightarrow \mathcal{N}(a_i) \ni \mathcal{D}(a_{i+1}) \\
& e = Call(m.a^*) \wedge m'.o^* \in \mathcal{A}(e) & \Rightarrow \mathcal{N}(a_{\|a^*\|}) \ni m' \\
& e = Call(m.a^*) \wedge (v^*.b).o^* \in \mathcal{A}(e) & \Rightarrow \mathcal{N}(b) \ni e \\
\text{OP}_\mathcal{N}\colon & e = Op(f.a.b) & \Rightarrow \mathcal{N}(a) \ni \mathcal{D}(b) \wedge \mathcal{N}(b) \ni e
\end{array}
$$

In all other cases, except Try, $\mathcal{N}$ is a left-to-right traversal of the AST as in $\text{OP}_\mathcal{N}\colon$. The definition of $\mathcal{N}$ for Try is deferred until the end of this subsection. There are several points worth noting:

– methods are added to the domain and codomain of $\mathcal{N}$ to highlight a method entry point;
– conditionals may give rise to multiple evaluation flow paths: if $\mathcal{S}(p) = \{true, false\}$ then for $e = If(p.a.b)$, $\mathcal{N}^{-1}(e) = \{a, b\}$.

The up function $\mathcal{U} : (Expr \cup Method) \to \mathcal{P}(Expr \cup Method)$ approximates both dynamic scope and the call graph. As for the $\mathcal{N}$ function, the definition of $\mathcal{U}$ relies on the function $\mathcal{A}$ defined in Subsection 3.2.

$$
\begin{array}{lll}
\text{METHOD}_\mathcal{U}\colon & m = Method(v^*.b) & \Rightarrow \mathcal{U}(b) \ni m \\
\text{CALL}_\mathcal{U}\colon & e = Call(m.a^*) \wedge m'.o^* \in \mathcal{A}(e) & \Rightarrow \mathcal{U}(m') \ni e \\
\text{OP}_\mathcal{U}\colon & e = Op(f.a.b) & \Rightarrow \mathcal{U}(a) = \mathcal{U}(b) = \{e\}
\end{array}
$$

In other cases, as in $\text{OP}_\mathcal{U}\colon$, $\mathcal{U}$ is the inverse of the tree structure induced by the AST.

Values In order to reflect the call-by-value evaluation strategy, the rules for $\mathcal{S}$ on values check whether their subterms may yield a value. E.g. if a or b has no possible value then $Op(f.a.b)$ has no possible value either.

$$
\begin{array}{lll}
\text{DATA}_\mathcal{S}\colon & e = Data(n) & \Rightarrow \mathcal{S}(e) \ni e \\
\text{NEW}_\mathcal{S}\colon & e = New(c.a^*) \wedge (\forall i \in [1, \|a^*\|], \mathcal{S}(a_i) \ne \emptyset) & \Rightarrow \mathcal{S}(e) \ni e \\
\text{VECTOR}_\mathcal{S}\colon & e = Vector(n.a) \wedge \mathcal{S}(n) \ne \emptyset \wedge \mathcal{S}(a) \ne \emptyset & \Rightarrow \mathcal{S}(e) \ni e \\
\text{OP}_\mathcal{S}\colon & e = Op(f.a.b) \wedge \mathcal{S}(a) \ne \emptyset \wedge \mathcal{S}(b) \ne \emptyset & \Rightarrow \mathcal{S}(e) \ni e
\end{array}
$$

Method call Since the dynamic types of the arguments can be found via $\mathcal{S}$, we can approximate the arguments and corresponding method activated on a call site by the function $\mathcal{A} : Call \rightarrow \mathcal{P}(Method \times Values^*)$ defined by the clause

$$\mathcal{A}(e) = \{m'.o^* | \forall 1 \leq i \leq \|a^*\|. \ \ o_i \in \mathcal{S}(a_i) \ \wedge \ m' = \mathcal{F}(e, Lookup(m, o^*), o^*)\}$$

if $e = Call(m.a^*)$.

Note that the definition of $\mathcal{A}$ reflects the call-by-value strategy since we need at least one value for each argument. With $\mathcal{A}$ we can define $\mathcal{S}$ both on method calls and variables.

CALL$_\mathcal{S}$: $e = Call(m.a^*) \ \wedge \ (v^*.b).o^* \in \mathcal{A}(e) \ \Rightarrow \ \mathcal{S}(v_i) \supseteq \mathcal{S}(o_i) \ (1 \leq i \leq \|a^*\|)$
$\qquad\quad e = Call(m.a^*) \ \wedge \ (v^*.b).o^* \in \mathcal{A}(e) \ \Rightarrow \ \mathcal{S}(e) \supseteq \mathcal{S}(b)$

The first rule transmits the values of arguments to formal parameters whereas the second rule transmits the values computed by the methods to the caller.

Variables Variables are approximated by following $\mathcal{N}^{-1}$ until reaching $Method$ or $SetVar$ nodes where their value has been initialised or modified. Upon reaching $Method$ and $SetVar$ nodes involving the variable to be approximated, the traversal is interrupted and, in particular, does not attempt to reach earlier $Method$ and $SetVar$ nodes involving v. In order to ensure termination of the analysis, the function $SearchVar$ specifying the traversal admits an extra argument that takes care of cycles in $\mathcal{N}^{-1}$-paths by keeping track of the expressions that have already been inspected.

GETVAR$_\mathcal{S}$: $e = GetVar(v) \ \Rightarrow \ \mathcal{S}(e) \supseteq SearchVar(v, e, \emptyset)$

where
$SearchVar(v, e, s) = \quad$ **if** $(e \in s)$ $\qquad\qquad\qquad\qquad$ **then** $\emptyset$
$\qquad\qquad\qquad\qquad\quad$ **else if** $(e = Method(v^*.b) \ \wedge \ v \in v^*)$ **then** $\mathcal{S}(v)$
$\qquad\qquad\qquad\qquad\quad$ **else if** $(e = SetVar(v.a))$ $\qquad\qquad$ **then** $\mathcal{S}(a)$
$\qquad\qquad\qquad\qquad\quad$ **else** $\qquad\qquad\qquad\qquad\qquad\qquad SearchVar_{aux}(v, e, s)$

$SearchVar_{aux}(v, e, s) = \bigcup_{e_i \in \mathcal{N}^{-1}(e)} SearchVar(v, e_i, s \cup \{e\})$

Example 1. The approximation of x in $x = 1; x = 2$ is 2.

Pointer equality Our analysis does not provide enough information to approximate pointer equality precisely. Indeed, approximations may only be used to detect inequality: if $\mathcal{S}(a) \cap \mathcal{S}(b) = \emptyset$, then a and b cannot be physically equal. In other cases, i.e. when $\mathcal{S}(a) \cap \mathcal{S}(b) \neq \emptyset$, one cannot conclude. While most static analyses adopt such an imprecise viewpoint, there exists several dedicated alias analyses, see e.g. [15,18], which provide precise approximations of pointer equality. Here we opt for an intermediate approach by letting approximations for pointer equality depend on an auxiliary function $\phi_=$, which acts as a (rudimentary) pointer analysis.

EQ$_\mathcal{S}$: $e = Eq(a.b) \ \Rightarrow \ \mathcal{S}(e) \supseteq DecideEq(a, b)$
where

$$DecideEq(a,b) = \textbf{if } \mathcal{S}(a) = \emptyset \ \vee \ \mathcal{S}(b) = \emptyset \qquad \textbf{then } \emptyset$$
$$\textbf{else if } \mathcal{S}(a) \cap \mathcal{S}(b) = \emptyset \qquad \textbf{then } \{false\}$$
$$\textbf{else if } \mathcal{S}(a) = \mathcal{S}(b) \ \wedge \ \Phi_=(a) \textbf{ then } \{true\}$$
$$\textbf{else} \qquad\qquad\qquad\qquad \{true, false\}$$

There are multiple implementations of the $\Phi_=$ function. The simple-minded approach referred above is to define $\Phi_=(x) = false$. In order to enhance the precision of the analysis, our prototype implementation of SRA (see Subsection 5.1) returns $true$ for $\Phi_=(e)$ whenever $\mathcal{S}(e) = \{o\}$ and there exists only one path starting from o and following $\mathcal{U}$. Our choice for such a rudimentary implementation is motivated by the following facts:

- the definition of $\Phi_=$ is accurate enough to derive the evaluator as a specific instance of our analysis;
- the definition of $\Phi_=$ is sufficient to illustrate how the choice of $\Phi_=$ influences the quality of approximations for field and vector access;
- more sophisticated implementations of $\Phi_=$ rely on techniques that are orthogonal to SRA and beyond the scope of this paper.

Example 2. The following example illustrates some of the subtleties involved with alias analysis.

```
class T {
static  main() {
    create() == create()
  }

static  create() {
    new T();
  }
}
```

The equality will be approximated to $false$ whenever the approximation function produces two different copies for the two invocations of `create`; this essentially corresponds to k-CFA for $k \geq 1$. If the approximation function merges the two invocations of `create` in a single copy, as done by 0-CFA, then the equality is approximated to $\{true, false\}$.

Example 3. The following example illustrates the usefulness of $\Phi_=$.

```
class T {
static   main() {
  let o = new T();
   o == o;
  }
}
```

Whereas standard analyses (setting $\Phi_=(x) = false$) approximate pointer equality by $\{true, false\}$, our approximation yields $\{true\}$ since $\Phi_=(o)$ holds.

Fields The rules for fields essentially follow the same pattern as those for variables and use an auxiliary function $SearchField$ to compute the approximations for field access. The function $SearchField$ takes as arguments an object o and a field f of this object, an expression e providing the AST upon which the traversal is to be made, and an accumulator s to keep track of nodes that have been visited previously so as to prevent loops. $SearchField$ specifies a traversal along $\mathcal{N}^{-1}$-paths until reaching corresponding New or $SetField$ nodes. Upon reaching such a node, say $SetField(t.f.a)$, one checks whether o is a possible value of the target expression, i.e $o \in \mathcal{S}(t)$; if so, then the new constraint $\mathcal{S}(a) \subseteq \mathcal{S}(e)$ is added. Then the search proceeds recursively.

GETFIELD$_S$: $e = GetField(o.f) \Rightarrow \mathcal{S}(e) \supseteq \bigcup_{o_i \in \mathcal{S}(o) \cap New} SearchField(o_i, f, e, \emptyset)$

where

$SearchField(o, f, e, s) =$
    if $(e \in s)$ **then** $\emptyset$
    else if $(e = SetField(t.f.a) \wedge o \in \mathcal{S}(t))$ **then** $\mathcal{S}(a) \cup SF_{cont}(t, o, f, e, s)$
    else if $(e = o = New(c.a^*))$ **then** $\mathcal{S}(a_{Index(f,c)}) \cup SF_{cont}(o, o, f, e, s)$
    else $SearchField_{aux}(o, f, e, s)$

$SF_{cont}(t, o, f, e, s)$ $=$ **if** $\Phi_=(t)$ **then** $\emptyset$
                                     else $SearchField_{aux}(o, f, e, s)$

$SearchField_{aux}(o, f, e, s) = \bigcup_{e_i \in \mathcal{N}^{-1}(e)} SearchField(o, f, e_i, s \cup \{e\})$

Note that, unlike for variables, one cannot systematically stop the search at nodes $SetField(t.f.a)$ satisfying $o \in \mathcal{S}(t)$ or at the creation node.

Example 4. Here is a simple example demonstrating why the search needs to proceed beyond creation nodes. Here the approximation function is supposed not to make copies of `create`.

```
class T {
field  x;
static main() {
        let o1 = create(1);
        o1.x = 2;
        let o2 = create(3);
        o1.x;
        }

static create(n) {
        new T(n);
        }
}
```

Beginning at `o1.x`, the only $\mathcal{N}^{-1}$-path reaches `new T(n)`. If we stop the search here, then one would obtain

$$\mathcal{S}(\texttt{o1.x}) = \mathcal{S}(\texttt{n}) = \{1, 3\}$$

which is incorrect: indeed the approximation is not conservative since o1.x evaluates to 2.

Now let us show that our rules are correct. Since there is a $\mathcal{U}$-fork from method `create`, we have to continue the search and can thus reach the o1.x=2 node. The approximation now becomes conservative:

$$\mathcal{S}(\texttt{o1.x}) = \{1, 2, 3\}$$

Example 5. The following example illustrates how $\phi_=$ may be used to good effect to improve the quality of approximations. If the approximation function duplicates the method `create` in the above example, the approximation of o1.x coincides with its approximation in the example below:

```
class T {
    field x;
    main() {
        let o1 = create1(1);
        o1.x = 2;
        let o2 = create2(3);
        o1.x;
    }
    create1(n) {
        new T(n);
    }
    create2(n) {
        new T(n);
    }
}
```

Beginning at o1.x, the only $\mathcal{N}^{-1}$-path bypasses the allocation done in `create2` (since this is not a value of o1) and reaches directly the o1.x = 2 node, yielding 2 as a result. The search stops at this node since there is no $\mathcal{U}$-fork starting from the method `create1`.

Conditionals and arithmetic comparisons The rules for conditionals are straightforward:

$$\begin{aligned}
\text{IF}_{\mathcal{S}} \colon e = If(p.a.b) \;\wedge\; true \in \mathcal{S}(p) &\;\Rightarrow\; \mathcal{S}(e) \supseteq \mathcal{S}(a)\\
e = If(p.a.b) \;\wedge\; false \in \mathcal{S}(p) &\;\Rightarrow\; \mathcal{S}(e) \supseteq \mathcal{S}(b)
\end{aligned}$$

Arithmetic comparisons are more subtle to handle. Since arithmetic operations are not reduced by $\mathcal{S}$, one needs to execute recursively all arithmetic operations pending before proceeding to the comparison. As for variables, termination of the analysis is ensured by an auxiliary function which detects cycles in the above mentioned process. In case a cycle is detected, then the execution of arithmetic operations is aborted and both *true* and *false* are returned.

COMP$_S$: $e = Comp(f.a.b) \wedge a' \in Collect(a) \wedge b' \in Collect(b) \Rightarrow S(e) \ni \underline{Comp}\ a'\ b'$
$e = Comp(f.a.b) \wedge CollectLoop(a) \qquad\qquad\qquad \Rightarrow S(e) \supseteq \{true, false\}$
$e = Comp(f.a.b) \wedge CollectLoop(b) \qquad\qquad\qquad \Rightarrow S(e) \supseteq \{true, false\}$

where

$$Collect(d) = \{n | Data(n) \in S(d)\} \cup$$
$$\{a'\underline{f}b' \mid a' \in Collect(a) \wedge b' \in Collect(b) \wedge Op(f.a.b) \in S(d)\}$$
$$CollectLoop(op) = S(op) \neq \emptyset \wedge CollectLoop(op, \emptyset)$$
$$CollectLoop(op, s) = \mathsf{let}\ op = Op(f.a.b),\ s' = s \cup \{op\}\ \mathsf{in}$$
$$op \in s \vee CollectLoop_{aux}(a, s') \vee CollectLoop_{aux}(b, s')$$
$$CollectLoop_{aux}(e, s) = \bigvee\nolimits_{(op=Op(f.a.b)) \in S(e)} CollectLoop(op, s)$$

In the above rules, *Collect* computes recursively a set of integers by applying the suitable operations. We underline function symbols to denote their implementation so that $3\underline{+}2$ is considered equal to 5. The function *CollectLoop* detects whether any loop arises in the process of computing the *Collect* operations.

Vectors The rules for vectors follow the same pattern as for fields and use an auxiliary function *SearchVector* to compute approximations for vector access. The function *SearchVector* takes as arguments a vector a and an index n in this vector, an expression e providing the AST upon which the traversal is to be made, and an accumulator s to keep track of nodes that have been visited previously so as to prevent loops. *SearchVector* specifies a traversal along $\mathcal{N}^{-1}$-paths until reaching corresponding *Vector* or *SetVector* nodes. Upon reaching such a node, say $SetVector(t.n'.b)$, one checks whether a is possible value of the target expression (i.e. $a \in S(t)$) and whether n and n' may yield the same value (by using rules provided by arithmetic comparisons); if so, then the new constraint $S(o) \subseteq S(e)$ is added. Then the search proceeds recursively depending of the same kind of check used in fields rules.

Finally, the rule for *Length* is straightforward:

LENGTH$_S$: $e = Length(a) \Rightarrow S(e) \supseteq \bigcup_{Vector(n_i, e_i) \in S(a)} S(n_i)$

Exceptions Exceptions in our language, as in Lisp, ML, Java or C++, have a dynamic scope: an expression $e = Try(a.g^*)$ can only catch exceptions raised during the execution of a. We therefore need to approximate the dynamic scope of a; to this end, we use $(\mathcal{U}^+)^{-1}(a) = (\mathcal{U}^{-1})^+(a)$. Besides, exceptions may be analysed by two different methods:

- starting from a *Try*, one might search for all the *Throws* in the evaluation flow. In this case, we start from an expression $e = Try(a.g^*)$ and follow $\mathcal{N}$ from $\mathcal{D}(a)$ until reaching a node $Throw(e')$. However, $Throw(e')$ may lie outside of the dynamic scope of a and/or there may be another *Try* between e and $Throw(e')$ which catches the latter;
- starting from a *Throw*, one might search for all the surrounding *Trys*. In this case, we start from an expression $e = Throw(a)$ and follow $\mathcal{U}$ until reaching

a node $Try(e'.g^*)$. However, several elements of $Throw$ can reach the same Try whereas only one is really activated.

In order to overcome the limitations of each method, the definition of $\mathcal{S}$ on exceptions uses both. More precisely, we start from a Try statement and search all the $Throw$ statements in the evaluation flow. When reaching such $Throw$ statements, we go back $\mathcal{U}$ to check whether we recover the original Try statement.

$\text{TRY}_{\mathcal{N}}: e = Try(a.g^*) \;\wedge\; \mathcal{S}(a) \neq \emptyset \qquad\qquad\qquad \Rightarrow\; \mathcal{N}(a) \ni e$

$e = Try(a.g^*) \;\wedge\; t.(m = v^*.b).o \in SearchThrow(a, g^*) \;\Rightarrow\; \mathcal{N}(t) \ni m \;\wedge\; \mathcal{N}(b) \ni e$

$\text{TRY}_{\mathcal{S}}: e = Try(a.g^*) \qquad\qquad\qquad\qquad\qquad\qquad \Rightarrow\; \mathcal{S}(e) \supseteq \mathcal{S}(a)$

$e = Try(a.g^*) \;\wedge\; t.(m = v^*.b).o \in SearchThrow(a, g^*) \;\Rightarrow\; \mathcal{S}(v_1) \ni o \;\wedge\; \mathcal{S}(e) \supseteq \mathcal{S}(b)$

where

$\overline{SearchThrow(a, g^*)} \quad = \bigcup_{t_i \in Thrown(\mathcal{D}(a), \emptyset), (o_j = New(c.a^*)) \in \mathcal{S}(t_i)} SBTry(g^*, t_i, o_j, c)$

$\overline{Thrown(e, s)} \qquad\quad = \textbf{if } (e \in s) \qquad\qquad\qquad\qquad \textbf{then } \emptyset$

$\qquad\qquad\qquad\qquad\quad \textbf{else if } (\mathcal{U}(e) = \{Throw(e)\}) \textbf{ then } \{e\}$

$\qquad\qquad\qquad\qquad\quad \textbf{else} \qquad\qquad\qquad \bigcup_{e_i \in \mathcal{N}(e)} Thrown(e_i, s \cup \{e\})$

$\overline{SBTry(g^*, t, o, c)} \quad = \{t.m_k.o \mid m_k \in SearchTry(g^*, c, t, \emptyset)\}$

$\overline{SearchTry(g^*, c, e, s)} = \textbf{if } (e \in s) \qquad\qquad\qquad\qquad \textbf{then } \emptyset$

$\qquad\qquad\qquad\qquad\quad \textbf{else if } (\mathcal{U}(e) = \{Try(e, g^*)\}) \textbf{ then } Caught(g^*, c)$

$\qquad\qquad\qquad\qquad\quad \textbf{else if } (\mathcal{U}(e) = \{Try(e, g'^*)\}) \textbf{ then}$

$\qquad\qquad\qquad\qquad\qquad\quad \textbf{if } Caught(g'^*, c) = \emptyset$

$\qquad\qquad\qquad\qquad\qquad\quad \textbf{then } SearchTry(g^*, c, \mathcal{U}(e)_1, s \cup \{e\})$

$\qquad\qquad\qquad\qquad\qquad\quad \textbf{else } \emptyset$

$\qquad\qquad\qquad\qquad\quad \textbf{else} \qquad\qquad\qquad \bigcup_{e_i \in \mathcal{U}(e)} SearchTry(g^*, c, e_i, s \cup \{e\})$

$\overline{Caught(g^*, c)} \qquad = \textbf{if } g^* = \emptyset \qquad\qquad\qquad\qquad \textbf{then } \emptyset$

$\qquad\qquad\qquad\qquad\quad \textbf{else if } g_1 = c'.m \;\wedge\; c \leq c' \textbf{ then } \{m\}$

$\qquad\qquad\qquad\qquad\quad \textbf{else} \qquad\qquad\qquad Caught(Tail(g^*), c)$

Note that:

- $SearchThrow(a.g^*)$ returns a set of $t.m.o$ where t is a thrown expression having a value o which is caught by the method m in g^*;
- the function $Thrown$ finds all exceptions raised during the continuation ($\mathcal{N}$) of an expression. $\mathcal{N}$ of the thrown expression is only defined when finding the corresponding guards;
- $SearchTry$ follows $\mathcal{U}$ until reaching a specific Try (given by its guards g^*); observe that the predicate $\mathcal{U}(e) = \{Try(e, g^*)\}$ is used to specify that we do not come from guards. $SearchTry$ returns $\emptyset$ when reaching a nested Try that has already caught the given exception value (given by its class c);
- the function $Caught$ simply verifies whether a list of guards can catch a specific class of exceptions. It is the only place where inheritance plays a role in the analysis (apart from the $Lookup$ function, which we leave unspecified);
- guards introduce local methods which, in order to ensure correctness, must always be copied when the surrounding method is copied.

4   Approximation Functions

The quality and efficiency of SRA rely on the approximation function. The purpose of this section is to define some standard approximation functions and exemplify their behaviour on some small examples.

4.1   Equivalence Classes of Expressions

In order to be correct, approximation functions are required to return copies that are α-equivalent to the original method. In order to enforce such a requirement, we are led to consider expressions up to α-conversion and to introduce the quotient set $\overline{Expr}$ of expressions up to α-conversion. Then one can define $\overline{\mathcal{S}} : \overline{Expr} \to \mathcal{P}(\overline{Value})$ by the clause

$$\overline{\mathcal{S}}(\overline{e}) = \{\overline{v} \mid v \in \mathcal{S}(e') \wedge e' \in \overline{e}\}$$

The auxiliary functions $\mathcal{N}$ and $\mathcal{U}$ can be merged the same way. Considering $\overline{Expr}$ and $\overline{\mathcal{S}}$ is useful for a number of purposes:

- *comparisons* between approximation functions may be formulated in terms of $\overline{\mathcal{S}}$. Formally, we define $\mathcal{F}_1 \leq \mathcal{F}_2$ iff

$$\forall \overline{e} \in \overline{Expr}.\ \overline{\mathcal{S}}_{\mathcal{F}_1}(\overline{e}) \subseteq \overline{\mathcal{S}}_{\mathcal{F}_2}(\overline{e})$$

- correct approximation functions are defined to be those functions $\mathcal{F}$ satisfying $\overline{\mathcal{F}(e, m, o^*)} = \overline{m}$ for every e, m and o^*. The *correctness* of SRA is then expressed as $\mathcal{F}_\infty \leq \mathcal{F}$ for all correct approximation functions $\mathcal{F}$.
- *synthetic results.* Most post-SRA analyses only require to know the possible values of the AST nodes of the original program. By considering $\overline{\mathcal{S}}$ instead of $\mathcal{S}$, it is possible to hide the AST nodes generated by the analysis.
- *termination criterion* of SRA. Our analysis terminates if only a finite number of copies is performed. One means to ensure that only a finite number of copies is performed is to define approximation functions as injections whose domains is the set of equivalence classes of calls and methods in the original program.

4.2   General Approximation Functions

We have already seen that the approximation function is $\mathcal{F}_0(e, m, o^*) = m$ for 0CFA-like analysis, and $\mathcal{F}_\infty(e, m, o^*) = Copy(m)$ for the evaluator. The purpose of this subsection is to illustrate the flexibility of our framework by defining several well-known approximation functions.

1CFA is a well-known family of approximations where a method is copied at most one for each class of call site. It may be formalized as any injection $1CFA : \overline{Call}, \overline{Method} \to Method$ such that $\overline{1CFA(\overline{e}, \overline{m})} = \overline{m}$. Then the associated approximation function is $\mathcal{F}_1(e, m, o^*) = 1CFA(\overline{e}, \overline{m})$. Intuitively, $\mathcal{F}_1$ unrolls one stage of the $\mathcal{F}_0$-call graph.

CFAn To unroll more than one stage, we have to choose between two directions: breadth and depth. The first direction will allow n copies for a call site, and is captured by the *CFAn* functions. Those may be any injection $CFAn : \overline{Call}, \overline{Method} \rightarrow Method^n$ such that $m_i \in CFAn(\overline{e}, \overline{m}) \Rightarrow \overline{m_i} = \overline{m}$ and $m_i, m_j \in CFAn(\overline{e}, \overline{m}) \wedge i \neq j \Rightarrow m_i \neq m_j$. The approximation function must have a (fair) strategy to choose between the methods, so we assume the existence of a function *ChooseBetween* enforcing fairness and set $\mathcal{F}_{breadth(n)}(e, m, o^*) = ChooseBetween(CFAn(\overline{e}, \overline{m}))$.

pCFA In order to capture unrolling, we use injections $pCFA : \overline{Call}^p, \overline{Method} \rightarrow Method$. The first argument of $pCFA$, computed by a function *CallStack*, is the sequence of (the equivalence classes) of the p method call nodes on top of the stack approximation; note that we don't collect a call node when we don't come from a method body. If a stack has less than p call nodes, we can safely complete the sequence with an arbitrary call node.

nCFAn We can combine both directions by considering injections of the form $pCFAn : \overline{Call}^p, \overline{Method} \rightarrow Method^n$ with

$$\mathcal{F}_{p,n}(e, m, o^*) = ChooseBetween(pCFAn(CallStack(p, e), \overline{m}))$$

While the $\mathcal{F}_{p,n}$ approximation functions allow to tune the precision of our analysis, they also have a main defect: for p or n greater than 1, they unroll recursive calls, whereas such unrolling does not yield any refinement in the analysis.

recCFA An alternative to the $\mathcal{F}_{p,n}$ functions is to merge methods only on recursion. This yields an approximation function

$$\mathcal{F}_r(e, m, o^*) = \textbf{if } \|m^* = SearchRec(m, e, \emptyset)\| > 0 \textbf{ then } m_1 \textbf{ else } Cache(e, \overline{m})$$

where

$$SearchRec(m, e, s) = \textbf{if } (e \in s) \qquad \textbf{then } \emptyset$$
$$\textbf{else if } (\overline{e} = \overline{m}) \textbf{ then } \{m\}$$
$$\textbf{else} \qquad \bigcup_{e_i \in \mathcal{U}(e)} SearchRec(m, e_i, s \cup \{e\})$$

Here *Cache* is any injection $Call, \overline{Method} \rightarrow Method$ such that $\overline{Cache(e, \overline{m})} = \overline{m}$. This function ensures that only a finite number of copies is made under a specific call site and thereby guarantees termination.

OOCFA An approximation function which is more accurate for object-oriented languages (single dispatch) will make copies for each different types of the target object

$$\mathcal{F}_{oo}(e, m, o^*) = \textbf{if } (o_1 = New(c.a^*)) \textbf{ then } OOCFA(\overline{m}, c) \textbf{ else } m$$

where *OOCFA* is any injection $\overline{Method}, Class \rightarrow Method$ such that

$$\overline{OOCFA(\overline{m}, c)} = \overline{m}$$

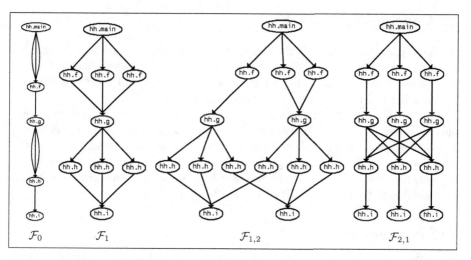

Fig. 2. DIFFERENCES BETWEEN SOME APPROXIMATION FUNCTIONS

This approximation function yields an analysis that bears some similarities with Agesen's Cartesian Product Algorithm (CPA) [1].

Example 6. In order to illustrate the differences between approximation functions, consider the (contrived) program:

```
class hh {
    static main() {
        f()*f()*f();
    }
    static f() {
        g();
    }
    static g() {
        h()+h()+h();
    }
    static h() {
        i();
    }
    static i() {
        1;
    }
}
```

The behaviour of various approximation functions is displayed in Figure 2.

5   Concluding Remarks

We have presented Static Reduction Analysis, a generic control-flow sensitive static analysis, for $\mathcal{L}_{ioe}$ an untyped, object-oriented language with exceptions. The salient feature of our analysis is to be AST-based, and not to require approximating run-time values such as environments.

Our work is an effort to provide a middle-ground between general (but complex) analyses à la k-CFA and simpler (but more specific) analyses à la Set-Based Analysis by allowing a simple definition of the analysis (the set AST nodes is the only domain) while allowing a parametrization of the analysis through approximation functions. The exact relationship between alternative proposals based on constrained types, see e.g. [17] remains to be unveiled.

5.1   Implementation

We have developed a prototype implementation of SRA in LISP. The prototype, which includes a graphical user interface, represents 3000 lines of code, including 500 lines for the SRA rules presented in this paper (in fact the implementation also deals with static methods without argument by adding a dummy argument, called hook, to method calls) and 100 lines for the different approximation functions.

The prototype supports all the functionalities described in this article (choice of the lookup function, choice of the approximation function), and also allows to preview the call-graphs (as PostScript files) of method calls in approximated programs. Thus far, the implementation has only been tested on small programs and no special provision has been made for efficiency. Further work is needed to achieve acceptable performance for large programs.

5.2   Future Work

We intend to pursue this work in two directions: first, we would like to establish formally the correctness of SRA and clarify its relationship with alternative proposals for polyvariant flow analysis, see e.g. [10,12,17]. Second, we would like to investigate applications of SRA, including partial evaluation and information flow.

In a different line of work, we are interested in adapting Static Reduction Analysis to the Java Card Virtual Machine as formalised in [2].

Acknowledgments We would like to thank the anonymous referees for suggesting improvements to the paper.

References

1. O. Agesen. The cartesian product algorithm: Simple and precise typing of parametric polymorphism. In W. Olthoff, editor, *Proceedings of ECOOP'95*, volume 952 of *Lecture Notes in Computer Science*, pages 2–26. Springer-Verlag, 1995. 359

2. G. Barthe, G. Dufay, L. Jakubiec, B. P. Serpette, S. Melo de Sousa, and S.-W. Yu. Formalisation of the Java Card Virtual Machine in Coq. In S. Drossopoulou, S. Eisenbach, B. Jacobs, G. T. Leavens, P. Müller, and A. Poetzsch-Heffter, editors, *Proceedings of FTfJP'00—ECOOP Workshop on Formal Techniques for Java Programs*, 2000. 360

3. C. Flanagan and M. Felleisen. Componential set-based analysis. In *Proceedings of PLDI'97*, volume 32 of *ACM SIGPLAN Notices*, pages 235–248. ACM Press, 1997. 346

4. J. Gosling, B. Joy, and G. Steele. *Java Language Specification*. The Java Series. Addison-Wesley, 1996. 344

5. D. P. Grove. *Effective Interprocedural Optimization of Object-Oriented Languages*. PhD thesis, University of Wahsington, 1998. 345

6. N. Heintze and D. McAllester. On the complexity of set-based analysis. In *Proceedings of ICFP'97*, pages 150–163. ACM Press, 1997. 346

7. S. Jagannathan and S. Weeks. A unified treatment of flow analysis in higher-order languages. In *Proceedings of POPL'95*, pages 393–407. ACM Press, 1995. 345

8. K. Leino and R. Joshi. A semantic approach to secure information flow. In J. Jeuring, editor, *Proceedings of MPC'98*, volume 1422 of *Lecture Notes in Computer Science*, pages 254–271, 1998. 345, 345

9. S. S. Muchnick and N. D. Jones, editors. *Program Flow Analysis: Theory and Applications*. Prenctice Hall, 1981. 344

10. F. Nielson and H. R. Nielson. Infinitary control flow analysis: a collecting semantics for closure analysis. In *Proceedings of POPL'97*, pages 332–345. ACM Press, 1997. 345, 360

11. F. Nielson, H. R. Nielson, and C. Hankin. *Principles of Program Analysis*. Springer-Verlag, 1999. 344

12. J. Palsberg and C. Pavlopoulou. From polyvariant flow information to intersection and union types. In *Proceedings of POPL'98*, pages 197–208. ACM Press, 1998. 360

13. J. Pleyvak. *Optimization of Object-Oriented and Concurrent Programs*. PhD thesis, University of Illinois at Urbana-Champaign, 1996. 345

14. B. Serpette. Approximations d'évaluateurs fonctionnels. In *Proceedings of WSA'92*, pages 79–90, 1992. 346, 347

15. M. Shapiro and S. Horwitz. Fast and accurate flow-insensitive points-to analysis. In *Proceedings of POPL'97*, pages 1–14. ACM Press, 1997. 351

16. O. Shivers. *Control-Flow Analysis of Higher-Order Languages*. PhD thesis, School of Computer Science, Carnegie Mellon University, May 1991. Also appears as Technical Report CMU-CS-91-145. 345, 345, 346, 347

17. S. F. Smith and T. Wang. Polyvariant flow analysis with constrained types. In G. Smolka, editor, *Proceedings of ESOP'00*, volume 1782 of *Lecture Notes in Computer Science*, pages 382–396. Springer-Verlag, 2000. 360, 360

18. B. Steensgard. Points-to analysis in almost linear time. In *Proceedings of POPL'96*, pages 32–41. ACM Press, 1996. 351

19. D. Volpano and G. Smith. Language Issues in Mobile Program Security. In G. Vigna, editor, *Mobile Agent Security*, volume 1419 of *Lecture Notes in Computer Science*, pages 25–43. Springer-Verlag, 1998. 345

An Abstract Interpretation Approach to Termination of Logic Programs

Roberta Gori

Dipartimento di Informatica, Università di Pisa, Corso Italia 40, 56125 Pisa, Italy
gori@di.unipi.it
Ph.: +39-050-887248   Fax: +39-050-887226

Abstract. In this paper we define a semantic foundation for an abstract interpretation approach to universal termination and we develop a new abstract domain useful for termination analysis. Based on this approximation we define a method which is able to detect classes of goals which universally terminate (with a fair selection rule). We also define a method which is able to characterize classes of programs and goals for which depth-first search is fair.

Keywords: *Abstract interpretation, Logic programming, Infinite derivations, Universal termination.*

1 Introduction

A lot of techniques have been proposed to approach various kinds of termination problems for logic programs (see [22] for a detailed survey). For logic programs there exist basically two notions of termination: universal termination and existential termination. In this paper we will be concerned with universal termination only.

The existing results on termination are either automatic methods for detecting non-termination or more theoretical approaches, providing manually verifiable criteria for non-termination. In particular, there exist correct and complete methods to prove universal termination (see, for example, [2]). Since termination is known to be undecidable, such methods are not effective. On the other hand, there are techniques providing sufficient decidable conditions. For example, to prove that a logic program terminates for a given goal it is sufficient to prove a strict decrease in some measure over a well founded domain on the sequence of procedure calls. Most of the termination analyses apply this approach but focus on different aspects on proving termination of programs. Several papers [23,10,38] tackle the problem of inferring norms and well founded orders. Other [45,38,11,43,7] define techniques for computing inter-argument relations. The analysis in [34] shows how to infer classes of terminating queries using approximations on Natural and on Boolean constraint domains. Based on this ideas, several systems for automatic termination analyses have been proposed, such as TermiWeb [13], TermiLog [30], cTI [35] and Mercury's termination analyzer [42].

M. Parigot and A. Voronkov (Eds.): LPAR 2000, LNAI 1955, pp. 362–380, 2000.
© Springer-Verlag Berlin Heidelberg 2000

These systems are very powerful and are able to automatically prove that classes of goals (for example, all the goals whose arguments are ground) do universally terminate. Of course these systems are not able to analyze all the programs. For example, since these systems are based on approximations on domains of symbolic norms, they have some problem in analyzing programs, where termination depends on the structure of terms.

Example 1. By using the systems in [30,35] we were not able to prove that the following program [31] always terminates.

$$P_1 : at(telaviv, mary)$$
$$at(jerusalem, mary)$$
$$at(x, fido) \leftarrow at(x, mary), near(x)$$
$$near(jerusalem)$$

Example 2. Consider now the program P_2.

$$P_2 : p(a, b)$$
$$p(a, f(x)) \leftarrow p(a, x)$$
$$q(f(x), y) \leftarrow p(x, y)$$

The analyses in [30,35] tells us that the query $p(f(b), y)$ terminates if y is bound, e.g. finite and ground. Note however that $p(f(b), y)$ does universally terminate for any y.

Since automatic verification of termination must be based on some notion of approximation, it seems reasonable to tackle the problem by abstract interpretation techniques. Abstract interpretation techniques have already been used for validating termination analysis methods [45,31,14]. Moreover, [14] shows that the semantics in [24] is suitable to deal with termination and that some norm-based methods can be viewed as abstractions of this semantics. We want to push forward this approach, by explicitly using abstract interpretation to systematically derive the "right" semantics to model termination and various effective abstractions modelling different abstract properties. In addition, abstract interpretation provides techniques to systematically combine different analyses in a powerful automatic system. As a matter of fact, we believe that most of the existing automatic methods [44,38,32,45,34] can be reconstructed as abstract interpretations of the "termination semantics" on suitable abstract domains. Such a reconstruction would allow one to easily combine the existing techniques and to extend them with new analyses based on new abstract domains, thus improving the precision of the approximate analysis.

The main contributions of this paper are the definition of a semantic foundation for the abstract interpretation approach to universal termination and the development of a new abstract domain, which leads to an analysis which solves the problem shown in Examples 1 and 2. The abstract domain is a variation of the well-known depth-k abstraction. The resulting approximate analysis often devises classes of (terminating) goals which are smaller than the ones one would obtain using the systems in [30,35,13,42] or in general with methods based on inter-argument relations (see [22] for a detailed survey). However, the approximation on the structure of terms can be improved by increasing the k in the abstraction. Hence, for example, in program P_1 it is sufficient to choose $k = 2$, to be able to prove that the goal $at(x, y)$ will always terminate and in P_2 it is sufficient to choose $k = 3$ to show that the goals $p(f(b), x), p(b, x), p(f(b), x), p(f(f(\_)), x)$ always terminate. In particular, we can prove an interesting result (see Corollary 1), showing that for any goal which does universally terminate, by increasing k we will eventually be able to prove it. As a final remark, we want to stress that the idea behind our analysis is that such an analysis should be combined with the existing ones, to improve the overall precision.

In order to reason on termination by using abstract interpretation, the first step is to define a suitable concrete semantics. Section 2 explains which properties such semantics has to enjoy and why the previously defined semantics are not suitable to our purposes. Section 3 introduces an adequate semantics (exact answers) which is able to model infinite derivations in a compositional way. This semantics is then used to derive computable approximations of non-terminating computations and effective conditions to reason about termination. Section 4 defines the depth-k abstraction, which yields an approximation of exact answers and therefore of the set of goals having at least one infinite derivation. This approximation is used for a termination analysis which is able to detect classes of goals which universally terminate (Section 5). Finally, in Section 6, the approximation is used for detecting a class of goals for which the depth-first search rule is fair.

2 Which Semantics for Infinite Derivations

Since we can not reason about possibly non-terminating computations without taking into account the infinite behavior of goals, we start by providing a semantics which faithfully models this observable. Moreover, in order to apply the abstract interpretation framework, we look for a semantics which is obtained as a fixpoint of a suitable operator. Furthermore, to reason in a modular way, we need this semantics to be compositional w.r.t. the syntactic operators. In particular, we want the semantics to be AND-compositional, i.e. we want to be able to infer the infinite behavior of a conjunctive goal by using the information on the infinite behavior of atomic goals only. These requirements have some consequences. First of all information about successful computations has to be collected. In fact, a conjunctive goal has an infinite computation (via a fair selection rule) if at least one of the atoms in the goal has an infinite computation and all the other atoms are successful. Moreover we also have to faithfully model *answers* of infinite and successful derivations. In fact, this allows us to understand whether a conjunctive goal has an infinite derivation or a finite failure due to the computation of incompatible substitutions. Following the terminology of [21], we aim at modelling *exact answers*, i.e. the set of substitutions computed by successful or non-terminating derivations.

Unfortunately all the semantics defined so far for modeling the infinite behavior are not adequate to our needs. For example, the semantics in [33,36,25,40,28] do not model exact answers, but their downward closure. The semantics in [29] is able to model many aspects of Prolog computations as well as the infinite behavior. However it is not able to model exact answers of infinite derivations. On the other hand, [21,37] introduce a categorical approach that allows them to model exact answers as the colimit of the ω-chain consisting of the iterates of a functor F. Anyway, this construction seems quite hard to fit into the abstract interpretation framework.

Intuitively, the main difficulty in modeling exact answers is due to the choice of a suitable domain. If the domain is too abstract, in a sense, we are not able to exclude instances of the exact answers from the computation of the fixpoint. Our intuition is that without any information on the number of steps necessary to compute a substitution, a greatest fixpoint semantics would deliver as answers instances of substitutions for non-terminating derivations, which would never be computed.

Our semantics models exact answers and is defined as the greatest fixpoint of a co-continuous operator on a quite simple domain (the order is the pointwise extension of subset inclusion). Clearly, this semantics is not effectively computable. Hence we can not use it as the base for the effective conditions we are looking for. Anyway, by following the usual approach of abstract interpretation, we use it as the collecting semantics to build more abstract approximate yet effective semantics. It is worth noting that our construction allows us to define approximate semantics which still distinguish between the results of successful and infinite computations, which is not the case for most of ad-hoc semantics proposed in the literature [33,36,25,40,21].

3   A Fixpoint Semantics Modeling Exact Answers

The reader is assumed to be familiar with the terminology and the basic results in the semantics of logic programs [1,33] and with the theory of abstract interpretation as presented in [18,19]. In the following $x, x_1, x_2, \ldots, y, y_1, y_2, \ldots$ denote variables, while by x and t tuples of distinct variables and of terms respectively, B and G denote (possible empty) conjunctions of atoms.

We want to define a semantics which models exact answers of infinite and successful derivations via a *fair* selection rule (the parallel selection rule [1]).

An important step is the choice of a domain which is able to precisely represent the behaviors we are interested in. In particular, the domain has to be able to represent finite and infinite substitutions computed by non-terminating computations. A natural choice is to use sequences of substitutions to represent possibly infinite substitutions. Given the goal G, we consider the sets of (possibly infinite) sequences $\vartheta_1 :: \vartheta_2 :: \ldots :: \vartheta_m :: \ldots$ of substitutions, increasingly more instantiated ($\forall i \; G\vartheta_i \leq G\vartheta_{i+1}$) representing the sequence of partial substitutions computed at each step by successful and infinite derivations for G. In particular, an infinite sequence $\vartheta_1 :: \vartheta_2 :: \ldots :: \vartheta_m :: \ldots$ represents the sequence of partial substitutions computed by an infinite derivation, while a finite

---

[1] Let $G = A_1, \ldots, A_n$. The parallel selection rule R, selects at the first step, some A_j. Then, in the next step it will select the atom A_{j+1} (of the goal G) up to A_n, then it will select A_1 and so on.

sequence $\vartheta_1 :: \vartheta_2 :: \ldots :: \vartheta_n$ represents the finite sequence of partial substitutions computed by a successful derivation.

Example 3. Consider the following program P_3

$$P_3 : p(x) \leftarrow p(x)$$
$$q(f(x)) \leftarrow q(x)$$
$$q(a)$$

The substitution computed by the infinite derivation for $p(x)$ may be represented by the infinite sequence $x/x_1 :: x/x_2 :: x/x_3 :: \ldots$, while the substitution computed by the infinite derivations for $q(x)$ may be represented by the infinite sequence $x/f(x_1) :: x/f(f(x_2)) :: \ldots :: x/f^n(x_n) :: \ldots$. Finally, the substitutions computed by the successful derivations for $q(x)$ may be represented by the finite sequences x/a, $x/f(x_1) :: x/f(a)$, $x/f(x_1) :: x/f(f(x_2)) :: x/f(f(a))$, $x/f(x_1) :: x/f(f(x_2)) :: x/f(f(f(x_3))) :: x/f(f(f(a)))$, $\ldots$

It is worth noting that this approach is already sufficient to avoid the difficulties of other approaches, which generally consider much more abstract domains made up of simple substitutions. The information on the number of steps needed to compute a substitution implicitly contained in our sequences is necessary to obtain a semantics modelling exact answers. Consider the previous example, with the substitutions $\{x/f(t)\}$, where t is any term, as possible substitutions computed for $p(x)$. It is easy to see that there is no way of not including the substitutions $\{x/f(t)\}$ in the computation of a greatest fixpoint semantics for $p(x)$, even if actually $\{x/f(t)\}$ is not a substitution computed by any successful or infinite derivation in P, for any t. Indeed, this is not the case in our approach as shown by the following examples.

Once this choice is made, a second problem arises with And-compositionality. In fact, it seems not possible to achieve And-compositionality without any information on the clauses used in the derivation. This is because the substitutions for a conjunctive goal, computed at a given step, depend on the selected atom. Since the selection rule we consider is the parallel selection rule (note however that the same problem would arise with any *non local rule*), the substitution computed at a given rewriting step depends on the form of the current goal in the concrete derivation and on the previously selected atoms. However, we do not have all this information in our abstract domain unless we consider the concrete derivation itself.

In order to solve this problem, we consider sequences of substitutions computed at specific steps of the rewriting process. These specific steps allow us to compose sequences of substitutions without any additional information on the concrete derivation they come from.

Let $d = G \rightarrow G_1 \rightarrow G_2 \rightarrow \ldots$, be a derivation of G via the parallel selection rule. Let $n_1 = \text{length}(G)$ be the number of steps necessary for all the atoms in G to be selected exactly once [2]. Let G_1 be the goal at step n_1. n_2 is the number of steps necessary for all the atoms in G_1 to be selected ($\text{length}(G_1)$). Then G_2 is the goal at step $n_1 + n_2$, and n_3 is the number of steps necessary for all of its

---

[2] $\text{length}(A_1, \ldots, A_m) = m$

atoms to be selected and so on. We can single out a sequence of intermediate steps $n_1, n_2, \ldots$ of the derivation d for a goal **G**. This sequence can be formally defined by the recursive definition, $n_1 = \text{length}(\mathbf{G})$, $n_{i+1} = \text{length}(\mathbf{G}_{\sum_{k \leq i} n_k})$. Our idea is to use this sequence of steps to refine our sequence of substitutions. We consider then sequences of substitutions $<\!::_{n_1} \vartheta_1 ::_{n_2} \vartheta_2 :: \ldots ::_{n_m} \vartheta_m :: \ldots >$, $(\mathbf{G}\vartheta_i \leq \mathbf{G}\vartheta_{i+1})$, labelled by step numbers, where ϑ_j is the substitution (restricted to the variables of **G**) computed by the derivation up to step $\Sigma_{k \leq j} n_k$. This choice is crucial to obtain an And-compositional fixpoint semantics modeling exact answers.

Thus, let **S** be the powerset of the set of finite and infinite sequences of substitutions $::_{n_1} \vartheta_1 ::_{n_2} \vartheta_2 ::_{n_i} \ldots ::_{n_m} \vartheta_m :: \ldots$, ordered by set inclusion. Let *Goals* be the set of all goals in the program P. Our domain is $\mathbb{A}_{inf} \subseteq [Goals \rightharpoonup \mathbf{S}]$, i.e. the domain of all the partial functions ordered by $\sqsubseteq$, the pointwise extension of the order in **S**. $(\mathbb{A}_{inf}, \sqsubseteq)$ is clearly a complete lattice.

Since we need to And-compose sequences of different lengths, we introduce a notion of *completed* version of a sequence w.r.t. a given length n (where possibly $n = \omega$).

Definition 1. $s_1 \in \mathbf{S}$ *is the completed version w.r.t.* n *of a sequence* $s_2 \in \mathbf{S}$, $s_2 =::_{n_1} \vartheta_1 :: \ldots ::_{n_m} \vartheta_m$, *if* $n \geq m$ *and* $s_1 = s_2 ::_0 \underbrace{\vartheta_m ::_0 \vartheta_m ::_0 \ldots}_{n-m}$

Our semantic operator $\mathcal{P}[[P]] : \mathbb{A}_{inf} \rightarrow \mathbb{A}_{inf}$ is defined as follows. For each generic atomic goal $p(\mathbf{x})$ we compute the new set of exact answers by considering each clauses in P defining the procedure p and each exact answer (according to the interpretation $I \in \mathbb{A}_{inf}$) for the conjunctive goal, body of the clause. For each clause $p(\mathbf{t}) \leftarrow p_1(\mathbf{t}_1), \ldots, p_n(\mathbf{t}_n)$ (renamed version w.r.t. **x** of a clause defining p in P), we obtain a new exact answer for $p(\mathbf{x})$ whose first substitution is $\vartheta = \text{mgu}(p(\mathbf{x}), p(\mathbf{t}))$ and the following sequence is the composition of ϑ with the substitutions forming a possible exact answer for the conjunctive goal $p_1(\mathbf{t}_1), \ldots, p_n(\mathbf{t}_n)$ according to I. It is worth noting that a possible exact answer for the conjunctive goal $p_1(\mathbf{t}_1), \ldots, p_n(\mathbf{t}_n)$ according to I is computed by And-composing and composing w.r.t. instantiation the exact answers (found in I) for the atomic generic goals $p_j(\mathbf{x}_j)$, $j = 1, \ldots, n$.

Definition 2. *Let* $I \in \mathbb{A}_{inf}$. $\mathcal{P}[[P]]I =$
$\lambda p(\mathbf{x}).\{<\!::_1 \vartheta ::_{n_1} \vartheta_1 \ldots ::_{n_m} \vartheta_m \ldots > \mid \exists p(\mathbf{t}) \leftarrow \mathbf{B}$ *a renamed version w.r.t.* **x**
$\qquad$ *of a clause in* P, $\mathbf{B} = p_1(\mathbf{t}_1), \ldots, p_n(\mathbf{t}_n)$,
$\qquad$ $\vartheta := \{\mathbf{x}/\mathbf{t}\}$, *for* $j = 1, \ldots, n, s_j \in I(p_j(\mathbf{x}_j))$,
$\qquad$ $<\!::_{n_1^j} \vartheta_1^j :: \ldots ::_{n_m^j} \vartheta_m^j :: \ldots >$ *is the completed*
$\qquad$ *version w.r.t.* $w = \max_{j=1,\ldots,n}(\text{length}(s_j))$
$\qquad$ *of the sequence* s_j, *for* $h = 1 \ldots w$, $n_h = \Sigma_j n_h^j$
$\qquad$ $\vartheta_h = \vartheta \cdot \text{mgu}(\mathbf{B}, (p_1(\mathbf{x}_1)\vartheta_h^1, \ldots, p_n(\mathbf{x}_n)\vartheta_h^n))_{|\mathbf{x}}\}$

Theorem 1. *[27]* $\mathcal{P}[[P]]$ *is monotone and co-continuous on* $\mathbb{A}_{inf}$.

We define the fixpoint semantics of P as $\text{gfp}(\mathcal{P}[[P]]) = \text{glb}(\{\ \mathcal{P}[[P]] \downarrow i \mid i < \omega\}) = \cap_{i < \omega} \mathcal{P}[[P]] \downarrow i$, where $\mathcal{P}[[P]] \downarrow i$ are the usual ordinal powers.

Example 4.

$$P_4: \quad q(a) \leftarrow p(x)$$
$$p(f(x)) \leftarrow p(x)$$

$\mathcal{P}[[P_4]] \downarrow 0 = \top^{inf} (q(x)) = \{ s \mid s \text{ is a sequence } \}$

$\mathcal{P}[[P_4]] \downarrow 0 = \top^{inf} (p(x)) = \{ s \mid s \text{ is a sequence } \}$

$\mathcal{P}[[P_4]] \downarrow 1 \; q(x) = \{ s \mid s \text{ is a sequence starting with } ::_1 x/a\}$

$\mathcal{P}[[P_4]] \downarrow 1 \; p(x) = \{ s \mid s \text{ is a sequence starting with } ::_1 \{x/f(x_1)\} \}$

$\mathcal{P}[[P_4]] \downarrow 2 \; q(x) = \{ s \mid s \text{ is a sequence starting with } ::_1 \{x/a\} ::_1 \{x/a\} \}$

$\mathcal{P}[[P_4]] \downarrow 2 \; p(x) = \{ s \mid s \text{ is a sequence starting with}$
$$::_1 \{x/f(x_1)\} ::_1 \{x/f(f(x_2))\} \}$$

$$\vdots$$

$gfp(\mathcal{P}[[P_4]]) \; q(x) = \{<::_1 \{x/a\} ::_1 \ldots ::_1 \{x/a\} ::_1 \ldots >\}$

$gfp(\mathcal{P}[[P_4]]) \; p(x) = \{<::_1 \{x/f(x_1)\} ::_1 \{x/f(f(x_2))\} ::_1 \ldots ::_1 \{x/f^n(x_n)\} :: \ldots >\}$

$$P_5: q(x) \leftarrow q(f(x))$$
$$p(f(x))) \leftarrow q(x)$$

$gfp(\mathcal{P}[[P_5]]) \; q(x) = \{<::_1 \{x/x_1\} ::_1 \ldots ::_1 \{x/x_n\} ::_1 \ldots >\}$

$gfp(\mathcal{P}[[P_5]]) \; p(x) = \{<::_1 \{x/f(x_1)\} ::_1 \ldots ::_1 \{x/f(x_n)\} ::_1 \ldots >\}$

$$P_6: t(a) \leftarrow p(x,y), q(x,y)$$
$$q(f(f(y)), f(y)) \leftarrow q(f(y), y)$$
$$p(f(y), f(y)) \leftarrow p(y,y)$$
$$p(a, a)$$

$gfp(\mathcal{P}[[P_6]]) \; t(x) = \{ \; \}$

$gfp(\mathcal{P}[[P_6]]) \; q(x,y) =$
$$\{<::_1 \{x/f(f(x_1)), y/f(x_1)\} ::_1 \ldots ::_1 \{x/f^{n+1}(x_n), y/f^n(x_n)\} \ldots >\}$$

$gfp(\mathcal{P}[[P_6]]) \; p(x,y) = \{<::_1 \{x/a, y/a\} >,$
$$<::_1 \{x/f(x_1), y/f(x_1)\} ::_1 \{x/f(a), y/f(a)\} >, \ldots$$
$$<::_1 \{x/f(x_1), y/f(x_1)\} ::_1 \ldots ::_1 \{x/f^n(x_n), y/f^n(x_n)\} \ldots >\}$$

$$P_7: even(s(s(x))) \leftarrow even(x).$$
$$even(0)$$

$gfp(\mathcal{P}[[P_7]]) \; even(x) = \{<::_1 \{x/0\} >, \; <::_1 \{x/s(s(x_1))\} ::_1 \{x/s(s(0))\} >, \ldots$
$$<::_1 \{x/s(s(x_1))\} ::_1 \{x/s(s(s(s(x_2))))\} ::_1 \ldots ::_1 \{x/s^{2n}(x_n)\} \ldots >\}$$

The derivation of the above fixpoint semantics deserves some interest by itself. In fact, we have systematically derived it by using abstract interpretation techniques, starting from a very concrete semantics introduced in [26], which extends with infinite SLD fair derivations the semantics in [16], we have defined a Galois insertion which does not lose any information on exact answers. It turns out that the abstraction is precise (complete): $gfp(\mathcal{P}[[P]])$ faithfully models the exact answers of infinite and successful derivations. Moreover the semantics is fully abstract as stated by the following theorem.

Theorem 2. *[27] Let* **G** *be a goal.* $gfp(\mathcal{P}[[P]]) = gfp(\mathcal{P}[[Q]])$ *iff every goal* **G** *has the same answers computed by infinite or successful derivations in the program* P *and in the program* Q.

The semantics is compositional w.r.t. *instantiation* and *And-composition* of atoms. That is, the behavior of $p(t)$ can be inferred from the behaviors of a generic atom $p(x)$, and we can infer the behavior of conjunctive goals from the information on the exact answers of atomic goals only. For a detailed definition of these operations see [27].

4 Approximate Semantics

Our fixpoint semantics is not effective, since it deals with infinite sequences of substitutions and needs to compute a greatest fixpoint. Hence, in this section, we introduce an effective approximation of $\mathsf{gfp}(\mathcal{P}[[P]])$.

4.1 The Depth-k Domain

The idea is to approximate an infinite set of exact answers by means of a $\mathsf{depth}(k)$ cut [41]. Terms are cut by replacing each sub-term rooted at depth k with a new fresh variable taken from a set W (disjoint from the set of program variables V). A $\mathsf{depth}(k)$ term represents a set of terms, obtained by instantiating the variables of W with any term built over V.

These operations define a function α_k on terms. We can extend α_k to substitutions to obtain abstract substitutions of the form $\alpha_k(\vartheta) = \{x/\alpha_k(t) \mid x/t \in \vartheta\}$. We assume that, for any binding in ϑ, cuts are performed by using distinct variables of W. We denote by Subst_k the set of substitutions $V \to T_k$, where T_k is the set of $\mathsf{depth}(k)$ terms.

4.2 The Abstraction

In order to make the approximation effectively computable, we have to get rid of the information on the number of steps and on the partial substitution computed at step n_i. The idea is to abstract a sequence of substitutions $<::_{n_1} \vartheta_1 ::_{n_2} \vartheta_2 :: \ldots ::_{n_m} \vartheta_m :: \ldots >$ for $\mathbf{G}$ with the abstract substitution $\alpha_k(\vartheta_s)$, where $\forall v > s$, $\alpha_k(\vartheta_s) = \alpha_k(\vartheta_v)$. It follows from bounded depth of terms in the $\mathsf{depth}(k)$ domain that such an s always exists. Moreover, since we have lost all the information about the number of steps, we have to find some way to distinguish between successful and infinite derivations. To this aim, we use the two symbols $\square$ and $\Diamond$.

Let $\mathcal{S}_k^\alpha$ be the domain of sets of pairs $< \vartheta, \circ >$, where $\vartheta \in \mathsf{Subst}_k$ and $\circ \in \{\square, \Diamond\}$, ordered by set inclusion. Consider now the abstract domain $\mathbb{A}_k^\alpha \subseteq [Goals \rightharpoonup \mathcal{S}_k^\alpha]$, the domain of all the partial functions ordered by $\sqsubseteq_\alpha$, the pointwise extension of the order in $\mathcal{S}_k^\alpha$. $(\mathbb{A}_k^\alpha, \sqsubseteq_\alpha)$ is a complete lattice.

The abstraction $\alpha^k : \mathbb{A}_{inf} \to \mathbb{A}_k^\alpha$ and the concretization $\gamma^k : \mathbb{A}_k^\alpha \to \mathbb{A}_{inf}$ are defined as follows.

$$\alpha^k(I) := \lambda G.\{< \alpha_k(\vartheta_h), \square > \mid t =<::_{n_1} \vartheta_1 ::_{n_2} \vartheta_2 :: \ldots ::_{n_m} \vartheta_m :: \ldots >\in I(G),$$
$$t \text{ is a finite sequence and}$$
$$\forall \, v > h, \ \alpha_k(\vartheta_h) = \alpha_k(\vartheta_v)\}\cup$$
$$\{< \alpha_k(\vartheta_h), \Diamond > \mid t =<::_{n_1} \vartheta_1 ::_{n_2} \vartheta_2 :: \ldots ::_{n_m} \vartheta_m :: \ldots >\in I(G),$$
$$t \text{ is an infinite sequence and}$$
$$\forall \, v > h, \ \alpha_k(\vartheta_h) = \alpha_k(\vartheta_v)\}$$

$$\gamma^k(I^a) :=$$
$$\lambda G.\{ \ s =<::_{n_1} \vartheta_1 :: \ldots ::_{n_m} \vartheta_m :: \ldots > \mid s \text{ is a finite sequence and}$$
$$< \vartheta, \square >\in I^a(G) \text{ and } \exists \, h, \ \vartheta = \alpha_k(\vartheta_h) \text{ and}$$
$$\forall \, v > h, \ \alpha_k(\vartheta_h) = \alpha_k(\vartheta_v)\}\cup$$
$$\{ \ s =<::_{n_1} \vartheta_1 :: \ldots ::_{n_m} \vartheta_m :: \ldots > \mid s \text{ is an infinite sequence and}$$
$$< \vartheta, \Diamond >\in I^a(G) \text{ and } \exists \, h, \ \vartheta = \alpha_k(\vartheta_h) \text{ and}$$
$$\forall \, v > h, \ \alpha_k(\vartheta_h) = \alpha_k(\vartheta_v)\}$$

Lemma 1. $\langle \alpha^k, \gamma^k \rangle : (\mathbb{A}_{inf}, \sqsubseteq) \rightleftharpoons (\mathbb{A}_k^\alpha, \subseteq)$ *is a Galois insertion.*

Following the abstract interpretation theory, we can derive the optimal abstraction $\mathcal{P}_k^\alpha[[P]] : \mathbb{A}_k^\alpha \to \mathbb{A}_k^\alpha$ of $\mathcal{P}[[P]]$ on the abstract domain $\mathbb{A}_k^\alpha$, defined as $\alpha_k \cdot \mathcal{P}[[P]] \cdot \gamma_k$.

Definition 3. *Let* $I \in \mathbb{A}_k^\alpha$. $\mathcal{P}_k^\alpha[[P]]I =$
$$\lambda p(\mathbf{x}).\{< \alpha_k(\vartheta), \square > \mid \exists p(\mathbf{t}) \leftarrow \mathbf{B} \ a \ renamed \ version \ w.r.t. \ \mathbf{x}$$
$$of \ a \ clause \ in \ \mathbf{P}, \ \mathbf{B} = p_1(\mathbf{t}_1), \ldots, p_n(\mathbf{t}_n),$$
$$\vartheta' := \{\mathbf{x}/\mathbf{t}\}, \ \exists < \vartheta^j, \square >\in I(p_j(\mathbf{x}_j))$$
$$\vartheta = \vartheta' \cdot mgu(\mathbf{B}, (p_1(\mathbf{x}_1)\vartheta^1, \ldots, p_n(\mathbf{x}_n)\vartheta^n))_{|\mathbf{x}}\}\cup$$
$$\{< \alpha_k(\vartheta), \Diamond > \mid \exists p(\mathbf{t}) \leftarrow \mathbf{B} \ a \ renamed \ version \ w.r.t. \ \mathbf{x}$$
$$of \ a \ clause \ in \ \mathbf{P}, \ \mathbf{B} = p_1(\mathbf{t}_1), \ldots, p_n(\mathbf{t}_n), \ n > 0,$$
$$and \ \vartheta' := \{\mathbf{x}/\mathbf{t}\}, \exists \bar{j} \in \{1, \ldots, n\}, \ < \vartheta^{\bar{j}}, \Diamond >\in I(p_{\bar{j}}(\mathbf{x}_{\bar{j}}))$$
$$and \ for \ j \in \{1, \ldots, n\}, \ j \neq \bar{j}$$
$$there \ exist \ < \vartheta^j, \circ >\in I(p_j(\mathbf{x}_j)), \ \circ \in \{\Diamond, \square\},$$
$$\vartheta = \vartheta' \cdot mgu(\mathbf{B}, (p_1(\mathbf{x}_1)\vartheta^1, \ldots, p_{\bar{j}}(\mathbf{x}_{\bar{j}})\vartheta^{\bar{j}}, \ldots, p_n(\mathbf{x}_n)\vartheta^n))_{|\mathbf{x}}\}$$

The semantics of P is the greatest fixpoint of the $\mathcal{P}_k^\alpha[[P]]$ operator, which is monotone on the *finite* domain $\mathbb{A}_{inf}^\alpha$.

Of course, our approximation $gfp(\mathcal{P}_k^\alpha[[P]])$ has lost precision w.r.t. the concrete semantics $gfp(\mathcal{P}[[P]])$. However, it still distinguishes between exact answers of successful and infinite derivations, i.e. the abstract operator does not necessarily compute the same answers for successful and infinite derivations, as shown by the results for programs P_6 and P_7 in the following example. As we have already pointed out, this is not the case for most of the concrete semantics proposed to model infinite computations [33,36,25,40,21].

Example 5. Consider the programs in Example 4. Let $w, w_1, w_2 \in W$ and let $k = 3$.

$\mathcal{P}_k^\alpha[[P_4]]\ (\top^a)\ (q(x)) = \{ < \{x/a\}, \square >, < \{x/a\}, \Diamond > \}$

$\mathcal{P}_k^\alpha[[P_4]]\ (\top^a)(p(x)) = \{ < \{x/f(a)\}, \square >, < \{x/f(x_1)\}, \square >,$
$\qquad < \{x/f(f(w))\}, \square >, < \{x/f(a)\}, \Diamond >, < \{x/f(x_1)\}, \Diamond >,$
$\qquad < \{x/f(f(w))\}, \Diamond >, \}$

$\vdots$

$gfp(\mathcal{P}_k^\alpha[[P_4]])\ (q(x)) = \{ < \{x/a\}, \square >, < \{x/a\}, \Diamond > \}$

$gfp(\mathcal{P}_k^\alpha[[P_4]])\ (p(x)) = \{ < \{x/f(f(w))\}, \square >, < \{x/f(f(w))\}, \Diamond > \}$

$gfp(\mathcal{P}_k^\alpha[[P_5]]\ )(q(x)) = \{ < \{x/a\}, \square >, < \{x/f(a)\}, \square >, < \{x/x_1\}, \square >,$
$\qquad < \{x/f(x_1)\}, \square >, < \{x/f(f(w))\}, \square >, < \{x/a\}, \Diamond >,$
$\qquad < \{x/f(a)\}, \Diamond >, < \{x/x_1\}, \Diamond >, < \{x/f(x_1)\}, \Diamond >,$
$\qquad < \{x/f(f(w))\}, \Diamond > \}$

$gfp(\mathcal{P}_k^\alpha[[P_5]]\ )(p(x)) = \{ < \{x/f(a)\}, \square >, < \{x/f(x_1)\}, \square >,$
$\qquad < \{x/f(f(w))\}, \square >, < \{x/f(a)\}, \Diamond >, < \{x/f(x_1)\}, \Diamond >,$
$\qquad < \{x/f(f(w))\}, \Diamond > \}$

$gfp(\mathcal{P}_k^\alpha[[P_6]])(t(x)) = \{ < \{x/a\}, \square >, < \{x/a\}, \Diamond > \}$

$gfp(\mathcal{P}_k^\alpha[[P_6]])(q(x,y)) = \{ < \{x/f(f(w_1)), y/f(f(w_2))\}, \square >,$
$\qquad < \{x/f(f(w_1)), y/f(f(w_2))\}, \Diamond > \}$

$gfp(\mathcal{P}_k^\alpha[[P_6]])(p(x,y)) = \{ < \{x/a, y/a\}, \square >, < \{x/f(a), y/f(a)\}, \square >,$
$\qquad < \{x/f(f(w_1)), y/f(f(w_2))\}, \square >$
$\qquad < \{x/f(f(w_1)), y/f(f(w_2))\}, \Diamond > \}$

$gfp(\mathcal{P}_2^\alpha[[P_7]])\ even(x) = \{ < \{x/0\}, \square >, < \{x/s(s(w))\}, \square > < \{x/s(s(w))\}, \Diamond > \}$

The abstract fixpoint semantics $gfp(\mathcal{P}_k^\alpha[[P]])$ correctly approximates $gfp(\mathcal{P}[[P]])$, i.e, for every $p(x)$, $gfp(\mathcal{P}[[P]])\ (p(x)) \subseteq \gamma^k(gfp(\mathcal{P}_k^\alpha[[P]]))\ (p(x))$ and is finitely computable.

Theorem 3. *[27] If $p(x)$ has an infinite or successful derivation in P, computing the sequence of substitutions s, then, for any k, $s \in \gamma^k(gfp(\mathcal{P}_k^\alpha[[P]]))(p(x))$.*

It is important to note that the abstract semantics is still compositional w.r.t. instantiation and And-composition. For a detailed definition of these abstract operations see [27].

We can use the information in $gfp(\mathcal{P}_k^\alpha[[P]])$ to give a correct upward approximation of the sets of goals which have at least an infinite derivation in P. By Theorem 3 and the results on compositionality, we can define the set Inf_P^k.

Definition 4. *Let*

$$Inf_P^k = \{ \mathbf{G} \mid \mathbf{G} = p_1(t_1), \ldots, p_n(t_n),$$
$$\exists \bar{i} \in \{1, \ldots, n\},\ < \sigma_{\bar{i}}, \Diamond > \in gfp(\mathcal{P}_k^\alpha[[P]])(p_{\bar{i}}(x_{\bar{i}}))\}\ and$$
$$for\ i = \{1, \ldots, n\},\ \exists < \sigma_i, \circ > \in gfp(\mathcal{P}_k^\alpha[[P]])(p_i(x_i))\ such\ that$$
$$\mathbf{G}\ and\ p_1(x_1)\sigma_1, \ldots, p_n(x_n)\sigma_n\ are\ unifiable\}$$

As a consequence of the correctness of the approximation $gfp(\mathcal{P}_k^\alpha[[P]])$, (Theorem 3), we can prove the following theorem.

Theorem 4. *If the goal* **G** *has an infinite derivation in* **P** *(via a fair selection rule), then, for all* k, **G** $\in$ Inf$_P^k$.

The following result tells us that, by increasing k, we get more precise results.

Theorem 5. *[27] If the goal* **G** *does not have an infinite derivation in the program* P *(via a fair selection rule), then there exists a* k, *such that* $\forall \bar{k} \geq$ k, **G** $\notin$ Inf$_P^{\bar{k}}$.

Let us now spend a few words on the complexity of the computation of the fix-point abstract semantics on which the termination analysis is based. The size of a depth-k term is bounded exponentially by k, where k is given and does not depend on the particular program. The complexity of the abstract unification is the same as the one of concrete unification, i.e. linear on the term sizes. Since the size of the depth-k terms is bounded, the complexity of the abstract unification is bounded exponentially by k. Let us now analyze the complexity of the computation of the abstract semantics. We want to find a bound for the number of iterations of the abstract fixpoint operator necessary to compute the abstract semantics. It is worth noting that the number of the different depth-k terms on the depth-k domain is bounded exponentially by k. Again, k is a given constant and does not depend on the particular program. Therefore, the number of iterations of the abstract fixpoint operator is exponential in k in the worst case. It is worth noting however that this case is very rare and that there exist techniques such as widening [20] which can be used to speed up convergency. Many static analyses as groundness, sharing, based on an abstract semantics whose computation requires (in the worst case) a number of iterations comparable to the one of our abstract semantics, have been successfully implemented in real analyzers such as CIAO [12] and China [6].

The two following sections show how our abstract semantics can be applied to study universal termination and to analyze the fairness of the depth-first strategy.

5   Universal Termination

Let us formally introduce the notion of universal termination.

Definition 5. *A logic program* P *and a goal* G *universally terminate w.r.t. a set of selection rules* S, *if every SLD-derivation of* P *and* G, *via any selection rule from* S, *is finite.*

The early approaches to the characterization of terminating programs focused on universal termination w.r.t. *all* selection rules [8,9]. Indeed, this is a strong property holding only for simple programs and goals. Therefore the following step was the study of universal termination w.r.t. specific selection rules, and, in particular Prolog's leftmost selection rule [3,4,5]. Recently, [39] has introduced the concept of ∃-Universal Termination, which is related to the existence of at least one selection rule for which every SLD-derivation is finite.

Definition 6. *[39] A program* P *and a goal* G ∃-*universally terminate iff there exists a selection rule* s *such that every SLD-derivation of* G *via* s *is finite.*

Since fair selection rules select any atom of a goal in a finite number of steps, they allow a conjunctive goal to fail if at least one of its atom fails. Therefore, the following result holds.

Theorem 6. *[39] A program* P *and a goal* G ∃-*universally terminate iff they universally terminate w.r.t. the set of fair selection rules.*

In [39] Ruggeri introduces a characterization of ∃-universal termination by means of a notion of fair-bounded programs and queries, which provides us with a sound and complete method for proving ∃-universal termination. Anyway this characterization is undecidable.

By using our effective approximation $\mathcal{P}_k^\alpha[[P]]$, we can define sufficient yet effective conditions. In fact, we can compute, for any given k, a superset of the goals which have at least an infinite derivation by a fair selection rule. Therefore, if, for a given k, a goal G (or, in general, a class of goals) has no infinite derivation according to Inf_P^k, we can conclude that the goal G (or the class of goals) ∃-universally terminates. As a consequence of Theorem 4, we can state the following corollary.

Corollary 1. *[27] Let* P *be a program and* $G = p_1(t_1), \ldots, p_n(t_n)$ *be a goal. The program* P *and the goal* G ∃-*universally terminate iff there exists a* k *such that* $G \notin \mathrm{Inf}_P^k$, *i.e., for all* $\sigma_1, \ldots, \sigma_n$ *such that*

- *for at least one* $\bar{\imath}$, $\bar{\imath} \in \{1, \ldots, n\}$, $< \sigma_{\bar{\imath}}, \Diamond > \in \mathrm{gfp}(\mathcal{P}_k^\alpha[[P]])(p_{\bar{\imath}}(x_{\bar{\imath}}))$,
- *for* $i = \{1, \ldots, n\}$, $< \sigma_i, \circ > \in \mathrm{gfp}(\mathcal{P}_k^\alpha[[P]])(p_i(x_i))$,

there exists no $\mathrm{mgu}(G, p_1(x_1)\sigma_1, \ldots, p_n(x_n)\sigma_n)$.

Since the depth-k domain is finite, once k is given, we can check that $G \notin \mathrm{Inf}_P^k$ in an effective way. Moreover, if G belongs to Inf_P^k, for a given k, yet it does ∃-universally terminate, we can always increase k, until we find a k for which the conditions of Theorem 1 are satisfied.

Example 6. Let P_8 be the following program.

$$P_8 : q(a) \leftarrow p(f(f(a))).$$
$$p(f(f(b))) \leftarrow p(x).$$

Let $k = 3$ and $w \in W$.

$$\mathrm{gfp}(\mathcal{P}_k^\alpha[[P_8]])(q(x)) = \{< \{x/a\}, \Diamond >, < \{x/a\}, \Box >\}$$
$$\mathrm{gfp}(\mathcal{P}_k^\alpha[[P_8]])(p(x)) = \{< \{x/f(f(w))\}, \Diamond >, < \{x/f(f(w))\}, \Box >\}$$

$q(x) \in \mathrm{Inf}_P^k$. However the goal $q(x)$ has a finite failure in P_8. Therefore, for $k = 3$, our analysis can not conclude that the goal $q(x)$ terminates.
We can then try to increase k to improve the precision of our analysis.
For $k = 4$,

$$\mathrm{gfp}(\mathcal{P}_k^\alpha[[P_8]])(q(x)) = \{\}$$
$$\mathrm{gfp}(\mathcal{P}_k^\alpha[[P_8]])(p(x)) = \{< \{x/f(f(b))\}, \Diamond >, < \{x/f(f(b))\}, \Box >\}$$

In this case we prove (by correctness of the approximation) that the goal $q(x)$ terminates. In fact, for $k = 4$, $q(x) \notin Inf_p^k$. It is worth noting that this is a very simple example, where the right k can be guessed just by looking at the program. Of course, this is not always the case for more complex programs.

An advantage of the proposed method is that we can deal with classes of goals, for which the termination can be formally proved just by computing an abstract greatest fixpoint. In the following examples we sometimes deal with classes of goals defined extensionally using natural language. It is worth noting that such classes could be equivalently described using elements of a suitable composition of abstract domains (for groundness, types and depth-k)[19].

Example 7. Consider the program P_1 of Example 1. Assume we want to prove that $at(x, y)$ always terminates. Let $k = 2$. It can be checked that

$$gfp(\mathcal{P}_k^\alpha[[P_1]])(at(x, y)) = \{< \{x/telaviv, y/fido\}, \square >,$$
$$< \{x/jerusalem, y/fido\}, \square >\}$$
$$gfp(\mathcal{P}_k^\alpha[[P_1]])(near(x)) = \{< \{x/jerusalem\}, \square >\}$$

Note that $at(x, y) \notin Inf_p^2$, since $Inf_p^2 = \emptyset$.

Consider now the program P_9. We assume that the language of P_9 contains also the constants a and b.

$$P_9 : reverse([x|x_s], y_s) \leftarrow reverse(x_s, z_s), append(z_s, [x], y_s).$$
$$reverse([\,], [\,]).$$
$$append([x|x_s], y_s, [x|z_s]) \leftarrow append(x_s, y_s, z_s)$$
$$append([\,], y_s, y_s).$$

Assume we want to prove that $reverse(x, y)$ terminates whenever its first argument is a list of ground elements whose length is less than 100 or its second argument is a list of ground elements whose length is less than 100.
For the sake of simplicity, we consider only elements of the form $< -, \Diamond >$ in $gfp(\mathcal{P}_k^\alpha[[P_9]])$, since the queries we are interested in are atomic. Let $k = 101$ and $w_1, w_2 \in W$.
$gfp(\mathcal{P}_k^\alpha[[P_9]])(append(x, y, z)) =$
$\{< \{x/[x_1, x_2, \ldots, x_{100}, w_1], y/y_{100}, z/[z_1, z_2, \ldots, z_{100}, w_2]\}, \Diamond > \ |$ where
$\qquad\qquad\qquad x_1, \ldots, x_{100},$ and $z_1, \ldots, z_{100}$ are
$\qquad\qquad\qquad$ variables in V or one of the constants $a, b\}$

$gfp(\mathcal{P}_k^\alpha[[P_9]])(reverse(x, y)) =$
$\{< \{x/[x_1, x_2, \ldots, x_{100}, w_1], y/[y_1, y_2, \ldots, y_{100}, w_2]\}, \Diamond > \ |$ where $x_1, \ldots, x_{100},$
$\qquad\qquad\qquad$ and $y_1, \ldots, y_{100}$ are variables in V
$\qquad\qquad\qquad$ or one of the constants $a, b \qquad\qquad \}$

Note that, any goal of the form $reverse(x, y)$, such that x is a list of ground elements whose length is less or equal 100 or y is a list of ground elements whose length is less or equal 100, does not belong to Inf_p^{101}, i.e., there exists no mgu between such goals and $reverse([x_1, x_2, \ldots, x_{100}, w_1], [y_1, y_2, \ldots, y_{100}, w_2])$.
Therefore, we may conclude that for our queries P_9 terminates.
On the other hand, for all k, $reverse(x, y) \in Inf_p^k$. Therefore, for any k, we can

not conclude that the goal reverse(x, y) terminates in P_9. Indeed this goal has an infinite derivation in P_9.

Let us now consider the program P_2 of Example 2. Again we consider only elements of the form $< -, \Diamond >$ in $gfp(\mathcal{P}_k^\alpha[[P_2]])$. Let $k = 4$ and $w \in W$.
$$gfp(\mathcal{P}_k^\alpha[[P_2]])(p(x,y)) = \{ \ < \{x/a, y/f(f(f(w)))\}, \Diamond > \}$$
$$gfp(\mathcal{P}_k^\alpha[[P_2]])(q(x,y)) = \{ \ < \{x/f(a), y/f(f(f(w)))\}, \Diamond > \}$$
which allow us to prove that the goals $q(t, y)$, where t does not unify with $f(a)$, terminate, while the goals $q(x, y), q(f(x), y), q(f(a), y)$ universally terminate only if y is a ground depth-4 term.

Finally, let us consider the program P_{10}.

$$P_{10} : odd(s(x)) : -even(x)$$
$$even(s(s(x))) : -even(x)$$
$$even(0).$$

Let $k = 4$ and $w \in W$.
$$gfp(\mathcal{P}_k^\alpha[[P_{10}]])(odd(x)) = \{ \ < \{x/s(0)\}, \Box >, < \{x/s(s(s(0)))\}, \Box >,$$
$$< \{x/s(s(s(s(w))))\}, \Box >, < \{x/s(s(s(s(w))))\}, \Diamond > \}$$

$$gfp(\mathcal{P}_k^\alpha[[P_{10}]])(even(x)) = \{ \ < \{x/0\}, \Box >, < \{x/s(s(0))\}, \Box >,$$
$$< \{x/s(s(s(s(w))))\}, \Box >, < \{x/s(s(s(s(w))))\}, \Diamond > \}$$

We can prove that the conjunctive goal even(x), odd(x) terminates whenever x is ground and its depth is less than 4. In this case, in fact, $even(x), odd(x) \notin Inf_P^4$. Note that, for any k, we can not prove that even(x), odd(x) terminates. Indeed this goal has an infinite derivation.

6   Non-safe Programs and Goals

The theory of logic programming tells us that SLD-resolution is sound and complete. As a consequence, given a program P, every SLD-tree for a goal **G** is a complete search space for finding an SLD-refutation of **G**. In the actual implementation of logic languages (e.g. Prolog), the critical choice is that of the tree-searching algorithm. Two basic tree-search strategies are: the breadth-first search, which visits the tree by levels, and the depth-first search, which visits the tree by branches. The former is a fair strategy, since it finds a success node if one exists, whereas the latter is incomplete, since success nodes can be missed if an infinite branch is visited first. Universal termination tells us that we can be independent from the search algorithm. This means that if the program P and the goal **G** universally terminate we can "safely" replace breadth-first search by depth-first search. This is not the only case where depth-first search is fair. For example, if a goal **G** does not universally terminate in P, yet it has no successful derivations, breadth-first search and depth-first search will not yield different results.

Hence it is useful to have an effective way to understand, for a given program, which are the goals for which the choice of the search strategy becomes relevant. To study this problem we introduce the following set.

Definition 7. *Let* P *be a program and* **G** *be a goal.*
Non − safe(P) = {**G**| **G** *has at least one infinite*
and one successful derivation in P }

If **G** is *non-safe* in P, then the depth-first search is in general non-equivalent to the breadth-first search. The set of *non-safe* goals of P are the goals which do not universally terminate, yet they do existentially terminate successfully via a fair selection rule and a breadth-first search rule.

Our goal is to find a *correct* upward approximation of the set of non-safe goals, i.e. if **G** is a non-safe goal in P then we want **G** to belong to our approximation. Since the set of non-safe goals contains goals which have at least a successful and an infinite derivation, we use our approximate semantics to single out the goals which have an infinite derivation. We also need information on goals which have a successful derivation. Of course we could use our approximate semantics as well to single out the goals which have a successful derivation. However, in order to improve the precision of the analysis we introduce an approximation on the depth-k domain of the answers computed by *successful derivations* obtained as a least fixpoint. The improvement in precision in this new approximation w.r.t. the information on exact answers of successful derivations is due to the fact that, since we are observing just the successful behavior (which is a finite behavior), we can use a least fixpoint computation. It is worth noting that a similar remark was already made by Cousot in [17]. Therefore we introduce the following abstract operator approximating computed answers, a domain $\tilde{A} \subseteq [Goals \rightarrow \tilde{S}]$, where $\tilde{S}$ is the set of sets of substitutions of Subst$_k$.

Definition 8. *[15]* $\widetilde{\mathcal{P}}_k^\alpha[[P]]I =$
$\lambda p(\mathbf{x}).\{\alpha_k(\vartheta) \mid \exists p(\mathbf{t}) \leftarrow \mathbf{B}$ *a renamed version w.r.t.* **x** *of a clause in* P,
    $\mathbf{B} = p_1(\mathbf{t}_1),\ldots,p_n(\mathbf{t}_n),\ \exists \vartheta_j \in I(p_j(\mathbf{x}_j)),$
    $\vartheta = \{\mathbf{x}/\mathbf{t}\} \cdot mgu(\mathbf{B}, (p_1(\mathbf{x}_1)\vartheta_1,\ldots,p_n(\mathbf{x}_n)\vartheta_n))_{|\mathbf{x}}\}$

The abstract fixpoint semantics is the lfp of the $\widetilde{\mathcal{P}}_k^\alpha[[P]]$ operator.

By And-compositionality, we can define the set of goals which have a successful derivation according to the information in lfp($\widetilde{\mathcal{P}}_k^\alpha[[P]]$).

Definition 9. Succ$_P^k =$
{**G** | **G** = $p_1(\mathbf{t}_1),\ldots,p_n(\mathbf{t}_n)$, *for* $i = \{1,\ldots,n\}$, $\exists \sigma_i \in$ lfp($\widetilde{\mathcal{P}}_k^\alpha[[P]]$)($p_i(\mathbf{x}_i)$)
    such that **G** *and* $p_1(\mathbf{x}_1)\sigma_1,\ldots,p_n(\mathbf{x}_n)\sigma_n$ *are unifiable*}

In [15], it was proved that this is a correct upward approximation of the computed answers of the program P. Therefore we can state the following result.

Theorem 7. *[15] If* **G** *has at least a successful derivation in the program* P, *then for any* k, **G** $\in$ Succ$_P^k$.

As a consequence of Theorems 4 and 7, we can state the following result.

Corollary 2. *[27] If* $\exists k$, *such that at least one of the following holds,* **G** $\notin$ Succ$_P^k$
or **G** $\notin$ Inf$_P^k$, *then a goal* **G** $\notin$ Non − safe(P).

Therefore, for a given k, $\mathrm{lfp}(\widetilde{\mathcal{P}}_k^\alpha[[P]])$ and $\mathrm{gfp}(\mathcal{P}_k^\alpha[[P]])$ allow us to define a correct approximation of the non-safe set of goals of a program P. Note that the conditions in Definitions 4 and 9 can effectively be checked.

Example 8. Consider the program P_{11}.

$$P_{11} : t(a) \leftarrow p(x,y), q(x,y)$$
$$q(f(f(y)), f(y)) \leftarrow q(f(y), y)$$
$$q(f(f(y)), a) \leftarrow q(y, y)$$
$$p(f(y), f(y)) \leftarrow p(y, y)$$
$$p(a, a)$$

Let $k = 3$ and $w_1, w_2 \in W$. For the sake of simplicity, we will consider only elements of the form $< \_, \Diamond > \in \mathrm{gfp}(\mathcal{P}_k^\alpha[[P_{11}]])$.

$\mathrm{gfp}(\mathcal{P}_k^\alpha[[P_{11}]]) \, t(x) = \quad \{ < \{x/a\}, \Diamond > \}$
$\mathrm{gfp}(\mathcal{P}_k^\alpha[[P_{11}]]) \, q(x,y) = \{ < \{x/f(f(w_1)), y/f(f(w_2))\}, \Diamond > \}$
$\mathrm{gfp}(\mathcal{P}_k^\alpha[[P_{11}]]) \, p((x,y) = \{ < \{x/f(f(w_1)), y/f(f(w_2))\}, \Diamond > \}$

$\mathrm{lfp}(\widetilde{\mathcal{P}}_k^\alpha[[P_{11}]]) \, t(x) = \quad \{ \, \}$
$\mathrm{lfp}(\widetilde{\mathcal{P}}_k^\alpha[[P_{11}]]) \, q(x,y) = \{ \, \}$
$\mathrm{lfp}(\widetilde{\mathcal{P}}_k^\alpha[[P_{11}]]) \, p(x,y) = \{ \{x/a, y/a\}, \{x/f(a), y/f(a)\},$
$$\{ x/f(f(w_1)), y/f(f(w_2)) \} \}$$

Looking at our approximations, we can conclude that $q(x,y)$ (and all of its instances) are safe for any ϑ. Indeed, $q(x,y)$ and all its instances do not belong to Succ_P^3. Therefore, for these goals, depth-first search is equivalent to breadth-first search. The same holds for $t(x)$ and all its instances, since $t(x)$ and all its instances do not belong to Succ_P^3. On the other side, $p(x,y)\eta$ is safe just for $\eta \in \{\{x/f(x_1), y/x_1\}, \{x/x_1, y/f(x_1)\}\}$ or whenever η is a grounding substitution for x or y, belonging to Subst_3. In this cases, in fact, $p(x,y)\eta$ does not belong to Inf_P^3.

Note that, for the goal $p(x,y)$, we can not infer that a depth-first search is equivalent to a breadth-first search. In fact, in order to find the successful derivations for $p(x,y)$ we need to use a breadth-first search.

7 Future Work

In this paper we have proposed an abstract interpretation approach to universal termination. We have provided a semantic foundation for such an approach and developed a new abstract domain useful for termination analysis. Following these guidelines, other methods can be derived simply by computing an abstraction of our concrete semantics. In particular, we believe that this framework could allow us to reconstruct most of the existing automatic methods such as [44,38,32,45,34] as abstractions of the "exact answers" semantics on a suitable abstract domain. Once the different methods have been reconstructed in our framework, we can use general results of abstract interpretation. In particular, by using the reduced product [19], an operation on abstract domains which allow us to formally define a new abstract domain as the optimal combination of abstract domains, it will be possible to combine the results of different approximations so that for each program and goal we can systematically obtain the most precise results. The resulting method can be viewed as a theoretical basis for the design of a refined system able to analyze termination of real Prolog programs. In order to achieve this goal, however, a further step needs to be performed, i.e. the abstract semantics has to be modified to take into account some Prolog features, such as control strategies and extra-logical features. We believe that it would not be very difficult, for example, to modify the abstract semantics in order to deal with Prolog selection rule. Then an analyzer able to deal with full Prolog programs can be implemented and compared on benchmarks with the existing systems [30,35,13,42].

Finally we would like to point out that even if the k in the abstraction must be "guessed", we think that useful heuristics can be devised so as to provide "good" k's. Anyway we remind that a coarse initial abstraction can be easily refined by incrementing such k's.

References

1. K. R. Apt. Introduction to Logic Programming. In J. van Leeuwen, editor, *Handbook of Theoretical Computer Science*, volume B: Formal Models and Semantics, pages 495–574. Elsevier and The MIT Press, 1990. 365
2. K. R. Apt. *From Logic Programming to Prolog*. Prentice Hall, 1997. 362
3. K. R. Apt and D. Pedreschi. Studies in Pure Prolog: Termination. In J. W. Lloyd, editor, *Computational Logic*, pages 150–176. Springer-Verlag, 1990. 372
4. K. R. Apt and D. Pedreschi. Reasoning about termination of pure PROLOG programs. *Information and Computation*, 106(1):109–157, 1993. 372
5. K. R. Apt and D. Pedreschi. Modular termination proofs for logic and pure Prolog programs. In G. Levi, editor, *Advances in Logic Programming Theory*, pages 183–229. Oxford University Press, 1994. 372
6. R. Bagnara. China. http://www.cs.unipr.it/ bagnara/China/. 372
7. F. Benoy and A. King. Inferring argument size relationships with clp(r). In *Proc. of the Sixth International Workshop on Logic Program Synthesis and Transformation (LOPSTR'96)*, 1996. 362
8. M. Bezem. Characterizang termination of logic programs with level mappings. In E.L. Lusk and R.A. Overbeek, editors, *Proc. of the North American Conference on Logic Programming*, pages 69–80. The MIT Press, 1989. 372
9. M. Bezem. Strong termination of logic programs. *Journal of Logic Programming*, 15(1 & 2):79–98, 1993. 372
10. A. Bossi, N. Cocco, and M. Fabris. Proving Termination of Logic Programs by Exploiting Term Properties. In S. Abramsky and T. S. E. Maibaum, editors, *Proc. TAPSOFT'91*, volume 494 of *Lecture Notes in Computer Science*, pages 153–180. Springer-Verlag, 1991. 362
11. A. Brodsky and Y. Sagiv. Inference of monotonicity constraints in datalog programs. In *Proceedings of the Eighth ACM SIGACT-SIGART-SIGMOND Symposium on Principles of Database Systems*, pages 190–199, 1989. 362
12. F. Bueno, M. Carro, M. Hermenegildo, P. López, and G. Puebla. Ciao prolog development system. http://www.clip.dia.fi.upm.es/Software/Ciao/. 372

13. M. Codish and C. Taboch. Termiweb. http://www.cs.bgu.ac.il/ taboch/TerminWeb/. 362, 363, 378

14. M. Codish and C. Taboch. A semantic basis for the termination analysis of logic programs. *Journal of Logic Programming*, 41(1):103–123, 1999. 363, 363

15. M. Comini, G. Levi, and M. C. Meo. A theory of observables for logic programs. *Information and Computation*, 1999. To appear. 376, 376, 376

16. M. Comini and M. C. Meo. Compositionality properties of *SLD*-derivations. *Theoretical Computer Science*, 211(1-2):275–309, 1999. 368

17. P. Cousot. Constructive design of a hierarchy of semantics of a transition system by abstract interpretation (Invited Paper). In S. Brookes and M. Mislove, editors, *Proc. of the 13th Internat. Symp. on Mathematical Foundations of Programming Semantics (MFPS '97)*, volume 6 of *Electronic Notes in Theoretical Computer Science*. Elsevier, Amsterdam, 1997. 376

18. P. Cousot and R. Cousot. Abstract Interpretation: A Unified Lattice Model for Static Analysis of Programs by Construction or Approximation of Fixpoints. In *Proc. Fourth ACM Symp. Principles of Programming Languages*, pages 238–252, 1977. 365

19. P. Cousot and R. Cousot. Systematic Design of Program Analysis Frameworks. In *Proc. Sixth ACM Symp. Principles of Programming Languages*, pages 269–282, 1979. 365, 374, 377

20. P. Cousot and R. Cousot. Comparing the Galois Connection and Widening/Narrowing Approaches to Abstract Interpretation. In M. Bruynooghe and M. Wirsing, editors, *Proc. of PLILP'92*, volume 631 of *Lecture Notes in Computer Science*, pages 269–295. Springer-Verlag, 1992. 372

21. F.S. de Boer, A. Di Pierro, and C. Palamidessi. Nondeterminism and Infinite Computations in Constraint Programming. *Theoretical Computer Science*, 151(1), 1995. 364, 365, 365, 370

22. S. Decorte and D. De Schreye. Termination of logic programs: the never-ending story. *Journal of Logic Programming*, 19/20:199–260, 1994. 362, 363

23. S. Decorte, D. De Schreye, and M. Fabris. Automatic Inference of Norms: a Missing Link in Automatic Termination Analysis. In D. Miller, editor, *Proc. 1993 Int'l Symposium on Logic Programming*, pages 420–436. The MIT Press, 1993. 362

24. M. Gabbrielli and R. Giacobazzi. Goal independency and call patterns in the analysis of logic programs. In J. Urban E. Deaton, D. Oppenheim and H. Berghel, editors, *Proceedings of the Ninth ACM Symposium on Applied Computing*, pages 394–399, Phoenix AZ, 1994. ACM Press. 363

25. W. G. Golson. Toward a declarative semantics for infinite objects in logic programming. *Journal of Logic Programming*, 5:151–164, 1988. 364, 365, 370

26. R. Gori. A fixpoint semantics for reasoning about finite failure. In H. Ganzinger, D. McAllester, and A. Voronkov, editors, *Proceedings of the 6th International Conference on Logic for Programming and Automated Reasoning*, volume 1705 of *LNAI*, pages 238–257. Springer, 1999. 368

27. R. Gori. *Reasoning about finite failure and infinite computations by abstract interpretation*. PhD thesis, Dipartimento di Matematica, Università di Siena, 1999. http://www.di.unipi.it/ gori/gori.html. 367, 368, 369, 371, 371, 372, 373, 376

28. Marta Z. Kwiatkowska. Infinite Behaviour and Fairness in Concurrent Constraint Programming. In J. W. de Bakker, W. P.de Roever, and G. Rozenberg, editors, *Semantics: Foundations and Applications*, volume 666 of *Lecture Notes in Computer Science*, pages 348–383, Beekbergen The Nederland, June 1992. REX Workshop, Springer Verlag. 364

29. B. Le Charlier, S. Rossi, and P. Van Hentenryck. An abstract interpretation framework which accurately handles prolog search-rule and the cut. In *Proc. of the 1994 International Logic Programming Symposium*. The MIT Press, 1994. 364

30. N. Lindenstrauss. TermiLog. http://www.cs.huji.ac.il/ talre/form.html. 362, 363, 363, 363, 378

31. N. Lindenstrauss and Y. Sagiv. Checking termination of queries to logic programs. http://www.cs.huji.ac.il/ naomil. 363, 363

32. N. Lindenstrauss and Y. Sagiv. Automatic termination analysis of logic programs. In L.Naish, editor, *Proc. of the Fourteenth International Conference on Logic Programming*, pages 63–67, 1997. 363, 377

33. J. W. Lloyd. *Foundations of Logic Programming*. Springer-Verlag, 1987. Second edition. 364, 365, 365, 370

34. F. Mesnard. Inferring left-terminating classes of queries for constraint logic programs. In M.J. Maher, editor, *Proc. of the Joint International Conference on Logic Programming*, pages 7–21, 1996. 362, 363, 377

35. F. Mesnard and U. Neumerkel. cti. http://www.complang.tuwien.ac.at/cti/. 362, 363, 363, 363, 378

36. M. A. NaitAbdallah. On the intepretation of infinite computations in logic programming. In J. Paredaens, editor, *Proc. of Automata, Languages and Programming*, volume 172, pages 374–381. Springer Verlag, 1984. 364, 365, 370

37. S. O. Nyström and B. Jonsson. Indeterminate Concurrent Constraint Programming: A Fixpoint Semantics for Non-Terminating Computations. In D. Miller, editor, *Proc. of the 1993 International Logic Programming Symposium*, Series on Logic Programming, pages 335–352. The MIT Press, 1993. 365

38. L. Plümer. *Termination Proofs for Logic Programs*, volume 446 of *Lecture Notes in Artificial Intelligence*. Springer-Verlag, 1990. 362, 362, 363, 377

39. S. Ruggeri. ∃-universal termination of logic programs. *Theoretical Computer Science*, 1999. To appear. 372, 373, 373, 373

40. V. A. Saraswat. *Concurrent Constraint Programming Languages*. PhD thesis, Carnegie-Mellon University, January 1989. 364, 365, 370

41. T. Sato and H. Tamaki. Enumeration of Success Patterns in Logic Programs. *Theoretical Computer Science*, 34:227–240, 1984. 369

42. C. Speirs, Z. Somogyi, and H. Sondergaard. Mercury's termination analyzer. http://www.cs.mu.oz.au/research/mercury/. 362, 363, 378

43. A. Van Gelder. Deriving Constraints Among Argument Sizes in Logic Programs. In *Proc. of the eleventh ACM Conference on Principles of Database Systems*, pages 47–60. ACM, 1990. 362

44. K. Verschaetse and D. De Schreye. Deriving Termination Proofs for Logic Programs, Using Abstract Procedures. In K. Furukawa, editor, *Proc. Eighth Int'l Conf. on Logic Programming*, pages 301–315. The MIT Press, 1991. 363, 377

45. K. Verschaetse and D. De Schreye. Deriving linear size relations by abstract interpretation. *New Generation Computing*, 13(2):117–154, 1995. 362, 363, 363, 377

Using an Abstract Representation to Specialize Functional Logic Programs[*]

Elvira Albert[1], Michael Hanus[2], and Germán Vidal[1]

[1] DSIC, UPV, Camino de Vera s/n, E-46022 Valencia, Spain
{ealbert,gvidal}@dsic.upv.es
[2] Institut für Informatik, CAU Kiel, Olshausenstr. 40, D-24098 Kiel, Germany
mh@informatik.uni-kiel.de

Abstract. This paper introduces a novel approach for the specialization of functional logic languages. We consider a maximally simplified abstract representation of programs (which still contains all the necessary information) and define a non-standard semantics for these programs. Both things mixed together allow us to design a simple and concise partial evaluation method for modern functional logic languages, avoiding several limitations of previous approaches. Moreover, since these languages can be automatically translated into the abstract representation, our technique is widely applicable. In order to assess the practicality of our approach, we have developed a partial evaluation tool for the multi-paradigm language Curry. The partial evaluator is written in Curry itself and has been tested on an extensive benchmark suite (even a meta-interpreter). To the best of our knowledge, this is the first purely declarative partial evaluator for a functional logic language.

1 Introduction

Partial evaluation (PE) is a source-to-source program transformation technique for specializing programs w.r.t. parts of their input (hence also called *program specialization*). PE has been studied, among others, in the context of functional programming (e.g., [9,21]), logic programming (e.g., [12,24]), and functional logic programming (e.g., [4,22]). While the aim of traditional partial evaluation is to specialize programs w.r.t. some known data, several PE techniques are able to go beyond this goal, achieving more powerful program optimizations. This is the case of a number of PE methods for functional programs (e.g., positive supercompilation [27]), logic programs (e.g., partial deduction [24]), and functional logic programs (e.g., narrowing-driven PE [4]). A common pattern of these techniques is that they are able to achieve optimizations regardless of whether known data are provided (e.g., they can eliminate some intermediate data structures, similarly to Wadler's deforestation [28]). In some sense, these techniques are stronger *theorem provers* than traditional PE approaches.

[*] This work has been partially supported by CICYT TIC 98-0445-C03-01, by Acción Integrada hispano-alemana HA1997-0073, and by the DFG under grant Ha 2457/1-1.

Recent proposals of multi-paradigm declarative languages amalgamate the most important features of functional, logic and concurrent programming (see [14] for a survey). The operational semantics of these languages is usually based on a combination of two different operational principles: narrowing and residuation [15]. The *residuation* principle is based on the idea of delaying function calls until they are ready for a deterministic evaluation (by rewriting). On the other hand, the *narrowing* mechanism allows the instantiation of variables in input expressions and, then, applies reduction steps to the function calls of the instantiated expression. Due to its optimality properties w.r.t. the length of derivations and the number of computed solutions, *needed narrowing* [6] is currently the best narrowing strategy for functional logic programs. The formulation of needed narrowing is based on the use of *definitional trees* [5], which define a strategy to evaluate functions by applying narrowing steps.

In this work, we are concerned with the PE of functional logic languages. The first approach to this topic was the narrowing-driven PE of [4], which considered functional logic languages with an operational semantics based solely on narrowing. Recently, [2] introduced an extension of this basic framework in order to consider also the residuation principle. Using the terminology of [13], the narrowing-driven PE methods of [2,4] are able to produce both *polyvariant* and *polygenetic* specializations, i.e., they can produce different specializations for the same function definition and can also combine distinct original function definitions into a comprehensive specialized function. This means that narrowing-driven PE has the same potential for specialization as *positive supercompilation* [27] and *conjunctive partial deduction* [10] (a comparison can be found in [4]).

Despite its power, the narrowing-driven approach to PE suffers from several limitations: (i) Firstly, in the context of *lazy* functional logic languages, expressions in *head normal form* (i.e., rooted by a constructor symbol) cannot be evaluated at PE time. This restriction is imposed because the *backpropagation* of bindings to the left-hand sides of residual rules can incorrectly restrict the domain of functions (see Example 2). (ii) Secondly, if one intends to develop a PE scheme for a realistic multi-paradigm declarative language, several high-level constructs have to be considered: higher-order functions, constraints, program annotations, calls to external functions, etc. A complex operational calculus is required to properly deal with these additional features of modern languages. It is well-known that a partial evaluator normally includes an interpreter of the language. Therefore, as the operational semantics becomes more elaborated, the associated PE techniques become (more powerful but) also increasingly more complex. (iii) Finally, an interesting application of PE is the generation of compilers and compiler generators [21]. For this purpose, the partial evaluator must be self-applicable, i.e., able to partially evaluate itself. This becomes difficult in the presence of high-level constructs such as those mentioned in (ii). As advised in [21], *it is essential to cut the language down to the bare bones* in order to achieve self-application.

In order to overcome the aforementioned problems, a promising approach successfully tested in other contexts (e.g., [7,25]) is to consider programs written in a maximally simplified programming language, into which programs written in a higher-level language can be automatically translated. Recently, [18] introduced an explicit representation of the structure of definitional trees (used to guide the needed narrowing strategy) in the rewrite rules. This provides more explicit control and leads to a calculus simpler than standard needed narrowing. Moreover, source programs can be automatically translated to the new representation.[1] In this work, we consider a very simple abstract representation of functional logic programs which is based on the one introduced in [18]. As opposed to [18], our abstract representation includes also information about the evaluation type of functions: *flexible* —which enables narrowing steps— or *rigid* —which forces delayed evaluation by rewriting. Then, we define a *non-standard* semantics which is specially well-suited to perform computations at PE time. This is a crucial difference with previous approaches [2,4], where the same mechanism is used both for program execution and for PE. The use of an abstract representation, together with the new calculus, allows us to design a simple and concise automatic PE method for modern functional logic languages, breaking the limitations of previous approaches.

Finally, since truly lazy functional logic languages can be automatically translated into the abstract representation (which still contains all the necessary information about programs), our technique is widely applicable. Following this scheme, partially evaluated programs will be also written in the abstract representation. Since existing compilers use a similar representation for intermediate code, this is not a restriction. Rather, our specialization process can be seen as an optimization phase (transparent to the user) performed during the compilation of the program. In order to assess the practicality of our approach, we have developed a PE tool for the multi-paradigm language Curry [19]. The partial evaluator is written in Curry itself and has been tested on an extensive set of benchmarks (even a meta-interpreter). To the best of our knowledge, this is the first purely declarative partial evaluator for a functional logic language.

The structure of this paper is as follows. After providing some preliminary definitions in Sect. 2, we present our approach for the PE of functional logic languages based on the use of an abstract representation in Sect. 3. We also discuss the limitations of using the standard semantics during PE and, then, introduce a more suitable semantics. Section 4 presents a fully automatic PE algorithm based on the previous ideas, and Sect. 5 shows some benchmarks performed with an implementation of the partial evaluator. Finally, Sect. 6 concludes and discusses some directions for future work. More details and missing proofs can be found in [3].

2 Preliminaries

In this section we recall, for the sake of completeness, some basic notions from term rewriting [11] and functional logic programming [14]. We consider a (*many-*

---

[1] Indeed, it constitutes the basis of a recent proposal for an standard intermediate language, FlatCurry, for the compilation of Curry programs [20].

sorted) *signature* Σ partitioned into a set $\mathcal{C}$ of *constructors* and a set $\mathcal{F}$ of (defined) *functions* or *operations*. We write $c/n \in \mathcal{C}$ and $f/n \in \mathcal{F}$ for n-ary constructor and operation symbols, respectively. There is at least one sort *Bool* containing the constructors `True` and `False`. The set of *terms* and *constructor terms* with *variables* (e.g., x, y, z) from $\mathcal{V}$ are denoted by $\mathcal{T}(\mathcal{C} \cup \mathcal{F}, \mathcal{V})$ and $\mathcal{T}(\mathcal{C}, \mathcal{V})$, respectively. The set of variables occurring in a term t is denoted by $\mathcal{V}ar(t)$. A term is *linear* if it does not contain multiple occurrences of any variable. We write $\overline{o_n}$ for the *sequence of objects* $o_1, \ldots, o_n$. We denote by $root(t)$ the symbol at the root of the term t. A *position* p in a term t is denoted by a sequence of natural numbers. Positions are ordered by: $u \leq v$, if $\exists w$ such that $u.w = v$. The *subterm* of t at position p is denoted by $t|_p$, and $t[s]_p$ is the result of *replacing the subterm* $t|_p$ by the term s.

We denote a *substitution* σ by $\{x_1 \mapsto t_1, \ldots, x_n \mapsto t_n\}$ with $\sigma(x_i) = t_i$ for $i = 1, \ldots, n$ (where $x_i \neq x_j$ if $i \neq j$), and $\sigma(x) = x$ for all other variables x. By abuse, $\mathcal{D}om(\sigma) = \{x \in \mathcal{V} \mid \sigma(x) \neq x\}$ is called the *domain* of σ. Also, $\mathcal{R}an(\theta) = \{\theta(x) \mid x \in \mathcal{D}om(\theta)\}$. A substitution σ is a *constructor* substitution, if $\sigma(x)$ is a constructor term $\forall x \in \mathcal{D}om(\sigma)$. The identity substitution is denoted by $\{ \}$. Given a substitution θ and a set $V \subseteq \mathcal{V}$, we denote the substitution obtained from θ by restricting its domain to V by $\theta_{\upharpoonright V}$. We write $\theta = \sigma \; [V]$ if $\theta_{\upharpoonright V} = \sigma_{\upharpoonright V}$, and $\theta \leq \sigma \; [V]$ denotes the existence of a substitution γ such that $\gamma \circ \theta = \sigma \; [V]$. A term t' is an *instance* of t if $\exists \sigma$ with $t' = \sigma(t)$.

A set of rewrite rules $l = r$ such that $l \notin \mathcal{V}$, and $\mathcal{V}ar(r) \subseteq \mathcal{V}ar(l)$ is called a *term rewriting system* (TRS). The terms l and r are called the *left-hand side* and the *right-hand side* of the rule, respectively. A *rewrite step* is an application of a rewrite rule to a term, i.e., $t \rightarrow_{p,R} s$ if there exists a position p in t, a rewrite rule $R = (l = r)$ and a substitution σ with $t|_p = \sigma(l)$ and $s = t[\sigma(r)]_p$. Given a relation $\rightarrow$, we denote by $\rightarrow^+$ the transitive closure of $\rightarrow$, and by $\rightarrow^*$ the transitive and reflexive closure of $\rightarrow$. A (constructor) *head normal form* is either a variable or a term rooted by a constructor symbol. To evaluate terms containing variables, narrowing nondeterministically instantiates the variables so that a rewrite step is possible. Formally, $t \rightsquigarrow_{p,R,\sigma} t'$ is a *narrowing step* if p is a non-variable position in t and $\sigma(t) \rightarrow_{p,R} t'$. We denote by $t_0 \rightsquigarrow^*_\sigma t_n$ a sequence of narrowing steps $t_0 \rightsquigarrow_{\sigma_1} \ldots \rightsquigarrow_{\sigma_n} t_n$ with $\sigma = \sigma_n \circ \cdots \circ \sigma_1$. (If $n = 0$ then $\sigma = \{ \}$.) In functional programming, one is interested in the computed *value* whereas logic programming emphasizes the different bindings (*answers*). In an integrated setting, given a narrowing derivation $t_0 \rightsquigarrow^*_\sigma t_n$, we say that t_n is the computed value and σ is the computed answer for t_0.

3   Using an Abstract Representation for PE

In this section, we present an appropriate abstract representation for modern functional logic languages. We also provide a non-standard operational semantics which is specially well-suited to perform computations during partial evaluation.

First, let us briefly recall the basis of the narrowing-driven approach to PE of [4]. Informally speaking, given a particular narrowing strategy $\rightsquigarrow$, the (parametric) notions of resultant and partial evaluation are defined as follows. A *resultant* is a program rule of the form: $\sigma(s) = t$ associated to a narrowing derivation: $s \leadsto_\sigma^+ t$. A *partial evaluation* for a term s in a program $\mathcal{R}$ is computed by constructing a finite (possibly incomplete) narrowing *tree* for this term, and then extracting the resultants associated to the root-to-leaf derivations of the tree. Depending on the considered class of programs (and the associated narrowing strategy), a PE might require a post-processing of renaming to recover the same class of programs. An intrinsic feature of the narrowing-driven approach is the use of the same operational mechanism for both execution and PE.

3.1 The Abstract Representation

Recent approaches to functional logic programming consider inductively sequential systems as programs and a combination of needed narrowing and residuation as operational semantics [15,19]. The precise mechanism (narrowing or residuation) for each function is specified by *evaluation annotations*, which are similar to coroutining declarations in Prolog, where the programmer specifies conditions under which a call is ready for a resolution step. Functions to be evaluated in a deterministic manner are declared as *rigid* (which forces deferred evaluation by rewriting), while functions providing for nondeterministic evaluation steps are declared as *flexible* (which enables narrowing steps).

Similarly to [18], we present an abstract representation for programs in which the definitional trees (used to guide the needed narrowing strategy) are made explicit by means of case constructs. Moreover, here we distinguish two kinds of case expressions in order to make also explicit the flexible/rigid evaluation annotations. In particular, we assume that all functions are defined by one rule whose left-hand side contains only variables as parameters and the right-hand side contains case expressions for pattern-matching. Thanks to this new representation, we can define a simple operational semantics, which will become essential to simplify the definition of the associated PE scheme. The syntax for programs in the abstract representation is summarized as follows:

$$
\begin{array}{llll}
\mathcal{R} ::= D_1 \dots D_m & t ::= v & & \text{(variable)} \\
D ::= f(v_1, \dots, v_n) = t & \quad\mid\quad c(t_1, \dots, t_n) & & \text{(constructor)} \\
& \quad\mid\quad f(t_1, \dots, t_n) & & \text{(function call)} \\
p ::= c(v_1, \dots, v_n) & \quad\mid\quad case \ t_0 \ of \ \{p_1 \to t_1; \dots; p_n \to t_n\} & & \text{(rigid case)} \\
& \quad\mid\quad fcase \ t_0 \ of \ \{p_1 \to t_1; \dots; p_n \to t_n\} & & \text{(flexible case)}
\end{array}
$$

where $\mathcal{R}$ denotes a program, D a function definition, p a pattern and t an arbitrary expression. A program $\mathcal{R}$ consists of a sequence of function definitions D such that the left-hand side is linear and has only variable arguments, i.e., pattern matching is compiled into case expressions. The right-hand side of each function definition is a term t composed by variables, constructors, function calls, and case expressions. The form of a case expression is: $(f)case \ t \ of \ \{c_1(\overline{x_{n_1}}) \to t_1, \dots, c_k(\overline{x_{n_k}}) \to t_k\}$, where t is a term, $c_1, \dots, c_k$ are different constructors of the type of t, and $t_1, \dots, t_k$ are terms (possibly containing case expressions).

The variables $\overline{x_{n_i}}$ are called *pattern variables* and are local variables which occur only in the corresponding subexpression t_i. The difference between *case* and *fcase* shows up when the argument t is a free variable: *case* suspends (which corresponds to residuation) whereas *fcase* nondeterministically binds this variable to the pattern in a branch of the case expression (which corresponds to narrowing). Functions defined only by *fcase* (resp. *case*) expressions are called *flexible* (resp. *rigid*). Thus, flexible functions act as generators (like predicates in logic programming) and rigid functions act as consumers. Concurrency is expressed by a built-in operator "&" which evaluates its two arguments concurrently. This operator can be defined by the rule: True & True = True and, hence, in the following we simply consider it as an ordinary function symbol.

Example 1. Consider the rules defining the (rigid) function " $\leqslant$ ":[2]

$$
\begin{aligned}
0 \leqslant n &= \text{True} \\
(\text{Succ } m) \leqslant 0 &= \text{False} \\
(\text{Succ } m) \leqslant (\text{Succ } n) &= m \leqslant n
\end{aligned}
$$

By using case expressions, they can be represented by the following rewrite rule:

$$
x \leqslant y = \text{ case } x \text{ of } \{0 \qquad \rightarrow \text{True}; \\
(\text{Succ } x_1) \rightarrow \text{case } y \text{ of } \{0 \rightarrow \text{False}; \\
(\text{Succ } y_1) \rightarrow x_1 \leqslant y_1\} \}
$$

Due to the presence of fresh pattern variables in the right-hand side of the rule, this is not a standard rewrite rule. Nevertheless, the reduction of a case expression binds these pattern variables so that they disappear during a concrete evaluation (see [18]).

3.2   The Residualizing Semantics

An automatic transformation from inductively sequential programs to programs using case expressions is introduced in [18]. They also provide an appropriate operational semantics for these programs: the LNT calculus (Lazy Narrowing with definitional Trees), which is equivalent to needed narrowing over inductively sequential programs. In this work, we consider functional logic languages with a more general operational principle, namely a combination of (needed) narrowing and residuation. Nevertheless, the translation method of [18] could be easily extended to cover programs containing evaluation annotations; namely, flexible (resp. rigid) functions are translated by using only *fcase* (resp. *case*) expressions. Moreover, the LNT calculus of [18] can be also extended to correctly evaluate *case/fcase* expressions. In the following, we refer to the LNT calculus to mean the LNT calculus of [18] extended to cope with *case/fcase* expressions (the formal definition can be found in [3]).

Unfortunately, by using the standard semantics during PE, we would have the same problems of previous approaches (see Sect. 1). In particular, one of the

---

[2] Although we consider in this work a first-order language, we use a curried notation in the examples (as is usual in functional languages).

main problems comes from the *backpropagation* of variable bindings to the left-hand sides of residual rules. In the context of lazy (call-by-name) functional logic languages, this can provoke an incorrect restriction on the domain of functions (regarding the ability to compute head normal forms) and, thus, the loss of correctness for the transformation whenever some term in head normal form is evaluated during PE. The following example illustrates this point.

Example 2. Consider the following program:

$$
\begin{array}{lll}
\text{isZero 0} & = & \text{True} \\
\text{nonEmptyList } (x : xs) & = & \text{True} \\
\text{foo x} & = & \text{isZero x} : [] \\
\end{array}
$$

Here we use "[]" and ":" as constructors of lists, and "0" and "Succ" to define natural numbers. Then, given the (unique) computation for foo y:

$$\text{foo y} \rightsquigarrow_{\{\}} \ (\text{isZero y}) : [] \ \rightsquigarrow_{\{y \mapsto 0\}} \ \text{True} : []$$

where (isZero y) : [] is in head normal form, we get the residual rule:

$$\text{foo 0} = \text{True} : []$$

However, the expression nonEmptyList (foo (Succ 0)) can be evaluated to True in the original program (reduced functions are underlined):

$$\text{nonEmptyList } (\underline{\text{foo}} \ (\text{Succ 0})) \rightsquigarrow_{\{\}} \underline{\text{nonEmptyList}} \ (\text{isZero (Succ 0)} : [])$$
$$\rightsquigarrow_{\{\}} \text{True}$$

whereas it is not possible if the residual rule for foo is used (together with the original definitions for isZero and nonEmptyList).

The restriction on forbidding the evaluation of head normal forms can drastically reduce the optimization power of the transformation in some cases. Therefore, we propose a *residualizing* version of the LNT calculus which allows us to avoid this restriction. In the new calculus, variable bindings are encoded by case expressions (and are considered "residual" code). The inference rules of the new calculus, RLNT (Residualizing LNT), can be seen in Fig. 1. Let us explain the inference rules defining the one-step relation $\Rightarrow$. We note that the symbols "[" and "]" in an expression like $[\![t]\!]$ are purely syntactical (i.e., they do not denote "the value of t"). Indeed, they are only used to *guide* the inference rules and, most importantly, to mark which part of an expression can be still evaluated (within the square brackets) and which part must be definitively residualized (not within the square brackets). Let us briefly describe the rules of the calculus:

HNF. The HNF (Head Normal Form) rules are used to evaluate terms in head normal form. If the expression is a variable or a constructor constant, the square brackets are removed and the evaluation process stops. Otherwise, the evaluation proceeds with the arguments. This evaluation can be made in a don't care nondeterministic manner. Note, though, that this source of nondeterminism can be easily avoided by considering a fixed selection rule, e.g., by selecting the leftmost argument which is not a constructor term.

HNF
$$[\![t]\!] \Rightarrow t \quad \text{if } t \in \mathcal{V} \text{ or } t = c() \text{ with } c/0 \in \mathcal{C}$$
$$[\![c(t_1, \ldots, t_n)]\!] \Rightarrow c([\![t_1]\!], \ldots, [\![t_n]\!])$$

Case-of-Case
$$[\![(f)\,case\;((f)\,case\;t\;of\;\{\overline{p_k \to t_k}\})\;of\;\{\overline{p'_j \to t'_j}\}]\!]$$
$$\Rightarrow [\![(f)\,case\;t\;of\;\{\overline{p_k \to (f)\,case\;t_k\;of\;\{\overline{p'_j \to t'_j}\}}\}]\!]$$

Case Function
$$[\![(f)\,case\;g(\overline{t_n})\;of\;\{\overline{p_k \to t'_k}\}]\!] \Rightarrow [\![(f)\,case\;\sigma(r)\;of\;\{\overline{p_k \to t'_k}\}]\!]$$
$$\text{if } g(\overline{x_n}) = r \in \mathcal{R} \text{ is a rule with fresh variables}$$
$$\text{and } \sigma = \{\overline{x_n \mapsto t_n}\}$$

Case Select
$$[\![(f)\,case\;c(\overline{t_n})\;of\;\{\overline{p_k \to t'_k}\}]\!] \Rightarrow [\![\sigma(t'_i)]\!] \quad \text{if } p_i = c(\overline{x_n}),\; c \in \mathcal{C},\; \sigma = \{\overline{x_n \mapsto t_n}\}$$

Case Guess
$$[\![(f)\,case\;x\;of\;\{\overline{p_k \to t_k}\}]\!] \Rightarrow (f)\,case\;x\;of\;\{\overline{p_k \to [\![\sigma_k(t_k)]\!]}\}$$
$$\text{if } \sigma_i = \{x \mapsto p_i\},\; i = 1, \ldots, k$$

Function Eval
$$[\![g(\overline{t_n})]\!] \Rightarrow [\![\sigma(r)]\!] \quad \text{if } g(\overline{x_n}) = r \in \mathcal{R} \text{ is a rule with fresh}$$
$$\text{variables and } \sigma = \{\overline{x_n \mapsto t_n}\}$$

Fig. 1. RLNT Calculus

Case-of-Case. This rule moves the outer case inside the branches of the inner one. Rigorously speaking, this rule can be expanded into four rules (with the different combinations for *case* and *fcase* expressions), but we keep the above (less formal) presentation for simplicity. Observe that the outer case expression may be duplicated several times, but each copy is now (possibly) scrutinizing a known value, and so the Case Select rule can be applied to eliminate some case constructs.

Case Function. This rule can be only applied when the argument of the case is operation-rooted. In this case, it allows the *unfolding* of the function call.

Case Guess. It represents the main difference w.r.t. the standard LNT calculus. In order to imitate the instantiation of variables in needed narrowing steps, this rule is defined in the standard LNT calculus as follows:

$$[\![fcase\;x\;of\;\{\overline{p_k \to t_k}\}]\!] \Rightarrow^{\sigma} [\![\sigma(t_i)]\!] \quad \text{if } \sigma = \{x \mapsto p_i\},\; i = 1, \ldots, k$$

However, in this case, we would inherit the limitations of previous approaches. Therefore, it has been modified in order not to backpropagate the bindings of variables. In particular, we "residualize" the case structure and continue with the evaluation of the different branches (by applying the corresponding substitution in order to propagate bindings forward in the computation). Note that, due to this modification, no distinction between flexible and rigid case expressions is needed in the RLNT calculus.

Function Eval. This rule performs the unfolding of a function call. As in proof procedures for logic programming, we assume that we take a program rule with fresh variables in each such evaluation step.

In contrast to the standard LNT calculus, the inference system of Fig. 1 is completely deterministic, i.e., there is no don't know nondeterminism involved in the computations. This means that only one derivation can be issued from a given term (thus, there is no need to introduce a notion of RLNT "tree").

Example 3. Consider the well-known function app to concatenate two lists:

$$\text{app x y} = \text{case x of } \{ \; [] \quad \rightarrow \text{y} \; ; \\ (\text{a} : \text{b}) \rightarrow \text{a} : (\text{app b y}) \; \}$$

Given the call app (app x y) z to concatenate three lists, we have the following (partial) derivation using the rules of the RLNT calculus:

⟦app (app x y) z⟧

⇒ ⟦case (app x y) of {[] → z; (a : b) → (a : app b z)}⟧

⇒ ⟦case (case x of {[] → y; (a′ : b′) → (a′ : app b′ y)})
       of {[] → z; (a : b) → (a : app b z)}⟧

⇒ ⟦case x of { []   → case y of {[] → z; (a : b) → (a : app b z)};
              (a′ : b′) → case (a′ : app b′ y) of {[] : z; (a : b) → (a : app b z)}⟧

⇒ case x of { []   → ⟦case y of {[] → z; (a : b) → (a : app b z)}⟧;
             (a′ : b′) → ⟦case (a′ : app b′ y) of {[] → z; (a : b) → (a : app b z)}⟧

⇒* case x of { []   → case y of {[] → z; (a : b) → (a : ⟦app b z⟧)};
              (a′ : b′) → ⟦case (a′ : app b′ y) of {[] → z; (a : b) → (a : app b z)}⟧

⇒* case x of { []   → case y of {[] → z; (a : b) → (a : ⟦app b z⟧)};
              (a′ : b′) → (a′ : ⟦app (app b′ y) z⟧)}

The resulting RLNT calculus shares many similarities with the driving mechanism of [27] and Wadler's deforestation [28] (although we obtained it independently by refining the original LNT calculus to avoid the backpropagation of bindings). The main differences w.r.t. the driving mechanism are that we include the Case-of-Case rule and that driving is defined also for if_then_else constructs (which can be expressed in our representation by means of case expressions). The main difference w.r.t. deforestation is revealed in the Case Guess rule, where the patterns p_i are substituted in the different branches, like in the driving transformation. Although it may seem only a slight difference, situations may arise during transformation in which our calculus (as well as the driving mechanism) takes advantage of the sharing between different arguments while deforestation may not (see [27]).

A common restriction in related program transformations is to forbid the unfolding of function calls using program rules whose right-hand side is not linear. This avoids the duplication of calls under an eager (call-by-value) semantics or under a lazy (call-by-name) semantics implementing the *sharing* of common variables. Since our computation model is based on a lazy semantics, which does not consider the sharing of variables, we cannot incur into the risk of duplicated computations. Nevertheless, if sharing is considered (as in, e.g., the language Curry), this restriction can be implemented by requiring right-linear program rules to apply the Case Function and Function Eval rules.

Regarding the PE of programs with *flexible/rigid* evaluation annotations, [2] introduced a special treatment in order to correctly infer the evaluation annotations for residual definitions. Within this approach, one is forced to split resultants by introducing several intermediate functions in order not to mix bindings which come from the evaluation of flexible and rigid functions. Moreover, to avoid the creation of a large number of intermediate functions, only the computation of a single needed narrowing step for suspended expressions is allowed. Now, by using case expressions (instead of functions defined by patterns as in [2]), we are able to proceed the specialization of suspended expressions beyond a single needed narrowing step without being forced to split the associated resultant (and hence without increasing the size of the residual program). This is justified by the fact that case constructs preserve the rigid or flexible nature of the functions which instantiate the variables.[3] The following example is taken from [2] and illustrates that the use of case constructs to represent function definitions simplifies the residual program.

Example 4. Consider a program and its PE for the term $f \; x \; (g \; y \; (h \; z))$, according to the technique introduced in [2]:

f 0 (Succ 0)	= 0	% flex	f′ 0 Y Z		= f′₁ Y Z	% flex
g 0 0	= (Succ 0)	% rigid	f′₁ (Succ 0) Z		= f′₂ Z	% rigid
h 0	= 0	% flex	f′₂ 0		= f′₃	% flex
			f′₃		= 0	% flex

where $f \; x \; (g \; y \; (h \; z))$ is renamed as $f' \; x \; y \; z$. The original program can be translated to our abstract representation as follows:

$$f \; x \; y \; = \; \mathtt{fcase} \; x \; \mathtt{of} \; \{0 \rightarrow \mathtt{fcase} \; y \; \mathtt{of} \; \{(\mathtt{Succ} \; 0) \rightarrow 0\}\}$$
$$g \; x \; y \; = \; \mathtt{case} \; x \; \mathtt{of} \; \{0 \rightarrow \mathtt{case} \; y \; \mathtt{of} \; \{0 \rightarrow (\mathtt{Succ} \; 0)\}\}$$
$$h \; x \; = \; \mathtt{fcase} \; x \; \mathtt{of} \; \{0 \rightarrow 0\}$$

The following PE for $f \; x \; (g \; y \; (h \; z))$, constructed by using the rules of the RLNT calculus, avoids the introduction of three intermediate rules and, thus, is notably simplified:

$$f' \; x \; y \; z = \mathtt{fcase} \; x \; \mathtt{of} \; \{0 \rightarrow \mathtt{case} \; y \; \mathtt{of} \; \{(\mathtt{Succ} \; 0) \rightarrow \mathtt{fcase} \; z \; \mathtt{of} \; \{0 \rightarrow 0\}\}\}$$

The next result establishes a precise equivalence between the standard semantics (the LNT calculus) and its residualizing version. In the following, we denote by $\Rightarrow_{Guess}$ the application of the following rule from the standard semantics:

$$[\![fcase \; x \; of \; \{\overline{p_k \rightarrow t_k}\}]\!] \; \Rightarrow^{\sigma}_{Guess} \; [\![\sigma(t_i)]\!] \quad \text{if } \sigma = \{x \mapsto p_i\}, \; i = 1, \ldots, k$$

Furthermore, we denote by $del_{sq}(t)$ the expression which results from t by deleting all the occurrences of "$[\![$" and "$]\!]$" (if any).

Theorem 1. *Let t be a term, $V \supseteq Var(t)$ a finite set of variables, d a constructor term, and $\mathcal{R}$ a program in the abstract representation. For each LNT*

---

[3] Indeed, the treatment for *case/fcase* expressions is the same in the RLNT calculus.

derivation $\llbracket t \rrbracket \overset{*}{\Rightarrow}{}^{\sigma} d$ *for* t *w.r.t.* $\mathcal{R}$ *computing the answer* σ, *there exists a* RLNT *derivation* $\llbracket t \rrbracket \Rightarrow^* t'$ *for* t *w.r.t.* $\mathcal{R}$ *such that there is a finite sequence* $\llbracket del_{sq}(t') \rrbracket \Rightarrow^{\sigma_1}_{Guess} \cdots \Rightarrow^{\sigma_n}_{Guess} d$, *where* $\sigma_n \circ \ldots \circ \sigma_1 = \sigma$ $[V]$, *and vice versa.*

Roughly speaking, for each (successful) LNT derivation from t to a constructor term d computing σ, there is a corresponding RLNT derivation from t to t' in which the computed substitution σ is encoded in t' by case expressions and can be obtained by a (finite) sequence of $\Rightarrow_{Guess}$ steps (deriving the same value d).

4   Control Issues for Partial Evaluation

Following [12], a simple *on-line* PE algorithm can proceed as follows. Given a term t and a program $\mathcal{R}$, we compute a finite (possibly incomplete) RLNT derivation $t \Rightarrow^+ s$ for t w.r.t. $\mathcal{R}$.[4] Then, this process is iteratively repeated for any subterm which occurs in the expression s and which is not *closed* w.r.t. the set of terms already evaluated. Informally, the *closedness* condition guarantees that each call which might occur during the execution of the residual program ·is covered by some program rule. If this process terminates, it computes a set of partially evaluated terms S such that the closedness condition is satisfied and, moreover, it uniquely determines the associated residual program.

First, we formalize the notion of closedness adjusted to our abstract representation.

Definition 1. *Let* S *be a set of terms and* t *be a term. We say that* t *is* S-*closed if* $closed(S, t)$ *holds, where the relation "closed" is defined inductively as follows:*

$$closed(S,t) = \begin{cases} true & \text{if } t \in \mathcal{V} \\ closed(S,t_1) \wedge \ldots \wedge closed(S,t_n) & \text{if } t = c(t_1, \ldots, t_n), \ c \in \mathcal{C} \\ closed(t') \wedge \bigwedge_{i \in \{1,\ldots,k\}} closed(t_i) & \text{if } t = (f)case \ t' \ of \ \{\overline{p_k \to t_k}\} \\ \bigwedge_{t' \in \mathcal{R}an(\theta)} closed(S,t') & \text{if } \exists s \in S \text{ such that } t = \theta(s) \end{cases}$$

A set of terms T *is* S-*closed, written* $closed(S, T)$, *if* $closed(S, t)$ *holds for all* $t \in T$.

According to this definition, variables are always closed, while an operation-rooted term is S-closed if it is an instance of some term in S and the terms in the matching substitution are recursively S-closed. On the other hand, for constructor-rooted terms and for case expressions, we have two nondeterministic ways to proceed: either by checking the closedness of their arguments or by proceeding as in the case of an operation-rooted term. For instance, a case expression such as *case* t *of* $\{p_1 \to t_1, \ldots, p_k \to t_k\}$ can be proved closed w.r.t. S either by checking that the set $\{t, t_1, \ldots, t_k\}$ is S-closed[5] or by testing whether the whole case expression is an instance of some term in S.

---

[4] Note that, since the RLNT calculus is deterministic, there is no branching. Thus, only a single derivation can be computed from a term.

[5] Patterns are not considered here since they are constructor terms and hence closed by definition.

Example 5. Let us consider the following set of terms:

$$S = \{\texttt{app a b}, \texttt{case (app a b) of} \{[] \rightarrow \texttt{z}; (\texttt{x} : \texttt{y}) \rightarrow (\texttt{app y z})\} \} \ .$$

The following expression $\texttt{case (app a' b') of} \{[] \rightarrow \texttt{z'}; (\texttt{x'} : \texttt{y'}) \rightarrow (\texttt{app y' z'})\}$ can be proved S-closed using the first element of the set (by checking that the subterms $\texttt{app a' b'}$ and $\texttt{app y' z'}$ are instances of $\texttt{app a b}$) or by testing that the whole expression is an instance of the second element of the set.

The PE algorithm outlined above involves two control issues: the so-called *local* control, which concerns the computation of partial evaluations for single terms, and the *global* control, which ensures the termination of the iterative process but still guaranteeing that the closedness condition is eventually reached. Following [12], we present a PE procedure which is parameterized by:

– An unfolding rule $\mathcal{U}$ (local control), which determines how to stop RLNT derivations. Formally, $\mathcal{U}$ is a (total) function from terms to terms such that, whenever $\mathcal{U}(s) = t$, then there exists a *finite* RLNT derivation $[\![s]\!] \Rightarrow^+ t$.

– An abstraction operator *abstract* (global control), which keeps the set of partially evaluated terms finite. It takes two sets of terms S and T (which represent the current partially evaluated terms and the terms to be added to this set, respectively) and returns a *safe* approximation of $S \cup T$. Here, by "safe" we mean that each term in $S \cup T$ is closed w.r.t. the result of $abstract(S, T)$.

Definition 2. *Let $\mathcal{R}$ be a program and T a finite set of expressions. We define the PE function $\mathcal{P}$ as follows:*

$$\mathcal{P}(\mathcal{R}, T) = S \quad \text{if } abstract(\{\}, T) \longmapsto^*_{\mathcal{P}} S \quad \text{and } S \longmapsto_{\mathcal{P}} S$$

where $\longmapsto_{\mathcal{P}}$ is defined as the smallest relation satisfying

$$\frac{S' = \{s' \mid s \in S \wedge \mathcal{U}(s) = s'\}}{S \longmapsto_{\mathcal{P}} abstract(S, S')}$$

We note that the function $\mathcal{P}$ does not compute a partially evaluated program, but a set of terms S from which a S-closed PE can be uniquely constructed using the unfolding rule $\mathcal{U}$. To be precise, for each term $s \in S$ with $\mathcal{U}(s) = t$, we produce a residual rule $s = t$. Moreover, in order to ensure that the residual program fulfills the syntax of our abstract representation, a renaming of the partially evaluated calls is necessary. This can be done by applying a standard post-processing renaming transformation. We do not present the details of this transformation here but refer to [3].

As for local control, a number of well-known techniques can be applied for ensuring the finiteness of RLNT derivations, e.g., depth-bounds, loop-checks, well-founded (or well-quasi) orderings (see, e.g., [8,23,26]). For instance, an unfolding rule based on the use of the homeomorphic embedding ordering has been proposed in [4].

As for global control, an abstraction operator should essentially distinguish the same cases as in the closedness definition. Intuitively, the reason is that the abstraction operator must first check whether a term is closed and, if not, try to add this term (or some of its subterms) to the set. Therefore, given a call $abstract(S, \{t\})$, an abstraction operator usually distinguishes three main cases depending on t:

- if t is constructor-rooted, it tries to add the arguments of t;
- if it is operation-rooted and is an instance of some term in S, it tries to add the terms in the matching substitution;
- otherwise (an operation-rooted term which is not an instance of any term in S), it is simply added to S (or *generalized* in order to keep the set S finite).

Our particular abstraction operator uses a *quasi-ordering*, namely the homeomorphic embedding relation $\trianglelefteq$ (see, e.g., [23]), to ensure termination and generalizes those calls which do not satisfy this ordering by using the *msg* (*most specific generalization*) between terms.[6]

As opposed to previous abstraction operators [4], here we need to give a special treatment to case expressions. Of course, if one considers the *case* symbol as an ordinary constructor symbol, the extension would be straightforward. Unfortunately, this will often provoke a serious loss of specialization, as the following example illustrates.

Example 6. Let us consider again the program app and the RLNT derivation of Example 3:

$[\![\text{app (app x y) z}]\!]$

$\Rightarrow^*$ $[\![\text{case (case x of } \{[] \rightarrow \text{y; } (\text{a}' : \text{b}') \rightarrow (\text{a}' : \text{app b}' \text{ y})\})$
$\qquad\qquad \text{of } \{[] \rightarrow \text{z; } (\text{a} : \text{b}) \rightarrow (\text{a} : \text{app b z})\}]\!]$

$\Rightarrow^*$ $\text{case x of } \{ \; [] \qquad\quad \rightarrow \text{case y of } \{[] \rightarrow \text{z; } (\text{a} : \text{b}) \rightarrow (\text{a} : [\![\text{app b z}]\!])\};$
$\qquad\qquad\quad (\text{a}' : \text{b}') \rightarrow (\text{a}' : [\![\text{app (app b}' \text{ y) z}]\!])\}$

If one considers an unfolding rule which stops the derivation at the intermediate case expression, then the abstraction operator will attempt to add only the operation-rooted subterms app b$'$ y and app b y to the set of terms to be specialized. This will prevent us from obtaining an efficient (recursive) residual function for the original term, since we will never reach again an expression containing app (app x y) z (see Example 7).

On the other hand, by treating case expressions as operation-rooted terms, the problem is not solved. For instance, if we consider that the unfolding rule returns the last term of the above derivation, then it is not convenient to add the whole term to the current set. Here, the best choice would be to treat the *case* symbol as a constructor symbol. Moreover, a similar situation arises when considering constructor-rooted terms, since the RLNT calculus has no restrictions to evaluate terms in head normal form.

---

[6] A *generalization* of the set of terms $S = \{t_1, \ldots, t_n\}$ is a pair $\langle t, \{\theta_1, \ldots, \theta_n\}\rangle$ such that, $\forall i \in \{1, \ldots, n\}$, $\theta_i(t) = t_i$. The pair $\langle t, \{\theta_1, \ldots, \theta_n\}\rangle$ is the *most specific generalization* of S, written $msg(S)$, if $\langle t, \{\theta_1, \ldots, \theta_n\}\rangle$ is a generalization and for every other generalization $\langle t', \{\theta'_1, \ldots, \theta'_n\}\rangle$ of S, t' is more general than t.

Luckily, the RLNT calculus gives us some leeway. The key idea is to take into account the position of the square brackets of the calculus: an expression within square brackets should be added to the set of partially evaluated terms (if possible), while expressions which are not within square brackets should be definitively residualized (i.e., ignored by the abstraction operator, except for operation-rooted terms).

Definition 3. *Given two finite sets of terms, T and S, we define:*[7]

$$abstract(S, T) = \begin{cases} S & \textit{if } T = \varnothing \\ abs(\dots abs(S, t_1), \dots, t_n) & \textit{if } T = \{t_1, \dots, t_n\}, n \geq 1 \end{cases}$$

The function $abs(S, t)$ distinguishes the following cases:

$$abs(S, t) = \begin{cases} S & \textit{if } t \in \mathcal{V} \\ abstract(S, \{t_1, \dots, t_n\}) & \textit{if } t = c(t_1, \dots, t_n), \ c \in \mathcal{C} \\ abstract(S, \{t', t_1, \dots, t_n\}) & \textit{if } t = (f)case \ t' \ of \ \{\overline{p_n \to t_n}\} \\ try\_add(S, t) & \textit{if } t = f(t_1, \dots, t_n), \ f \in \mathcal{F} \\ try\_add(S, t') & \textit{if } t = [\![t']\!] \end{cases}$$

Finally, the function $try\_add(S, t)$ is defined as follows:

$$try\_add(S, t) = \begin{cases} abstract(S \setminus \{s\}, \{s'\} \cup \mathcal{R}an(\theta_1) \cup \mathcal{R}an(\theta_2)) \\ \quad \textit{if } \exists s \in S. \ root(s) = root(t) \ and \ s \trianglelefteq t, \\ \quad \textit{where } \langle s', \{\theta_1, \theta_2\} \rangle = msg(\{s, t\}) \\ S \cup \{t\} \qquad \textit{otherwise} \end{cases}$$

Let us informally explain this definition. Given a set of terms S, in order to add a new term t, the abstraction operator *abs* distinguishes the following cases:

- variables are disregarded;
- if t is rooted by a constructor symbol or by a case symbol, then it recursively inspects the arguments;
- if t is rooted by a defined function symbol or it is enclosed within square brackets, then the abstraction operator tries to add it to S with *try_add* (even if it is constructor-rooted or a case expression). Now, if t does not embed any *comparable* (i.e., with the same root symbol) term in S, then t is simply added to S. Otherwise, if t embeds some comparable term of S, say s, then the *msg* of s and t is computed, say $\langle s', \{\theta_1, \theta_2\} \rangle$, and it finally attempts to add s' as well as the terms in θ_1 and θ_2 to the set resulting from removing s from S.

Let us consider an example to illustrate the complete PE process.

Example 7. Consider the program $\mathcal{R}_{\text{app}}$ which contains the rule defining the function app. In order to compute $\mathcal{P}(\mathcal{R}_{\text{app}}, \{\text{app (app x y) z}\})$, we start with:

$$S_0 = abstract(\{\}, \{\text{app (app x y) z}\}) = \{\text{app (app x y) z}\}$$

---

[7] The particular order in which the elements of T are added to S by *abstract* cannot affect correctness but can degrade the effectiveness of the algorithm. A more precise treatment can be easily given by using sequences instead of sets of terms.

For the first iteration, we assume that:

$\mathcal{U}($app (app x y) z$) =$

    case x of $\{$ [] $\rightarrow$ case y of $\{$[] $\rightarrow$ z; (a : b) $\rightarrow$ (a : [app b z])$\})$;
                     (a$'$: b$'$) $\rightarrow$ (a$'$: [app (app b$'$ y) z])$\}$

(see derivation in Example 3). Then, we compute:

$S_1 = abstract(S_0, \{\mathcal{U}($app (app x y) z$)\}) = \{$app (app x y) z$)$, app b z$\}$

For the next iteration, we assume that:

$\mathcal{U}($app b z$) =$ case b of $\{$[] $\rightarrow$ z; (c : d) $\rightarrow$ c : [app d z] $\}$

Therefore, $abstract(S_1, \{\mathcal{U}($app b z$)\}) = S_1$ and the process finishes. The associated residual rules are (after renaming the original expression by dapp x y z):

    dapp x y z = case x of $\{$ [] $\rightarrow$ case y of $\{$[] $\rightarrow$ z;
                                    (a : b) $\rightarrow$ (a : app b z)$\}$;
                   (a$'$: b$'$) $\rightarrow$ (a$'$: dapp b$'$ y z)$\}$
    app b z   = case b of $\{$ [] $\rightarrow$ z; (c : d) $\rightarrow$ (c : app d z)$\}$

Note that the optimized function dapp is able to concatenate three lists by traversing the first list only once, which is not possible in the original program.

The following proposition states that the operator *abstract* of Def. 3 is *safe*.

Proposition 1. *Given two finite sets of terms, T and S, if $S' = abstract(S, T)$, then for all $t \in (S \cup T)$, t is closed with respect to S'.*

Finally, we establish the termination of the complete PE process:

Theorem 2. *Let $\mathcal{R}$ be a program and S a finite set of terms. The computation of $\mathcal{P}(\mathcal{R}, S)$ terminates using a finite unfolding rule and the abstraction operator of Def. 3.*

5 Experimental Evaluation

In order to assess the practicality of the ideas presented in this work, the implementation of a partial evaluator for the multi-paradigm declarative language Curry has been undertaken.[8] Curry [19] integrates features from logic (logic variables, partial data structures, built-in search), functional (higher-order functions, demand-driven evaluation) and concurrent programming (concurrent evaluation of constraints with synchronization on logical variables). Furthermore, Curry is a complete programming language which is able to implement distributed applications (e.g. Internet servers [16]) or graphical user interfaces at a high-level [17]. In order to develop an effective PE tool for Curry, one has to extend the basic

---

[8] It is publicly available at http://www.dsic.upv.es/users/elp/soft.html.

Benchmark	mix	original	specialized	speedup
allones	470	430	290	1.48
double_app	510	370	320	1.16
double_flip	750	550	400	1.37
kmp	1440	730	35	20.9
length_app	690	310	290	1.07

Table 1. Benchmark results

PE scheme to cover all high-level features. This extension becomes impractical within previous frameworks for the PE of functional logic languages due to the complexity of the resulting semantics. By using an abstract representation and translating high-level programs to this notation (see [20]), the extension becomes simple and effective. A detailed description of the concrete manner in which each feature is treated can be found in [3]. Moreover, as opposed to previous partial evaluators for Curry (e.g., INDY [1]), it is completely written in Curry. To the best of our knowledge, this is the first purely declarative partial evaluator for a functional logic language.

Firstly, we have benchmarked several examples which are typical from partial deduction and from the literature of functional program transformations. Table 1 shows the results obtained from some selected benchmarks (a complete description can be found, e.g., in [4]). For each benchmark, we show the specialization time including the reading and writing of programs (column mix), the timings for the original and specialized programs (columns original and specialized), and the speedups achieved (column speedup). Times are expressed in milliseconds and are the average of 10 executions on a Sun Ultra-10. Runtime input goals were chosen to give a reasonably long overall time. All benchmarks have been specialized w.r.t. function calls containing no static data, except for the kmp example (what explains the larger speedup produced). Speedups are similar to those obtained by previous partial evaluators, e.g., INDY [1]. Indeed, these benchmarks were used in [4] to illustrate the power of the narrowing-driven approach (and are not affected by the discussed limitations). This indicates that our new scheme for PE is a conservative extension of previous approaches on comparable examples. Note, though, that our partial evaluator is applicable to a wider class of programs (including higher-order, constraints, several built-in's, etc), while INDY is not.

Secondly, we have considered the PE of the collection of programs in the Curry library (see http://www.informatik.uni-kiel.de/~curry). Here, our interest was to check the ability of the partial evaluator to deal with realistic programs which make extensive use of all the features of the Curry language. Our partial evaluator has been successfully applied to all the examples producing in some cases significant improvements. We refer to [3] for the source code of some benchmarks. Finally, we have also considered the PE of a meta-interpreter w.r.t. a source program. Although the partial evaluator successfully specialized it, regarding improvement in efficiency, the results were not so satisfactory. To improve this situation, we plan to develop a binding-time analysis to determine, for each expression, whether it can be definitively evaluated at PE time (hence, it should not be generalized by the abstraction operator) or whether this decision must be taken online. This kind of (*off-line*) analysis would be also useful to reduce specialization times.

Altogether, the experimental evaluation is encouraging and gives a good impression of the specialization achieved by our partial evaluator.

6 Conclusions

In this work, we introduce a novel approach for the PE of truly lazy functional logic languages. The new scheme is carefully designed for an abstract representation in which high-level programs can be automatically translated. We have shown how a non-standard (residualizing) semantics can avoid several limitations of previous frameworks. The implementation of a fully automatic PE tool for the language Curry has been undertaken and tested on an extensive benchmark suite. To the best of our knowledge, this is the first purely declarative partial evaluator for a functional logic language. Moreover, since Curry is an extension of both logic and (lazy) functional languages, we think that our PE scheme can be easily adapted to other declarative languages.

From the experimental results, we conclude that our partial evaluator is indeed suitable for "real" Curry programs. Anyway, there is still room for further improvements. For instance, although self-application is already (theoretically) possible, the definition of a precise binding-time analysis seems mandatory to achieve an *effective* self-applicable partial evaluator. On the other hand, we have not considered a formal treatment to measuring the *effectiveness* of our partial evaluator. Another promising direction for future work is the development of abstract criteria to formally measure the potential benefit of our PE algorithm.

References

1. E. Albert, M. Alpuente, M. Falaschi, and G. Vidal. INDY User's Manual. Technical report, UPV, 1998. Available from URL:
   http://www.dsic.upv.es/users/elp/papers.html. 396, 396
2. E. Albert, M. Alpuente, M. Hanus, and G. Vidal. A Partial Evaluation Framework for Curry Programs. In *Proc. of the 6th Int'l Conf. on Logic for Programming and Automated Reasoning, LPAR'99*, pages 376–395. Springer LNAI 1705, 1999. 382, 382, 383, 390, 390, 390, 390
3. E. Albert, M. Hanus, and G. Vidal. Using an Abstract Representation to Specialize Functional Logic Programs. Technical report, UPV, 2000. Available from URL:
   http://www.dsic.upv.es/users/elp/papers.html. 383, 386, 392, 396, 396
4. M. Alpuente, M. Falaschi, and G. Vidal. Partial Evaluation of Functional Logic Programs. *ACM Transactions on Programming Languages and Systems*, 20(4):768–844, 1998. 381, 381, 382, 382, 382, 383, 384, 392, 393, 396, 396
5. S. Antoy. Definitional trees. In *Proc. of the 3rd Int'l Conference on Algebraic and Logic Programming, ALP'92*, pages 143–157. Springer LNCS 632, 1992. 382
6. S. Antoy, R. Echahed, and M. Hanus. A Needed Narrowing Strategy. *Journal of the ACM*, 2000 (to appear). Previous version in *Proc. of POPL'94*, pages 268–279. 382

7. A. Bondorf. A Self-Applicable Partial Evaluator for Term Rewriting Systems. In *Proc. of TAPSOFT'89*, pages 81–95. Springer LNCS 352, 1989. 382
8. M. Bruynooghe, D. De Schreye, and B. Martens. A General Criterion for Avoiding Infinite Unfolding. *New Generation Computing*, 11(1):47–79, 1992. 392
9. C. Consel and O. Danvy. Tutorial notes on Partial Evaluation. In *Proc. ACM Symp. on Principles of Programming Languages*, pages 493–501, 1993. 381
10. D. De Schreye, R. Glück, J. Jørgensen, M. Leuschel, B. Martens, and M.H. Sørensen. Conjunctive Partial Deduction: Foundations, Control, Algorihtms, and Experiments. *Journal of Logic Programming*, 41(2&3):231–277, 1999. 382
11. N. Dershowitz and J.-P. Jouannaud. Rewrite Systems. In J. van Leeuwen, editor, *Handbook of Theoretical Computer Science*, volume B: Formal Models and Semantics, pages 243–320. Elsevier, Amsterdam, 1990. 383
12. J. Gallagher. Tutorial on Specialisation of Logic Programs. In *Proc. of Partial Evaluation and Semantics-Based Program Manipulation*, pages 88–98. ACM, New York, 1993. 381, 391, 392
13. R. Glück and M.H. Sørensen. A Roadmap to Metacomputation by Supercompilation. In *Partial Evaluation. Int'l Dagstuhl Seminar*, pages 137–160. Springer LNCS 1110, 1996. 382
14. M. Hanus. The Integration of Functions into Logic Programming: From Theory to Practice. *Journal of Logic Programming*, 19&20:583–628, 1994. 382, 383
15. M. Hanus. A unified computation model for functional and logic programming. In *Proc. of POPL'97*, pages 80–93. ACM, New York, 1997. 382, 385
16. M. Hanus. Distributed Programming in a Multi-Paradigm Declarative Language. In *Proc. of PPDP'99*, pages 376–395. Springer LNCS 1702, 1999. 395
17. M. Hanus. A Functional Logic Programming Approach to Graphical User Interfaces. In *Int'l Workshop on Practical Aspects of Declarative Languages*, pages 47–62. Springer LNCS 1753, 2000. 395
18. M. Hanus and C. Prehofer. Higher-Order Narrowing with Definitional Trees. *Journal of Functional Programming*, 9(1):33–75, 1999. 383, 383, 383, 385, 386, 386, 386, 386, 386
19. M. Hanus (ed.). Curry: An Integrated Functional Logic Language. Available at http://www.informatik.uni-kiel.de/~curry, 2000. 383, 385, 395
20. M. Hanus, S. Antoy, J. Koj, P. Niederau, R. Sadre, and F. Steiner. PAKCS 1.2: User Manual. Available at http://www.informatik.uni-kiel.de/~pakcs, 2000. 383, 396
21. N.D. Jones, C.K. Gomard, and P. Sestoft. *Partial Evaluation and Automatic Program Generation*. Prentice-Hall, Englewood Cliffs, NJ, 1993. 381, 382, 382
22. Laura Lafave. *A Constraint-based Partial Evaluator for Functional Logic Programs and its Application*. PhD thesis, Department of Computer Science, University of Bristol, 1998. 381
23. M. Leuschel. On the Power of Homeomorphic Embedding for Online Termination. In G. Levi, editor, *Proc. of SAS'98*, pages 230–245. Springer LNCS 1503, 1998. 392, 393
24. J.W. Lloyd and J.C. Shepherdson. Partial Evaluation in Logic Programming. *Journal of Logic Programming*, 11:217–242, 1991. 381, 381
25. A.P. Nemytykh, V.A. Pinchuk, and V.F. Turchin. A Self-Applicable Supercompiler. In *Proc. of Dagstuhl Sem. on Part. Evaluation*, pages 322–337. Springer LNCS 1110, 1996. 382
26. M.H. Sørensen and R. Glück. An Algorithm of Generalization in Positive Supercompilation. In *Proc. of ILPS'95*, pages 465–479. MIT Press, 1995. 392
27. M.H. Sørensen, R. Glück, and N.D. Jones. A Positive Supercompiler. *Journal of Functional Programming*, 6(6):811–838, 1996. 381, 382, 389, 389
28. P.L. Wadler. Deforestation: Transforming programs to eliminate trees. *Theoretical Computer Science*, 73:231–248, 1990. 381, 389

Binding-Time Analysis by Constraint Solving
A Modular and Higher-Order Approach for Mercury

Wim Vanhoof*

Department of Computer Science, K.U.Leuven, Belgium
wimvh@cs.kuleuven.ac.be

Abstract. In this paper we present a binding-time analysis for the logic programming language Mercury. Binding-time analysis is a key analysis needed to perform off-line program specialisation. Our analysis deals with the higher-order aspects of Mercury, and is formulated by means of constraint normalisation. This allows (at least part of) the analysis to be performed on a modular basis.

1 Introduction

Mercury is a recently introduced logic programming language, comprising many features needed for modern software engineering practice: polymorphism, type-classes and a strong module system are some examples of the means available to the programmer to design and build modular programs that employ a lot of abstraction and reuse of general components.

Employing abstraction and generality, however, imposes a penalty on the efficiency of the resulting program due to the presence of for example procedure calls and tests for which the input is (partially) known at compile-time. To overcome this performance problem, the Mercury compiler performs several optimizations on the original source code. Although most of these optimizations are implemented as different processes during compilation, some of them are instances of the more general framework of partial evaluation. Examples are inlining of procedure bodies, higher-order call specialisation, specialisation of type-info's [4]. A problem shared by these optimizations is knowing at what points in the code enough information is present to apply the optimization under consideration. Currently, this problem is solved mostly by using some heuristics that are hard coded in the analysis.

A more general approach is the use of binding-time analysis (BTA) to perform a thorough dataflow analysis, and propagate information through the program about what variables are definitely bound to a value at compile-time, independent of the program's runtime input. Such information can be used by a so-called off-line program specialiser to partially evaluate certain parts of the code w.r.t. some given input. An advantage of the more general approach is that the results of BTA can be shared by more than one optimization. As such, different optimizations do not need to redo the analysis, additional optimizations can more

---

* Supported by the IWT, Belgium.

M. Parigot and A. Voronkov (Eds.): LPAR 2000, LNAI 1955, pp. 399–416, 2000.

easily be plugged in, and precision improvements of the BTA have an impact on a broader scope of optimizations. The results of BTA can be shown as annotations on the original source code and provide as such excellent feedback to the user, enabling a better understanding of why several optimizations are (not) performed.

In a logic programming setting, work on partial evaluation has mainly concentrated on on-line specialisation (where the specialisation process is controlled by the concrete input rather than by a previous analysis [6,11]). Consequently, little attention has been paid to off-line specialisation and BTA [9,12,2]. In previous work [16] we have defined a completely automatic BTA for a subset of the logic programming language Mercury. The current work reformulates and extends our previous work extensively: in contrast with [16], our analysis now deals with the higher-order aspects of Mercury, and its formulation by constraint solving allows it to be performed (at least partially) on the same modular basis as compilation.

Binding-time analysis has been studied thoroughly before in the context of functional languages (e.g. [1,7,5]). In this context, the actual binding-time analysis is usually preceded by a flow (or closure) analysis to determine the higher-order control flow in a program (e.g. [8,13]). In [1], such a flow analysis is combined with a monovariant binding-time analysis by constraint solving. The work in [7,5] describes a polyvariant BTA by deriving binding-time descriptions that are polymorphic w.r.t. the binding-times of a function's arguments. The described analysis deals with a monomorphic lambda-calculus like language. Techniques exist [10] to deal with more involved type systems and partially static data structures.

This work adapts and generalises ideas from BTA of functional programs in several ways. Basing our domain of binding-times on the type system of Mercury enables us to handle type polymorphism, represent partially static data structures and propagate closure information as part of the binding-time information. Hence our analysis does not require a separate closure analysis, but computes this information using a single set of constraints. Our analysis is polyvariant: it computes different binding-times for the variables in a predicate depending on the binding-times (including the closure information) of the predicate's input arguments. In this work, we develop the basic machinery for such a binding-time analysis for the logic programming language Mercury. Since Mercury is a language specifically tuned towards use in large scale applications, programs written in Mercury usually consists of a number of modules. Hence, performing (at least a large part of) the analysis one module at a time is crucial for such applications. We are closely collaborating with the Mercury developers in Melbourne to implement a BTA based on the presented material in a version of the Mercury compiler which should enable to perform some large scale experiments.

The remainder of the paper is organised as follows: In Sect. 2 we describe the technique of BTA by constraint solving for a first order subset of Mercury, emphasising on the modular approach. In Sect. 3, we alter the technique to deal with the higher-order aspects of the Mercury language.

2   BTA by Constraint Solving: A First-order Setting

Binding-time analysis is about knowing *what* runtime values will be known already at compile-time, without being interested in the values themselves. In order to be useful when complex data structures are involved, the analysis must deal with values that are *partially* known at compile-time. First, we describe a suitable domain for representing such knowledge that was originally introduced in [16]. The analysis itself is presented afterwards.

2.1   A Precise Domain of Binding-Times

In a statically typed language like Mercury, the set of possible values a program variable can have at runtime can be described at compile time by a finite type graph. Consider for example the definition of a type list(T), denoting a polymorphic list, where such a list is either the empty list (denoted by []) or a cons (denoted by [|]) of a value of type T and a list of T. A possible type graph for this type is denoted in Fig. 1.

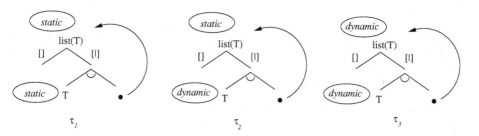

Fig. 1. Labelled type graphs for list(T).

A suitable domain of binding-times is obtained by associating the label *static* or *dynamic* to every or-node in the type graph, the idea being that if a node is labelled *dynamic* respectively *static*, the corresponding subterm(s) in the value described by the binding-time are free, respectively bound to a functor. Also in Fig. 1, τ_1, τ_2 and τ_3 represent such associations denoting, respectively a value that is completely known at compile-time, a value for which the skeleton of the list is definitely known at compile-time but the elements are not, and a value that is possibly completely unknown at compile-time.

We first introduce the necessary concepts and notation. The basic Mercury entity our analysis deals with is a compilation unit; that is a Mercury module M together with the definitions from the interfaces of the modules that are imported by M. Throughout the paper, we often simply refer to such a unit as a "module", though. Let $\mathcal{T}_V$, $\mathcal{T}_C$ and $\mathcal{T}_F$ denote the sets of respectively type variables, type constructors and type functors for a given module.

In this section, we consider only first-order types: a type variable or a type constructor applied to a number of types

Definition 1. *Let $\mathcal{MT}$ denote the set of all possible types for a program P:*

$$\mathcal{MT} ::= V \in \mathcal{T}_V \mid$$
$$\gamma(t_1, \ldots, t_n) \ with \ \gamma/n \in \mathcal{T}_C \ and \ t_i \in \mathcal{MT}, \forall i \in \{1 \ldots n\}$$

Each type constructor $\gamma/n \in \mathcal{T}_C$ is defined by a type definition in P. The set of all these definitions is denoted by $\mathcal{TD}ef$.

Definition 2. *A type constructor is defined as follows, with the constraint that all type variables occurring in the right hand side also occur in the left hand side.*

$$\mathcal{TD}ef ::= \gamma(V_1, \ldots, V_n) \longrightarrow c_1(t_{1_1}, \ldots, t_{1_{m_1}}) ; \ldots ; c_l(t_{l_1}, \ldots, t_{l_{m_l}}).$$
$$where \ \gamma/n \in \mathcal{T}_C, c_i \in \mathcal{T}_F, V_i \in \mathcal{T}_V \ and \ t_{i_j} \in \mathcal{MT}$$

A *type substitution* is defined as a mapping $TSubst : \mathcal{T}_V \mapsto \mathcal{MT}$. To ease notation, we define the following shorthand notation to denote the set of constructors of a type and a specific subtype of a type:. For a type $t \in \mathcal{MT}$, define $C_t = \{c_1/m_1, \ldots, c_l/m_l\} \subseteq \mathcal{T}_F$ and $ST_t^{c_i,k} = t_{i,k}\theta$ if $t = \gamma(V_1, \ldots, V_n)\theta$ with $\theta \in TSubst$ and $\gamma(V_1, \ldots, V_n) \longrightarrow c_1(t_{1,1}, \ldots, t_{1,m_1}) ; \ldots ; c_l(t_{l,1}, \ldots, t_{l,m_l}) \in \mathcal{TD}ef$.

Example 1. The type `list(T)` introduced above is formally defined by the following definition:

```
:- type list(T) ---> [] ; [T | list(T)].
```

Given the type $list(T)$, $C_{list(T)} = \{[]/0, [|]/2\}$ and $ST_{list(T)}^{[|],1} = T$ and $ST_{list(T)}^{[|],2} = list(T)$.

In what follows, we represent a *sequence* over a set S by $\langle S \rangle$; $\langle \rangle$ denotes the empty sequence, and $\langle d_1, \ldots, d_n \rangle \bullet \langle e_1, \ldots, e_m \rangle$ denotes the concatenation of the two sequences, resulting in the new sequence $\langle d_1, \ldots, d_n, e_1, \ldots, e_m \rangle$. For any sequence S, S_i denotes the i'th element of S, if it exists.

We will often need to refer to subvalues of values, according to a type tree. Since a node in a type tree is uniquely defined by a path from the root towards it, we define the notion of a *type path* as a sequence over $\mathcal{T}_F \times \mathbb{N}$. The set of all type paths is denoted by $TPath$. Type paths are used to select (type) nodes in a type graph: For a type t, $t^{\langle \rangle} = t$ and $t^{\langle (c,k) \rangle \bullet \delta}$ denotes $t_{c,k}^\delta$ if $c \in C_t$ and $t_{c,k} = ST_t^{c,k}$.

Given the notion of type paths, we can define a (possibly infinite) type tree in a straightforward way:

Definition 3. *For a type $t \in \mathcal{MT}$, the first order type tree of t, $\mathcal{L}_t \in 2^{TPath}$ is recursively defined as:*

- $\langle \rangle \in \mathcal{L}_t$
- *if $t = \gamma(t_1, \ldots, t_n)$ with $\gamma \in \mathcal{T}_C$ then $\forall c/n \in C_t, k \in \{1, \ldots, n\}, \langle (c,k) \rangle \bullet \delta \in \mathcal{L}_t$ where $\delta \in \mathcal{L}_{t_{c,k}}$ and $t_{c,k} = ST_t^{c,k}$.*

Recursive type definitions correspond to infinite type trees. However, to consider finite type *graphs*, it is sufficient to impose an equivalence relation on the set of type paths for a given type, turning it into a finite set. We consider two type paths δ, δ' to be *equivalent* w.r.t a type t if and only if $\delta = \delta' \bullet \gamma$ (for some $\gamma \in TPath$) and $t^\delta = t^{\delta'}$. Let $\equiv$ denote the transitive closure of the *equivalent* relation and $\mathcal{L}_t^\equiv$ denotes a type tree modulo $\equiv$.

Example 2. For $list(T)$ as defined in Example 1, $\mathcal{L}_{list(T)}^\equiv = \{\langle\rangle, \langle([\,|\,], 1)\rangle\}$, since the path referring to the recursive occurrence of $list(T)$, $\langle([\,|\,], 2\rangle$ is equivalent with $\langle\rangle$.

If we introduce the domain $\mathcal{B} = \{static, dynamic\}$, we can define a binding-time for a type t by a labelling of $\mathcal{L}_t^\equiv$'s nodes as follows:

Definition 4. *A binding-time for a type* $t \in \mathcal{MT}$ *is a function* $\tau : \mathcal{L}_t^\equiv \mapsto \mathcal{B}$ *such that* $\forall \delta \in dom(\tau) : \tau(\delta) = $ dynamic *implies that* $\tau(\delta') = $ dynamic *for all* $\delta' \in dom(\tau), \delta' = \delta \bullet \epsilon$. *The set of all binding-times (regardless their type) is denoted by* $\mathcal{BT}$.

We impose the ordering $dynamic > static$ on $\mathcal{B}$ which induces an ordering on $\mathcal{BT}$: $\tau_1 \geq \tau_2$ if and only if $\tau_1(\delta) \geq \tau_2(\delta)$ for all $\delta \in TPath$. In Fig. 1, for example, $\tau_3 > \tau_2 > \tau_1$.

2.2 Generating a Constraint System

The task of BTA can be described as follows: given binding-times for a predicate's input arguments, compute binding-times for the predicate's remaining variables. In [16], we formulated BTA for a subset of Mercury as a top-down, call dependent analysis. Extending such a call dependent analysis to deal with multi-module programs is nontrivial and often leads to duplication of analysis efforts. In this work, we reformulate BTA in a constraint-solving setting, enabling an efficient modular approach: First, the data flow inside a predicate is examined resulting in a number of constraints on the variables' binding-times. These constraints are created without taking the call pattern of interest into account. Next, the least solution to this constraint system, combined with the call pattern of interest, provides a binding-time for each variable that is both correct and as much static as possible.

First, we show how to translate a module to a constraint system. Such a module can be considered as a set of *procedures*, which are obtained by translating the original predicates and functions to superhomogeneous form [15]. The definition of a procedure is converted to a single clause: the arguments in the head of the clause and in predicate calls in the body are distinct variables, explicit unifications are generated for these variables in the body goal, and complex unifications are broken down into several simpler ones. A goal is either an atom or a number of goals connected by *conjunction, disjunction, if then else* or *not*. An atom is either a unification or a procedure call.

Definition 5.

$Proc ::= p(H_1, \ldots, H_n) : -G$ with $p \in \mathcal{T}_P, G \in Goal, H_1, \ldots, H_n \in Var$
$Goal ::= Atom \mid not\ G \mid (G_1, G_2) \mid (G_1; G_2) \mid if\ G_1\ then\ G_2\ else\ G_3$
    with $G, G_1, G_2, G_3 \in Goal$
$Atom ::= X = f(Y^1, \ldots, Y^n) \mid X = Y \mid p(Y^1, \ldots, Y^n)$
    with $X, Y, Y^1, \ldots, Y^n \in Var, f \in \mathcal{T}_F$ and $p \in \mathcal{T}_P$

An essential part of compiling a Mercury module is *mode analysis* [15]. During this analysis, unifications are split into four different types, according their data flow: *test* denoted by $X == Y$ (where X and Y are both input and of atomic type - that is having a type tree consisting of only the root node), *assignment* denoted by $X := Y$ (where Y is input, X is output), *construction* denoted by $X \Leftarrow f(Y^1, \ldots, Y^n)$ (where X is output, $Y^1, \ldots, Y^n$ input and $f \in \mathcal{T}_F$) and *deconstruction* denoted by $Y \Rightarrow f(X^1, \ldots, X^n)$ (where Y is input, $X^1, \ldots, X^n$ output and $f \in \mathcal{T}_F$). For a procedure p, $\mathcal{A}rg(p)$ denotes the sequence $\langle H_1, \ldots, H_n \rangle$ of p's formal arguments. Each atom in a procedure's body is uniquely identified by a *program point* (PP).

Example 3. Consider the predicate $append(list(T) :: in, list(T) :: in, list(T) :: out)$ where for each argument position $t :: in$ or $t :: out$ denotes that the corresponding argument is of type t and is an input, respectively output argument. The predicate is defined as follows, where each atom is subscribed with a natural number, representing its program point.

$\{append(X, Y, Z)\}_0 :- \quad \{X \Rightarrow []\}_1, \{Z := Y\}_2;$
$\{X \Rightarrow [E \mid Es]\}_3, \{append(Es, Y, R)\}_4, \{Z \Leftarrow [E|R]\}_5.$

Variables can be initialised in different branches of a disjunction (or if-then-else). To improve precision, we associate different binding-times to such variables, one for each occurrence in such a branch. Notationally we distinguish between such occurrences by subscribing a variable with the program point (represented by a natural number) where it got initialised. In the *append* example, Z_2 denotes the occurrence of Z in the first branch of the disjunction, whereas Z_5 denotes the occurrence of Z in the second branch. We identify the head of a procedure with program point "0" and use this program point to denote input arguments and (final occurrences) of output arguments. The set of all occurrences of variables is denoted by Var_{pp}.

When computing *the* binding-time of a variable, it is mandatory to take the right (sub)set of its occurrences into account. We therefore introduce the following notions: An *execution path* in a predicate p is a sequence of program points in p denoting a possibly nonfailing derivation of the atoms associated to the program points. Two program points p_1 and p_2 *share an execution path* in p if there exists an execution path S in p such that both p_1 and p_2 are elements of S. We define the function $init : Var \times PP \mapsto 2^{Var_{pp}}$ as $init(X, pp) = \{X_{pp'} \mid X$ got initialised at pp' and pp and pp' share an execution path$\}$.

Example 4. In the *append* example, the only two execution paths are $\langle 0, 1, 2 \rangle$ and $\langle 0, 3, 4, 5 \rangle$. We have, for example, that $init(Y, 2) = \{Y_0\}$, $init(X, 1) = \{X_0\}$, $init(E, 5) = \{E_3\}$ and $init(R, 5) = \{R_4\}$.

The binding-times that can be associated to (an occurrence of) a variable are constrained by the binding-times of those variables it is related to through a data flow relation. To express data flow relations in terms of binding-times, we introduce the notion of a *binding-time constraint*.

Definition 6. *A* binding-time constraint *is denoted as* $X_{pp}^{\delta} \geq S$, *where* $X_{pp} \in Var_{pp}$, $\delta \in TPath$ *and* $S = $ static $\mid$ dynamic $\mid Y_{pp'}^{\gamma}$ *(with* $Y_{pp'} \in Var_{pp}$ *and* $\gamma \in TPath$). *The set of all binding-time constraints is denoted by* $\mathcal{B}_C$.

Due to the well-modedness of Mercury procedures, it is possible to trace the dataflow back to the procedure's input arguments, and consequently to express the binding-time constraints on a variable in function of the binding-times of the procedure's (input) arguments. Binding-time constraints in this format are said to be in *normal form*.

Definition 7. *A binding-time constraint* $X_{pp}^{\delta} \geq S$ *is in* normal form *if its right-hand side* S *is either* static, dynamic *or* Y_0^{γ} *with* Y *being an input argument.*

In particular, we are often interested in the normalised binding-time constraints on a procedure's output arguments. For a procedure p, we denote with p's *normal form* a set of binding-time constraints, where the left-hand side is an (output) argument of p. We start by defining how to translate a procedure into a set of binding-time constraints. A unification can straightforwardly be translated to a set of binding-time constraints on the involved variables, whereas translating a call to a set of binding-time constraints requires the called predicate's normal form. If we denote with μ a function $Proc \mapsto 2^{\mathcal{B}_C}$ such that for a procedure q, $\mu(q)$ denotes q's normal form, we can define for any atom A its associated set of binding-time constraints as $\mathcal{A}_{\mu}(A)$, where $\mathcal{A}$ is defined as follows:

Definition 8. *Consider an atom at program point* pp. $\mathcal{A}_{\mu}(A)$ *is as follows, depending on the type of atom:*

$$\mathcal{A}_{\mu}(X == Y) \qquad = \{\}$$

$$\mathcal{A}_{\mu}(X := Y) \qquad = \{X_{pp} \geq Y_{pp_Y} \mid Y_{pp_Y} \in init(Y, pp)\}$$

$$\mathcal{A}_{\mu}(X \Leftarrow f(Y^1, \ldots, Y^n)) = \begin{cases} \{X_{pp} \geq \text{static}\} \cup \\ \{X_{pp}^{\langle\langle f, 1 \rangle\rangle} \geq Y_{pp_{Y^1}}^1 \mid Y_{pp_{Y^1}}^1 \in init(Y^1, pp)\} \\ \cup \ldots \cup \\ \{X_{pp}^{\langle\langle f, n \rangle\rangle} \geq Y_{pp_{Y^n}}^n \mid Y_{pp_{Y^n}}^n \in init(Y^n, pp)\} \end{cases}$$

$$\mathcal{A}_{\mu}(Y \Rightarrow f(X^1, \ldots, X^n)) = \begin{cases} \{X_{pp}^1 \geq Y_{pp_Y}^{\langle\langle f, 1 \rangle\rangle} \mid Y_{pp_Y} \in init(Y, pp)\} \\ \cup \ldots \cup \\ \{X_{pp}^n \geq Y_{pp_Y}^{\langle\langle f, n \rangle\rangle} \mid Y_{pp_Y} \in init(Y, pp)\} \end{cases}$$

$$\mathcal{A}_{\mu}(p(X^1, \ldots, X^n)) \qquad = \mathcal{C}_{\mu(p)}(p(X^1, \ldots, X^n))$$

where $\mathcal{C}_N(p(X^1, \dots, X^n))$ is defined as

$$\{X_{pp}^{i^\delta} \geq X_{pp_j}^{j^\gamma} \mid X_{pp_j}^j \in init(X^j, pp) \text{ and } F_0^{i^\delta} \geq F_0^{j^\gamma} \in N\}$$
$$\cup \{X_{pp}^{i^\delta} \geq c \mid F_0^{i^\delta} \geq c \in N \text{ with } c = \text{static } or \text{ dynamic}\}$$
$$where \langle F^1, \dots F^n \rangle = \mathcal{A}rg(p)$$

Due to the well-modedness of procedures, the set $init(Y, pp)$ in Definition 8 contains program points that precede the unification in the procedure's body. Naturally, an atom constrains only the binding-time(s) of its output variable(s) in order to comply with the *congruence* [8] condition: if a (subvalue of) a variable is constructed using (a subvalue of) another variable, then the corresponding subvalue in the former's binding-time should be at least as dynamic as the corresponding part in the latter's binding-time. Note that a construction with a constant, $X \Leftarrow c$, leads to a (superfluous) constraint $X \geq static$ whereas a deconstruction with a constant $X \Rightarrow c$ leads to no constraints at all. The function $\mathcal{C}_{\mu(p)}$ maps the normal form of p's binding-time constraints – expressed in terms of the formal arguments F_0^i – to the actual arguments of the call to p: F_0^i in a left-hand side occurrence is replaced by X_{pp}^i; F_0^i in a right-hand side occurrence is replaced by $X_{pp_i}^i$ with pp_i the program point(s) where X^i gets initialised.

Now, we are in a position to define the set of constraints associated to a procedure:

Definition 9. *Given $\mu : Proc \mapsto 2^{\mathcal{B}c}$ such that $\mu(q)$ denotes q's normal form, we define for a procedure $p(F^1, \dots, F^n) : -G$*

$$BTC_p(\mu) = \mathcal{F} \cup \bigcup_{atom \ A \in G} \mathcal{A}_\mu(A)$$

where $\mathcal{F} = \{F_0^i \geq F_{pp}^i \mid F_{pp}^i \in init(F^i, 0) \text{ and } F^i \text{ an output argument}\}$

Note that in the above definition, $\mathcal{F}$ links the final binding-time of p's output arguments to each occurrence of that argument in p's body where it is initialised.

Example 5. If we consider $\mu(append) = \{\}$ for the *append* procedure from Example 3, $BTC_{app}(\mu)$ is defined as:

$$Z_0 \geq Z_2 \qquad Z_2 \geq Y_0 \qquad E_3 \geq X_0^{\langle\langle([]], 1\rangle\rangle} \qquad Z_5^{\langle\langle([]], 1\rangle\rangle} \geq E_3$$
$$Z_0 \geq Z_5 \qquad \qquad \qquad \qquad Es_3 \geq X_0 \qquad \qquad Z_5 \geq R_4$$

Bringing a set of binding-time constraints to normal form can be achieved by repeatedly unfolding the variable in the right-hand side, replacing it with the right-hand sides of the constraints that exist on that value. Unfolding is complicated by the use of subvalues, since a variable may need to be unfolded w.r.t. constraints that define a subvalue as well as a supervalue of the former. If we consider two subvalues of a variable, say X^δ and X^γ, we know that one of them is a subvalue of the other if either δ is an extension of γ or vice versa.

Definition 10. *We define* $ext : TPath \times TPath \mapsto TPath \times TPath$ *as follows:*

$$ext(\gamma, \delta) = (\langle\rangle, \eta) \ if \ \gamma = \delta \bullet \eta$$
$$ext(\gamma, \delta) = (\eta, \langle\rangle) \ if \ \gamma \bullet \eta = \delta$$

and undefined otherwise.

Note that if $ext(\gamma, \delta) = (\alpha, \beta)$ then $\gamma \bullet \alpha = \delta \bullet \beta$. Unfolding a constraint $X_{ppX}^{\gamma} \geq Y_{ppY}^{\delta}$ w.r.t. a set of constraints results in a set of new constraints on (possible subvalues of) X_{ppX}^{γ}, with as right hand sides the appropriate subvalues of the right hand sides of the constraints that were used for unfolding. To denote a subvalue of a constraint's right-hand side S, we use the notation $S^{\bullet\eta}$. If S denotes the variable X_{pp}^{γ}, then $S^{\bullet\eta}$ equals $X_{pp}^{\gamma\bullet\eta}$. Otherwise, if S denotes the constant *static* or *dynamic*, $S^{\bullet\eta}$ simply equals S.

$$project(X_{ppX}^{\gamma} \geq Y_{ppY}^{\delta}, \mathcal{D}) = \{ \ X_{ppX}^{\gamma\bullet\eta_1} \geq S'^{\bullet\eta_2} \mid Y_{ppY}^{\delta'} \geq S' \in \mathcal{D} $$
$$and \ ext(\delta, \delta') = (\eta_1, \eta_2)\}$$

Note that unfolding results in an empty set when the right-hand side that is unfolded is a variable on which no constraints are defined. Normalising a constraint set then consists of repeatedly selecting a constraint that is not yet in normal form and replacing it with the constraints resulting from unfolding, until a fixed point is reached.

Definition 11. *We define* $compr_p : 2^{\mathcal{B}c} \mapsto 2^{\mathcal{B}c}$ *as follows:*

$$compr_p(\mathcal{D}) = \mathcal{D} \setminus \{X_{ppX}^{\gamma} \geq Y_{ppY}^{\delta} \mid X_{ppX}^{\gamma} \geq Y_{ppY}^{\delta} \in \mathcal{D} \ and \ pp_Y \neq 0\}$$
$$\cup \ project(X_{ppX}^{\gamma} \geq Y_{ppY}^{\delta}, \mathcal{D})$$

and use the classical notation to apply $compr_p$ *a number of times starting with a set* $\mathcal{D}$: $compact_p(\mathcal{D}) = compr_p \uparrow \omega$ *where*

$$compr_p \uparrow 0 = \mathcal{D}$$
$$compr_p \uparrow i = compr_p(compr_p \uparrow (i-1)) \ \forall i > 0$$

Computing the normal form of the predicates defined in a module M is done through a fixed point computation. Note that we compute, for each procedure p, a set of binding-time constraints that is larger than p's normal form, as we are interested in the normalised binding-time constraints on all local variables of p. The result of the process for a module M is denoted by a function $\mu_M : Proc \mapsto 2^{\mathcal{B}c}$, mapping a procedure to a set of normalised binding-time constraints. For such a function μ_M, we denote with $\mu_{M_{|o}}$ the restricted mapping, where for each p, $\mu_{M_{|o}}(p) \subseteq \mu_M(p)$ and $\mu_{M_{|o}}(p)$ denotes p's normal form. When the analysis for the module M starts, we assume that all procedures defined in other modules have been analysed before, the result recorded in a function μ_I. If they are not, μ_I is initialised with *dynamic* for all the output arguments.[1] Construction of μ_M

---

[1] See [3] for a discussion of how a inter module analysis can cope with cyclic dependencies between modules.

is then defined as a fixed point iteration over an operator $T_\mu : (Proc \mapsto 2^{\mathcal{B}c}) \mapsto (Proc \mapsto 2^{\mathcal{B}c})$ defined as follows:

$$T_\mu(S) = S \setminus \{(p, N) \mid (p, N) \in S \text{ and } p \text{ defined in } M\}$$
$$\cup \{(p, compact_p(BTC_p(S_{|o})))\}$$

The analysis starts from an initial mapping where each predicate of M is mapped onto an empty set of constraints. During each iteration round, the constraints from a procedure's body are generated and normalised. This process is repeated until a fixed point is reached.

Definition 12. *For a module M, the result of constraint generation is the mapping $\mu_M = T_\mu \uparrow \omega$, where $T_\mu \uparrow \omega$ is the fixed point of*

$$T_\mu \uparrow 0 = \mu_I \cup \{(p, \{\}) \mid p \in Proc \text{ defined in } M\}$$
$$T_\mu \uparrow k = T_\mu(T_\mu \uparrow (k-1)) \; \forall k \geq 1$$

Example 6. Let M_{app} denote a module consisting only of the definition of *append*. The result of constraint generation, for the *append* procedure, is the following set of binding-time constraints:

$$Z_0 \geq Y_0 \qquad\qquad Z_2 \geq Y_0 \quad E_3 \geq X_0^{\langle([|],1)\rangle} \quad R_4 \geq Y_0 \qquad Z_5^{\langle([|],1)\rangle} \geq X_0^{\langle([|],1)\rangle}$$
$$Z_0^{\langle([|],1)\rangle} \geq X_0^{\langle([|],1)\rangle} \qquad\qquad E_{s3} \geq X_0 \qquad R_4 \geq X_0^{\langle([|],1)\rangle} \quad Z_5 \geq Y_0$$

It can be proven that the fixed point $\mu_M = T_\mu \uparrow \omega$ is reached after a finite number of steps. Informally, since binding-time constraints reflect the data flow relations in a *well-moded* procedure, no circular dependencies exist in the set and hence unfolding is finite and results in a finite set of binding-time constraints in normal form. New constraints are incorporated in each iteration round, but due to monotonicity of *compact* and the fact that the number of possible binding-time constraints is finite for a procedure, the iteration process results in a fixed point.

2.3   Solving the Cconstraint System

Given a set of binding-time constraints in normal form for a procedure and a call pattern of interest for that procedure, it is straightforward to compute the binding-time of an occurrence of a variable in that procedure. Let us formally define a call pattern ($Callp$) for a procedure p/n as a sequence over $\mathcal{B}T$ of length n. To compute binding-times, we define the function $\beta : Proc \times Callp \times 2^{\mathcal{B}c} \times Var_{pp} \mapsto \mathcal{B}T$ as follows:

Definition 13. *We define $\beta : Proc \times Callp \times 2^{\mathcal{B}c} \times \mathcal{B} \mapsto \mathcal{B}T$ as follows: For a program variable X of type t_X that is initialised at program point pp,*

$$\beta(p, \pi, \mathcal{D}, X_{pp}) = \text{ if } \quad X = Arg(p)_i \text{ and input then } \pi_i$$
$$\text{else } \{(\delta, \text{static}) \mid \delta \text{ a } TPath \in t_X\}$$
$$\sqcup \{(\gamma \bullet \eta, c) \mid X_{pp}^\gamma \geq c \in \mathcal{D}, c \in \{\text{static}, \text{dynamic}\} \text{ and}$$
$$\eta \text{ a } TPath \in t_X^\gamma\}$$
$$\sqcup \{(\gamma \bullet \eta, \tau(\eta)) \mid X_{pp}^\gamma \geq Y_0^{i^\delta} \in \mathcal{D}, \eta \text{ a } TPath \in t_X^\gamma \text{ and}$$
$$\tau = \pi_i^\delta\}$$

Note that computing binding-times using β is a cheap process, as it merely consists of looking up some binding-times from the call pattern and combining (subvalues of) these using the least upperbound operator on $\mathcal{BT}$. The resulting binding-time is the *least dynamic* one satisfying the constraints from $\mathcal{D}$.

Example 7. Reconsider $\mu_{M_{app}}$ from Example 6 and the binding-times as depicted in Fig. 1. Given the call pattern $\pi = \langle \tau_2, \tau_1, \bot \rangle$,

$$\beta(app, \pi, \mu_{M_{app}}, Z_0) = \{(\langle\rangle, static), (\langle\langle[|], 1\rangle\rangle, static)\}$$
$$\sqcup \{(\langle\rangle, static), (\langle\langle[|], 1\rangle\rangle, static)\}$$
$$\text{from } Z_0 \geq Y_0$$
$$\text{and } \beta(p, \pi, \mathcal{D}, Y_0) = \{(\langle\rangle, static), (\langle\langle[|], 1\rangle\rangle, static)\}$$
$$\sqcup \{(\langle\langle[|], 1\rangle\rangle, dynamic)\}$$
$$\text{from } Z_0^{\langle\langle[|], 1\rangle\rangle} \geq X_0^{\langle\langle[|], 1\rangle\rangle}$$
$$\text{and } \beta(p, \pi, \mathcal{D}, X_0) = \{(\langle\rangle, static), (\langle\langle[|], 1\rangle\rangle, dynamic)\}$$
$$= \{(\langle\rangle, static), (\langle\langle[|], 1\rangle\rangle, dynamic)\} = \tau_2$$

In this section, we have described BTA for a first-order subset of Mercury as a 2-phase process. The first, and computationally most involved phase – the computation of a set of normalised binding-time constraints – is performed *independent* of any call pattern of interest. Hence, this phase needs to be run only once for each module; its results can be recorded and used when analysing other modules that import this one. It is only the second phase – computing binding-times w.r.t. a call pattern – that needs to be repeated for every call and call pattern of interest.

Example 8. Consider a module M_{rev}, importing the module M_{app} and consisting of the definition of reverse: $rev(list(T) :: in, list(T) :: out)$.[2]

$$rev(X, R) : -\ X \Rightarrow [], R \Leftarrow [];$$
$$X \Rightarrow [E|Es], rev(Es, Rs), X' \Leftarrow [E], append(Rs, X', R).$$

When construction $\mu_{M_{rev}}$, the constraints for the call to *append* are renamings of $\mu_{M_{app}}(append)$ (See Example 6) w.r.t. the mapping $\{X \mapsto Rs, Y \mapsto X', Z \mapsto R\}$:

$$R^{\langle\langle[|], 1\rangle\rangle} \geq Rs^{\langle\langle[|], 1\rangle\rangle} \qquad\qquad R \geq X'$$

resulting (without re-analysing *append* itself) in the following constraints for *rev*:

$$E \geq X^{\langle\langle[|], 1\rangle\rangle} \qquad X'^{\langle\langle[|], 1\rangle\rangle} \geq X^{\langle\langle[|], 1\rangle\rangle} \qquad R \geq static$$
$$Es \geq X \qquad\qquad X' \geq static \qquad\qquad R^{\langle\langle[|], 1\rangle\rangle} \geq X^{\langle\langle[|], 1\rangle\rangle}$$

---

[2] In the remaining examples, we associate only one occurrence with each variable and leave out the program point subscription in order to ease notation.

3   BTA in a Higher-Order Setting

Mercury is a higher-order language in which *closures* can be created, passed as arguments of predicate calls, and in turn be called themselves. In this section, we reconsider the defined BTA in such a higher-order context.

To deal with the higher-order issues in Mercury, it suffices to extend the definition of superhomogeneous form (see Definition 5) with two new kinds of atoms: A *higher-order unification*, $X \Leftarrow p(V^1, \ldots, V^k)$ with $p \in \mathcal{T}_P$ and $V^1, \ldots, V^k \in Var$, constructs a closure from a predicate p where $V^1, \ldots, V^k$ are the curried arguments. Closures are called using a *higher-order call* $X(V^{k+1}, \ldots, V^n)$ where $V^{k+1}, \ldots, V^n \in Var$ are the closure's arguments. To express higher-order binding-times, we also consider a *higher-order type* to be included in the set of types ($\mathcal{MT}$). A higher-order type is a type definition of a predicate like $p(t_1, \ldots, t_n)$ with $t_i \in \mathcal{MT}$. When constructing a type tree, higher-order types are considered leaf nodes (the argument types are thus not taken into account in the type tree).

3.1   Closure Information

The basic problem when analysing a procedure involving higher-order calls, is that the control flow in the procedure is determined by the value of the higher-order variables. Consequently, without knowing (an approximation of) these values, it is impossible to compute meaningful data dependencies between the procedure's variables. Consider the following example:

Example 9. The *map* predicate converts the first list of type T to a new list of type T using the predicate provided in the second argument.

$$: -pred\ map(list(T), pred(T, T), list(T)).$$
$$: -mode\ map(in, in(pred(in, out)), out)\ is\ det.$$
$$map(L_1, P, L_0) : -L_1 \Rightarrow [], L_0 \Leftarrow [];$$
$$L_1 \Rightarrow [E_1 | Es_1], P(E_1, E_2), map(Es_1, P, Es_2), L_0 \Leftarrow [E_2 | Es_2].$$

In this example, we know from the mode declaration that E_1 is input and E_2 is output from the call to P. To compute a meaningful binding-time for E_2 (and all variables depending on E_2), we need to consider the dependencies that most likely exist between E_2 and E_1. Obviously, this requires information on the possible closures P can be bound to. Without this information, we can only approximate E_2's binding-time by *dynamic*, resulting in the following constraints for *map*:

$$L_0 \geq static \qquad L_0 \geq Es_2 \qquad L_0^{\langle\langle[]], 1\rangle\rangle} \geq Es_2$$
$$E_1 \geq L_1^{\langle\langle[]], 1\rangle\rangle} \qquad Es_1 \geq L_1 \qquad E_2 \geq dynamic$$

However, if there is in the module a call to *map* that binds P for example to the predicate *rev* from above, then we can create a more precise set of constraints for *map w.r.t. the fact that $P = rev$*, and the call $P(E_1, E_2)$, can, during constraint generation, be treated as a first-order call $rev(E_1, E_2)$, resulting in the following extra constraints (which are a renaming from $\mu_{M_{rev}}(rev)$, see Example 8):

$$E_2 \geq static \qquad E_2^{\langle([],1)\rangle} \geq E_1^{\langle([],1)\rangle}$$

Normalising and incorporating the constraints for the call to map now results in the following set of constraints:

$$E_1 \geq L_1^{\langle([],1)\rangle} \qquad E_2^{\langle([],1)\rangle} \geq L_1^{\langle([],1),([],1)\rangle} \qquad L_0 \geq static$$
$$Es_1 \geq L_1 \qquad E_2 \geq static \qquad L_0^{\langle([],1)\rangle} \geq static$$
$$L_0^{\langle([],1),([],1)\rangle} \geq L_1^{\langle([],1),([],1)\rangle}$$

Informally, the constraints on map's output argument L_0 express that in the least solution, the binding-times of the elements of the output lists will be the same as the binding-times of the elements of the input lists.

3.2 Higher-Order BTA

The general idea behind performing the analysis in a higher-order setting is to associate different sets of constraints to the same predicate, depending on the closures occurring in its call pattern.

To make this information available during BTA, we extend the notion of a binding-time to include, for higher-order variables, a set of closures. This set approximates the specific closure bound to the variable at runtime. For analysis purposes, we denote such a closure with $p(\tau_1, \ldots, \tau_k)$ where p is a procedure and $\tau_1, \ldots, \tau_k$ are binding-times for the curried arguments. The set of all such closures is denoted by $Clos$. In what follows, we extend the necessary definitions from Sect. 2. First, we extend the domain B by an explicit bottom element $\perp$ and elements $static(S)$ with $S \in 2^{Clos}$. The domain is ordered by the lattice in Fig. 2.

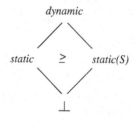

Fig. 2. The ordering on B

The ordering between elements $static(S_1)$ and $static(S_2)$ in turn is determined by the lattice $\{2^{Clos}, \supseteq\}$, that is: $static(S_1) \geq static(S_2)$ if and only if $S_1 \supseteq S_2$. Given this notion of B, the definition of a binding-time remains unchanged.

In the higher-order BTA, we associate a set of binding-time constraints to a call/call pattern pair and consequently denote the result of analysis by a function $\mu^c : Proc \times Callp \mapsto 2^{\mathcal{B}c}$. Like in the first-order case, the entries in μ^c contain the set of normalised binding-time constraints of the involved procedure/call pattern pair and we denote with $\mu^c_{|_o}$ the restriction of these sets to the procedure's normal form (under the involved call pattern).

Computing the set of binding-time constraints associated to a procedure p w.r.t. call pattern π requires the normal forms of the procedures called in the body of p. Hence, we denote this set with $BTC_p^\pi(\mu^c)$, which is defined like BTC_p in the first-order case, except for the handling of procedure calls.

Like before, handling a first-order call involves renaming the called predicate's normal form, which now depends on that call's call pattern. Let $q(X^1, \ldots, X^n)$ denote a call in procedure p's body at program point pp. The constraints associated with this call are $\mathcal{C}_{\mu^c(q,\langle \tau_1,\ldots,\tau_n \rangle)}(q(X^1, \ldots, X^n))$ where $\mathcal{C}$ is defined as in Definition 8 and $\langle \tau_1, \ldots, \tau_n \rangle$ denotes the call pattern of the call, which is computed using the set of binding-time constraints already associated with p and π: $\forall i : \tau_i = \beta(p, \pi, \mu^c(p, \pi), X^i_{pp_i})$.

Handling a higher-order call $P(X^{k+1}, \ldots, X^n)$ is a bit more complicated. First of all, the binding-time of P is computed in the current environment (that is, the procedure p and call pattern π). If this binding-time turns out to be *dynamic* (indicating that P is not bound to a set of closures at specialisation-time), the binding-times of the call's output arguments are simply approximated by *dynamic*, by introducing the following constraints for the call:

$$\{X^i_{pp} \geq dynamic \,|\, X^i \text{ is an output argument}\}$$

If, on the other hand, the binding-time of P turns out to be $static(S) - S$ denoting the set of closures P can be bound to, the higher-order call can be treated as a number of first-order calls (one for each closure in S), and its set of associated constraints is the union of the (renamed) normal forms associated to each of the derived first-order calls:

$$\bigcup_{q(\tau_1,\ldots,\tau_k)\in S} \mathcal{C}_{\mu^c_{|_o}(q,\langle \tau_1,\ldots,\tau_n \rangle))}(q(X^{k+1}, \ldots, X^n))$$

Note that the involved call pattern $\langle \tau_1, \ldots, \tau_n \rangle$ is a combination of the binding-times from the closure and the binding-times of the remaining arguments (computed as in the first-order case). The constraints associated to $q(\langle \tau_1, \ldots, \tau_n \rangle)$ are renamed with respect to the sequence of variables

$$\langle P^{\langle (pred,1) \rangle}, \ldots, P^{\langle (pred,k) \rangle}, X^{k+1}, \ldots, X^n \rangle$$

where $P^{\langle (pred,j) \rangle}$ denotes a new binding-time variable that refers to the j-th binding-time from the closure associated to P.

Example 10. Consider a higher-order call $P(L_2, L_3)$ and suppose we know that P is bound to a closure $append(\tau_1)$ (the *append* predicate from Example 3 with its first argument curried). This call introduces the following constraints, by appropriate renaming of *append*'s normal form:

$$L_3 \geq L_2 \qquad L_3^{\langle([|],1)\rangle} \geq P^{\langle(pred,1)\rangle}$$

3.3   Towards Maximum Modularity

Like in the first-order case, when analysing a module M, we start from an initial table μ_I^c, containing the analysis results for the imported modules. We return to the issue of modularity further on. An entry for a call and call pattern (p, π) of interest is added to the table, by computing $BTC_p^\pi(\mu_I^c)$ resulting in a new table in which (empty) entries are added for all call/call pattern pairs for which, during computation of $BTC_p^\pi(\mu_I^c)$, the normal form needed to be retrieved but was not yet available. The set of normalised binding-times is then (re)computed for each entry in the table belonging to a procedure in M, possibly requiring to add new entries to the table. This process is repeated until a fixed point is reached.

It is important to note that, when constructing the table, a new entry for p w.r.t. call pattern π should only be created if there is not yet an entry for (p, π') in the table, where π' is a call pattern of which the *higher-order parts* (i.e. the involved closures) are equal to those of π. For constraint generation, only these higher-order parts are significant, since the set of generated constraints does not depend on the first-order parts of the call pattern.

The analysis of a module starts from $(p_i, \langle dynamic, \ldots, dynamic\rangle)$ for every exported predicate p_i. Using $\langle dynamic, \ldots, dynamic\rangle$ as a call pattern ensures that no (higher-order) information from outside the module is assumed and thus the result of analysis, μ_M^c, can be used when analysing other modules that import M. Higher-order information that is *local* to a module, however, is used if BTA deals with higher-order unification. Therefore, we extend the notion of a binding-time constraint so that its right-hand side now can include $static(p(Y^1, \ldots, Y^k))$, as we associate the constraints

$$\{X \geq static(p(Y_{pp_1}^1, \ldots, Y_{pp_k}^k)) \mid Y_{pp_1}^1 \in init(Y^1, pp), \ldots, Y_{pp_k}^k \in init(Y^k, pp)\}$$

with such a higher-order unification $X \Leftarrow p(Y^1, \ldots, Y^k)$. The minimal binding-time for X satisfying this constraint says that X is bound to a closure created from the predicate p and binding-times for $Y^1, \ldots, Y^k$.

The right-hand side of such a constraint is "constant", and it can readily be used by β to compute a binding-time for X. All β needs to do is to compute the binding-times $\tau_1, \ldots, \tau_k$ for $Y^1, \ldots, Y^k$ (in the same environment) and guarantee that if the binding-time for X turns out to be $static(S)$ (by evaluating possibly other constraints on X), that the closure $p(\tau_1, \ldots, \tau_k) \in S$. Note however that while such a constraint can be considered to be in normal form (as it can readily be used by β), it can only be considered as part of the normal form of the procedure it is part of, if the arguments of the closure construction are arguments of that procedure. If not, such a constraint can not be directly renamed into a caller's environment.

Consider the following example, defining a predicate $lrev(list(list(T))$ :: $in, list(list(T))$ :: $out)$ that reverses each of the lists in its first argument:

$$lrev(L_1, L_2) : -P \Leftarrow rev, map(L_1, P, L_2).$$

When analysing $lrev$ w.r.t. the call pattern $\langle dynamic, dynamic \rangle$, the constraint $P \geq static(rev())$ is derived for $lrev$ in a first iteration round. In a next round, this constraint is used to derive $\langle dynamic, static(\{rev()\}), dynamic \rangle$ as call pattern for the call to map. If map and rev are defined in the same module as $lrev$, a specific version of the constraints for map is created, leading to the following set of normalised constraints for $lrev$ (the constraints with only $static$ in the right-hand side are removed):

$$\{L_2^{\langle ([|], 1), ([|], 1) \rangle} \geq L_1^{\langle ([|], 1), ([|], 1) \rangle}\}$$

Informally, this constraint implies that in the least solution the binding-times of the elements of the output lists will be the same as the binding-times of the elements of the input lists. Note that this set of constraints is independent of the call pattern of $lrev$, as this does not contain any higher-order information.

If map and/or rev are defined in a different module (say M'), then there are several possibilities to follow. A first one is to rename the constraints associated to a version of map w.r.t. a call pattern that is more dynamic. Note that this option is always available, as map – being an exported predicate of M' – will at least be analysed w.r.t. $\langle dynamic, \ldots, dynamic \rangle$, the result recorded in μ_f^c. While correct, this option will not result in useful constraints for predicates using map.

A more interesting option is to make sure that when analysing M, enough information is available such that map can be (re)analysed w.r.t. this new call pattern, outside the scope of the earlier analysis of M'. Note that doing so, might require to (re)analyse all procedures involving higher-order arguments imported by map and other procedures imported therein. Recent work [3] indicates how such a call dependent analysis can still be performed on a modular basis, re-analysing one module at a time and propagating analysis results – possibly triggering reanalysis of other modules, until "enough" precision is obtained.

4   Discussion

In this work, we have rephrased a BTA for Mercury [16] using constraint normalisation, an approach that allows – in contrast with our earlier work – to perform (a large part of) the analysis on a modular basis. When no higher-order control flow is involved, constraint normalisation can be performed one module at a time, bottom-up in the module graph. To obtain maximal precision with predicates that do involve higher-order flow, it can be necessary to re-analyse the predicate with respect to specific closure information in the predicate's call pattern. Dealing with modular programs is mandatory to apply program analysis tools on real-world programs. Recently, analysis of modular programs gained attention in a logic programming setting. For example, [14] discusses some – mainly practical – issues in the analysis of modular programs.

The current work formalises the ideas presented in [17] and extends the analysis to deal with the higher-order concepts of Mercury. In our analysis, closures are encapsulated in the notion of a binding-time, and our analysis does not require a separate closure analysis. Closure information from the call pattern is incorporated during constraint generation and normalisation. The analysis is polyvariant in this respect, as different such call patterns result in different constraint sets for the same predicate. Computing concrete binding-times is straightforward given binding-times for a predicate's input arguments, as it merely consists of computing the least solution of the predicate's associated normal form. The analysis is also polyvariant in this respect: different call patterns result in a different least solution of the same constraint set.

We are closely collaborating with the Mercury developers in Melbourne to implement the described BTA in a version of the Mercury compiler. To that extent, the tight interaction between the binding-time analysis and a concrete specialiser (ensuring, for example, that binding-times of a call's output arguments are treated as *dynamic* when the specialiser would not unfold the call) needs to be modeled. This can be achieved by adding constraints to the system that model the conditions under which atoms are evaluated at specialisation-time. Topics of further work are to perform some large scale experiments with the analysis, and couple it with the partial evaluation mechanisms of the compiler.

Acknowledgments

The author likes to thank Maurice Bruynooghe for fruitful discussions on the subject, and anonymous referees for their valuable feedback. Special thanks go to Zoltan Somogyi for providing the opportunity of working together with the Mercury team, and to Tyson Dowd, Fergus Henderson, David Jeffrey, David Overton, Peter Ross, Mark Brown and Anthony Senyard for explaining over and over again several implementation issues of the Melbourne Mercury compiler, but perhaps most of all for providing an environment which made the author's visit to Australia an unforgettable experience.

References

1. A. Bondorf and J. Jørgensen. Efficient analyses for realistic off-line partial evaluation. *Journal of Functional Programming*, 3(3):315–346, July 93. 400, 400
2. Maurice Bruynooghe, Michael Leuschel, and Kostis Sagonas. A polyvariant binding-time analysis for off-line partial deduction. In C. Hankin, editor, *Programming Languages and Systems, Proc. of ESOP'98, part of ETAPS'98*, pages 27–41, Lisbon, Portugal, 1998. Springer-Verlag. LNCS 1381. 400
3. F. Bueno, M. de la Banda, M. Hermenegildo, K. Marriott, G. Puebla, and P. Stuckey. A model for inter-module analysis and optimizing compilation. In *Preproceedings of LOPSTR 2000*, 2000. 407, 414

4. Tyson Dowd, Zoltan Somogyi, Fergus Henderson, Thomas Conway, and David Jeffery. Run time type information in Mercury. In *Proceedings of the International Conference on the Principles and Practice of Declarative Programming*, volume 1702 of *Lecture Notes in Computer Science*, pages 224–243. Springer-Verlag, 1999. 399

5. D. Dussart, F. Henglein, and C. Mossin. Polymorphic recursion and subtype qualifications: Polymorphic binding-time analysis in polynomial time. In A. Mycroft, editor, *International Static Analysis Symposium, Glasgow, Scotland, September 1995, (Lecture Notes in Computer Science, vol. 983)*, pages 118–136. Berlin: Springer-Verlag, 1995. 400, 400

6. J. Gallagher. Specialisation of logic programs: A tutorial. In *Proceedings PEPM'93, ACM SIGPLAN Symposium on Partial Evaluation and Semantics-Based Program Manipulation*, pages 88–98, Copenhagen, June 1993. ACM Press. 400

7. F. Henglein and C. Mossin. Polymorphic binding-time analysis. *Lecture Notes in Computer Science*, 788, 1994. 400, 400

8. N. D. Jones, C. K. Gomard, and P. Sestoft. *Partial Evaluation and Automatic Program Generation*. Prentice Hall, 1993. 400, 406

9. J. Jørgensen and M. Leuschel. Efficiently generating efficient generating extensions in Prolog. In O. Danvy, R. Glück, and P. Thiemann, editors, *Proceedings Dagstuhl Seminar on Partial Evaluation*, pages 238–262, Schloss Dagstuhl, Germany, 1996. Springer-Verlag, LNCS 1110. 400

10. J. Launchbury. Dependent sums express separation of binding times. In K. Davis and J. Hughes, editors, *Functional Programming*, pages 238 – 253. Springer-Verlag, 1989. 400

11. M. Leuschel, B. Martens, and D. De Schreye. Controlling generalisation and polyvariance in partial deduction of normal logic programs. *ACM Transactions on Programming Languages and Systems*, 20(1), 1998. 400

12. T. Mogensen and A. Bondorf. Logimix: A self-applicable partial evaluator for Prolog. In K.-K. Lau and T. Clement, editors, *Proceedings LOPSTR'92*, pages 214–227. Springer-Verlag, Workshops in Computing Series, 1993. 400

13. Jens Palsberg. Closure analysis in constraint form. *ACM Transactions on Programming Languages and Systems*, 17(1):47–62, January 1995. 400

14. G. Puebla and M. Hermenegildo. Some issues in analysis and specialization of modular Ciao-Prolog programs. Las Cruces, 1999. Proceedings of the Workshop on Optimization and Implementation of Declarative Languages. In Electronic Notes in Theoretical Computer Science, Volume 30 Issue No.2, Elsevier Science. 414

15. Zoltan Somogyi, Fergus Henderson, and Thomas Conway. The execution algorithm of Mercury, an efficient purely declarative logic programming language. *Journal of Logic Programming*, 29(1–3):17–64, October–November 1996. 403, 404

16. W. Vanhoof and M. Bruynooghe. Binding-time analysis for Mercury. In D. De Schreye, editor, *16th International Conference on Logic Programming*, pages 500 – 514. MIT Press, 1999. 400, 400, 401, 403, 414

17. W. Vanhoof and M. Bruynooghe. Towards modular binding-time analysis for first-order Mercury. Las Cruces, 1999. Proceedings of the Workshop on Optimization and Implementation of Declarative Languages. In Electronic Notes in Theoretical Computer Science, Volume 30 Issue No.2, Elsevier Science. 415

Efficient Evaluation Methods for Guarded Logics and Datalog LITE

Erich Grädel

Mathematische Grundlagen der Informatik, RWTH Aachen
graedel@informatik.rwth-aachen.de

Abstract

Guarded logics are fragments of first-order logic, fixed point logic or second-order logic in which all quantifiers are relativised by guard formulae in an appropriate way. Semantically, this means that such logics can simultaneously refer to a collection of elements only if all these elements are 'close together' (e.g. coexist in some atomic fact). Guarded logics are powerful generalizations of modal logics (such as propositional multi-modal logic or the modal μ-calculus) that retain and, to a certain extent, explain their good algorithmic and model-theoretic properties.

In this talk, I will survey the recent research on guarded logics. I will also present a guarded variant of Datalog, called Datalog LITE, which is semantically equivalent to the alternation-free portion of guarded fixed point logic. The main focus of the talk will be on model checking (or equivalently, query evaluation) algorithms for guarded logics. While the complexity of evaluating arbitrary guarded fixed point formulae is closely related to the model checking problem for the modal μ-calculus (for which no polynomial-time algorithms are known up to now), there are interesting fragments that admit efficient, in fact linear time, evaluation algorithms. In particular this is the case for the guarded fragment of first-order logic and for Datalog LITE.

M. Parigot and A. Voronkov (Eds.): LPAR 2000, LNAI 1955, pp. 417–417, 2000.
© Springer-Verlag Berlin Heidelberg 2000

On the Alternation-Free Horn μ-Calculus

Jean-Marc Talbot

Max-Planck Institut für Informatik
Stuhlsatzenhausweg 85 - 66123 Saarbrücken - Germany

Abstract. The Horn μ-calculus is a formalism extending logic programs by specifying for each predicate symbol with which (greatest or least) fix-point semantics, its denotation has to be computed. When restricted to a particular class of logic programs called uniform, the Horn μ-calculus provides a syntactic extension for Rabin tree automata. However, it has been shown [1] that the denotation of the Horn μ-calculus restricted to a uniform program remains a regular set of trees and that moreover, the emptiness of the denotation of a predicate p is a DEXPTIME-complete problem (in the size of the program). In [3], these results have been extended to uniform programs that may contain both existential and universal quantifications on the variables occurring in the body of "clauses": considering this extension, the denotation of a program remains a regular set of trees, but the best known algorithm for testing the emptiness of the denotation of a predicate is doubly-exponential in the size of the program.

In this paper, we consider uniform logic programs with both kinds of quantification in the body. But we add to the Horn μ-calculus a limitation on the way the fix-point semantics is specified for predicates. This restriction is close to the one defining the alternation-free fragment of the μ-calculus. Therefore, we name this fragment of the Horn μ-calculus the *alternation-free* fragment. We devise for it an algorithm which performs the emptiness test for the denotation of a predicate in single-exponential time in the size of the program.

To obtain this result, we develop a constructive approach based on a new kind of tree automata running on finite and infinite trees, called *monotonous tree automata*. These automata are defined by means of a family of finite and complete lattices. The acceptance condition for monotonous tree automata is based on the ordering relations of the lattices.

1   Introduction

The Horn μ-calculus [1] is a formalism extending logic programs by means of fix-point operators. In the classical framework of logic programming, the ground [1] semantics of a program is usually expressed in terms of a least fix-point computation. However, for some different purposes, the semantics related to the greatest fix-point (also called reactive semantics) may also be of some interest [11]. The idea of the Horn μ-calculus is to integrate this semantical point in the program itself by specifying for each predicate symbol whether its semantics has to be computed as a least or a greatest fix-point.

---

[1] In this paper, we consider the semantics of programs over the complete Herbrand base, that is atoms built over finite and infinite trees.

M. Parigot and A. Voronkov (Eds.): LPAR 2000, LNAI 1955, pp. 418–435, 2000.

Semantics for a logic program can alternatively be expressed in terms of ground proof trees. A ground atom belongs to the least fix-point semantics if there exists a finite ground proof tree rooted by this atom. Since the number of predicate symbols is finite, this finiteness condition over the proof tree can be rephrased as "for any predicate p, along every branch, one encounters only finitely many atoms defined with p". On the other hand, an atom belongs to the greatest fix-point semantics if it exists a ground (possibly infinite) proof tree rooted by this atom. Thus, in this case predicate symbols can occur freely infinitely often in atoms along each branch. The Horn μ-calculus captures such notions by associating with each predicate symbol a positive integer, called priority. An atom is then "accepted" by the program if there exists a proof tree for it such that the priorities of predicates in the atoms along each branch satisfy the *parity condition* [7]: the maximal priority occurring infinitely often along each branch is even. Roughly speaking, the semantics for predicate symbols with an odd arity is computed as a least fix-point and a greatest fix-point is used for predicate symbols with an even arity.

When this Horn μ-calculus is restricted to a class of logic programs called uniform programs [8], then its semantics coincide with the notion of regularity *à la* Rabin [14]: the set of grounds atoms accepted by the program is a regular set of trees. This result has extended other results concerning the least [108108] and the greatest [2626] fix-point semantics for uniform logic programs obtained in the area of set-based analysis. Hence, the Horn μ-calculus restricted to uniform programs can be viewed as a particular technique for performing (set-based) relaxation of the Horn μ-calculus for arbitrary programs.

As we have said above, uniform logic programs are related to set-based analysis; they are more specifically connected with some classes of set constraints called the definite [9] and the co-definite class [4]. To be fully precise, uniform logic programs correspond to the definite and the co-definite classes extended with some set description called quantified set expressions in [10] and membership expressions in [5656]. Quantified set expressions (or equivalently, membership expressions) are in fact intentional description of sets of the form $\{x|\Psi(x)\}$; they are interpreted as the set of all trees τ which satisfy the given property Ψ, *i.e.* as the set of trees τ such that $\Psi(\tau)$ holds. This property Ψ is presented as an existentially quantified conjunction of atoms. So, rephrased in the logic programming framework, such an expression would correspond to the clause $p(x) \leftarrow \Psi(x)$. In [16181618], the membership expressions have been extended by allowing variables to be also universally quantified. The aim of this extension was to provide a uniform view of the definite and the co-definite classes of set constraints. So, one may wonder whether this extension can be carried over the Horn μ-calculus for uniform programs. This would lead to consider "clauses"[2] of the from $p(t) \leftarrow \Psi$ where the local variables in Ψ (the ones which do not occur in $p(t)$) can be quantified existentially or universally.

The Horn μ-calculus restricted to uniform programs with both quantification kinds in the body has been considered by Charatonik, Niwiński and Podelski in [3] for a model-checking purpose. It is shown in that paper that this extension preserves the regularity,

---

[2] The word *clause* is used here in a not fully proper way: universal quantification on variables in the body part leads to formulas that are no longer Horn clauses.

in the sense that the denotation of the program is a regular set of trees. Furthermore, Charatonik and *al.* have proposed an algorithm performing the emptiness test for the denotation of some predicate; this algorithm runs in doubly-exponential time in the size of the program, whereas the best known lower-bound for the complexity of this problem is DEXPTIME.

As in [3], this paper is based on logic programs for which variables occurring in the body of clauses can be quantified universally or existentially together with a function that assigns to predicate symbols a priority (*i.e.* a natural number). However, we will consider only a fragment of this calculus; this fragment is defined by a syntactic restriction based on priorities. It is quite similar to the restriction defining from the μ-calculus its alternation-free fragment. Therefore, we call this fragment the *alternation-free Horn μ-calculus*. Roughly speaking the restriction for the alternation-free μ-calculus requires that the computation of the semantics of a formula can be achieved layer-by-layer, each of those layers representing an homogeneous block of fix-point quantifiers. In a similar way, for the alternation-free Horn μ-calculus, the semantics of any predicate p can not depend on the semantics of some predicate q having a priority strictly greater than the one of p: the denotation of a predicate can not depend on the denotation of predicates belonging to upper layers. For this alternation-free fragment, we rephrase its semantics by means of fix-point operators based on this idea of layers.

We show here that for the alternation-free Horn μ-calculus restricted to uniform programs with both quantification kinds in bodies, deciding whether the denotation of a predicate p is empty (that is, deciding whether there exists a tree τ such that $p(\tau)$ belongs to the denotation of the program) can be achieved in single-exponential time in the size of the program improving the result from [3] for this specific fragment. To obtain this result, we have designed a new kind of tree automaton, called *monotonous tree automaton*.

Those automata are based on a family of finite and complete lattices $(S_i, \preceq^i)$. A state is a tuple of the S_i's. The notion of runs coincides with the classical one but the acceptance condition of those runs is based on the family of orderings $(\preceq^0, \ldots, \preceq^k)$ on which lattices are defined. We show that for a monotonous tree automaton the emptiness of the accepted language can be checked in polynomial time in the size of the automaton. For a program from the alternation-free Horn μ-calculus (P, Ω), we fix a general shape of the monotonous automata we have to consider. Then, we design some fix-point operators defined over those automata that mimic the fix-point semantics of (P, Ω). Our algorithm yields a monotonous tree automaton that recognizes the semantics of (P, Ω).

The paper is organized as follows: in Section 2, we give the syntactic definition for the Horn μ-calculus as well as its semantics in terms of proof trees. We also present there the alternation-free fragment together with its alternative semantics in terms of fix-points. Section 3 is devoted to the presentation of the generalized uniform programs we consider. Those programs allow in *projection clauses* variables occurring in the body to be either universally or existentially quantified. This section settles also the main result of this paper that is the emptiness problem for the alternation-free Horn μ-calculus when restricted to those so-extended uniform programs is DEXPTIME-complete. The rest of the paper that is Section 4 addresses the method we used to obtain this result: we present there monotonous tree automata and give some basic results about them like the regularity of the recognized language and the complexity for the emptiness problem. Then, we present an algorithm that given a uniform program from the alternation-free Horn μ-calculus computes an equivalent monotonous tree automaton: the semantics of program and the language accepted by the automaton coincide.

Due to lack of space, most of the proofs are not in this paper but can be found in the long version [17].

2   The Horn μ-Calculus

We consider a finite ranked signature Σ. We denote T^*_Σ, the set of all finite and infinite trees generated over Σ. For a tree τ, $dom(\tau)$ will denote its tree domain; for any position d in $dom(\tau)$, $\tau(d)$ denotes the function symbol from Σ labeling τ at position d and $\tau[d]$ denotes the sub-tree of τ rooted at position d.

We consider also a finite set of monadic [3] predicate symbols Pred; we will denote HB*, the complete Herbrand base generated over Pred and T^*_Σ, i.e. the set of all ground atoms $p(\tau)$ with $p \in$ Pred and $\tau \in T^*_\Sigma$.

For a term t or a first-order formula Ψ, Var(t) and Var(Ψ) will denote the set of all free variables respectively in t and in Ψ. Assuming that in the formula Ψ, two different quantifications always address two different variables, we denote Var$_\exists(\Psi)$ and Var$_\forall(\Psi)$ the set of all variables that are respectively existentially and universally quantified in Ψ.

2.1   Definitions

A *generalized clause* (or simply, a clause in the rest of the paper) is a first-order formula that generalizes Horn clauses by allowing universal quantification on the variables appearing in the body part of the clause. More formally, a generalized clause is an implication

$$p(t) \leftarrow \Psi \qquad \text{with } \Psi ::= p'(t') \mid \Psi \wedge \Psi \mid \exists y \Psi \mid \forall y \Psi \mid true$$

where t and t' s are first-order terms and p, p' belong to Pred. We assume wlog that the free variables of the formula Φ must occur in the head of the clause, that is in the term t.

Definition 1. *A Horn μ-program (P, Ω) is given by a set P of generalized clauses and a mapping Ω from* Pred *to the set of natural numbers.*

For a predicate p, $\Omega(p)$ is the *priority* of p and $\max(\Omega(P))$ is the maximal priority in the range of Ω over the predicates of P. For a clause c, $\Omega(c)$ denotes the priority of the predicate symbol occurring in the head of c.

The semantics for a Horn μ-program is given by means of ground proof trees. A proof tree for a ground atom $p(\tau)$ is a (possibly) infinite and infinitely-branching tree rooted in a node labeled by $p(\tau)$. Moreover, for any node n in the proof tree, labeled by some ground atom $p'(\tau')$, there exists a clause $p'(t) \leftarrow \Psi$ and a substitution $\sigma : Var(t) \mapsto T^*_\Sigma$ such that:

---

[3] For a matter of simplicity, we shall consider only monadic predicate symbols. The notions presented in this section extend naturally to predicate symbols with arbitrary arity.

- $\sigma(p'(t)) = p'(\tau')$ and,
- the sons of the node n are labeled by elements from the set $Sons(n, p'(t) \leftarrow \Psi)$ in such a way that for all elements l in $Sons(n, p'(t) \leftarrow \Psi)$, there is exactly one son of n labeled by l.

This set $Sons(n, p'(t) \leftarrow \Psi)$ is a set of ground atoms and is a minimal model over the universe HB^* of the formula Ψ under the substitution σ, i.e. $Sons(n, p'(t) \leftarrow \Psi), \sigma \models \Psi$.

Note that leaves in a proof tree correspond necessarily to the case where the clause used to compute $Sons(n, p'(t) \leftarrow \Psi)$ is a fact, i.e. $\Psi = true$ In this case, $Sons(n, p'(t) \leftarrow \Psi)$ is empty.

For an infinite path π in a proof tree, we denote $Inf(\pi)$ the set of all priorities that occur infinitely often along the path π. We say that a proof tree accepts the atom $p(\tau)$ if it is rooted in a node labeled by $p(\tau)$ and for all paths π starting from the root, the maximal element of $Inf(\pi)$ is even. A Horn μ-program (P, Ω) accepts a ground atom $p(\tau)$ if there exists a proof tree which accepts the atom $p(\tau)$.

We denote $[\![(P, \Omega)]\!]$ (resp. $[\![(P, \Omega), p]\!]$) the set of all ground atoms (resp. of all ground atoms for the predicate symbol p) accepted by the Horn μ-program (P, Ω).

2.2   The Alternation-Free Fragment of the Horn μ-Calculus

We give here a syntactic restriction for Horn μ-calculus as we presented it above. This restriction limits the dependency on predicates according to their respective priorities. We say that a predicate p depends on a predicate q if there exists a clause $p(t) \leftarrow \Psi$ such that either q occurs in Ψ or some predicate r occurs in Ψ and r depends on q.

Definition 2 (Alternation-free Horn μ-calculus). *A Horn μ-program is said to be alternation-free if for any predicates p, q, if p depends on q then the priority of q is smaller or equal to the priority of p i.e. $\Omega(q) \leq \Omega(p)$*

Note that "classical" logic programs with respect to their usual least fix-point semantics as well as to their reactive greatest fix-point semantics fall into this alternation-free fragment: for a logic program P, the least fix-point semantics of P is exactly the denotation of the alternation-free Horn μ-program (P, Ω_l) where $\Omega_l(p) = 1$ for any predicate p in P whereas the greatest semantics is the denotation of (P, Ω_g) where $\Omega_g(p) = 0$ for any p in P.

We give for the alternation-free Horn μ-calculus a semantics in terms of fix-points. This latter is however equivalent to the ground proof tree semantics. For simplicity, we will assume from now on that the range of the function Ω is an interval over $\mathbb{N}$ of the form $[0 \ldots \max(\Omega(P))]$.

Let $T^i_{(P,\Omega)} : \mathsf{HB}^* \mapsto \mathsf{HB}^*$ be an operator defined for any integer i in the range of $\Omega(P)$ as follows:

$$T^i_{(P,\Omega)}(S) = (S \smallsetminus \mathsf{HB}^*_i) \cup \left\{ \sigma(p(t)) \left| \begin{array}{l} \text{there exists } p(t) \leftarrow \Psi \text{ in } P \text{ such that } \Omega(p) = i \\ \text{there exists a substitution } \sigma : Var(t) \mapsto T^*_\Sigma \\ \text{such that } S, \sigma \models \Psi \end{array} \right. \right\}$$

where HB_i^* the set of all ground atoms defined for predicates p with priority i.
Let us consider a set A such that $A \cap \mathsf{HB}_i^* = \varnothing$. Now, it is easy to see that $L_A = (\{A \cup S \mid S \subseteq \mathsf{HB}_i^*\}, \subseteq)$ is a complete lattice and that $T_{(P,\Omega)}^i$ is monotonic over this lattice. So, by Knaster-Tarski'theorem, $T_{(P,\Omega)}^i$ admits a least and a greatest fix-points over this lattice respectively denoted $\mathit{lfp}_A(T_{(P,\Omega)}^i)$ and $\mathit{gfp}_A(T_{(P,\Omega)}^i)$.

In order to define the fix-point semantics for some alternation-free Horn μ-program (P, Ω), we introduce a family $(F_{(P,\Omega)}^i)$ of sets of ground atoms for each i in $\Omega(P)$ as:

$$F_{(P,\Omega)}^0 = \mathit{gfp}_\varnothing(T_{(P,\Omega)}^0)$$

$$F_{(P,\Omega)}^i = \mathit{lfp}_{F_{(P,\Omega)}^{i-1}}(T_{(P,\Omega)}^i) \qquad \text{if } i \text{ is odd}$$

$$F_{(P,\Omega)}^i = \mathit{gfp}_{F_{(P,\Omega)}^{i-1}}(T_{(P,\Omega)}^i) \qquad \text{if } i \text{ is even}$$

The fix-point semantics of (P, Ω) is simply the set $F_{(P,\Omega)}^n$ where n is $\max(\Omega(P))$. As claimed earlier, the fix-point and the proof tree semantics coincide as stated in the following theorem:

Theorem 1. *Let (P, Ω) be an alternation-free Horn μ-program with $\Omega(P) = [1 \ldots n]$. Then $[\![(P, \Omega)]\!] = F_{(P,\Omega)}^n$.*

Example 1. Let us consider (P, Ω), the alternation-free Horn μ-program given by

$$P = \left\{ \begin{array}{lll} p_0(f(x,y)) \leftarrow p_0(x), p_0(y) & p_0(a) & p_0(b) \\ p_1(f(x,y)) \leftarrow p_1(x), p_0(y) & p_1(a) & p_1(c) \\ p_2(x) \leftarrow \exists y\, p_1(f(x,y)) \wedge p_0(x) \end{array} \right\} \text{ and for all } i, \Omega(p_i) = i.$$

The set $F_{(P,\Omega)}^0$ is exactly the denotation of the predicate p_0, that is the set of all ground atoms $p_0(\tau)$ where τ is a finite or an infinite tree built over the binary function symbol f and the two constants a and b.

The set $F_{(P,\Omega)}^1$ contains $F_{(P,\Omega)}^0$ and the denotation of the predicate p_1. The denotation of this latter is the set of all ground atoms $p_1(\tau')$; these trees τ' consist of a finite left backbone of f symbols and terminated either with a or with c. Each of those f symbols from the backbone has as right son a finite or infinite tree built over f, a and b. The trees τ' can be depicted as

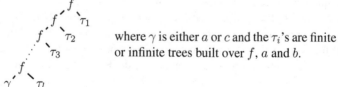

where γ is either a or c and the τ_i's are finite or infinite trees built over f, a and b.

Finally, the set $F_{(P,\Omega)}^2$ is the denotation of (P, Ω); it contains the set $F_{(P,\Omega)}^1$ and the denotation of p_2, that is the set of all ground atoms $p_2(\tau'')$, where τ'' is a finite left backbone of f terminated with the constant a and each right son of the f symbols from the backbone is a finite or infinite tree built over f, a and b. Hence, a tree τ'' is similar to the tree τ' depicted above except that for τ'', γ is necessarily the constant a.

3   Horn μ-Calculus for Uniform Programs

Uniform logic programs aimed to express the set-based analysis in terms of logic programs. They are the basis of the results from [1313] where it is shown that the Horn μ-calculus restricted to uniform programs is decidable. In this section, we investigate the case of the alternation-free Horn μ-calculus restricted to uniform programs that are extended with both existential and universal quantification in the body of clauses.

Definition 3. *A* uniform Horn μ-program *is given as a pair* (P, Ω) *where* Ω *is a priority function and* P *is a set of clauses (defined over a set* **Pred** *of monadic predicate symbols) of the following forms:*

$$p(f(x_1, \ldots, x_m)) \leftarrow p_1(x_1), \ldots, p_m(x_m)$$

$$p(x) \leftarrow \Psi$$

where: - $\Psi ::= p'(t') \mid \Psi \wedge \Psi \mid \exists y \Psi \mid \forall y \Psi \mid true$
         - Ψ *contains at most* x *as a free variable,* i.e. $\text{Var}(\Psi) \subseteq \{x\}$.

The first kind of clause is called *automaton clause* as a reference the transition rule in a classical tree automaton. The second kind is called *projection clause*. From now on, the expression *uniform program* relates to a program of the form given in Definition 3.

It should be noticed that the alternation-free Horn μ-program given in Example 1 is uniform.

We have already mentioned the regularity aspect of the Horn μ-calculus restricted to uniform programs. Actually, this calculus can be viewed as a syntactic extension for the definition of Rabin tree automata. It turns out that Rabin tree automata, or to be more precise *parity tree automata*, correspond to uniform programs with clauses of the form $p(f(x_1, \ldots, x_m)) \leftarrow p_1(x_1), \ldots, p_n(x_n)$. Additionally, when uniform clauses of the form $p(x) \leftarrow q(x), r(x)$ are considered then those programs correspond to *alternating parity tree automata* [13].

As we have said before, we consider only programs based on a restriction of the Horn μ-calculus, namely the alternation-free fragment. Therefore, one may wonder whether there is an existing class of tree automata for which the alternation-free fragment of the Horn μ-calculus restricted to uniform programs is a generalization.

In [12] Muller and *al.* have introduced the notion of *weak-alternating tree automata*. They are based on a Büchi acceptance condition, that is on a set F such that a run is accepted iff for every path, the set of states occurring infinitely often intersects F. The main feature of weak-alternating tree automata are the following requirements:

- there exists a partition $(Q_0, \ldots, Q_n)$ of the set of states Q such that for all i, either $Q_i \subseteq F$ (Q_i is said to be *accepting*) or $Q_i \cap F = \varnothing$ (Q_i is said to be *rejecting*).
- the partition $(Q_0, \ldots, Q_n)$ is equipped with a partial ordering relation $\leq$,
- and, rephrased in our "uniform program" setting, for every transition rule in the automaton $p(f(x_1, \ldots, x_m)) \leftarrow p_1(x_1), \ldots, p_m(x_m)$ or $p(x) \leftarrow p_1(x), \ldots, p_m(x)$, for every p_i, if p belongs to Q_j and p_i belongs to Q_k, then $Q_k \leq Q_j$.

The encoding of the weak-alternating tree automata into the alternating-free fragment of the Horn μ-calculus is straightforward: it is sufficient to define a priority function Ω compatible with the ordering $\leq$ and such that $\Omega(q)$ is even if q belongs to an accepting set Q_i and $\Omega(q)$ is odd otherwise.

The main result of this paper can be stated as:

Theorem 2. *Let (P, Ω) be a uniform and alternation-free Horn μ-program. Deciding whether $[\![(P, \Omega), p]\!]$ is empty is* DEXPTIME-*complete.*

The DEXPTIME-hardness follows from [15]. To prove the completeness we develop a constructive approach: we consider in the next section a class of tree automata for finite and infinite trees, called monotonous tree automata. The main ingredient of these automata is a family of finite and complete lattices. States are defined component-wise as the product of elements of these lattices. Orderings for lattices are lifted to states and then, to runs. The acceptance condition is then expressed in terms of those orderings overs runs. Later on, we will give an algorithm that builds from a uniform and alternation-free Horn μ-program (P, Ω) a monotonous tree automaton $\mathcal{A}$. The basic idea is simply that the semantics of (P, Ω) coincide with the language accepted by $\mathcal{A}$.

As uniform and alternation-free Horn μ-programs extend weak-alternating tree automata, our method provides a new technique to check emptiness for those automata. Furthermore, emptiness problem for weak-alternating tree automata being DEXPTIME-hard, our method is theoretically optimal.

4   From Alternation-Free Horn μ-Programs to Tree Automata

4.1   Monotonous Tree Automata

We consider now a special kind of tree automata running on both finite and infinite trees.

Definition 4. *A monotonous tree automaton $\mathcal{A}$ is a triple $(\Sigma, (\mathcal{O}_i)_{i \in \{0,\ldots,k\}}, \Delta)$ where Σ is a finite signature and $(\mathcal{O}_i)_{i \in \{0,\ldots,k\}}$ is a family of complete finite lattices, i.e. each of the $\mathcal{O}_i = (S_i, \preceq^i)$ is a complete lattice over a finite set S_i. A state in $\mathcal{A}$ is a k-tuple $(s_0, \ldots, s_k)$ such each of the s_i's belongs to S_i. We denote Q the set of all states in $\mathcal{A}$. We assume each of the orderings $\preceq^i$ to be lifted in a component-wise way on states, i.e. $q \preceq^i q'$ iff for s_i, s_i' the ith components of respectively q and q', $s_i \preceq^i s_i'$ holds. Δ is a set of transition rules $f(q_1, \ldots, q_n) \to q$ with $q_1, \ldots, q_n, q$ in Q and f in Σ, and moreover,*

- *Δ is deterministic: there exists at most one transition rule $f(q_1, \ldots, q_n) \to q$ for any $f, q_1, \ldots, q_n$.*
- *Δ is complete: there exists at least one transition rule $f(q_1, \ldots, q_n) \to q$ for any $f, q_1, \ldots, q_n$.*
- *the rules from the set Δ have the* monotonicity *property: for any two transition rules $f(q_1, \ldots, q_m) \to q$ and $f(q_1', \ldots, q_m') \to q'$, for any integer i in $\{0, \ldots, k\}$, if for all $j \leq i$ and for all $l \in 1..m$, $q_l \preceq^j q_l'$, then $q \preceq^i q'$.*

The *rank* of the automaton $\mathcal{A}$ given above is denoted $rank(\mathcal{A})$ and is equal to k; we denote $\mathcal{F}_\preceq$ the family of orderings $(\preceq^i)_{i \in \{0,...,k\}}$.

We define for monotonous tree automata a notion of run that actually coincide with the classical one: a *run* $r : dom(\tau) \mapsto Q$ for a tree τ in an automaton $\mathcal{A}$ is a mapping from the tree domain of τ to the set of states of the automaton. Moreover, this mapping r satisfies for any position d in $dom(\tau)$, labeled with a function symbol f (for arity m) and having $d.1,...,d.m$ as child positions, that $f(r(d.1), \ldots, r(d.m)) \rightarrow r(d)$ is a transition rule of $\mathcal{A}$.

Note that since a monotonous tree automaton has to be (ascending) deterministic and complete, any finite tree τ admits a unique run in this automaton. Of course, this property is not true for infinite trees.

The acceptance condition is based on the family of orderings $\mathcal{F}_\preceq = (\preceq^0, \ldots, \preceq^k)$. We first extend each of those orderings $\preceq^i$ over runs as follows: for two runs r and r' (for a tree τ), $r \preceq^i r'$ iff for any position d in the tree domain of τ, $r(d) \preceq^i r'(d)$.

We say that the set of all 0-accepting runs for a tree τ is the set of all runs for τ. Recursively, for $0 \le i \le k$, we say that r is a $(i + 1)$-accepting run iff r is a minimal (resp. maximal) i-accepting run in the sense of $\preceq^i$ if i is odd (resp. even).

Definition 5 (Acceptance condition). *A run* r *for the tree* τ *in the automaton* $\mathcal{A}$ *is said to be* accepting *if* r *is a* $(k + 1)$-accepting run.

Theorem 3. *For all trees* τ, *there exists a unique accepting run for* τ *in* $\mathcal{A}$.

Proof. *We consider for any tree* τ *and for* i *in* $\{0, \ldots, k+1\}$ *the set* $\mathcal{R}_i^{\tau, \mathcal{A}}$ *of* i-accepting *runs for* τ *in* $\mathcal{A}$. *We show that* $\mathcal{R}_0^{\tau, \mathcal{A}}$ *is not empty and that for* i *in* $\{0, \ldots, k\}$, $\mathcal{R}_i^{\tau, \mathcal{A}}$ *admits some minimal and maximal elements in the sense of* $\preceq^i$. *Thus, for* i *in* $\{0, \ldots, k+1\}$, *none of the* $\mathcal{R}_i^{\tau, \mathcal{A}}$'s *is empty. Therefore, any tree* τ *has at least one accepting run in* $\mathcal{A}$. *Finally, using the definition of states and of* i-acceptance, *it is easy to see that* $\mathcal{R}_{k+1}^{\tau, \mathcal{A}}$ *can be at most a singleton. See [17] for the details.*

From now on, we denote $r_\tau^\mathcal{A}$ the unique accepting run for the tree τ in $\mathcal{A}$. A state q is *reachable* in the automaton $\mathcal{A}$ if there exists a tree τ such that for the unique accepting run $r_\tau^\mathcal{A}$, $r_\tau^\mathcal{A}(\epsilon) = q$. For a state q, we define $\mathcal{L}(\mathcal{A}, q)$ the language accepted by the automaton $\mathcal{A}$ in this state q as the set of trees $\{\tau \in T_\Sigma^* \mid r_\tau^\mathcal{A}(\epsilon) = q\}$.

Theorem 4. *The language* $\mathcal{L}(\mathcal{A}, q)$ *is a regular set of trees.*

Proof. *By reduction to* SkS (k *being the maximal arity in* Σ): *for the language* $\mathcal{L}(\mathcal{A}, q)$, *we construct a* SkS-formula $\psi_q^\mathcal{A}$ *such that the full* k-ary *tree containing a tree* τ *is a model of* $\psi_q^\mathcal{A}$ *iff* $\tau \in \mathcal{L}(\mathcal{A}, q)$. *See [17] for the detailed proof.*

Note that the constructive proof for showing the regularity of $\mathcal{L}(\mathcal{A}, q)$ can easily be adapted to show that reachability for a state can be encoded into SkS as well. Unfortunately, this would provide a quite high-complexity algorithm whereas as stated in the next theorem, reachability can be tested efficiently.

For the automaton $\mathcal{A}$, we define the size of $\mathcal{A}$ as $(|\mathcal{O}_0| \times \ldots \times |\mathcal{O}_k|)^{c_\Sigma}$, where $|\mathcal{O}_i|$ is the cardinality of the lattice $\mathcal{O}_i$ and c_Σ is a constant depending only on the signature Σ.

Theorem 5.

(i) *Reachability for a state q can be decided in polynomial time in the size of the automaton $\mathcal{A}$.*
(ii) *The emptiness of the language $\mathcal{L}(\mathcal{A}, q)$ can be decided in polynomial time in the size of $\mathcal{A}$.*

Proof. See [17] for the proof of (i). For (ii), checking emptiness for $\mathcal{L}(\mathcal{A}, q)$ simply amounts to check whether the state q is reachable. This can be achieved in polynomial time due to (i).

To conclude this section, let us say a few words about the expressiveness of monotonous tree automata. We have shown in Theorem 4 that the accepted language (for a fixed final state, and so, for a finite set of final states) is a Rabin regular tree language. On one hand, we will see in the next section how monotonous tree automata can be used to accept the same language as the ones defined by a uniform and alternation-free Horn μ-program. On the other hand, we have already said that these programs extend weak-alternating tree automata; furthermore, it is known that weak-alternating tree automata accept exactly languages that can be defined by Büchi tree automata whose complement is also a Büchi tree automaton (often denoted as Büchi $\cap$ co-Büchi). Therefore, monotonous tree automata are at least as expressive as Büchi $\cap$ co-Büchi. It is not yet known whether this inclusion is strict or not.

When one considers only signatures restricted to unary function symbols, then tree automata correspond to word automata. In this case, Theorem 4 claims that monotonous automata accept languages which are regular sets of words, *i.e.* ω-regular languages. On the other hand, it is know that weak-alternating word automata accept also exactly the ω-regular languages. Hence, when restricted to words (that is, unary function symbols), monotonous tree automata accept exactly the regular languages.

One should also notice that the class of monotonous tree automata is closed under complementation. For an automaton $\mathcal{A}$ whose states are in Q, let us consider $F \subseteq Q$ a set of final states. We define the language $\mathcal{L}(\mathcal{A}, F)$ as $\bigcup_{q \in F} \mathcal{L}(\mathcal{A}, q)$. We know already that this language is a regular set of trees. Due to the uniqueness of the accepting run for any tree τ, the complement of this language, $\overline{\mathcal{L}(\mathcal{A}, F)}$ is simply $\mathcal{L}(\mathcal{A}, Q \setminus F)$. Hence, given a monotonous tree automaton $\mathcal{A}$ together with a set of final states F, one can construct an automaton recognizing the complemented language in linear time.

4.2 Fix-Point Operators for Automata

In this section, we present a construction that given a uniform and alternation-free Horn μ-program computes a monotonous tree automaton such that the language recognized by the automaton coincides with the semantics of the program. The construction is based on an instantiation (depending on the program (P, Ω)) of the definition we gave for monotonous tree automata in the previous section: to be a little bit more explicit, states of these automata will be seen as sets of predicate symbols occurring in (P, Ω) and the family of orderings defining the acceptance condition will be expressed in terms of set inclusions taking the priority of predicates into account.

Let us consider (P, Ω), a uniform and alternation-free Horn μ-program for which the set of predicates is Pred. Pred_i is the set of all predicates in Pred with priority i; $\mathsf{Pred}_i = \{p \in \mathsf{Pred} \mid \Omega(p) = i\}$. We instantiate the notion of automata given in the previous section. This yields a finite family of automata, denoted $\mathcal{F}_{(P,\Omega)}$, satisfying:

- Σ is the set of function symbols over which P is defined,
- the family of finite and complete lattices is given by $\mathcal{O}_{i \in \{0,\dots,\max(\Omega(P))\}}$, where for each i, $\mathcal{O}_i = (\wp(\mathsf{Pred}_i), \subseteq)$. $\wp(\mathsf{Pred}_i)$ denotes the set of all subsets of Pred_i. A state in these automata is a tuple of sets of predicate symbols such that the ith component is a set that contains only predicate symbols of priority i. The ordering of the lattice $\mathcal{O}_i$ is simply the inclusion relation.

For convenience, we will regard a state simply as a unique set of predicate symbols. Since components of a tuple representing a state are pair-wise disjoint, the set of all predicate symbols occurring in a tuple corresponds to a unique state and vice-versa. This view simply imposes that for the family of orderings $(\preceq^0, \dots, \preceq^n)$, each ordering $\preceq^i$ as to be defined as: $q \preceq^i q'$ iff $q \cap \mathsf{Pred}_i \subseteq q' \cap \mathsf{Pred}_i$ for any states q, q' viewed as sets of predicates.

The construction is defined for this instantiation by means of fix-point operators tranforming an automaton from $\mathcal{F}_{(P,\Omega)}$ into another one. These operators are defined for each clause: the definition of an operator depends on the priority of the predicate occurring in the head and the nature ("automaton"/"projection") of its clause.

Let us first sketch the basic idea of our approach: the semantics for a Horn μ-program (P, Ω) associates with each of its predicates a set of trees. This can be viewed the other way round: the semantics can be expressed as a unique mapping $\xi_{(P,\Omega)}$ from the set of trees to sets of predicates as a kind of characteristic function: for instance, for some tree τ, $\xi_{(P,\Omega)}(\tau)$ could be $\{p, q\}$ which stands for "both $p(\tau)$ and $q(\tau)$ belong to $[\![(P, \Omega)]\!]$ and for no other predicate r, $r(\tau)$ belongs to $[\![(P, \Omega)]\!]$".

Note that the mapping $\xi_{(P,\Omega)}$ defines also for a fixed tree τ a unique mapping from the tree domain of τ to sets of predicate symbols simply by associating to each position d in τ the value of $\xi_{(P,\Omega)}(\tau[d])$ (recall that $\tau[d]$ denotes the subtree of τ at position d). Therefore, $\xi_{(P,\Omega)}$ for a fixed tree τ is of the same kind as a run for the tree τ in a monotonous tree automaton from $\mathcal{F}_{(P,\Omega)}$. What we show here is that for $\xi_{(P,\Omega)}$, there exists a monotonous tree automaton $\mathcal{A}_{(P,\Omega)}$ such that for any tree τ, $\xi_{(P,\Omega)}$ over the tree domain of τ is exactly the accepting run for τ in $\mathcal{A}_{(P,\Omega)}$. Furthermore, we give an algorithm that computes this automaton $\mathcal{A}_{(P,\Omega)}$.

We have described so far the shape of the tree automata we have to consider (namely the family $\mathcal{F}_{(P,\Omega)}$) and gave briefly the intuition of what the construction should yield. We are now going to explicitly formalize this construction. Roughly speaking, this construction is an alternating fix-point computation that mimics somehow the fix-points semantics for alternation-free Horn μ-programs given in Section 2.2.

To make the link clearer, we need to rephrase the definition for the operators $T^i_{(P,\Omega)}$ in terms of contributions of each clause that is taken into account by this operator. For $T^i_{(P,\Omega)}$, these clauses will be the ones for which the predicate in their head has priority i. Let us denote C^i the set of such clauses. Hence, for any set of atoms S, one have

$$T^i_{(P,\Omega)}(S) = \bigcup_{c \in C^i} T^i_{(c,\Omega)}(S)$$

where $T^i_{(c,\Omega)}$ is defined as $T^i_{(P,\Omega)}$ for a program having c as a unique clause.

Our construction will mimic this rephrased definition: we will define for each priority i, an operator $T^i_{\mathcal{F}}$ over tree automata from $\mathcal{F}_{(P,\Omega)}$ as the "union" of individual operators $T_{\mathcal{F},c}$ defined for all clauses with an head of priority i. Thus, as fix-point operators introduced for the semantics transform a set of atoms into another set of atoms, the operators $T_{\mathcal{F},c}$ and $T^i_{\mathcal{F}}$ will transform a tree automaton from $\mathcal{F}_{(P,\Omega)}$ into another tree automaton.

We start with defining the "union" operator $\sqcup$ for monotonous tree automata,

Definition 6. *For $\mathcal{A}, \mathcal{A}'$ in $\mathcal{F}_{(P,\Omega)}$ having respectively $\Delta_{\mathcal{A}}$ and $\Delta_{\mathcal{A}'}$ as sets of transition rules, the monotonous tree automaton $\mathcal{A} \sqcup \mathcal{A}'$ has, as transition rules, the set*

$$\{ lhs \to q \cup q' \mid lhs \to q \text{ in } \Delta_{\mathcal{A}}, \ lhs \to q' \text{ in } \Delta_{\mathcal{A}'} \}$$

For each clause c from P we will introduce an operator $T_{\mathcal{F},c} : \mathcal{F}_{(P,\Omega)} \mapsto \mathcal{F}_{(P,\Omega)}$ and we define $T^i_{\mathcal{F}}$, the tree automata operator as,

$$T^i_{\mathcal{F}}(\mathcal{A}) = \bigsqcup_{c \in C^i} T_{\mathcal{F},c}(\mathcal{A})$$

Now for clauses c from P, the operators $T_{\mathcal{F},c}$ are defined in a generic way by distinguishing clauses according the priority of the predicate in their head and according to their nature (automaton *vs.* projection).

Let us start with the case where c is an automaton clause.

Definition 7. *Let $\mathcal{A}$ be a tree automaton from $\mathcal{F}_{(P,\Omega)}$ with $\Delta_{\mathcal{A}}$ as set of transition rules. Let $c = p(f(x_1, \ldots, x_m)) \leftarrow p_1(x_1), \ldots, p_m(x_m)$ be an automaton clause from P. The operator $T_{\mathcal{F},c}$ is defined according to the parity of $\Omega(p)$ as*

- *if $\Omega(p)$ is odd: the tree automaton $T_{\mathcal{F},c}(\mathcal{A})$ has for transition rules the set*

$$\left\{ lhs \to q' \ \middle| \ \begin{array}{l} lhs \to q \in \Delta_{\mathcal{A}} \text{ and} \\ q' = \begin{cases} q \cup \{p\} & \text{if } lhs = f(q_1 \ldots, q_m) \text{ and for all } i : p_i \in q_i \\ q & \text{otherwise} \end{cases} \end{array} \right\}$$

- *if $\Omega(p)$ is even: the tree automaton $T_{\mathcal{F},c}(\mathcal{A})$ has for transition rules the set*

$$\left\{ lhs \to q' \ \middle| \ \begin{array}{l} lhs \to q \in \Delta_{\mathcal{A}} \text{ and} \\ q' = \begin{cases} q \smallsetminus \{p\} & \text{if } lhs \neq f(q_1 \ldots, q_m) \text{ or exists } i : p_i \in q_i \\ q & \text{otherwise} \end{cases} \end{array} \right\}$$

Defining $T_{\mathcal{F},c}$ for projection clauses is more involved and requires some auxiliary notions. We first point out an algebraic view of monotonous tree automata. The second step consists of a formalization of the different processing of quantified and free variables in the formulas from the body of clauses we have to consider.

We consider the finite algebra $A_{\mathcal{A}}$ defined by a monotonous tree automaton $\mathcal{A}$; the carrier of $A_{\mathcal{A}}$ is the set of all states of $\mathcal{A}$ and function symbols are interpreted according to transition rules: the function symbol f is interpreted in the algebra $A_{\mathcal{A}}$ by a function $f^{A_{\mathcal{A}}}$ such that for any tuple of states $q_1, \ldots, q_m$, $f^{A_{\mathcal{A}}}(q_1, \ldots, q_m) = q$ iff $f(q_1, \ldots, q_m) \to q$ is a transition rule in $\mathcal{A}$. Note that we use here the fact that $\mathcal{A}$ is ascending deterministic and complete.

This algebra $A_{\mathcal{A}}$ can be extended to a unique structure $R_{\mathcal{A}}$ simply by interpreting the monadic predicate p from the program as the set of all states that contain this predicate. Formally, the semantics of p in $R_{\mathcal{A}}$ is $\{q \in Q \mid p \in q\}$.

The main idea to address projection clauses is to consider this finite structure $R_{\mathcal{A}}$ for interpreting formulas in the body of clauses.

But this very simple and natural idea requires an additional technical point: for a clause $p(x) \leftarrow \Psi(x)$, we want to interpret the formula $\Psi(x)$ in the finite structure of states $R_{\mathcal{A}}$. However, for correctness, we have to consider the variable x ranging over arbitrary states, whereas for the quantified variables that may occur in $\Psi(x)$, we must consider them as ranging only over reachable states. To model formally this requirement, we will introduce a new formula $\tilde{\Psi}$ for each formula Ψ as follows:

- if Ψ is an atom $p(t)$ then $\tilde{\Psi} = p(t)$.
- if $\Psi = \Psi_1 \wedge \Psi_2$ (resp. $\Psi = \Psi_1 \vee \Psi_2$) then $\tilde{\Psi} = \tilde{\Psi}_1 \wedge \tilde{\Psi}_2$ (resp. $\tilde{\Psi} = \tilde{\Psi}_1 \vee \tilde{\Psi}_2$).
- if $\Psi = \exists y\, \Psi'$ then $\tilde{\Psi} = \exists y\, p_{\top}(y) \wedge \tilde{\Psi}'$,
- if $\Psi = \forall y\, \Psi'$ then $\tilde{\Psi} = \forall y\, p_{\top}(y) \Rightarrow \tilde{\Psi}'$.

The predicate symbol $p_{\top}$ is a new predicate symbol. Intuitively, from the Herbrand semantics point of view, its semantics is the set of all trees. One can imagine this predicate defined as $\forall x\, p_{\top}(x)$ when Herbrand structures are considered. For any quantified variable y in the formula Ψ, the corresponding variable y in the formula $\tilde{\Psi}$ will be *guarded* by an atom $p_{\top}(y)$.

As we said above, due to the particular interpretation of $p_{\top}$ in Herbrand structure,

Remark 1. For any Herbrand structure R_{HB^*} defining semantics for the predicate symbols and such that the semantics of $p_{\top}$ is the set of all trees, $R_{HB^*}, [x/\tau] \models \Psi(x)$ holds iff $R_{HB^*}, [x/\tau] \models \tilde{\Psi}(x)$ holds.

However, things are going to be different in the automaton structure $R_{\mathcal{A}}$. In this structure $R_{\mathcal{A}}$, we fix the semantics of the predicate $p_{\top}$ to the set of all reachable states in $\mathcal{A}$. Hence, this will ensure in a formal way that for the formula Ψ, one considers quantified variables instantiated only with reachable states, whereas the free variable of this formula may take arbitrary states for values.

Definition 8. *Let $\mathcal{A}$ be a tree automaton from $\mathcal{F}_{(P,\Omega)}$ with $\Delta_{\mathcal{A}}$ as set of transition rules. Let $c = p(x) \leftarrow \Psi(x)$ be a projection clause from P. The operator $T_{\mathcal{F},c}$ is defined according to the parity of $\Omega(p)$ as*

– *if $\Omega(p)$ is odd: the tree automaton $T_{\mathcal{F},c}(\mathcal{A})$ has for transition rules the set*

$$\left\{ lhs \to q' \,\middle|\, \begin{array}{l} lhs \to q \in \Delta_\mathcal{A} \text{ and} \\ q' = \begin{cases} q \cup \{p\} & \text{if } \mathsf{R}_\mathcal{A}, [x/q] \models \tilde{\Psi}(x) \\ q & \text{otherwise} \end{cases} \end{array} \right\}$$

– *if $\Omega(p)$ is even: the tree automaton $T_{\mathcal{F},c}(\mathcal{A})$ has for transition rules the set*

$$\left\{ lhs \to q' \,\middle|\, \begin{array}{l} lhs \to q \in \Delta_\mathcal{A} \text{ and} \\ q' = \begin{cases} q \smallsetminus \{p\} & \text{if } \mathsf{R}_\mathcal{A}, [x/q] \not\models \tilde{\Psi}(x) \\ q & \text{otherwise} \end{cases} \end{array} \right\}$$

Before carrying on the presentation of our algorithm, it may be worth to clarify a point about the operators $T_{\mathcal{F}}^i$ and $T_{\mathcal{F},c}$. We have claim earlier that those operators are defined over $\mathcal{F}_{(P,\Omega)}$. When it is quite clear that, due to their respective definition, $T_{\mathcal{F}}^i$ and $T_{\mathcal{F},c}$ associates with an automaton in $\mathcal{F}_{(P,\Omega)}$ a tree automaton defined over the same signature and the same set of state which is both (ascending) deterministic and complete, the monotonicity property is less straightforward for the resulting automaton. However, this is obvious that if the monotonicity property holds for $\mathcal{A}$ and $\mathcal{A}'$, then it holds for $\mathcal{A} \sqcup \mathcal{A}'$. This is also true for any of the $T_{\mathcal{F},c}$ operators as claimed in the next proposition

Proposition 1. *Let $\mathcal{A}$ be an automaton from $\mathcal{F}_{(P,\Omega)}$, then both $T_{\mathcal{F},c}(\mathcal{A})$ and $T_{\mathcal{F}}^i(\mathcal{A})$ satisfy the monotonicity property.*

Proof. The detailed proof for $T_{\mathcal{F},c}$ can be found in [17]. Taking into account the definition for $T_{\mathcal{F}}^i$ in terms of $\sqcup$ yields the proof.

We have yet introduced all the material needed to describe the algorithm. As we have said earlier, this latter simply mimics the computation of the fix-point semantics for (P, Ω) in terms of tree automata.

Let us consider an automaton $\mathcal{A}^\perp$ that belongs $\mathcal{F}_{(P,\Omega)}$ and that we call *initial*. This initial automaton is the one from $\mathcal{F}_{(P,\Omega)}$ having the right-hand side of its transition rules equals to a particular state denoted $q_{\mathcal{F}_\preceq}$. This state $q_{\mathcal{F}_\preceq}$ is the state satisfying: for all predicate symbols p, $p \in q_{\mathcal{F}_\preceq}$ iff $\Omega(p)$ is even.
The family of automata $(\mathcal{A}_{(P,\Omega)}^i)_{i \in \Omega(P)}$ is given by:

$$\mathcal{A}_{(P,\Omega)}^0 = (T_{\mathcal{F}}^0)^*(\mathcal{A}^\perp) \qquad \text{and} \qquad \mathcal{A}_{(P,\Omega)}^i = (T_{\mathcal{F}}^i)^*(\mathcal{A}_{(P,\Omega)}^{i-1}) \quad \text{for } 0 < i$$

where $(T_{\mathcal{F}}^i)^*(\mathcal{A})$ is the unique automaton $\mathcal{A}'$ for which there exists a least integer m such that $\mathcal{A}' = \underbrace{T_{\mathcal{F}}^i \circ \ldots \circ T_{\mathcal{F}}^i}_{m \text{ times}}(\mathcal{A})$ and $T_{\mathcal{F}}^i(\mathcal{A}') = \mathcal{A}'$.

The fact that such an integer m exists follows directly from the definitions of $T_{\mathcal{F},c}$ and $\sqcup$ and from the definition for the initial automaton $\mathcal{A}^\perp$.

The output of the algorithm is the tree automaton $\mathcal{A}_{(P,\Omega)}^n$ where $n = \max(\Omega(P))$. It should be noticed that the computation of $\mathcal{A}_{(P,\Omega)}^n$ <u>is not</u> the computation of a model of (P, Ω) on a finite algebra (or, finite pre-interpretation): performing the computation of a structure (that is, of an interpretation of predicate symbols) over a finite algebra would imply that the considered algebra is fixed all along the computation whereas the structure evolves. Contrary to this latter, in our approach, the finite algebra we consider changes during the computation according to the automata whereas the structure is fixed for a given algebra.

The correctness of our approach, that is the equivalence between the uniform and alternation-free Horn μ-program (P, Ω) and the tree automaton $\mathcal{A}^n_{(P,\Omega)}$ can be phrased as

Theorem 6. *A ground atom $p(\tau)$ belongs to $[\![(P, \Omega)]\!]$ iff $p \in r_\tau^{\mathcal{A}^n_{(P,\Omega)}}(\epsilon)$ for $n = \max(\Omega(P))$.*

Proof. See [17] for the proof.

Example 2. We illustrate our algorithm with the uniform and alternation-free Horn μ-program (P, Ω) given in Example 1. For conciseness, we identify a tree automaton with its set of transition rules and use a meta-representation for states occurring in the left-hand sides of transition rules. A state is represented as a pair $[X, Y]$ such that $X, Y \subseteq \mathsf{Pred}$ and $X \cap Y = \varnothing$. The pair $[X, Y]$ represents all states that contain X and that do not contain Y. For instance, $[\{p_0\}, \{p_1\}]$ represents the states $\{p_0\}$ and $\{p_0, p_2\}$. As an extension, the (meta-)transition rule $g([\{p_0\}, \{p_1\}]) \rightarrow q$ represents the two transition rules $g(\{p_0\}) \rightarrow q$ and $g(\{p_0, p_2\}) \rightarrow q$. Finally, we simply refer to a predicate symbol with its index, *i.e.* 2 stands for p_2.

We compute the family of automata $(\mathcal{A}^i_{(P,\Omega)})_{i \in \{0,1,2\}}$. The automaton $\mathcal{A}^2_{(P,\Omega)}$ is equivalent to the denotation of (P, Ω) as stated in Theorem 6.

The initial automaton $\mathcal{A}^\perp$ is given by

$$\left\{ \begin{array}{ccc} a \rightarrow \{0,2\} & b \rightarrow \{0,2\} & c \rightarrow \{0,2\} \\ g([\varnothing, \varnothing]) \rightarrow \{0,2\} & f([\varnothing, \varnothing], [\varnothing, \varnothing]) \rightarrow \{0,2\} \end{array} \right\}$$

For $c_1 = p_0(a)$ and $c_2 = p_0(b)$, $T_{\mathcal{F},c_1}(\mathcal{A}^\perp)$ and $T_{\mathcal{F},c_2}(\mathcal{A}^\perp)$ are equal respectively

to
$$\left\{ \begin{array}{c} a \rightarrow \{0,2\} \\ b \rightarrow \{2\} \\ c \rightarrow \{2\} \\ g([\varnothing, \varnothing]) \rightarrow \{2\} \\ f([\varnothing, \varnothing], [\varnothing, \varnothing]) \rightarrow \{2\} \end{array} \right\}$$
and to
$$\left\{ \begin{array}{c} a \rightarrow \{2\} \\ b \rightarrow \{0,2\} \\ c \rightarrow \{2\} \\ g([\varnothing, \varnothing]) \rightarrow \{2\} \\ f([\varnothing, \varnothing], [\varnothing, \varnothing]) \rightarrow \{2\} \end{array} \right\}.$$

And for $c_3 = p_0(f(x,y)) \leftarrow p_0(x), p_0(y)$,

$$T_{\mathcal{F},c_3}(\mathcal{A}^\perp) = \left\{ \begin{array}{cccc} a \rightarrow \{2\} & b \rightarrow \{2\} & c \rightarrow \{2\} & g([\varnothing, \varnothing]) \rightarrow \{2\} \\ f([\{0\}, \varnothing], [\{0\}, \varnothing]) \rightarrow \{0,2\} & & f([\varnothing, \{0\}], [\{0\}, \varnothing]) \rightarrow \{2\} \\ f([\{0\}, \varnothing], [\varnothing, \{0\}]) \rightarrow \{2\} & & f([\varnothing, \{0\}], [\varnothing, \{0\}]) \rightarrow \{2\} \end{array} \right\}$$

So, for $T_{\mathcal{F}}^0(\mathcal{A}^\perp) = T_{\mathcal{F},c_1}(\mathcal{A}^\perp) \sqcup T_{\mathcal{F},c_2}(\mathcal{A}^\perp) \sqcup T_{\mathcal{F},c_3}(\mathcal{A}^\perp)$, we have

$$\left\{ \begin{array}{cccc} a \rightarrow \{0,2\} & b \rightarrow \{0,2\} & c \rightarrow \{2\} & g([\varnothing, \varnothing]) \rightarrow \{2\} \\ f([\{0\}, \varnothing], [\{0\}, \varnothing]) \rightarrow \{0,2\} & & f([\varnothing, \{0\}], [\{0\}, \varnothing]) \rightarrow \{2\} \\ f([\{0\}, \varnothing], [\varnothing, \{0\}]) \rightarrow \{2\} & & f([\varnothing, \{0\}], [\varnothing, \{0\}]) \rightarrow \{2\} \end{array} \right\}$$

It is easy to see that $T_{\mathcal{F}}^0(T_{\mathcal{F}}^0(A^\perp)) = T_{\mathcal{F}}^0(A^\perp)$. Therefore, $A_{(P,\Omega)}^0 = T_{\mathcal{F}}^0(A^\perp)$. The next computed automaton is then $T_{\mathcal{F}}^1(A_{(P,\Omega)}^0)$, which is equal to

$$
\left\{
\begin{array}{ll}
a \to \{0,1,2\} \quad b \to \{0,2\} \quad c \to \{1,2\} \quad g([\varnothing,\varnothing]) \to \{2\} \\
f([\{0,1\},\varnothing],[\{0\},\varnothing]) \to \{0,1,2\} \quad f([\{0\},\{1\}],[\{0\},\varnothing]) \to \{0,2\} \\
f([\{1\},\{0\}],[\{0\},\varnothing]) \to \{1,2\} \quad f([\varnothing,\{0,1\}],[\{0\},\varnothing]) \to \{2\} \\
f([\{0\},\varnothing],[\varnothing,\{0\}]) \to \{2\} \quad f([\varnothing,\{0\}],[\varnothing,\{0\}]) \to \{2\}
\end{array}
\right\}
$$

Once again it is easy to see that $T_{\mathcal{F}}^1(T_{\mathcal{F}}^1(A_{(P,\Omega)}^0)) = T_{\mathcal{F}}^1(A_{(P,\Omega)}^0)$. So, $A_{(P,\Omega)}^1 = T_{\mathcal{F}}^1(A_{(P,\Omega)}^0)$. In this automaton, all the states in the right-hand side of the transition rules are reachable (and of course, only those ones): the states $\{0,1,2\}$, $\{0,2\}$, $\{1,2\}$ and $\{2\}$ are reachable since they label the root of the unique (and thus, accepting) run for respectively a, b, c and $g(a,a)$. Now, for computing the automaton $T_{\mathcal{F}}^2(A_{(P,\Omega)}^1)$, one has to check for each state q in the right-hand side of transition rules whether, in the structure induced by $A_{(P,\Omega)}^1$, the valuation $[x/q]$ renders the formula $\exists y\, (y \in \top \wedge p_1(f(x,y))) \wedge p_0(x)$ true or not. This formula is falsified by any state q which doesn't contain $\{0\}$. Hence, by definition of $T_{\mathcal{F}}^2$, $T_{\mathcal{F}}^2(A_{(P,\Omega)}^1)$ will contain the following (meta-)transition rules

$$
\left\{
\begin{array}{ll}
c \to \{1\} \quad g([\varnothing,\varnothing]) \to \varnothing \\
f([\{1\},\{0\}],[\{0\},\varnothing]) \to \{1\} \quad f([\varnothing,\{0,1\}],[\{0\},\varnothing]) \to \varnothing \\
f([\{0\},\varnothing],[\varnothing,\{0\}]) \to \varnothing \quad f([\varnothing,\{0\}],[\varnothing,\{0\}]) \to \varnothing
\end{array}
\right\}
$$

Now for the other rules, let us start with $q = \{0,1,2\}$ (that is the rules for a and for $f([\{0,1\},\varnothing],[\{0\},\varnothing])$). It is possible to find a reachable state q' such that $[x/q,y/q'] \models p_1(f(x,y))$. For instance due to the rule $f(\{0,1,2\},\{0,1,2\}) \to \{0,1,2\}$ in $A_{(P,\Omega)}^1$ whose right-hand side contains 1, the state $\{0,1,2\}$ is a proper choice for q'. So, $T_{\mathcal{F}}^2(A_{(P,\Omega)}^1)$ contains the unchanged rules

$$
a \to \{0,1,2\} \quad f([\{0,1\},\varnothing],[\{0\},\varnothing]) \to \{0,1,2\}
$$

On the other hand, for the remaining rules with $\{0,2\}$ as right-hand side, one can check that for $q = \{0,2\}$, it is not possible to find a reachable state q' such that $[x/q,y/q'] \models p_1(f(x,y))$. Therefore, $T_{\mathcal{F}}^2(A_{(P,\Omega)}^1)$ contains the rules

$$
b \to \{0\} \quad f([\{0\},\{1\}],[\{0\},\varnothing]) \to \{0\}
$$

It is easy to check that $T_{\mathcal{F}}^2(T_{\mathcal{F}}^2(A_{(P,\Omega)}^1)) = T_{\mathcal{F}}^2(A_{(P,\Omega)}^1)$. So, $A_{(P,\Omega)}^2 = T_{\mathcal{F}}^2(A_{(P,\Omega)}^1)$. Let us present the relevant part of the automaton $A_{(P,\Omega)}^2$, that is restricted over states occurring in the right-hand side of transition rules.

$$
\begin{array}{llll}
 & a \to \{0,1,2\} & b \to \{0\} & c \to \{1\} \\
g(\varnothing) \to \varnothing & g(\{0\}) \to \varnothing & g(\{1\}) \to \varnothing & g(\{0,1,2\}) \to \varnothing \\
f(\varnothing,\varnothing) \to \varnothing & f(\varnothing,\{0\}) \to \varnothing & f(\varnothing,\{1\}) \to \varnothing & f(\varnothing,\{0,1,2\}) \to \varnothing \\
f(\{0\},\varnothing) \to \varnothing & f(\{0\},\{0\}) \to \{0\} & f(\{0\},\{1\}) \to \varnothing & f(\{0\},\{0,1,2\}) \to \{0\} \\
f(\{1\},\varnothing) \to \varnothing & f(\{1\},\{0\}) \to \{1\} & f(\{1\},\{1\}) \to \varnothing & f(\{1\},\{0,1,2\}) \to \{1\} \\
 & f(\{0,1,2\},\varnothing) \to \varnothing & f(\{0,1,2\},\{0\}) \to \{0,1,2\} & \\
f(\{0,1,2\},\{1\}) \to \varnothing & & f(\{0,1,2\},\{0,1,2\}) \to \{0,1,2\} &
\end{array}
$$

One can notice for this automaton that if a tree τ contains one occurrence of the function symbol g, then any run r for τ will satisfy $r(\epsilon) = \varnothing$. So, τ does not belong to the denotation of any of the predicates p_0, p_1 and p_2.

Now let us consider f^ω the infinite binary tree with all its nodes labeled by f. This tree f^ω admits three runs in $\mathcal{A}^2_{(P,\Omega)}$. The first run r_1 associates with each node the state $\varnothing$, the second run r_2 the state $\{0\}$ and the last run r_3 the state $\{0,1,2\}$. r_1 can not be the accepted run since r_2 and r_3 are greater than r_1 in the sense of $\preceq^0$. Finally, r_2 is the accepted run since it is smaller in the sense of $\preceq^1$ than r_3. So, the tree f^ω belongs to the denotation of the predicate p_0 but not to the denotation of p_1 and p_2.

Finally, let us consider the two trees τ_1 and τ_2: τ_1 is a finite left backbone of f terminated with the constant c and each f from the backbone has f^ω as right son. The tree τ_2 is similar to τ_1 except that the backbone ends up with the constant a. Thus, the accepted run r_{τ_1} for τ_1 associates with each node outside of the backbone the state $\{0\}$ and because of the two rules $c \rightarrow \{1\}$ and $f(\{1\}, \{0\}) \rightarrow \{1\}$, with each node from the backbone the state $\{1\}$. Therefore, τ_1 belongs only to the denotation of p_1. The accepted run r_{τ_2} for τ_2 is similar to r_{τ_1} outside of the backbone and, due to the rules $a \rightarrow \{0,1,2\}$ and $f(\{0,1,2\}, \{0\}) \rightarrow \{0,1,2\}$, associates the state $\{0,1,2\}$ with the root of τ_2. Therefore, the tree τ_2 belongs to the denotation of p_0, p_1 and p_2.

The complexity for the construction of $\mathcal{A}^n_{(P,\Omega)}$ can be roughly estimated: the size for each automaton in the family $\mathcal{F}_{(P,\Omega)}$ is single-exponential in the size of P. Moreover, the basic operations $T_{\mathcal{F},c}$ and $\sqcup$ can be performed in polynomial time in the size of the automaton. Thus, computing $T_{\mathcal{F}}^i$ can be achieved by a polynomial-time algorithm in the size of the automaton. For each step from $\mathcal{A}^i_{(P,\Omega)}$ to $\mathcal{A}^{i+1}_{(P,\Omega)}$, the operator $T_{\mathcal{F}}^i$ has to be iterated. However, taking into account the definition for the basic operations and the initial automaton, we can claim that the number of iterations is bounded by a single-exponential in the size of P. So, globally, the automaton $\mathcal{A}^n_{(P,\Omega)}$ can be computed with an algorithm running in single-exponential time in the size of P.

By Theorem 6, testing whether $[\![(P, \Omega), p]\!]$ is empty simply amounts to search for reachability in the automaton $\mathcal{A}^n_{(P,\Omega)}$ for a state containing p. Then, by combining the complexity for the construction of $\mathcal{A}^n_{(P,\Omega)}$ and Theorem 5, the result claimed earlier in Theorem 2 follows.

Acknowledgments The author thanks Viorica Sofronie-Stokkermans and Witold Charatonik for discussions and reading. The author is grateful to Sophie Tison for her careful reading of a preliminary version of this paper. Finally, the author is deeply grateful to Damian Niwiński who gave him a crucial idea for the emptiness test of monotonous tree automata.

References

1. W. Charatonik, D. McAllester, D. Niwiński, A. Podelski, and I. Walukiewicz. The Horn Mu-Calculus. In *Proceedings of the 13th IEEE Symposium on Logic in Computer Science*, pages 58–69, 1998. 418, 418, 424

2. W. Charatonik, D. McAllester, and A. Podelski. The Greatest Fixed Point of the τ_P abstraction is Regular. Seminar on Applications of Tree Automata in Rewriting, Logic and Programming, oct 1997. 419

3. W. Charatonik, D. Niwiński, and A. Podelski. Model checking for uniforms programs. Draft, 1998. 418, 419, 420, 420, 424

4. W. Charatonik and A. Podelski. Co-definite Set Constraints. In *Proceedings of the 9th International Conference on Rewriting Techniques and Applications*, LNCS, 1998. 419

5. P. Devienne, J-M. Talbot, and S. Tison. Solving Classes of Set Constraints with Tree Automata. In G. Smolka, editor, *Proceedings of the 3rd International Conference on Principles and Practice of Constraint Programming*, LNCS 1330, pages 62–76, oct 1997. 419

6. P. Devienne, J-M. Talbot, and S. Tison. Co-definite Set Constraints with Membership Expressions. In Joxan Jaffar, editor, *Proceedings of the 1998 Joint International Conference and Symposium on Logic Programming*, pages 25–39. MIT-Press, jun 1998. 419, 419

7. E. A. Emerson and C. S. Jutla. Tree Automata, Mu-Calculus and Determinacy. In IEEE, editor, *Proceedings of the 32nd Annual Symposium on Foundations of Computer Science*, pages 368–377. IEEE Computer Society Press, October 1991. 419

8. T. Frühwirth, E. Shapiro, M.Y. Vardi, and E. Yardeni. Logic Programs as Types for Logic Programs. In *Proceedings of the 6th IEEE Symposium on Logic in Computer Science*, pages 300–309, jun 1991. 419, 419

9. N. Heintze and J. Jaffar. A Decision Procedure for a Class of Herbrand Set Constraints. In *Proceedings of the 5th IEEE Symposium on Logic in Computer Science*, pages 42–51, jun 1990. 419

10. N. Heintze and J. Jaffar. A Finite Presentation Theorem for Approximating Logic Programs. In *Proceedings of the 17th ACM SIGPLAN-SIGACT Symposium on Principles of Programming Languages*, pages 197–209, jan 1990. 419, 419

11. J. Lloyd. *Foundations of Logic Programming*. Springer-Verlag, 1987. 418

12. D. E. Muller, A. Saoudi, and P. E. Schupp. Weak Alternating Automata Give a Simple Explanation of Why Most Temporal and Dynamic Logics are Decidable in Exponential Time. In *Proceedings of the Third IEEE Symposium on Logic in Computer Science*, pages 422–427. IEEE Computer Society, jul 1988. 424

13. D. E. Muller and P. E. Schupp. Alternating Automata on Infinite Trees. *Theoretical Computer Science*, 54(2–3):267–276, 1987. 424

14. M.O. Rabin. Decidability of Second-order Theories and Automata on Infinite Trees. In *Transactions of American Mathematical Society*, volume 141, pages 1–35, 1969. 419

15. H. Seidl. Haskell Overloading is DEXPTIME-complete. *Information Processing Letter*, 52:57–60, 1994. 425

16. J-M. Talbot. *Contraintes Ensemblistes Définies et Co-définies : Extensions et Applications*. PhD thesis, Université des Sciences et Technologies de Lille, jul 1998. 419

17. J-M Talbot. On the alternation-free horn mu-calculus. Technical report, Max-Planck-Institut für Informatik, 2000. Long version of LPAR'00. 421, 426, 426, 427, 431, 432

18. J-M. Talbot, P. Devienne, and S. Tison. Generalized Definite Set Constraints. *CONSTRAINTS - An International Journal*, 5(1):161–202, 2000. 419

The Boundary between Decidable and Undecidable Fragments of the Fluent Calculus

Steffen Hölldobler [1] and Dietrich Kuske [2] *

[1] Institut für Künstliche Intelligenz, TU Dresden, D-01062 Dresden, Germany
[2] Institut für Algebra, TU Dresden, D-01062 Dresden, Germany

Abstract. We consider entailment problems in the fluent calculus as they arise in reasoning about actions. Taking into account various fragments of the fluent calculus we formally show decidability results, establish their complexity, and prove undecidability results. Thus we draw a boundary between decidable and undecidable fragments of the fluent calculus.

1 Introduction

Intelligent agents need to reason about the state of the world, the actions that they can perform and the effects that are achieved by executing actions. To elaborate on the question whether there exists a sequence of actions such that a given goal can be achieved by executing this sequence in the current state is one of the most important tasks an intelligent agent has to perform. From a logical point of view this amounts in solving an entailment problem as already laid down in [18]. Likewise, many other problems in reasoning about actions can be formalized as entailment problems in a suitable logic.

There is a variety of proposals for reasoning about actions, notably the situation [19,16], the fluent [11,30] and the event calculus [13,26], approaches based on the linear connection method [1,2], linear logic [8,17], transaction logic [3], temporal action logics [4], action languages [7], the features and fluent approach [25], etc. We opt for the fluent calculus because it is a logic with standard semantics and a sound and complete calculus, and many of the problems in reasoning about actions like the frame and the ramification problem can be dealt with in a representationally as well as computationally adequate way. There are also many extensions of the fluent calculus for reasoning about non–deterministic actions [31], specificity [10], sensing actions [32], etc., where the fluent calculus is at least equivalent and sometimes superior to alternative approaches.

Central to the fluent calculus is the idea to model world states by multisets of fluents and to represent these multisets as so–called state terms using an AC1–theory. The use of multisets instead of sets, or, equivalently, the use of

---

* new address: University of Leicester, Department of Mathematics and Computer Science, University Road, Leicester LE1 7RH, UK

M. Parigot and A. Voronkov (Eds.): LPAR 2000, LNAI 1955, pp. 436–450, 2000.

an AC1–theory instead of an ACI1–theory allows for an elegant solution of the frame problem [9]. Whereas in a multiset of fluents the elements are viewed as resources and actions produce and consume resources, one sometimes likes to view resources as properties, which do or do not hold in a state. This view can be supported in the fluent calculus as well by additionally requiring that each fluent occurs at most once in a multiset while retaining the abovementioned solution of the frame problem (see e.g. [28]).

In this paper we are concerned with the question of how the boundary between decidable and undecidable fragments of the fluent calculus looks like. To answer this question we consider various fragments of the fluent calculus and prove that the entailment problems in these fragments are either decidable or undecidable. In the former case we are additionally interested in the complexity of the decision procedure. In particular, we establish the following results:

1a. We consider a monadic second order fragment of the fluent calculus with restricted state update axioms and finitely many fluent constants. The corresponding entailment problem is shown to be decidable using the decidability of Presburger arithmetic and of the monadic second order theory of labeled trees. If we additionally assume that fluents may occur at most once in a state term then this result corresponds to a similar result achieved for the situation calculus in [29].

1b. We show that the entailment problem from 1a. cannot be solved in elementary time even if fluents may occur at most once in a state term. This solves an open problem posed in [29].

1c. We show that the entailment procedure used in 1a. can be modified to become elementary, if we consider a first order version of the fluent calculus with restricted state update axioms and finitely many fluent constants.

2a. We consider a first order fragment of the fluent calculus with unrestricted state update axioms and finitely many fluent constants. The corresponding entailment problem is shown to be undecidable by reducing it to the acceptance problem of two–counter machines.

2b. We show that the entailment problem is decidable if we additionally assume that fluents may occur at most once in a state term.

2c. We show that the entailment problem is again undecidable if we additionally assume to have two unary function symbols mapping fluents onto fluents.

The paper is organized as follows. In Section 2 we define the basic structure underlying the fluent calculus. The language of the fluent calculus as well as the various fragments and entailment problems are specified in Section 3. The monadic fragment together with the decidability result and the complexity considerations are presented in Section 4. The undecidability results are proved in Section 5. Finally, a brief discussion of our results in Section 6 concludes this paper.

2   The Structures

We consider a two–sorted version of the fluent calculus by understanding the fluents as state constants and the actions as functions. So let ACT and FL be two finite sets of *actions* and *fluents*, resp. An (ACT, FL)–structure is a two–sorted structure

$$\mathcal{M} = (\text{SIT}^{\mathcal{M}}, \text{ST}^{\mathcal{M}}; (\text{do}_a^{\mathcal{M}})_{a \in \text{ACT}}, \text{state}^{\mathcal{M}}, \circ^{\mathcal{M}}, 0^{\mathcal{M}}, s_I^{\mathcal{M}}, (F^{\mathcal{M}})_{F \in \text{FL}})$$

where $\text{do}_a^{\mathcal{M}} : \text{SIT}^{\mathcal{M}} \to \text{SIT}^{\mathcal{M}}$, $\circ^{\mathcal{M}} : \text{ST}^{\mathcal{M}} \times \text{ST}^{\mathcal{M}} \to \text{ST}^{\mathcal{M}}$ and $\text{state}^{\mathcal{M}} : \text{SIT}^{\mathcal{M}} \to \text{ST}^{\mathcal{M}}$ are functions, $0^{\mathcal{M}}, F^{\mathcal{M}} \in \text{ST}^{\mathcal{M}}$ and $s_I^{\mathcal{M}} \in \text{SIT}^{\mathcal{M}}$ are constants such that

1. situations have unique names, i.e. there are no action $a \in \text{ACT}$ and situation $s \in \text{SIT}^{\mathcal{M}}$ such that $\text{do}_a^{\mathcal{M}}(s) = s_I^{\mathcal{M}}$, and $\text{do}_{a_1}^{\mathcal{M}}(s_1) = \text{do}_{a_2}^{\mathcal{M}}(s_2)$ implies $a_1 = a_2$ and $s_1 = s_2$ for any actions $a_1, a_2 \in \text{ACT}$ and any situations $s_1, s_2 \in \text{SIT}^{\mathcal{M}}$,
2. any situation is reachable from the initial situation $s_I^{\mathcal{M}}$, i.e., for all $s \in \text{SIT}^{\mathcal{M}}$ there exist finite sequences $s_i \in \text{SIT}^{\mathcal{M}}$ and $a_i \in \text{ACT}$ such that $s_0 = s_I$, $s_{i+1} = \text{do}_{a_i}^{\mathcal{M}}(s_i)$ and $s_n = s$ for some $n \in \mathbb{N}$,
3. the structure $(\text{ST}^{\mathcal{M}}; \circ^{\mathcal{M}}, 0^{\mathcal{M}})$ is isomorphic to the algebra of finite multisets over $\{F^{\mathcal{M}} \mid F \in \text{FL}\}$ and $F^{\mathcal{M}} \neq G^{\mathcal{M}}$ for $F, G \in \text{FL}$ with $F \neq G$ together with multiset union as operation.

We consider total functions $\text{do}_a^{\mathcal{M}}$. In other words, in this model any action can be executed in any situation. This might contradict the intuition that some conditions have to be satisfied to be able to execute a particular action. Indeed, one could use partial functions instead which would considerably complicate the presentation. So we only remark that all we are doing can be done with partial actions, too. Finally, let $\text{MS} = (\text{ST}^{\mathcal{M}}; \circ^{\mathcal{M}}, 0^{\mathcal{M}}, (F^{\mathcal{M}})_{F \in \text{FL}})$.

Note that by the first two requirements on the situations and the functions $\text{do}_a^{\mathcal{M}}$, they form an infinite tree. More precisely, write $s \sqsubseteq s'$ iff the situation s' can be reached from the situation s by finitely many applications of functions $\text{do}_a^{\mathcal{M}}$. Then the first two restrictions ensure that $\sqsubseteq$ is a partial order and that $(\text{SIT}^{\mathcal{M}}; \sqsubseteq)$ is a tree. Because any situation is reachable from the initial situation in a unique way, we can identify the situation $\text{do}_{a_n}^{\mathcal{M}}(\text{do}_{a_{n-1}}^{\mathcal{M}}(\text{do}_{a_{n-2}}^{\mathcal{M}}(\cdots(s_I^{\mathcal{M}})\cdots)))$ with the word $a_1 a_2 \ldots a_n$ over the alphabet ACT. In other words, we can always assume that the underlying set of situations is the set $\text{ACT}^\star$ of finite words over the alphabet ACT, that the functions $\text{do}_a^{\mathcal{M}}$ map $w \in \text{SIT}^{\mathcal{M}}$ to $wa \in \text{SIT}^{\mathcal{M}}$, and that the initial situation $s_I^{\mathcal{M}}$ is the empty word ε.

Furthermore, the third requirement (explicitly) ensures that the algebra of states $\text{ST}^{\mathcal{M}}$ together with the function $\circ^{\mathcal{M}}$ and the constant $0^{\mathcal{M}}$ is isomorphic to the set of finite multisets over the set FL together with multiset union as operation. Therefore, without loss of generality, we can always assume that these two sets are equal.

An (ACT, FL)–structure will be called *canonical tree structure* if the set of situations equals $\text{ACT}^\star$, the functions $\text{do}_a^{\mathcal{M}}$ are the extension of the argument by the letter a, the initial situation is the empty word, and the set of states is the set of finite multisets over FL (where $\circ^{\mathcal{M}}$ is the multiset union and $0^{\mathcal{M}}$ denotes the empty multiset).

3   The Language

We will use a language that allows to make first order statements on states and monadic second order statements on situations. More formally, we have variables $x_1, x_2, \ldots$ ranging over states as well as finitely many state constants $0, F$ for $F \in$ FL that are of sort ST. There are elementary variables $s_1, s_2, \ldots$ ranging over situations and a situation constant s_I that are of sort SIT. In addition, we allow set variables $S_1, S_2, \ldots$ ranging over sets of situations.

Terms of sort ST and SIT are defined in the usual way where terms of sort ST use only the constants $0, F \in$ FL, the function symbol $\circ$ and state variables x_i. $\circ$ is assumed to be an AC1–symbol written infix with 0 being its unit element. *State formulas* are built up from equations of the form $t_1 = t_2$ by the logical connectives $\wedge, \vee, \neg$ etc. and quantification over states, where t_i are terms of sort ST. E.g., $(\exists x_1)\, [x_1 \circ F_1 = x_2 \wedge \neg(x_1 = F_2)]$ is a state formula with one free state variable x_2.

In many applications, in particular in those where we require that each fluent $F \in$ FL occurs at most once in a state s, it is convenient to introduce a macro $holds(F, s) := (\exists x)\, \text{state}(s) = F \circ x$.

Next, we describe what we mean by an instance of the fluent calculus. It consists of the finite sets ACT of actions, FL of fluents and $\mathcal{F} = \mathcal{F}_I \cup \mathcal{F}_{su}$ of axioms of the following form:

The set $\mathcal{F}_I$: The set $\mathcal{F}_I = \{\varphi(\text{state}(s_I))\}$ describes the initial state and contains one axiom where φ is a state formula with one free variable. E.g., if in a simple scenario we have some partial knowledge about the initial state, viz. that an agent is known to have three quarters (q), a cookie (c) but no tea t then

$$\mathcal{F}_I = \{(\exists x)\, [\text{state}(s_I) = q \circ q \circ q \circ c \circ x \wedge \neg(\exists x')\, x = t \circ x']\}. \tag{1}$$

The set $\mathcal{F}_{su}$ *of state update axioms:* For any action $a \in$ ACT, we have several state update axioms of the form $\Delta_a = (\forall s)\, \delta_a(\text{state}(s), \text{state}(\text{do}_a(s))))$, where δ_a is a state formula with two free variables. A *restricted state update axiom* is a state update axiom Δ_a as above where δ_a is only a Boolean combination of formulas $\varphi(\text{state}(s))$ and $\varphi(\text{state}(\text{do}_a(s)))$, and where φ is a state formula with one free variable. E.g., with the help of the axiom

$$(\exists x)\, \text{state}(s) = q \circ q \circ q \circ x \rightarrow (\exists x')\, state(\text{do}_{get\_tea}(s)) = t \circ x'$$

we can describe the preconditions as well as the positive effect of an action *get_tea*, viz. that the agent needs to have three quarters in some situation s in order to get a cup of tea.

If in the instance (ACT, FL, $\mathcal{F}$) of the fluent calculus each state update axiom is restricted, then this is an *instance of the restricted fluent calculus.*

Let T_1, T_2 be terms of sort SIT and φ a state formula with one free state variable. Then $T_1 = T_2$, $\varphi(\text{state}(T_1))$ and $T_1 \in S$ are atomic formulas, where S is a set variable. Formulas are built up from atomic formulas by the logical connectives $\land, \lor, \neg$ etc. and quantification over situations or sets of situations. The satisfaction relation $\mathcal{M} \models \varphi$ between (ACT, FL)–structures $\mathcal{M}$ and monadic formulas φ is defined canonically. If $\mathcal{F}$ is a set of formulas and φ is a formula, we write $\mathcal{F} \models \varphi$ iff for any (ACT, FL)–structure $\mathcal{M}$, we have $\mathcal{M} \models \mathcal{F} \to \varphi$.

Monadic queries: A *monadic query* is a formula without free variables. This query language is quite expressive. E.g., we can express properties like "every maximal path in the tree of situations originating in a situation, in which fluent F_1 holds, contains a situation in which fluent F_2 holds":

$$(\forall s, S) \, [[holds(F_1, s) \land s \in S \land (\forall s_1) \, (s_1 \in S \to \bigvee_{a \in \text{ACT}} \text{do}_a(s_1) \in S)]$$
$$\to (\exists s_2)(s_2 \in S \land holds(F_2, s_2))]$$

The *monadic entailment problem* consists of an instance (ACT, FL, $\mathcal{F}$) of the restricted fluent calculus and a monadic query Q, and is the question, whether $\mathcal{F} \models Q$ holds? We are going to show that this problem is decidable, but that it is not elementary decidable (i.e. the time complexity cannot be described by a function using addition, multiplication, and exponentiation).

State queries: Formulas of the form $(\exists s) \, \varphi(\text{state}(s))$, where φ is a state formula with one free variable, are called *state queries*. Thus, any state query is a monadic query. E.g., $(\exists s) \, holds(t, s)$ is a state query asking for a situation s in which an agent holds a cup of tea. We will show that the restriction of the monadic entailment problem to state queries is decidable in elementary time.

Unrestricted state update axioms: Recall that restricted state update axioms are Boolean combinations of formulas of the form $\varphi(\text{state}(s))$ and $\varphi(\text{state}(\text{do}_a(s)))$, universally quantified over all situations s. E.g., the formula

$$(\exists x) \, \text{state}(s) = q \circ q \circ q \circ x \to \text{state}(\text{do}_{get\_tea}(s)) \circ q \circ q \circ q = \text{state}(s) \circ t \quad (2)$$

is not a restricted state update axiom because it directly relates a state with its successor state. Here, executing the *get_tea* action results in the consumption of three quarters and the production of a cup of tea, whereas all other fluents are preserved. We can now answer queries like $(\exists s, x_1) \, \text{state}(s) = t \circ x_1$: Using (1) and (2) we conclude that $(\exists x)\text{state}(\text{do}_{get\_tea}(s_I)) = c \circ t \circ x \land \neg(\exists x') \, x = t \circ x'$. Due to lack of space we cannot give an extrended introduction into the fluent calculus and must refer the interested reader to the literature (e.g. [30]).

We will show that the abovementioned entailment problem is undecidable even for state queries if we allow unrestricted state update axioms, but consider arbitrary instances of the fluent calculus. However, if we additionally assume that each fluent occurs at most once in each state term, then the entailment problem consisting of an instance of the fluent calculus and state queries becomes decidable. This problem becomes again undecidable, if we allow to define fluents with the help of unary function symbols.

4 The Monadic Entailment Problem

4.1 Decidability of the Monadic Entailment Problem

To show that the monadic entailment problem is decidable, we show that one can decide whether there exists an (ACT, FL)–structure $\mathcal{M}$ with $\mathcal{M} \models \mathcal{F} \cup \{\neg Q\}$. Recall that any (ACT, FL)–structure is isomorphic to a canonical tree structure $\mathcal{M}$. Furthermore, a canonical tree structure over ACT and FL is uniquely given by the function $\text{state}^{\mathcal{M}}$ that maps words over ACT to finite multisets over FL. Note that the range of the function $\text{state}^{\mathcal{M}}$ can be infinite, i.e., the set $\{\text{state}^{\mathcal{M}}(w) \mid w \in \text{SIT}^{\mathcal{M}}\}$ can contain infinitely many states. The announced decidability result relies on the fact that we can reduce this infinite range to a finite one.

Let the *size* of a formula be its length and $k \in \mathbb{N}$ be some nonnegative integer. Furthermore, let SF_k denote the set of all state formulas of size at most k, with one free variable x_1 and which use at most the variable names $x_1, \dots, x_k$. SF_k and its powerset are finite. A set $M \subseteq \text{SF}_k$ is *consistent* if there exists a state $x \in \text{ST}^{\mathcal{M}}$ such that $\mathcal{M}\mathcal{S} \models \varphi(x)$ for any $\varphi \in M$.

Lemma 4.1. *There exists an algorithm solving the following problem in time* $O(\exp^3(cf(f+k+d+2)^k))$ *for some constants c and d: Given a finite set* FL *with $f = |\text{FL}|$, an integer k and $M \subseteq \text{SF}_k$, is M consistent?*

Proof. The formulas in SF_k are state formulas with one free variable x_1. Let $\varphi = \bigwedge M$. Then one has to decide whether $\psi = \exists x_1 \varphi$ holds in the structure $\mathcal{M}\mathcal{S}$. One can obviously consider a multiset $x \in \text{ST}^{\mathcal{M}}$ as a function $\text{FL} \to \mathbb{N}$ or, equivalently, as a tuple of nonnegative integers $x^1, \dots x^f$, where $f = |\text{FL}|$ is the number of fluents and x^i denotes the number of occurrences of the i th fluent in the multiset x. We inductively define a first order formula ψ' in the language of the structure $(\mathbb{N}; +, 0, 1)$ that is equivalent to ψ: First, let t be a state term and let $1 \le i \le f$. If $t = F_i$ (where F_i is the i th fluent), let $t^i = 1$, if $t = F^j$ with $i \ne j$, set $t^i = 0$, if $t = x_\ell$ is a variable, set $t^i = x^i_\ell$, and if $t = 0$ let $t^i = 0$. Furthermore, $(t_1 \circ t_2)^i = t^i_1 + t^i_2$. Thus, we have defined a term over the signature $(+, 0, 1)$ from a term over the signature $(\circ, 0, \text{FL})$.

Now let $\alpha = (t_1 = t_2)$ be an equation of two state terms. Then $\alpha' = \bigwedge_{1 \le i \le f}(t^i_1 = t^i_2)$, which is a formula in the language of the structure $(\mathbb{N}; +, 0, 1)$. If $\alpha = \exists x_\ell \beta$ is a state formula, let $\alpha' = \exists x^1_\ell \exists x^2_\ell \dots \exists x^f_\ell \beta'$. Finally, $(\neg \beta)' = \neg(\beta')$ and $(\beta_1 \theta \beta_2)' = \beta'_1 \theta \beta'_2$ where θ is any binary logical connector.

One can easily check, and the explanation above should give an intuition, that $\mathcal{M}\mathcal{S} \models \psi$ iff $(\mathbb{N}; +, 0, 1) \models \psi'$. By [23], the latter can be checked effectively.

Next we prove the complexity bound: Let n denote the number of state formulas in SF_k. Let d denote the number of symbols that can occur in formulas. Recall that any state formula in SF_k uses at most the variables $x_1, \dots, x_k$. Hence, $f + k + d$ symbols can occur in a state formula from SF_k. Because these state formulas have length at most k, we get $n \le (f + k + d)^k$.

M has at most n elements. Hence, $\psi = (\exists x_1) \bigwedge M$ has size $O(kn)$. Since in the construction of ψ' from ψ, we have essentially replaced each variable occurrence by f variable occurrences, the formula ψ' has length $O(fkn)$. By [22], the validity of ψ' in $(\mathbb{N}; +, 0, 1)$ can be checked in time $\exp^3(c'fkn)$ for some constant c'. Thus, we obtain an upper bound for the deterministic time complexity of $\exp^3(fkc(f + k + d)^k) = O(\exp^3(c(f + k + d + 2)^k))$. $\qquad\square$

We only gave an upper bound for the complexity of the consistency test. The proof of this bound is based on the result by Oppen that the validity of a first order formula in $(\mathbb{N}; +, 0, 1)$ can be checked in triple exponential time. A lower bound for this validity check was proved by Fischer and Rabin [6]: Any *nondeterministic* algorithm needs at least double exponential time. Hence the upper bound above can be improved at most by one exponentiation.

For a state $x \in \mathrm{ST}^{\mathcal{M}}$, let $\mathrm{Th}_k(x)$ denote the set of state formulas in SF_k that are satisfied by x, i.e., $\mathrm{Th}_k(x) = \{\varphi \in \mathrm{SF}_k \mid \mathcal{MS} \models \varphi(x)\}$. A set $M \subseteq \mathrm{SF}_k$ is *maximally consistent* if there exists a state $x \in \mathrm{ST}^{\mathcal{M}}$ with $M = \mathrm{Th}_k(x)$. We use Lemma 4.1 to show that a maximally consistent superset of a given set $M \subseteq \mathrm{SF}_k$ can be computed effectively:

Lemma 4.2. *There exists an algorithm of elementary time complexity computing a function* Red *with the following property: Given a finite set* FL *, an integer k and $M \subseteq \mathrm{SF}_k$ it computes a maximally consistent set $\mathrm{Red}_k(M)$ containing M if M is consistent, and $\emptyset$ if M is inconsistent.*

Proof. Let $n = |\mathrm{SF}_k|$ and $f = |\mathrm{FL}|$. To compute a maximally consistent superset of $M \subseteq \mathrm{SF}_k$, do the following: Let $\mathrm{SF}_k = \{\varphi_1, \ldots, \varphi_n\}$ and $M_1 = M$. For $1 \le i \le n$, set $M_{i+1} = M_i \cup \{\varphi_i\}$ if this set is consistent, and $M_{i+1} = M_i$ otherwise. Thus, we have to check the consistency of n sets of state formulas from SF_k. Each such test can be done in time $O(\exp^3(cf(f + k + d + 2)^k))$ by Lemma 4.1. In the proof of Lemma 4.1 we showed that $n \le (k + f + d)^k$. Hence, the computation of a maximally consistent superset of M can be done in time $O((k + f + d)^k \cdot \exp^3(c(f + k + d + 2)^k))$ which is elementary in f and k. $\qquad\square$

The function Red can be used to check consistency. From [6], we obtain a minimal nondeterministic time complexity doubly exponential in $(f + k)^k$.

Let $\mathcal{M}$ be some canonical tree structure. We define a new function λ : $\mathrm{SIT}^{\mathcal{M}} \to \mathcal{P}(\mathrm{SF}_k)$ by $\lambda(w) = \mathrm{Th}_k(\mathrm{state}(w)) = \{\varphi \in \mathrm{SF}_k \mid \mathcal{MS} \models \varphi(\mathrm{state}^{\mathcal{M}}(w))\}$ for any $w \in \mathrm{SIT}^{\mathcal{M}}$. The structure $\mathrm{Th}_k(\mathcal{M}) = (\mathrm{SIT}^{\mathcal{M}}; (\mathrm{do}_a^{\mathcal{M}})_{a \in \mathrm{ACT}}, s_I^{\mathcal{M}}, \lambda)$ is again a labeled tree. But now the labeling assumes only finitely many values, namely maximally consistent subsets of SF_k.

Similarly to the structure, we convert formulas: Let ψ be a monadic formula. Recall that the building blocks of these formulas are formulas of the form $\varphi(\mathrm{state}(T))$ where φ is a state formula with one free variable and T is a term of sort situation. Let $k \in \mathbb{N}$ such that any such building block occurring in ψ belongs to SF_k (since ψ is finite, such an integer exists). Now replace any building block $\varphi(\mathrm{state}(T))$ in ψ by $\varphi \in \lambda(T)$. The formula $\varphi \in \lambda(T)$ is equivalent to $\bigwedge \lambda(T) = M$ where the disjunction runs over all $M = \mathrm{Th}_k(w)$ for some $w \in \mathrm{ACT}^\star$ with $\varphi \in M$. Hence the result $\mathrm{Red}_k(\psi)$ of these replacement is a formula of the monadic second order logic for labeled trees where the labels are maximally consistent subsets of SF_k.

More formally, we define $Red_k(\psi)$ by structural induction on ψ: If $\psi = (T_1 = T_2)$ where T_1 and T_2 are situation terms, then $Red_k(\psi) = (T_1 = T_2)$. Similarly, if $\psi = (T \in S)$ where T is a term of sort SIT and S is a set variable, we define $Red_k(\psi) = (T \in S)$. If $\psi = \varphi(\text{state}(T))$ where φ is a state formula with one free variable and T is a term of sort SIT, we set $Red_k(\psi) = (\varphi \in \lambda(T))$. For formulas ψ_1 and ψ_2, a variable s_i of sort SIT and a set variable S_j, we proceed by setting $Red_k(\neg\psi_1) = \neg Red_k(\psi_1)$, $Red_k(\psi_1 \wedge \psi_2) = Red_k(\psi_1) \wedge Red_k(\psi_2)$, $Red_k((\exists s_i)\ \psi_1) = (\exists s_i)\ Red_k(\psi_1)$, and similarly $Red_k((\exists S_j)\ \psi_1) = (\exists S_j)\ Red_k(\psi_1)$.

Lemma 4.3. *Let ψ be a monadic formula and $k \in \mathbb{N}$ such that any state formula occurring in ψ belongs to SF_k. Let $\mathcal{M}$ be a canonical tree structure. Then $\mathcal{M} \models \psi$ iff $Th_k(\mathcal{M}) \models Red_k(\psi)$.*

Proof. The lemma is shown by structural induction on the formula ψ. In the following, let T_1, T_2 be situation terms, S a set variable, and φ a state formula with one free variable. Then, for any interpretation of the variables, we obviously obtain $\mathcal{M} \models (T_1 = T_2)$ iff $Th_k(\mathcal{M}) \models (T_1 = T_2)$ since the underlying trees $ACT^\star$ and the functions do_a of the two structures coincide. Similarly, $\mathcal{M} \models (T_1 \in S)$ iff $Th_k(\mathcal{M}) \models (T_1 \in S)$. The only nontrivial case is a formula of the form $\varphi(\text{state}(T_1))$: Let $w \in ACT^\star$ be the situation denoted by the term T_1 (under the variable interpretation in consideration). Then $\mathcal{M} \models \varphi(\text{state}(T_1))$ iff $\mathcal{M} \models \varphi(\text{state}^{\mathcal{M}}(w))$. But this is equivalent to $\varphi \in \lambda(w)$, i.e., to $Th_k(\mathcal{M}) \models \varphi \in \lambda(T_1)$. Now the induction proceeds straightforwardly. $\square$

Note that in the structure $Red_k(\mathcal{M})$ any set $\lambda(w)$ for $w \in SIT^{\mathcal{M}}$ is maximally consistent. Next we show that conversely any mapping λ that assumes only maximally consistent sets stems from some structure $Red_k(\mathcal{M})$:

Lemma 4.4. *Let $k \in \mathbb{N}$ and $\lambda : ACT^\star \to \mathcal{P}(SF_k)$ with $\lambda(w)$ maximally consistent for any $w \in ACT^\star$. Then there exists an (ACT, FL)–structure $\mathcal{M}$ with $Th_k(\mathcal{M}) \cong (ACT^\star; (do_a)_{a\in ACT}, \varepsilon, \lambda)$.*

Proof. Let $w \in ACT^\star$ and let $M = \lambda(w)$. M is maximally consistent. The consistency of M implies that there exists an $x \in ST^{\mathcal{M}}$ such that $\mathcal{M}S \models \varphi(x)$ for any $\varphi \in M$. Let $\text{state}^{\mathcal{M}}(w) = x$. Since M is maximally consistent, we get $M = \{\varphi \in SF_k \mid \mathcal{M}S \models \varphi(\text{state}^{\mathcal{M}}(w))\}$ and, therefore, $Th_k(\mathcal{M}) \cong (ACT^\star; (do_a)_{a\in ACT}, \varepsilon, \lambda)$. $\square$

Now we can describe the decision procedure for the restricted fluent calculus and monadic queries: Let $(ACT, FL, \mathcal{F})$ be an instance of the restricted fluent calculus and Q a monadic query. Now proceed as follows:

Step 1: Compute $k \in \mathbb{N}$ such that any state formula that is a subformula of some formula in $\mathcal{F} \cup \{\neg Q\}$ belongs to SF_k.

Step 2: Compute $\alpha = Red_k(\bigwedge \mathcal{F} \wedge \neg Q)$.

Step 3:   Compute Σ_k as the set of maximally consistent subsets of SF_k.

Step 4:   Check whether there exists a Σ_k-labeled tree $\mathcal{T} = (\text{ACT}^\star, \text{do}_a, \varepsilon, \lambda)$ such that $\mathcal{T} \models \alpha$.

The first two steps are recursive (take, for instance, the maximal size of a formula in $\mathcal{F} \cup \{\neg Q\}$ as k). The third step can be performed effectively by Lemma 4.1. The decidability of the fourth step, i.e., the existence of a Σ_k-labeled tree $\mathcal{T}$ that satisfies α is Rabin's Theorem [24]. Thus, we have

Theorem 4.5. *The monadic entailment problem is decidable.*

Remark 4.6. Note that the proof of the theorem above uses the decidability of two theories: The first one is the first order theory of the natural numbers with addition known as "Presburger arithmetic" in mathematical logic. The second decidable theory is the monadic second order theory of labeled trees, known as Rabin's Theorem. We had to investigate the interplay between these two theories using ideas going back to [5] and [27].

4.2   Complexity of the Monadic Entailment Problem

We already analyzed the complexity of the function Red that is used in our decision procedure (Lemma 4.2). Although it is of pretty high time complexity, the most important source of complexity in the monadic entailment problem is the fourth step. Using a very simple instance of the fluent calculus, we show that the complexity of the monadic entailment problem is non–elementary: So let $\text{ACT} = \{a\}$ be a singleton and $\text{FL} = \emptyset$ be the empty set. The (restricted) state update axiom Δ_a is of the trivial form $\forall s(0 = 0)$. There is just one (ACT, FL)–structure, which is isomorphic to $\mathcal{N} = (\mathbb{N}, \{\emptyset\}; +1, \circ^{\mathcal{M}}, \emptyset, 0, ())$. Any monadic second order sentence on the structure $(\mathbb{N}; +1)$ can be considered as an equivalent monadic query for this particular structure. The validity of monadic second order sentences in $(\mathbb{N}; +1)$ is known to be of non–elementary complexity [20] even if one restricts the set quantification to finite sets. Thus, we obtain

Theorem 4.7. *The monadic entailment problem is not elementary decidable.*

4.3   State Queries

In this section we are going to show that the entailment problem for state queries is elementary. Since the source of the non–elementary complexity in our decision procedure is its fourth step, we will only alter this one. From the third step we know the set Σ_k of all maximally consistent subsets of SF_k. This set will be the vertex set of a directed graph. Before we construct the edges, recall that a restricted state update axiom is an axiom Δ_a of the form $(\forall s)\, \delta_a(s)$ where $\delta_a(s)$ is a Boolean combination of state formulas of the form $\varphi(\text{state}(s))$ and $\varphi(\text{state}(\text{do}_a(s)))$. We may replace $\text{state}(s)$ by x_1 and $\text{state}(\text{do}_a(s))$ by x_2 in the formula δ_a. The result, denoted δ_a', is a state formula with two free variables x_1 and x_2. Furthermore, let $(\exists s)\, \psi(\text{state}(s))$ denote the state query. Then ψ is a state formula with one free variable. Now we define a graph in three steps:

Step 4.1: Let $V_0 \subseteq \Sigma_k$ be the set of maximally consistent subsets M of SF_k with $\psi \notin M$.

Step 4.2: Let E_0 denote the set of all triples $(M, a, N) \in V_0 \times \mathrm{ACT} \times V_0$ for which there exist finite multisets X and Y over FL with $\mathrm{MS} \models \bigwedge_{\varphi \in M} \varphi(X) \wedge \delta'_a(X, Y) \wedge \bigwedge_{\varphi \in N} \varphi(Y)$.

Step 4.3: From the graph (V_0, E_0), erase all vertices M together with all incident edges (M, b, M') and (M', b, M) for which there is an action $a \in \mathrm{ACT}$, but no $N \in \mathrm{SF}_k$ with (M, a, N). Repeat this deletion of vertices and edges until any remaining vertex has an a-successor for any action $a \in \mathrm{ACT}$.

Lemma 4.8. *Let* $(\mathrm{ACT}, \mathrm{FL}, \mathcal{F})$ *be an instance of the restricted fluent calculus and let* $Q = (\exists s)\ \psi(\mathrm{state}(s))$ *be a state query. Let* (V, E) *denote the graph obtained from steps 1–3, 4.1–4.3. Let* $\varphi(\mathrm{state}(s_I))$ *be the only element of* $\mathcal{F}_I$. *Then* $\mathcal{F} \not\models Q$ *iff in the graph* (V, E) *there is a vertex* M *with* $\varphi \in M$.

Proof. Suppose $M \in V$ with $\varphi \in M$. Let E' be some subset of E such that for any $N \in V$ and $a \in \mathrm{ACT}$, there is a unique node $N' \in V$ with $(N, a, N') \in E'$. Now, for any $u \in \mathrm{ACT}^\star$, there is a unique path in (V, E') starting in M with edge label u. Let $\lambda(u)$ denote the target node of this path. Then $\lambda(u) \in V \subseteq \Sigma_k$. Hence, by Lemma 4.4, there exists an $(\mathrm{ACT}, \mathrm{FL})$-structure $\mathcal{M}$ with $\mathrm{Th}_k(\mathcal{M}) \cong (\mathrm{ACT}^\star; (\mathrm{do}_a^{\mathcal{I}})_{a \in \mathrm{ACT}}, \varepsilon, \lambda) = \mathcal{I}$. By Lemma 4.3, $\mathcal{M} \models \mathcal{F}$ because $\mathcal{I} \models \mathrm{Red}_k(\mathcal{F})$. Since $\lambda(u) \in V \subseteq V_0$, there is no node u in the structure $\mathcal{I}$ satisfying $\psi \in \lambda(u)$. Hence, $\mathcal{I}$ does not satisfy $(\exists s)\ \psi \in \lambda(s)$. But this formula equals $\mathrm{Red}_k(Q)$. Hence, by Lemma 4.3, $\mathcal{M} \models \neg Q$, i.e., $\mathcal{F} \not\models Q$ which proves one implication.

Conversely, suppose $\mathcal{F} \not\models Q$. Then there exists an $(\mathrm{ACT}, \mathrm{FL})$-structure $\mathcal{M}$ with $\mathcal{M} \models \mathcal{F} \cup \{\neg Q\}$. Let $\mathcal{I} = (\mathrm{ACT}^\star; (\mathrm{do}_a^{\mathcal{I}})_{a \in \mathrm{ACT}}, \varepsilon, \lambda) = \mathrm{Th}_k(\mathcal{M})$ and define $V_1 = \{\lambda(u) \mid u \in \mathrm{ACT}^\star\}$. Since $\mathcal{M} \models \neg Q$, there is no $u \in \mathrm{ACT}^\star$ with $\psi \in \lambda(u)$, i.e., $V_1 \subseteq V_0$. Furthermore, for $u \in \mathrm{ACT}^\star$, we have $(\lambda(u), a, \lambda(ua)) \in E_0$ by Lemma 4.3 because $\mathcal{M}$ satisfies the state update axioms. Thus, any node $M \in V_1$ has at least one a-successor in V_1. This implies $V_1 \subseteq V$. But now $\mathcal{M} \models \varphi(\mathrm{state}(s_I))$ ensures $\varphi \in \lambda(\varepsilon)$. Since $M = \lambda(\varepsilon)$ belongs to $V_1 \subseteq V$, we showed the second implication, too. □

Theorem 4.9. *The entailment problem for state queries in the restricted fluent calculus is elementary decidable.*

Proof. Using Lemma 4.8 we obtain the decidability because all steps are effectively computable. By Lemma 4.1, the only problematic steps 3 and 4.2 have an elementary complexity. □

5   The Undecidability of the Unrestricted Fluent Calculus

Recall that in the unrestricted fluent calculus state update axioms have the form $(\forall s)\ \varphi(\mathrm{state}(s), \mathrm{state}(\mathrm{do}_a(s)))$. More precisely, we will consider state update axioms of the form $(\forall s)\,[\psi(\mathrm{state}(s)) \rightarrow \mathrm{state}(\mathrm{do}_a(s)) \circ \vartheta^- = \mathrm{state}(s) \circ \vartheta^+]$, where ϑ^- and ϑ^+ are ground state terms denoting the direct negative and positive effects of action do_a respectively. Given an instance $(\mathrm{ACT}, \mathrm{FL}, \mathcal{F})$ of the fluent calculus we are interested in the question whether it entails state queries of the form $(\exists s)\,\mathrm{state}(s) = t$, where t is a ground state term. This problem is shown to be undecidable by reducing it to the undecidable [21,14] problem whether a configuration of a two–counter machine is accepted.

5.1   Two–Counter Machines

For integers i, j with $i \leq j$ let $[i, j]$ denote the set $\{i, i+1, \ldots, j\}$. A *deterministic two–counter machine* is given by an integer $m > 0$ and a mapping

$$\tau : [1, m] \rightarrow \{+\} \times \{1, 2\} \times [0, m] \cup \{-\} \times \{1, 2\} \times [0, m]^2.$$

In the sequel, let (m, τ) be a fixed deterministic two-counter machine. A *configuration* is a triple $(i, p, q) \in [0, m] \times \mathbb{N}^2$. If the machine is in configuration (i, p, q) then its *successor configuration* is

$$
\begin{array}{lll}
(j, p+1, q) & \text{if } \tau(i) = (+, 1, j) & \text{(increment first counter)}, \\
(j, p-1, q) & \text{if } \tau(i) = (-, 1, j, k) \text{ and } p > 0 & \text{(decrement first counter)}, \\
(k, p, q) & \text{if } \tau(i) = (-, 1, j, k) \text{ and } p = 0 & \text{(test first counter)},
\end{array}
$$

Incrementing, decrementing, and testing of the second counter are dealt with likewise. Let $\Rightarrow$ denote the successor relation on configurations. A configuration (i, p, q) is *accepted* by the machine if, starting from (i, p, q), the machine eventually reaches the configuration $(0, 0, 0)$.

5.2   Encoding Two–Counter Machines

Let $\mathrm{ACT} = \{a\}$, $\mathrm{FL} = \{c_0, c_1, \ldots, c_m, d, e\}$, $F^0 = 0$ and $F^{n+1} = F^n \circ F$ for $n \geq 0$, and $\mathrm{do} = \mathrm{do}_a$. Intuitively, the action a denotes one computation step in the two–counter machine (m, τ). A configuration (i, p, q) is encoded into the state term $c_i \circ d^p \circ e^q$ and, consequently, $(0, 0, 0)$ is encoded into $c_0 \circ 0 \circ 0$, which is equivalent to c_0. The state update axioms $\mathcal{F}_{su}$ encode the successor relation of the two–counter machine. To this aim, let $\mathcal{F}_{su}$ consist of the following axioms for $1 \leq i \leq m$:

$$(\forall s)\,[holds(c_0, s) \rightarrow \mathrm{state}(\mathrm{do}(s)) = \mathrm{state}(s)]$$

if $\tau(i) = (+, 1, j)$:
$$(\forall s)\,[holds(c_i, s) \rightarrow \mathrm{state}(\mathrm{do}(s)) \circ c_i = \mathrm{state}(s) \circ d \circ c_j]$$

if $\tau(i) = (-, 1, j, k)$:
$$(\forall s)\,[holds(c_i, s) \wedge holds(d, s) \rightarrow \mathrm{state}(\mathrm{do}(s)) \circ c_i \circ d = \mathrm{state}(s) \circ c_j]$$
$$(\forall s)\,[holds(c_i, s) \wedge \neg holds(d, s) \rightarrow \mathrm{state}(\mathrm{do}(s)) \circ c_i = \mathrm{state}(s) \circ c_k]$$

In case $\tau = (+, 2, j)$ or $\tau = (-, 2, j, k)$ we add state update axioms similar to the preceding ones, where each occurrence of d is replaced by e. Recall that $holds(F, s)$ is an abbreviation for the formula $(\exists x)\ \text{state}(s) = F \circ x$. Hence, the state update axioms we have defined are indeed of the form $(\forall s)\ \delta(\text{state}(s), \text{state}(\text{do}(s)))$.

We define a *progression operator* $\rightsquigarrow$ on states: For $t, t' \in \text{ST}^{\mathcal{M}}$ we write $t \rightsquigarrow t'$ if there exists a state update axiom $(\forall s)\ \delta(\text{state}(s), \text{state}(\text{do}(s))) \in \mathcal{F}_{su}$ with $\mathcal{MS} \models \delta(t, t')$. Let $\overset{*}{\rightsquigarrow}$ denote the transitive and reflexive closure of $\rightsquigarrow$.

5.3   Soundness and Completeness of the Encoding

We start by showing that there is a one–to–one correspondence between the successor relation on configurations of two–counter machines and applications of the progression operator in the fluent calculus.

Lemma 5.1. $(i, p, q) \Rightarrow (i', p', q')$ *iff* $c_i \circ d^p \circ e^q \rightsquigarrow c_{i'} \circ d^{p'} \circ e^{q'}$.

Proof. Suppose $(i, p, q) \Rightarrow (i', p', q')$. We show $c_i \circ d^p \circ e^q \rightsquigarrow c_{i'} \circ d^{p'} \circ e^{q'}$ by case analysis depending on the form of $\tau(i)$. If $\tau(i) = (+, 1, j)$, we get $i' = j$, $p' = p + 1$, and $q' = q$. Therefore, we find a state update axiom $(\forall s)\ \delta(\text{state}(s), \text{state}(\text{do}(s))) \in \mathcal{F}_{su}$ with $\delta(x_1, x_2) = ((\exists x)\ x_1 = x \circ c_i \rightarrow x_1 \circ c_i = x_2 \circ d \circ c_j)$. Now let $t = c_i \circ d^p \circ e^q$ and $t' = c_j \circ d^{p+1} \circ e^q$. Then one can easily check that $\mathcal{MS} \models \delta(t, t')$, i.e., that $t \rightsquigarrow t'$. The other cases follow similarly.

Conversely, suppose $c_i \circ d^p \circ e^q \rightsquigarrow c_{i'} \circ d^{p'} \circ e^{q'}$. Again, we distinguish several cases depending on $\tau(i)$. If $\tau(i) = (+, 1, j)$, then only the state update axiom $(\forall s)\ \delta_a(\text{state}(s), \text{state}(\text{do}(s))) \in \mathcal{F}_{su}$ with $\delta(x_1, x_2) = ((\exists x) x_1 = x \circ c_i \rightarrow x_1 \circ c_i = x_2 \circ d \circ c_j)$ is applicable. Hence, $\mathcal{MS} \models \delta(c_i \circ d^p \circ e^q, c_{i'} \circ d^{p'} \circ e^{q'})$ which implies $i' = j$, $p' = p + 1$ and $q' = q$. Consequently, $(i, p, q) \Rightarrow (j, p+1, q) = (i', p', q')$, which concludes this case. The other cases follow similarly. □

Theorem 5.2. *Let* (m, τ) *be a two–counter machine,* (i, p, q) *a configuration and* $\mathcal{F} = \mathcal{F}_{su} \cup \{\text{state}(s_I) = c_i \circ d^p \circ e^q\}$. *Then,* (i, p, q) *is accepted by* (m, τ) *iff* $\mathcal{F} \models (\exists s)\ \text{state}(s) = c_0$.

Proof. Let $\mathcal{M}$ be an (ACT, FL)–structure that satisfies $\mathcal{F}$. Because there is only one action, the set of situations of $\mathcal{M}$ can be identified with the natural numbers, 0 being the initial situation and $\text{do}_a^{\mathcal{M}}$ the successor function. Furthermore, for any situation s, we have $\text{state}^{\mathcal{M}}(s) \rightsquigarrow \text{state}^{\mathcal{M}}(s + 1)$ by the definition of the state update axioms. Furthermore, $\text{state}(s_I) = c_i \circ d^p \circ e^q$.

Now remember that (i, p, q) is accepted by (m, τ) iff there exists $n \in \mathbb{N}$ with $(i, p, q) \overset{n}{\Rightarrow} (0, 0, 0)$. By induction on n using Lemma 5.1 we learn for all $n \in \mathbb{N}$ that $(i, p, q) \overset{n}{\Rightarrow} (0, 0, 0)$ iff $c_i \circ d^p \circ e^q \overset{n}{\rightsquigarrow} c_0$. Thus, (i, p, q) is accepted by (m, τ) iff $\mathcal{M} \models \text{state}(s) = c_0$ for some $s \in \text{ACT}^*$. Because $\mathcal{M}$ is the only (ACT, FL)–structure that satisfies $\mathcal{F}$, we conclude that (i, p, q) is accepted iff $\mathcal{F} \models (\exists s)\text{state}(s) = c_0$. □

5.4   Undecidability and Decidability Results

From Theorem 5.2 and the fact that there is a deterministic two–counter machine for which the set of accepted configurations is not recursive [21,14] we obtain

Theorem 5.3. *The entailment problem in the fluent calculus is undecidable.*

In some applications of the fluent calculus it is preferable to restrict state terms such that each fluent occurs at most once in each state term. Formally, this can be modeled by requiring that the structure $\mathcal{M}$ satisfies the set of axioms $\mathcal{F}_{mset} = \{\bigwedge_{F \in \mathrm{FL}} \neg(\exists s, x) \text{ state}(s) = F \circ F \circ x\}$, where s, x are variables of sort SIT and ST respectively.

Theorem 5.4. *Given an instance* $(\mathrm{ACT}, \mathrm{FL}, \mathcal{F})$ *of the fluent calculus and a monadic query* Q, *it is decidable whether* $\mathcal{F} \cup \mathcal{F}_{mset} \models Q$.

Proof. Because FL is finite and $\mathcal{M} \models \mathcal{F}_{mset}$ the set of states is finite. Consequently, each state update axiom is equivalent to set of restricted state update axioms. The result is obtained by an application of Theorem 4.5. $\square$

Remark 5.5. Theorem 5.4 corresponds to a result obtained in [29] for the situation calculus. In [29] an open problem is discussed, viz. whether the entailment problem is non–elementary. Theorem 4.7 answers this question positively, because the proof of Theorem 4.7 remains unchanged even if we require that the structure $\mathcal{M}$ satisfies $\mathcal{F}_{mset}$. A restricted version of Theorem 5.4 where state instead of monadic queries are considered was formally proved in [12].

The core of the proof of Theorem 5.3 was the encoding of a counter in a multiset x as the number of occurences of d in x. This encoding is impossible if each fluent may occur at most once in a state ($\mathcal{M} \models \mathcal{F}_{mset}$). If there are two unary function symbols $f_d, f_e : \mathrm{FL} \to \mathrm{FL}$, we can encode two counters again where, e.g., the value 3 of the first counter is encoded by $f_d(f_d(b))$. Then by a proof similar to the one of Theorem 5.3, we obtain:

Theorem 5.6. *Suppose the language contains two unary function symbols from* FL $\to$ FL. *Then, given an instance* $(\mathrm{ACT}, \mathrm{FL}, \mathcal{F})$ *of the fluent calculus and a state query* Q, *it is undecidable whether* $\mathcal{F} \cup \mathcal{F}_{mset} \models Q$.

6   Discussion

In this paper we have drawn a boundary between decidable und undecidable fragments of the fluent calculus by establishing a relation between the fluent calculus and known results in logic, complexity and automata theory.

Independently, it was shown in [15] that the entailment problem in the fluent calculus is undecidable by reducing it to an undecidable model checking problem in Petri nets. The same paper also describes a decidable first–order fragment of the fluent calculus. In this fragment fluents may occur more than once in a state, but syntactic constraints on the state update axioms ensure decidability.

We have focussed our attention on the core of the fluent calculus. It remains to extend the boundary to extensions of the fluent calculus capable of solving advanced problems like the ramification problem or of dealing with more complex actions like non–deterministic actions or continuous change.

Acknowledgements: This research was inspired by many discussions within the DFG–funded postgraduate programme ("Graduiertenkolleg") *Specification of Discrete Processes and Systems of Processes by Operational Models and Logics* bringing together researchers from Mathematics and Computer Science.

References

1. W. Bibel. A deductive solution for plan generation. *New Generation Computing*, 4:115–132, 1986. 436
2. W. Bibel. Let's plan it deductively! *Artificial Intelligence*, 103(1-2):183–208, 1998. 436
3. M.C.A.J. Bonner and M. Kifer. Transaction logic programming. *Theoretical Computer Science*, 133:205–265, 1994. 436
4. P. Doherty, J. Gustafsson, L. Karlsson, and J. Kvanström. Tal: Temporal action logics language specification and tutorial. *Electronic Transactions on Artificial Intelligence*, 2(3-4):273–306, 1998. 436
5. S. Feferman and R.L. Vaught. The first order properties of algebraic systems. *Fund. Math.*, 47:57–103, 1959. 444
6. M.J. Fischer and M.O. Rabin. Super–exponential complexity of Presburger arithmetic. In *Complexity of Computation, SIAM-AMS Proc. Vol. VII*, pages 27–41. AMS, 1974. 442, 442
7. M. Gelfond and V. Lifschitz. Action languages. *Electronic Transactions on Artificial Intelligence*, 2(3-4):193–210, 1998. 436
8. J. Y. Girard. Linear logic. *Journal of Theoretical Computer Science*, 50(1):1 – 102, 1987. 436
9. S. Hölldobler. On deductive planning and the frame problem. In A. Voronkov, editor, *Proceedings of the Conference on Logic Programming and Automated Reasoning*, pages 13–29. Springer, LNCS, 1992. 437
10. S. Hölldobler. Descriptions in the fluent calculus. In *Proceedings of the International Conference on Artificial Intelligence*, volume III, pages 1311–1317, 2000. 436
11. S. Hölldobler and J. Schneeberger. A new deductive approach to planning. *New Generation Computing*, 8:225–244, 1990. 436
12. S. Hölldobler and H.-P. Störr. Solving the entailment problem in the fluent calculus with binary decision diagrams. In *Proceedings of the First International Conference on Computational Logic*, pages 747–761, 2000. 448
13. R.A. Kowalski and M. Sergot. A logic-based calculus of events. *New Generation Computing*, 4:67–95, 1986. 436
14. J. Lambek. How to program an infinite abacus. *Canad. Math. Bull.*, 4:295–302, 1961. 446, 448

15. H. Lehmann and M. Leuschel. Decidability results for the propositional fluent calculus. In *Proceedings of First International Conference on Computational Logic*, pages 762–776, 2000. 448

16. H. Levesque, F. Pirri, and R. Reiter. Foundations for a calculus of situations. *Electronic Transactions on Artificial Intelligence*, 2(3-4):159–192, 1998. 436

17. M. Masseron, C. Tollu, and J. Vauzielles. Generating plans in linear logic. In *Foundations of Software Technology and Theoretical Computer Science*, pages 63–75. Springer, LNCS *472*, 1990. 436

18. J. McCarthy. Situations and actions and causal laws. Stanford Artificial Intelligence Project: Memo 2, 1963. 436

19. J. McCarthy and P. J. Hayes. Some philosophical problems from the standpoint of Artificial Intelligence. In B. Meltzer and D. Michie, editors, *Machine Intelligence 4*, pages 463 – 502. Edinburgh University Press, 1969. 436

20. A. R. Meyer. Weak monadic second order theory of one successor is not elementary recursive. In *Proc. Logic Colloquium*, volume 453 of *Lecture Notes in Mathematics*, pages 132–154. Springer, 1975. 444

21. M. L. Minsky. Recursive unsolvability of Post's problem of "tag" and other topics in theory of Turing machines. *The Annals of Mathematics*, 74(3):437–455, 1961. 446, 448

22. D. C. Oppen. A $2^{2^{2^{cn}}}$ upper bound on the complexity of Presburger arithmetic. *J. Comp. System Sci.*, 16:323–332, 1978. 442

23. M. Presburger. Über die Vollständigkeit eines gewissen Systems der Arithmetik ganzer Zahlen, in welchem die Addition als einzige Operation hervortritt. *Sprawozdanie z 1 Kongresu Matematyków Krajow Slowiańskich, Ksiaznica Atlas*, pages 92–10, 1930. 441

24. M. O. Rabin. Decidability of second-order theories and automata on infinite trees. *Trans. Amer. Math. Soc.*, 141:1–35, 1969. 444

25. E. Sandewall. *Features and Fluents. The Representation of Knowledge about Dynamical Systems*. Oxford University Press, 1994. 436

26. M. Shanahan. *Solving the Frame Problem: A Mathematical Investigation of the Common Sense Law of Inertia*. MIT Press, 1997. 436

27. S. Shelah. The monadic theory of order. *Annals of Mathematics*, 102:379–419, 1975. 444

28. H.-P. Störr and M. Thielscher. A new equational foundation for the fluent calculus. In *Proceedings of the First International Conference on Computational Logic*, pages 733–746, 2000. 437

29. E. Ternovskaia. Automata theory for reasoning about actions. In *Proceedings of the International Joint Conference on Artificial Intelligence*, pages 153–158, 1999. 437, 437, 448, 448

30. M. Thielscher. Introduction to the fluent calculus. *Electronic Transactions on Artificial Intelligence*, 2(3-4):179–192, 1998. 436, 440

31. Michael Thielscher. Nondeterministic actions in the fluent calculus: Disjunctive state update axioms. In S. Hölldobler, editor, *Intellectics and Computational Logic*. Kluwer Academic, 2000. 436

32. Michael Thielscher. Representing the knowledge of a robot. In *Proceedings of the International Conference on Principles of Knowledge Representation and Reasoning (KR)*, pages 109–120, 2000. 436

Solving Planning Problems by Partial Deduction

Helko Lehmann and Michael Leuschel

Department of Electronics and Computer Science
University of Southampton
Highfield, Southampton, SO17 1BJ, UK
{mal,hel99r}@ecs.soton.ac.uk

Abstract. We develop an abstract partial deduction method capable of solving planning problems in the Fluent Calculus. To this end, we extend "classical" partial deduction to accommodate both, equational theories and regular type information. We show that our new method is actually complete for conjunctive planning problems in the propositional Fluent Calculus. Furthermore, we believe that our approach can also be used for more complex systems, e.g., in cases where completeness can not be guaranteed due to general undecidability.

1 Introduction

One of the most widely used computational logic based formalism to reason about action and change is the situation calculus. In the situation calculus a situation of the world is represented by the sequence of actions $a_1, a_2, \ldots, a_k$ that have been performed since some initial situation s_0. Syntactically, a situation is represented by a term $do(a_k, do(a_{k-1}, \ldots, do(a_1, s_0) \ldots))$. There is no explicit representation of what properties hold in any particular situation: this information has to be derived using rules which define which properties are *initiated* and which ones are *terminated* by any particular action a_i.

The fluent calculus ($\mathcal{FC}$) "extends" the situation calculus by adding *explicit state* representations: every situation is assigned a *multi-set* of so called *fluents*. Every action a not only produces a new situation $do(a, \ldots)$ but also modifies this multi-set of fluents. This enables the fluent calculus to solve the (representational and inferential) frame problem in a simple and elegant way [10]. The fluent calculus can also more easily handle partial state descriptions and provides a solution to the explanation problem.

The multi-sets of fluents of $\mathcal{FC}$ are represented using an extended equational theory (EUNA, for extended unique name assumptions [11]) with an associated extended unification ($AC1$) which treats $\circ$ as a commutative and associative function symbol and 1° as the neutral element wrt $\circ$ (i.e., for any s, $s \circ 1^\circ =_{AC1} 1^\circ \circ s =_{AC1} s$). Syntactically, the empty multi-set is thus 1° and a multi-set of k fluents is thus represented as a term of the form $f_1 \circ \ldots \circ f_k$. This allows for a natural encoding of resources (à la linear logic) and it has the advantages that adding and removing a fluent f to a multi-set M can be very easily expressed using $AC1$ unification: $Add =_{AC1} M \circ f$ and $Del \circ f =_{AC1} M$.

In this light, the underlying execution mechanism of $\mathcal{FC}$ is the so-called SLDE–resolution, which extends ordinary SLD by adding support for an underlying equational theory, in this case $AC1$. An alternative way of implementing

M. Parigot and A. Voronkov (Eds.): LPAR 2000, LNAI 1955, pp. 451–468, 2000.

$\mathcal{FC}$ would be to implement the equational theory as a rewrite system and then use narrowing. Unfortunately, both of these approaches are useless for many interesting applications of $\mathcal{FC}$. For example, both SLDE and narrowing, cannot solve the so-called *conjunctive planning problem*, which consists of finding a plan which transforms an initial situation i so as to arrive at a situation which contains (at least) all the fluents of some goal situation g, e.g. [3]. Indeed, for but the most trivial examples, both SLDE and narrowing will loop if no plan exists, and will (due to depth-first exploration) often fail to find a plan if one exists.

Part of the problem is the lack of detection of infinite failure (see [2]), but another problem is the incapability of producing a finite representation of infinitely many computed answers. In general, of course, these two problems are undecidable and so is the conjunctive planning problem. However, the conjunctive planning problem is not very different from the so called *coverability problem* in Petri nets: is it possible to fire a sequence of Petri net transitions so as to arrive at a marking which covers some goal marking. This problem *is* decidable (e.g., using the Karp-Miller procedure [12]) and in [13] it has been shown that a fragment of the fluent calculus has strong relations to Petri nets.

So, one might try to apply algorithms from the Petri net theory to $\mathcal{FC}$ in order to tackle the conjunctive planning problem. However, there is also a logic programming based approach which *can* solve these problems as well and which scales up to any extension expressible as a logic program. This approach is thus a more natural candidate, as it can not only handle the fragment of $\mathcal{FC}$ described in [13], but any $\mathcal{FC}$ domain which can be represented as a definite logic program. (although it will no longer be a decision procedure). Indeed, from [16] we know that *partial deduction* can be successfully applied to solve coverability problems of Petri nets. Thus, the idea of this paper is to apply partial deduction to fluent calculus specifications in order to decide the conjunctive planning problem for an interesting class of $\mathcal{FC}$ specifications (and to provide a useful procedure for more general $\mathcal{FC}$'s). There are several problems that still need to be solved in order for this approach to work:

- $\mathcal{FC}$ relies on $AC1$-unification, but unification under equational theory is not directly supported by partial deduction as used in [16] and one would have to apply partial deduction to a meta-interpreter implementing $AC1$-unification. Although this is theoretically feasible, this is still problematic in practice for efficiency and precision reasons. A more promising approach is to extend partial deduction so that it can handle an equational theory.

- Another problem lies with an inherent limitation of "classical" partial deduction, which relies on a rather crude domain for expressing calls: in essence a term represents all its instances. This was sufficient for handling Petri nets in [16] (where a term such as $[0,s(X)]$ represents all Petri net markings with no tokens in place 1 and at least 1 token in place 2), but is not sufficient to handle $\mathcal{FC}$ whose state expressions are more involved. For example, given a $\mathcal{FC}$ specification with two fluents f_1 and f_2, it is impossible to represent a state which has one or more f_1's but no f_2's. Indeed, in "classical" partial deduction, a term such as $f_1 \circ X$ represents all its instances, and thus also represents states which contain f_2's. To solve this we propose to use so called *abstract partial deduction* [14] with an abstract domain based upon regular types [24], and extend it to cope with equational theories.

Although in this paper we are mainly interested in applying partial deduction to the $\mathcal{FC}$ based upon $AC1$, we present the generalised partial deduction independently of the particular equational theory. However, the use of this general method in practice relies on an efficient unification procedure. If such a procedure can not be provided and/or one wishes to specialise the underlying equational theory, other approaches, e.g., based on narrowing [8] [1], should be considered. The reason we extend classical partial deduction for SLDE–resolution rather than building on top of [1], is that we actually do not wish to modify the underlying equational theory. As we will see later in the paper, this leads to a simpler theory with simpler correctness results, and also results in a tighter link with classical partial deduction used in [16]. This also means that it is more straightforward to integrate abstract domains as described in [14] (no abstract specialisation exists as of yet for narrowing-based approaches).

In the remainder of the paper, we thus develop a partial deduction method which considers both, equational theories and regular type information. The method will then enable us to solve conjunctive planning problems in the *simple Fluent Calculus*. In particular, we show that our method is actually complete for conjunctive planning problems in the *propositional Fluent Calculus*. We believe that our approach can also be used for more complex systems, without changing much of the algorithm, e.g., in cases where completeness can not be guaranteed due to general undecidability.

2   The Simple Fluent Calculus

The Fluent Calculus $\mathcal{FC}$ is a method for representation and reasoning about action and change [10]. In contrast to the Situation Calculus, states of the world are represented explicitly by terms of sort St. The solution of the frame problem in $\mathcal{FC}$ relies heavily on the use of the equational theory $AC1$ which defines $(St, \circ)$ to be a commutative monoid:

$$\forall (x, y, z : St). \, (x \circ y) \circ z =_{AC1} x \circ (y \circ z)$$
$$\forall (x, y : St). \, x \circ y =_{AC1} y \circ x \qquad\qquad (AC1)$$
$$\forall (x : St). \, x \circ 1^\circ =_{AC1} x$$

The operation $\circ$ is used to compose states by combining atomic elements, called *fluents*, which represent elementary propositions. Although in general the Fluent Calculus can be seen as an extension of the Situation Calculus [23], we restrict ourselves here to $\mathcal{FC}$ domains as introduced in [10], since they can be represented as definite logic programs. We call such $\mathcal{FC}$ domains *simple*. In simple $\mathcal{FC}$ domains actions are defined using the predicate $\mathtt{action}(\mathcal{C}(\vec{x}),\ \mathcal{A}(\vec{x}),\ \mathcal{E}(\vec{x}))$ where $\mathcal{C}(\vec{x})$, $\mathcal{E}(\vec{x})$ are terms of sort St which might depend on variables in $\vec{x}$ and $\mathcal{A}(\vec{x})$ is a term of sort A which has the parameters $\vec{x}$, where the sort A represents the *actions*. Intuitively, executing an action $\mathcal{A}(\vec{x})\theta$ will consume the fluents in $\mathcal{C}(\vec{x})\theta$ and produce the fluents in $\mathcal{E}(\vec{x})\theta$. If all fluents appearing in terms of type St in predicates `action` of a simple $\mathcal{FC}$ domain are constants, we call the domain *propositional*.

Example 1. (propositional $\mathcal{FC}$ domain) Let Σ_p be the following propositional $\mathcal{FC}$ domain with the fluents $f_1, \ldots, f_5$ and the actions $a_1, \ldots, a_6$.

```
action(f₁,a₁,f₂).        action(f₃,a₄,f₂).
action(f₁,a₂,f₄).        action(f₄,a₅,f₅ ∘ f₅).
action(f₂,a₃,f₃ ∘ f₃).    action(f₅,a₆,f₄).
```

□

Example 2. (simple $\mathcal{FC}$ domain) Let Σ_s be the simple $\mathcal{FC}$ domain with the fluents f_1, f_4, f_5, the actions a_2, a_5, a_6 as defined for Σ_p in example 1 and the following predicates for $a_1, a_3(X), a_4(X)$:

```
action(f₁,a₁,f₂(0)).
action(f₂,a₃(X),f₃(foo(X))∘f₃(foo(foo(X)))).
action(f₃(X),a₄(X),f₂(X)).
```

□

The *conjunctive planning problem (CPP)* consists of deciding whether there is a finite sequence of actions such that its execution in a given initial state leads to a state which contains at least, i.e. *covers*, certain goal properties. The initial state and the goal properties are given as conjunctions of fluents, which can be represented as terms of sort St.

In the following, we consider the initial state to be completely known and represented by a ground term St_{init} of sort St.

To describe the execution of action sequences, we define the following predicate which describes all pairs of states, such that the second state can be reached from the first state by executing a finite sequence of actions:

```
reachable(S,S).
reachable(C∘V,T) ← action(C,A,E) ∧ reachable(V∘E,T).
```

Note that, in order to keep the representation simple, we do not keep explicitly track of the action sequence. Furthermore, since we propose to use program transformation techniques to solve the CPP, we do not encode the goal in the definition of `reachable/2`.[1] Also note that, for this interpreter to work correctly, it is important that ∘ is treated as a commutative and associative function symbol (e.g., $(f \circ g) \circ h$ should unify with $g \circ V$ with unifier $\{V/f \circ h\}$).

To specify and reason about equalities in a standard logical programming environment like Prolog, the particular underlying equational theory (e.g., $AC1$) has to be expressed by appropriate axioms. These axioms often cause trouble, e.g., if the solution to a unification problem is infinite, but it has been shown that equational theories can be successfully built into the unification procedure [20]. To allow a general treatment, SLD–resolution has been extended to SLDE–resolution which uses a universal unification procedure based on the proper ties common to all equational theories [9,7]. In contrast to other techniques, SLDE–resolution allows to cut down the often tremendous search space by merging equation solving and standard resolution steps. (Narrowing [8] is an efficient approach to solve certain equational theories, and can be integrated as part of the unification into SLDE.)

---

[1] This is in contrast to [10] where containment of goal properties is encoded in the program.

3   SLDE–Resolution

Formally, simple Fluent Calculus domains are (definite) E-programs (P, E), i.e. logic programs P with an equational theory E, [9,7]. An equational theory E is defined as a set of universally closed formulas of the form $\forall (s =_E t)$ for some predicate $=_E$ complemented by the standard axioms of equality.[2] Consequently, if $E = \emptyset$ we obtain the *standard equational theory*, i.e. only syntactically identical terms are considered to be equal. In simple Fluent Calculus domains E is given by $AC1$. Many other equational theories have been investigated, see e.g. [21] for a review.

An E*–unification problem* consists of terms s, t and the question whether there exists a substitution σ with $\mathrm{Dom}(\sigma) \subseteq Vars(s) \cup Vars(t)$, s.t. $s\sigma =_E t\sigma$. If such a substitution σ exists s and t are called E*–unifiable* with E*–unifier* σ. For example, the terms $V \circ a$ and $a \circ b$ are $AC1$–unifiable with $\{V/b\}$.

A term s is an E*–instance* of a term t, denoted $s \leq_E t$, iff there is a substitution σ with $s =_E \sigma t$. Similarly, $\theta \leq_E \sigma$, for substitutions θ, σ, iff for all terms s: $s\theta \leq_E s\sigma$.

Let $U_E(s, t)$ denote the set of all E–unifiers of the terms s and t. Then, $U \subseteq U_E(s, t)$ is called *complete* if for all $\theta \in U_E(s, t)$ there exists $\sigma \in U$ and a substitution λ s.t. $\forall x \in Vars(s) \cup Vars(t)$: $x\theta =_E x\sigma\lambda$. If U is complete and for all $\theta, \sigma \in U$, $\theta \leq_E \sigma$ implies $\theta = \sigma$, then it is called *minimal*. Correspondingly, an unification algorithm is called *complete* (*minimal*) if, for arbitrary terms s, t, it computes a complete (minimal) set of E–unifiers.

Note that minimal sets of E–unifiers are always unique if they exist. Hence, we denote the minimal E–unifier of s and t by $\mu U_E(s, t)$. We call a substitution in $\mu U_E(s, t)$ a *most general E–unifier* (*mgeu*) of s and t.

Based upon this, one can define the concepts of SLDE–resolution, SLDE–derivations, SLDE–refutations and computed answers in the classical way. One can also define SLDE–trees, where the only difference with SLD–trees is that resolution with a clause can lead to more than one child (as $\mu U_E(s, t)$ may contain more than one substitution)!

SLDE–trees are guaranteed to be finitely branching if the equational theory E is *finitary*, i.e. if the complete set of E unifiers $U_E(s, t)$ is finite. For example, it is well known that the equational theory $AC1$ is finitary.

In [10] it has been shown that SLDE–resolution is sound and complete for CPP in simple $\mathcal{FC}$ domains, i.e. every solution of a CPP is entailed by SLDE–resolution. However, even for propositional $\mathcal{FC}$ domains the SLDE–tree may contain infinite derivations and consequently, the search for a plan may not terminate.

Example 3. (Ex. 1 cont'd) If we repeatedly apply the actions a_3 and a_4 in alternation, we obtain an infinite derivation:

$\leftarrow \underline{\texttt{reachable}}(f_2, S)$

$\leftarrow \underline{\texttt{action}}(f_2, A, E) \wedge \texttt{reachable}(1^\circ \circ E, S) \quad \{A/a_3, E/f_3 \circ f_3\}$

$\leftarrow \underline{\texttt{reachable}}(f_3 \circ f_3, S)$

---

[2] These are reflexivity, symmetry, transitivity and substitutivity for all function and predicate symbols, respectively.

$\leftarrow \underline{\text{action}}(f_3, A', E') \ \land \ \text{reachable}(f_3 \circ E', S)$ $\{A'/a_4, E'/f_2\}$
$\leftarrow \underline{\text{reachable}}(f_3 \circ f_2, S)$
$\leftarrow \underline{\text{action}}(f_2, A'', E'') \ \land \ \text{reachable}(f_3 \circ E'', S)$ $\{A''/a_3, E''/f_3 \circ f_3\}$
... $\square$

To enable for solving the CPP or similar problems despite of potentially infinite SLDE–derivations we propose to use partial deduction techniques. To this end, we extend the partial deduction method used in [16,15] to fit SLDE–resolution. Furthermore, we allow conjuncts to carry additional type information, thereby enabling for more precise specialisations.

4   A Partial Deduction Algorithm for E–Programs

The general idea of partial deduction of ordinary logic programs [18] is to construct, given a query $\leftarrow Q'$ of interest, a finite number of finite but possibly incomplete SLD–trees which "cover" the possibly infinite SLD–tree for $P \cup \{\leftarrow Q'\}$ (and thus also all SLD–trees for all instances of $\leftarrow Q'$). The derivation steps in these SLD–trees are the computations which have been pre-evaluated and the clauses of the specialised program are then extracted by constructing one specialised clause (called *resultant*) per branch.

While the initial motivation for partial deduction was program specialisation, one can also use partial deduction as a *top-down flow analysis* of the program under consideration. Indeed, partial deduction will unfold the initial query of interest until it spots a dangerous growth, at which point it will generalise the offending calls and restart the unfolding from the thus obtained more general call. Provided a suitably refined control technique is used (e.g., [17,4]), one can guarantee termination as well as a precise flow analysis. As was shown in [15] such a partial deduction approach is powerful enough to provide a decision procedure for coverability problems for (reset) Petri nets and bears resemblance to the Karp–Miller procedure [12]. In the context of the CPP, the initial query of interest would be `reachable(init,goal)`, where `init` and `goal` are the initial and the goal state respectively and one would hope to obtain as a result a flow analysis from which it is clear whether the CPP has a solution.

Unfortunately, it has been demonstrated in [15] that "classical" partial deduction techniques may not be precise enough if state descriptions are complex. Similar problems occur if states are represented using non-empty equational theories, since abstractions just based on the "instance-of" relation and the associated most specific generalisation (*msg*) may be too crude (c.f., also [14]).

Example 4. (Ex. 1 cont'd) The *msg* of the atoms `reachable`$(f_3 \circ f_3, S)$ and `reachable`$(f_3 \circ f_3 \circ f_3, S)$ is `reachable`$(f_3 \circ f_3 \circ X, S)$. This is quite unsatisfactory, as X can represent *any* term, i.e., also terms containing other fluents such as f_4. In the context of CPP this means that any action can potentially be executed from $f_3 \circ f_3 \circ X$, and we have thrown away too much information for the generalisation to be useful. For example, if our goal state is f_4, we would not be able to prove that we cannot solve the CPP from the initial state $f_3 \circ f_3$. $\square$

In classical partial deduction there is no way of overcoming this problem, due to its inherent limitation that a call must represent all its instances (the same holds for narrowing-based partial evaluation [1]). Fortunately, this restriction has been lifted, e.g., in the abstract partial deduction framework of [14]. In essence, [14] extends partial deduction and conjunctive partial deduction [4] by working on *abstract conjunctions* on which *abstract unfolding* and *abstract resolution* operations are defined:

- An abstract conjunction is linked to the concrete domain of "ordinary" conjunctions via a concretisation function γ. In contrast to classical partial deduction, γ can be much more refined than the "instance-of" relation. For example, an abstract conjunction can be a couple (Q, τ) consisting of a concrete conjunction Q and some type[3] information τ, and $\gamma((Q, \tau))$ would be all the instances of Q which respect the type information τ. We could thus disallow f_4 to be an instance of X in Ex. 4.
- An abstract unfolding operation maps an abstract conjunction A to a set of *concrete* resultants $H_i \leftarrow B_i$, which have to be totally correct for all possible calls in $\gamma(A)$.
- For each such resultant $H_i \leftarrow B_i$ the abstract resolution will produce an *abstract* conjunction Q_i approximating all the possible resolvent goals which can occur after resolving an element of $\gamma(A)$ with $H_i \leftarrow B_i$.

It is to this framework, suitably adapted to cope with SLDE–resolution, that we turn to remedy our problems. We will actually only consider abstract atoms consisting of a concrete atom together with some type information. The latter will be represented by canonical (deterministic) regular unary logic (RUL) programs [24,5]. To use a RUL program R in an SLDE–setting it must be ensured that every type t defined by R is "E–closed", i.e. if some term is of type t then all E–equivalent terms are of type t as well.

Definition 1. *A canonical regular unary clause is a clause of the form*

$$t_0(f(X_1, \ldots, X_n)) \leftarrow t_1(X_1) \wedge \ldots t_n(X_n)$$

where $n \geq 0$ and $X_1, \ldots, X_n$ are distinct variables. A canonical regular unary logic (RUL) program is a program R where R is a finite set of regular unary clauses, in which no two different clause heads have a common instance.

Let E be an equational theory. R is called E–closed if the least Herbrand model of R and the least Herbrand model of (R, E) are identical.

The set of ground terms r such that $R \models t(r)$ is denoted by $\tau_R(t)$. A ground term r is of type t in R iff $r \in \tau_R(t)$. Given a (possibly non-ground) conjunction T, we write $R \models \forall(T)$ iff for all ground instances T' of T, $R \cup \{\leftarrow T'\}$ has an SLD–refutation.

So, to solve Ex. 4 one could use the following (E–closed) RUL program, representing all states using just the fluent f_3, and give the variable X in Ex. 4 the type t_3:

$$t_3(1^\circ). \qquad\qquad t_3(f_3). \qquad\qquad t_3(X \circ Y) \leftarrow t_3(X) \wedge t_3(Y).$$

---

[3] A type is simply a decidable set of terms closed under substitution.

Given two canonical *RUL*-programs R_1, R_2, there exist efficient procedures for checking inclusion, computing the intersection and computing an upper bound using well known algorithms on corresponding automata [24]. Because of our definition, we can simply re-use the first two procedures to efficiently decide inclusion and compute the intersection of E–closed *RUL* programs. Furthermore, the intersection of two E–closed *RUL* programs is an E–closed *RUL* program. Given two *RUL* programs R_1, R_2 and two types t_1, t_2, we will denote by $(R_1, t_1) \cap (R_2, t_2)$ the couple (R_3, t_3) obtained by the latter procedure (i.e., we have $\tau_{R_3}(t_3) = \tau_{R_1}(t_1) \cap \tau_{R_2}(t_2)$). We will not make use of the upper bound and provide our own generalization mechanism.

Given some *RUL* program R, a *type conjunction (in R)* is simply a conjunction of the form $t_1(X_1) \wedge \ldots \wedge t_n(X_n)$, where all the X_i are variables (not necessarily distinct) and all the t_i are defined in R. We also define the notation $types_T(X) = \{t_j \mid t_j(X) \in T\}$ (where we allow $\in$ to be applied to conjunctions). E.g., $types_{t(X) \wedge t'(Z)}(Z) = \{t'\}$.

We now define the abstract domain used to instantiate the framework of [14]:

Definition 2. *We define the RULE domain* $(\mathcal{AQ}, \gamma, E)$ *to consist of an equational theory* E, *abstract conjunctions of the form* $\langle Q, T, R \rangle \in \mathcal{AQ}$ *where* Q *is a concrete conjunction,* R *an* E–*closed RUL program, and* T *a type conjunction in* R *such that* $T = t_1(X_1) \wedge \ldots t_n(X_n)$, *where* $Vars(Q) = \{X_1, \ldots, X_n\}$.[4] *The concretisation function* γ *is defined by:* $\gamma(\langle Q, T, R \rangle) = \{Q\theta \mid R \models \forall(T\theta)\}$.

We now define simplification and projection operations for type conjunctions. This will allow us to apply substitutions to abstract conjunctions as well as to define an (abstract) unfolding operation. As the above definition requires every variable to have exactly one type, the type of a variable Z occurring in a substitution such as $\{X/Z, Y/Z\}$ has to be determined by type intersection.

Definition 3. *Let* R *be some RUL program. The relation* $\leadsto_R$ *which maps type conjunctions to type conjunctions is defined as follows:*
  - $t_1 \wedge t_2 \leadsto_R s_1 \wedge s_2$ *if* $t_1 \leadsto_R s_1, t_2 \leadsto_R s_2, s_1 \neq fail$, *and* $s_2 \neq fail$
  - $t(X) \leadsto_R t(X)$ *if* X *is a variable*
  - $t(c) \leadsto_R true$ *if* c *is a constant with* $c \in \tau_R(t)$
  - $t(f(r_1, \ldots, r_n)) \leadsto_R s_1 \wedge \ldots \wedge s_n$ *if* $t(f(X_1, \ldots, X_n)) \leftarrow t_1(X_1) \wedge \ldots \wedge t_n(X_n) \in R$ *and* $t_i(r_i) \leadsto_R s_i$
  - $t(r) \leadsto_R fail$ *otherwise*

We define a projection *which projects a type conjunction* T *in the context of a RUL program on a concrete conjunction* Q, *resulting in new abstract conjunction:*
$proj(Q, T, R) = \langle Q, S', R' \rangle$, *where* $T \leadsto_R S$ *and*
  - $S' = S, R' = \emptyset$ *if* $S = fail$ *or* $Vars(Q) = \emptyset$,
  - *otherwise* $S' = t_1(X_1) \wedge \ldots \wedge t_n(X_n)$ *where* $Vars(Q) = \{X_1, \ldots, X_n\}$, $types_S(X_i) = \{t_{i_1}, \ldots, t_{i_k}\}$, $(R_i, t_i) = (R, t_{i_1}) \cap \ldots \cap (R, t_{i_k})$. *In this case* $R' = R_1 \cup \ldots \cup R_n$.

---

[4] Note that when writing, e.g., $Vars(Q) = \{X_1, \ldots, X_n\}$ all X_i are assumed to be distinct.

We now define applying substitutions on abstract conjunctions: $\langle Q, T, R \rangle \theta = proj(Q\theta, T, R).$

For example, using the *RUL* program R above, we have $t_3(f_3 \circ Z \circ Z) \rightsquigarrow_R$ $true \land t_3(Z) \land t_3(Z)$. We would thus have for $\theta = \{X/(f_3 \circ Z \circ Z)\}$ that $\langle p(X), t_3(X), R \rangle \theta = \langle p(f_3 \circ Z \circ Z), t_3(Z), R \rangle.$

To extend the notion of instantiation preorder to abstract conjunctions the subset relation between types has to be considered:

Definition 4. *Let* $A = \langle Q, T, R \rangle$, $A' = \langle Q', T', R' \rangle$ *be abstract conjunctions in the RULE domain* $(\mathcal{AQ}, \gamma, E)$. *We call* A' *a RULE–instance of* A, *denoted by* $A' \leq_{RULE} A$ *iff*

1. *there exists a substitution* θ *such that* $A\theta = \langle Q', T'', R'' \rangle$ *and*
2. *for all* $X \in Vars(Q')$ *with* $types_{T'}(X) = \{t'\}$ *and* $types_{T''}(X) = \{t''\}$, *we have* $\tau_{R'}(t') \subseteq \tau_{R''}(t'')$.

We define $<_{RULE}$ *and* $=_{RULE}$ *accordingly.*

In the above example, $\langle p(f_3 \circ Z \circ Z), t_3(Z), R \rangle <_{RULE} \langle p(X), t_3(X), R \rangle.$

Definition 5. *An* unfolding rule *is a function which, given a definite* E*–program* (P, E) *and a goal* $\leftarrow Q$, *returns a non-trivial[5] and possibly incomplete SLDE–tree for* (P, E) *and* $\leftarrow Q$.

Let τ *be a finite (possibly incomplete) SLDE–tree for* (P, E), $\leftarrow Q$. *Let* $\leftarrow G_1, \ldots, \leftarrow G_m$ *be the goals in the leaves of the non-failing branches of* τ. *Let* $\theta_1, \ldots, \theta_n$ *be the computed answer substitutions of the SLDE–derivations from* $\leftarrow Q$ *to* $\leftarrow G_1, \ldots, \leftarrow G_n$, *respectively. Then the set of* SLDE–resultants, $resultants(\tau)$, *is defined to be the set of clauses* $\{Q\theta_1 \leftarrow G_1, \ldots, Q\theta_n \leftarrow G_n\}$.

We can now define an *abstract unfolding* and an *abstract resolution* in the *RULE* domain. When a conjunction of the *RULE* domain is unfolded, the information concerning the types of variables can be used to reduce the number of resultants. Additionally, we will use Def. 3 to determine the types of leaf conjunctions.

Definition 6. *Let* (P, E) *be a definite* E*–program,* $\langle Q, T, R \rangle$ *an abstract conjunction in the RULE domain* $(\mathcal{AQ}, \gamma, E)$, U *an unfolding rule. We define the abstract unfolding and resolution operations* $aunf(.)$, $ares(.)$ *as follows:*

- $aunf(\langle Q, T, R \rangle) = \{Q\theta \leftarrow B \mid Q\theta \leftarrow B \in resultants(U(Q)) \land T\theta \not\rightsquigarrow_R fail\}$
- $ares(\langle Q, T, R \rangle) = \{proj(B, T\theta, R) \mid Q\theta \leftarrow B \in aunf(\langle Q, T, R \rangle)\}$

The following is a generic algorithm for abstract partial deduction, which structures the abstract conjunctions to be specialised in a *global tree* (see, e.g., [17]), and is parametrised by a covering test *covered*, a *whistle* detecting potential infinite loops, an a generalisation operation *abstract* and a function *partition* to separate conjunctions into sub-conjunctions.

---

[5] A trivial SLDE–tree has a single node where no literal has been selected for resolution.

Algorithm 4.1 (*generic partial deduction algorithm*)
Input: a definite E–program (P, E), an abstract conjunction A in $(\mathcal{AQ}, \gamma, E)$.
Output: a set of abstract conjunctions $\mathcal{A}$, a specialised program P, a global tree λ
Initialisation: $\lambda :=$ a "global" tree with a single unmarked node, labelled by A
repeat
    pick an unmarked or abstracted leaf node L in λ
    if $covered(L, \lambda)$ **then**   mark L as processed
    else
       if $whistle(L, \lambda) = \mathbf{T}$ **then**
          mark L as abstracted
          $label(L) := abstract(L, \lambda)$
       else
          mark L as processed
          for all $A \in ares(label(L))$ **do**
             for all $A' \in partition(A)$ **do**
               add a new unmarked child C of L to λ
               $label(C) := A'$
until all nodes are processed
output $\mathcal{A} := \{label(A) \mid A \in \lambda\}$, $P := \{aunf(A) \mid A \in \mathcal{A}\}$, and λ

Algorithm 4.2 (a partial deduction algorithm for the Fluent Calculus) We define a particular instance of the above algorithm as follows:

Unfolding used by $aunf(.)$: Let $\langle Q, T, R \rangle$ be an abstract conjunction in the *RULE* domain and (P, E) be a definite E–program. We define $U(Q)$ to be the maximal SLDE–tree τ such that every predicate p is selected at most once in every branch of τ.

covered Let L be a node labelled by an abstract conjunction in the *RULE* domain $(\mathcal{AQ}, \gamma, E)$ and λ a tree labeled by elements of $\mathcal{AQ}$. Then we define $covered(L, \lambda)$ as true iff there is an ancestor L' of L such that $label(L') =_{RULE} label(L)$.

whistle We extend the well-established homeomorphic embedding relation [22], to take regular types and the $AC1$ equational theory into account. To simplify the presentation, we use $\circ$ as a variable arity functor to represent terms of sort St and disallow the use of 1° and nesting of $\circ$ (e.g., we represent $a \circ ((b \circ 1^\circ) \circ c)$ by $\circ(a, b, c)$ and $1^\circ \circ 1^\circ$ by $\circ()$).

Definition 7. *Let* $A = \langle Q, T, R \rangle$, $A' = \langle Q', T', R' \rangle$ *be abstract conjunctions in the RULE domain* $(\mathcal{AQ}, \gamma, E)$. *For the purposes of this definition we suppose that* $\wedge$ *is handled by* E *as an associative and commutative function symbol. We say that* A *is homeomorphically embedded in* A', $A \trianglelefteq_E A'$, *iff* $Q \trianglelefteq_E Q'$ *where* $\trianglelefteq_E$ *on expressions is inductively defined as follows:*

1. $X \trianglelefteq_E Y$ *if* X, Y *variables with* $\tau_R(types_T(X)) \subseteq \tau_{R'}(types_{T'}(Y))$
2. $r \trianglelefteq_E Y$ *for all variables* Y *and ground terms* r, *with* $r \in \tau_R(types_{T'}(Y))$
3. $r \trianglelefteq_E f(s_1, \ldots, s_n)$ *if* $f \neq \circ$ *and* $r \trianglelefteq_E s_i$ *for some* $1 \leq i \leq n$
4. $f(r_1, \ldots, r_n) \trianglelefteq_E f(s_1, \ldots, s_n)$ *if* $f \neq \circ$ *and* $\forall i \in \{1, \ldots, n\} : r_i \trianglelefteq_E s_i$.
5. $\circ(r_1, \ldots, r_m) \trianglelefteq_E \circ(s_1, \ldots, s_n)$ *if there exists a permutation* $s'_1, \ldots, s'_n$ *of* $s_1, \ldots, s_n$ *such that* $\forall i \in \{1, \ldots, m\} : r_i \trianglelefteq_E s'_i$.

Note that for point 3. we may have $n = 0$, and for point 5. m, n can be 0. Intuitively, $s \trianglelefteq_E t$, means that we can obtain s from t by "striking out" certain sub-terms and by using the equational theory to re-write s and t. E.g., we have $f(0) \circ g \trianglelefteq_E g \circ h \circ f(s(0))$. In general, of course, $\trianglelefteq_E$ will be quite expensive to compute ([19]). However, one can introduce a lot of optimisations to obtain an efficient implementation (sorting fluents and defining a normal form for terms; one can also always use the classical homeomorphic embedding which ignores E).

The homeomorphic relation is a well-quasi order (provided $\subseteq$ is a well-quasi order on the possible regular types of variables; see below) and can thus be used to ensure termination of program specialisation techniques [22]. We use $\trianglelefteq_E$ as follows. Let L be a node labelled by an abstract conjunction in the $RULE$ domain $(\mathcal{AQ}, \gamma, E)$ and λ a tree labeled by elements of $\mathcal{AQ}$. We define $whistle_{\trianglelefteq_E}(L, \lambda) = $ T iff L is not marked as abstracted and there is an ancestor L' of L such that $label(L') \trianglelefteq_E label(L)$.

abstract To ensure that abstractions of types may occur only finitely often, we require the use of a well founded type system.

Definition 8. *Let E be an equational theory and $\mathcal{T}$ a set of tuples (R, t) where R is a RUL program and t a predicate defined in R. We call $\mathcal{T}$ a well founded type system iff there is no infinite sequence $(R_1, t_1), (R_2, t_2), \ldots$ of elements of $\mathcal{T}$ such that $\tau_{t_i}(R_i) \subset \tau_{t_{i+1}}(R_{i+1})$ for all $i \geq 1$.*

Definition 9. *Let E be an equational theory, $\mathcal{T}$ be a well-founded type system and $\mathcal{A}$ a set of abstract conjunctions in $(\mathcal{AQ}, \gamma, E)$. The abstract conjunction $M = \langle Q, T, R \rangle$ is called a RULE–generalisation of $\mathcal{A}$ wrt $\mathcal{T}$ iff*
   1. for all $t(X) \in T$ we have $(t, R) \in \mathcal{T}$,
   2. for all $A \in \mathcal{A}$, $A \leq_{RULE} M$.
Furthermore, M is called a most specific RULE–generalisation of $\mathcal{A}$ wrt $\mathcal{T}$, denoted by $M \in msg_{\mathcal{T}}(\mathcal{A})$, iff there exists no M' such that conditions 1, 2 hold for M' and $M' <_{RULE} M$.

For example, $\mathcal{A} = \{\langle f_3, true, \emptyset \rangle, \langle f_3 \circ f_3, true, \emptyset \rangle\}$ and using the single type defined by the RUL program for t_3 we get $msg_{\mathcal{T}}(\mathcal{A}) = \{\langle f_3 \circ X, t_3(X), R \rangle\}$.

Again, for other equational theories than $(AC1)$ and more complicated type systems a most specific generalisation might be difficult to compute (and may not be unique). To accelerate convergence (and to simplify our completeness proof for CPP later on), we actually choose an element $M' = \langle Q, T, R \rangle$ of $msg_{\mathcal{T}}(\mathcal{A})$ and then remove the maximum number of subterms from Q so that the resulting abstract conjunction is still more general than M' (in the sense of $\leq_{RULE}$). We will denote the result by $nmsg_{\mathcal{T}}(\mathcal{A})$. For example, we would instead of using M' $= \langle f_3 \circ X, t_3(X), R \rangle$ use the more general $nmsg_{\mathcal{T}}(\mathcal{A}) = \langle X, t_3(X), R \rangle$. This loses some precision, but convergence is accelerated, and actually no vital information for the CPP is lost!

Let L be a node labelled by an abstract conjunction in the $RULE$ domain $(\mathcal{AQ}, \gamma, E)$ and λ a tree labeled by elements of $\mathcal{AQ}$. Let $\mathcal{L}$ denote the set of all ancestors of L in λ such that $L' \in \mathcal{L}$ iff $label(L') \trianglelefteq_E label(L)$. Furthermore, let $\mathcal{A}$ denote the set of labels of $\mathcal{L}$. Then we define $abstract(L, \lambda) = nmsg_{\mathcal{T}}(\mathcal{A})$.

partition Let $A = \langle Q, T, R \rangle$ be an abstract conjunction in $(\mathcal{AQ}, \gamma, E)$ and $\mathcal{T}$ a well-founded type system. We define $partition(A) = atoms(nmsg_\mathcal{T}(\{A\}))$.

Example 5. (Ex. 1 cont'd) Additionally to the actions of Ex. 1 and the domain independent `reachable/2`, let the initial state be defined as $St_{init} = f_1$.

In this example every abstract conjunction $C \in \mathcal{AQ}$ will be of the form $\langle \texttt{reachable}(u,v), T, R \rangle$ where $v = V_{f_1} \circ \ldots \circ V_{f_5}$ and $\forall 1 \le i \le 5 : t_{f_i}(V_{f_i}) \in T$ (representing that $f_1, \ldots, f_5$ may occur arbitrarily often in the final state). Furthermore, u is of sort St where $U \in Vars(u) \Rightarrow \exists 1 \le i \le 5$ s.t. $t_{f_i}(U) \in T$. Finally, R consists of predicates (t_{f_i}) for each fluent f_i:

$$t_{f_i}(1^\circ). \qquad\qquad t_{f_i}(f_i). \qquad\qquad t_{f_i}(X \circ Y) \leftarrow t_{f_i}(X) \wedge t_{f_i}(Y).$$

We define the type system as $\mathcal{T} = \{(R, t_{f_i}) \mid 1 \le i \le 5\}$. Then, the following tree is generated by our partial deduction algorithm with input Σ_p and initial abstract conjunction $\langle \texttt{reachable}(St_{init}, V_{f_1} \circ \ldots \circ V_{f_5}), \{t_{f_1}(V_{f_1}) \wedge \ldots \wedge t_{f_5}(V_{f_5})\}, R \rangle$:

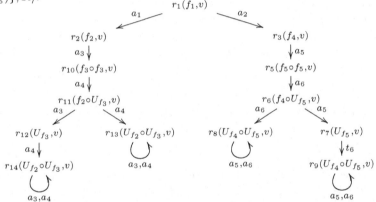

To simplify the picture the *RUL* programs and type conjunctions have not been represented. The *RUL* programs do not change in this example and the type information has been depicted as follows: v represents $V_{f_1} \circ \ldots \circ V_{f_5}$ and the type conjunction T_j for each node $r_j(u,v)$ contains atoms $t_{f_i}(V_{f_i})$, $i = 1, \ldots, 5$, and $t_{f_i}(U_{f_i})$ if they are used. t_{f_i} is defined by the corresponding *RUL* program (t_{f_i}). Finally, $r_j(u,v)$ is the jth node with label $\langle \texttt{reachable}(u,v), T_j, R \rangle$.

For example, we can conclude from the tree that every fluent can be generated arbitrarily often. But, e.g., it is impossible to reach a state containing both, f_2 and f_4. □

Example 6. (Ex. 2 cont'd) Additionally to the actions defined in Ex. 2 and the domain independent `reachable/2`, let the initial state be defined as $St_{init} = f_1$.

In this example every abstract conjunction $C \in \mathcal{AQ}$ will be of the form $\langle \texttt{reachable}(u,v), T, R \rangle$ where again $v = V_{f_1} \circ \ldots \circ V_{f_5}$ and $t_{f_i}(V_{f_i}) \in T$. Also, u is of sort St where for all $U \in Vars(u)$ either $\exists 1 \le i \le 5$ s.t. $t_{f_i}(U) \in T$ or

$t_{foo}(U) \in T$. R contains (t_{f_i}) of Ex. 5 for $i = 1, 4, 5$, and for $i = 2, 3$ and t_{foo}, respectively:

$t_{f_i}(1°)$.

$t_{f_i}(f_i(X)) :\!- t_{foo}(X)$.

$t_{f_i}(Y \circ X) :\!- t_{f_i}(Y), \ t_{f_i}(X)$.

$t_{foo}(0)$.

$t_{foo}(foo(X)) :\!- t_{foo}(X)$.

We define the type system $T = \{(R, t_{f_i}) \mid 1 \le i \le 5\} \cup \{(R, t_{foo})\}$. Then, the following tree is generated by our partial deduction algorithm with input Σ_s and initial abstract conjunction $\langle \texttt{reachable}(St_{init}, V_{f_1} \circ \ldots \circ V_{f_5}), \{t_{f_1}(V_{f_1}) \wedge \ldots \wedge t_{f_5}(V_{f_5})\}, R \rangle$:

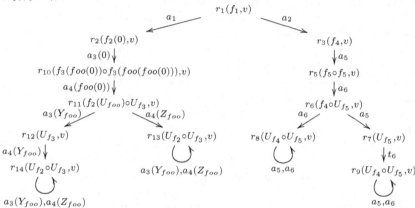

Again, to simplify the picture the RUL programs and type conjunctions have not been represented. RUL programs do not change in this example and the type information has been depicted as follows: v represents $V_{f_1} \circ \ldots \circ V_{f_5}$ and the type conjunction T_j for each node $r_j(u, v)$ contains atoms $t_{f_i}(V_{f_i})$, $i = 1, \ldots, 5$, and $t_{f_i}(U_{f_i})$, $t_{foo}(U_{foo})$ if they are used. t_{f_i}, t_{foo} is defined by the corresponding RUL programs (t_{f_i}) and (t_{foo}). Finally, $r_j(u, v)$ is the jth node with label $\langle \texttt{reachable}(u, v), T_j, R \rangle$.

For example, we can conclude from the tree that it is possible to generate a state containing arbitrary many instances of the fluent f_2. But we cannot conclude whether we can generate a state containing arbitrary many copies of one particular instance of f_2.   □

5   Completeness wrt. $\mathcal{FC}_{PL}$

In [13] it has been shown that Petri net algorithms can be used to decide temporal properties of propositional Fluent Calculus domains. In particular, to every propositional $\mathcal{FC}$ domain with completely defined initial state exists a bisimilar Petri net. Furthermore, the conjunctive planning problem for the propositional $\mathcal{FC}$ can be expressed as a formula in the temporal logic CTL (CTL respects bisimulation). The same formula is known to describe *coverability* properties of Petri nets. Coverability problems can be decided using the Karp-Miller tree [12]. This tree can also be generated by the partial deduction algorithm 4.1 using the instantiations of section 4. By doing so, we show that the proposed partial deduction method is complete wrt. conjunctive planning problems.

Theorem 1. *Let Σ be a propositional $\mathcal{FC}$ domain, Δ_Σ the RULE domain defined as in example 5 and St_{init} some ground term of sort St. Then the partial deduction algorithm applied to Σ, Δ_Σ and $A = \langle\texttt{reachable}(St_{init}, V_{f_1} \circ \ldots \circ V_{f_n}),$ $\{t_{f_1}(V_{f_1}) \wedge \ldots \wedge t_{f_n}(V_{f_n})\}, R\rangle$ will produce a global tree λ which is isomorphic to a Karp-Miller coverability tree of the corresponding Petri net Π.*

6   Conclusion

We have presented a generic and a more specific abstract partial deduction method for equational logic programs, based upon an abstract domain with regular types. This is one of the first full instantiations of the framework in [14] (see also the independently developed [6]). The main motivation was to obtain a useful method for tackling the conjunctive planning problem in the fluent calculus, stimulated by earlier success of partial deduction for solving coverability problems in the Petri net area. We were able to prove that our more specific method *is* a decision procedure for the conjunctive planning problem in the propositional fluent calculus. However, the method can also be applied to more expressive fragments of the fluent calculus or extended to cope with other formalisms such as process algebras (where, contrary to Petri nets, type information is also vital), and we believe that it will be able to provide useful results in that setting. Finally, the methods can of course also be used to *specialise* fluent calculus descriptions, and can also be applied to "ordinary" logic programs, where the additional precision of the regular types should pay off in terms of improved specialisation. In the future we hope to produce a full-fledged implementation to test these claims.

References

1. M. Alpuente, M. Falaschi, and G. Vidal. Partial evaluation of functional logic programs. *ACM Trans. Program. Lang. Syst.*, 20(4):768–844, 1998. 453, 453, 457
2. M. Bruynooghe, H. Vandecasteele, D. A. de Waal, and M. Denecker. Detecting unsolvable queries for definite logic programs. In C. Palamidessi, H. Glaser, and K. Meinke, editors, *Proceedings of ALP/PLILP'98*, LNCS 1490, pages 118–133. Springer, 1998. 452
3. D. Chapman. Planning for conjunctive goals. *AIJ*, 32(3):333–377, 1985. 452
4. D. De Schreye, R. Glück, J. Jørgensen, M. Leuschel, B. Martens, and M. H. Sørensen. Conjunctive partial deduction: Foundations, control, algorithms and experiments. *J. Logic Program.*, 41(2 & 3):231–277, 1999. 456, 457
5. J.P. Gallagher and D. A. de Waal. Fast and precise regular approximations of logic programs. In P. Van Hentenryck, editor, *Proceedings of ICLP'94*, pages 599–613. The MIT Press, 1994. 457
6. J. P. Gallagher and J. C. Peralta. Using regular approximations for generalisation durign partial evaluation. In *Proceedings of PEPM'00*, pages 44–51. ACM Press. 464
7. J. H. Gallier and S. Raatz. Extending SLD resolution to equational horn clauses using E-unification. *J. Logic Program.*, 6(1-2):3–43, 1989. 454, 455
8. M. Hanus. The integration of functions into logic programming. *J. Logic Program.*, 19 & 20:583–628, May 1994. 453, 454
9. S. Hölldobler. *Foundations of Equational Logic Programming*, LNAI 353. Springer, 1989. 454, 455

10. S. Hölldobler and J. Schneeberger. A new deductive approach to planning. *New Gen. Comput.*, 8:225–244, 1990. 451, 453, 453, 454, 455

11. S. Hölldobler and M. Thielscher. Computing change and specificity with equational logic programs. *Annals of Mathematics and Artificial Intelligence*, 14:99–133, 1995. 451

12. R. M. Karp and R. E. Miller. Parallel program schemata. *Journal of Computer and System Sciences*, 3:147–195, 1969. 452, 456, 463, 466

13. H. Lehmann and M. Leuschel. Decidability results for the propositional fluent calculus. In J. Lloyd et al., editor, *Proceedings of CL'2000*, LNAI 1861, pages 762–776, London, UK, 2000. Springer. 452, 452, 463, 466

14. M. Leuschel. Program specialisation and abstract interpretation reconciled. In Joxan Jaffar, editor, *Proceedings of JICSLP'98*, pages 220–234, Manchester, UK, 1998. MIT Press. 453, 453, 456, 457, 457, 458, 464

15. M. Leuschel and H. Lehmann. Coverability of reset Petri nets and other well-structured transition systems by partial deduction. In J. Lloyd et al., editor, *Proceedings of CL'2000*, LNAI 1861, pages 101–115, London, UK, 2000. Springer. 456, 456, 456

16. M. Leuschel and H. Lehmann. Solving coverability problems of Petri nets by partial deduction. In *Proceedings of PPDP'2000*. ACM Press, 2000. To appear. 452, 452, 452, 453, 456

17. M. Leuschel, B. Martens, and D. De Schreye. Controlling generalisation and polyvariance in partial deduction of normal logic programs. *ACM Trans. Program. Lang. Syst.*, 20(1):208–258, 1998. 456, 459

18. J. W. Lloyd and J. C. Shepherdson. Partial evaluation in logic programming. *J. Logic Program.*, 11(3&4):217–242, 1991. 456

19. R. Marlet. *Vers une Formalisation de l'Évaluation Partielle*. PhD thesis, Université de Nice - Sophia Antipolis, 1994. 461

20. G. Plotkin. Building in equational theories. In B. Meltzer and D. Michie, editors, *Machine Intelligence*, number 7, pages 73–90, Edinburgh, Scotland, 1972. Edinburgh University Press. 454

21. J. H. Siekmann. Unification theory. *Journal of Symbolic Computation*, 7(3–4):207–274, 1989. 455

22. M. H. Sørensen and R. Glück. An algorithm of generalization in positive supercompilation. In J. Lloyd, editor, *Proceedings of ILPS'95*, pages 465–479, Portland, USA, 1995. MIT Press. 460, 461

23. M. Thielscher. From Situation Calculus to Fluent Calculus: State update axioms as a solution to the inferential frame problem. *AIJ*, 111(1–2):277–299, 1999. 453

24. E. Yardeni and E. Shapiro. A type system for logic programs. *J. Logic Program.*, 10(2):125–154, 1990. 453, 457, 458

A   Completeness of Partial Deduction

Definition 10. *A Petri net Π is a tuple (S, T, F, M_0) consisting of a finite set of places S, a finite set of transitions T with $S \cap T = \emptyset$ and a flow relation F which is a function from $(S \times T) \cup (T \times S)$ to $\mathbb{N}$. A marking m for Π is a mapping $S \mapsto \mathbb{N}$. M_0 is a marking called* initial.

A transition $t \in T$ is enabled *in a marking M iff $\forall s \in S : M(s) \geq F(s, t)$. An enabled transition can be fired, resulting in a new marking M' defined by $\forall s : M'(s) = M(s) - F(s, t) + F(t, s)$. We will denote this by $M[t\rangle M'$. By $M[t_1, \ldots, t_k\rangle M'$ we denote the fact that for some intermediate markings $M_1, \ldots, M_{k-1}$ we have $M[t_1\rangle M_1, \ldots, M_{k-1}[t_k\rangle M'$.*

We define the reachability tree $RT(\Pi)$ inductively as follows: Let M_0 be the label of the root node. For every node n of $RT(\Pi)$ labelled by some marking M and for every transition t which is enabled in M, add a node n' labelled M' such that $M[t\rangle M'$ and add an arc from n to n' labelled t. The set of all labels of $RT(\Pi)$ is called the reachability set *of Π, denoted $RS(\Pi)$.*

For convenience, we denote $M \geq M'$ iff $M(s) \geq M'(s)$ for all places $s \in S$. We also introduce *pseudo-markings*, which are functions from S to $\mathbb{N} \cup \{\omega\}$ where we also define $\forall n \in \mathbb{N} : \omega > n$ and $\omega + n = \omega - n = \omega + \omega = \omega$. Using this we also extend the notation $M_{k-1}[t_1, \ldots, t_k\rangle M'$ for such markings.

Many interesting properties of Petri nets can be investigated using the so-called *Karp-Miller tree* resulting of the following algorithm[6], first defined in [12]. The Karp-Miller tree is a finite abstraction of the set of reachable markings $RS(\Pi)$ with which we can decide whether it is possible to "cover" some arbitrary marking M' (i.e., $\exists M'' \in RT(\Pi) \mid M'' \geq M'$) simply by checking whether a node in the tree covers M'.

Algorithm A.1 (*Karp–Miller–Tree*)

Input: a Petri net $\Pi = (S, T, F, M_0)$
Output: a tree $KM(\Pi)$ of nodes labelled by pseudo-markings
Initialisation: set $U := \{node(r, M_0)\}$ of unprocessed nodes
while $U \neq \emptyset$
    select some $(k, M) \in U$;
    $U := U \setminus \{(k, M)\}$;
    if there is no ancestor node (k_1, M_1) of (k, M) with $M = M_1$ **then**
        $M_2 = M$;
        for all ancestors (k_1, M_1) of (k, M) such that $M_1 < M$ **do**
            for all places $p \in S$ such that $M_1(p) < M(p)$ **do** $M_2(p) = \omega$;
        $M := M_2$;
        for every transition t such that $M[t\rangle M'$ **do**
            create node (k', M');
            create arc labelled t from (k, M) to (k', M');
            $U := U \cup (k', M')$;

We will now formally prove that the algorithm 4.1 with the instantiation of section 4.2 can be used to decide coverability problems. For this we need to establish a link between pseudo-markings in the Karp-Miller tree and abstract conjunctions produced by partial deduction.

Let Σ be some propositional $\mathcal{FC}$ domain. Let $F_\Sigma = \{f_1, \ldots, f_n\}$ be the fluents and $A_\Sigma = \{a_1, \ldots, a_m\}$ the actions defined in Σ. We define $|t, f|$ as the number of occurrences of $f \in F_\Sigma$ in the ground term t. According to [13], the Petri net (S, T, F, M_0) corresponding to a propositional $\mathcal{FC}$ domain Σ is given by associating an unique place $S(f) \in S$ to each $f \in F_\Sigma$. Every clause $\texttt{action}(\mathcal{C}, a, \mathcal{E})$ in Σ with $a \in A_\Sigma$ is associated a transition $T(a) \in T$. The flow relation is defined by $F(T(a), S(f)) = |\mathcal{C}, f|$ and $F(S(f), T(a)) = |\mathcal{E}, f|$ for every $f \in F_\Sigma$ and $a \in A_\Sigma$. Let Δ_Σ be the *RULE* domain $(\mathcal{AQ}, \gamma, AC1)$ where every abstract conjunction $C \in \mathcal{AQ}$ is of the form $\langle \texttt{reachable}(u, v), H, R \rangle$ where $v = V_{f_1} \circ \ldots \circ V_{f_n}$, u are terms of sort St and for all variables $U_f \in Vars(u)$, where $f \in F_\Sigma$, $t_f(U) \in H$ and for all $V_f \in Vars(v)$, where $f \in F_\Sigma$, $t_f(V) \in H$.

---

[6] The algorithm presented here differs slightly from the original.

R consists of the predicates (t_f) for each fluent $f \in F_\Sigma$ as defined in example 5 for (t_{f_i}). Then, we define the pseudo-marking C^μ for each $f \in F_\Sigma$:

$$C^\mu(f) = \begin{cases} \omega & \text{if } X \in Vars(u) \wedge t_f(X) \in H \\ |u, f| & \text{otherwise} \end{cases}$$

Accordingly, the initial marking corresponding to some St_{init} is given by

$$\langle \mathtt{reachable}(St_{init}, V_{f_1} \circ \ldots \circ V_{f_n}), \{t_{f_1}(V_{f_1}) \wedge \ldots \wedge t_{f_n}(V_{f_n})\}, R \rangle^\mu$$

Additionally, we associate with every pseudo-marking M and RUL program R as defined above an abstract conjunction $M^\alpha = \langle \mathtt{reachable}(u, v), H, R \rangle$ s.t. for every fluent $f \in F_\Sigma$, the term u contains f exactly $M(S(f))$ times if $M(S(f)) \neq \omega$. For every $f \in F_\Sigma$ with $M(S(f)) = \omega$, u contains a variable X and H a type declaration $t_f(X)$, and $v = V_{f_1} \circ \ldots \circ V_{f_n}$ with $t_{f_i}(V_{f_i}) \in H$ for all $1 \leq i \leq n$.

To prove that the tree generated by our PD algorithm is isomorphic to the Karp-Miller tree, we use the following propositions establishing links between the algorithms 4.2 and A.1.

Lemma 1. *Let L_1, L_2 be some nodes of the tree λ which is labelled by abstract conjunctions of Δ_Σ. Let $C_1 = \langle \mathtt{reachable}(u_1, v), T_1, R \rangle = label(L_1)$ and $C_2 = \langle \mathtt{reachable}(u_2, v), T_2, R \rangle = label(L_2)$. Then $C_1 =_{RULE} C_2$ iff $C_1{}^\mu = C_2{}^\mu$.*
Proof. This follows using the mappings $\_^\alpha$ and $\_^\mu$ between markings and abstract conjunctions as defined above and the fact that $(C^\mu)^\alpha =_{RULE} C$ for markings $C = \langle \mathtt{reachable}(u, v), T, R \rangle$: from the definition, $C_1 =_{RULE} C_2$ iff for all fluents f holds either 1. the number of f in u_1 and u_2 must be equal, or 2. there are variables X in u_1 and Y in u_2 s.t. $t_f(X) \in T_1$ and $t_f(Y) \in T_2$. $\square$

Lemma 2. *Let L be some node of the tree λ which is labelled by abstract conjunctions of Δ_Σ. Let $C = \langle \mathtt{reachable}(u, v), T, R \rangle = label(L)$ and $C_0, C_1, \ldots, C_n$ is the sequence of labels of the ancestors of L in λ where C_0 is the label of the root node. whistle$(L, \lambda) = T$ iff there is some L_k labelled C_k, $0 \leq k \leq n$, with $C_k{}^\mu \leq C^\mu$.*
Proof. According to the definition, *whistle* returns T iff there is some ancestor L_k of L labelled C_k s.t. $C_k \trianglelefteq_E C$. If u does not contain a variable of type t_f for fluent $f \in F_\Sigma$ and $C_k \trianglelefteq_E C$, then by case 5 of the definition of $\trianglelefteq_E$ follows $C_k{}^\mu(S(f)) \leq C^\mu(S(f))$. Otherwise, by case 1 follows $C_k{}^\mu(S(f)) \leq C^\mu(S(f))$ if C_k contains a variable of type t_f as well, or by case 2, $C_k{}^\mu(S(f)) \leq C^\mu(S(f))$ if C_k contains any number of copies of f. Note that due to the use of $nmsg$ a label C' in λ may never contain both, copies of f and a variable of type t_f. Now, let $C_k = \langle \mathtt{reachable}(u_k, v_k), T_k, R \rangle$ be an abstract conjunction in Δ_Σ and $C_k{}^\mu(S(f)) \leq C^\mu(S(f))$ for all fluents f. If C_k contains a variable of type t_f, case 1 of $\trianglelefteq_E$ applies iff C contains an appropriate variable, i.e. iff $C^\mu(S(f)) = \omega$. Otherwise, if C_k contains copies of f then case 2 applies iff $C^\mu(S(f)) = \omega$ and case 5 applies iff $C_k{}^\mu(S(f)) \leq C^\mu(S(f)) \neq \omega$. $\square$

Lemma 3. *Let L be some node of the tree λ which is labelled by abstract conjunctions of Δ_Σ. Let $C = \langle \mathtt{reachable}(u, v), T, R \rangle = label(L)$ and $\{C_1, \ldots, C_n\}$ is the sequence of labels of ancestors of L in λ s.t. $C_i \trianglelefteq_E C$ for all $1 \leq i \leq n$. Let $\mathcal{T}$ consist of all pairs (t, R) s.t. t is a predicate in R. $C' = abstract(L, \lambda)$ iff $C'^\mu = M'$ and M' is defined as follows: if for some fluent f there exists an ancestor C_k of C in λ s.t. $C_k{}^\mu < C^\mu$ and $C_k{}^\mu(f) < C^\mu(f)$, $M'(f) = \omega$, otherwise $M'(f) = C^\mu(f)$.*

Proof. Note that $\mathcal{T}$ is a finite set s.t. for any two $(R, t_1), (R, t_2) \in \mathcal{T}$, $\tau_R(t_1) \cap \tau_R(t_2) = \emptyset$. Furthermore, every $\tau_R(t)$ with $(R, t) \in \mathcal{T}$ consists only of terms constructed by combining copies of one particular fluent $f \in F_\Sigma$ using $\circ$. Hence, from the definition of *abstract* and $nmsg_T$ follows that C' contains a variable of type t_f iff there is some ancestor L_k of L labelled C_k s.t. $C_k \trianglelefteq_E C$ and $C_k{}^\mu(f) < C^\mu(f)$: on one hand, condition 2 in definition 9 ensures that C and C_k are both instances of $C' = \langle \text{reachable}(u', v), T', R \rangle$. This can only be the case if $C_k{}^\mu(f) < C'^\mu(f)$ and $C^\mu(f) \leq C'^\mu(f)$ and hence, u' must contain a variable of type t_f representing ω. Furthermore, from definition of $nmsg_T$, a fluent f must not occur in u' if u' contains a variable of type t_f. On the other hand, from definition 9 follows, that u' must not contain a variable of type t_f if $C_k{}^\mu(f) = C^\mu(f)$ for all ancestors with label C_k and $C_k{}^\mu < C^\mu$. In this case, since $C \trianglelefteq_E C'$, the same number of copies of fluent f occurs in u' as in u. $\qquad\square$

Lemma 4. *Let L_1, L_2 be some nodes of the tree λ which is labelled by abstract conjunctions of Δ_Σ. Let $C_1 = \langle \text{reachable}(u_1, v), T_1, R \rangle = label(L_1)$ and $C_2 = \langle \text{reachable}(u_2, v), T_2, R \rangle = label(L_2)$. $C_2 \in partition(ares(aunf(L_1)))$ iff there is an action $\mathcal{A}$ s.t. $C_1{}^\mu [T(\mathcal{A})\rangle C_2{}^\mu$.*

Proof. The procedures $ares()$ and $aunf()$ can be simplified, since a variable may never have two or more types, conjunctions of types do not have to be computed. $ares()$ and $aunf()$ unfold and ensure type of variables, only. According to the used unfolding rule an atom $\text{reachable}(u_1, v)$ is unfolded s.t. every occuring predicate is unfolded once, i.e. into the subgoals $\text{action}(\mathcal{C}, \mathcal{A}, \mathcal{E})$ where $\mathcal{C}$, $\mathcal{A}$, $\mathcal{E}$ are ground and $\text{reachable}(u_1', v)$ where $u_1' =_{AC1} V \circ \mathcal{C}$ and $u_1' =_{AC1} V \circ \mathcal{E}$. According to the $AC1$ unification, if $u_1 =_{AC1} V \circ \mathcal{C}$ either $|\mathcal{C}, f| \leq |u_1, f|$ or there is a variable of type t_f in u_1. Consequently, $u_1 =_{AC1} V \circ \mathcal{C}$ iff $T(\mathcal{A})$ is enabled in $C_1{}^\mu$. Furthermore, if u_1 does not contain a variable of type t_f, it holds $|u_1', f| = |u_1, f| - |\mathcal{C}, f| + |\mathcal{E}, f|$. Otherwise, the codomain of any $mgeu$ for u_1 and $V \circ \mathcal{C}$ must contain a variable X s.t. $t_f(X)$. Let T_1' be the set of such type declarations. Then, with $C_1' = \langle \text{reachable}(u_1', v), T_1', R \rangle$, it follows $C_1{}^\mu [T(\mathcal{A})\rangle C_1'{}^\mu$. However, u_1' may contain copies of a fluent f even if there is a variable X in u_1' with $t_f(X) \in T$. Using the partition function with $nmsg_T$, u_2 is defined as u_1' where such additional copies are removed. By this it is ensured that for every marking M with $C_1{}^\mu [T(\mathcal{A})\rangle M$, $M^\alpha =_{RULE} C_2$. $\qquad\square$

Proof. (**theorem 1**) Per definition $U = \{node(r, M_0)\}$ where $M_0 = A^\mu$. Now, we show the correspondence between each step in algorithm A.1 and algorithm 4.1. First, both algorithms terminate if no unprocessed nodes remain. Second, in algorithm 4.1 a selected node L is marked processed if $covered(L, \lambda)$ is true. Let (k, M) be the selected node by algorithm A.1 with $M = label(L)^\mu$. According to lemma 1, $covered(L, \lambda)$ iff there is an ancestor node (k_1, M_1) with $M = M_1$. In this case (k, M) is marked processed by algorithm A.1 (i.e. removed from the list of unprocessed nodes). Third, algorithm 4.1 calls $abstract(L, \lambda)$ if $whistle(L, \lambda) = \text{T}$. Using lemma 2 $whistle(L, \lambda) = \text{T}$ iff there is some ancestor L_k of L s.t. $label(L_k)^\mu \leq label(L)^\mu$. In algorithm A.1, abstraction is performed for every ancestor (k_1, M_1) of (k, M) with $M_1 < M$. Since the case $M_1 = M$ and $label(L_k) =_{RULE} label(L)$, respectively, has already been checked, it remains to be shown, that $C' = abstract(L, \lambda)$ iff $C'^\mu = M'$ and M' is defined as follows: if for some fluent f there exists an ancestor L_k of L in λ s.t. $label(L_k)^\mu < label(L)^\mu$ and $label(L_k)^\mu(f) < label(L)^\mu(f)$, $M(f) = \omega$, otherwise $M(f) = label(L)^\mu(f)$. This has been shown in lemma 3. Finally, from lemma 4 follows that $C_2 \in partition(ares(aunf(L_1)))$ iff there is an action $\mathcal{A}$ s.t. $C_1{}^\mu [T(\mathcal{A})\rangle C_2{}^\mu$. $\qquad\square$

A Kripkean Semantics for Dynamic Logic Programming

Ján Šefránek

Institute of Informatics, Comenius University
811 03 Bratislava, Slovakia
e-mail: sefranek@fmph.uniba.sk

Keywords: knowledge representation and reasoning, nonmonotonic reasoning, knowledge evolution, updates, dynamic logic programming, stable model, Kripke structure, dynamic Kripke structure

Abstract. The main goal of the paper is to propose a tool for a semantic specification of program updates (in the context of dynamic logic programming paradigm). A notion of Kripke structure $\mathcal{K}_P$ associated with a generalized logic program P is introduced. It is shown that some paths in $\mathcal{K}_P$ specify stable models of P and vice versa, to each stable model of P corresponds a path in $\mathcal{K}_P$. An operation on Kripke structures is defined: for Kripke structures $\mathcal{K}_P$ and $\mathcal{K}_U$ associated with P (the original program) and U (the updating program), respectively, a Kripke structure $\mathcal{K}_{P \oplus U}$ is constructed. $\mathcal{K}_{P \oplus U}$ specifies (in a reasonable sense) a set of updates of P by U. There is a variety of possibilities for a selection of an updated program.

1 Introduction

Knowledge evolution is a problem of crucial importance from the non-monotonic reasoning point of view. In fact, the non-monotony of reasoning is only a symptom of the evolution of knowledge.[1]

A formalization of some essential features of knowledge evolution was proposed recently in [3], see also the predecessors [14,16,2,10,11]. Knowledge bases (KB) are represented in [3] by generalized logic programs which allow default negation also in heads of the rules. As a consequence, both insertions and deletions may be specified by the rules of a program. The basic situation is as follows. A program P (the initial program) is given. P is updated by another program U (the updating program). A new program $P \oplus U$ (the updated program) is the result of the update. This situation is generalized in [3] to sequences of program updates $P \oplus U_1 \oplus \cdots \oplus U_n$. The paradigm of dynamic logic programming provides an appropriate tool for a representation of dynamically changing knowledge (dynamic knowledge bases).

---

[1] "...non-monotonic behaviour ...is a *symptom*, rather than the essence of non-standard inference" according to [20].

M. Parigot and A. Voronkov (Eds.): LPAR 2000, LNAI 1955, pp. 469–486, 2000.
© Springer-Verlag Berlin Heidelberg 2000

The approach of [3] is based on this basic decision: an update KB' of one knowledge base KB by another knowledge base U should not just depend on the semantics of the knowledge bases KB and U but it should also depend on their syntax (the dependencies among literals are encoded in the syntax). The decision is implemented via a syntactic transformation. First, the set of propositional letters is extended. For each propositional letter a quintuple of new propositional letters is introduced. Second, the updated program $P \oplus U$ contains for each original clause from P and U a modified clause in the extended language. $P \oplus U$ also contains for each original propositional letter six new clauses.

The main goal of this paper is to investigate semantic foundations of dynamic logic programming paradigm. For each generalized logic program P an associated Kripke structure $\mathcal{K}_P$ is defined. Dependencies among literals are encoded in the accessibility relation of the Kripke structure. We can specify updated logic programs using a new Kripke structure $\mathcal{K}_{P \oplus U}$. $\mathcal{K}_{P \oplus U}$ is the result of an operation on Kripke structures $\mathcal{K}_P$ and $\mathcal{K}_U$, associated with an original program P and an updating program U, respectively. There is no need for an extended language and for some new types of clauses when the updated programs are created.

Updated programs are not specified by the operation in a unique way. It is not a drawback, it is a basic general property of updates. In this paper we propose some simple, "cautious" approaches to the updated program selection. In a next paper we investigate the problem more thoroughly. The approach of [3] will be discussed from the viewpoint of possible-world semantics in a more detail in the forthcoming paper, too. The main goals of this paper are:

- the introduction of the Kripkean semantics,
- a demonstration that the semantics is useful for stable models identification (computation),
- and that there is an operation on Kripke structures which can be used as a basis for a specification of updates of generalized logic programs.

The paper is structured as follows. The problem is introduced, motivated, and the preliminary technicalities are sketched in the Sections 2 – 4. The kernel of the paper: Section 5 is devoted to Kripke structures associated with given generalized logic programs. It is proved that stable models are encoded in Kripke structures (a method of stable models computation is implicit in this encoding). A construction of the Kripke structure $\mathcal{K}_{P \oplus U}$ is introduced in Section 6. The construction is defined over given Kripke structures $\mathcal{K}_U$ and $\mathcal{K}_P$ associated with programs U and P, respectively. Finally, $\mathcal{K}_{P \oplus U}$ is presented as a tool for a semantic specification of an update of P by U in Section 7 . Some results concerning the correctness of the specification are proved.

2   Interpretation Updates and Dynamic Logic Programs

The so called interpretation update approach emphasizes the role of a semantics in updating: A KB' is considered to be an update of KB by U if the set of models of KB' coincides with the set of updated models of KB. We may express it as $Mod(KB') = Update_U(Mod(KB))$, where $Mod(X)$ is the set of (relevant)[2] models of X and $Update_U(M)$ is an update of a set M of models. The update is determined by the program U, more precisely by a set of (relevant) models of U.

The goal (and a strength) of the interpretation update is an abstraction from the superficial syntactic features when specifying updates. Unfortunately, it is impossible to respect dependencies among literals, to account for justifications, using the interpretation update (and using the traditional AGM-postulates, [1,8], too, see [21]). This is the reason why the interpretation update is refused in [10], and then also in [3]. The fact that $Update_U(KB)$ should not just depend on the interpretations of KB and U is illustrated by a simple example:

Example 1 ([3]) Let P be a program: *innocent $\leftarrow$ not found_guilty*.

Consider the stable model semantics [9] as the representation of the program meaning. The meaning of P is $Mod(P) = \{\{innocent\}\}$, the only stable model of P is $S = \{innocent\}$.

If P is updated by $U = \{found\_guilty \leftarrow\}$, then according to the interpretation update approach we should insert *found_guilty* into S, i.e.

$$Update(Mod(P)) = \{\{innocent, found\_guilty\}\}.$$

Of course, $\{innocent, found\_guilty\}$ is not the intended semantic characterization of the update of P by U. □

Therefore, it is decided to base the updated program $P \oplus U$ on a syntactic transformation, see [3].

3 Motivation

Our next goal is to propose a new semantics of a generalized logic program. An important feature of the semantics should be an ability to handle and to record the dependencies among literals, the justifications.

Example 2 (Continuation of the Example 1) In a sense, *innocent* is justified (in P) by *not found_guilty*. This justification is uprooted by the updating program U. It seems that dependencies, justifications, arguments are important from the semantic point of view. We propose a Kripkean semantics in order to provide a semantic characterization of the dependencies, justifications, arguments. The justifications are represented (encoded) by the accessibility relation (between interpretations).

The graphs GP and GU of the Figure 1 visualize the relevant parts of the Kripke structures associated with programs P and U, respectively. The nodes of the graphs (the possible worlds) represent (partial) interpretations. An accessibility relation is defined on the interpretations as follows. A partial interpretation

---

[2] For example, the relevant models may be the stable models.

M is accessible from another partial interpretation M', if the body of a rule of the given program is satisfied in M' and both the body and the head of the rule are satisfied in M (M is justified by M').

The graph GC provides a semantic characterization of the update of P by U. It is constructed over the graph GU. Some parts of the graph GP may be – in general – connected to GU, but in our example it is impossible: no edge of GP can be appended to $u1$ (no edge of GP is compatible with $found\_guilty$).

Therefore, $GU = GC$ and the stable model of the updated program should be the same as the stable model of the updating program. Of course, $innocent$ is not true in GC. $\square$

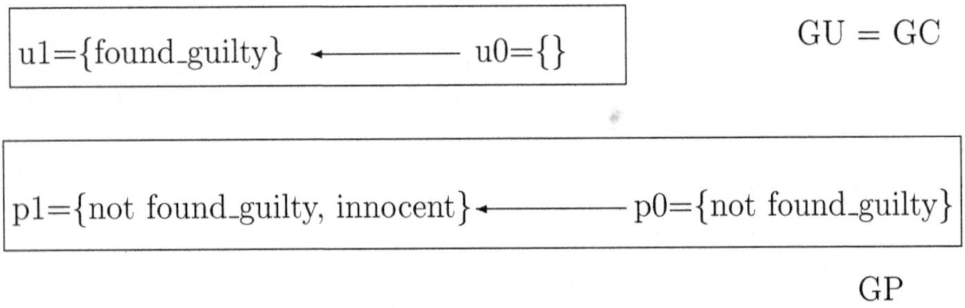

u1={found_guilty} ◀──────── u0={}

$$GU = GC$$

p1={not found_guilty, innocent} ◀──────── p0={not found_guilty}

$$GP$$

Fig. 1. The node $p0$ represents the interpretation $\{not\ found\_guilty\}$, $p1 = \{not\ found\_guilty, innocent\}$, $u0 = \emptyset$, $u1 = \{found\_guilty\}$. The edges $(p0, p1)$ and $(u0, u1)$ represent the dependencies among literals (the second member of a pair is justified by the first member). The update is determined by U, therefore the graph associated with the update (GC) is constructed over the graph associated with the program U (GU). Some parts of the graph associated with P (GP) may be – in general – connected to GU, but in our case it is impossible: no edge of GP can be put before $u0$, similarly, no edge can be appended to $u1$ (no edge of GP is compatible with $found\_guilty$).

The example shows that there is a possibility of an adequate semantic treatment of dependencies among literals. Moreover, the semantics enables to identify and to compute stable models and it enables also to connect relevant parts of one Kripke structure to another. This "connectivity" serves as a basis for updates specification in terms of a purely semantic construction.

We are going to the details.

4   Preliminaries

Consider a finite set of propositional symbols $\mathcal{L}$. The set $\mathcal{L}_{not}$ is defined as $\mathcal{L} \cup \{not\ A : A \in \mathcal{L}\}$. A member of $\mathcal{L}_{not}$ is called $literal$. We will denote the set $\{not\ A : A \in \mathcal{L}\}$ by $\mathcal{D}$ (defaults, assumptions).

A *generalized clause* is a formula c of the form $L \leftarrow L_1, \ldots, L_k$, where L, L_i are literals. We will denote L also by $head(c)$ and the conjunction $L_1, \ldots, L_k$ by $body(c)$. A set of generalized clauses is called a *generalized logic program*. In the following, whenever we use "clause" or "program" we mean "generalized clause" and "generalized logic program", respectively.

For each $A \in \mathcal{L}$, A and $not\ A$ are called *conflicting literals*. A set of literals is *consistent*, if it does not contain a pair of conflicting literals. *Partial interpretation* (of a language $\mathcal{L}_{not}$) is a consistent subset of $\mathcal{L}_{not}$. *Total interpretation* is a partial interpretation $\mathcal{I}$ such that for each $A \in \mathcal{L}$ either $A \in \mathcal{I}$ or $not\ A \in \mathcal{I}$. We are interested in sets of propositional symbols determined by programs. By $\mathcal{L}^P$ we denote the set of all propositional symbols used in the program P. A partial interpretation of a *program* P is a consistent subset of $\mathcal{L}^P_{not}$. The set of all partial interpretations of P we denote by Int_P. Each inconsistent set of literals we denote by $w_\perp$.

A literal L is *satisfied* in a partial interpretation $\mathcal{I}$ if $L \in \mathcal{I}$. A clause $L \leftarrow L_1, \ldots, L_k$ is satisfied in a partial interpretation $\mathcal{I}$ if L is satisfied in $\mathcal{I}$ whenever each L_i is satisfied in $\mathcal{I}$. A partial interpretation $\mathcal{I}$ is a *model* of a program P if each clause $c \in P$ is satisfied in $\mathcal{I}$. Notice that propositional generalized logic programs can be treated as Horn theories: each literal $not\ A$ can be considered as a new propositional symbol (if $not\ A \in \mathcal{L}$ it has to be renamed). The least model of the Horn theory H we denote by $Least(H)$.

Definition 3 (Stable model, [3]) Let P be a generalized logic program and S be an interpretation of P. It is said that S is a stable model of P iff $S = Least(P \cup S^-)$, where $S^- = \{not\ A : not\ A \in S\}$. □

We will visualize Kripke structures as graphs. If e is an edge (w_i, w_{i+1}) of a graph G, the node w_i is called the *source* of e and w_{i+1} the *target* of e. A sequence σ of edges $(w_0, w_1), (w_1, w_2), \ldots, (w_{n-1}, w_n)$ is called a path, w_0 we denote also by $begin(\sigma)$ and w_n by $end(\sigma)$.

5   Kripke Structure Associated with a Program

A notion of Kripke structure associated with a program is defined in this Section. Moreover, it is shown that some distinguished paths in the defined structure represent stable models of logic programs and, conversely, for each stable model there is a distinguished path in the Kripke structure.

The basic idea of our approach was illustrated in the Example 2. A more complicated example is presented below.

Example 4 ([17]) Let P be

$$p \leftarrow not\ q, r$$
$$q \leftarrow not\ p$$
$$r \leftarrow not\ s$$
$$s \leftarrow not\ p.$$

A fragment of the $\mathcal{K}_P$ is depicted in the Figure 2. The nodes are partial interpretations. We distinguish two kinds of edges – ρ_1, and ρ_2.

Consider $(w1, w2)$, an example of an ρ_1-edge, where $w1 = \{not\ p\}$ and $w_2 = \{not\ p, q, s\}$. There are two clauses with the body satisfied in $w1$. Consequences of these clauses are appended to $w1$, the possible world $w2$ is the result of this operation.

Finally, a motivation for ρ_2. There is no total interpretation u such that $(w_2, u) \in \rho_1$, i.e. no clause is applicable to the partial interpretation $w_2 = \{not\ p, q, s\}$ (except of $q \leftarrow not\ p$ and $s \leftarrow not\ p$, but they do not change the possible world $w2$). It means, that P does not enable to justify the truth of r (if we suppose $w2$). Therefore, we may assume by default that r is not true (w.r.t. P and $w2$). The ρ_2-edge from $w2$ to $w3$ represents a completion of $\{not\ p, q, s\}$ by $not\ r$.

□

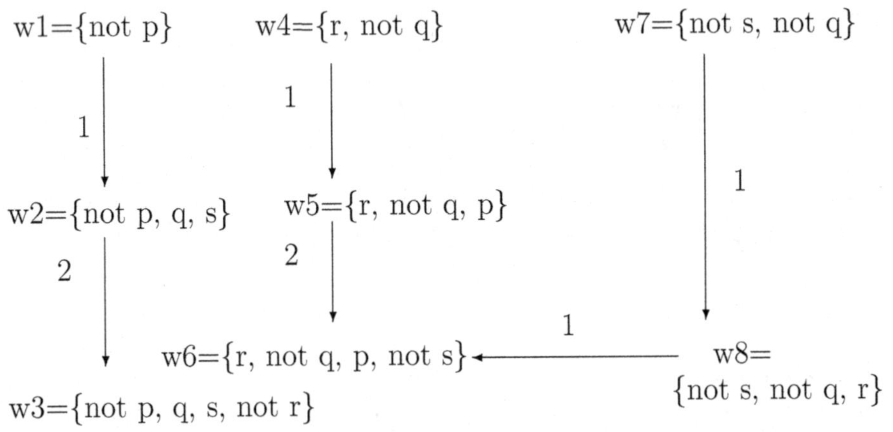

w1={not p}   w4={r, not q}   w7={not s, not q}

1 (w1→w2)   1 (w4→w5)   1 (w7→w8)

w2={not p, q, s}   w5={r, not q, p}

2 (w2→w3)   2 (w5→w6)

w6={r, not q, p, not s} ←—— 1 —— w8={not s, not q, r}

w3={not p, q, s, not r}

Fig. 2. A fragment of $\mathcal{K}_P$. An edge labeled by i is a ρ_i-edge.

Let us summarize: A ρ_1-edge corresponds to an application of a clause to a partial interpretation. A clause c is applicable to a partial interpretation w if $w \models body(c)$. In general, for each $c \in P$: if w is a model of $body(c)$, then $head(c) \in w'$ for some w' such that $w \subseteq w'$ and $(w, w') \in \rho_1$. Intuitively, (w, w') represents a step in a computation bottom-up.

If an atom A is not computed (bottom-up), we assume that $not\ A$ holds. The relation ρ_2 represents a completion (by default negations) of partial interpretations that cannot be changed by some clauses of P.

Now we are ready to define a Kripke structure $\mathcal{K}_P$ associated with P.

Definition 5 Let P be a program. A Kripke structure $\mathcal{K}_P$ associated with P is a pair (W, ρ), where:

- $W = Int_P \cup \{w_\perp\}$, W is called the set of possible worlds, Int_P is the set of all partial interpretations of P, $w_\perp$ is the representative of the set of all inconsistent sets of literals,
- ρ is a binary relation on $W \times W$, it is called the accessibility relation and it is composed of two relations: $\rho = \rho_1 \cup \rho_2$, where
    1. the accessibility relation ρ_1 contains the set of all pairs (w, w') such that $w' = w \cup \{head(c_i) : i = 1, \ldots, k\}$, where $c_1, \ldots, c_k$ are (not necessary all) clauses from P such that $w \models body(c_i)$,
    2. if w is not a total interpretation and for no $u \neq w$ there is an edge $(w, u) \in \rho_1$, then $(w, w') \in \rho_2$, where $w' = w \cup \{not\ A : A \notin w\}$.

$\square$

Of course, $\mathcal{K}_P$ may be viewed as a graph.

Definition 6 ρ-path is a sequence σ of edges $(w_0, w_1), (w_1, w_2), \ldots, (w_{n-1}, w_n)$ in $\mathcal{K}_P$ such that each $(w_i, w_{i+1}) \in \rho$.

We say that this σ is *rooted* in w_0 (also w_0-rooted). If there is no ρ-edge (w_n, w) in $\mathcal{K}_P$ such that $w \neq w_n$, we say that σ is *terminated* in w_n (also: w_n is a terminal node of $\mathcal{K}_P$). $\square$

Sometimes we denote paths by the shorthand $\langle w_0, w_1, w_2, \ldots, w_{n-1}, w_n \rangle$. Similarly, a ρ_1-path could be defined.

We have seen that Kripke structures are appropriate for recording justifications (of interpretations by another interpretations). The justifications have to be non-circular. There are two kinds of basic assumptions – facts (with empty interpretation as the justification, edges to facts are $\emptyset$-rooted) and default negations (subsets of $\mathcal{D}$), called non-monotonic assumptions in TMS [6]: if there is no evidence against, we assume $not\ A$ (where A is an atom). Therefore, the Kripke structure $\mathcal{K}_P$ associated with a program P enables to identify (and to compute) the stable models of P.

Example 7 Let us return to the Example 4 (and to the Figure 2)

There is no fact in P, hence there is no $\emptyset$-rooted path in $\mathcal{K}_P$. As a consequence, relevant paths are only those rooted in some w such that $\emptyset \neq w \subseteq \mathcal{D}$ (only defaults can be assumed). There is a $\{not\ s, not\ q\}$-rooted ρ-path terminated in a stable model $\{not\ s, r, not\ q, p\}$ and a $\{not\ p\}$-rooted (similarly, also a $\{not\ p, not\ r\}$-rooted) ρ-path terminated in another stable model $\{p, not\ q, not\ s, r\}$. $\square$

Now we are ready to state conditions for stable models in terms of nodes and paths in $\mathcal{K}_P$.

Definition 8 Let P be a program, σ be an acyclic ρ-path $\langle w_0, w_1, \ldots, w_n \rangle$ from $\mathcal{K}_P$. We say that σ is *correctly rooted*, if

- either $w_0 = \emptyset$
- or $\emptyset \neq w_0 \subseteq \mathcal{D}$. $\square$

Theorem 9 *Let P be a program, $\mathcal{K}_P$ be the Kripke structure associated with P, $\sigma = (w_0, w_1), (w_1, w_2), \ldots, (w_{n-1}, w_n)$ be an acyclic ρ-path in $\mathcal{K}_P$ terminated in a total interpretation w_n.*

If σ is correctly rooted, then w_n is a stable model of P.

Proof Sketch:
Let P be a generalized logic program. Let P' be $P \cup \{not\ A \leftarrow: not\ A \in w_n^-\}$. Consider P' as a definite program (each literal $not\ A$ is a new propositional letter) with integrity constraints of the form $\leftarrow A, not\ A$ for each propositional symbol $A \in \mathcal{L}^P$.

According to [3], see also the Definition 3: w_n is a stable model of P iff $w_n = Least(P \cup w_n^-)$, where $w_n^- = \{not\ A : not\ A \in w_n\}$.

We assume that $\sigma = \langle w_0, w_1, \ldots, w_{n-1}, w_n \rangle$ is correctly rooted and w_n is a total interpretation. If $(w_{n-1}, w_n) \in \rho_1$ it is straightforward to show that $w_n = Least(P \cup w_n^-)$. Otherwise, notice that $w^* = w_0 \cup (w_n \setminus w_{n-1}) \subseteq w_n^-$ and $\langle w^*, (w_1 \cup w^*), \ldots, (w_{n-1} \cup w^*) \rangle$ is a correctly rooted acyclic ρ_1-path terminated in w_n. It means, $Least(P') = w_n$. Clearly, integrity constrains are satisfied in w_n. Finally, $Least(P') = Least(P \cup w_n^-)$. □

Theorem 10 *Let S be a stable model of a generalized logic program P and $\mathcal{K}_P$ be a Kripke structure associated with P.*

There is a correctly rooted and acyclic ρ-path $\sigma = \langle w_0, \ldots, w_n, S \rangle$ in $\mathcal{K}_P$ terminated in S.

Proof Sketch:
We again use $S = Least(P \cup S^-)$. We can construct a correctly rooted (in S^-) ρ-path terminated in S both if $S^- = \emptyset$ and if $S^- \neq \emptyset$. □

Fact 11 *Let $P, \mathcal{K}_P$ be as in the Theorem 10. If $(\mathcal{D}, w_\perp) \notin \rho_1$, then $\mathcal{D}$ is the only stable model of P.*

Proof: First, $\mathcal{D}$ is a stable model of P: Let $\mathcal{D}' \neq \emptyset$ be a proper subset of $\mathcal{D}$. Then $\langle \mathcal{D}', \mathcal{D} \rangle$ is a correctly rooted ρ-path terminated in the total interpretation $\mathcal{D}$.

Let $\langle w_0, \ldots, w_n \rangle$ be a correctly rooted ρ-path terminated in a total interpretation $w_n \neq \mathcal{D}$. Hence, $A \in w_n$ for at least one atom A. Of course, there is an atom A, a rule $A \leftarrow L_1, \ldots, L_k$, and a correctly rooted ρ-path $\langle u_0, \ldots, u_m \rangle$ such that $u_m = w_n$ and $u_0 \models L_1, \ldots, L_k$, where $u_0 \subset \mathcal{D}$. Therefore, $\mathcal{D} \models L_1, \ldots, L_k$ and $(\mathcal{D}, w_\perp) \in \rho_1$. It means, $\mathcal{D}$ is the only stable model of P. □

Fact 12 *Let P and $\mathcal{K}_P$ be as in the Theorem 10. If $\sigma = \langle w_0, w_1, \ldots, w_n \rangle$ is a ρ-path in $\mathcal{K}_P$, terminated in $w_n \neq w_\perp$, then w_n is a model of P.*

If M is a model of P, then there is a ρ-path in $\mathcal{K}_P$ terminated in M.

Proof: If $c \in P$ and $w_i \models body(c)$ for some w_i, then $head(c) \in w_{i+1}$.

M is not an isolated node: If $M = \mathcal{D}$, we can use the edge $(\mathcal{D}', \mathcal{D})$ from the proof of the Fact 11. If $M \neq \mathcal{D}$ and $w = M \setminus \mathcal{D}$, then there is a path σ in $\mathcal{K}_P$ such that $begin(\sigma) = w$ and $end(\sigma) = M$.

M is a terminal node: $(M, w_\perp) \notin \rho_1$, otherwise there is a clause $c \in P$ which is not true in M. $\square$

6   Updated Kripke Structures

We are going to construct a Kripke structure $\mathcal{K}_{P\oplus U}$ over two Kripke structures, over $\mathcal{K}_P$ (let us recall that it specifies the semantics of an original program P) and over $\mathcal{K}_U$ (specifying the semantics of an updating program U). We intend to use the structure $\mathcal{K}_{P\oplus U}$ as a semantic specification of an updated program.

First we motivate definitions of some notions needed for the construction of $\mathcal{K}_{P\oplus U}$. The concept called *continuation node* is the most important one.

We assume that the nodes of $\mathcal{K}_{P\oplus U}$ are the (partial) interpretations of the language $\mathcal{L}^{P\cup U}$.

Example 13 ([3]) Let $P = \{s \leftarrow not\ t; a \leftarrow t; t \leftarrow\}$ be given. We assume that P is updated by $U = \{not\ t \leftarrow p; p \leftarrow\}$. The relevant parts of $\mathcal{K}_P$ and $\mathcal{K}_U$ are illustrated on the Figure 3. We construct $\mathcal{K}_{P\oplus U}$ over $\mathcal{K}_U$, the update is dominated by $\mathcal{K}_U$. If P can consistently add something to U, it should be accepted. Hence, some paths from $\mathcal{K}_P$ may be connected to $\mathcal{K}_U$.

Consider possible worlds from $\mathcal{K}_P$: $w1 = \emptyset$, $w2 = \{t\}$, $w3 = \{t, a\}$, $w4 = \{t, a, not\ s\}$, $w5 = \{not\ t\}$, $w6 = \{not\ t, s\}$. Similarly, the relevant possible worlds from $\mathcal{K}_U$ are: $u1 = \emptyset$, $u2 = \{p\}$, $u3 = \{p, not\ t\}$.

An important decision should be made: Which paths of $\mathcal{K}_P$ may be connected to which nodes of $\mathcal{K}_U$?

Above all, the nodes of $\mathcal{K}_U$ which terminate ρ_1-paths are the reasonable continuation nodes. If we connect a path of $\mathcal{K}_P$ to an intermediate node of a ρ_1-path of $\mathcal{K}_U$, then some information of U could be lost. On the other hand, the acceptance of default assumptions should be postponed until all ρ_1-paths of $\mathcal{K}_{P\oplus U}$ are constructed.

Let us summarize, we have a first example of continuation nodes – the terminal nodes of ρ_1-paths.

Now we proceed to the connection of relevant paths to the continuation nodes. A path σ of $\mathcal{K}_P$ may be connected to a continuation node w of $\mathcal{K}_U$, if $begin(\sigma)$ is compatible – in a sense – with w.

In our simple example, the only relevant continuation node is $u3$. If we connect the path $\langle w1, w2, w3, w4 \rangle$ to the continuation node $u3 = \{p, not\ t\}$, the first edge $(w1, w2)$ leads to $w_\perp$ – the node $w2 = \{t\}$ contradicts the node $u3$.

On the contrary, the path $(w5, w6)$ may be connected successfully to the node $u3$. The node $w5$ is compatible with the node $u3$: $w5 \subseteq u3$, it means that every literal satisfied in $u3$ is satisfied in $w5$, too. Moreover, $w6$ and $u3$ are consistent.

Therefore, the path of $\mathcal{K}_{P\oplus U}$ could be $\sigma = \langle u1, u2, u3, w, w' \rangle$, where $w = u3 \cup w6$ (notice that $u3 = u3 \cup w5$) and $w' = w \cup \{not\ a\}$. The edge (u_3, w) we obtain by connecting (w_5, w_6) to $u3$. The last edge, (w, w') is a ρ_2-edge. This completion is made w.r.t. the language $\mathcal{L}_{not}^{P\cup U}$. The relevant part of $\mathcal{K}_{P\oplus U}$ is on the Figure 3.

478   Ján Šefránek

The path σ is correctly rooted and it is terminated by the total interpretation w'. We can consider a correctly rooted path from $\mathcal{K}_{P\oplus U}$ which terminates in a total interpretation to be a basis for a semantic specification of updated programs $P \oplus U$.

By the way, $w' = \{p, not\ t, s, not\ a\}$ is the only stable model (modulo irrelevant literals) of the updated program $P \oplus U$, as defined in [3]. $\square$

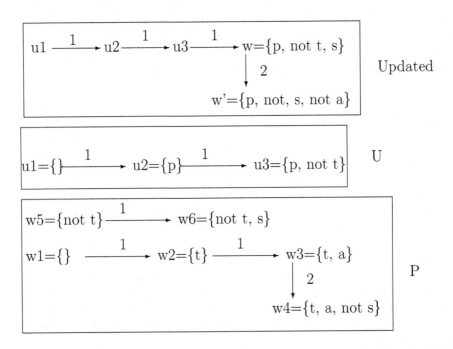

Fig. 3. The relevant parts of $\mathcal{K}_P$ and $\mathcal{K}_U$ from the Example 13. The edges are labeled as in the Figure 2. The edge $(w5, w6)$ from $\mathcal{K}_P$ is connected to the path $\langle u1, u2, u3 \rangle$ from $\mathcal{K}_U$ and the path is completed by the edge (w, w'). The resulting path from $\mathcal{K}_{P\oplus U}$ is $(u1, u2), (u2, u3), (u3, w), (w, w')$, where $w = u3 \cup w6$ and $w' = w \cup \{not\ a\} = \{p, not\ t, s, not\ a\}$.

The example motivates our first decision about continuation nodes: Each terminal node of a ρ_1-path from $\mathcal{K}_U$ is a continuation node. Let w be a continuation node. We may connect a path σ from $\mathcal{K}_P$ to w, if all formulae satisfied in w are satisfied also in $begin(\sigma)$ and if a consistency criterion is satisfied. The node w can be considered as a justification of the connected path.

Now we extend our idea of continuation nodes: It is acceptable to connect some paths of $\mathcal{K}_P$ before some nodes of $\mathcal{K}_U$: Possible continuation nodes are also $w_0 = \emptyset$ and $\emptyset \neq w_0 \subseteq \mathcal{D}$, if there is in $\mathcal{K}_U$ no ρ_1-path rooted in w_0.

We are now ready to present a series of definitions.

Definition 14 Let $\mathcal{K}_U$ be a Kripke structure associated with an update program U.

Continuation nodes of $\mathcal{K}_U$ are

(i) all nodes terminated a ρ_1-path
(ii) $\emptyset$ or w such that $\emptyset \neq w \subseteq \mathcal{D}$, if they are not the source of a ρ_1-edge.

□

Definition 15 The path $\sigma = \langle w_0, w_1, \ldots, w_n \rangle$ from $\mathcal{K}_P$ may be connected to a node w from $\mathcal{K}_U$ iff $w_0 \subseteq w$ and $w \cup w_1$ is consistent. □

Definition 16 Let $\sigma = \langle u_0, \ldots, u_n \rangle$ be a ρ-path and w be a node.

Then *connect σ to w* is a partial operation as follows: if σ may be connected to w, then $(w, u_1 \cup w), \ldots, (u_{n-1} \cup w, u_n \cup w)$ is a ρ-path. If for some $i > 1$ holds that $w \cup u_i$ is inconsistent, it is replaced by $w_\perp$ and the rest of the path is removed. □

Definition 17 Let $\mathcal{K}_P$ and $\mathcal{K}_U$ be the Kripke structures associated with non-empty programs P and U, respectively.

We construct $\mathcal{K}_{P \oplus U}$ as follows:

1. each ρ_1-edge from $\mathcal{K}_U$ is an ρ_1-edge of $\mathcal{K}_{P \oplus U}$,
2. for each continuation node w from $\mathcal{K}_U$ and each ρ_1-path $\sigma = \langle u_0, u_1, \ldots, u_n \rangle$ from $\mathcal{K}_P$: connect σ to w,
3. introduce new ρ_2-edges whenever it is possible.

□

7 Updated Programs Specification

In this Section we present some useful properties of $\mathcal{K}_{P \oplus U}$ and then we sketch some simple methods of updated programs construction.

7.1 Good Worlds and the Stability Condition

First, we introduce a definition in order to simplify the description of $\mathcal{K}_{P \oplus U}$. By analogy to the results of Section 5, correctly rooted ρ-paths terminated in a total interpretation from $\mathcal{K}_{P \oplus U}$ deserve a special interest. We will use them as a basis for a specification of $P \oplus U$.

Definition 18 (Good worlds) Let a Kripke structure $\mathcal{K}_{P \oplus U}$ be given. Let σ be a correctly rooted ρ-path from $\mathcal{K}_{P \oplus U}$ terminated in a total interpretation w.

We say that σ is a *distinguished ρ-path* and w is a *good world*. □

Now it can be said that we will use distinguished ρ-paths and good worlds as a tool for a specification of $P \oplus U$. We accept a cautious strategy in this paper: for each distinguished ρ-path σ (and the corresponding good world w) from $\mathcal{K}_{P \oplus U}$ we are aiming at specifying a program Π such that w is the only stable model of Π. It means, we consider $\mathcal{K}_{P \oplus U}$ as the specification of a variety of updates.

Our next goal is to define a criterion of a reasonable update of P by U. Updated programs specified by $\mathcal{K}_{P \oplus U}$ should satisfy the criterion. The criterion is called the stability condition. It provides a natural characterization of what to accept (or what to reject) from the original program P, if a model M of the updating program U is given. The model M represents an (alternative) belief set dominating the update.

The results of this Subsection – Fact 23, Theorem 24, and Consequence 26 show that

- stability condition and good worlds agree, in a sense,
- both concepts (stability condition, good worlds) enable to specify updated programs compatible with U,
- good worlds are stable models of the updated programs.

A crucial issue is what to accept and what to reject from the original program P, if the updating program U is given. Next example motivates why sometimes the defaults from U override facts from P.

Example 19 Let P be $\{a \leftarrow; b \leftarrow a\}$ and U be $\{not\ b \leftarrow c; c \leftarrow not\ a; a \leftarrow not\ c\}$.

U specifies an intuitively acceptable update of P: a new propositional symbol c is introduced, the meaning of c is the opposite to the meaning of a, and c is a condition for $not\ b$ (while a – according to P – is a condition for b). Notice that no path of $\mathcal{K}_U$ is rooted in $\emptyset$ and the stable models of U are based on some default assumptions.

The relevant parts of $\mathcal{K}_P$, $\mathcal{K}_U$, and $\mathcal{K}_{P \oplus U}$ are illustrated on the Figure 4. The continuation nodes of $\mathcal{K}_U$ are $w2$ and $w4$. The ρ-path $\langle u0, u1, u2 \rangle$ from $\mathcal{K}_P$ may not be connected to $w2$, the edge $(u0, u1)$ leads immediately to the $w_{\perp}$ ($w2 \cup u1$ is not consistent). If we connect the path to the node $w4$ we get $w = \{not\ c, a, b\}$ (a redundant cycle $(w4, w4) = (w4, w4 \cup u1)$ is removed).

Let us summarize – we have two ρ-paths terminated in a total interpretation in $\mathcal{K}_{P \oplus U}$: $\langle w3, w4, w \rangle$ and $\langle w0, w1, w2 \rangle$. The total interpretation w respects the facts from P, but the total interpretation $w2$ does not respect them – it prefers the default assumptions of U.

Our attitude here is a cautious one: we allow both interpretations to determine an updated program $P \oplus U$. $\square$

The example 19 shows that sometimes it is justified to reject some facts of P. Let us suppose that a literal L holds in a stable model S of the updating program U and $L' \leftarrow$ is a fact of the original program P, where L and L' are conflicting literals. The fact is rejected, if we accept the belief set S.

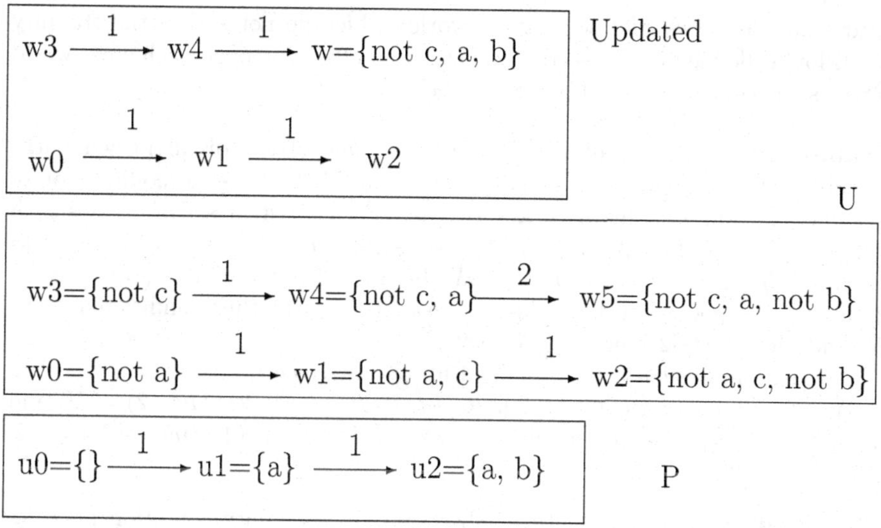

$$w3 \xrightarrow{\;1\;} w4 \xrightarrow{\;1\;} w=\{not\ c,\ a,\ b\}$$

$$w0 \xrightarrow{\;1\;} w1 \xrightarrow{\;1\;} w2$$

Updated

U

$$w3=\{not\ c\} \xrightarrow{\;1\;} w4=\{not\ c,\ a\} \xrightarrow{\;2\;} w5=\{not\ c,\ a,\ not\ b\}$$

$$w0=\{not\ a\} \xrightarrow{\;1\;} w1=\{not\ a,\ c\} \xrightarrow{\;1\;} w2=\{not\ a,\ c,\ not\ b\}$$

$$u0=\{\} \xrightarrow{\;1\;} u1=\{a\} \xrightarrow{\;1\;} u2=\{a,\ b\}$$

P

Fig. 4. A fragment of graphs from the Example 19. The relevant parts of $\mathcal{K}_{P\oplus U}$ are the same as of $\mathcal{K}_U$ with the only exception – the node $w = \{not\ c, a, b\}$ instead of $w5$ and $(w4, w) \in \rho_1$.

Definition 20 Let M be an interpretation of an updating program U, and L, L' be conflicting literals. Let P be an original program.

- $Rejected(M) = \{c \in P : (\exists c' \in U)\ ((head(c), head(c')\ \text{are conflicting literals}$ and $M \models body(c')\} \cup \{(L \leftarrow) \in P : L' \in M\}$
- $Residue(M) = U \cup (P \setminus Rejected(M))$
- $Defaults(M) = \{not\ A : (\forall c \in Residue(M))\ (head(c) = A \Rightarrow M \not\models body(c))\}$, where A is an atom.

□

Our definition of rejected clauses slightly differs from that of [3]. The basic difference is that in [3] facts from P are not rejected when they are in conflict with a stable model S.[3] Similarly, our definition of defaults is different: we define defaults with respect to the $Residue(M)$, while in [3] they are defined w.r.t. $P \cup U$.

Definition 21 (Stability condition) Let programs P, U be given. Let w be a possible world from $\mathcal{K}_{P\oplus U}$. We say that w satisfies the *stability condition*, if holds

$$w = Least(Residue(w) \cup Defaults(w)).\ \square$$

---

[3] From this point of view, the approach of [10,11] is similar to our approach. On the other hand, $Rejected(M)$ may be defined in a distinct way also in our setting. A more detailed comparison and an analysis of some possibilities will be presented in the forthcoming paper.

Next example shows that some good worlds which do not satisfy the stability condition as defined in [3][4] satisfy our Definition 21. Moreover, each good world satisfies the condition (see Theorem 24 below).

Example 22 Let us recall the Example 19. One of the distinguished paths terminates in the good world $w2 = \{not\ a, c, not\ b\}$. Consider a modification of $Residue(w2)$ and $Defaults(w2)$. Let $\Pi \subseteq P$ be a consistent set of clauses such that $(a \leftarrow) \in \Pi$. Let Δ be $\{not\ A : \forall c \in (\Pi \cup U)\ (head(c) = A \Rightarrow w2 \not\models body(c))\}$. Then $w2 \neq Least(\Pi \cup U \cup \Delta)$, because of $not\ A \notin Least(\Pi \cup U \cup \Delta)$. It means, the good world $w2$ does not satisfy the stability condition for the modified $Residue(w2)$ and $Defaults(w2)$.

Notice that $Residue(w2)$ as defined in [3] contains $a \leftarrow$.

According to our Definition 20: $Rejected(w2) = P$, $Residue(w2) = U$, and $Defaults(w2) = \{not\ a, not\ b\}$, hence $Least(Residue(w2) \cup Defaults(w2)) = w2$. □

We proceed to the results of this Subsection. The stability condition provides an important criterion: Each possible world w satisfying this condition respect the information of the updating program U, w is a model of U. Moreover, w is a stable model of $Residue(w)$, where $Residue(w)$ can be viewed as a natural updated program.

Fact 23 *Let P, U be programs. If a possible world w from $\mathcal{K}_{P \oplus U}$ satisfies the stability condition, then*

- *w is a model of U*
- *w is a stable model of $Residue(w)$.*

Proof Sketch: It is straightforward to show that w is a model of U: $w = Least(Residue(w) \cup Defaults(w)) = Least(U \cup (P \setminus Rejected(w)) \cup Defaults(w))$. If $not\ A \in Defaults(w)$, then $A \notin w$, therefore $not\ A \in w^-$, i.e.

$$Least(Residue(w) \cup Defaults(w)) \subseteq Least(Residue(w) \cup w^-).$$

Let us suppose that $not\ A \in w^-$ and there is no clause $c \in Residue(w)$ such that $head(c) = not\ A$ and $w \models body(c)$. Therefore, for each clause $c' \in Residue(w)$ holds that if $head(c') = A$, then $w \not\models body(c')$ (otherwise $A \in w$). Hence, it holds that

$$Least(Residue(w) \cup w^-) \subseteq Least(Residue(w) \cup Defaults(w)).\square$$

Now we demonstrate the important role of distinguished paths and good worlds for updated programs specification. Good worlds and worlds satisfying the stability condition coincide.

Theorem 24 *Let P, U be given. Then w_n is a good world from $\mathcal{K}_{P \oplus U}$ iff w_n satisfies the stability condition.*

---

[4] The term "stability condition" is not used in [3].

Proof Sketch:

$\Rightarrow$

We assume a correctly rooted ρ-path $\sigma = \langle w_0, w_1, \ldots, w_n \rangle$ terminated in w_n. If $(w_{n-1}, w_n) \in \rho_1$, then $Defaults(w_n) = w_0$.

Otherwise, $Defaults(w_n) = w_0 \cup (w_n \setminus w_{n-1})$ and in both cases we have a "computation bottom-up" starting in w_0 and terminated in w_n, i.e.

$$w_n = Least(Residue(w_n) \cup Defaults(w_n)).$$

$\Leftarrow$

$w_n = Least(Residue(w_n) \cup Defaults(w_n))$ is assumed. According to the Fact 23, w_n is a stable model of the $Residue(w_n)$. It means, there is a correctly rooted ρ-path σ in $\mathcal{K}_{Residue(w_n)}$ terminated in w_n (the Theorem 10). Lemma 25 shows that w_n is a good world also w.r.t. $\mathcal{K}_{P \oplus U}$. $\square$

Lemma 25 *Let P and U be programs and w_n be a total interpretation from $\mathcal{K}_{P \oplus U}$.*

If $\sigma = \langle w_0, \ldots, w_n \rangle$ is a correctly rooted ρ-path from $\mathcal{K}_{Residue(w_n)}$ which is terminated in w_n, then there is a correctly rooted ρ-path σ' in $\mathcal{K}_{P \oplus U}$ which is terminated in w_n.

Proof Sketch: If $(w_i, w_{i+1}) \in \sigma$ and there are clauses $c \in U$ and $d \in P$ such that $w_i \models body(c)$, $w_i \models body(d)$, and $head(c), head(d) \in w_{i+1}$, $head(c) \neq head(d)$, then there is a path $\langle w_i, w', w_{i+1} \rangle$, where $w' = w_i \cup \{L \in w_{i+1} : \exists c \in U \ (head(c) = L)\}$.

By repeating this construction we get a path from $\mathcal{K}_{P \oplus U}$ which is correctly rooted and terminated in w_n. $\square$

Finally, the next straightforward consequence shows that good worlds from $\mathcal{K}_{P \oplus U}$ have reasonable properties from the viewpoint of updated programs specification.

Consequence 26 *Let P, U be programs and w be a good world of $\mathcal{K}_{P \oplus U}$. Then*

- *w is a model of U,*
- *w is a stable model of $Residue(w)$.*

It is time to specify $P \oplus U$ (using distinguished ρ-paths and good worlds).

7.2   Updated Programs

In general, each (non-trivial) update may be realized in different ways. (Moreover, we accept the stable-model semantics, therefore it is natural to allow more results of an update.)

The most simple possibility is to consider $Residue(w)$ as an updated program (for any good world w).

A further possible specification of an updated program: $\mathcal{K}_{P \oplus U}$ determines a set $\mathcal{S}$ of programs[5] as follows. Each distinguished ρ-path σ determines one program Π from the set.

The construction of Π: Let a distinguished ρ-path $\sigma = \langle w_0, \dots, w_n \rangle$ be given. For each edge $(w_i, w_{i+1}) \in \rho_1 \cup \rho_2$ let $w_i = \{L_1, \dots, L_m\}$ and $w_{i+1} \setminus w_i = \{L'_1, \dots, L'_k\}$. We put $L'_j \leftarrow L_1, \dots, L_m$ into Π for each $j = 1, \dots, k$.

The good world $end(\sigma)$ of σ is the (only) stable model of Π:

Fact 27 *Let Π be constructed from $\mathcal{K}_{P \oplus U}$ over a distinguished ρ-path σ as above.*

Then the good world $end(\sigma)$ of σ is the (only) stable model of Π.

Proof Sketch: First, $end(\sigma)$ is a stable model of Π: it is a good world and a terminal of a correctly rooted path from $\mathcal{K}_\Pi$. Second, it is the only total interpretation of $\mathcal{K}_\Pi$ which terminates a correctly rooted ρ-path. $\square$

Π introduced above is a member of a family of representatives of $P \oplus U$ in a sense.

Of course, there are more sophisticated possibilities how to construct $P \oplus U$. A special attention deserves an idea of partial evaluation of P with respect to the continuation nodes of $\mathcal{K}_U$, see [12].

All presented proposals for a specification of an updated program on the basis of $\mathcal{K}_{P \oplus U}$ are cautious, they select one of the possible alternatives. Skeptical solutions will be discussed in a forthcoming paper.

Remark 28 Our approach can be expressed also in terms of stable model (answer set) programming paradigm [15,13,17]. Consider a model w of U. It can be said that the model represents the information of U (from a point of view). The model can be viewed as a basis of a constraint satisfaction process and the rules of P can be viewed as constraints. Some of the constraints are not applicable to w (w does not satisfy the constraints), they are rejected. The rest of the constraints is applicable and may be added to the rules from U. The application of the constraints results in some modifications of w (the solutions of the constraint satisfaction process).

8   Conclusions

The approach presented in this paper shows that updates of programs may be specified in a purely semantic frame. The approach is very simple, it does not need an extension of the language and/or of the program(s). There is a variety of syntactic implementations of given semantic specification. In this paper some straightforward constructions are proposed.

The main contributions of the paper may be summarized as follows:

– a semantic treatment of justifications in terms of Kripke structures,

---

[5] We may say that $\mathcal{S}$ is a family of representatives for $P \oplus U$.

- a characterization of stable models in terms of Kripke structures,
- a semantic (and sensitive w.r.t. justifications) characterization of generalized logic programs revisions.

A forthcoming paper will be devoted to a more thorough comparison of the approach of [3] and of the approach presented here. Further, more sophisticated possibilities of $P \oplus U$ specification in terms of $\mathcal{K}_{P \oplus U}$ will be investigated. Similarly for an extension to the case of dynamic program updates specification by $\mathcal{K}_{\oplus \{P_s : s \in S\}}$ (some priorities have to be assigned to the edges of the Kripke structures).

Also the topic of inconsistent generalized logic programs and their revisions (their use in dynamic logic programming) devotes an interest.

Another open problem is a compilation of stable model computing in the spirit of [4], see also [5]. The off-line part of the computation provides a construction of the Kripke structure associated with the given program. The on-line part consists in identifying the stable models in the Kripke structure.

Our approach uses an old idea of TMS, [6] (and a formal reconstruction of TMS by Elkan, [7]). Updates must respect dependencies among literals. Justifications of believed facts are important parts of knowledge bases. Argumentation must not be a circular one. There are some basic assumptions of each argumentation (justification) – axioms (facts) and default assumptions.

Last, some remarks about dynamic Kripke structures (DKS): The concept was introduced and studied in [18,19]. The basic idea about DKS consisted in some transformations of possible worlds. A possibility to modify dynamically the accessibility relation was proposed in [19]. Now, in the present paper the dynamics is implicit in the operation on Kripke structures. Hence, a generalization of the DKS concept (and its applications to the study of knowledge evolution, of hypothetical, nonmonotonic reasoning) is a goal of our research in the future.

Acknowledgments I would like to thank anonymous referees for their comments. Thanks to Štefan Baloc, Damas Gruska, and Ivan Strohner for the remarks to an earlier version of the paper. The work was partially supported by Slovak agency VEGA under the grant 1/7654/20.

References

1. Alchourrón, C., Makinson, D., Gärdensfors, P. *On the logic of theory change. Partial meet contraction and revision functions.* Journal of Symbolic Logic, 50:510-530 (1985)   471
2. Alferes, J.J., Pereira,L.M. *Update-programs can update programs.* LNAI 11126, Springer 1996   469
3. Alferes, J.J., Leite, J.A., Pereira, L.M., Przymusinska, H., Przymusinski, T.C. *Dynamic Logic Programming.* Proc. KR'98, 1998   469, 469, 469, 470, 470, 471, 471, 471, 473, 476, 477, 478, 481, 481, 481, 482, 482, 482, 485
4. Cadoli, M., Donini, F.M., Schaerf, M. *Is intractability of non-monotonic reasoning a real drawback?* Artificial Intelligence 88, 1-2, 215-251   485

5. Cadoli, M., Donini, F.M., Liberatore, P., Schaerf, M. *Space Efficiency of Propositional Knowledge Representation Formalisms*. Journal of Artificial Intelligence Research 13 (2000), 1-31   485
6. Doyle, J. *A Truth Maintenance System*. AI Journal 12 (1979),231-272   475, 485
7. Elkan, C. *A Rational Reconstruction of Nonmonotonic Truth Maintenance Systems*. AI Journal 43 (1990) 219-234   485
8. Gärdenfors, P., Rott. H. *Belief Revision*. In D. Gabbay, C. Hogger, J. Robinson: Handbook of Logic in Artificial Intelligence and Logic Programming, vol. 4, Epistemic and Temporal Reasoning, 35-132, 1995   471
9. Gelfond, M., Lifschitz, V. *The Stable Model Semantics for Logic Programming*. Proc. 5th ICLP, MIT Press, 1988, 1070-1080   471
10. Leite, J., Pereira, L. *Generalizing Updates: from models to programs*. In LNAI 1471, 1997   469, 471, 481
11. Leite, J., Pereira, L. *Iterated Logic Programs Updates*. In Proc. of JICSLP98   469, 481
12. Lifschitz, V., Turner, H. *Splitting a Logic Program*. Proc. of the 11th Int. Conf. on Logic Programming, 1994, 23-37   484
13. Lifschitz, V. *Answer set planning*. Proc. of ICLP, 1999   484
14. Marek, W., Truszczynski, M. *Revision Programming*. Theoretical Computer Science, 190 (1998), 241-277   469
15. Marek, W., Truszczynski, M. *Stable models and an alternative logic programming paradigm*. In The Logic Programming Paradigm: a 25-Year Perspective, 375-398, Springer 1999   484
16. Przymusinski, T., Turner, H. *Update by inference rules*. The Journal of Logic Programming, 1997   469
17. Niemelä, I. *Logic Programs with Stable Model Semantics as a Constraint Programming Paradigm*. Workshop on computational aspects of nonmonotonic reasoning, Trento, 1998   473, 484
18. Šefránek, J. *Dynamic Kripke Structures*. Proc. of CAEPIA'97, Malaga, Spain   485
19. Šefránek, J. *Knowledge,Belief, Revisions, and a Semantics of Non-Monotonic Reasoning*. Proc. LPNMR'99, Springer 1999   485, 485
20. J. Van Benthem, *Semantic Parallels in Natural Language and Computation*, in: Logic Colloquium '87, eds. Ebbinghaus H.-D. et al., 1989, 331-375, North Holland, Amsterdam   469
21. Witteveen, C., Brewka, G. *Skeptical reason maintenance and belief revision*. Artificial Intelligence 61 (1993), 1-36   471

Author index